Handbook of Wood Chemistry and Wood Composites

SECOND EDITION

Edited by **Roger M. Rowell**

CRC Press
Taylor & Francis Group
Boca Raton London New York

CRC Press is an imprint of the
Taylor & Francis Group, an **informa** business

CRC Press
Taylor & Francis Group
6000 Broken Sound Parkway NW, Suite 300
Boca Raton, FL 33487-2742

First issued in paperback 2021

© 2013 by Taylor & Francis Group, LLC
CRC Press is an imprint of Taylor & Francis Group, an Informa business

No claim to original U.S. Government works

ISBN 13: 978-1-03-209916-3 (pbk)
ISBN 13: 978-1-4398-5380-1 (hbk)

Library of Congress Cataloging-in-Publication Data

Handbook of wood chemistry and wood composites / editor, Roger M. Rowell. -- 2nd ed.
    p. cm.
  "A CRC title."
  Includes bibliographical references and index.
  ISBN 978-1-4398-5380-1 (hardcover : alk. paper)
  1. Wood--Chemistry--Handbooks, manuals, etc. 2. Engineered wood--Handbooks, manuals, etc. I. Rowell, Roger M.

TS932.H36 2013
620.1'2--dc23   2012025068

Visit the Taylor & Francis Web site at
http://www.taylorandfrancis.com

and the CRC Press Web site at
http://www.crcpress.com

For a teacher,
success is not measured in the number of students who conform to your way
of thinking; it is measured in the number of students in whom you have confirmed
the excitement of the learning process.

Roger M. Rowell

Contents

PART I   Structure and Chemistry

PART II   Properties

Hōryū Temple in Nara, Japan is over 1300 years old and is one of the oldest remaining wooden structures in the world.

Preface

Wood has played a major role throughout human history. The earliest humans used wood to make shelters, cook food, construct tools, build boats, and make weapons. There are human marks on a climbing pole that were made over 300,000 years ago. We have found wood in the Egyptian pyramids, Chinese temples and tombs, and ancient ships that attest to the use of wood by earlier societies. Collectively, society learnt very early about the great advantages of using a resource that was widely distributed, multifunctional, strong, easy to work, aesthetic, sustainable, and renewable. Wood has been used by people for centuries as a building material and we have accepted its limitations in use, such as instability toward moisture, and degradations due to microorganisms, termites, fire, and ultraviolet radiation. We must accept that wood was designed by nature to perform in a wet environment, and nature is programmed to recycle wood to carbon dioxide and water using the chemistries of decay, thermal, ultraviolet, and moisture degradations. By accepting these limitations, however, we also limit our expectations of performance, which, ultimately, limits our ability to accept new concepts for improved biomaterials.

We are concerned about issues dealing with the environment, sustainability, recycling, energy, sequestering carbon, and the depletion of our natural resources with a growing world population. Wood is receiving a fresh look by scientists, politicians, and economists because of its unique properties, aesthetics, availability, abundance, and perhaps most important of all, its renewability. We will not, however, be able to realize the full potential that wood and wood products can play in our "modern society" as a material and chemical feed stock until we understand its chemistry and materials properties and performance. That understanding holds the key to its utilization. Wood will not reach its highest use potential until we fully describe it, understand the mechanisms that control its performance properties, and finally, are able to manipulate those properties to give us the desired performance we seek.

The purpose of this book is to present the latest concepts in wood chemistry and wood composites as understood by the various authors who have written the chapters. I thank them for their time and effort in the preparation of this book. The book is an update of two earlier books, *Chemistry of Solid Wood, Advances in Chemistry Series*, 207, American Chemical Society, Washington, DC, 1984, which is long out of print and the first edition of this CRC handbook published in 2005.

Acknowledgments

The editor thanks all authors who have contributed to these series of books over the past 30 years: Von Byrd, Daniel Caulfield, Ellis Cowling, L. Emilio Cruz-Barba, Ferencz Denes, W. Dale Ellis, William Feist, Irving Goldstein, Richard Gray, James Han, Danid N.-S. Hon, Rodney Jacobson, T. Kent Kirk, Susan LeVan, Sorin Manolache, John Meyer, Regis Miller, Darrel Nicholas, Russel Parham, Alan Preston, Jeffrey Rowell, Fred Shafizadeh, Chris Skaar, R.V. Subramanian, R. Sam Williams, and Eugene Zavarin.

Editor

Roger M. Rowell retired as a senior technical pioneering scientist after 41 years at the USDA (the United States Department of Agriculture) Forest Service, Forest Products Laboratory, Madison, Wisconsin. He also retired, at the same time, from the University of Wisconsin, Madison, after 35 years and is now a professor emeritus. He is currently a guest professor at Ecobuild in Stockholm, Sweden. He has worked on projects for the United Nations and with many universities, institutes, and companies around the world on chemical modification, sustainable development, and bio-based composites. Dr. Rowell has taught courses in wood chemistry in several countries and has presented many lectures at international and national scientific meetings. His research specialties are in the areas of carbohydrate chemistry, chemical modification of lignocellulosics for property enhancement, water quality, and sustainable materials. He has been a visiting scholar in Japan, Taiwan, Sweden, Mexico, the United Kingdom, New Zealand, and China. He is a fellow in the International Academy of Wood Science and the American Chemical Society, Division of Cellulose, Paper and Textiles.

He received a BS in chemistry and mathematics at Southwestern College in Winfield, Kansas, MS in biochemistry at Purdue University in West Lafayette, Indiana, and PhD in biochemistry at Purdue University. He has edited 11 books, and has over 350 publications and 24 patents.

He is married to his wonderful wife, Judith (who has spent many hours helping with this book), and his family includes three grown-up sons, their wives, and eight grandchildren.

Contributors

Ingeborga Andersone
Department of Wood Chemistry
Latvian State Institute of Wood Chemistry
Riga, Latvia

Bruno Andersons
Laboratory of Wood Protection Emission from
   Wood-Based Products
Latvian State Institute of Wood Chemistry
Riga, Latvia

Georg Avramidis
Department of Science and Technology
HAWK University of Applied Sciences
University of Göttingen
Göttingen, Germany

Marius Catalin Barbu
University Transilvania of Brasov
Brasov, Romania

Lars Bergland
Wallenberg Wood Science Center
Royal Institute of Technology
Stockholm, Sweden

Craig M. Clemons
Forest Products Laboratory
U.S. Department of Agriculture
Madison, Wisconsin

Mark A. Dietenberger
Forest Products Laboratory
U.S. Department of Agriculture
Madison, Wisconsin

Philip D. Evans
Centre for Advanced Wood Processing
University of British Columbia
Vancouver, British Columbia, Canada

Charles R. Frihart
Forest Products Laboratory
U.S. Department of Agriculture
Madison, Wisconsin

Rebecca E. Ibach
Forest Products Laboratory
U.S. Department of Agriculture
Madison, Wisconsin

Mark A. Irle
Ecole Superieure du Bois
Nantes, France

Joseph Jakes
Forest Products Laboratory
U.S. Department of Agriculture
Madison, Wisconsin

Holger Militz
Department of Wood Biology and
   Wood Technology
Georg-August-University Göttingen
Göttingen, Germany

Jouko Peltonen
Center for Functional Materials
Abo Akademi University
Turku, Finland

Roger Pettersen
Forest Products Laboratory
U.S. Department of Agriculture
Madison, Wisconsin

David Plackett
Department of Chemical and Biochemical
   Engineering
Technical University of Denmark
Lyngby, Denmark

Roman Reh
Department of Mechanical Technology
   of Wood
Technical University in Zvolen
Zvolen, Slovakia

Christopher D. Risbrudt
Forest Products Laboratory
U.S. Department of Agriculture
Madison, Wisconsin

Roger M. Rowell
Biological Systems Engineering
University of Wisconsin
Madison, Wisconsin

B. Kristoffer Segerholm
EcoBuild Institute Excellence Centre
SP Technical Research Institute of Sweden
Stockholm, Sweden

Mandla A. Tshabalala
Forest Products Laboratory
U.S. Department of Agriculture
Madison, Wisconsin

Mark R. VanLandingham
Army Research Laboratory
Adelphi, Maryland

Wolfgang Viöl
HAWK University of Applied Sciences
    and Arts
Göttingen, Germany

Shaoxia Wang
Center for Functional Materials
Abo Akademi University
Turku, Finland

Alex C. Wiedenhoeft
Forest Products Laboratory
U.S. Department of Agriculture
Madison, Wisconsin

Jerrold E. Winandy
Department of Bio-Based Products and
    Biosystems Engineering
University of Minnesota
St. Paul, Minnesota

1 Wood and Society

Christopher D. Risbrudt

CONTENT

Forests, and the wood they produce, have played an important role in human activity since before recorded history. Indeed, one of the first major innovations was utilizing fire, fueled by wood, for cooking and heating. It is very likely that early hominids used wood fires for cooking, as long as 1.5 million years ago (Clark and Harris 1985). Clear evidence of this use of wood exists from sites 400,000 years old (Sauer 1962). Since this ancient beginning, the uses of wood, and the value of the forest, have expanded dramatically, as the populations of humans and their economies grew. Wood was used in myriad products, such as agricultural implements and tools, shelters and houses, bridges, road surfaces, ships and boats, arrows and bows, spears, shoes, wheelbarrows, wagons, ladders, and thousands of others. Other important products that forests provided were food, in the form of berries, nuts, fruits, and wild animals, and of course, fuel. Wood was the most important material in early human economies, and though other materials have grown in importance, wood used for solid products, fiber, composites, and chemicals is still the largest single type of raw material input by weight—with the one exception of crushed stone, sand, and gravel—into today's economy (Haynes 2003).

Wood is still the major source of cooking and heating fuel for most of the world. In 2010, world production of fuelwood and charcoal totaled 1,860,403,792 cubic meters. This represents over 58% of the world's consumption of wood. About 43% of this fuelwood consumption occurs in Asia, while Africa consumes 31%. The United States consumes only 1% of the world's total of fuelwood and charcoal (Food and Agriculture Organization [FAO] of the United Nations 2011). Total world consumption of roundwood, which includes fuelwood, charcoal, and industrial wood, amounted to 3275 million m³ in 2010 (FAO of the UN 2011).

Owing to the large volumes of wood consumed around the world, and wood's critical role in heating and cooking, the sustainability of this valuable material is of special concern. As noted earlier, more than half of the wood consumed in the world is used for fuel. Although the majority of wood is used for this purpose, the economics of transportation limits the distances fuelwood can be moved. Hence, very little fuelwood is traded internationally (Steierer 2011). This means the sustainability of wood, and fuelwood in particular, must be evaluated and protected through management, locally.

Besides producing fuelwood and wood for construction and other uses, forests have always been an important part of the American cultural landscape, playing a key role in the social, economic, and spiritual life of the country. However, as the American population and economy grew, forests were removed to make way for farms, cities, and roadways. After the first European settlements in North America, forests were often viewed as an obstacle to farming and travel. Huge acreages were cleared in the nineteenth century to make way for field, pastures, cities, and industry. In 1800, total cropland area in the United States extended across 20 million acres. By 1850, this had grown to 76 million acres, with pasture and hayland at perhaps twice that amount. Most of this farmland expansion was at the expense of forests (MacCleery 1996). The amount of cropland in the United States peaked in 1932, at about 361 million acres (USDA National Agricultural Statistical Service 2003). However,

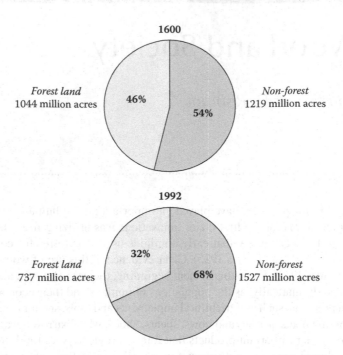

FIGURE 1.1 U.S. forest area. Forests as a percent of total U.S. land area, 1600–1992.

while much forest land has been converted to other uses, the net area of forest land has remained relatively stable since the 1920s (Alig et al. 2003). As shown in Figure 1.1, about 70% of the original amount of forested land still remains as forest, although much of it is likely modified from its structure and composition in 1600. Since 1932, however, as farmed land acreage decreased, forest area in the United States has been increasing. Forests have been the beneficiary of the conversion from animal power to mechanical power in farming. An estimated 20 million acres of grain fields and pastures were no longer needed when gasoline tractors replaced horses and mules. As agricultural productivity per acre increased, as a result of plant breeding, fertilizers, and pesticides, forests have reclaimed many acres back from farm fields.

Wood has remained an important substance throughout history because of its unique and useful properties. Wood is recyclable, renewable, and biodegradable. Many species are shock resistant, bendable, and stable (although all wood changes dimensions as it loses or gains moisture). Density among species varies greatly; the balsa popular with model-airplane builders can weigh as little as 6 lb/ft^3, while some tropical hardwoods weigh more than 70 lb. Wood and lignin can be converted to many useful industrial chemicals, such as ethanol and plastics. Wood can be treated to resist decay, and with proper construction techniques, and stains or paint, wood buildings can last hundreds of years. The oldest surviving wood structure is an Asian temple, built in the seventh century. Today, wood is used in tools, paper, buildings, bridges, guardrails, railroad ties, posts, poles, mulches, furniture, packaging, and thousands of other products.

Wood's versatility makes many wood products recyclable. Perhaps the earliest and simplest recycling was the burning of used wood for heat, whether in a wood stove, fireplace, or furnace. New technologies are improving the efficiency with which used or scrap wood can generate electricity and heat. The paper and paperboard industry has recycled paper to augment virgin wood pulp for decades. At first, recycled paper generally found its way into newsprint and other low-grade products, but recent advances in recycling technology permit used paper to go into the manufacture of higher quality papers, where appearance, texture, and consistency are important. Other products, whether railroad ties or structural timbers from 60-year-old buildings, find second lives as lumber. Affluent consumers, especially in the United States, have long been willing to pay a premium for the aesthetics of using 100-year-old barn siding as interior paneling. More recently, entrepreneurs have recognized the availability and potential value of millions of board feet of high-quality lumber in World War II-era buildings sitting on closed (or soon-to-be-closed) military bases.

Other recycling opportunities for wood include the manufacture of wood fiber and plastic composite materials, where wood fibers improve the strength-to-weight ratio over that of plastic alone—a performance characteristic that has strong appeal in the automotive industry among others.

The ability of forests to regenerate on abandoned farmland—or after a destructive forest fire—testifies to the renewable nature of the wood resource. Though consumption of wood outpaces growth in some parts of the world, in the United States, trees have been growing and producing wood faster than they have been harvested since the early 1950s. By 1970, U.S. forestlands produced more than 20 billion ft^3 of wood, some 5 billion more than the harvest (MacCleery 1996). Much wood for U.S. home construction and the nation's paper industry comes from plantations—mostly southern pine—in the southeastern United States. Those plantations depend on the resource's renewability, with trees harvested and new ones planted and then harvested 15–30 years later. The major inputs are abundant: rain, sun, airborne carbon, and soil nutrients.

Wood's chemical makeup is largely carbon, hydrogen, and oxygen, arranged as cellulose, hemicellulose, and lignin. As such, wood presents an appetizing feast for a variety of fungi species that can metabolize either the sugar-like celluloses or the more complex lignin. With the help of these fungi in the presence of air and water, wood rots, or in environmental terms, is biologically degraded.

Wood is renewable, recyclable, and biodegradable—characteristics generally accepted as good for the environment. At the beginning of the twenty-first century, however, the most environmentally friendly aspect of wood may be its role in carbon sequestration. Growing trees soak up great quantities of carbon from atmospheric carbon dioxide (CO_2), widely regarded as a greenhouse gas that traps heat and affects global climates. Indeed, dry wood is roughly half carbon by weight. Each cubic foot of wood can contain between 11 and 20 lb of carbon. A single tree can easily contain a ton or more of carbon. In addition to the carbon in the tree above ground, significant carbon is locked in the roots and soil. With roughly one-third of the U.S. land area, or some 747 million acres, in forestland, the nation's forests hold more than 50 billion tons of carbon out of the atmosphere (U.S. Environmental Protection Agency 2002).

Forests hold unique significance in the environments that Americans value (see Table 1.1). While the United States is the world's largest importer *and* exporter of wood products, the forests of America are also highly regarded for their recreational, aesthetic, spiritual, and natural values. Forests are valued for providing wildlife and fish habitat, clean water, and clean air. Forests are further shown to be important sinks of carbon, slowing and ameliorating global warming. These concerns are all melded into the concept of sustainability. This concept, expressed through worldwide concern in the 1992 Rio de Janeiro agreement, has resulted in a multinational effort to measure forest sustainability. The Montreal Process, resulting from the Rio agreement, lists seven criteria and 67 indicators of factors and conditions that can help in the judgment of sustainability. The criteria are:

1. Conservation of biological diversity
2. Maintenance of productive capacity
3. Maintenance of forest ecosystem health and vitality
4. Maintenance of soil and water resources
5. Maintenance of forest contribution to global carbon cycles
6. Maintenance and enhancement of long-term multiple socioeconomic benefits to meet the needs of societies
7. Legal, institutional, and economic framework; capacity to measure and monitor changes; and capacity to conduct and apply research and development for forest conservation and sustainable management

TABLE 1.1
Indicator Variables for Outputs from FS- and BLM-Administered
Lands Suitable for an Ecosystem Market Basket

Indicator Variable

Carbon storage
Ecosystem health
Fire risk to life and property
Fish
Game
Minerals
Passive-use values
    Existence of salmon
    Existence of other threatened and endangered species
    Existence of unroaded areas
Range
Recreation
    Access (roads)
    Access to riparian areas
Science integrity
Soil productivity
Special forest and range products
Timber
Visibility
Water quality

Source: Haynes, R.W. An analysis of the timber situation in the United States: 1952–2050.
           General Technical Report PNW-GTR-560. USDA Forest Service, Portland, OR, 2003.

The United States has just completed its first assessment of these criteria and indicators (USDA Forest Service 2004).

Not surprisingly, the assessment produces mixed results in almost every category. While the total area of forest in the United States has remained stable for the past 80 years, the location of forests is changing and the nature of forests and how they are used is changing. Though much forest habitat has remained stable over recent decades, some forest plants, birds, and other animals are at risk of extinction. Net growth on timberland continues to exceed harvest removals.

The downside is that the net gain can represent overcrowding, which increases risk of wildfire and of susceptibility to insects and disease, thus being serious threats to the forest ecosystems.

Healthy forests function well as sources of water for towns and cities, especially in the western United States. The first sustainability assessment indicates however that at least 10% of the forested counties in each region have areas where forest conditions have deteriorated and water and soil quality have been compromised through reduced oxygen levels or higher sediment, dissolved salts, or acidity.

The assessment of long-term socioeconomic benefits to society reveals the interrelatedness and interdependence of many indicators. Recreational use of the nation's forests is increasing. At the same time, the increased demand for wood and wood products has not led to increased harvest of U.S. trees. It appears that increased wood imports and increased recovery of paper for recycling have enabled Americans to use more wood and paper without cutting more trees from America's forests. The global implications—economic, social, political, and environmental—can only be guessed at.

Since the Rio agreement, governmental and consumer concerns over the sustainability—the "greenness"—of products manufactured and consumed has only increased. In this regard, the renewability, recyclability, and biodegradability of wood products are receiving renewed appreciation. For example, wood-framed homes contain 16–17% less embodied energy (resulting from manufacturing and transportation), and a 26–31% lower carbon footprint than steel or concrete buildings (Bowyer et al. 2005).

In spite of being one of the oldest materials used by man, we are still discovering new properties and uses for wood. Nanotechnology—the study and control of the properties of materials at the billionth of a meter scale—represent the latest advance in wood materials science. Wood has a significant amount of material that can be reduced to nano size, with highly desirable properties. Wood fibers 4–6 nm in diameter and 200–300 nm long are extremely strong, can be made into clear products, and can be combined with other materials into composites. Significant research is underway to determine if the promise of this material can be captured economically (Moon et al. 2011).

The United States, through government agencies, nongovernmental organizations, and institutions, industry, and academia, conducts extensive research regarding forests, wood, recycling, and related topics. Nonetheless, the national debate about the proper care and nurture of forests appears in many cases to be rooted in emotion and politics rather than science. The debate can be shrill at times, which reflects perhaps the intensity with which our culture regards forests. Perhaps the greatest challenge to ensuring the long-term sustainability of healthy forests that provide our society with valued resources, recreation, and aesthetic opportunities as well as environmentally vital carbon sequestration lies not in science but in finding a way to sit down at the table together, agree on some goals, and the continued need to explore alternative ways to achieve those goals.

REFERENCES

Alig, R.J., Pantinga, A.J., Ahn, S.E., and Kline, J.D. Land use changes involving forestry in the United States: 1952 to 1997, with projections to 2050. General Technical Report PNW-GTR-587. Portland, OR: U.S. Department of Agriculture, Forest Service, Pacific Northwest Research Station. 92pp., 2003.

Bowyer, J. et al. Life cycle environmental performance of renewable materials in the context of building construction. Consortium for Research on Renewable Industrial Materials, 2005.

Clark, J.D. and Harris, J.W.K. Fire and its roles in early hominid lifeways. *The African Archaeological Review* 3: 3–27, 1985.

Food and Agriculture Organization of the United Nations. *FAO Statistical Data 2011 Forestry Data*, http:. faostat.fao.org.

Haynes, R.W. An analysis of the timber situation in the United States: 1952–2050. General Technical Report PNW-GTR-560. USDA Forest Service, Portland, OR, 2003.

MacCleery, D.W. American forests: A history of resiliency and recovery, revised 1996. Forest History Society Issues Series, Forest History Society, 14, 47, 1996.

Moon, R., Martini, A., Narin, J., Simonsen, J., and Youngblood, J. Cellulose nanomaterials review: Structure, properties and nanocomposites. *Society of Chemical Reviews* 40: 3941–3994, 2011.

Sauer, C.O. Fire and early man. *Paideuma* 7: 399–407, 1962.

Steierer. *Highlights on Wood Fuel: 2004–2009*, FAOSTAT-ForesSTAT. FAO Forestry Dept., Rome, 2011.

U.S. Environmental Protection Agency. Inventory of U.S. greenhouse gas emissions and sinks: 1990–2000, EPA 430-R-02-003, April, 6–8, 2002.

USDA National Agricultural Statistical Service. *Historical Track Records*. Washington, DC, 209pp., 2003.

USDA Forest Service. National report on sustainable forests—2003. FS-766, February, 139, 2004.

Part I

Structure and Chemistry

2 Structure and Function of Wood

Alex C. Wiedenhoeft

CONTENTS

2.1 INTRODUCTION

Wood is a complex biological structure, a composite of many cell types and chemistries acting together to serve the needs of a living plant. Attempting to understand wood in the context of wood technology, we have often overlooked the basic fact that wood evolved over the course of millions of years to serve three main functions in plants—conduction of water from the roots to the leaves, mechanical support of the plant body, and storage and synthesis of biochemicals. There is no property of wood—physical, mechanical, chemical, biological, or technological—not fundamentally derived from the fact that wood is formed to meet the needs of the living tree. To accomplish any of these functions, wood must have cells that are designed and interconnected in ways sufficient to perform them. These three functions have influenced the evolution of approximately 20,000 different species of woody plants, each with unique properties, uses, and capabilities, in both plant and human contexts. Understanding the basic requirements dictated by these three functions and identifying the structures in wood that perform them allow insight to the realm of wood as a composite material itself, and as a component of composite wood products (Hoadley 2000, Barnett and Jeronimidis 2003). The objective of this chapter is to review the basic biological structure of wood and provide a basis for interpreting its properties in an engineering context. By understanding the function of wood in the living tree, we can better understand the strengths and limitations it presents as a material.

The component parts of wood must be defined and delimited at a variety of spatial scales, and those parts related to the form and function of the plant. For this reason, this chapter explains the structure of wood at decreasing scales and in ways that demonstrate the biological rationale for a plant to produce wood with such features. This background will permit the reader to understand the biological bases for the properties presented in subsequent chapters, and to access directly the primary literature in wood structure.

Although shrubs and many vines form wood, the remainder of this chapter will focus on wood from trees, which are the predominant source of wood for commercial and engineering applications and provide examples of virtually all features that merit discussion.

2.2 TREES

A tree has two main domains, the shoot and the roots. Roots are the subterranean structures responsible for water and mineral nutrient uptake, mechanical anchoring of the shoot, and the storage of biochemicals. The shoot is made up of the trunk or bole, branches, and leaves (Raven et al. 1999). The remainder of the chapter will be concerned with the trunk of the tree.

If one cuts down a tree and looks at the stump, several gross observations can be made. The trunk is composed of various materials present in concentric bands. From the outside of the tree to the inside are outer bark, inner bark, vascular cambium, sapwood, heartwood, and the pith (Figure 2.1).

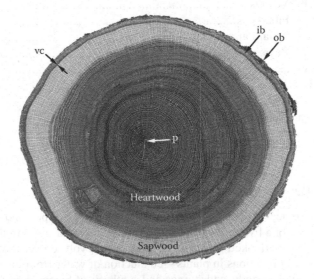

FIGURE 2.1 Macroscopic view of a transverse section of a *Quercus alba* trunk. Beginning at the outside of the tree is the outer bark (ob). Next is the inner bark (ib) and then the vascular cambium (vc), which is too narrow to be seen at this magnification. Interior to the vascular cambium is the sapwood, which is easily differentiated from the heartwood that lies the interior. At the center of the trunk is the pith (p), which is barely discernible in the center of the heartwood.

Outer bark provides mechanical protection to the softer inner bark and also helps to limit evaporative water loss. Inner bark is the tissue through which sugars produced by photosynthesis (photosynthate) are translocated within the tree. The vascular cambium is the layer between the bark and the wood that produces both these tissues each year. The sapwood is the active, "living" wood that conducts the water (or sap) from the roots to the leaves. It has not yet accumulated the often-colored chemicals that set apart the nonconductive heartwood found as a core of darker-colored wood in the middle of most trees. The pith, at the center of the trunk, is the remnant of early growth before wood was formed.

2.3 SOFTWOODS AND HARDWOODS

Despite what one might think based on the names, not all softwoods have soft, lightweight wood, nor do all hardwoods have hard, heavy wood. To define them botanically, softwoods are those woods that come from gymnosperms (mostly conifers), and hardwoods are woods that come from angiosperms (flowering plants). In the temperate portion of the northern hemisphere, softwoods are generally needle-leaved evergreen trees such as pine (*Pinus*) and spruce (*Picea*), whereas hardwoods are typically broadleaf, deciduous trees such as maple (*Acer*), birch (*Betula*), and oak (*Quercus*). Softwoods and hardwoods not only differ in terms of the types of trees from which they are derived but they also differ in terms of their component cells. Softwoods have a simpler basic structure than do hardwoods because they have only two cell types and relatively little variation in structure within these cell types. Hardwoods have greater structural complexity because they have both a greater number of basic cell types and a far greater degree of variability within the cell types. The most important distinction between softwoods and hardwoods is that hardwoods have a characteristic type of cell called a vessel element (or pore) which softwoods lack (Figure 2.2). An important cellular similarity between softwoods and hardwoods is most cells are dead at maturity, even in the sapwood. Cells alive at functional maturity in wood are typically limited to the sapwood, are known as parenchyma cells, and can be found in both softwoods and hardwoods.

2.4 SAPWOOD AND HEARTWOOD

In both softwoods and hardwoods, the wood in the trunk of the tree is typically divided into two functionally distinct zones, sapwood and heartwood. Sapwood is located adjacent to the bark and represents the actively conducting portion of the stem, in which parenchyma cells are still alive and metabolically active. Heartwood is found to the interior of the sapwood, is decommissioned sapwood, and in many species, can be easily distinguished thanks to its darker color (Figure 2.1).

   In the living tree, sapwood is responsible not only for the conduction of sap but also for storage and synthesis of biochemicals. An important storage function is the long-term storage of photosynthate. Carbon necessary to form a new flush of leaves or needles must be stored somewhere in the tree, and parenchyma cells of the sapwood are often where this material is stored. The primary storage forms of photosynthate are starch and lipids. Starch grains are stored in parenchyma cells and can be easily seen with a microscope (Figure 2.2a). The starch content of sapwood can have important ramifications in the wood industry. For example, in the tropical tree ceiba (*Ceiba pentandra*), an abundance of starch can lead to growth of anaerobic bacteria that produce ill-smelling compounds, making the wood commercially unusable (Chudnoff 1984). In southern yellow pines of the United States, and in other softwood timbers, a high starch content encourages growth of sap-stain fungi (Figure 2.2b) that, though they do not affect the strength of the wood, can nonetheless decrease the lumber value for aesthetic reasons (Simpson 1991).

   Parenchyma cells of the sapwood are also the agents of heartwood formation, as biochemicals must be actively synthesized and translocated by living cells (Hillis 1996). Heartwood functions in long-term storage of biochemicals of many varieties depending on the species in question, and these chemicals are what impart color to the heartwood. Heartwood chemicals are known collectively as

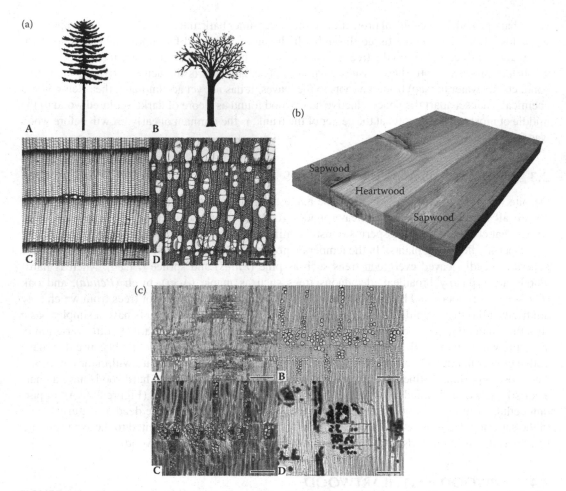

FIGURE 2.2 (a) (A) The general form of a generic softwood tree. (B) The general form of a generic hardwood tree. (C) Transverse section of *Pseudotsuga menziesii,* a typical softwood; the 13 round white spaces are resin canals. (D) Transverse section of *Betula allegheniensis,* a typical hardwood; the many large, round white structures are vessels or pores, the characteristic feature of a hardwood. Scale bars = 780 μm. (b) Sap-stain in the sapwood of a softwood. Note that the heartwood does not exhibit the darker color of the fungal stain, but the sapwood does. This is a dramatic demonstration of the storage function in wood, as the sapwood contents encourage fungal growth, and the heartwood contents prevent it. Sap-stain does not alter the mechanical properties of wood, but can be considered a defect in some products. (c) Light micrographs of starch grains in the parenchyma cells of *Hevea brasiliensis.* (A) Transmitted light microscopy of unstained wood, radial section. The many small spherical objects are starch grains. (B) same as (A), but at higher magnification. Individual starch grains can easily be resolved. (C) same image as (B), but illuminated with polarized light. The cell walls and starch grains are birefringent. The cross- or x-shaped pattern of light in the starch grains are characteristic of amylose. (D) Another radial section, stained with I_2KI; the starch grains are dark brown or purple and contrast strongly with the background. Scale bars = (A) 390 μm, (B)–(D) 98 μm.

extractives because they are not structural components of the wood itself, and can be extracted from the wood using solvents (Hillis 1987). In the past, heartwood was thought to be a disposal site for harmful byproducts of cellular metabolism, the so-called secondary metabolites. This led to the concept of the heartwood as a dumping ground for chemicals that, to a greater or lesser degree, would harm living cells if not sequestered in a safe place. We now know that extractives are a normal part of the plant's system of protecting its wood. Extractives are accumulated by parenchyma cells either at a distinct heartwood–sapwood boundary (Type 1 or *Robinia*-type heartwood formation) or in the aging sapwood tissue (Type 2 or *Juglans*-type heartwood formation) (Magel 2000; Burtin et al. 1998) and are then exuded through pits into adjacent cells (Hillis 1996). In this way, dead cells can become occluded or infiltrated with extractives despite the fact that these cells lack the ability to synthesize or accumulate these compounds on their own.

Extractives are responsible for imparting several larger-scale characteristics to wood (Hillis 1987). For example, extractives provide natural durability to timbers that have a resistance to decay fungi. In the case of a wood like teak (*Tectona grandis*), known for its stability and water resistance, these properties are conferred in large part by the waxes and oils formed and deposited in the heartwood. Many woods valued for their colors, such as mahogany (*Swietenia mahagoni*), African blackwood (*Diospyros melanoxylon*), Brazilian rosewood (*Dalbergia nigra*), and others, owe their value to the type and quantity of extractives in the heartwood. For these species, the sapwood has little or no value, because the desirable properties are imparted by heartwood extractives. Gharu wood, or eagle wood (*Aquilaria malaccensis*), has been driven to endangered status due to human harvest of the wood to extract valuable resins used in perfume making (Lagenheim 2003). Sandalwood (*Santalum spicatum*), a wood famed for its use in incenses and perfumes, is valuable only if the heartwood is rich with the desired scented extractives.

2.5 AXIAL AND RADIAL SYSTEMS

More detailed inquiry into the structure of wood shows that wood is composed of discrete cells connected and interconnected in an intricate and predictable fashion to form an integrated system continuous from root to twig. The cells of wood are typically many times longer than wide and are specifically oriented in two separate systems: the axial system and the radial system. Cells of the axial system have their long axes running parallel to the long axis of the organ (up and down the trunk). Cells of the radial system are elongated perpendicularly to the long axis of the organ and are oriented like radii in a circle or spokes in a bicycle wheel, from the pith to the bark. In the trunk of a tree, the axial system runs up and down, functions in long-distance sap movement, and provides the bulk of the mechanical strength of the tree. The radial system runs in a pith-to-bark direction, provides lateral transport for biochemicals, and in many cases, performs a large fraction of the storage function in wood. These two systems are interpenetrating and interconnected, and their presence is a defining characteristic of wood as a tissue.

2.6 PLANES OF SECTION

Although wood can be cut in any direction for examination, the organization and interrelationship between the axial and radial systems give rise to three main perspectives from which wood should be viewed to glean the most information. These three perspectives are the transverse plane of section (the cross section), the radial plane of section, and the tangential plane of section. Radial and tangential sections are referred to as longitudinal sections because they extend parallel to the axial system (along the grain).

The transverse plane of section is the face exposed when a tree is cut down. Looking down at the stump one sees the transverse section (as in Figure 2.3h); cutting a board across the grain exposes the transverse section. The transverse plane of section provides information about features that vary both in the pith to bark direction (called the radial direction) and also those that vary in the circumferential direction (called the tangential direction). It does not provide information about variations up and down the trunk.

The radial plane of section runs in a pith-to-bark direction (Figure 2.3a, top), and is parallel to the axial system, thus it provides information about longitudinal changes in the stem and from pith to bark along the radial system. To describe it geometrically, it is parallel to the radius of a cylinder,

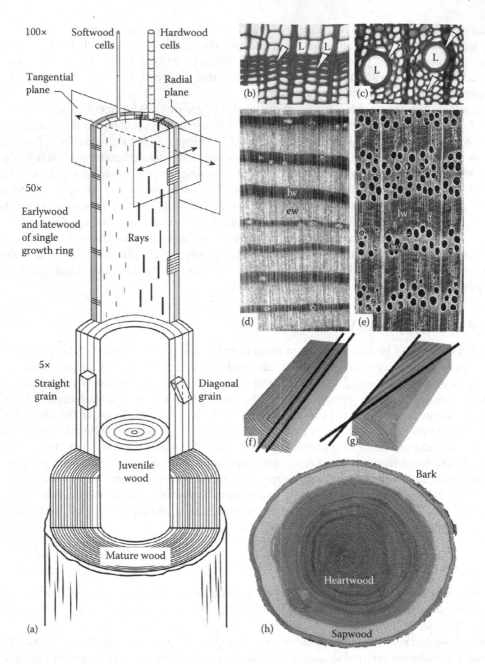

FIGURE 2.3 (a) Illustration of a cut-away tree at various magnifications, corresponding roughly with the images to its right; at the top, at an approximate magnification of 100×, a softwood cell and several hardwood cells are illustrated, to give a sense of scale between the two; one tier lower, at an approximate magnification of 50×, is a single growth ring of a softwood (left) and a hardwood (right), and an indication of the radial and tangential planes; the next tier, at approximately 5× magnification, illustrates many growth rings together and how one might produce a straight-grained rather than a diagonal-grained board; the lowest tier includes an illustration of the relative position of juvenile and mature wood in the tree, at 1× magnification. (b, c) Light microscopic views of the lumina (L) and cell walls (arrowheads) of a softwood (b) and a hardwood (c). (d, e) Hand-lens views of growth rings, each composed of earlywood (ew) and latewood (lw), in a softwood (d) and a hardwood (e). (f) A straight-grained board; Note that the line along the edge of the board is parallel to the line along the grain of the board. (g) A diagonal-grained board; Note that the two lines are markedly not parallel; this board has a slope of grain of about 1 in 7. (h) The gross anatomy of a tree trunk, showing bark, sapwood, and heartwood.

and extending up and down the length of the cylinder. In a practical sense, it is the face or plane exposed when a log is split exactly from pith to bark. It does not provide any information about features that vary in the tangential direction.

The tangential plane is at a right angle to the radial plane (Figure 2.3a, top). Geometrically, it is parallel to any tangent line that would touch the cylinder, and it extends along the length of the cylinder. One way in which the tangential plane would be exposed is if the bark were peeled from a log; the exposed face is the tangential plane. The tangential plane of section does not provide any information about features that vary in the radial direction, but it does provide information about the tangential dimensions of features.

All three planes of section are important to the proper observation of wood, and only by looking at each can a holistic and accurate understanding of the three-dimensional structure of wood be gleaned. The three planes of section are determined by the structure of wood and the way in which the cells in wood are arrayed. The topology of wood and the distribution of the cells are accomplished by a specific part of the tree stem.

2.7   VASCULAR CAMBIUM

The axial and radial systems and their component cells are derived from the vascular cambium, which is a thin layer of cells between the inner bark and the wood (Figures 2.1, 2.4). It produces, by means of many cell divisions, wood (or secondary xylem) to the inside and bark (or secondary phloem) to the outside, both of which are vascular conducting tissues (Larson 1994). As the vascular cambium adds cells to the layers of wood and bark around a tree, the girth of the tree increases, and thus the total surface area of the vascular cambium itself must increase; this is accomplished by cell division as well.

The axial and radial systems are generated in the vascular cambium by two component cells: fusiform initials and ray initials. Fusiform initials, named to describe their long, slender shape, give rise to cells of the axial system, and ray initials give rise to the radial system. In most cases, the radial system in the wood is continuous into the inner bark, through the vascular cambium (Sauter 2000). In this way wood, the water-conducting tissue, remains connected to the inner bark, the photosynthate-conducting tissue. They are interdependent tissues because living cells in wood require photosynthate for respiration and cell growth and the inner bark requires water in which to dissolve and transport the photosynthate. The vascular cambium is an integral feature that not only gives rise to these tissue systems but also links them so that they may function in the living tree.

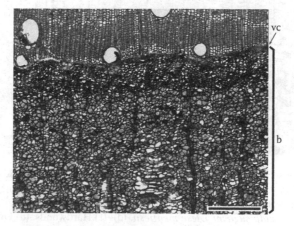

FIGURE 2.4   Light microscopic view of the vascular cambium. Transverse section showing vascular cambium (vc) and bark (b) in *Croton macrobothrys*. The tissue above the vascular cambium is wood. Scale bar = 390 μm.

2.8 GROWTH RINGS

Wood is produced by the vascular cambium, one layer of cell division at a time, but we know from general experience that in many woods, large groups of cells are produced more or less together in time, and these groups act together to serve the tree. These cohorts of cells produced together over a discrete time interval are known as growth increments or growth rings. Cells formed at the beginning of the growth increment are called earlywood cells, and cells formed in the latter portion of the growth increment are called latewood cells (Figure 2.3d,e). Springwood and summerwood were terms formerly used to refer earlywood and latewood, respectively, but their use is no longer recommended (IAWA 1989).

In parts of the world with annual seasonality in temperature and/or precipitation, or in trees with annual cycles in growth, the growth increments are often called annual rings. In parts of the world without such seasonality, and in many evergreen plants (e.g., many trees in the tropics), growth rings are not distinct in the wood. The absence of obvious growth increments does not necessarily imply continuous cambial growth; research has uncovered several characteristics whereby wood structure can be correlated with seasonality changes in some tropical species (Worbes 1995, 1999, Callado et al. 2001).

Woods that form distinct growth rings show one of three fundamental patterns within a growth ring: no change in cell pattern across the ring; a gradual reduction of the inner diameter of conducting elements from earlywood to latewood; or a sudden and distinct change in the inner diameter of the conducting elements across the ring (Figure 2.5). These patterns appear in both softwoods and hardwoods but differ in each because of the distinct anatomical differences between the two.

Softwoods (nonporous woods, woods without vessels) can exhibit any of these three general patterns. Some softwoods such as western red-cedar (*Thuja plicata*), northern white-cedar (*Thuja occidentalis*), and species of spruce (*Picea*) and true fir (*Abies*) have growth increments that undergo a gradual transition from the thin-walled wide-lumined earlywood cells to the thicker-walled, narrower-lumined latewood cells (Figure 2.5b). Other woods undergo an abrupt transition

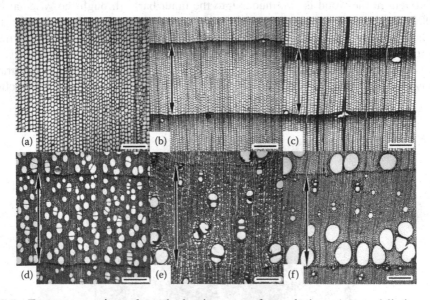

FIGURE 2.5 Transverse sections of woods showing types of growth rings. Arrows delimit growth rings, when present. (a)–(c) Softwoods. (a) No transition within the growth ring (growth ring absent) in *Podocarpus imbricata*. (b) Gradual transition from earlywood to latewood in *Picea glauca*. (c) Abrupt transition from earlywood to latewood in *Pseudotsuga menziesii menziesii*. (d)–(f) Hardwoods. (d) Diffuse–porous wood (no transition) in *Acer saccharum*. (e) Semi-ring-porous wood (gradual transition) in *Diospyros virginiana*. (f) Ring–porous wood (abrupt transition) in *Fraxinus americana*. Scale bars = 300 μm.

from earlywood to latewood, such as southern yellow pine (*Pinus*), larch (*Larix*), Douglas-fir (*Pseudotsuga menziesii*), baldcypress (*Taxodium distichum*), and redwood (*Sequoia sempervirens*) (Figure 2.5c). Because most softwoods are native to the north temperate regions, growth rings are clearly evident. Only in species such as araucaria (*Araucaria*) and some podocarps (*Podocarpus*) does one find no transition within the growth ring (Figure 2.5a). Some authors report this state as growth rings being absent or only barely evident (Phillips 1948, Kukachka 1960).

Hardwoods (porous woods, woods with vessels) also exhibit the three patterns of change within a growth ring, and this is referred to as porosity. In diffuse-porous woods, vessels either do not markedly differ in size and distribution from earlywood to latewood, or the change in size and distribution is gradual and no clear distinction between earlywood and latewood can be found (Figure 2.5d). Maple (*Acer*), birch (*Betula*), aspen/cottonwood (*Populus*), and yellow-poplar (*Liriodendron tulipifera*) are diffuse-porous species.

This pattern is in contrast to ring-porous woods wherein the transition from earlywood to latewood is abrupt, with vessel diameters decreasing substantially (often by an order of magnitude or more); this change in vessel size is often accompanied by a change in the pattern of vessel distribution as well. This creates a ring pattern of large earlywood vessels around the inner portion of the growth increment, and then denser, more fibrous tissue in the latewood, as is found in hackberry (*Celtis occidentalis*), white ash (*Fraxinus americana*), and northern red oak (*Quercus rubra*) (Figure 2.5f).

Sometimes the vessel size and distribution pattern falls more or less between these two definitions, and this condition is referred to as semi-ring-porous (Figure 2.5e). Black walnut (*Juglans nigra*) is a temperate-zone semi-ring-porous wood. Although most tropical hardwoods are diffuse-porous, the best-known commercial exceptions to this are the Spanish-cedars (*Cedrela* spp.) and teak (*Tectona grandis*), which are generally semi-ring-porous and ring-porous, respectively.

Few distinctly ring-porous species grow in the tropics and comparatively few grow in the southern hemisphere. In genera that span temperate and tropical zones, it is common to have ring-porous species in the temperate zone and diffuse-porous species in the tropics. The oaks (*Quercus*), ashes (*Fraxinus*), and hackberries (*Celtis*) native to the tropics are diffuse-porous, whereas their temperate congeners are ring-porous. Numerous detailed texts provide more information on growth increments in wood, a few of which are of particular note (Panshin and deZeeuw 1980, Dickison 2000, Carlquist 2001, Schweingruber 2007).

2.9 CELLS IN WOOD

Understanding a growth ring in greater detail requires some familiarity with the structure, function, and variability of cells that make up the ring. A living plant cell consists of two primary domains: the protoplast and the cell wall. The protoplast is the sum of the living contents that are bounded by the cell membrane. The cell wall is a nonliving, largely carbohydrate matrix extruded by the protoplast to the exterior of the cell membrane. The plant cell wall protects the protoplast from osmotic lysis and often provides mechanical support to the plant at large (Raven et al. 1999, Dickison 2000, Evert 2006).

For cells in wood, the situation is somewhat more complicated than this highly generalized case. In many cases in wood, the ultimate function of the cell is borne solely by the cell wall. This means that many mature wood cells not only do not require their protoplasts, but indeed must completely remove their protoplasts prior to achieving functional maturity. For this reason, a common convention in wood literature is to refer to a cell wall without a protoplast as a cell, and it will be observed throughout the remainder of the chapter.

In the case of a mature cell in wood in which there is no protoplast, the open portion of the cell where the protoplast would have existed is known as the lumen (plural: lumina). Thus, in most cells in wood there are two domains; the cell wall and the lumen (Figure 2.3b,c). The lumen is a critical component of many cells, whether in the context of the amount of space available for water conduction or in the context of a ratio between the width of the lumen and the thickness of the cell wall. The lumen has no structure per se, as it is the void space in the interior of the cell. Thus, wood is a substance that has two basic domains; air space (mostly in the lumina of the cells) and the cell walls of the component cells, a fact that will be discussed later when speaking of wood technology.

2.10 CELL WALLS

Cell walls in wood impart the majority of the properties discussed in later chapters. Unlike the lumen, which is a void space, the cell wall itself is a highly regular structure, from one cell type to another, between species, and even when comparing softwoods and hardwoods. The cell wall consists of three main regions: the middle lamella, the primary wall, and the secondary wall (Figure 2.6). In each region, the cell wall has three major components: cellulose microfibrils (with characteristic distributions and organization), hemicelluloses, and a matrix or encrusting material, typically pectin in primary walls and lignin in secondary walls (Panshin and deZeeuw 1980). In a general sense, cellulose can be understood as a long string-like molecule with high tensile strength; microfibrils are collections of cellulose molecules into even longer, stronger thread-like macromolecules. Lignin is a brittle matrix material. The hemicelluloses are smaller, branched molecules thought to help link the lignin and cellulose into a unified whole in each layer of the cell wall.

To understand these wall layers and their interrelationships, it is necessary to remember that plant cells generally do not exist singly in nature; instead they are adjacent to many other cells, and this association of thousands of cells, taken together, forms an organ, such as a leaf. Each of the individual cells must adhere to one another in a coherent way to ensure that the cells can act as a unified whole. This

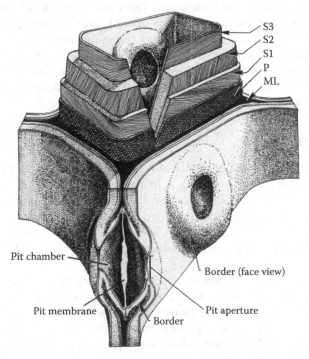

FIGURE 2.6 Cut-away drawing of the cell wall, including the structural details of a bordered pit. The various layers of the cell wall are detailed at the top of the drawing, beginning with the middle lamella (ML). The next layer is the primary wall (P), and on the surface of this layer the random orientation of the cellulose microfibrils is detailed. Interior to the primary wall is the secondary wall in its three layers: S1, S2, and S3. The microfibril angle of each layer is illustrated, as well as the relative thickness of the layers. The lower portion of the illustration shows bordered pits in both sectional and face view.

means they must be interconnected to permit the movement of biochemicals (such as photosynthate, hormones, cell-signaling agents) and water. This adhesion is provided by the middle lamella, the layer of cell wall material between two or more cells, a part of which is contributed by each of the individual cells (Figure 2.6). This layer is the outermost layer of the cell wall continuum and in a nonwoody organ, is rich in pectin. In the case of wood, the middle lamella is lignified.

The layer interior to the middle lamella is the primary wall (Figure 2.6). The primary wall is characterized by a largely random orientation of cellulose microfibrils; like thin threads wound round a balloon in no particular order, where any microfibril angle from 0° to 90° relative to the long axis of the cell may be present. In cells of wood, the primary wall is thin and is generally indistinguishable from the middle lamella. For this reason, the term compound middle lamella is used to denote the primary cell wall of a cell, the middle lamella, and the primary cell wall of the adjacent cell (Kerr and Bailey 1934). Even when viewed with transmission electron microscopy, the compound middle lamella often cannot be separated unequivocally into its component layers.

The remaining cell wall domain, found in virtually all cells in wood (and in many cells in nonwoody plants or plant parts), is the secondary cell wall. The secondary cell wall is composed of three layers (Figure 2.6), distinguished mainly on the basis of the different angle that the helically oriented microfibrils make with the long axis of the cell (Frey-Wyssling 1976, Abe and Funada 2005). As cell wall layers are deposited, the lumen volume is progressively reduced. The first-formed secondary cell wall layer is the S_1 (Figure 2.6), which is adjacent to the compound middle lamella (or technically, the primary wall). This layer is thin and characterized by a large microfibril angle. That is to say, the angle between the mean microfibril direction and the long axis of the cell is large (50–70°).

The next wall layer is arguably the most important cell wall layer in determining the properties of the cell and, thus, the wood properties at a macroscopic level (Panshin and deZeeuw 1980). This layer, formed interior to the S_1 layer, is the S_2 layer (Figure 2.6). As the thickest secondary cell wall layer it makes the greatest contribution to the overall properties of the cell wall. It is characterized by a lower lignin percentage and a low microfibril angle (5–30°). The microfibril angle of the S_2 layer of the wall has a strong but not fully understood relationship with wood properties at a macroscopic level (Kretschmann et al. 1998, Sheng-zuo et al. 2006), and this is an area of active research.

Interior to the S_2 layer is the relatively thin S_3 layer (Figure 2.6). The microfibril angle of the S_3 layer is relatively high and similar to the S_1 (>70°). The S_3 layer has the lowest percentage of lignin of any of the secondary wall layers, which is likely in part related to the movement of sap within the tree. For more detail on these wall components and information on transpiration and the role of the cell wall, see any college-level plant physiology textbook (e.g., Pallardy 2008, Taiz and Zeiger 2010).

2.11   PITS

Any discussion of cell walls in wood must be accompanied by a discussion of the ways in which cell walls are modified to allow intercellular communication and transport in the living plant. These wall modifications, called pit-pairs (or more commonly just pits), are thin areas in the cell walls between two cells and are a critical aspect of wood structure too often overlooked in wood technological treatments. Pits have three domains: the pit membrane, the pit aperture, and the pit chamber. The pit membrane (Figure 2.6) is the thin semi-porous remnant of the primary wall; it is a carbohydrate, a critical fact for biologists accustomed to thinking of phospholipid cell or organelle membranes. The pit aperture is the opening or hole leading into the open area of the pit, which is called the pit chamber (Figure 2.6). The type, number, size, and relative proportion of pits can be characteristic of certain types of wood and furthermore can directly affect how wood behaves in a variety of situations, such as how wood interacts with surface coatings (DeMeijer et al. 1998, Rijkaert et al. 2001).

Pits of predictable types occur between different types of cells. For two adjacent cells, pits will form in the wall of each cell separately but in a coordinated location so that the pitting of one cell will match up with the pitting of the adjacent cell (thus a pit-pair). When this coordination is lacking and a pit is formed only in one of the two cells, it is called a blind pit. Blind pits are fairly rare in wood. Understanding the type of pit can permit one to determine what type of cell is being examined in the absence of other information. It can also allow one to make a prediction about how the cell might behave, particularly in contexts that involve fluid flow. Pits occur in three varieties: bordered, simple, and half-bordered (Raven et al. 1999, Evert 2006).

Bordered pits are thus named because the secondary wall overarches the pit chamber and the aperture is generally smaller or differently shaped than the pit chamber, or both. The portion of the cell wall overarching the pit chamber is called the border (Figures 2.6, 2.7a,d). When seen in face view, bordered pits often are round in appearance and look somewhat like a doughnut (Figure 2.6). When seen in sectional view, the pit often looks like a pair of V's with the open ends of the V's facing each other (Figure 2.7a,d). In this case, the long stems of the V represent the borders, the secondary walls that are overarching the pit chamber. Bordered pits always occur between two conducting cells, and sometimes between other cells, typically those with thick cell walls. The structure and function of bordered pits, particularly those in softwoods (see following section), are much-studied and considered to be well-suited to the safe and efficient conduction of sap. The status of the bordered pit (whether it is open or closed) has great importance in the field of wood preservation and can affect wood finishing and adhesive bonding.

Simple pits lack any sort of border (Figure 2.7c,f). The pit chamber is straight-walled, and the pits are uniform in size and shape in each of the partner cells. Simple pits are typical between parenchyma cells and in face view, merely look like clear areas in the walls.

Half-bordered pits occur between a conducting cell and a parenchyma cell. In this case, each cell forms the kind of pit that would be typical of its type (bordered in the case of a conducting cell and simple in the case of a parenchyma cell) and thus half of the pit pair is simple and half is bordered (Figure 2.7b,e). In the living tree, these pits are of great importance because they represent the communication between conducting cells and biochemically active parenchyma cells.

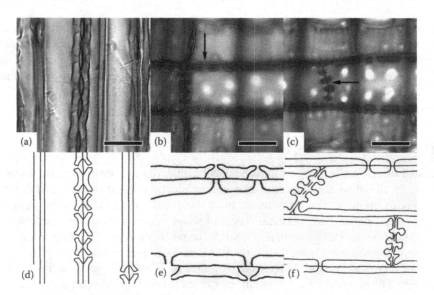

FIGURE 2.7 Light micrographs and sketches of the three types of pits. (a,d) Longitudinal section of bordered pits in *Xanthocyparis vietnamensis*; the pits look like a vertical stack of thick-walled letter vs. (b,e) Half-bordered pits in *Pseudotsuga menziesii*; the arrow shows one half-bordered pit. (c,f) Simple pits on an end wall in *Pseudotsuga menziesii*; the arrow indicates one of five simple pits on the end wall. Scale bars = 20 μm.

2.12   MICROSCOPIC STRUCTURE OF SOFTWOODS AND HARDWOODS

As discussed previously, the fundamental differences between woods are founded on the types, sizes, proportions, pits, and arrangements of different cells that comprise the wood. These fine details of structure can affect the use of a wood.

2.12.1   SOFTWOODS

The structure of a typical softwood is relatively simple. The axial or vertical system is composed mostly of axial tracheids, and the radial or horizontal system is made of rays, which are composed mostly of ray parenchyma cells.

2.12.1.1   Tracheids

Tracheids are long cells, often more than 100 times longer (1–10 mm) than wide, and comprise over 90% of the volume of softwoods. They serve both the conductive and mechanical needs of softwoods. On the transverse view (Figure 2.8a), tracheids appear as square or slightly rectangular cells in radial rows. Within one growth ring they are typically thin-walled in the earlywood and thicker-walled in the latewood. For water to flow between tracheids, it must pass through circular bordered pits that are concentrated in the long, tapered ends of the cells. Tracheids overlap with adjacent cells across both the top and bottom 20–30% of their length. Water flow thus must take a slightly zigzag path as it goes from one cell to the next through the pits. Because the pits have a pit membrane, resistance to flow is substantial. The resistance of the pit membrane coupled with the narrow diameter of the lumina makes tracheids relatively inefficient conduits compared with the conducting cells

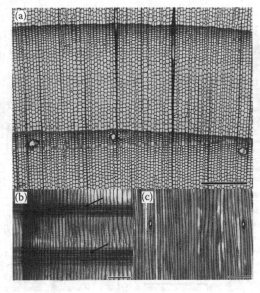

FIGURE 2.8   Microscopic structure of *Picea glauca*, a typical softwood. (a) Transverse section, scale bar = 390 μm; the bulk of the wood is made of tracheids, the small rectangles of various thicknesses; the three large, round structures are resin canals and their associated cells; the dark lines running from the top to the bottom of the photo are the ray cells of the rays. (b) Radial section showing two rays (arrows) running from left to right; each cell in the ray is a ray cell, and they are low, rectangular cells; the rays begin on the right in the earlywood (thin-walled tracheids) and continue into and through the latewood (thick-walled tracheids) and into the earlywood of the next growth ring, on the left side of the photo; scale bar = 195 μm. (c) Tangential section; rays seen in end view, mostly only one cell wide; two rays are fusiform rays; there are radial resin canals embedded in the rays, causing them to bulge; scale bar = 195 μm.

of hardwoods. Detailed treatments of the structure of wood in relation to its conductive functions can be found in the literature (Zimmermann 1983, Pallardy 2008).

2.12.1.2 Axial Parenchyma and Resin Canal Complexes

Another cell type sometimes present in softwoods is axial parenchyma. Axial parenchyma cells are similar in size and shape to ray parenchyma cells, but they are vertically oriented and stacked one on top of the other to form a parenchyma strand. In transverse section they often look like axial tracheids but can be differentiated when they contain dark-colored organic substances in their lumina. In the radial or tangential section they appear as long strands of cells generally containing dark-colored substances. Axial parenchyma is most common in redwood, juniper, cypress, baldcypress, and some species of *Podocarpus* but rarely, if ever, makes up even 1% of the volume of a block of wood. Axial parenchyma strands are generally absent in pine, spruce, larch, hemlock, and species of *Araucaria* and *Agathis*.

In species of pine, spruce, Douglas-fir, and larch, structures commonly called resin ducts or resin canals are present axially (Figure 2.9) and radially (Figure 2.9c). These structures are voids or spaces in the wood and are not cells. Specialized parenchyma cells that function in resin production surround resin canals. When referring to the resin canal and all the associated parenchyma cells, the correct term is axial or radial resin canal complex (Wiedenhoeft and Miller 2002). In pine, resin canal complexes are often visible on the transverse section to the naked eye, but they are much smaller in spruce, larch, and Douglas-fir, and a hand lens is needed to see them. Radial resin canal complexes are embedded in specialized rays called fusiform rays (Figures 2.8c, 2.9c). These rays are typically taller and wider than normal rays. Resin canal complexes are absent in the normal wood of other softwoods, but some species can form large tangential clusters of traumatic axial resin canals in response to substantial injury.

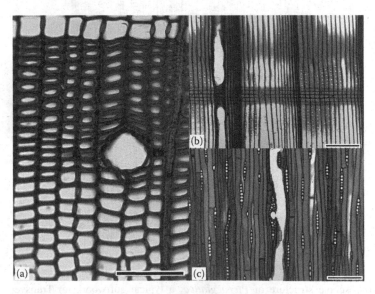

FIGURE 2.9 Resin canal complexes in *Pseudotsuga menziesii*. (a) Transverse section showing a single axial resin canal complex. In this view the tangential and radial diameters of the canal can be measured accurately. Scale bar = 100 μm. (b) Radial section showing an axial resin canal complex embedded in the latewood. It is crossed by a ray that also extends into the earlywood on either side of the latewood. Scale bar = 195 μm. (c) Tangential section showing the anastomosis between an axial and a radial resin canal complex. The fusiform ray bearing the radial resin canal complex is in contact with the axial resin canal complex. Scale bar = 195 μm.

2.12.1.3 Rays

The other cells in Figure 2.8a that are barely visible and appear as dark lines running in a top-to-bottom direction are ray parenchyma cells. Ray parenchyma cells are rectangular prisms or brick-shaped cells. Typically they are approximately 15–20 μm high × 10–20 μm wide × 150–250 μm long in the radial or horizontal direction (Figure 2.8b). These brick-like cells form the rays, which function primarily in synthesis, storage, and lateral transport of biochemicals and, to a lesser degree, water. In radial view or section (Figure 2.8b), the rays look like brick walls and the ray parenchyma cells are sometimes filled with dark-colored substances. In tangential section (Figure 2.8c), the rays are stacks of ray parenchyma cells one on top of the other forming a ray one cell in width, called a uniseriate ray.

When ray parenchyma cells intersect with axial tracheids, specialized pits are formed to connect the axial and radial systems. The area of contact between the tracheid wall and the wall of the ray parenchyma cells is called a cross-field. The type, shape, size, and number of pits in the cross-field are generally consistent within a species and can be diagnostic for wood identification.

Species that have axial and radial resin canal complexes also have ray tracheids, which are specialized tracheids that normally are situated at the margins of the rays. Ray tracheids have bordered pits like axial tracheids but are much shorter and narrower. Ray tracheids also occur in a few species that do not have resin canals. Alaska yellow-cedar (*Chamaecyparis nootkatensis*), hemlock (*Tsuga*), and rarely some species of true fir (*Abies*) have ray tracheids. Additional details regarding the microscopic structure of softwoods can be found in the literature (Phillips 1948, Kukachka 1960, Panshin and deZeeuw 1980, IAWA 2004).

2.12.2 HARDWOODS

The structure of a typical hardwood is more complicated than that of a softwood. The axial system is composed of fibrous elements of various kinds, vessel elements in various sizes and arrangements, and axial parenchyma in various patterns and abundance. As in softwoods, rays comprise the radial system and are composed of ray parenchyma cells, but hardwoods show greater variety in cell sizes and shapes.

2.12.2.1 Vessels

Vessel elements are the specialized water-conducting cells of hardwoods. They are stacked one on top of another to form vessels. Where the ends of the vessel elements come in contact with one another, a hole (with no membrane; it is not a pit) is formed called a perforation plate. If there are no obstructions across the perforation plate, it is called a simple perforation plate. If bars are present, the perforation plate is called a scalariform perforation plate. Thus hardwoods have perforated tracheary elements (vessels elements) for water conduction, whereas softwoods have imperforate tracheary elements (tracheids). On the transverse section, vessels appear as large openings and are often referred to as pores (Figure 2.2d).

Vessel diameters may be small (<30 μm) or quite large (>300 μm), but typically range from 50 to 200 μm. Vessels are much shorter than tracheids and range from 100 to 1200 μm (0.1–1.2 mm). Vessels can be arranged in various patterns. As mentioned under the discussion of growth rings, if the vessels are of the same size and more or less scattered throughout the growth ring, the wood is called diffuse-porous (Figure 2.5d). If the earlywood vessels are much larger than the latewood vessels, the wood is called ring porous (Figure 2.5f). Vessels can also be arranged in a tangential or oblique arrangement, in radial arrangement, in clusters, or in many combinations of these types (IAWA 1989). In addition, individual vessels may occur alone (solitary) or in pairs or radial multiples of up to five or more vessels in a row.

Where vessel elements come in contact with each other, intervessel pits are formed. These pits are bordered, range in size from 2 to >16 μm in height, and are arranged on the vessel walls in three basic ways. The most common arrangement is alternate, where the pits are offset by half the diameter of a pit from one row to the next. In the opposite arrangement, the pits are in files with their apertures aligned vertically and horizontally. In the scalariform arrangement, the pits are much wider than high. Combinations of these arrangements can also be observed in some species. Where vessel elements come in contact with ray cells, often half-bordered pits are formed called vessel–ray pits. These pits can be the same size and shape as the intervessel pits or much larger.

2.12.2.2   Fibers

Fibers in hardwoods provide mechanical support to the wood. They are shorter than softwood tracheids (0.2–1.2 mm), average about half the width of softwood tracheids, but are usually 2–10 times longer than vessel elements of the same wood (Figure 2.10b). The thickness of the fiber cell wall is a major factor governing density and mechanical strength of hardwood timbers. Species with thin-walled fibers, such as cottonwood (*Populus deltoides*), basswood (*Tilia americana*), ceiba, and balsa (*Ochroma pyramidale*), have low density and strength; species with thick-walled fibers, such as hard maple, black locust (*Robinia pseudoacacia*), ipe (*Tabebuia serratifolia*), and bulletwood (*Manilkara bidentata*), have high density and strength. Pits between fibers are generally inconspicuous and may be simple or bordered. In some woods such as oak (*Quercus*) and meranti/lauan (*Shorea*), vascular or vasicentric tracheids are present, especially near or surrounding the vessels. These specialized fibrous elements in hardwoods typically have bordered pits, are thin walled, and are shorter than the fibers of the species; they should not be confused with the tracheids in softwoods, which are much longer.

2.12.2.3   Axial Parenchyma

Axial parenchyma in softwoods is absent or only occasionally present as scattered cells, but hardwoods have a wide variety of axial parenchyma patterns (Figure 2.11). The axial parenchyma cells in hardwoods and softwoods are roughly the same size and shape, and they also function in the same manner. The difference comes in the abundance and specific patterns in hardwoods. Two major types of axial parenchyma are found in hardwoods. Paratracheal parenchyma is associated with the vessels, whereas apotracheal parenchyma is not associated with the vessels. Paratracheal parenchyma is further divided into vasicentric (surrounding the vessels, Figure 2.11a), aliform (surrounding the vessel and with wing-like extensions, Figure 2.11c), and confluent (several connecting patches of paratracheal parenchyma sometimes forming a band, Figure 2.11e). Apotracheal parenchyma is divided into diffuse (scattered), diffuse-in-aggregate (short bands, Figure 2.11b), and banded, whether at the beginning or end of the growth ring (marginal, Figure 2.11f) or within a growth ring (Figure 2.11d). Every wood has a particular pattern of axial parenchyma, which is more or less consistent from specimen to specimen, and these cell patterns are important in traditional hardwood identification.

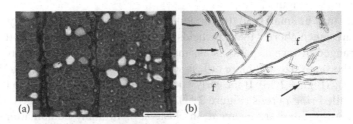

FIGURE 2.10   Fibers in *Quercus rubra*. (a) Transverse section showing thick-walled, narrow-lumined fibers; three rays are passing vertically through the photo, and there are a number of axial parenchyma cells, the thin-walled, wide-lumined cells, in the photo; scale bar = 50 μm. (b) Macerated wood; there are several fibers (f), two of which are marked; also easily observed are parenchyma cells (arrows), both individually and in small groups; note the thin walls and small rectangular shape compared to the fibers; scale bar = 300 μm.

FIGURE 2.11 Transverse sections of various woods showing a range of hardwood axial parenchyma patterns. (a, c, e) are woods with paratracheal types of parenchyma. (a) Vasicentric parenchyma in *Enterolobium maximum;* note that two vessels in the middle of the view are connected by parenchyma, which is the feature also shown in (e); the other vessels in the image present vasicentric parenchyma only. (c) Aliform parenchyma in *Afzelia africana;* the parenchyma cells are the light-colored, thin-walled cells, and are easily visible. (e) Confluent parenchyma in *Afzelia cuazensis.* (b, d, f) are woods with apotracheal types of parenchyma. (b) Diffuse-in-aggregate parenchyma in *Dalbergia stevensonii.* (d) Banded parenchyma in *Micropholis guyanensis.* (f) Marginal parenchyma in *Juglans nigra,* in this case, the parenchyma cells are darker in color, and they delimit the growth rings (arrows). Scale bars = 780 μm.

2.12.2.4   Rays

The rays in hardwoods are structurally more diverse than those found in softwoods. In some species such as willow (*Salix*), cottonwood, and koa (*Acacia koa*), the rays are exclusively uniseriate and are much like softwood rays. In hardwoods, most species have rays more than one cell wide. In oak and hard maple, the rays are two-sized, uniseriate and more than eight cells wide (Figure 2.12a). In most species, rays are 1–5 cells wide and <1 mm high (Figure 2.12b). Rays in hardwoods are composed of ray parenchyma cells that are either procumbent or upright. As the name implies, procumbent ray cells are horizontally elongated and are similar in shape and size to softwood ray parenchyma cells (Figure 2.12c). Upright ray cells have their long axis oriented axially (Figure 2.12d). Upright ray cells are generally shorter than procumbent cells, and sometimes are nearly square. Rays that have only one type of ray cell, typically only procumbent cells, are called homocellular rays. Those that have procumbent and upright cells are called heterocellular rays. The number of rows of upright ray cells, when present, varies from one to many and can be diagnostic in wood identification.

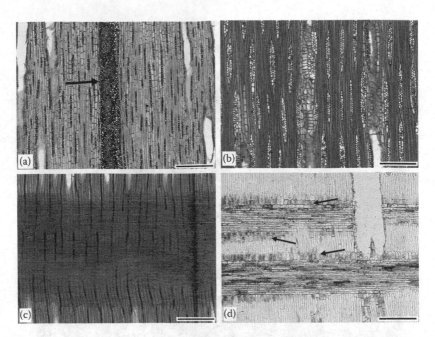

FIGURE 2.12 Rays in longitudinal sections. (a) and (b) Show tangential sections, scale bars = 300 μm. (a) *Quercus falcata* showing a wide multiseriate ray (arrow) and many uniseriate rays. (b) *Swietenia macrophylla* showing numerous rays ranging from 1 to 4 cells wide; note that in this wood the rays are arranged roughly in rows from side to side. (c) and (d) Show radial sections, scale bars = 200 μm. (c) Homocellular ray in *Tilia americana;* all cells in the ray are procumbent cells; they are longer radially than they are tall. (d) Two heterocellular rays in *Khaya ivorensis;* the central portion of the ray is composed of procumbent cells, but the margins of the ray, both top and bottom, have two rows of upright cells (arrows), which are as tall as or taller than they are wide.

The great diversity of hardwood anatomy is treated in many sources throughout the literature (Metcalfe and Chalk 1950, 1979, 1987, Panshin and deZeeuw 1980, IAWA 1989, Gregory 1994, Cutler and Gregory 1998, Dickison 2000, Carlquist 2001, Schweingruber et al. 2006).

2.13 WOOD TECHNOLOGY

Though briefly discussing each kind of cell in isolation is necessary, the beauty and complexity of wood are found in the interrelationship between many cells at a much larger scale. The macroscopic properties of wood such as density, hardness, bending strength, and others are properties derived from the cells that make up the wood. Such larger-scale properties are based on chemical and anatomical details of wood structure (Panshin and deZeeuw 1980, Barnett and Jeronimidis 2003).

2.14 MOISTURE RELATIONS

The cell wall is largely made up of cellulose and hemicellulose, and the hydroxyl groups on these chemicals make the cell wall hygroscopic. Lignin, the agent cementing cells together and rigidifying the cell wall, is a comparatively hydrophobic molecule. This means that the cell walls in wood have a great affinity for water, but the ability of the walls to take up water is limited in part by the presence of lignin. Moisture in wood has a strong effect on wood properties, and wood–water relations greatly affect the industrial use of wood in wood products.

Often it is useful to know how much water is contained in the tree or a piece of wood. This relationship is called moisture content and is the weight of water in the wood expressed as a percentage of the weight of wood with no water (oven-dry weight). Although it is somewhat oversimplified, water exists in wood in two forms, free water and bound water. Free water is the liquid water within the lumina of the cells. Bound water is the water that is adsorbed to the cellulose and hemicellulose in the cell wall. Free water is only found when all sites for the adsorption of water in the cell wall are filled; this point is called the fiber saturation point (FSP). All water added to wood after the FSP has been reached exists as free water.

Wood of a freshly cut tree is said to be green and the moisture content of green wood can be over 100%; in such cases the mass of water in the wood is greater than the mass of the dried cells. In softwoods the moisture content of the sapwood is typically much higher than that of the heartwood, but in hardwoods, the difference may not be as great and in a few cases the heartwood has a higher moisture content than the sapwood.

When drying from the green condition to FSP (commonly reported as 25–30% moisture content for many species) only free water is lost, and thus no change in the cell-wall volumes has occurred. Once the wood is dried below the FSP, bound water is removed from the cell walls and shrinkage of the wood begins. Some of the shrinkage that occurs from green to dry is irreversible; no amount of rewetting can swell the wood back to its original dimensions. After this process of irreversible shrinkage has occurred, however, shrinkage and swelling is reversible from 0% moisture content up to FSP. Controlling the rate at which bound water is removed from green wood is the subject of entire fields of research. By properly controlling the rate at which wood dries, drying defects such as cracking, checking, honeycombing, and collapse can be minimized (Hillis 1996).

2.15 DENSITY

Density (or specific gravity) is one of the most important physical properties of wood (Desch and Dinwoodie 1996, Bowyer et al. 2003). Density is the mass of wood divided by the volume of wood at a given moisture content. Thus, units for density are typically expressed as grams per cubic centimeter (g cm^{-3}) or kilograms per cubic meter (kg m^{-3}). When density values are reported in the literature, the moisture content of the wood must also be given. Usually density values are listed as air-dry, which means 12% moisture content in North America and Europe, but sometimes means 15% moisture content in tropical countries. Specific gravity is the density of the sample normalized to the density of water—the ratio of the density of wood to the density of water. Since 1 cm^3 of water weighs 1 g, density in g/cm^3 is numerically equivalent to specific gravity. Density in kg/m^3 must be divided by 1000 to get the same numerical value as specific gravity. Because specific gravity is a ratio, it does not have units.

Basic specific gravity = density of wood (oven-dry wt/volume when green) ÷ density of water

Specific gravity can be determined at any moisture content, but typically it is based on mass when oven-dry and volume when green; this known as the basic specific gravity, and is generally the standard used throughout the world. The most important reason for measuring basic specific gravity is repeatability. The mass of wood (and the moisture therein) can be determined at any moisture content, but conditioning the wood to a given moisture content consistently is difficult. The oven-dry mass (0% moisture content) is easy to obtain on a consistent basis. Green volume is also relatively easy to determine using water displacement method, as a sample can be large or small and nearly any shape. Thus basic specific gravity can be determined as follows:

Basic specific gravity = oven-dry mass ÷ mass of displaced water

Basic specific gravity is the scientifically preferred metric for comparing between woods; it is the oven-dry mass of the wood divided by the volume of the wood at the time the tree was felled. This basis for comparison avoids the changes in volume associated with wood drying. When consulting the literature or reporting wood density or specific gravity, it is necessary to establish under what conditions the values were obtained.

Wood structure determines the wood specific gravity; softwoods in which latewood is abundant (Figure 2.3d) in proportion to earlywood have higher specific gravity (e.g., 0.59 specific gravity in longleaf pine, *Pinus palustris*). The reverse is true when there is more earlywood than latewood

(Figure 2.5b) (e.g., 0.35 specific gravity in eastern white pine, *Pinus strobus*). To say it another way, specific gravity increases as the proportion of cells with thick cell walls increases. In hardwoods, specific gravity is dependent not only on fiber wall thickness, but also on the amount of void space occupied by vessels and parenchyma. In balsa, vessels are large (typically >250 μm in tangential diameter) and there is an abundance of axial and ray parenchyma. Fibers that are present are thin walled, and the specific gravity may be <0.20. In dense woods, the fibers are thick walled, lumina are virtually absent, and fibers are abundant in relation to vessels and parenchyma. Some tropical hardwoods have specific gravities >1.0.

2.16 JUVENILE WOOD AND REACTION WOOD

Two key examples of the biology of the tree affecting wood quality can be seen in the formation of juvenile wood and reaction wood (Barnett and Jeronimidis 2003). They are grouped together because they share several common cellular, chemical, and tree physiological characteristics, and each may or may not be present in a certain piece of wood.

Juvenile wood is the first-formed wood of the young tree—the rings closest to the pith (Figure 2.3a, bottom). Juvenile wood in softwoods is in part characterized by the production of axial tracheids that have a higher microfibril angle in the S2 wall layer (Larson et al. 2001). A higher microfibril angle in the S2 is correlated with drastic longitudinal shrinkage of the cells when the wood is dried for human use, resulting in a piece of wood that has a tendency to exhibit normal deformations, such as cup and radial checks, and develop anomalous ones, like bow, spring, and cross-grain checking. The morphology of the cells themselves is often altered so that the cells, instead of being long and straight, are often shorter and angled, twisted, or bent. Juvenile wood is thought to afford the young tree mechanical advantages in that stage of growth, but the mechanism of formation is still not fully understood (Barnett and Bonham 2004).

Reaction wood is similar to juvenile wood in several respects but is formed by the tree for different reasons. Most any tree of any age will form reaction wood when the woody organ (whether a twig, branch, or the trunk) is deflected from the vertical by more than 1° or 2°. This means that all nonvertical branches form considerable quantities of reaction wood. The type of reaction wood formed by a tree differs in softwoods and hardwoods. In softwoods, reaction wood is formed on the underside of the leaning organ and is called compression wood (Figure 2.13a) (Timmel 1986). In hardwoods, reaction wood forms on the top side of the leaning organ and is called tension wood (Figure 2.13b) (Desch and Dinwoodie 1996, Bowyer et al. 2003). In compression wood, tracheids are shorter, misshapened cells with a large S_2 microfibril angle and high lignin content (Timmel 1986). They also take on a distinctly rounded outline (Figure 2.13c). In tension wood, the fibers fail to form a proper secondary wall and instead form a highly cellulosic wall layer called the G layer, or gelatinous layer (Figure 2.13d). As mentioned earlier, many features of juvenile wood and reaction wood are similar and analogies and differences between the two are being explored (Xu et al. 2011).

2.17 WOOD IDENTIFICATION

The identification of wood can be of critical importance to the primary and secondary wood using industry, government agencies, museums, law enforcement, and scientists in the fields of botany, ecology, anthropology, forestry, and wood technology. Wood identification is the recognition of characteristic cell patterns and wood features and is generally accurate only to the generic level. Because woods of different species from the same genus often have different properties and perform differently under various conditions, serious problems can develop if species or genera are mixed during the manufacturing process and in use. Because foreign woods are imported to the U.S. market, both buyers and sellers must have access to correct identifications and information about properties and uses, and to ensure compliance with laws governing timber imports.

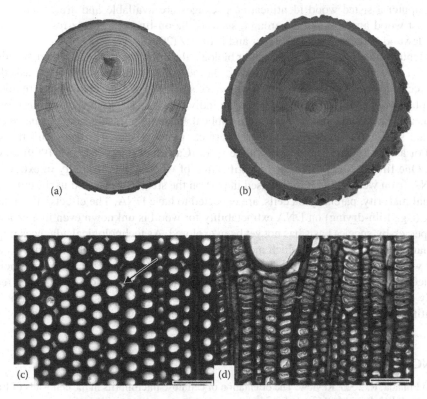

FIGURE 2.13 Macroscopic and microscopic views of reaction wood in a softwood and a hardwood. (a) Compression wood in *Pinus* sp.; note that the pith is not in the center of the trunk, and the growth rings are much wider in the compression wood zone. (b) Tension wood in *Juglans nigra;* the pith is nearly centered in the trunk, but the growth rings are wider in the tension wood zone. (c) Transverse section of compression wood in *Picea engelmannu;* the trachelds are thick-walled and round in outline, giving rise to prominent intercellular spaces in the cell corners (arrow). (d) Tension wood fibers showing prominent gelatinous layers in *Croton gossypiifolius;* the gelatinous layers in the fibers are most pronounced across the top of the image on either side of and just below the vessel; the fibers in the lower half of the image show thinner gelatinous layers. Scale bars = 50 μm.

Lumber graders, furniture workers, those working in the industry, and hobbyists often identify wood without laboratory tools. Features often used are color, odor, grain patterns, density, and hardness. With experience, these features can be used to identify many different woods, but the accuracy of the identification is dependent on the experience of the person and the quality of the unknown wood. If the unknown wood specimen is atypical, decayed, or small, often the identification is incorrect. Examining woods, especially hardwoods, with a 10× to 20× hand lens, greatly improves the accuracy of the identification (Panshin and deZeeuw 1980, Hoadley 1990, Brunner et al. 1994, CITES 2002, Richter and Oelker 2002, Wiedenhoeft 2011). Some foresters and wood technologists armed with a hand lens and sharp knife can accurately identify lumber in the field by examining cell patterns on the transverse surface.

The accepted technique for scientifically rigorous, accurate identifications requires that wood be sectioned and examined with a light microscope. With the light microscope, even with only a 10× objective, many more features are available for use in making a determination. Equally important as the light microscope in wood identification is the reference collection of correctly identified specimens to which unknown samples can be compared (Wheeler and Baas 1998). If a reference collection is not available, books of photomicrographs or books or journal articles with anatomical descriptions and dichotomous keys can be used (Miles 1978, Schweingruber 1978, Core et al. 1979, Gregory 1980, Ilic 1991, Miller and Détienne 2001, Ogata et al. 2008). In addition to these resources, several computer-assisted wood identification packages are available and are suitable for people with a robust wood anatomical background, such as the on-line searchable resource InsideWood (http://insidewood.lib.ncsu.edu/) or Richter and Dallwitz (2000).

Wood identification by means of molecular biological techniques is an active and promising field of research (Lowe and Cross 2011). Depending on the scale of the question being asked, various molecular techniques can be applied to determine the species (DNA barcoding), the region or stand of origin (phylogeographic methods), or even the individual tree (DNA fingerprinting) from which a wood was derived. With increased research, public databasing of results, and advanced analytical methods, the substantial population-biological effects previously expected to limit the statistical likelihood of a robust identification for routine work (Canadian Forest Service 1999) may well be overcome. One limit to the DNA-based identification of wood is the difficulty in extracting high-quality DNA from wood. This limit is based in part on the structure of wood itself; only cells alive at functional maturity, parenchyma cells, are expected to have DNA. The effects of standard wood processing (e.g., kiln-drying) on DNA extractability for wood is unknown even in a general sense, and on a species-by-species basis has not yet been explored. As technological advances improve the quality, quantity, and speed with which molecular data can be collected, the difficulty and cost of molecular wood identification will decrease. We can reasonably expect that at some point in the future molecular tools will be employed in routine identification of wood and that such techniques will greatly increase the specificity and accuracy of identification. For the present, routine scientific wood identification still depends on microscopic evaluation of wood anatomical features.

REFERENCES

Abe, H. and Funada, R. 2005. Review: The orientation of cellulose microfibrils in the cell walls of tracheids in conifers. *IAWA Journal*. 26(2): 161–174.

Barnett, J.R. and Bonham, V.A. 2004. Cellulose microfibril angle in the cell wall of wood fibres. *Biological Reviews*. 79: 461–472.

Barnett, J.R. and Jeronimidis, G. 2003. *Wood Quality and Its Biological Basis*. Blackwell Publishing Ltd., Boca Raton, FL. 225pp.

Bowyer, J., Shmulsky, R., and Haygreen, J.G. 2003. *Forest Products and Wood Science: An Introduction*. 4th ed. Iowa City, IA: Iowa State Press. 554pp.

Brunner, M., Kucera, L.J., and Zürcher, E. 1994. Major timber trees of Guyana: A lens key. *Tropenbos Series* 10. Wageningen, The Netherlands: The Tropenbos Foundation. 224pp.

Burtin, P., Jay-Allemand, C., Charpentier, J., and Janin, G. 1998. Natural wood colouring process in Juglans sp. (*J. nigra, J. regia,* and hybrid *J. nigra* 23 × *J. regia*) depends on native phenolic compounds accumulated in the transition zone between sapwood and heartwood. *Trees*. 12(5): 258–264.

Callado, C.H., da Silva Neto, S.J., Scarano, F.R., and Costa, C.G. 2001. Periodicity of growth rings in some flood-prone trees of the Atlantic rain forest in Rio de Janeiro. *Brazil. Trees*. 15: 492–497.

Canadian Forest Service, Pacific Forestry Centre. 1999. Combating tree theft using DNA technology. *Breakout Session Consensus*. Victoria, BC.

Carlquist, S. 2001. *Comparative Wood Anatomy*. 2nd ed. Berlin: Springer. 448pp.

Chudnoff, M. 1984. Tropical timbers of the world. *Agric. Handb*. 607. Madison, WI: U.S. Department of Agriculture, Forest Service, Forest Products Laboratory. 464pp.

CITES. 2002. *CITES Identification Guide—Tropical Woods*. Wildlife Enforcement and Intelligence Division, Enforcement Branch, Environment Canada.

Core, H.A., Côte, W.A., and Day, A.C. 1979. *Wood Structure and Identification*. 2nd ed. Syracuse, NY: Syracuse University Press. 182pp.

Cutler, D.F. and Gregory, M. 1998. *Anatomy of the Dicotyledons*. 2nd ed. New York: NY: Oxford University Press. 304pp. Vol. IV.

DeMeijer, M., Thurich, K., and Militz, H. 1998. Comparative study on penetration characteristics of modern wood coatings. *Wood Science and Technology*. 32: 347–365.

Desch, H.E. and Dinwoodie, J.M. 1996. *Timber Structure, Properties, Conversion and Use*. 7th ed. London, UK: Macmillan Press. 306pp.

Dickison, W. 2000. *Integrative Plant Anatomy*. New York, NY: Academic Press. 533pp.

Evert, R.F. 2006. *Esau's Plant Anatomy: Meristems, Cells, and Tissues of the Plant Body: Their Structure, Function, and Development.* 3rd ed. John Wiley & Sons, Inc., Hoboken, New Jersey. 601pp.

Frey-Wyssling, A. 1976. The plant cell wall. In: *Handbuch der Pflanzenanatomie*, Band 3, Teil 4. Abt. Cytologie, 3rd rev. ed. Gebruder Borntraeger, Berlin.

Gregory, M. 1980. Wood identification: An annotated bibliography. *IAWA Bull.* 1. ISSN: New Series. 1(1): 3–41.

Gregory, M. 1994. Bibliography of systematic wood anatomy of dicotyledons. *IAWA Journal.* Supplement 1. 265pp.

Hillis, W.E. 1987. *Heartwood and Tree Exudates.* Springer-Verlag, Berlin. 240pp.

Hillis, W.E. 1996. Formation of robinetin crystals in vessels of Intsia species. *IAWA Journal.* 17(4): 405–419.

Hoadley, R.B. 1990 *Identifying Wood: Accurate Results with Simple Tools.* Newtown, CT: Taunton Press. 223pp.

Hoadley, R.B. 2000. *Understanding Wood: A Craftsman's Guide to Wood Technology.* 2nd ed. Newtown, CT: Taunton Press. 280pp.

IAWA Committee. 1989. IAWA list of microscopic features for hardwood identification. In: Wheeler, E.A.; Baas, P.; Gasson, P., eds. *IAWA Bull. 10.* ISSN: New Series 10(3): 219–332.

IAWA Committee. 2004. IAWA list of microscopic features of softwood identification. In: Richter, H.G.; Grosser, D.; Heinz, I.; Gasson, P., eds. *IAWA Journal.* 25(1): 1–70.

Ilic, J. 1991. *CSIRO Atlas of Hardwoods.* Bathurst, Australia: Crawford House Press. 525pp.

Kerr, T. and Bailey, W. 1934. The cambium and its derivative tissues. X. Structure, optical properties and chemical composition of the so-called middle lamella. *J. Arnold Arb.* 15: 327–349.

Kretschmann, D.E., Alden, H.A., and Verrill, S. 1998. Variations of microfibril angle in loblolly pine: Comparison of iodine crystallization and X-ray diffraction techniques. In: Butterfield, B.G., ed. *Microfibril Angle in Wood.* New Zealand: University of Canterbury. 157–176.

Kukachka, B.F. 1960. Identification of coniferous woods. *Tappi Journal.* 43(11): 887–896.

Lagenheim, J.H. 2003. *Plant Resins: Chemistry, Evolution, Ecology, and Ethnobotany.* Portland, OR: Timber Press. 448pp.

Larson, P.R. 1994. *The Vascular Cambium, Development and Structure.* Berlin: Springer-Verlag. 725pp.

Larson, P.R., Kretschmann, D.E., Clark, A., III and Isenbrands, J.G. 2001. Formation and properties of juvenile wood in southern pines: A synopsis. Gen. Tech. Rep. FPL–GTR–129. Madison, WI: U.S. Department of Agriculture, Forest Service, Forest Products Laboratory.

Lowe, A.J. and Cross, H.B. 2011. The application of DNA methods to timber tracking and origin verification. *IAWA Journal.* 32(2): 251–262.

Magel, E.A. 2000. Biochemistry and physiology of heartwood formation. In: Savidge, R.A., Barnett, J.R., Napier, R. Cell and Molecular Biology of Wood Formation. Eds. *BIOS Scientific*, Oxford. 363–376.

Metcalfe, C.R. and Chalk, L. 1950. *Anatomy of the Dicotyledons.* Oxford, UK: Clarendon Press. 2 vols. 1,500pp.

Metcalfe, C.R. and Chalk, L. 1979. *Anatomy of the Dicotyledons.* 2nd ed. New York: Oxford University Press. Vol. I. 276pp.

Metcalfe, C.R. and Chalk, L. 1987. *Anatomy of the Dicotyledons.* 2nd ed. New York: Oxford University Press. Vol. III. 224pp.

Miles, A. 1978. *Photomicrographs of world woods. Building Research Establishment.* Princes Risborough Laboratory; London: Her Majesty's Stationery Office. 233pp.

Miller, R.B. and Détienne, P. 2001. Major timber trees of Guyana: Wood anatomy. *Tropenbos Series* 20. Wageningen, The Netherlands: The Tropenbos Foundation. 218pp.

Ogata, K., Fujii T., Abe, H., and Baas, P. 2008. *Identification of the Timbers of Southeast Asia and the Western Pacific.* Shiga-ken, Japan: Kaiseisha Press. 400pp.

Pallardy, S.G. 2008. *Physiology of Woody Plants.* Third edition. Burlington, MA: Elsevier. Chapters 11, 12.

Panshin, A.J. and deZeeuw, C. 1980. *Textbook of Wood Technology.* 4th ed. New York: McGraw-Hill. 722pp.

Phillips, E.W.J. 1948. Identification of softwoods by microscopic structure. *Forest Products Res. Bull.* 22. 64pp.

Raven, P., Evert, R., and Eichhorn, S. 1999. *Biology of Plants.* 6th ed. New York, NY: W.H. Freeman & Company. 944pp.

Richter, H.G. and Dallwitz, M.J. 2000 onwards. Commercial timbers: Descriptions, illustrations, identification, and information retrieval. In English, French, German, Portuguese, and Spanish. Version: 16th April 2006. http://delta-intkey.com.

Richter, H.G. and Oelker, M. 2002 onwards. MacroHOLZdata Commercial timbers: Descriptions, illustrations, identification, and information retrieval. In English and German. Version: October 2002.

Rijkaert, V., Stevens, M., de Meijer, M., and Militz, H. 2001. Quantitative assessment of the penetration of water-borne and solvent-borne wood coatings in Scots pine sapwood. *Holz als Roh-und Werkstoff.* 59: 278–287.

Sauter, J.J. 2000. Photosynthate allocation to the vascular cambium: facts and problems. In: Savidge, R.A., Barnett, J.R., and Napier, R. *Cell and Molecular Biology of Wood Formation.* Eds. BIOS Scientific, Oxford. 71–83.

Sheng-zuo, F., Wen-zhong, Y., and Xiang-xiang, F. 2006. Variation of microfibril angle and its correlation to wood properties in poplars. *Journal of Forestry Research*. 15(4): 261–267.

Schweingruber, F. 1978. Microscopic wood anatomy. *Birmensdorf*: Swiss Federal Institute for Foreign Research. 800pp.

Schweingruber, F.H., Borner A., and Schulze E.-D. 2006. *Atlas of Woody Plant Stems. Evolution, Structure, and Environmental Modifications*. Berlin: Springer-Verlag. 229pp.

Schweingruber, F.H. 2007. *Wood Structure and Environment*. Berlin: Springer-Verlag. 279pp.

Simpson, W.T., ed. 1991. Dry kiln operator's manual. *Agric. Handb.* AH-188. Madison, WI: U.S. Department of Agriculture, Forest Service, Forest Products Laboratory. 274pp.

Taiz, L. and Zeiger, E. 2010. *Plant Physiology*, Fifth Edition. Sunderland, MA: Sinauer.782pp.

Timmel, T.E. 1986. *Compression Wood in Gymnosperms*. Heidelberg: Springer. 2,150pp.

Wheeler, E.A. and Baas, P. 1998. Wood identification: A review. *IAWA Journal*. 19(3): 241–264.

Wiedenhoeft, A.C. and Miller, R.B. 2002. Brief comments on the nomenclature of softwood axial resin canals and their associated cells. *IAWA Journal*. 23(3): 299–303.

Wiedenhoeft, A.C. 2011. *Identification of Central American Woods. Identificatiòn de las Especies Maderables de Centroamérica*. Madison, WI: Forest Products Society. 167pp.

Worbes, M. 1995. How to measure growth dynamics in tropical trees: A review. *IAWA Journal*. 16(4): 337–351.

Worbes, M. 1999. Annual growth rings, rainfall-dependent growth and long-term growth patterns of tropical trees in the Capar Forest Reserve in Venezuela. *Journal of Ecology*. 87: 391–403.

Xu, P., Liu, H., Donaldson, L.A., and Zhang, Y. 2011. Mechanical performance and cellulose microfibrils in wood with high S_2 microfibril angles. *Journal of Materials Science*. 46(2): 534–540.

Zimmermann, M.H. 1983. *Xylem Structure and the Ascent of Sap*. New York: Springer-Verlag. 143pp.

3 Cell Wall Chemistry

Roger M. Rowell, Roger Pettersen, and Mandla A. Tshabalala

CONTENTS

Wood is best defined as a three-dimensional biopolymer composite composed of an interconnected network of cellulose, hemicelluloses and lignin with minor amounts of extractives, and inorganics. The major chemical component of a living tree is water, but on a dry weight basis, all wood cell walls consist mainly of sugar-based polymers (carbohydrates, 65–75%) that are combined with lignin (18–35%). Overall, dry wood has an elemental composition of about 50% carbon, 6% hydrogen, 44% oxygen, and trace amounts of inorganics. Simple chemical analysis can distinguish between hardwoods (angiosperms) and softwoods (gymnosperms) but such techniques cannot be used to identify individual tree species because of the variation within each species and the similarities among species. In general, the coniferous species (softwoods) have a higher cellulose content (40–45%), higher lignin (26–34%), and lower pentosan (7–14%) content as compared to deciduous species (hardwoods) (cellulose 38–49%, lignin 23–30%, and pentosans 19–26%). Table 3.1 shows a summary of the carbohydrates, lignin, and ash content of hardwoods and softwoods in the United States (Pettersen 1984).

A complete chemical analysis accounts for all the components of wood. Vast amounts of data are available on the chemical composition of wood. All tables in this chapter summarize data for wood species in North America (Pettersen 1984).

3.1   CARBOHYDRATE POLYMERS

3.1.1   HOLOCELLULOSE

The major carbohydrate portion of wood is composed of cellulose and hemicellulose polymers with minor amounts of other sugar polymers such as starch and pectin (Stamm 1964). The combination

TABLE 3.1

Summary of Carbohydrate, Lignin, and Ash Compositions for U.S. Hardwoods and Softwoods

Species	Holocellulose	α-cellulose	Pentosans	Klason Lignin	Ash
Hardwoods	71.7 ± 5.7	45.4 ± 3.5	19.3 ± 2.2	23.0 ± 3.0	0.5 ± 0.3
Softwoods	64.5 ± 4.6	43.7 ± 2.6	9.8 ± 2.2	28.8 ± 2.6	0.3 ± 0.1

Source: Adapted from Pettersen, R.C. 1984. *The Chemistry of Solid Wood*, Advances in Chemistry Series 20, Chapter 2, pp. 57–126, Washington, DC: ACS.

of cellulose (40–45%) and the hemicelluloses (15–25%) are called holocellulose and usually accounts for 65–70% of the wood dry weight. These polymers are made up of simple sugars, mainly, D-glucose, D-mannose, D-galactose, D-xylose, L-arabinose, D-glucuronic acid, and lesser amounts of other sugars such as L-rhamnose and D-fucose. These polymers are rich in hydroxyl groups that are responsible for moisture sorption through hydrogen bonding (see Chapter 4).

3.1.2 CELLULOSE

Cellulose is the most abundant organic chemical on the face of the earth. It is a glucan polymer of D-glucopyranose units, which are linked together by β-(1 → 4)-glucosidic bonds. Actually the building block for cellulose is cellobiose since the repeating unit in cellulose is a two sugar unit (Figure 3.1).

The number of glucose units in a cellulose molecule is referred to as the degree of polymerization (DP). Goring and Timell (1962) determined the average DP for native celluloses from several sources using a nitration isolation procedure that minimized depolymerization and maximized yield. These molecular weight determinations, done by light-scattering experiments, indicate the wood cellulose has an average DP of at least 9000–10,000 and possibly as high as 15,000. An average DP of 10,000 would mean a linear chain length of approximately 5 μm in wood. This would mean an approximate molecular weight for cellulose ranging from about 10,000 to 150,000. Figure 3.2 shows a partial structure of cellulose.

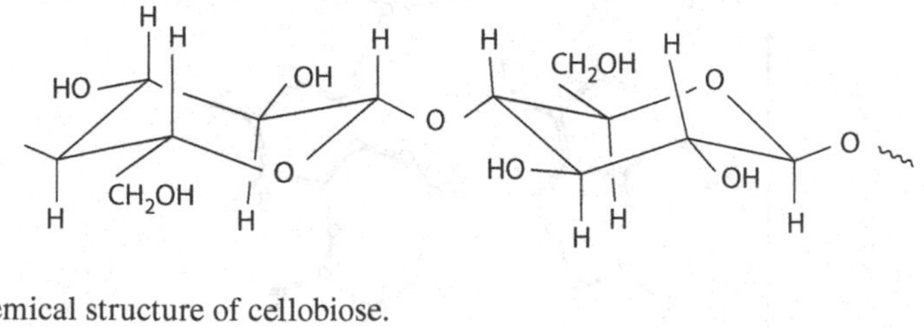

FIGURE 3.1 Chemical structure of cellobiose.

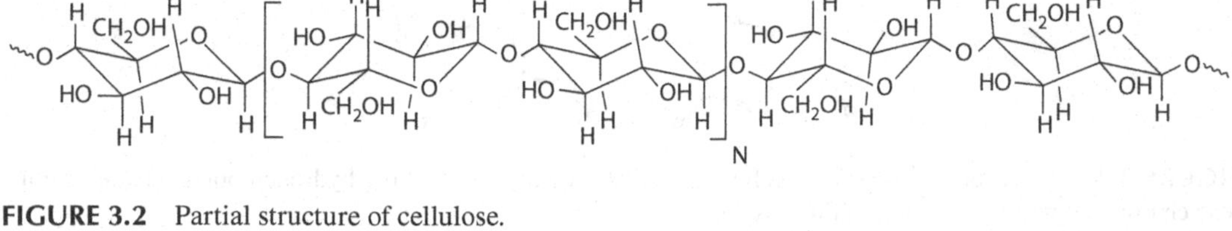

FIGURE 3.2 Partial structure of cellulose.

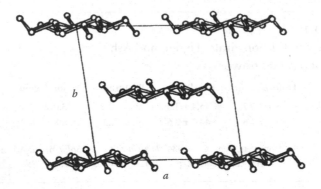

FIGURE 3.3   Axial projection of the crystal structure of cellulose I.

Cellulose molecules are randomly oriented and have a tendency to form intra- and intermolecular hydrogen bonds. As the packing density of cellulose increases, crystalline regions are formed. Most wood-derived cellulose is highly crystalline and may contain as much as 65% crystalline regions. The remaining portion has a lower packing density and is referred to as amorphous cellulose. X-ray diffraction experiments indicate crystalline cellulose (*Valonia ventricosa*) has a space group symmetry $a = 16.34$ Å and $b = 15.72$ Å (Figure 3.3, Gardner and Blackwell 1974). The distance of one repeating unit, that is, one cellobiose unit is $c = 10.38$ Å (Figure 3.4). The unit cell contains eight cellobiose moieties. The molecular chains pack in layers that are held together by weak van der Waals' forces. The layers consist of parallel chains of anhydroglucopyranose units and the chains are held together by intermolecular hydrogen bonds. There are also intramolecular hydrogen bonds between the atoms of adjacent glucose residues (Figure 3.4). This structure is referred to as cellulose I or native cellulose.

There are several types of cellulose in wood: crystalline and noncrystalline (as described above) and accessible and nonaccessible. Accessible and nonaccessible celluloses refer to the availability of the cellulose to water, microorganisms, and so on. The surfaces of crystalline cellulose are accessible

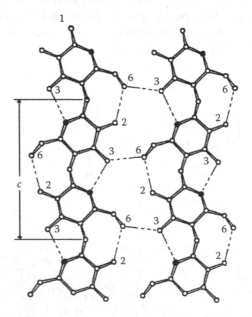

FIGURE 3.4   Planar projection of two cellulose chains showing some of the hydrogen bond between cellulose chains and within a single cellulose chain.

but the rest of the crystalline cellulose is nonaccessible. Most of the noncrystalline cellulose is accessible but part of the noncrystalline cellulose is so covered with both hemicelluloses and lignin that it becomes nonaccessible. Concepts of accessible and nonaccessible cellulose are very important in moisture sorption, pulping, chemical modification, extractions, and interactions with microorganisms.

Cellulose II is another important type of cellulose used for making cellulose derivatives. It is not found in nature as cellulose II. Cellulose II is obtained by mercerization and regeneration of native cellulose. Mercerization is treatment of cellulose I with strong alkali. Regeneration is a treatment with carbon disulfide to form a soluble xanthate derivative. The derivative is converted back to cellulose and reprecipitated as cellulose II. Cellulose II has space group $a = 8.01$ Å, $b = 9.04$ Å (Figure 3.3), and $c = 10.36$ Å (Figure 3.4).

There is also cellulose III which is treatment of cellulose I with liquid ammonia at about $-80°C$ followed by evaporation of the ammonia. Alkali treatment of cellulose III gives cellulose II. Cellulose IV is formed by heating cellulose III in glycerol at 260°C.

Another type of cellulose (based on the method of extraction from wood) often referred to in the literature is Cross and Bevan cellulose. It consists largely of cellulose I but also contains some hemicellulose. It is obtained by chlorination of wood meal, followed by washing with aqueous solutions of 3% sulfur dioxide (SO_2) and 2% sodium sulfite ($NaSO_3$).

Finally, there is another cellulose referred to as Kürschner cellulose also based on the method of isolation. Kürschner cellulose is obtained by refluxing wood meal three times for 1 h with a 1:4 (v/v) mixture of nitric acid and ethyl alcohol. The water washed and dried cellulose is referred to as Kürschner cellulose which also contains some hemicelluloses. This method of cellulose isolation is not often used because it destroys some of the cellulose and the nitric acid–ethanol mixture is potentially explosive.

Cellulose I is insoluble in most solvents including strong alkali. Alkali will make the cellulose to swell but not dissolve it. Cellulose dissolves in strong acids such as 72% sulfuric acid, 41% hydrochloric acid, and 85% phosphoric acid but degradation occurs rapidly. It is difficult to isolate cellulose from wood in a pure form because it is intimately associated with lignin and hemicellulose. The analytical method for isolating cellulose is given in the analytical procedures section of this chapter.

3.1.3 HEMICELLULOSES

In general, the hemicellulose fraction of woods consists of a collection of polysaccharide polymers with a lower DP than cellulose (average DP of 100–200) and containing mainly the sugars D-xylopyranose, D-glucopyranose, D-galactopyranose, L-arabinofuranose, D-mannopyranose, D-glucopyranosyluronic acid, and D-galactopyranosyluronic acid with minor amounts of other sugars. The structure of hemicelluloses can be understood by first considering the conformation of the monomer units. There are three entries under each monomer in Figure 3.5. In each entry, the letter designation D and L refer to a standard configuration for the two optical isomers of glyceraldehydes, the simplest carbohydrate, and designate the conformation of the hydroxyl group at carbon four for pentoses (xylose and arabinose) and carbon five for hexoses (glucose, galactose, and mannose). The Greek letters α and β refer to the configuration of the hydroxyl group on carbon one. The two configurations are called anomers. The first entry is a shortened form of the sugar name. The second entry indicates the ring structure. Furanose refers to a five-membered ring and pyranose refers to a six-membered ring. The six-membered ring is usually in a chair conformation. The third entry is an abbreviation commonly used for a sugar residue in a polysaccharide (Whistler et al. 1962, Timell 1964, 1965, Whistler and Richards 1970, Jones et al. 1979).

Hemicelluloses are intimately associated with cellulose and contribute to the structural component on the tree. Some hemicelluloses are present in very large amounts when the tree is under stress, that is, compression wood where the wood has a higher D-galactose content as well as a

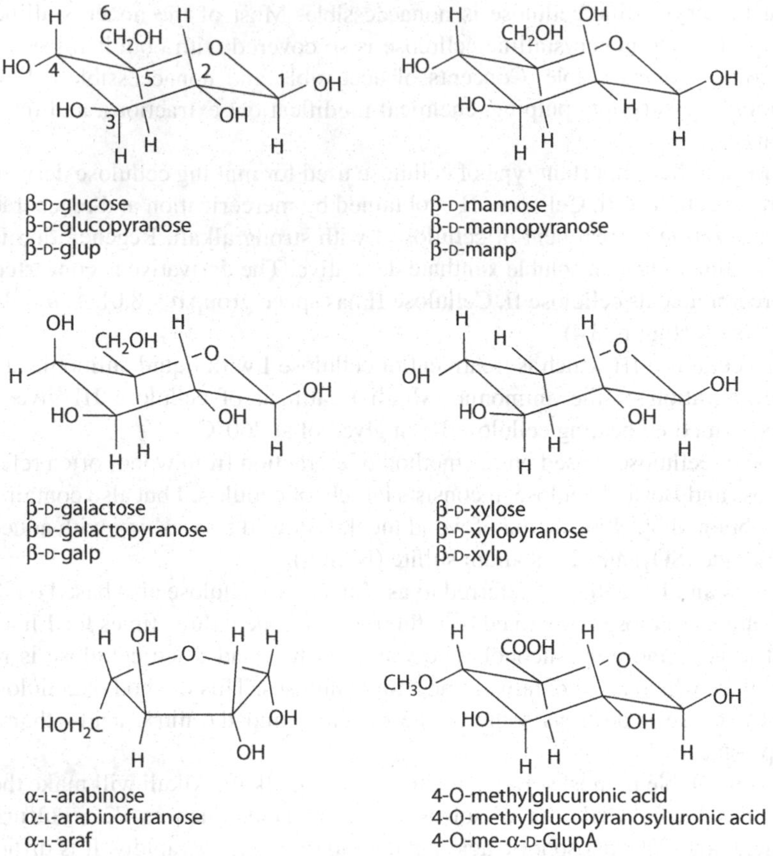

FIGURE 3.5 Sugar monomer components of wood hemicellulose.

higher lignin content (Timell 1982). They usually contain a backbone consisting of one repeating sugar unit linked β-(1 → 4) with branch points (1 → 2), (1 → 3), and/or (1 → 6).

Hemicelluloses usually consist of more than one type of sugar unit and are sometimes referred to by the sugars they contain, for example, galactoglucomannan, arabinoglucuronoxylan, arabinogalactan, glucuronoxylan, glucomannan, and so on. The hemicelluloses also contain acetyl- and methyl-substituted groups. Hemicelluloses are soluble in alkali and are easily hydrolyzed by acids. A gradient elution at varying alkali concentrations can be used for a crude fractionation of the hemicelluloses from wood. The hemicelluloses can then be precipitated from the alkaline solution by acidification using acetic acid. Further treatment with a neutral organic solvent such as ethyl alcohol to the neutralized solution results in a more complete precipitation (Sjöström 1981). The detailed structures of most wood hemicelluloses have not been determined. Only the ratio of sugars these polysaccharides contain have been studied.

3.1.3.1 Hardwood Hemicelluloses

Figure 3.6 shows a partial structure of O-acetyl-4-O-methyl-glucuronoxylan from a hardwood. This class of hemicelluloses is usually referred to as glucuronoxylans. This polysaccharide contains a xylan backbone of D-xylopyranose units linked β-(1 → 4) with acetyl groups at C-2 or C-3 of the xylose units on an average of 7 acetyls per 10 xylose units (Sjöström 1981). The xylan is substituted with side chains of 4-O-methylglucuronic acid units linked to the xylan backbone α-(1 → 2) with an average frequency of approximately 1 uronic acid group per 10 xylose units. The side chains are quite short.

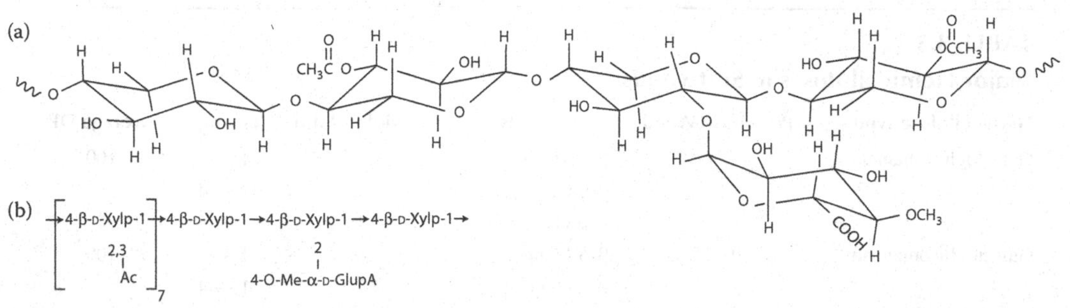

(a)

(b)

FIGURE 3.6 Partial molecular structure (a) and structure representation (b) of O-acetyl-4-O-methyl-lucuronoxylan.

TABLE 3.2
Major Hemicelluloses in Hardwoods

Hemicellulose Type	Percent in Wood	Units	Molar Ratio	Linkage	DP
Glucuronoxylan	15–30	β-D-Xylp	10	1 → 4	200
		4-O-Me-α-D-GlupA	1	1 → 2	
		Acetyl	7		
Glucomannan	2–5	β-D-Manp	1–2	1 → 4	200
		β-D-Glup	1	1 → 4	

Hardwoods also contain 2–5% of a glucomannan composed of β-D-glucopyranose and β-D-mannopyranose units linked (1 → 4). The glucose:mannose ratio varies between 1:2 and 1:1 depending on the wood species. Table 3.2 shows the major hemicelluloses found in hardwoods.

3.1.3.2 Softwood Hemicelluloses

Table 3.3 shows the major hemicelluloses from softwoods. One of the main hemicelluloses from softwoods contains a backbone polymer of D-galactose, D-glucose, and D-mannose (Sjöström 1981). The galactoglucomannan is the principal hemicellulose (ca. 20%) with a linear or possibly slightly branched chain with β-(1 → 4) linkages (Figure 3.7). Glucose and mannose make up the backbone polymer with branches containing galactose. There are two fractions of these polymers differing by their galactose content. The low galactose fraction has a ratio of galactose:glucose:mannose of about 0.1:1:4 while the high galactose fraction has a ratio of 1:1:3. The D-galactopyranose units are linked as a single unit side chain by α-(1 → 6) bonds. The 2- and 3-positions of the backbone polymer have acetyl groups substituted on them an average of 3–4 hexose units.

Another major hemicellulose polymer in softwoods (5–10%) is an arabinoglucuronoxylan consisting of a backbone of β-(1 → 4) xylopyranose units with a (1 → 2) branches of D-glucopyranosyl-uronic acid on the average of every 2–10 xylose units and α-(1 → 3) branches of L-arabinofuranose on the average of every 1.3 xylose units (Figure 3.8).

Another hemicellulose that is found mainly in the heartwood of larches is an arabinogalactan. Its backbone is a β-(1 → 3)-linked D-galactopyranose polymer with almost every unit having a branch attached to carbon 6 of β-D-galactopyranose residues. In some cases this side chain is β-L-arabinofuranose linked (1 → 3) or β-D-arabinopyranose linked (1 → 6).

TABLE 3.3

Major Hemicelluloses in Softwoods

Hemicellulose Type	Percent in Wood	Units	Molar Ratio	Linkage	Average DP
Galactoglucomannan	5–8	β-D-Manp	3	1 → 4	100
		β-D-Glup	1	1 → 4	
		α-D-Galp	1	1 → 6	
Galactoglucomannan	10–15	β-D-Manp	4	1 → 4	100
		β-D-Glup	1	1 → 4	
		α-D-Galp	0.1	1 → 6	
		Acetyl	1		
Arabinoglucuronoxylan	7–10	β-D-Xylp	10	1 → 4	100
		4-O-Me-α-D-GlupA	2	1 → 2	
		α-L-Araf	1.3	1 → 2	
Arabinogalactan	5–35	β-D-Galp	6	1 → 4	200
(Larch wood)				1 → 6	
		α-L-Araf	2–3	1 → 6	
		β-D-Arap	1–3	1 → 3	
		β-D-GlupA	Trace	1 → 6	

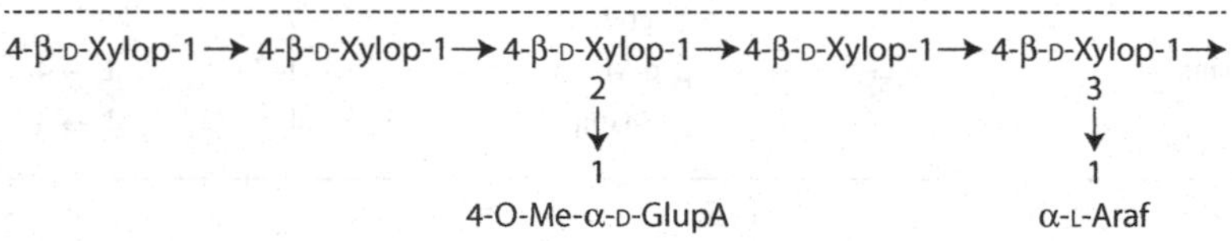

FIGURE 3.7   Partial structure of a softwood arabino 4-O-methylglucuronoxylan.

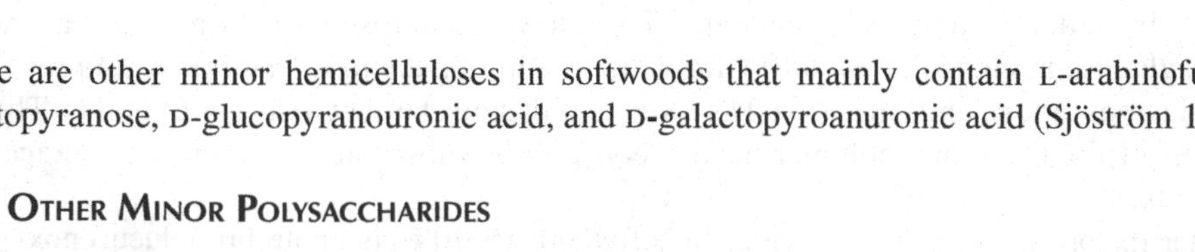

FIGURE 3.8   Partial structure of a softwood O-acetyl-galacto-glucomannan.

There are other minor hemicelluloses in softwoods that mainly contain L-arabinofuranose, D-galactopyranose, D-glucopyranouronic acid, and D-galactopyroanuronic acid (Sjöström 1981).

3.1.4   OTHER MINOR POLYSACCHARIDES

Both softwoods and hardwoods contain small amounts of pectins, starch, and proteins. Pectin is a polysaccharide polymer made up of repeating units of D-galacturonic acid linked α-(1 → 4). Pectin is found in the membranes in the boarded pits between wood cells and in the middle lamella. Degradation of this membrane by microorganisms increases permeability of wood to water-based treatment chemicals such as fire retardants and wood preservatives. Pectins are found in high concentration in the parenchyma cell walls in the inner bark where they may act as a binder. L-Arabinofuranose and D-galactopyranose are often found as a minor part of the pectic substance. Pectin is also found as the methyl ester.

Starch is the principal reserve polysaccharide in plants. Small amount of starch can also be found in the wood cell wall. Starch normally occurs as granules and is composed of D-glucopyranose units linked α-(1 → 4) (amylose) or α-(1 → 4) with branches about every 25 glucopyranosyl units at α-(1 → 6) (amylopectin). Amylose occurs as a helix structure in the solid state due to the α-configuration in the polymer. Amylopectin is highly branched.

3.2 LIGNIN

Lignins are amorphous, highly complex, mainly aromatic, polymers of phenylpropane units (Figure 3.9) that are considered to be an encrusting substance. The three-dimensional polymer is made up of C–O–C and C–C linkages. The precursors of lignin biosynthesis are p-coumaryl alcohol (Figure 3.9, Structure 1), coniferyl alcohol (Figure 3.9, Structure 2), and sinapyl alcohol (Figure 3.9, Structure 3). Structure 1 is a minor precursor of both softwood and hardwood lignins, Structure 2 is the predominate precursor of softwood lignin, and Structures 2 and 3 are both precursors of hardwood lignin (Adler 1977).

Softwood lignin has a methoxyl content of 15–16% while hardwood lignin has a methoxyl content of 21%. Lignin does not have a single repeating unit like cellulose of the hemicelluloses but consists of a complex arrangement of substituted phenolic units.

Lignins can be classified in several ways but they are usually divided according to their structural elements (Sjöström 1981). All wood lignins consist mainly of three basic building blocks of guaiacyl, syringyl, and p-hydroxyphenyl moieties, although other aromatic type units also exist in many different types of woods. There is a wide variation of structures within different wood species. The lignin content of hardwoods is usually in the range of 18–25%, where the lignin content of softwoods varies between 25% and 35%. The phenylpropane can be substituted at the α, β, or γ positions into various combinations linked together both by ether and carbon to carbon linkages (Sakakibara 1991).

Lignins from softwoods are mainly a polymerization product of coniferyl alcohol and are called "guaiacyl lignin." Hardwood lignins are mainly "syringyl–guaiacyl lignin" as they are a copolymer of coniferyl and sinapyl alcohols. The ratio of these two varies in different lignins from a ratio of 4:1 to 1:2 (Sarkanen and Ludwig 1971). A proposed structure for a hardwood lignin (*Fagus sylvatica* L.) is shown in Figure 3.10 (Adler 1977).

Lignins found in woods contain significant amounts of constituents other than guaiacyl- and syringylpropane units (Sarkanen and Ludwig 1971). Lignin is distributed throughout the secondary cell wall with the highest concentration in the middle lamella. Because of the difference in the volume of middle lamella to secondary cell wall, about 70% of the lignin is located in the cell wall.

Lignin can be isolated from wood in several ways. The so-called Klason lignin is obtained after hydrolyzing the polysaccharides with 72% sulfuric acid. It is highly condensed and does not truly

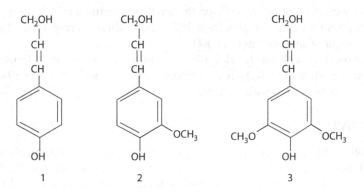

FIGURE 3.9 Chemical structures of lignin precursors. 1 = p-coumaryl alcohol, 2 = coniferyl alcohol, and 3 = sinapyl alcohol.

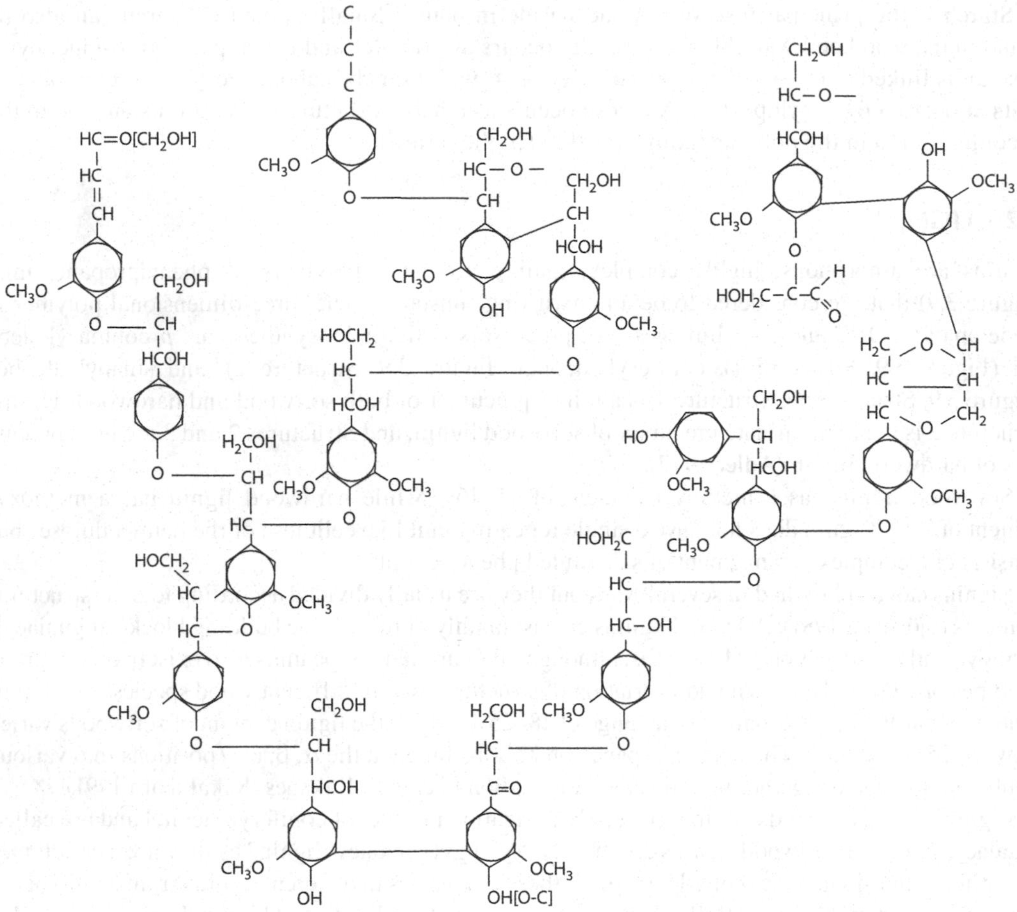

FIGURE 3.10   Partial structure of a softwood lignin.

represent the lignin in its native state in the wood. The polysaccharides can be removed using enzymes to give an "enzyme lignin" that is much closer to a native lignin than Klason lignin. "Milled wood lignin" or Björkman lignin can be isolated by using a vibratory ball mill on fine wood flour and then extracted with suitable organic solvents (Björkman 1956, 1957). Approximately, 30–50% of the native lignin is isolated using this procedure. This is a tedious procedure but does isolate a lignin closer to a native lignin.

The molecular weight of lignin depends on the method of extraction. Klason lignin, since it is highly condensed, has molecular weights from 260 to 50 million (Goring 1962). Björkman lignin has a molecular weight of approximately 11,000.

Lignins are associated with the hemicelluloses forming, in some cases, lignin–carbohydrate complexes that are resistant to hydrolysis even under pulping conditions (Obst 1982). There is no evidence that lignin is associated with cellulose.

3.3   EXTRACTIVES

As the name implies, extractives (also referred to as natural products) are chemicals in the wood that can be extracted using several solvents. In some cases, the extractives are classified by the solvent used to extract them, for example, water-soluble or toluene–ethanol or ether-soluble extractives. Hundreds of extractives have been identified and in some cases, their role in the tree is well understood. In other cases, it is not clear why they are present (Rowe 1989). Extractives, such as pine

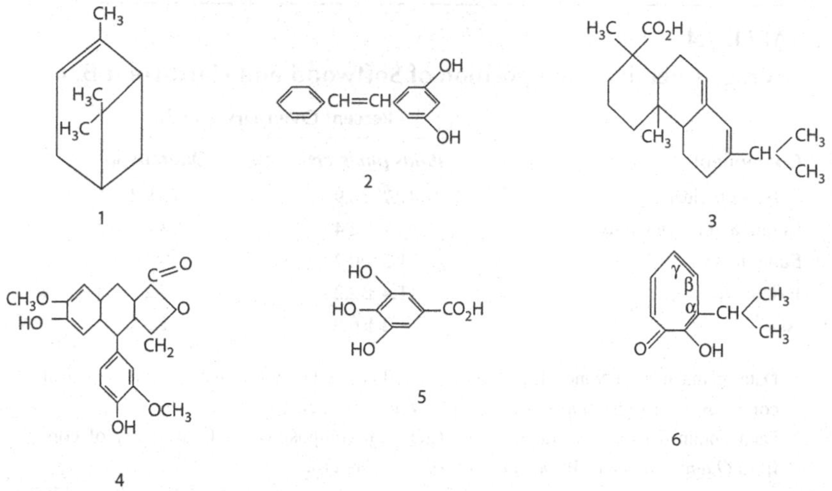

FIGURE 3.11 Chemical structures of some of the extractives in wood. 1 = abietic acid, 2 = α-pinene, 3 = pinosylvin, 4 = pinoresinol, 5 = gallic acid, and 6 = α-, β-, and γ-thujaplicin.

pitch and resins, have been used for centuries to waterproof wooden boats, used in torches and as a binder. They have also found application in medicine, cosmetics, and as a preservative (Hillis 1989). Some of the extractives in wood are precursors to other chemicals, some are in response to wounds, and some act as a defense mechanism.

The extractives are a group of cell wall chemicals mainly consisting of fats, fatty acids, fatty alcohols, phenols, terpenes, steroids, resin acids, rosin, waxes, and many other minor organic compounds. These chemicals exist as monomers, dimers, and polymers. In general, softwoods have higher extractives content than hardwoods and most of the extractives in both softwoods and hardwoods are located in the heartwood and some are responsible for the color, smell, and durability. The qualitative difference in extractive content from species to species is the basis of chemotaxonomy (taxonomy based on chemical constituents).

Resins and fats are made up of resin acid and fatty acids, respectively. Fatty acids are esters with alcohols such as glycerol and mainly occur in sapwood. Resin acids have a free carboxylic acid function are mainly found in heartwood (Kai 1991). Abietic acid (Figure 3.11-1) is a common type of resin acid.

The most common terpene in softwoods is α-pinene (Figure 3.11-2) and other similar chemical structures. One of the most important polyphenols is pinosylvin (Figure 3.11-3) which is very toxic and found in pine heartwood. Lignans are a combination of two phenylpropane units and are common in softwoods (Gottlieb and Yoshida 1989). Conidendrin (Figure 3.11-4) is found in spruce and hemlock. Tannins in wood can be classified into three classes: gallotannins, ellagitannins, and condensed tannins (Hemingway 1989, Porter 1989). Gallotannins are polymeric esters of gallic acid (Figure 3.11-5) and are usually associated with sugars (Haslam 1989). Tropolones are responsible for the durability of cedar wood. α-, β-, and γ-Thujaplicin (Figure 3.11-6) are examples of this class of extractives (Kollmann and Côté 1968).

3.4 BARK

Bark is a very complex tissue that is composed of two principal zones: the inner bark and the outer bark. The outer bark, which is sometimes referred to as rhytidome is also known as the periderm, and is made up of three layers: the phellem (cork cells), phellogen (cork cambium), and the phellogen (cork skin). The thickness of the periderm varies greatly between and within species and with the age of the bark. The inner bark, which is referred to as the phloem or bast is complex in structure

TABLE 3.4
Average Chemical Composition of Softwood and Hardwood Bark

| | Percent Oven-Dry Weight | |
Component	Pinus pinaster[a]	Quercus suber[b]
Polysaccharides	41.7 ± 0.9	19.9 ± 2.6
Lignin and polyphenols	43.7 ± 2.4	23.0 ± 0.5
Suberin	1.5 ± 0.2	39.4 ± 1.7
Extractives	11.4 ± 2.2	14.2 ± 1.1
Ash	1.2 ± 0.6	1.2 ± 0.2

[a] Data obtained from Nunes, E., Quilhó, T., and Pereira, H. 1996. Anatomy and chemical composition of *Pinus pinaster* bark. *IAWA Journal* 17(2): 141–149.

[b] Data obtained from Pereira, H. 1988. Chemical composition and variability of cork from *Quercus suber* L. *Wood Sci. Technol.* 22: 211–218.

and is composed of several types of cells including sieve tubes, fiber cells, albuminose cells, companion cells, parenchyma cells, ideoblasts, and lactifers. Not all cell types occur in every bark. The bark is divided from the wood or xylem by the vascular cambium layer (Sandved et al. 1992).

The chemical composition of bark is equally complex, and varies between and within species, and also between the inner and outer bark (Toman et al. 1976). Proximate chemical analysis of bark from different species indicates that the chemical constituents of bark can be classified into four major groups: polysaccharides (cellulose, hemicellulose, and pectic materials), lignin and polyphenols, hydroxy acid complexes (suberin), and extractives (fats, oils, phytosterols, resin acids, waxes, tannins, terpenes, phlobaphenes, and flavonoids). Table 3.4 illustrates the variability of the chemical composition of bark between softwood and hardwood species, *Pinus pinaster* and *Quercus suber*, respectively.

3.4.1 EXTRACTIVES

The extractives content of bark is quite high compared to wood, but values reported in the literature can be very different even for the same species. These apparent differences depend on the method of extraction. For example, McGinnis and Parikh (1975) reported 19.9% extractives for loblolly pine bark using, petroleum ether, benzene, ethanol, and cold and hot water. Labosky (1979) extracted loblolly pine bark with hexane, benzene, ethyl ether, ethanol, water, and 1% sodium hydroxide (NaOH) and reported 27.5% extractives.

The analysis methods developed for wood cannot be used for bark directly. There are many compounds in bark that are not found in wood which interfere with these analysis methods. For example, the presence of suberin in bark tends to limit access of delignification reagents to the lignin in the bark, and therefore may lead to a holocellulose that is not pure enough to permit fractionation of individual bark polysaccharides. Suberin, polyflavonoids, and other high molecular weight condensed tannins can also complicate analysis of bark lignin, resulting in false high values of lignin content in bark.

Owing to the interference of the extractives in polysaccharide and lignin analysis, procedures for elucidation of the chemical composition of bark begin with an extraction protocol that consists of sequential extraction solvents of increasing polarity. A common protocol begins with a diethyl ether extraction step that yields waxes, fatty acids, fats, resin acids, phytosterols, and terpenes. This is followed by an ethyl alcohol step that yields condensed tannins, flavonoids, and phenolics. The third step uses hot water, and yields condensed tannins, and water-soluble carbohydrates. To release phenolic acids, hemicelluloses, and suberin monomers from the residue from the third step, 1% aqueous NaOH is used (Holloway and Deas 1973, Kolattukudy 1984).

The extract fractions from the above-mentioned steps are then subjected to further workup to separate each into easy-to-analyze mixtures of compounds. For example, partitioning the diethyl ether fraction against aqueous sodium bicarbonate separates the fatty acids and resin acids from the neutral components, tannins, terpenes, and flavonoids. The neutral fraction is then saponified to give the alcohols and salts of fatty acids, dicarboxylic, hydroxy-fatty, and ferulic acids. Ethanol extraction followed by hot water extraction of the insoluble ether fraction yields soluble simple sugars and condensed tannins. Sodium hydroxide extraction of the insoluble residue gives soluble suberin monomers, phenolic acids, and hemicelluloses. Sulfuric acid treatment of the insoluble fraction yields lignin (Chang and Mitchell 1955, Hemingway 1981, Laks 1991).

3.4.1.1   Chemical Composition of Extractives

The waxes in bark are esters of high molecular weight long-chain monohydroxy alcohol fatty acids. A lot of research has been done on softwood waxes but very little on hardwood waxes. At one time, hardwood waxes were produced commercially for polishes, lubricants, additives to concrete, carbon papers and fertilizers, and in fruit coatings (Hemingway 1981).

Terpenes are a condensation of two or more 5-carbon isoprene (2-methy-1,3-butadiene) units in a linear or cyclic structure. They can also contain various functional groups. The most common of the monoterpenes are α- and β-pinenes found in firs and pines. Birch bark can contain up to 25% of the total dry weight (Seshadri and Vedantham 1971).

Flavonoids are a group of compounds based on a 15-carbon, hydroxylated tricyclic unit (Laks 1991). They are often found as glycosides. Many tree barks are rich in mono- and polyflavonoids (Hergert 1960, 1962). Their function seems to be as an antioxidant, pigment, and growth regulator (Laks 1991).

Hydrolysable and condensed tannins are also major extractives from bark. The hydrolysable tannins are esters of carboxylic acids and sugars that are easily hydrolyzed to give benzoic acid derivatives and sugars. Over 20 different hydrolysable tannins have been isolated from oaks (Nonaka et al. 1985).

The condensed tannins are a group of polymers based on a hydroxylated C-15 flavonoid monomer unit. Low DP tannins are soluble in polar solvents whereas the high DP tannins are soluble in dilute alkali solutions (Hemingway et al. 1983). It is difficult to isolate pure fractions of tannins and the structure can be altered by the extraction procedure.

Free sugars are also extracted from bark. Hot water extraction yields about 5% free sugar fraction mainly composed of glucose and fructose and this amount varies depending on the growing season. For example, the free-sugar content is low in early spring and increases during the growing season reaching a maximum in the fall (Laks 1991). Other minor free sugars found in bark include galactose, xylose, mannose, and sucrose. Hydrolysis of the hot water extract of bark yields more free sugars. The most abundant one being arabinose. These sugars are tied up as glycosides, or are tied up in the hemicelluloses. Other sugars released during hydrolysis are glucose, fructose, galactose, xylose, mannose, and rhamnose.

3.4.2   HEMICELLULOSES

The hemicellulose content of different barks varies from 9.3% for *Quercus robur* to 23.1% for *Fagus sylvatica* (Dietrichs et al. 1978). The main hemicellulose in conifer barks is a galactoglucomannan and arabino-4-O-methyl-glucuronoxylan in deciduous barks. In general, bark xylans and glucomannans are similar to ones found in wood. Other hemicelluloses that have been isolated from barks include 4-O-methy-glucuronoxylans, glucomannans, O-acetyl-galactoglucomannan, and O-acetyl-4-O-methyl-glucuronoxylan (Painter and Purves 1960, Jiang and Timell 1972, Dietrichs 1975). In the xylans, the xylose units are connected β-(1 $\rightarrow$ 4) and the glucuronic acid groups are attached to the xylan backbone α-(1 $\rightarrow$ 2). The ratio of xylose to GluU is 10 to 1 with a DP of between 171 and 234 (Mian and Timell 1960). Glucomannans from deciduous barks contain mannose and glucose units in a ratio of about 1:1 to 1.4:1 (Timell 1961b). The mannans from the barks of aspen and willow, galactose

TABLE 3.5
Sugars Present in Hydrolysates of Some Tree Barks

Species	Glu	Man	Gal	Xyl	Ara	Rha	UrA	Ac
Abies amabilis	37.4	8.0	1.6	3.2	3.2	—	5.6	0.8
Picea abies	36.6	6.5	1.3	4.8	1.8	0.3	—	—
Picea engelmannii	35.7	2.9	2.4	3.8	3.3	—	8.0	0.5
Pinus contoria								
Inner bark	40.9	2.5	4.3	3.7	10.6	—	9.9	0.2
Outer bark	26.8	2.5	4.2	3.4	5.5		7.7	0.8
Pinus sylvestris	30.2	5.4	2.4	5.8	2.1	0.3	—	—
Pinus taeda								
Inner bark	21.3	2.5	3.1	2.1	5.6	0.3	4.6	—
Outer bark	15.8	2.6	2.5	3.8	1.8	0.1	2.1	—
Betula papyrifera								
Inner bark	28.0	0.2	1.0	21.0	2.7	—	2.2	—
Fagus sylvatica	29.7	0.2	3.1	20.1	3.1	1.2	—	—
Quercus robur	32.3	0.5	1.3	16.4	2.0	0.5	—	—

Source: Adapted from Fengel, D. and Wegener, G. 1984. *Wood: Chemistry, Ultrastructure and Reactions.* W. de. Gruyter, Berlin.

units were found as side chains. The ratio of manose:glucose:galactose was 1.3:1:0.5 with an average DP of 30–50 (Timell 1961b).

Arabinans have been reported in the barks of aspen, spruce, and pine (Painter and Purves 1960). The backbone is α-(1 → 5) arabinofuranose units and in the case of pine, the average DP is 95 (Timell 1961b). A group of galacturonic acid polymers has been isolated from birch. One is a galacturonic acid backbone α-(1 → 4) with arabinose side chains in a ratio of galacturonic acid to arabinose of 9:1 and another one consisting of galacturonic acid, arabinose, and galactose in a ratio of 7:3:1. Small amounts of glucose, xylose, and rhamnose were also found in these polymers (Mian and Timell 1960, Timell 1961b).

A pectic substance has been isolated from barks, which contains either galactose alone or galactose and arabinose units (Toman et al. 1976). The pure galactan is water-soluble and consists of 33 β-(1 → 4)-linked galactose units with side chains at C6 of the backbone. A highly branched arabinogalactan was found in the bark of spruce bark with a ratio of galactose to arabinose of 10:1 (Painter and Purves 1960).

In almost all cases, the hemicelluloses found in bark are similar to those found in wood with some variations in the composition.

Table 3.5 shows the sugars present after hydrolysis of the polysaccharides in bark.

3.4.3 CELLULOSE

The cellulose content of barks ranges from 16% to 41% depending on the method of extraction. In unextracted bark, the cellulose content was between 20.2% for pine and 32.6% for oak (Dietrichs et al. 1978). The high extractive content, especially of suberin, requires harsh conditions to isolate the cellulose so the cellulose content is usually low and the cellulose is degraded during the isolation process. The outer bark usually contains less cellulose than the inner bark (Harun and Labosky 1985).

Timell (1961a,b) and Mian and Timell (1960), found a number average DP for bark cellulose of 125 (*Betula papyrifera*) to 700 (*Pinus contorta*) and a weight average of 4000 (*Abies amabilis, Populus grandidentata*) to 6900 (*Pinus contorta*). Bark cellulose has the same type of crystalline lattice (cellulose I) as normal wood but the degree of crystallinity is less.

3.4.4 LIGNIN

As with other analyses involving bark components, literature values for lignin content can vary depending on the method of extraction (Kurth and Smith 1954, Higuchi et al. 1967). Bark contains high contents of condensed and hydrolysable tannins and sulfuric acid insoluble suberin that can give false high values of lignin content. For example, the Klason lignin from *Pinus taeda* bark is 46.0% when including both lignin and condensed tannins but only 20.4% when the bark is first extracted with alkali (McGinnis and Parikh 1975). Other researchers have found lignin contents from 38% to 58% (Labosky 1979). The elemental composition and functional group content of bark lignins are similar to the lignin from the wood of the same species (Sarkanen and Hergert 1971, Hemingway 1981). There is less lignin in the inner bark as compared to the outer bark.

There is a lower ratio of OCH_3 groups in aspen bark than in the wood and a higher ratio of phenolic OH groups to OCH_3 (Clermont 1970). There are more guaiacyl units in deciduous bark and more *p*-hydroxyphenyl units in coniferous bark as compared to the wood of the same species (Andersson et al. 1973). While there are some differences in the ratio of components, no structural difference have been found between most bark lignins and the corresponding wood.

3.4.5 INORGANICS AND pH

Bark is generally higher in inorganics than normal wood. The inorganic (ash) content can be as high as 13% and, in general, the inner bark contains more inorganics as compared to the outer bark (Young 1971, Choong et al. 1976, Hattula and Johanson 1978, Harder and Einspahr 1980). For example, the outer bark of willow contains 11.5% ash, the inner bark 13.1% compared to 0.9% in sapwood; sweetgum outer 10.4%, inner 12.8%, sapwood 0.5%; red oak outer 8.9%, inner 11.1%, sapwood 0.9%; and, ash, outer 12.3%, inner 12.1%, sapwood 0.9%. The major inorganic elements in bark are Na, K, Ca, Mg, Mn, Zn, and P (Choong et al. 1976). There is more Na, K, Mg, Mn, Zn, and P in sapwood than in bark and more Ca in bark than in sapwood.

In general, the pH of bark is lower than normal wood due to the higher inorganic content of bark compared to normal wood. For example, Martin and Gray (1971) reported the pH values of southern pines ranging from about 3.1 to 3.8 with an average of 3.4 to 3.5 compared to a pH of 4.4 to 4.6 for sapwood. The outer bark has a lower pH than the inner bark presumably due to a higher content of Ca in the outer bark (Volz 1971). The pH of bark decreases slightly with the age of the tree.

3.5 INORGANICS

The inorganic content of a wood is usually referred to its ash content which is an approximate measure of the mineral salts and other inorganic matter in the fiber after combustion at a temperature of $575 \pm 25°C$. The inorganic content can be quite high in woods containing large amounts of silica, however, in most cases, the inorganic content is less than 0.5% (Browning 1967). This small amount of inorganic material contains a wide variety of elements (Ellis 1965, Young and Guinn 1966). Ca, Mg, and K make up 80% of the ash in wood. These elements probably existed as oxalates, carbonates, and sulfates in the wood or bound to carboxyl groups in pectic materials (Hon and Shiraishi 1991). Other elements present are Na, Si, B, Mn, Fe, Mo, Cu, Zn, Ag, Al, Ba, Co, Cr, Ni, Pb, Rb, Sr, Ti, Au, Ga, In, La, Li, Sn, V, and Zr (Ellis 1965). Some of these are essential for wood growth. Inorganic ions are absorbed into the tree through the roots and transported throughout the tree. Young and Guinn (1966) determined the distribution of 12 inorganic elements in various part of a tree (roots, bark, wood, and leaves) and concluded that both the total inorganic content and concentration of each element varied widely within and between species. The inorganic content varies depending on the environmental conditions in which the tree lives. See Tables 3.11 through 3.13 for a partial list of the inorganic content of some woods.

Saka and Goring (1983) studied the distribution of inorganics from the pith to the outer ring of black spruce (*Picea mariana* Mill) using energy-dispersive x-ray analysis (EDXA). They found 15 different elements including Na, Mg, Al, S, K, Ca, Fe, Ni, Cu, Zn, and Pb. They also found that the inorganic content was higher in early wood as compared to late wood.

The pH of wood varies from 4.2 (*Pinus sylvestris*) to 5.3 (*Fagus sylvatica*) with an average of approximately 4.7.

3.6 DISTRIBUTION IN CELL WALL

The content of cell wall components depends on the tree species and where in the tree the sample is taken. Softwoods are different from hardwoods, heartwood from sapwood and latewood from springwood. Table 3.6 shows the cell wall polysaccharides in earlywood compared to latewood (Saka 1991). Latewood contains more glucomannans as compared to earlywood but earlywood contains more glucuronoarabinoxylans. Heartwood contains more extractives than sapwood and as sapwood is transformed into heartwood, aspiration of the bordered pits takes place in softwoods and encrustation of pit membranes with the formation of tyloses occurs in hardwoods. Springwood contains more lignin than summerwood.

Figure 3.12 shows the distribution of components across the cell wall of scotch pine. The middle lamella and primary wall is mainly composed of lignin (84%) with lesser amounts of hemicelluloses (13.3%) and even less cellulose (0.7%). The S_1 layer is composed of 51.7% lignin, 30.0% cellulose, and 18.3% hemicelluloses. The S_2 layer is composed of 15.1% lignin, 54.3% cellulose, and 30.6% hemicelluloses. The S_3 layer has little or no lignin, 13% cellulose, and 87% hemicelluloses. The content of xylan is lowest in the S_2 layer and higher in the S_1 and S_3 layers. The concentration of galactoglucomannan is higher in the S_2 than the S_1 or S_3 layers. On a percentage basis, the middle lamella and primary wall contain the highest concentration of lignin but there is more lignin in the S_2 because it is a much thicker layer as compared to the middle lamella and primary wall. The lignin in the S_2 layer is evenly distributed throughout the layer.

The angle of the cellulose microfibrils in the various cell wall layers, in relation to the fiber axis is known as the fibril angle. It is one of the most important structural parameters determining mechanical properties of wood. For normal wood, the microfibril angle of the cellulose in the S_2 layer is 14–19°. It is because this angle is so low in the thick S_2 layer that wood does not swell or shrink (0.1–0.3%) in the longitudinal direction.

Figure 3.13 shows more detail of the cell wall. It shows possible lignin–carbohydrate bonding and very little, if any, contact between lignin and cellulose.

TABLE 3.6
Cell Wall Polysaccharides in Earlywood and Latewood in Pine

Cell Wall Component	Earlywood	Latewood
	%	
Cellulose	56.7	56.2
Galactan	3.4	3.1
Glucomannan	20.3	24.8
Arabinan	1.0	1.8
Glucuronoarabinoxylan	18.6	14.1

Source: Adapted from Saka, S. 1991. *Wood and Cellulosic Chemistry*, Chapter 2, pp. 59–88, New York, NY: Marcel Dekker, Inc.

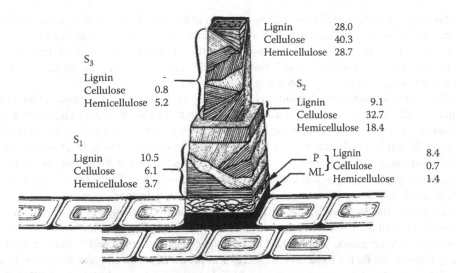

Lignin   28.0
Cellulose   40.3
Hemicellulose   28.7

S_3
Lignin   -
Cellulose   0.8
Hemicellulose   5.2

S_2
Lignin   9.1
Cellulose   32.7
Hemicellulose   18.4

S_1
Lignin   10.5
Cellulose   6.1
Hemicellulose   3.7

P ⎫
ML ⎭
Lignin   8.4
Cellulose   0.7
Hemicellulose   1.4

FIGURE 3.12   Chemical composition of the cell wall of scots pine.

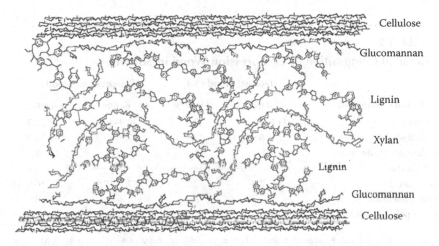

Cellulose

Glucomannan

Lignin

Xylan

Lignin

Glucomannan

Cellulose

FIGURE 3.13   Schematic of the wood cell wall.

A further discussion of the distribution of the hemicelluloses in the cell wall can be found in Chapter 15. Strength properties of wood are related to the distribution of hemicelluloses in the cell wall (Chapter 11).

3.7   JUVENILE WOOD AND REACTION WOOD

Juvenile wood is the wood that develops in the early stages of tree growth. It physical properties are described in the chapter. Juvenile wood cells are shorter, have smaller cell diameter, larger microfibril angle (up to 55°) and have a high content of compression wood as compared to mature wood. Juvenile wood has a lower density and strength than mature wood. Juvenile wood has less cellulose, more hemicelluloses, and lignin compared to mature wood. There is a gradual increase in cellulose content as the cells mature and a gradual decrease in hemicellulose content. The lignin content decreases more rapidly as the cell mature.

Normal wood growth is erect and vertical. When a tree is forced out of this pattern either by wind or gravitational forces, abnormal woody tissue is formed in different parts of the tree to compensate for the abnormal growing conditions. The wood cells that are formed when softwoods and hardwoods are out of vertical are called reaction wood since these cells are reacting to the stressful conditions. In softwoods, irregular cells develop on the underside of a stem or branch and are referred to as compression wood. In hardwoods, irregular cells develop on the upper side of a stem or branch and are referred to as tension wood.

Table 3.7 shows the chemical composition of softwood compression wood (Panshin and de Zeeuw 1980, Timell 1982). Compression wood has higher lignin content and lower cellulose content as compared to normal wood. The cellulose in the S_2 layer has a lower degree of crystallization than normal wood and the lignin is largely concentrated in the S_2 layer as compared to normal wood. Forty percent of the lignin is in the outer zone of the S_2 layer and an additional 40% is uniformly distributed over the remaining part of the S_2 layer (Panshin and de Zeeuw 1980). There are more galactoglucomannans in normal wood and more $1 \rightarrow 3$ linked glucans and galactans in compression wood. The microfibril angle in the modified S_2 layer in compression wood is quite high (44–47°) and have more rounded tracheids that are 10–40% shorter than normal tracheids. Compression wood is weaker than normal wood and lower elastic properties. The reduced cellulose content and high microfibril angle is probably responsible for the reduction in mechanical properties (Panshin and de Zeeuw 1980).

Table 3.8 shows the chemical composition of hardwood tension wood (Schwerin 1958). Tension wood has lower lignin content and higher cellulose content as compared to normal wood. There is

TABLE 3.7
Chemical Composition of Compression Wood in Softwoods

Cell Wall Component	Normal Wood (%)		Compression Wood (%)	
	Range	Average	Range	Average
Lignin	24.2–33.3	28.8	30.9–40.9	37.7
Cellulose	37.7–60.6	44.6	27.3–53.7	34.9
Galactoglucomannan	—	18	—	9
1,3-Glucan	—	Trace	—	2
Galactan	1.0–3.8	2.2	7.1–12.9	10.0
Glucuronoarabinoxylan	—	8	—	8
Other polysaccharides	—	2	—	2

Source: Adapted from Panshin, A.J. and de Zeeuw, D. 1980. *Textbook of Wood Technology*, McGraw-Hill, New York.

TABLE 3.8
Chemical Composition of Tension Wood in Hardwoods

Cell Wall Component	Normal Wood (%)	Compression Wood (%)
Lignin	29	14
Cellulose	44	57
Pentosans	15	11
Acetyl	3	2
Galactosans	2	7

Source: Adapted from Schwerin, G. 1958. *Holzforschung* 12: 43–48.

lower content of pentosans (xylans) and acetyls than in normal wood and more galactosans in tension wood. There is no S_3 layer in tension wood but rather what is known as a G layer or gelatinous layer. This layer is approximately 98% cellulose. The cellulose in the G layer is highly crystalline with a microfibril angle of only 5% and contains very little hemicelluloses or lignin. The G layer is as thick or thicker than the S_2 layer in normal wood and contains about the same quantity of inorganics. Tension wood has lower mechanical properties as compared to normal wood (Panshin and De Zeeuw 1980). For example, compression parallel and perpendicular to the grain, modulus of elasticity in bending, modulus of rupture in static bending, and longitudinal shear are all reduced in tension wood as compared to normal wood.

3.8 ANALYTICAL PROCEDURES

Chemical composition varies from species to species and within different parts of the same wood species. Chemical composition also varies within woods from different geographic locations, ages, climates, and soil conditions.

There are hundreds of reports on the chemical composition of wood material. In reviewing this vast amount of data, it becomes apparent that the analytical procedures used, in many cases, are different from lab to lab and a complete description of what procedure was used in the analysis is unclear. For example, many descriptions do not describe if the samples were preextracted with some solvent before analysis. Others do not follow a published procedure so comparison of data is not possible. The following section is composed of standard procedures used in many laboratories to determine the chemical components of the wood cell wall. Tables 3.9 through 3.14 provide summaries of various types of chemical composition of hardwoods and softwoods in the United States

TABLE 3.9
Methoxyl Content of Some Common Hardwoods and Softwoods

Type of Wood	Methoxy Content (%)
Hardwoods	
Balsa	5.68
Basswood	6.00
Yellow birch	6.07
Shellbark hickory	5.63
Sugar maple	7.25
Mesquite	5.55
Tanoak	5.74
Softwoods	
Incense cedar	6.24
Alaska cedar	5.25
Douglas fir	4.95
Western larch	5.03
Longleaf pine	5.05
Western white pine	4.56
Redwood	5.21
White spruce	5.30

Source: Adapted from Moore, W. and Johnson, D. 1967. *Procedures for the Chemical Analysis of Wood and Wood Products.* USDA, Forest Service, Forest Products Laboratory.

TABLE 3.10

Acetyl Content of Some Common Hardwoods and Softwoods

Type of Wood	Acetyl Content (%)
Hardwoods	
Aspen	3.4
Balsa	4.2
Basswood	4.2
Beech	3.9
Yellow birch	3.3
White birch	3.1
Paper birch	4.4
American elm	3.9
Shellbark hickory	1.8
Red maple	3.8
Sugar maple	3.2
Mesquite	1.5
Overcup oak	2.8
Southern red oak	3.3
Tanoak	3.8
Softwoods	
Eastern white-cedar	1.1
Incense-cedar	0.7
Western red-cedar	0.5
Alaska-cedar	1.1
Douglas-fir	0.7
Balsam fir	1.5
Eastern hemlock	1.7
Western hemlock	1.2
Western larch	0.5
Jack pine	1.2
Loblolly pine	1.1
Longleaf pine	0.6
Western white pine	0.7
Redwood	0.8
White spruce	1.3
Tamarack	1.5

Source: Adapted from Moore, W. and Johnson, D. 1967. *Procedures for the Chemical Analysis of Wood and Wood Products.* USDA, Forest Service, Forest Products Laboratory.

and the world. These data have been collected from the analytical laboratories of the USDA, Forest Service, Forest Products Laboratory from 1927 to 1968.

3.8.1 SAMPLING PROCEDURES

In reporting the chemical content of a wood, it is very important to report as much information about the samples as possible. Since the chemical content of a given species may vary depending upon the growing conditions, harvesting times of the year, and so on, it is critical to report these conditions along with the chemical analysis. It is also important to report the exact analytical

TABLE 3.11
Chemical Composition of North American Hardwoods and Softwoods

Botanical Name	Common Name	Holo cellulose	α-cellulose	Pentosans	Klason Lignin	1% NaOH	Solubility Hot Water	EtOH/Benzene	Ether	Ash
				Hardwoods						
Acer macrophyllum	Bigleaf maple	—	46.0	22.0	25.0	18.0	2.0	3.0	0.7	0.5
Acer negundo	Boxelder	—	45.0	20.0	30.0	10.0	—	—	0.4	—
Acer rubrum	Red maple	77.0	47.0	18.0	21.0	16.0	3.0	2.0	0.7	0.4
Acer saccharinum	Silver maple	—	42.0	19.0	21.0	21.0	4.0	3.0	0.6	—
Acer saccharum	Sugar maple	—	45.0	17.0	22.0	15.0	3.0	3.0	0.5	0.2
Alnus rubra	Red alder	74.0	44.0	20.0	24.0	16.0	3.0	2.0	0.5	0.3
Arbutus menziesii	Pacific madrone	—	44.0	23.0	21.0	23.0	5.0	7.0	0.4	0.7
Betula alleghaniensis	Yellow birch	73.0	47.0	23.0	21.0	16.0	2.0	2.0	1.2	0.7
Betula nigra	River birch	—	41.0	23.0	21.0	21.0	4.0	2.0	0.5	—
Betula papyrifera	Paper birch	78.0	45.0	23.0	18.0	17.0	2.0	3.0	1.4	0.3
Carya cordiformis	Bitternut hickory	—	44.0	19.0	25.0	16.0	5.0	4.0	0.5	—
Carya glabra	Pignut hickory	71.0	49.0	17.0	24.0	17.0	5.0	4.0	0.4	0.8
Carya ovata	Shagbark hickory	71.0	48.0	18.0	21.0	18.0	5.0	3.0	0.4	0.6
Carya pallida	Sand hickory	69.0	50.0	17.0	23.0	18.0	7.0	4.0	0.4	1.0
Carya tomentosa	Mockernut hickory	71.0	48.0	18.0	21.0	17.0	5.0	4.0	0.4	0.6
Celtis laevigata	Sugarberry	—	40.0	22.0	21.0	23.0	6.0	3.0	0.3	—
Eucalyptus gigantea	—	72.0	49.0	14.0	22.0	16.0	7.0	4.0	0.3	0.2
Fagus grandifolia	American beech	77.0	49.0	20.0	22.0	14.0	2.0	2.0	0.8	0.4
Fraxinus americana	White ash	—	41.0	15.0	26.0	16.0	7.0	5.0	0.5	—
Fraxinus pennsylvanica	Green ash	—	40.0	18.0	26.0	19.0	7.0	5.0	0.4	—
Gleditsia triacanthos	Honey locust	—	52.0	22.0	21.0	19.0	—	—	0.4	—
Laguncularia racemosa	White mangrove	—	40.0	19.0	23.0	29.0	15.0	6.0	2.1	—
Liquidambar styraciflua	Sweetgum	—	46.0	20.0	21.0	15.0	3.0	2.0	0.7	0.3
Liriodendron tulipifera	Yellow-poplar	—	45.0	19.0	20.0	17.0	2.0	1.0	0.2	1.0

continued

TABLE 3.11 (continued)

Chemical Composition of North American Hardwoods and Softwoods

Botanical Name	Common Name	Holo cellulose	α-cellulose	Pento sans	Klason Lignin	Solubility 1% NaOH	Hot Water	EtOH/Benzene	Ether	Ash
Lithocarpus densiflorus	Tanoak	71.0	46.0	20.0	19.0	20.0	5.0	3.0	0.4	0.7
Melaleuca quinquenervia	Cajeput	—	43.0	19.0	27.0	21.0	4.0	2.0	0.5	—
Nyssa aquatica	Water tupelo	—	45.0	16.0	24.0	16.0	4.0	3.0	0.6	0.6
Nyssa sylvatica	Black tupelo	72.0	45.0	17.0	27.0	15.0	3.0	2.0	0.4	0.5
Populus alba	White poplar	—	52.0	23.0	16.0	20.0	4.0	5.0	0.9	—
Populus deltoides	Eastern cottonwood	—	47.0	18.0	23.0	15.0	2.0	2.0	0.8	0.4
Populus tremuloides	Quaking aspen	78.0	49.0	19.0	19.0	18.0	3.0	3.0	1.2	0.4
Populus trichocarpa	Black cottonwood	—	49.0	19.0	21.0	18.0	3.0	3.0	0.7	0.5
Prunus serotina	Black cherry	85.0	45.0	20.0	21.0	18.0	4.0	5.0	0.9	0.1
Quercus alba	White oak	67.0	47.0	20.0	27.0	19.0	6.0	3.0	0.5	0.4
Quercus coccinea	Scarlet oak	63.0	46.0	18.0	28.0	20.0	6.0	3.0	0.4	—
Quercus douglasii	Blue oak	59.0	40.0	22.0	27.0	23.0	11.0	5.0	1.4	1.4
Quercus falcata	Southern red oak	69.0	42.0	20.0	25.0	17.0	6.0	4.0	0.3	0.4
Quercus kelloggii	California black oak	60.0	37.0	23.0	26.0	26.0	10.0	5.0	1.5	0.4
Quercus lobata	Valley oak	70.0	43.0	19.0	19.0	23.0	5.0	7.0	1.0	0.9
Quercus lyrata	Overcup oak	—	40.0	18.0	28.0	24.0	9.0	5.0	1.2	0.3
Quercus marilandica	Blackjack oak	—	44.0	20.0	26.0	15.0	5.0	4.0	0.6	—
Quercus prinus	Chestnut oak	76.0	47.0	19.0	24.0	21.0	7.0	5.0	0.6	0.4
Quercus rubra	Northern red oak	69.0	46.0	22.0	24.0	22.0	6.0	5.0	1.2	0.4
Quercus stellata	Post oak	—	41.0	18.0	24.0	21.0	8.0	4.0	0.5	1.2
Quercus velutina	Black oak	71.0	48.0	20.0	24.0	18.0	6.0	5.0	0.2	0.2
Salix nigra	Black willow	—	46.0	19.0	21.0	19.0	4.0	2.0	0.6	—
Tilia heterophylla	Basswood	77.0	48.0	17.0	20.0	20.0	2.0	4.0	2.1	0.7
Ulmus americana	American elm	73.0	50.0	17.0	22.0	16.0	3.0	2.0	0.5	0.4
Ulmus crassifolia	Cedar elm	—	50.0	19.0	27.0	14.0	—	—	0.3	—

Softwoods

Abies amabilis	Pacific silver fir	—	44.0	10.0	29.0	11.0	3.0	3.0	0.7	0.4
Abies balsamea	Balsam fir	—	42.0	11.0	29.0	11.0	4.0	3.0	1.0	0.4
Abies concolor	White fir	66.0	49.0	6.0	28.0	13.0	5.0	2.0	0.3	0.4
Abies lasiocarpa	Subalpine fir	67.0	46.0	9.0	29.0	12.0	3.0	3.0	0.6	0.5
Abies procera	Noble fir	61.0	43.0	9.0	29.0	10.0	2.0	3.0	0.6	0.4
Chamaecyparis thyoides	Atlantic white cedar	—	41.0	9.0	33.0	16.0	3.0	6.0	2.4	—
Juniperus deppeana	Alligator juniper	57.0	40.0	5.0	34.0	16.0	3.0	7.0	2.4	0.3
Larix laricina	Tamarack	64.0	44.0	8.0	26.0	14.0	7.0	3.0	0.9	0.3
Larix occidentalis	Western larch	65.0	48.0	9.0	27.0	16.0	6.0	2.0	0.8	0.4
Libocedrus decurrens	Incense cedar	56.0	37.0	12.0	34.0	9.0	3.0	3.0	0.8	0.3
Picea engelmannii	Engelmann spruce	69.0	45.0	10.0	28.0	11.0	2.0	2.0	1.1	0.2
Picea glauca	White spruce	—	43.0	13.0	29.0	12.0	3.0	2.0	1.1	0.3
Picea mariana	Black spruce	—	43.0	12.0	27.0	11.0	3.0	2.0	1.0	0.3
Picea sitchensis	Sitka spruce	—	45.0	7.0	27.0	12.0	4.0	4.0	0.7	—
Pinus attenuata	Knobcone pine	—	47.0	14.0	27.0	11.0	3.0	1.0	—	0.2
Pinus banksiana	Jack pine	66.0	43.0	13.0	27.0	13.0	3.0	5.0	3.0	0.3
Pinus clausa	Sand pine	—	44.0	11.0	27.0	12.0	2.0	3.0	1.0	0.4
Pinus contorta	Lodgepole pine	68.0	45.0	10.0	26.0	13.0	4.0	3.0	1.6	0.3
Pinus echinata	Shortleaf pine	69.0	45.0	12.0	28.0	12.0	2.0	4.0	2.9	0.4
Pinus elliottii	Slash pine	64.0	46.0	11.0	27.0	13.0	3.0	4.0	3.3	0.2
Pinus monticola	Western white pine	69.0	43.0	9.0	25.0	13.0	4.0	4.0	2.3	0.2
Pinus palustris	Longleaf pine	—	44.0	12.0	30.0	12.0	3.0	4.0	1.4	—
Pinus ponderosa	Ponderosa pine	68.0	41.0	9.0	26.0	16.0	4.0	5.0	5.5	0.5
Pinus resinosa	Red pine	71.0	47.0	10.0	26.0	13.0	4.0	4.0	2.5	—
Pinus sabiniana	Digger pine	—	46.0	11.0	27.0	12.0	3.0	1.0	—	0.2
Pinus strobes	Eastern white pine	68.0	45.0	8.0	27.0	15.0	4.0	6.0	3.2	0.2
Pinus sylvestris.	Scotch or scots pine	—	47.0	11.0	28.0	—	1.0	—	1.6	0.2
Pinus taeda	Loblolly pine	68.0	45.0	12.0	27.0	11.0	2.0	3.0	2.0	—
Pseudotsuga menziesii	Douglas fir	66.0	45.0	8.0	27.0	13.0	4.0	4.0	1.3	0.2

continued

TABLE 3.11 (continued)
Chemical Composition of North American Hardwoods and Softwoods

Botanical Name	Common Name	Holo cellulose	α-cellulose	Pento sans	Klason Lignin	1% NaOH	Solubility Hot Water	EtOH/Benzene	Ether	Ash
Sequoia sempervirens	Redwood old growth	55.0	43.0	7.0	33.0	19.0	9.0	10.0	0.8	0.1
	Redwood second growth	61.0	46.0	7.0	33.0	14.0	5.0	<1.0	0.1	0.1
Taxodium distichum	Bald cypress	—	41.0	12.0	33.0	13.0	4.0	5.0	1.5	—
Thuja occidentalis	Northern white cedar	59.0	44.0	14.0	30.0	13.0	5.0	6.0	1.4	0.5
Thuja plicata	Western red cedar	—	38.0	9.0	32.0	21.0	11.0	14.0	2.5	0.3
Tsuga canadensis	Eastern hemlock	—	41.0	9.0	33.0	13.0	4.0	3.0	0.5	0.5
Tsuga heterophylla	Western hemlock	67.0	42.0	9.0	29.0	14.0	4.0	4.0	0.5	0.4
Tsuga mertensiana	Mountain hemlock	60.0	43.0	7.0	27.0	12.0	5.0	5.0	0.9	0.5

Source: Adapted from Petersen, R.C. 1984. *The Chemistry of Solid Wood*, Advances in Chemistry Series 20, Chapter 2, pp. 57–126, Washington, DC: ACS.

TABLE 3.12
Polysaccharide Content of Some North American Woods

Scientific Name	Common Name	Glu	Xyl	Gal	Arab	Mann	Uronic	Acetyl	Lignin	Ash
	Hardwoods									
Acer rubrum	Red maple	46	19	0.6	0.5	2.4	3.5	3.8	24	0.2
Acer saccharum	Sugar maple	52	15	<0.1	0.8	2.3	4.4	2.9	23	0.3
Betula alleghaniensis	Yellow birch	47	20	0.9	0.6	3.6	4.2	3.3	21	0.3
Betula papyrifera	White birch	43	26	0.6	0.5	1.8	4.6	4.4	19	0.2
Fagus grandifolia	Beech	46	19	1.2	0.5	2.1	4.8	3.9	22	0.4
Liquidambar styraciflua	Sweetgum	39	18	0.8	0.3	3.1	—	—	24	0.2
Platanus occidentalis	Sycamore	43	15	2.2	0.6	2.0	5.1	5.5	23	0.7
Populus deltoides	Eastern cottonwood	47	15	1.4	0.6	2.9	4.8	3.1	24	0.8
Populus tremuloides	Quaking aspen	49	17	2.0	0.5	2.1	4.3	3.7	21	0.4
Quercus falcata	Southern red oak	41	19	1.2	0.4	2.0	4.5	3.3	24	0.8
Ulmus Americana	White elm	52	12	0.9	0.6	2.4	3.6	3.9	24	0.3
	Softwoods									
Abies balsamea	Balsam fir	46	6.4	1.0	0.5	12	3.4	1.5	29	0.2
Gingo biloba	Ginko	40	4.9	3.5	1.6	10	4.6	1.3	33	1.1
Juniperus communis	Juniper	41	6.9	3.0	1.0	9.1	5.4	2.2	31	0.3
Larix decidua	Larch	46	6.3	2.0	2.5	11	4.8	1.4	26	0.2
Larix laricina	Tamarack	46	4.3	2.3	1.0	13	2.9	1.5	29	0.2
Picea abies	Norway spruce	43	7.4	2.3	1.4	9.5	5.3	1.2	29	0.5
Picea glauca	White spruce	45	9.1	1.2	1.5	11	3.6	1.3	27	0.3
Picea mariana	Black spruce	44	6.0	2.0	1.5	9.4	5.1	1.3	30	0.3
Picea rubens	Red spruce	44	6.2	2.2	1.4	12	4.7	1.4	28	0.3
Pinus banksiana	Jack pine	46	7.1	1.4	1.4	10	3.9	1.2	29	0.2
Pinus radiata	Radiata pine	42	6.5	2.8	2.7	12	2.5	1.9	27	0.2

continued

TABLE 3.12   (continued)
Polysaccharide Content of Some North American Woods

Scientific Name	Common Name	Glu	Xyl	Gal	Arab	Mann	Uronic	Acetyl	Lignin	Ash
Pinus resinosa	Red pine	42	9.3	1.8	2.4	7.4	6.0	1.2	29	0.4
Pinus rigida	Pitch pine	47	6.6	1.4	1.3	9.8	4.0	1.2	28	0.4
Pinus strobus	Eastern white pine	45	6.0	1.4	2.0	11	4.0	1.2	29	0.2
Pinus sylvestris	Scots pine	44	7.6	3.1	1.6	10	5.6	1.3	27	0.4
Pinus taeda	Loblolly pine	45	6.8	2.3	1.7	11	3.8	1.1	28	0.3
Pseudotsuga menziesii	Douglas-fir	44	2.8	4.7	2.7	11	2.8	0.8	32	0.4
Thuja occidentalis	Northern white cedar	43	10.0	1.4	1.2	8.0	4.2	1.1	31	0.2
Tsuga canadensis	Eastern hemlock	44	5.3	1.2	0.6	11	3.3	1.7	33	0.2

Source: Adapted from Pettersen, R.C. 1984. *The Chemistry of Solid Wood*, Advances in Chemistry Series 20, Chapter 2, pp. 57–126, Washington, DC: ACS.

TABLE 3.13

Chemical Composition of Selected Hardwoods from the Southeastern United States (Percent of Oven-Dry Wood)

| Scientific Name | Common Name | Carbohydrates | | Components of Hemicelluloses | | | | Lignin | Total Ext | Ash |
		Cell	Total Hemi	Gluco mann	AcGlu UrXyl	Arab Gal	Pectin			
Acer rubrum	Red maple	40.7	30.4	3.5	23.5	1.6	1.9	23.3	5.3	0.3
Aesculus octandra	Yellow buckeye	40.6	25.8	3.6	18.6	1.0	2.6	30.0	3.1	0.5
Carya glabra	Pignut hickory	46.2	26.7	1.1	22.1	1.2	2.3	23.2	3.4	0.6
Carya illinoensis	Pecan	38.7	30.2	1.6	24.7	1.6	2.3	23.2	3.4	0.6
Carya tomentosa	Mockernut	43.5	27.7	1.5	21.5	1.3	3.5	23.6	5.0	0.4
Cornus florida	Flowering dogwood	35.8	35.4	3.4	27.2	1.0	5.0	21.8	4.6	0.3
Fagus grandifolia	American beech	35.0	29.4	2.7	23.5	1.3	1.8	30.9	3.4	0.4
Fraxinus Americana	White ash	39.5	29.1	3.8	22.1	1.4	1.9	24.8	6.3	0.3
Gordonia lasianthus	Loblolly-bay	43.8	29.1	4.1	22.1	1.1	1.8	21.5	5.2	—
Liquidambar styraciflua	Sweetgum	40.8	30.7	3.2	21.4	1.3	4.9	22.4	5.9	0.2
Liriodendron tulipifera	Yellow poplar	39.1	28.0	4.9	20.1	0.7	2.4	30.3	2.4	0.3
Magnolia virginiana	Sweetbay	44.2	37.7	4.3	20.2	1.6	1.6	24.1	3.9	0.2
Nyssa aquatica	Water tupelo	45.9	24.0	3.5	18.6	0.8	1.1	25.1	4.7	0.4
Nyssa sylvatica	Black tupelo	42.6	27.3	3.6	18.0	1.0	4.8	26.6	2.9	0.6
Oxydendrum arboreum	Sourwood	40.7	34.6	1.3	31.9	1.0	0.4	20.8	3.6	0.3
Persea borbonia	Redbay	45.6	25.6	1.0	23.2	0.9	0.5	23.6	5.0	0.2
Platanus occidentalis	Sycamore	43.0	27.2	2.3	22.3	1.4	1.2	25.3	4.4	0.1
Populus deltoids	Eastern cottonwood	45.5	26.6	4.4	16.8	1.6	1.8	25.9	2.4	0.6
Quercus alba	White oak	44.7	28.4	3.1	21.0	1.6	2.7	24.6	5.3	0.2
Quercus coccinea	Scarlet oak	43.2	29.2	2.3	23.3	1.4	2.2	20.9	6.6	0.1
Quercus falcata	Southern red oak	43.5	24.2	1.7	18.6	1.7	2.2	23.6	9.6	0.5
Quercus ilicifolia	Scrub oak	37.6	27.5	1.0	22.3	1.8	2.4	26.4	8.0	0.5
Quercus marilandica	Blackjack oak	33.8	28.2	2.0	21.0	2.3	2.9	30.1	6.6	1.3
Quercus nigra	Water oak	44.6	34.8	3.0	28.9	2.2	0.7	19.1	4.3	0.3
Quercus prinus	Chestnut oak	40.8	29.9	2.9	23.8	1.8	1.4	22.3	6.6	0.4

continued

TABLE 3.13   (continued)
Chemical Composition of Selected Hardwoods from the Southeastern United States (Percent of Oven-Dry Wood)

		Carbohydrates		Components of Hemicelluloses						
Scientific Name	Common Name	Cell	Total Hemi	Gluco mann	AcGlu UrXyl	Arab Gal	Pectin	Lignin	Total Ext	Ash
Quercus rubra	Northern red oak	42.2	33.1	3.3	26.6	1.6	1.6	20.2	4.4	0.2
Quercus stellata	Post oak	37.7	29.9	2.6	23.0	2.0	2.3	26.1	5.8	0.5
Quercus velutina	Black oak	39.6	28.4	1.9	23.2	1.1	1.9	25.3	6.3	0.5
Quercus virginiana	Live oak	38.1	22.9	1.0	18.3	1.7	1.9	25.3	13.2	0.6
Sassafras albidum	Sassafras	45.0	35.1	4.0	30.4	0.9	<0.1	17.4	2.4	0.2
Ulmus Americana	American elm	42.6	26.9	4.6	19.9	0.8	1.6	27.8	1.9	0.8

Source: Adapted from Pettersen, R.C. 1984. *The Chemistry of Solid Wood*, Advances in Chemistry Series 20, Chapter 2, pp. 57–126, Washington, DC: ACS.
Note: Cell = Cellulose.

TABLE 3.14

Elemental Composition of Some Woods

Scientific Name	Common Name	Ca ppt	K ppt	Mg ppt	P ppt	Mn ppt	Fe ppm	Cu ppm	Zn ppm	Na ppm	Cl ppm
Abies balsamea	Balsam fir	0.8	0.8	0.27	—	0.13	13	17	11	18	—
Acer rubrum	Red maple	0.8	0.7	0.12	0.03	0.07	11	5	29	5	18
Betula papyrifera	White birch	0.7	0.3	0.18	0.15	0.03	10	4	28	9	10
Fraxinus Americana	White ash	0.3	2.6	1.8	0.01	—	—	—	—	31	—
Liquidambar styraciflua	Sweetgum	0.55	0.3	0.34	0.15	0.08	—	—	19	81	—
Picea rubens	Red spruce	0.8	0.2	0.07	0.05	0.14	14	4	8	8	0.3
Pinus strobes	Eastern white pine	0.2	0.3	0.07	—	0.03	10	5	11	9	19
Populus deltoids	Eastern cottonwood	0.9	2.3	0.29	—	0.2	100	—	30	940	—
Populus tremuloides	Quaking aspen	1.1	1.2	0.27	0.10	0.03	12	7	17	5	—
Quercus alba	White oak	0.5	1.2	0.31	—	<0.01	—	—	—	21	15
Quercus falcate	Southern red oak	0.3	0.6	0.03	0.02	0.01	30	73	38	44	—
Tilia Americana	Basswood	0.1	2.8	0.35	—	—	—	—	—	63	38
Tsuga Canadensis	Eastern hemlock	1.0	0.4	0.11	0.12	0.15	6	5	2	6	—

Source: Adapted from Pettersen, R.C. 1984. *The Chemistry of Solid Wood*, Advances in Chemistry Series 20, Chapter 2, pp. 57–126, Washington, DC: ACS.

conditions and procedures used. This way, it may be possible to reproduce the results by other workers in different laboratories. Without this information, it is not possible to compare data from different laboratories.

The following information should accompany each chemical analysis:

1. Source of the wood:
   a. Geographic location
   b. Part of the tree sampled
   c. Date sample was taken
2. Sampling:
   a. Different anatomical parts
   b. Degree of biological deterioration, if any
   c. Sample size
   d. Drying method applied
3. Analytical procedure used
4. Calculations and reporting technique

All the above-mentioned criteria could contribute in one way or another toward variations in chemical analyses.

3.8.2   EXTRACTION

3.8.2.1   Scope and Summary

Wood materials = Extractives + holocellulose + lignin + inorganics (ash).

This method describes a procedure for extraction of wood for further analysis, such as holocellulose, hemicellulose, cellulose, and lignin analysis.

Neutral solvents, water, toluene or ethanol, or combinations of solvents are employed to remove extractives in wood. However, other solvents ranging from diethyl ether to 1% NaOH, and so on, could be applied according to the nature of extractives and sample type, that is, bark, leaves, and so on.

3.8.2.2   Sample Preparation

It is highly recommended to have a fresh sample. If not, keep the sample frozen or in a refrigerator to avoid fungal attack. Peel off the bark from the stem and separate the sample into component parts. Dry samples are oven dried for 24 h (usually at 105°C) prior to milling. Wet samples can be milled while frozen in order to prevent oxidation or other undesirable chemical reactions. Samples are ground to pass 40 mesh (0.40 mm) using a Wiley Mill.

3.8.2.3   Apparatus

Buchner funnel
Extraction thimbles
Extraction apparatus, extraction flask, 500 mL, Soxhlet extraction tube
Heating device, heating mantle or equivalent
Boiling chips, glass beads, boilers, or any inert granules for taming boiling action
Chemical fume hood
Vacuum oven

3.8.2.4   Reagents and Materials

Ethanol (ethyl alcohol), 200 proof
Toluene, reagent grade
Toluene–ethanol mixture, mix one volume of ethanol and one volume of toluene

3.8.2.5 Procedures

Weigh 2–3 g of samples into covered preweighed extraction thimbles. Place the thimbles in a vacuum oven not exceeding 45°C for 24 h, or to a constant weight. Cool the thimbles in a desiccator for 1 h and weigh. Then place the thimbles in Soxhlet extraction units. Place 200 mL of the toluene:ethanol mixture in a 500 mL round bottom flask with several boiling chips to prevent bumping. Carry out the extraction in a well-ventilated chemical fume hood for 2 h, keeping the liquid boiling so that siphoning from the extractor is no less than four times per hour. After extraction with the toluene:ethanol mixture, take the thimbles from the extractors, drain the excess solvent, and wash the samples with ethanol. Place them in the vacuum oven over night at temperatures not exceeding 45°C for 24 h. When dry, remove them to a desiccator for an hour and weigh. Generally, the extraction is complete at this stage; however, the extractability depends upon the matrix of the sample and the nature of extractives. Second and the third extractions with different polarity of solvents may be necessary. Browning (1967) suggests 4 h of successive extraction with 95% alcohol, however, two successive extractions, 4 h with ethanol followed with distilled water for 1 h can also be done. Pettersen (1984) extracted pine sample with acetone/water, followed by the toluene/ethanol mixture.

3.8.3 Ash Content (ASTM D-1102-84)

3.8.3.1 Scope

The ash content of fiber is defined as the residue remaining after ignition at 575 ± 25°C for 3 h, or longer if necessary to burn off all the carbon. It is a measure of mineral salts in the fiber, but it is not necessarily quantitatively equal to them.

3.8.3.2 Sample Preparation

Obtain a representative sample of the fiber, preferably ground to pass a 40-mesh screen. Weigh, to 5 mg or less, a specimen of about 5 g of moisture-free wood for ashing, preferably in duplicate. If the moisture in the sample is unknown, determine it by drying a corresponding specimen to constant weight in a vacuum oven at 105 ± 3°C.

3.8.3.3 Apparatus

Crucible: A platinum crucible or dish with lid or cover is recommended. If platinum is not available, silica may be used. Analytical balance having a sensitivity of 0.1 mg. Electric muffle furnace adjusted to maintain a temperature of 575 ± 25°C.

3.8.3.4 Procedure

Carefully clean the empty crucible and cover, and ignite them to constant weight in a muffle furnace at 575 ± 25°C. After ignition, cool slightly and place in a desiccator. When cooled to room temperature, weigh the crucible and cover on the analytical balance.

Place all, or as much as practicable, of the weighed specimen in the crucible. Burn the sample directly over a low flame of a Bunsen burner (or preferably on the hearth of the furnace) until it is well carbonized, taking care not to blow portions of the ash from the crucible. If a sample tends to flare up or lose ash during charring, the crucible should be covered, or at least partially covered during this step. If the crucible is too small to hold the entire specimen, gently burn the portion added and add more as the flame subsides. Continue heating with the burner only as long as the residue burns with a flame. Place the crucible in the furnace at 575 ± 25°C for a period of at least 3 h, or longer if needed, to burn off all the carbon. When ignition is complete, as indicated by the absence of black particles, remove the crucible from the furnace, replace the cover and allow the crucible to cool somewhat. Then place in a desiccator and cool to room temperature. Reweigh the ash and calculate the percentage based on the moisture-free weight of the fiber.

3.8.3.5 Report

Report the ash as a percentage of the moisture-free wood to two significant figures, or to only one significant figure if the ash is less than 0.1%.

3.8.3.6 Precision

The results of duplicate determinations should be suspected if they differ by more than 0.5 mg. Since the ignition temperature affects the weight of the ash, only values obtained at $575 \pm 25°C$ should be reported as being in accordance with this method. Porcelain crucibles can also be used in most cases for the determination of ash. Special precautions are required in the use of platinum crucibles. There can be significant losses in sodium, calcium, irons, and copper at temperatures over 600°C.

3.8.4 PREPARATION OF HOLOCELLULOSE (CHLORITE HOLOCELLULOSE)

3.8.4.1 Scope

Holocellulose is defined as a water-insoluble carbohydrate fraction of wood materials. According to Browning (1967) there are three ways of preparing holocellulose and their modified methods: (1) chlorination method, (2) modified chlorination method, and (3) chlorine dioxide and chlorite method. The standard purity of holocellulose is checked following lignin analysis.

3.8.4.2 Sample Preparation

The sample should be extractive and moisture free and prepared after Procedure 9.2. If Procedure 9.2 is skipped for some reason, the weight of the extractives should be accounted for in the calculation of holocellulose.

3.8.4.3 Apparatus

Buchner funnel
250 mL Erlenmeyer flasks
25 mL Erlenmeyer flasks
Water bath
Filter paper
Chemical fume hood

3.8.4.4 Reagents

Acetic acid, reagent grade
Sodium chlorite, $NaClO_2$, technical grade, 80%

3.8.4.5 Procedure

To 2.5 g of sample, add 80 mL of hot distilled water, 0.5 mL acetic acid, and 1 g of sodium chlorite in a 250 mL Erlenmeyer flask. An optional 25 mL Erlenmeyer flask is inverted in the neck of the reaction flask. The mixture is heated in a water bath at 70°C. After 60 min, 0.5 mL of acetic acid and 1 g of sodium chlorite are added. After each succeeding hour, fresh portions of 0.5 mL acetic acid and 1 g sodium chlorite are added with shaking. The delignification process degrades some of the polysaccharides and the application of excess chloriting should be avoided. Continued reaction will remove more lignin but hemicellulose will also be lost (Rowell 1980).

Addition of 0.5 mL acetic acid and 1 g of sodium chlorite is repeated until the wood sample is completely separated from lignin. It usually takes 6–8 h of chloriting and the sample can be left without further addition of acetic acid and sodium chlorite in the water bath for over night. At the end of 24 h of reaction, cool the sample and filter the holocellulose on filter paper using a Buchner funnel until the yellow color (the color of holocellulose is white) and the odor of chlorine dioxide is removed. If the weight of the holocellulose is desired, filter the holocellulose on a tarred fritted disc glass thimble, wash with acetone, vacuum-oven dry at 105°C for 24 h, place in a desiccator for an hour, and weigh. The holocellulose should not contain any lignin and the lignin content of holocellulose should be determined and subtracted from the weight of the prepared holocellulose.

3.8.5　Preparation of α-cellulose (Determination of Hemicelluloses)

3.8.5.1　Scope

The preparation of α-cellulose is a continuous procedure from Procedure 9.4. The term hemicellulose is defined as the cell wall components that are readily hydrolyzed by hot dilute mineral acids, hot dilute alkalies, or cold 5% NaOH.

3.8.5.2　Principle of Method

Extractive-free, lignin-free holocellulose is treated with NaOH and then with acetic acid, with the residue defined as α-cellulose. The soluble fraction represents the hemicellulose content.

3.8.5.3　Apparatus

A thermostat or other constant-temperature device will be required that will maintain a temperature of $20 \pm 0.1°C$ in a container large enough to hold a row of at least three 250 mL beakers kept in an upright position at all times.

Filtering crucibles of Alundum or fritted glass thimbles of medium porosity.

3.8.5.4　Reagents

　Sodium hydroxide solution, NaOH, 17.5 and 8.3%
　Acetic acid, 10% solution

3.8.5.5　Procedure

Weigh out about 2 g of vacuum-oven-dried holocellulose and place into a 250 mL glass beaker provided with a glass cover. Add 10 mL of 17.5% NaOH solution to the holocellulose in a 250 mL beaker, cover with a watch glass, and maintain at 20°C in a water bath. Manipulate the holocellulose lightly with a glass rod with a flat end so that entire specimen becomes soaked with the NaOH solution. After the addition of the first portion of 17.5% NaOH solution to the specimen, at 5 min intervals, add 5 mL more of the NaOH solution, and thoroughly stir the mixture with the glass rod. Continue this procedure until the NaOH is consumed. Allow the mixture to stand at 20°C for 30 min making the total time for NaOH treatment 45 min.

Add 33 mL of distilled water at 20°C to the mixture. Thoroughly mix the contents of the beaker and allow to stand at 20°C for 1 h before filtering.

Filter the cellulose with the aid of suction into the tarred, alkali-resistant Alundum or fritted-glass crucible of medium porosity. Transfer all the holocellulose residue to the crucible, and wash with 100 mL of 8.3% NaOH solution at 20°C. After the NaOH wash solution has passed through the residue in the crucible, continue the washing at 20°C with distilled water, making certain that all particles have been transferred from the 250 mL beaker to the crucible. Washing the sample in the crucible is facilitated by releasing the suction, filling the crucible to within 6 mm of the top with water, carefully breaking up the cellulose mat with a glass rod so as to separate any lumps present, and again applying suction. Repeat this step twice.

Pour 15 mL of 10% acetic acid at room temperature into the crucible, drawing the acid into the cellulose by suction but, while the cellulose is still covered with acid, release the suction. Subject the cellulose to the acid treatment for 3 min from the time the suction is released; then apply suction to draw off the acetic acid. Without releasing the suction, fill the crucible almost to the top with distilled water at 20°C and allow to drain completely. Repeat the washing until the cellulose residue is free of acid as indicated by litmus paper. Give the cellulose a final washing by drawing, by suction, an additional 250 mL of distilled water through the cellulose in the crucible. Dry the crucible on the bottom and sides with a cloth and place it overnight in an oven to dry at 105°C. Cool the crucible and weighing bottle in a desiccator for 1 h before weighing.

3.8.5.6   Calculation and Report

Calculate the percentage of α-cellulose on the basis of the oven-dry holocellulose sample, as follows:

$$\alpha\text{-cellulose } (\%) = (W_2/W_1) \times 100$$

where

   W_2 = weight of the oven-dry α-cellulose residue, and
   W_1 = weight of the original oven-dry holocellulose sample.

3.8.6   PREPARATION OF KLASON LIGNIN

3.8.6.1   Scope

Klason lignin gives a quantitative measure of the acid-insoluble lignin and is not suitable for the study of lignin structures and some other lignins such as cellulolytic enzyme lignin, or Björkman (milled-wood lignin) should be prepared (Sjöström 1981) for the study of lignin structure. This procedure is a modified version of ASTM D-1106-84. The lignin isolated using this procedure is also called sulfuric acid lignin.

3.8.6.2   Apparatus

   Autoclave
   Buchner funnel
   100 mL centrifuge tube, Pyrex 8240
   Desiccator
   Glass rods
   Water bath
   Glass fiber
   Filter paper, Whatman Cat No. 1827-021, 934-AH
   Glass micro-filter, 2.1 cm

3.8.6.3   Reagent

   Sulfuric acid, H_2SO_4, 72 and 4% by volume
   Fucose, 24.125 % in 4% H_2SO_4 [w/w]

3.8.6.4   Procedure

Prepare samples by Procedure 9.2 and dry the sample at 45°C in a vacuum oven overnight. Accurately weigh out approximately 200 mg of ground-vacuum-dried sample, into a 100 mL centrifuge tube. To the sample in a 100 mL centrifuge tube, add 1 mL of 72% (w/w) H_2SO_4 for each 100 mg of sample. Stir and disperse the mixture thoroughly with a glass rod twice, then incubate the tubes in a water bath at 30°C for 60 min. Add 56 mL of distilled water. This results in a 4% solution for the secondary hydrolysis. Add 1 mL fucose internal standard (this procedure is required only if five sugars are to be analyzed by HPLC as a part of the analysis). Autoclave at 121°C, 15 psi, for 60 min. Remove the samples from the autoclave and filter off the lignin, with glass fiber filters (filters were rinsed into crucibles, dried, and tarred) in crucibles using suction, keeping the solution hot. Wash the residue thoroughly with hot water and dry at 105°C overnight. Move to a desiccator, and let it sit an hour and weigh. Calculate Klason lignin content from weights.

3.8.6.5   Additional Information

Condensation reactions involving protein can cause artificially high Klason lignin measurements when tissues containing significant protein contents are analyzed. A nitrogen determination can be done to indicate possible protein content.

3.8.7   Determination of Methoxyl Groups

3.8.7.1   Scope

Methoxyl groups ($-OCH_3$) are present in the lignin and lignin derivatives as side chains of aromatic phenylpropanes and in the polysaccharides mainly as methoxy uronic acids. Methoxyl content is determined using ASTM, D-1166-84.

3.8.7.2   Principle of Method

In the original method, methyl iodide was absorbed in an alcoholic solution of silver nitrate. The solution was diluted with water, acidified with nitric acid, and boiled. The silver iodide was removed by filtration, washed, and weighed in the manner usual for halide determinations. A volumetric modification is based on absorption of the methyl iodide in a known volume of standard silver nitrate solution and titration of the unused silver nitrate with standard potassium thiocyanate solution (ferric alum indicator solution). In this procedure, the methyl iodide is collected in an acetic acid solution of potassium acetate containing bromine.

$$CH_3I + Br_2 \rightarrow CH_3Br + IBr$$

$$IBr + 2Br_2 + 3H_2O \rightarrow HIO_3 + 5HBr$$

The excess bromine is destroyed by addition of acid, and the iodate equivalent of the original methoxyl content is determined by titration with sodium thiosulfate of the iodine liberated in the reaction:

$$HIO_3 + 5HI \rightarrow 3I_2 + 3H_2O$$

One methoxyl group is equivalent to six atoms of iodine and, consequently, a favorable analytical factor is obtained.

3.8.7.3   Sample Preparation

The sample is dried, ground, and extracted accordingly prior to analysis.

3.8.7.4   Apparatus

> Reaction flask
> Heat source
> Vertical air-cooled condenser
> Scrubber
> Absorption vessels

3.8.7.5   Reagents

> Bromine, liquid.
> Cadmium sulfate solution—Dissolve 67.2 g of $CdSO_4 \cdot 4H_2O$ in 1 L of water.
> Carbon dioxide gas.
> Formic acid, 90%.
> Hydroiodic acid.
> Phenol.
> Potassium acetate solution in acetic acid. Anhydrous potassium acetate (100 g) is dissolved in
>   1 L of glacial acetic acid.
> Potassium iodide solution—Dissolve 100 g of KI in water and dilute to 1 L.
> Sodium acetate solution—Dissolve 415 g of sodium acetate trihydrate in water and dilute
>   to 1 L.

Sodium thiosulfate solution (0.1 N)-Dissolve 25 g of $Na_2S_2O_3 \cdot H_2O$ in 200 mL of water and dilute to 1 L.

Starch indicator solution (10 g/L).

Sulfuric acid—Mix one volume of H_2SO_4 (sp gr 1.84) with nine volumes of water.

3.8.7.6   Procedure

Weigh the sample, about 100 mg of wood or 50 mg of lignin and place in the reaction flask. Place in the reaction flask 15 mL of HI, 7 g of phenol, and a boiling tube. Place in the scrubber a mixture of equal volumes of $CdSO_4$ solution and $Na_2S_2O_3$. The volume of solution should be adjusted so that the inlet tube of the scrubber is covered to a depth of about 4 mm. Adjust the flow of CO_2 to about 60 bubbles per minute through the scrubber. Heat the flask and adjust the rate of heating so that the vapors of the boiling HI rise about 100 mm into the condenser. Heat the flask under these conditions for 30–45 min, or longer if necessary, to remove methoxyl-containing or other interfering substances which are usually present in the reagents.

Let the distilling flask cool below 100°C. In the meantime, add to 20 mL of the potassium acetate solution, about 0.6 mL of bromine, and mix. Add approximately 15 mL of the mixture to the first receiver and 5 mL to the second, and attach the receiver to the apparatus. Seal the ground-glass joint with a small drop of water from a glass rod.

Remove the distilling flask and introduce the test specimen. Immediately reconnect the flask and seal the ground-glass joint with a drop of molten phenol from a glass rod. Bring the contents of the flask to reaction temperature while passing a uniform stream of CO_2 through the apparatus.

Adjust the rate of heating so that the vapors of the boiling HI rise about 100 mL into the condenser. Continue the heating for a time sufficient to complete the reaction and sweep out the apparatus. Usually, not more than 50 min are required.

Wash the contents of both receivers into a 250 mL Erlenmeyer flask that contains 15 mL of sodium acetate solution. Dilute with water to approximately 125 mL and add 6 drops of formic acid. Rotate the flask until the color of the bromine is discharged, then add 12 more drops of formic acid and allow the solution to stand for 1–2 min. Add 10 mL of KI solution and 10 mL of H_2SO_4, and titrate the liberated iodine with $Na_2S_2O_3$ solution, adding 1 mL of starch indicator solution just before the end point is reached, continuing the titration to the disappearance of the blue color.

3.8.7.7   Calculation and Report

$$\text{Methoxyl, } \% = (VN \times 31.030 \times 100)/(G \times 1000 \times 6) = (VN/G) \times 0.517$$

where

V = milliliters of $Na_2S_2O_3$ solution required for the titration
N = normality of $Na_2S_2O_3$ solution
G = grams of moisture free sample

Table 3.9 shows the methoxyl content of some common hardwoods and softwoods.

3.8.8   DETERMINATION OF ACETYL BY GAS–LIQUID CHROMATOGRAPHY

3.8.8.1   Scope

The acetyl and formyl groups that are in the polysaccharide portion can be determined in one of three ways: (1) acid hydrolysis; sample is hydrolyzed to form acetic acid, (2) saponification; acetyl groups are split from polysaccharides with hot alkaline solution and acidified to form acetic acid, or (3) trans-esterificaion; sample is treated with methanol in acid or alkaline solution to form methyl acetate. Acetic acid and methyl acetate are analyzed by gas chromatography.

The procedure presented here is saponification and acetyl determined by gas chromatography.

$$CH_3COOR + NaOH \rightarrow CH_3COONa + ROH$$

$$CH_3COONa + H+ \rightarrow CH_3COOH$$

3.8.8.2   Reagents

2% formic acid. Dilute 2 mL of 90% formic acid to 900 mL of deionized H_2O.

Internal standard stock solution: Weigh 25.18 g of 99+% propionic acid in 500 mL volumetric flask, make to volume with 2% formic acid.

Internal standard solution: Pipette 10 mL stock solution into a 200 mL volumetric flask, make to volume with deionized water.

Acetic acid standard solution: Weigh 100 mg 99.7% glacial acetic acid into a 100 mL volumetric flask, make to volume with deionized water.

NaOH solution 1 N: Weigh 4 g sodium hydroxide, dissolve in 100 mL deionized water.

3.8.8.3   Sample Preparation

The amount of sample is based on the approximate acetyl content: Acetyl content (AC) 0–10%, 50 mg; AC 10%, 25 mg; AC 15%, 20 mg; AC 20%, 15 mg; AC 25%, 10 mg. Weigh an oven-dried sample in a long handled weighing tube and transfer it to an acetyl digestion flask and add boiling chips. Pipette 2 mL 1 N NaOH solution to wash down the neck of the flask. Connect the reaction flask to a water-cooled reflux condenser and reflux for 1 h. Cool the reaction flask to room temperature and pipette 1 mL of propionic acid (internal standard) into a 10 mL volumetric flask. Quantitatively transfer the liquid from the reaction flask to the volumetric flask. Wash the reaction flask and the solid residue with several portions of distilled water. Add 0.2 mL of 85% phosphoric acid and make to volume with distilled water. This solution may be filtered through a small plug of glass wool to remove solid particles. Analyze the sample by GLC and determine the average ratio. Milligrams of acetic acid are determined from the calibration curve.

3.8.8.4   Gas Chromatography

Column: Supelco 60/80 Carbopack C/0.3% carbowax 20 M/0.1% H_3PO_4–3 ft 1/4 in. O.D. and 4 mm I.D.; oven temperature 120°C; injection port 150°C; F.I.D. 175°C; nitrogen 20 mL/min.

The ratio of the area is determined by dividing the area of the acetic acid by the area of the propionic acid (internal standard). The average of the ratios is used to determine mg/mL of acetic acid from the calibration curve.

Preparation of a calibration curve: Pipette 1, 2, 4, 6, and 8 mL of standard acetic acid solution into 10 mL volumetric flasks. Pipette 1 mL of propionic acid internal standard into each sample, then add 0.2 mL 85% phosphoric acid. Make to volume with distilled water. Analyze each solution three times by GLC. Calculate the ratios by dividing the area of the acetic acid by the area of the propionic acid (internal standard). Plot the average ratios against milligrams per milliliter of acetic acid. Standard and sample solutions can be stored in the refrigerator for at least 1 week.

3.8.8.5   Reporting

Report the average, standard deviation, and precision of each sample. The results may be reported as percent acetic acid or as percent acetyl:

% Acetic acid = mg/mL acetic acid found × 10 mL % 100
                             sample weight in mg

% Acetyl = % acetic acid × 0.7172

REFERENCES

Adler, E. 1977. Lignin chemistry: Past, present and future. *Wood Sci. Technology* 11: 169–218.

Andersson, A., Erickson, M., Fridh, H., and Miksche, G.E. 1973. Gas chromatographic analysis of lignin oxidation products. XI. Structure of the bark lignins of hardwoods and softwoods. *Holzforschung* 27: 189–193.

ASTM, 1984. Standard method for methoxyl content in pulp and wood, D-1166-84.

ASTM, 1984. Standard method for lignin in wood, D-1106-84.

ASTM, 1984. Standard method for ash content in wood and wood-based materials, D-1102-84.

Björkman, A. 1956. Studies on finely divided wood. Part 1: Extraction of lignin with neutral solvents. *Svensk Papperstid.* 59: 477–485.

Björkman, A. 1957. Studies on finely divided wood. Part 2: The properties of lignins extracted with neutral solvents from softwoods and hardwoods. *Svensk Papperstid.* 60: 158–169.

Browning, G.L. 1967. *Methods in Wood Chemistry*, Vol. 2, New York, NY: Wiley Interscience.

Chang, Y.-P. and Mitchell, R.L. 1955. Chemical composition of common North American pulpwood parks, *Tappi* 38(5): 315.

Choong, E.T, Abdullah, G., and Kowalczuk, J. 1976. Louisiana State University, Wood Utilization Notes, No. 29.

Clermont, L.P. 1970. Study of lignin from stone cells of aspen poplar inner bark. *Tappi* 53(1): 52–57.

Dietrichs, H.H. 1975. Polysaccharides in bark. *Holz als Roh- und Werkstoff.* 33: 13–20.

Dietrichs, H.H., Graves, K., Behrenwsdorf, D., and Sinner, M. 1978. Studies on the carbohydrates in tree barks. *Holzforschung* 32(2): 60–67.

Ellis, E.L. 1965. *Cellular Ultrastructure of Woody Plants*. W.A. Côté Jr. (ed), New York: Syracuse University Press, pp. 181–189.

Fengel, D. and Wegener, G. 1984. *Wood: Chemistry, Ultrastructure and Reactions*. W. de. Gruyter, Berlin.

Gardner, K.H. and Blackwell, J. 1974. The hydrogen bonding in naïve cellulose. *Biochim. Biophys. Acta* 343: 232–237.

Goring, D.A.I. 1962. The physical chemistry of lignin. *Pure Appl. Chem.* 5: 233–254.

Goring, D.A.I and Timell, T.E. 1962. Molecular weight of native celluloses. *Tappi* 45(6): 454–460.

Gottlieb, O.R. and Yoshida, M. 1989. Lignins. In: Rowe, J.W. (ed), *Natural Products of Woody Plants*. I. Chapter 7.3, pp. 439–511, New York, NY: Springer-Verlag.

Harder, M.L. and Einspahr, D.W. 1980. Levels of some essential metals in bark. *Tappi* 63(2): 110.

Harun, J. and Labosky, P. 1985. Chemical constituents of five northeastern barks. *Wood and Fiber* 17(2): 274.

Haslam, E. 1989. Gallic acid derivatives and hydrolysable tannins. In: Rowe, J.W. (ed), *Natural Products of Woody Plants*. I. Chapter 7.2, pp. 399–438, New York, NY: Springer-Verlag.

Hattula, T. and Johanson, M. 1978. Determination of some of the trace elements in bark by neutron activation analysis and high resolution spectroscopy. *Radiochem. Radioanal. Letters* 32(1–2): 35.

Hemingway, R.W. 1981. Bark: Its chemistry and prospects for chemical utilization. In: Goldstein, I.S. Ed. *Organic Chemicals and Biomass*. Chapter 10, pp. 189–248, Boca Raton, FL: CRC Press.

Hemingway, R.W. 1989. Biflavonoids and proanthocyanidins. In: Rowe, J.W. (ed), *Natural Products of Woody Plants*. I. Chapter 7.6, pp. 571–650, New York, NY: Springer-Verlag.

Hemingway, R.W., Karchesy, J.J., McGraw, G.W., and Wielesek, R.A. 1983. Heterogeneity of interflavaniod bond located in loblolly pine bark procyanidins. *Phytochem.* 22: 275.

Hergert, H.T. 1960. Chemical composition of tannins and polyphenolics from conifer wood and bark. *Forest Products Journal* 10: 610–617.

Hergert, H.T. 1962. Economic importance of flavonoid compounds: Wood and bark. In: Geissmann, T.A. (ed), *The Chemistry of Flavonoid Compounds*, Chapter 17, New York, NY: MacMillan.

Higuchi, T., Ito, Y., Shimada, M., and Kawamura, I. 1967. Chemical properties of bark lignins. *Cellulose Chem. Technol.* 1: 585–595.

Hillis, W.E. 1989. Historical uses of extractives and exudates. In: Rowe, J.W. (ed), *Natural Products of Woody Plants*. I. Chapter 1.1, pp. 1–12, New York, NY: Springer-Verlag.

Holloway, P.J. and Deas, A.H.B. 1973. Epoxyoctadeconic acids in plant cutins and suberins. *Phytochemistry* 12: 1721.

Hon, D.N.-S. and Shiraishi, N. 1991. *Wood and Cellulosic Chemistry*. New York, NY: Marcel Dekker, Inc.

Jiang, K.S. and Timell, T.E. 1972. Polysaccharides in the bark of aspen (*Populus tremuloides*), III. The constitution of a galactoglucomann. *Cellulose Chem. Technol.* 6: 503–505.

Jones, R.W., Krull, J.H., Blessin, C.W., and Inglett, G. E. 1979. Neutral sugars of hemicellulose fractions of pith from stalks of selected woods, *Cereal Chem.* 56(5): 441.

Kai, Y. 1991. Chemistry of extractives. In: Hon, D.N.-S. and Shiraishi, N. (ed), *Wood and Cellulosic Chemistry*, Chapter 6, pp. 215–255, New York, NY: Marcel Dekker, Inc.

Kolattukudy, P.E. 1984. Biochemistry and function of cutin and suberin. *Canadian J. Botany* 62(12): 2918.

Kollmann, F.P. and Côté, W.A. Jr. 1968. *Principles of Wood Science and Technology*, New York, NY: Springer-Verlag.

Kurth, E.F. and Smith, J.E. 1954. The chemical nature of the lignin of Douglas-fir bark. *Pulp Paper. Mag. Canada.* 55: 125.

Labosky, P. 1979. Chemical constituents of four southern pine parks. *Wood Science* 12(2): 80–85.

Laks, P.E. 1991. Chemistry of bark. In: Hon, D.N.-S. and Shiraishi, N. (ed), *Wood and Cellulosic Chemistry*, Chapter 7, pp. 257–330, New York, NY: Marcel Dekker, Inc.

Martin, R.E. and Gray, G.R. 1971. pH of southern pine barks. *Forest Products Journal* 21(3): 49–52.

McGinnis, G.D. and Parikh, S. 1975. The chemical constituents of loblolly pine. *Wood Science* 7(4): 295–297.

Mian, A.J. and Timell, T.E. 1960. Isolation and characterization of a cellulose from the inner bark of white birch (*Betula papyrifera*). *Can. Journal Chemistry* 38: 1191–1198.

Moore, W. and Johnson, D. 1967. *Procedures for the chemical analysis of wood and wood products*, USDA, Forest Service, Forest Products Laboratory.

Nonaka, G., Nishimura, H., and Nishioka, I. 1985. Tannins and related compounds: Part 26. Isolation and structures of stenophyllanins A, B, and C, novel tannins from *Quercus stenophylla. J. Chem. Soc. Perkin Trans.* I. 163.

Nunes, E., Quilhó, T., and Pereira, H. 1996. Anatomy and chemical composition of *Pinus Pinaster* bark. *IAWA Journal* 17(2): 141–149.

Obst, J.R. 1982. Guaiacyl and syringyl lignin composition in hardwood cell components. *Holzforschung* 36(3): 143–153.

Painter, T.J. and Purves, C.B. 1960. Polysaccharides in the inner bark of white spruce. *Tappi* 43: 729–736.

Panshin, A.J. and de Zeeuw, D. 1980. *Textbook of Wood Technology*, McGraw-Hill, New York.

Pereira, H. 1988. Chemical composition and variability of cork from *Quercus suber L. Wood Sci. Technol.* 22: 211–218.

Pettersen, R.C. 1984. The chemical composition of wood. In: Rowell, R. M. (ed), *The Chemistry of Solid Wood*, Advances in Chemistry Series 20, Chapter 2, pp. 57–126, Washington, DC: ACS.

Porter, L.J. 1989. Condensed tannins. In: Rowe, J.W. (ed), *Natural Products of Woody Plants*. I Chapter 7.7, pp. 651–688, New York, NY: Springer-Verlag.

Rowe, J.W. (ed), 1989. *Natural Products of Woody Plants*, I and II. New York, NY: Springer-Verlag.

Rowell, R.M. 1980. Distribution of reacted chemicals in southern pine modified with methyl isocyanate. *Wood Sci.* 13(2): 102–110.

Saka, S. 1991. Chemical composition and distribution. In: Hon, D.N.-S. and Shiraishi, N. (eds), *Wood and Cellulosic Chemistry*, Chapter 2, pp. 59–88, New York, NY: Marcel Dekker, Inc.

Saka, S. and Goring. D.A.I. 1983. The distribution of inorganic constituents in black spruce wood as determed by TEM-EDXA. *Mokuzai Gakkaishi* 29: 648.

Sakakihara, A. 1991. Chemistry of lignin. In: Hon, D.N.-S. and Shiraishi, N. (ed), *Wood and Cellulosic Chemistry*, Chapter 4, pp. 113–175, New York, NY: Marcel Dekker, Inc.

Sarkanen, K.V. and Hergert, H.L. 1971. Classification and distribution, In: Sarkanen, K.V. and Ludwig, C.H. (eds), *Lignins, Occurrence, Formation, Structure and Reactions*, New York, NY: Wiley-Interscience.

Sarkanen, K.V. and Ludwig, C.H. (eds), 1971. *Lignins: Occurrence, Formation, Structure and Reactions*. New York, NY: Wiley-Interscience.

Schwerin, G. 1958. The chemistry of reaction wood. II. The polysaccharides of *Eucalyptus goniocalyx* and *Pinus radiate. Holzforschung* 12: 43–48.

Sandved, K.B., Prance, G.T., and Prance, A.E. 1992. *Bark: The Formation, Characteristics, and Uses of Bark Around the World*. Portland, OR: Timber Press.

Seshadri, T.R. and Vedantham, T.N.C. 1971. Chemical examination of the barks and heartwood of *Betula* species of American origin. *Phytochemistry* 10: 897.

Sjöström, E. 1981. *Wood Polysaccharides, Wood Chemistry, Fundamentals and Applications*, Chapter 3, pp. 51–67, New York, NY: Academic Press.

Stamm, A.J. 1964. *Wood and Cellulose Science*, New York, NY: The Ronald Press Co.

Timell, T.E. 1961a. Characterization of celluloses from the bark of gymnosperms. *Svensk Papperestid.* 64: 685–688.

Timell, T.E. 1961b. Isolation of polysaccharides from the bark of gymnosperms. *Svensk Papperstid.* 64: 651–661.

Timell, T.E. 1964. Wood hemicelluloses. Part 1. *Advances in Carbohydrate Chemistry.* 19: 247–302.

Timell, T.E. 1965. Wood hemicelluloses. Part 2. *Advances in Carbohydrate Chemistry.* 20: 409–483.

Timell, T.E. 1982. Recent progress in the chemistry and topochemistry of compression wood. *Wood Sci. Technology* 16: 83–122.

Toman, R., Karacsonyi, S., and Kubackova, M. 1976. Studies on pectin present in the bark of white willow (*Salic alba*), structure of the acidic and neutral oligosaccharides obtained by partial acid hydrolysis. *Cellulose Chem. Technol.* 10: 561.

Volz, K.R. 1971. Influence of inorganic content on the pH of bark. *Holz-Zentralbl.* 97: 1783.

Whistler, R.L. and Richards, E.L. 1970. Hemicelluloses. In: W. Pigman and D. Horton (eds), *The Carbohydrates*, Second edition Vol. 2A, pp. 447–469, New York, NY: Academic Press.

Whistler, R.L., Wolfrom, M.L., and BeMiller, J.N. (eds). 1962–1980. *Methods in Carbohydrate Chemistry*. Vols. 1–6, New York, NY: Academic Press.

Young, H.E. 1971. Preliminary estimates of bark percentages and chemical elements of bark percentages and chemical elements in complete trees of eight species in Main. *Forest Products Journal* 21(5): 56–59.

Young, H.E. and Guinn, V.P. 1966. Chemical elements in complete mature trees of seven species in Maine. *Tappi* 49(5): 190–197.

Part II

Properties

4 Moisture Properties

Roger M. Rowell

CONTENTS

Wood was designed by nature over millions of years to perform in a wet environment. The wood structure is formed in a water-saturated environment in the living tree and the water in the living tree keeps the wood elastic and able to withstand environmental strain such as high wind loads. We cut down a tree, dry the wood, and mainly use it in its dry state. But, wood in use remains a hygroscopic resource. Wood's dimensions, mechanical, elastic, and thermal properties depend on the moisture content. Wood is also anisotropic which means that its properties vary according to its growing direction [longitudinal (vertical or length direction), tangential (parallel to annual growth rings), and radial (perpendicular to the annual growth rings)]. The mechanical properties depend very much on both moisture content and growing direction.

4.1 MOISTURE CONTENT OF GREEN WOOD

Wood is composed of a complex capillary network through which transport occurs by capillarity, pressure permeability, and diffusion. Moisture exists in wood as both liquid moisture in the cell voids or lumens (free water) and as moisture in the cell wall (bound water). The moisture content in green wood is defined as the total amount of free and bound water in the living tree. This is the maximum moisture content that can exist in a living tree. The moisture content of green wood varies from species to species and depends on the specific gravity. The living tree holds much water in its cells. A southern pine log, 5 m long and 0.5 m in diameter, for example, may weigh as much as 1000 kg and contain ~47% or 0.46 m^3 water. Lumen volume decreases as the specific gravity increases so the green moisture content decreases with increasing specific gravity. The maximum moisture content M_{max} can be calculated by

$$M_{max} = \frac{100(1.54 - G_{bsg})}{1.54\,G_{bsg}}$$

where G_{bsg} is the basic specific gravity based on oven-dry weight and green volume and 1.54 is the specific gravity of the wood cell wall.

Using this equation, the maximum possible moisture content of green wood would be 267% with a basic specific gravity of 0.3 and the minimum possible moisture content would be 44% with a basic specific gravity of 0.9. The density of wood substance is approximately 1.5 g/cm^3. A species with a density of 0.5 g/cm^3 (based on oven-dry weight and volume) has a void volume of 66.6%. Each 100 kg of the totally dry wood occupies 0.2 m^3 and contains 0.067 m^3 wood substance and 0.133 m^3 void volume. The density of most woods fall between 320 and 720 kg/m^3 with a range of 160 kg/m^3 for balsa and 1040 kg/m^3 for some imported hardwoods.

Table 4.1 shows some average moisture contents of green heartwood and sapwood of some common United States wood species. In some cases, the green moisture content is highest in heartwood and in others the sapwood is highest in moisture content.

TABLE 4.1
Average Moisture Contents of Green Wood

Species	Moisture Content	
	Heartwood	Sapwood
Aspen	95	113
Basswood	81	133
Beech	55	72
Birch, Paper	89	72
Cedar, Incense	40	213
Cottonwood, Eastern	162	146
Douglas-fir, Costal	37	115
Elm, American	95	92
Fir, Balsam	88	173
Hemlock, Western	85	170
Maple, Sugar	65	72
Oak, Red	83	75
Pine, Longleaf	31	106
Pine, Ponderosa	40	148
Pine, Sugar	98	219
Poplar, Yellow	83	106
Redwood, Old growth	86	210
Spruce, Sitka	41	142
Sweetgum	79	137
Sycamore, American	114	130
Walnut, Black	90	73

Source: Adapted from Wood Handbook—Wood as an engineering material. 1999. Gen. Tech. Rep. FPL-GTE-113, Madison, WI: U.S. Department of Agriculture, Forest Service, Forest Products Laboratory, 463pp.

4.2   FIBER SATURATION POINT

As water is lost in green wood, there is no change in the volume of the wood until it reaches the Fiber Saturation Point (FSP). The FSP is defined as the moisture content of the cell wall when there is no free water in the voids and the cell walls are saturated with water. This point is called the fiber saturation point (FSP), and ranges from 20 to 50 wt.% gain depending on the wood species (Feist and Tarkow 1967). As moisture is removed below the FSP, the wood volume starts to shrink. As stated before, wood is anisotropic so the shrinkage in wood is different in all three growing directions. Figure 4.1 shows the change in wood shape as a cross section of a log is dried below the FSP. It can be seen that depending on where the piece of wood is located in the log, the wood will not only get smaller due to the loss of water but will become distorted due to the anisotropic properties of wood. As will be discussed later, tangential shrinkage is about twice that of radial shrinkage and longitudinal, in most woods, is almost zero.

The shrinkage of wood upon drying depends on several variables including specific gravity, rate of drying, and the size of the piece. As can be seen in Figure 4.1, the piece of wood cut from the center, left, and right middle is distorted the least (quarter sawing). While quarter sawing is somewhat wasteful, it does result in minimum distortion in the cut lumber.

Table 4.2 shows the average shrinkage values for some common United States woods. It can be seen that most radial shrinkage values are less that about 6%, most tangential shrinkage values less than 10% and most volumetric shrinkage values are less than 15%.

To determine the approximate volumetric shrinkage that would occur at a moisture content greater than oven-dry but less than the FSP, the approximate volumetric value at a given moisture can be calculated using the following formula using the date in Table 4.2:

$$S_m = S_o \cdot \frac{30 - M}{30}$$

where

S_m is the volumetric shrinkage at a given moisture content
S_o is the total volumetric shrinkage
M is the moisture content

Table 4.2 does not include any longitudinal shrinkage (shrinkage parallel to the grain) since is it usually less than 0.2% for almost all United States species. If a piece of wood is cut near the center of a tree that contains a large amount of juvenile wood or a piece containing reaction wood, the longitudinal shrinkage from green to oven-dry can be as high as 2%.

The size of the cell cavities remain almost the same size during the loss of water in the cell wall (Tiemann 1944). The thickness of the cell wall decreases in proportion to the moisture lost below

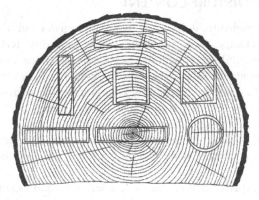

FIGURE 4.1   Shrinkage and distortion of wood upon drying.

TABLE 4.2

Average Radial, Tangential, and Volumetric Shrinkage

	Shrinkage from Green to Over-Dry Moisture Content		
Species	Radial	Tangential	Volumetric
Aspen	3.5	6.7	11.5
Basswood	6.6	9.3	15.8
Beech	5.5	11.9	17.2
Birch, Paper	6.3	8.6	16.2
Cedar, Incense	3.3	5.2	7.7
Cottonwood, Eastern	3.9	9.2	13.9
Douglas-fir, Costal	4.8	7.6	12.4
Elm, American	4.2	9.5	14.6
Fir, Balsam	2.9	6.9	11.2
Hemlock, Western	4.2	7.8	12.4
Maple, Sugar	4.8	9.9	14.7
Oak, Red	4.7	11.3	16.1
Pine, Longleaf	5.1	7.5	12.2
Pine, Ponderosa	3.9	6.2	9.7
Pine, Sugar	2.9	5.6	7.9
Redwood, Old growth	2.6	4.4	6.8
Spruce, Sitka	4.3	7.5	11.5
Sweetgum	5.3	10.2	15.8
Sycamore, American	5.0	8.4	14.1
Walnut, Black	5.5	7.8	12.8

Source: Adapted from Wood Handbook—Wood as an engineering material. 1999. Gen. Tech. Rep. FPL-GTE-113, Madison, WI: U.S. Department of Agriculture, Forest Service, Forest Products Laboratory, 463pp.

the FSP but the size of the cell lumen remains approximately constant. If this relationship is constant, the volumetric shrinkage V_s of a wood with a water soak specific gravity SP_w can be calculated as (Stamm and Loughborough 1942):

$$V_s = (M) (SP_w) \quad \text{or} \quad M = V_s / SP_w$$

This ratio should be the approximate FSP for most woods. The value of M for 107 hardwood species was 27 and for 52 softwood species the value was 26 (Stamm and Loughborough 1942).

4.3 EQUILIBRIUM MOISTURE CONTENT

As the green wood loses moisture, it does not change dimensions until the FSP is reached and after that the dimensions change respective of the relative humidity (RH) of the wood surroundings. When the wood is in equilibrium with the surrounding RH, the wood is defined as being at its equilibrium moisture content (EMC). The moisture content of wood is a dynamic property in that the moisture content of wood is constantly changing as the surrounding moisture content changes. When the wood stays at one RH for long periods of time, the wood will reach an equilibrium moisture content. Test results show that, for small pieces of wood at a constant RH, the EMC is reached in about 14 days. Larger wood members may take several weeks to reach its EMC. Wax and extractive content of wood can have a large effect on the length of time it takes for wood to reach its EMC.

Table 4.3 shows EMC experimental values for southern pine and aspen at 30%, 65%, 80%, and 90% RH.

TABLE 4.3
Equilibrium Moisture Content of Pine and Aspen

| | Equilibrium Moisture Content at | | | |
Species	30% RH	65% RH	80% RH	90% RH
Southern pine	5.8	12.0	16.3	21.7
Aspen	4.9	11.1	15.6	21.5

4.4 SORPTION ISOTHERMS

A sorption isotherm for wood is defined as the sorption of water in wood at a defined temperature. Figure 4.2 shows the sorption isotherm for Douglas-fir at 32°C (Spalt 1958). It is a plot of moisture content (M%) vs. relative vapor pressure (h = relative humidity/100); note that as moisture is lost from green wood (IN DES—initial descending), it follows a different curve than both the rewetting curve (ADS—adsorbing) and the second redrying (SEC DES—secondary descending) curve. The second, third, fourth, and all subsequent redrying and the second, third, and subsequent rewetting will approximately follow the SEC SEC and ADS curves. The difference between these curves is referred to as sorption hysteresis for wood (Skaar 1984). Understanding of the difference in moisture content from wet to dry and from dry to wet is very important when it comes to mechanical properties of wood that will be discussed later. The adsorbing (A) curve is always lower than the desorbing (D) curve and the A/D ratio generally ranges between 0.8 and 0.9 depending on the relative humidity and wood species (Okoh and Skaar 1980). Most mechanical properties of wood are very dependent on moisture content and Figure 4.2 shows that at a given relative humidity of say 60% (h = 0.6), the moisture content going from wet to dry is approximately 13% while the same wood going from dry to wet would have a moisture content of approximately 10%. This difference of 3% can make a major difference in mechanical properties of the wood. Wax and extractive content of wood can have a large effect on the sorption isotherm.

4.4.1 EFFECT OF TEMPERATURE ON SORPTION AND DESORPTION OF WATER

Sorption isotherms for wood decrease with increasing temperature at temperatures about 0°C. Figure 4.3 shows the effect of temperature on the sorption isotherm of wood (Skaar 1984). The

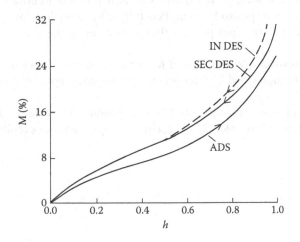

FIGURE 4.2   Sorption isotherm for wood.

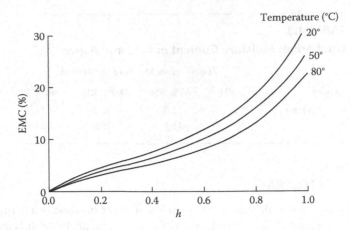

FIGURE 4.3   Effect of temperature on sorption isotherm.

desorption isotherm follows the same trend. Because temperature effects sorption and desorption of moisture in wood, EMC values must be determined at a single constant temperature.

4.5   SWELLING OF DRY WOOD IN WATER

Theoretically, it is possible to have absolutely dry wood. That is wood with a zero moisture content. In actual fact, wood with zero moisture content has never been achieved. When we talk about "oven dry" (dried above 100°C) wood, the actual moisture content is less than 1% but not zero. There is a small amount of water that is so tightly bound to wood that it is impossible to remove. Because of this, there is no such thing as the volume of zero moisture content wood. However, the amount of water in oven dry wood is so small, that the water volume is considered to be negligible.

Water content of wood is measured using the Karl Fisher titration method (Kollmann and Hockele 1962). A solution of pyridine, sulfur dioxide, and iodine in methanol are used. The reagents react with water according to the following equation (Figure 4.4). The end point of the titration may be determined either colorimetrically (free iodine). The method is best used on small samples with a moisture content less than 10%.

For wood to swell from the dry state, water or some other swelling agent must enter the cell wall. Entry may result from mass flow or diffusion of water vapor into the cell lumens and diffusion from there into the cell wall, or from a diffusion of bound water entirely within the cell wall. In most cases, both processes probably occur. Penetration by mass flow followed by diffusion into the cell wall is a much more rapid process than either vapor-phase or bound-water diffusion (Banks 1973).

Caulfield (1978) proposed a "zipper" model for the movement of water into the wood structure. That is, water moves through wood by forcing cell wall polymers apart as it moves deeper into the wood structure.

Although wood is a porous material (60–70% void volume), its permeability or flow of liquids under pressure, is extremely variable. This is due to the highly anisotropic shape and arrangement

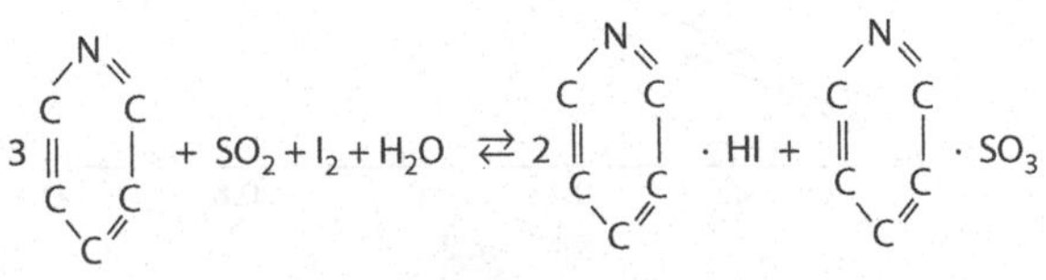

FIGURE 4.4   Reaction of the Karl Fisher reagents with water.

of the component cells and to the variable condition of the microscopic channels between cells. Wood is much more permeable in the longitudinal direction than in the radial or tangential directions. Because of this anisotropy, longitudinal flow paths are of major importance in the wetting of wood exposed to the weather (Miller and Boxall 1984). Wood is a hydroscopic resource which means that the hydroxyl groups in the cell wall polymers are attracted to and hydrogen bond with environmental moisture. As water is added to the cell wall, wood volume increases nearly proportionally to the volume of water added (Stamm 1964). Swelling of the wood continues until the cell reaches the FSP and water, beyond the FSP, is free water in the void structure and does not contribute to further swelling. This process is reversible, and wood shrinks as it loses moisture below the FSP. This dimensional instability has restricted wood from many applications where movement of a material due to changes in moisture content can not be tolerated.

The amount of swelling that occurs in wood because of hygroscopic expansion is dependent on the density of the wood (Stamm 1964). The percent volumetric swelling, V, is a function of the dry density d (in g/cm^3), and the FSP (K_{fsp}, in cm^3/g), so that

$$V = K_{fsp}d$$

This equation determines the approximate volumetric swelling of wood going from an oven-dry state to the FSP, and the approximate volumetric shrinkage of wood going from the FSP to oven-dry. Deviations from this relationship usually occur in species high in natural extractives.

All of the cell wall polymers (cellulose, the hemicelluloses, and lignin) are hydroscopic. The order of hydroscopicity is hemicellulose (HEMI) > cellulose > lignin (K LIG) as is shown in Figure 4.4 (Skaar 1984). The holocellulose isotherm is a combination of the hemicelluloses and cellulose polymers. The sorption of moisture by each cell wall polymer not only depends on its hydrophilic nature but also accessibility of water to the polymer's hydroxyl groups. Most, if not all, of the hydroxyl sites in the hemicelluloses are accessible to moisture and the same is probably true of the lignin. The noncrystalline portion of cellulose (~40%) and the surfaces of the crystallites are accessible to moisture but the crystalline part (~60%) is not (Stamm 1964, Sumi et al. 1964).

4.6   DISTRIBUTION OF MOISTURE

According to the Dent sorption theory, water is added to the cell wall polymers in monolayers (Figure 4.5) (Dent 1977). This theory is based on a modification of the BET model by Brunauer et al. (1938) that was based on a Langmuir model of 1918. Figure 4.5 shows the wood surface containing primary sorption sites (vertical lines), some of these sites are occupied by primary water molecules (dark circles) and some occupied by secondary water molecules (open circles). The primary sites are high-energy sites such as on hydroxyl groups and the secondary sites have lower energy.

The model also permits more than one layer of water on any particular sorption site. This means that when liquid water or water vapor comes into contact with wood, it does not concentrate in one spot but spreads out over the entire cell wall structure at equilibrium. Uneven swelling does occur when one part of the wood is wetter than another until equilibrium is reached.

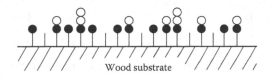

FIGURE 4.5   Schematic drawing of water molecules on the surface of wood.

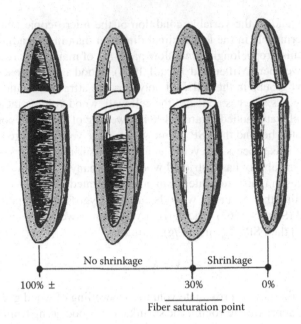

FIGURE 4.6   Swelling of the wood cell wall.

4.7   SWELLING AND SHRINKING OF WOOD

When green wood starts to lose moisture, it remains the same size until it reaches the fiber saturation point (Figure 4.6). Below the fiber saturation point, the wood cell wall starts to shrink and continues to shrink until dry. While the cell wall changes dimensions, the lumen stays the same size as was stated before.

This is demonstrated in Figure 4.7. The figure shows possible changes in the cell as the water content increases (Tiemann 1944). Wood at dimension of Figure 4.7a can stay the same size with the lumen getting smaller (Figure 4.7b) as moisture is sorbed, or the cell wall alone can swell to accommodate the water (Figure 4.7c) or both the cell wall and the lumen increase in volume (Figure 4.7d) (Skaar 1984). Experimental data show that Figure 4.7c takes place but some exotic woods do show some changes in lumen size as moisture is reduced.

4.8   MEASURING SWELLING

There are several ways to measure swelling resulting from interaction with water or other solvents in wood. Some report volumetric swelling which is a combination of radial, tangential, and longitudinal

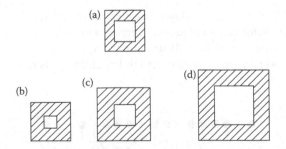

FIGURE 4.7   Theoretical changes in the cell wall and lumen dimensions. (a) Normal dry cell, (b) cell where lumen is smaller and wall is thicker, (c) cell wall thicker but lumen stays the same size, and (d) cell where both the cell wall and the lumen are larger.

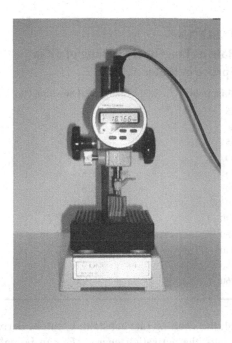

FIGURE 4.8   Flat bed micrometer for measuring wood dimensions.

swelling and some report just tangential swelling. Since tangential swelling is usually about twice that of radial, tangential swelling alone is representative of the swelling characteristics of each species.

In measuring the rate and extent of swelling in wood using a flat bed micrometer, it is best to cut specimens thin in the longitudinal direction for fast penetration of moisture into the wood with maximum tangential length. It is critical that the tangential grain runs parallel to the cut top and bottom edge. If the tangential grain is not parallel to the top and bottom edge, the specimen will go out of square when it swells and will not be measured accurately in the flat-bed micrometer (Figure 4.8). The micrometer can be connected to a computer so that measurements can be taken accurately and quickly.

4.9   RATE OF WATER SORPTION AND ACTIVATION ENERGY

The rate of swelling of Sitka spruce wood at 23°C is shown in Table 4.4. The specimen size was small (2.5 cm^2) but still it takes some time for water to penetrate into the wood structure. Table 4.4 shows that more than half of the final tangential swelling occurs in the first 15 min of liquid water contact with the wood. Since the swelling rate depends of specimen size, the rate of swelling would be faster with a thinner specimen (in the longitudinal direction) but the extent of swelling at equilibrium would be the same. Tangential swelling was determined as follows:

$$\text{Tangential swelling} = \frac{(\text{swollen dimension} - \text{oven dry dimension})}{\text{oven dry dimension}}$$

The rate of swelling of wood in water or other solvents is dependent on several factors (Banks and West 1989). Hydrogen bonding ability, molecular size of the reagent, extractives content, temperature, and specimen size. In the case of water, there is an initial induction period due to the diffusion of the water into the cell wall structure and then the water penetrates cell wall capillaries and moves from lumen to lumen in the fiber direction.

TABLE 4.4
Rate of Tangential Swelling of Sitka
Spruce in Water at 23°C

Time (min)	Tangential Swelling (%)
15	5.3
30	7.0
45	7.8
60	8.1
75	8.4
90	8.5
105	8.6
120	8.7
180	8.8
240	8.8
480	9.0
960	9.0

Using the data in Table 4.4, the swelling rate constant, k, can be derived from the slope of a plot of $\ln k$ vs $1/T$. From these data, the activation energy, E_a can be calculated from the standard Arrhenius equation:

$$k = Ae^{-E_a/RT}$$

where

A = constant
R = gas constant
T = temperature (°K)

Table 4.5 shows the maximum tangential swelling of wood in water, the swelling rate (k) and the activation energy of swelling (Mantanis et al. 1994a) for Sitka spruce, Douglas fir, and sugar

TABLE 4.5
Rate of Swelling, Maximum Tangential Swelling, and Activation
Energy of Wood in Water

Temperature (°C)	Sitka Spruce	Douglas Fir	Sugar Maple
	Maximum Tangential Swelling (%)		
23	6.5	7.6	10.0
40	6.9	7.7	10.8
60	6.9	7.7	11.5
80	7.2	8.0	11.5
100	7.3	8.1	12.3
	Swelling Rate, k		
23	1.3	0.3	0.3
40	3.3	0.9	1.1
60	6.5	1.8	3.6
80	10.8	4.7	6.6
100	14.8	6.9	7.6
Activation energy, E_a (kJ/mol)	32.2	38.9	47.6

TABLE 4.6
Rate of Swelling and Activation Energy for Unextracted and Extracted Woods

	Sitka Spruce		Douglas Fir		Sugar Maple	
	Unextracted	Extracted	Unextracted	Extracted	Unextracted	Extracted
Maximum Tangential Swelling (%) at T (°C)						
23	6.5	7.4	7.6	9.0	10.0	10.7
40	6.9	7.5	7.7	9.1	10.8	11.6
60	6.9	7.7	7.7	9.1	11.5	11.4
80	7.2	7.7	8.0	9.1	11.5	11.2
100	7.3	7.7	8.1	9.2	12.3	11.3
Swelling Rate, k at T (°C)						
23	1.3	7.2	0.3	0.2	0.3	0.9
40	3.3	14.9	0.9	1.2	1.1	4.0
60	6.5	25.6	1.8	1.4	3.6	7.1
80	10.8	32.4	4.7	3.2	6.6	16.7
100	14.8	36.2	6.9	3.4	7.6	20.7
Activation energy, E_a (kJ/mol)	32.2	23.3	38.9	41.5	47.6	42.3

maple. Of these three species, Sitka spruce swells the fastest and has the lowest activation energy. The rate of swelling increased with an increase in temperature (see Table 4.6). The higher density hardwood had greater activation energies as compared to softwoods which Stamm had found earlier (1964).

The presents of extractives also has a great effect on the rate of swelling of wood in water and other liquids. Stamm and Loughborough (1942) discuss two types of extractives in wood: extractives deposited in the coarse capillary structure and extractives deposited in the cell wall structure. The extractives deposited in the cell wall structure have a great influence on the rate of swelling. The effect of extractives on swelling rate as a function of temperature is shown in Table 4.6. Extractives were removed using 80% ethanol in water for 2 h at room temperature. In the case of Sitka spruce, the swelling rate greatly increases with the removal of extractives but the removal of extractives from Douglas fir and sugar maple has less effect.

4.10   CELL WALL ELASTIC LIMIT

The orientation of the cell wall polymers in the S_1 layer in with determines the extent of swelling of wood in water or other solvent. Swelling from the dry state to the water saturation state continues until the cross banding of the cell wall polymers in the S_1 layer restricts further swelling. This point is defined as the elastic limit of the cell wall. In many cases, data on swelling in solvents other than water, is given relative to the swelling in water so that the value of water is set at value of 1, 10, or 100 and all other solvents are reported above or below this value. Some solvents do swell wood greater than water due to several factors including extractive removal, plasticization, softening or solubilization of one or more of the cell wall polymers (usually lignin or hemicelluloses) (see Section 4.19).

4.11   SWELLING PRESSURE

The maximum swelling pressure of wood has been measured in wood (Tarkow and Turner 1958). It was measured by inserting wooden dowels into steel restraining rings equipped with a strain gage.

FIGURE 4.9   Evidence of holes cut in rock to split the rock (Egypt 2001).

The wooden dowels of different densities were then wetted with liquid water and the swelling pressure measured as a function of density. Extrapolation of the curve of swelling pressure vs density to the density of the cell wall (1.5) gave value of 91 MPa (13,200 psi). A theoretical value based on an osmotic pressure theory, gave a theoretical value of 158 MPa at room temperature. The difference between the actual and the theoretical values was thought to be due to hydroelastic factors. These factors are associated with the fact that restrained wood has a lower moisture content as compared to un-restrained wood.

The best example of using the swelling pressure of wood to practical use goes back to the Egyptians. On a trip to Egypt in 2001, the author saw the evidence first hand. On the edge of a shear granite cliff, holes about 15 cm long, 5 cm wide, and 10 cm deep had been chiseled into the rock. Dry wooden wedges the size of the cavities were driven into these holes. Water was then poured onto the dry wood and allowed to swell. The swelling pressure caused the granite to split down the chain of holes resulting in a giant obelisk or other large building blocks. Figure 4.9 shows the series of holes that were chiseled into a large rock and Figure 4.10 shows the result of the splitting process.

FIGURE 4.10   Rock split by using the swelling pressure of wood (Egypt 2001).

4.12   EFFECTS OF MOISTURE CYCLES

Drying and rewetting causes an increase in both the rate of swelling and the extent of swelling. The degradation and extraction of hemicelluloses and extractives as well as some degradation of the cell wall structure during wetting, drying and rewetting cycles results in more accessibility of water to the cell wall. This is especially evident in hot or boiling water extraction of wood where significant amounts of cell wall polymers can be lost. A similar effect occurs with wood is exposed to high relative humidity in repeated cycles. Even though no cell wall polymers are extracted, repeated humidity cycles result in a slight increase in moisture content with each cycle.

4.13   EFFECT ON VIBRATIONAL PROPERTIES

Moisture has a great effect on vibrational properties of wood. Because wood is a viscoelastic material, vibrational properties are highly dependent on the elasticity of, as well as the internal friction within, the cell wall polymers and matrix. One way of studying the viscoelastic properties of wood is through vibrational analysis. A simple harmonic stress results in a phase difference between stress and strain. The ratio of dynamic Young's modulus (E') to specific gravity (γ), that is, E'/γ = specific modulus and internal friction (tangent of the phase angle, tan δ) measurements can be used to study the visoelastic nature of wood. The E'/γ is related to sound velocity and tan δ to sound absorption or damping within the wood.

The sorption of water molecules between the wood cell wall polymers acts as a plasticizer to loosen the cell wall microstructure. This affects the tone quality of wooden musical instruments because as moisture content increases, the acoustic properties of wood, such as specific dynamic Young's modulus and internal friction are reduced or dulled (James 1964, Sasaki et al. 1988, Yano et al. 1993). The decrease in cohesive forces in the cell wall also enhances the deformation of wooden parts under stress.

In practical terms, as the moisture content of a wooden musical instrument, such as an oboe, clarinet, or recorder, increases, the quality of the sound, the brightness of the tone and the separation of sound between notes decreases.

4.14   EFFECT ON BIOLOGICAL PROPERTIES

The ability of microorganisms to attack wood depends on the moisture content of the cell wall. The old saying, "dry wood does not rot" is basically true. Ten-thousand year old wood from tombs in China that has remained dry through the years is essentially the same wood today as it was when it was first used (Rowell and Barbour 1990). Termites may seem to attack dry wood but, in fact, they bring their own moisture to the wood. White-rot fungi need the least water to attack, brown-rot fungi require more, and soft-rot fungi require the highest. But all of them require moisture at or near the fiber saturation point to be able to degrade wood (see Chapter 5).

4.15   EFFECTS ON INSULATION AND ELECTRICAL PROPERTIES

Thermal and electrical conductivity is very low in dry wood. Early log cabins were warmer in the winter and cooler in the summer due to the insulation properties of wood. Thermal conductivity increases with increasing moisture content. Heat transmission through dry wood is slow but heat transfer is much faster in moist wood using the water as the heat conduit.

The electrical resistance of wood is extremely sensitive to the woods moisture content. Moisture meters that determine the moisture content of wood are based on this sensitivity. There is also a strong increase in resistance with a decrease in wood temperature. Moisture meters measure the resistance between pairs of pin electrodes driven into the wood to various depths. The meter is calibrated by using data obtained on a given species at room temperature. Meter readings are less reliable at moisture contents above about 25% and read high when used on hot wood and vice versa on cold wood.

TABLE 4.7
Wet and Dry Strength and Stiffness of Different Soft Woods

Wood Species	MOR 12% kPa	MOR Wet kPa	Difference (%)	MOE 12% MPa	MOE Wet MPa	Difference (%)
Douglas fir	85,000	53,000	−38	13,400	10,800	−19
Loblolly pine	88,000	50,000	−43	12,300	9,700	−21
Sitka spruce	65,000	34,000	−48	9,900	7,900	−20

Source: Adapted from Wood Handbook—Wood as an engineering material. 1999. Gen. Tech. Rep. FPL-GTE-113, Madison, WI: U.S. Department of Agriculture, Forest Service, Forest Products Laboratory, 463pp.

4.16 EFFECTS ON STRENGTH PROPERTIES

Changes in moisture content of the wood cell wall below the fiber saturation point have a major effect on the mechanical properties of wood (Table 4.7; Winandy and Rowell 1984). Mechanical properties change very little at moisture contents above the fiber saturation point. Mechanical properties increase with decreasing moisture content with compression parallel to the grain being the most affected (see Chapter 11).

4.17 WATER REPELLENCY AND DIMENSIONAL STABILITY

The terms water repellency and dimensional stability are often used interchangeably as if they were the same. They are very different concepts. Water repellency is a rate phenomenon and dimensional stability is an equilibrium phenomenon (Rowell and Banks 1985). Confusion over these two concepts has led to some product failures in service costing contractors or owners considerable money.

A water-repellent treatment is one that prevents or slows down the rate moisture or liquid water is taken up by the wood. Examples of water repellents include coating, surface applied oils, or lumen filling. A dimensional stability treatment is one that reduces or prevents swelling in wood no matter how long it is in contact with moisture or liquid water. Examples of dimensional stability treatments include bulking the cell wall with polyethylene glycol, penetrating polymers or bonded cell wall chemicals, or cross-linking cell wall polymers (Rowell and Youngs 1981).

An attractive force exists between a solid and a liquid in contact with it. The net value of this force is governed by the relative magnitudes of the cohesive forces within the liquid and the adhesive forces generated between the liquid and solid. Where the adhesion of liquid to solid is equal to or greater than the cohesion of the liquid, a drop of liquid in contact with the solid spreads spontaneously. That is, the angle between solid and liquid at the solid/liquid/air interface, termed the "contact angle" is zero. If the liquid/solid adhesion is less than the liquid cohesion, an applied liquid droplet does not spread but stands on the surface making a finite contact angle with it (see

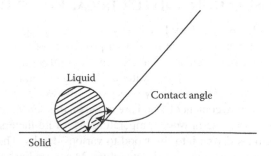

FIGURE 4.11   Contact angle of a water drop on wood.

Figure 4.11). The magnitude of the contact angle increases as the magnitude of the adhesive forces relative to the cohesion of the liquid decreases (Adam 1963). These relationships are expressed algebraically as the Young equation:

$$\gamma_S = \gamma_{SL} + \gamma_L \cos\theta$$

where

γ_S = surface tension of the solid
γ_{SL} = liquid/solid interfacial tension
γ_L = surface tension of the liquid
θ = contact angle between liquid and solid

Wood is a capillary porous medium. The pore structure is defined by the lumina of the cells and the cell wall openings (pits) interconnecting them. The primary routes for liquid penetration into wood are by these two routes. Except in the case where the contact angle is equal to 90° ($\cos\theta = 0$), any liquid contained in a cylindrical capillary of uniform bore has a curved surface. The pressure difference (Pc), often called the capillary pressure, across this curved surface is given by the following relationship, derived from the Kelvin equation:

$$Pc = \frac{-2\,\gamma_L \cos\theta}{r}$$

where

γ_L = surface tension of the liquid
θ = contact angle between liquid and solid
r = capillary radius

The pressure gradient set up by the pressure difference acts in the sense that liquid is forced into the capillary spontaneously for values of θ less than 90°. Conversely, where θ is greater than 90°, external pressure larger than Pc must be applied to force liquid in to the capillary. Although wood structure departs significantly from a simple cylindrical capillary model, the general principles of capillary penetration into its structure hold and the magnitude of Pc remains functionally related to the cosine of the contact angle.

In systems involving water as the liquid phase, surface forming contact angles less than 90° are said to be hydrophilic, whereas those giving a contact angle greater than 90° are said to be hydrophobic or water repellent.

The chemistry of surfaces giving rise to these properties is fairly well understood. Those surfaces presenting polar functional groups, especially those capable of forming hydrogen bonds with water, tend to be hydrophilic. In contrast, surfaces consisting of nonpolar moieties tend to be strongly hydrophobic.

Water-repellent effectiveness (WRE) is also measured as a time-dependent function of swelling as given

$$WRE = \frac{Dc - Dt}{Dc} \times 100$$

where

Dc = swelling (or weight of water uptake) of control during exposure in water for "t" minutes
Dt = swelling (or weight of water uptake) of treated specimen for the same "t" time

For a given exposure time, considerable variation in WRE may result from variations in specimen geometry and in permeability to water of the wood species. To ensure good reproducibility of swelling or water uptake tests, conditions must be closely specified and carefully adhered to.

As was stated before, water repellents are applied to wood principally to prevent or reduce the rate of liquid water flow into the cellular structure and do not significantly alter the dimensions or water sorption observed at equilibrium. Usually the treatments involve the deposition of a thin layer of a hydrophobic substance onto external and to some extent, internal cell lumen surfaces of wood. The measured WRE varies between 0% and 100% depending on the time the test specimens are exposed to water. In some cases, the time to reach equilibrium may be weeks, months or even years but eventually, maximum swelling will be reached at equilibrium. Generally, the equilibrium moisture content is not altered by the water-repellent treatments.

Water-repellent treatments, such as impregnation with a wax, may fail due to a failure of the bond formed between the wax and the wood cell wall. This is usually due to degradation of the wood (Banks and Voulgaridis 1980). Wood flooring that has been treated with methyl methacrylate or a similar monomer and polymerized *in situ* also give a very high WRE. Moisture is physically blocked from entering lumens and penetration of the water must proceed by wicking through the cell wall. Moisture pickup can take a very long time so this type of treatment can be confused as a treatment for dimensional stability.

Table 4.8 shows the results of three resin/wax treatments on WRE, contact angle, and the time it takes for the contact angle to fall to 90°. The data indicate that the hydrophobic effect (contact angle) only partly controls the water resistance of specimens treated with the resin/wax systems.

In contrast to water repellency which is a measure of the rate of moisture pickup, dimensional stability is the measure of the extent of swelling resulting from ultimate moisture pickup. A variety of terms have been used to describe the degree of dimensional stability given to wood by various treatments. The most common term, antishrink efficiency or ASE, is misleading because it seems to apply to shrinking and not to swelling. The abbreviation is acceptable because ASE might also stand for antiswelling efficiency; this usage requires a statement whether the values were determined during swelling or shrinking, and, in addition, because both are possible, whether they were obtained in liquid water or water vapor.

Changes in wood dimensions are a result of moisture uptake can be measured as a single-dimensional component, that is, usually tangential alone, but can also be measured volumetrically taking into account all three-dimensional changes. Calculations for dimensional stability are

$$S = \frac{V_2 - V_1}{V_1} \times 100$$

TABLE 4.8
Water-Repellent Effectiveness, Contact Angle, and Time for Contact Angle to Fall to 90°

Treating Solution	Concentration (%)	WRE (%)	Contact Angle (θ)	Time for θ to Fall to 90 (Min)
Alkyd resin	10	78	130	425
Paraffin wax	0.5			
Hydrocarbon resin	10	64	131	416
Paraffin wax	0.5			
Rosin ester	10	68	136	460
Paraffin wax	0.5			

Source: Adapted from Rowell, R.M. and Banks, W.B. 1985. Water repellency and dimensional stability of wood. USDA Forest Service General Technical Report FPL 50. Forest Products Laboratory, Madison, WI, 24pp.

where

S = Volumetric swelling coefficient
V_2 = Wood volume after humidity conditioning or wetting in water
V_1 = Wood volume of ovendried wood before conditioning or wetting

Then,

$$\text{ASE} = \frac{S_2 - S_1}{S_1} \times 100$$

where

ASE = Antishrink efficiency or reduction in swelling resulting from a treatment
S_2 = Volumetric swelling coefficient of ovendried after treatment
S_1 = Volumetric swelling coefficient of ovendried before treatment

The test conditions for determining the efficiency of a treatment to reduce dimensional changes depend on the treatment as well as the application of the treated product. For a water-leachable treatment, such as polyethylene glycol, a test method based on water vapor is usually used. Humidity tests are also applied to products intended for indoor use. Nonleachable treatments and products intended for exterior use are usually tested in liquid water. For changes in wood dimensions resulting from changes in humidity, the test must be continued long enough to ensure that final equilibrium swelling has been reached.

For a series of humidity cycles, the specimens are placed back and forth in the different humidity conditions and the swelling coefficients determined. The cycles can be repeated many times to show the efficiency of the treatment when exposed repeatedly to extremes of humidity.

If a liquid water test is used on a leachable chemical treatment, then a single water swelling test may be run. Because the treating chemicals are leached out during the swelling test, cyclic water tests cannot be done.

Nonleachable treatments are usually tested for dimensional stability by water soaking. The swelling coefficients are determined on the first swelling, again on the first drying, and again on the second wetting. A series of water soaking and drying cycles give the best indication of the durability of the treatment (Rowell and Ellis 1978). This repeated water soaking and ovendrying test is very severe and may result in specimen checking and splitting. It should only be used to determine the effectiveness of treatments for outdoor applications under the most harsh conditions.

Since determination of dimensional stability is based on a comparison between an untreated and a treated specimen, it is critical that the treated specimen come from the same source as the control. Usually, specimens are cut from the same board taking every other specimen for treatment and the other as controls. To illustrate this point, the S value for southern pine earlywood is 6–9, while southern pine latewood is 17–20. The average swelling coefficient for sample, therefore, depends on the proportion of latewood and earlywood. If a control is used to compare with a treated sample that differs in percent of latewood, then the values obtained for ASE are nearly useless.

Treatments that have been used for improving dimensional stability include lumen filling (methyl methacrylate), non-cell wall bonded-leachable (polyethylene glycol), non-cell wall bonded-nonleachable (phenol–formaldehyde) and cell wall bonded (reaction with acetic anhydride). Table 4.9 shows the ASE values for these treatments. More information on the chemistry of these treatments is found in Chapter 15.

TABLE 4.9
Antiswell and Antishrink Efficiency of Wood Treatments

Treatment[a]	Antiswell Efficiency 1st Wetting	Antishrink Efficiency 1st Drying	Antiswell Efficiency 2nd Wetting	Antishrink Efficiency 2nd Drying
Methyl methacrylate 65 WPG	15	20	13	20
Polyethylene glycol	85	15	10	10
Phenol-formaldehyde	83	80	82	81
Acetic anhydride 21 WPG	82	85	83	85

[a] WPG = weight percent gain.

Complete dimensional stability of wood, that is, ASE of 100, has only been accomplished through the process of petrification. Chemical treatments are unlikely ever to achieve this level of stabilization.

4.18 SWELLING IN WOOD COMPOSITES

The swelling that occurs in wood composites, such as flake-, particle-, or fiberboard, is much greater than in the wood itself (see Chapter 10). This is due to the release of compressive forces as well as normal wood swelling. The compressive forces are a result of the physical compression of the wood elements during pressing of the board. This type of swelling is known as irreversible swelling and is the release of the compressive forces during the first wetting of the composites. It is known as irreversible swelling because it is not reversible upon redrying. Reversible swelling also occurs during wetting which is the swelling of wood elements as a result of moisture pickup. It is known as reversible swelling because the wood shrinks again as a result of redrying.

Most of the irreversible swelling occurs early in the wetting of a composite but the rate can be slow due to the presence of waxes or other chemicals added to retard the rate of moisture pickup.

4.19 SWELLING IN LIQUIDS OTHER THAN WATER

The maximum swelling of wood in various organic liquids is mainly influenced by three solvent properties: the solvent basicity, the molar size, and the hydrogen bonding capacity of the liquid. In addition to these are the extractives content, temperature, and specimen size. An increase in molecule size not only decreases the swelling rate but the swelling equilibrium is also decreased due to the larger molecule slow diffusion into the fine capillary structure of the wood. Swelling of wood in organic liquids is closely associated with the swelling of cellulose alone (Stamm 1935, 1964, Stamm and Tarkow 1950, West 1988, Banks and West 1989, Mantanis et al. 1994a,b).

Table 4.10 shows the maximum tangential swelling of Sitka spruce, Douglas fir and sugar maple at 23°C (Mantanis et al. 1994b). The three different woods swell to different extents n the same organic liquid. For example, 2-methyl pyridine swells sugar maple almost twice as much as it does Douglas fir. Sugar maple also swells greater than Douglas fir or Sitka spruce in acetic acid, pyrrole, propionic acid, benzyl alcohol and 2,6-dimethyl pyridine.

Table 4.11 shows the decreasing volumetric swelling coefficient order of pine sapwood in various liquids at two different temperatures (Rowell 1984). Several organic liquids swell wood greater than water. It is believed that this is due to a partial softening and solubilization of the lignin in the cell wall that allows the wood to swell to a larger volume than in water (Stamm 1964). The largest swelling occurs in methyl isocyanates at 120°C but this is due to a noncatalyzed reaction of the isocyanates with the wood cell wall hydroxyl groups forming a urethane bond. The reacted wood is much

TABLE 4.10
Maximum Tangential Swelling of Wood at 23°C, 100 days

Solvent	Sitka Spruce	Douglas Fir	Sugar Maple
Water	8.4	8.8	10.6
n-Butylamine	14.5	16.4	19.3
Dimethyl sulfoxide	13.9	13.7	14.5
Pyridine	12.2	11.4	13.3
Dimethyl formamide	11.8	11.5	13.2
Formamide	11.2	9.6	16.8
2-Methylpyridine	10.8	11.6	11.9
Diethylamine	10.1	10.4	11.0
Ethylene glycol	9.5	9.1	10.4
Acetic acid	8.7	7.6	10.2
Methyl alcohol	8.2	7.3	8.7
Pyrrole	7.6	6.9	12.0
Butyrolacetone	7.2	7.5	9.6
Ethyl alcohol	7.0	6.3	6.9
Propionic acid	6.4	6.3	10.0
Acetone	5.0	4.6	7.1
Dioxane	5.7	7.5	8.4
Furfural	5.5	5.4	7.6
Methyl acetate	5.0	4.7	5.6
Propyl alcohol	4.9	5.1	5.3
Nitromethane	4.5	4.0	5.4
2-Butanone	4.3	4.1	5.1
Benzyl alcohol	2.9	2.6	8.9
Ethyl acetate	2.6	2.7	3.7
Propyl acetate	2.2	1.4	3.1
Ethylene dichloride	2.1	2.1	4.6
Toluene	1.6	1.5	1.5
Isopropyl ether	1.5	1.3	1.0
Chloroform	1.4	1.5	3.9
Dibutylamine	1.3	1.0	0.8
2,6-Dimethyl pyridine	1.3	1.0	8.3
Carbon tetrachloride	1.2	1.3	1.1
Benzaldehyde	1.0	0.9	2.1
Benzyl benzoate	1.0	0.9	1.0
Piperidine	0.9	0.2	4.3
Octane	0.9	0.7	0.6
Butylaldehyde	0.7	0.7	1.2
Nitrobenzene	0.5	0.4	0.7
Quinoline	0.4	0.3	0.6

larger than the green volume resulting from cell wall rupture due to excess bonded chemical in the cell wall (see Chapter 15). Acetic anhydride also reacts with cell wall hydroxyl groups at 120°C but there is no rupture of the cell wall occurring with this reaction.

Table 4.11 also shows at a high temperature (120°C) for 1 h versus room temperature (25°C) for 24 h in various liquids. Solvents such as n-butyl amine, dimethyl sulfoxide, dimethylformanide, pyridine, formaldehyde, cellosolve, and methyl cellosolve swell wood to the same extent at both the high and low temperature. Piperidine, aniline, 1,4-dioxane, and butylene oxide swell wood to a

TABLE 4.11
Volumetric Swelling Coefficients for Southern Pine
Sapwood in Various Solvents

Solvent	V 120°C, 1 h	V 25°C, 24 h
Methyl isocyanate	52.6[a]	5.1
n-Butylamine	15.5	15.2
Piperidine	13.3	0
Dimethyl solfoxide	13.3	11.7
Dimethyl formamide	12.8	12.5
Pyridine	11.3	13.1
Formaldehyde	12.3	12.3
Acetic anhydride	12.3[a]	1.5
Acetic acid	11.1	8.8
Aniline	11.0	0.5
Cellosolve	10.6	10.2
Methyl cellosolve	10.3	10.0
WATER	10.0	10.0
Methyl alcohol	9.0	9.3
Epichlorohydrin	6.9	5.9
Acrolein	6.7	7.0
1,4-Dioxane	6.5	0.6
Tetrahydrofuran	5.4	7.2
Propylene oxide	5.2	5.0
Acetone	5.1	5.6
Diethylamine	5.0	11.0
Acrylonitrile	4.6	4.5
Butylene oxide	4.1	0.7
Dichloromethane	3.8	3.3
Methyl ethyl ketone	3.6	5.0
N-Methyl aniline	2.6	0.8
Ethyl acetate	2.4	4.2
Cyclohexanone	2.3	0.5
N-Methylpiperidine	2.2	1.6
4-Methyl-2-pentanone	0.4	1.5
N,N-Dimethylaniline	0.3	0.5
Xylenes	0.1	0.2
Cyclohexane	0.1	0.1
Triethylamine	−0.1	2.1
Hexanes	−0.2	0.2

[a] Reaction occurred.

much greater extent at the high temperature as compared to the low temperature. Pyridine, tetrahydrofuran, and diethylamine swell wood to a greater extent at the low temperature as compared to the high temperature. Liquids such as trithylamine and hexanes cause a slight shrinkage of the wood at the high temperature.

Table 4.12 shows the effects of the extractives on the maximum tangential swelling of Sitka spruce and sugar maple at 23°C in various organic liquids (Mantanis et al. 1995a). Swelling is usually larger in the extracted wood. In the case of sugar maple in 2,6-dimethyl pyridine and piperidine, the swelling is much greater in the extracted wood as compared to the unextracted wood.

TABLE 4.12

Maximum Tangential Swelling of Wood at 23°C, 100 days

Solvent	Sitka Spruce		Sugar Maple	
	Unextracted	Extracted	Unextracted	Extracted
Water	5.9	6.3	9.5	10.6
n-Butylamine	11.1	13.3	17.7	19.6
Dimethyl sulfoxide	8.6	9.1	15.7	16.0
Pyridine	8.4	8.6	12.3	14.3
Dimethyl formamide	8.0	8.2	12.5	14.4
Formamide	8.1	8.0	13.3	13.9
2-Methylpyridine	7.8	10.8	11.6	13.2
Diethylamine	7.9	8.6	11.1	11.6
Ethylene glycol	6.9	7.1	10.0	10.8
Acetic acid	5.6	5.7	9.4	11.4
Methyl alcohol	5.8	6.1	9.2	10.2
Butyrolacetone	5.0	6.5	8.3	10.0
Ethyl alcohol	5.0	5.6	8.2	9.7
Propionic acid	4.1	5.3	7.9	9.5
Acetone	4.6	4.7	6.4	8.0
Dioxane	5.3	6.7	8.3	9.9
Furfural	4.0	5.7	7.6	9.4
Methyl acetate	3.7	4.2	5.2	6.9
Propyl alcohol	4.3	4.4	5.6	6.7
Nitromethane	2.7	2.8	5.3	6.6
2-Butanone	4.2	4.2	5.2	7.0
Ethyl acetate	2.8	3.1	4.0	6.1
Propyl acetate	1.1	2.0	2.3	4.6
Ethylene dichloride	2.2	2.1	3.3	4.6
Toluene	1.2	1.3	1.5	2.3
Isopropyl ether	1.7	1.9	1.4	1.8
Chloroform	2.6	3.4	4.3	6.4
Dibutylamine	0.8	0.8	0.5	0.5
2,6-Dimethyl pyridine	1.1	1.9	1.7	10.0
Carbon tetrachloride	1.1	1.5	1.4	1.7
Benzaldehyde	1.6	2.2	1.7	8.5
Benzyl benzoate	1.4	1.5	1.0	1.6
Piperidine	1.3	2.3	1.2	10.6
Octane	1.1	1.2	0.7	0.8
Butylaldehyde	0.4	0.6	1.0	2.2
Nitrobenzene	1.8	1.9	1.7	3.7

Table 4.13 shows the effect of the organic liquids basicity, hydrogen bonding potential, molar volume, and molecular weight on wood swelling (Mantanis et al. 1995b). Nayer (1948) and Stamm (1964) had speculated that only hydrogen bonding ability alone could explain the relative swelling of wood in different organic liquids. There are, however, several significant exception to this correlation including di-n-butyl amine, tri-n-butyl amine, benzaldehyde (Mantanis et al. 1994b). Swelling of wood in organic liquids can be predicted much more accurately using the four parameters of basicity, hydrogen bonding ability, molar volume, and molecular weight.

TABLE 4.13

Properties of Swelling Solvents Affecting Wood Swelling

Solvent	Basicity[a] (kcal/mol)	Hydrogen Bonding[b]	Molar Volume[c] (cm³)	Molecular Weight
Butyl amine	57.0	16.8	98.80	73.10
Pyridine	33.1	18.1	80.40	79.10
Dimethyl sulfoxide	29.8	7.7	71.00	78.10
Dimethyl formamide	26.6	11.7	77.00	73.10
Acetic acid	25.0	9.7	57.10	60.00
Formaldehyde	24.0	21.5	39.90	45.04
Ethyl alcohol	20.0	18.7	58.50	46.07
Methyl alcohol	19.0	18.7	40.70	32.04
Ethylene glycol	18.0	20.6	55.80	62.10
Water	18.0	39.0	18.05	18.02
Propyl alcohol	18.0	18.7	75.00	60.10
Propyl acetate	17.4	8.6	115.7	102.10
Ethyl acetate	17.1	8.4	98.50	88.10
Acetone	17.0	9.7	74.00	58.10
Methyl acetate	16.5	8.4	79.70	74.10
Dioxane	14.8	9.7	85.70	88.10
Chloroform	7.0	1.5	80.70	120.40
Nitrobenzene	4.4	2.8	102.30	123.10
Carbon tetrachloride	3.0	0.0	97.10	153.80
Toluene	3.0	4.5	106.40	92.10
Octane	0.0	0.0	162.50	114.20

[a] Gutmann (1976).

[b] Wave number shift × 10—Crowley et al. (1966), Gordy (1939)

[c] Robertson (1964), Lide (1993).

REFERENCES

Adam, N.K. 1963. Water proofing and water repellency. In: J.L. Moilliet, ed. *Principles of Water Repellency*. London, Elsevier.

Banks, W.B. 1973. Water uptake by Scots pie and its restriction by the use of water repellents. *Wood Sci. and Tech.* 7: 271–284.

Banks, W.B. and Voulgaridis, E. 1980. The performance of water repellents in the control of moisture absorption by wood exposed to the weather. *Records of the Annual Convention, British Wood Preservation Association;* June 24–27, Cambridge. British Wood Preservation Association 43–53.

Banks, W.B. and West, H. 1989. A chemical kinetics approach to the process of wood swelling. *Proc. Tenth Cellulose Conf.*, C. Schuerch, ed. John Wiley & Sons, New York, NY 1215.

Brunauer, S., Emmett, P.H., and Teller, E.J. 1938. Adsorption of gases in multi molecular Layers, *J. Am. Chem. Soc.* 60: 309–319.

Caulfield, D.F. 1978. The effect of cellulose on the structure of water. In *Fibre–Water Interactions in Paper-Making*. Clowes and Sons, Ltd., London, England.

Crowley, J.D., Teague, G.S., and Lowe, J.W. 1966. A three-dimensional approach to solubility. *J. Paint Technol.* 38: 269–280.

Dent, R.W. 1977. A multiplayer theory for gas sorption. Part 1: Sorption of a single gas. *Text. Res. J.* 47(2): 145–152.

Feist, W.C. and Tarkow, H. 1967. A new procedure for measuring fiber saturation points. *Forest Prod. J.* 17(10): 65–68.

Gordy, W.J. 1939. Spectroscopic comparison of the proton-attracting properties of liquids. *J. Phys. Chem.*7: 93–101.

Gutmann, V. 1976. Empirical parameters for donor and acceptor properties of solvents. *Electrochimica Acta* 21: 661–670.

James, W.L. 1964. Vibration, static strength, and elastic properties of clear Douglas fir at various levels of moisture content. *Forest Prod. J.* 14(9): 409–413.

Kollmann, H.D. and Hockele, G. 1962. Kritischer Vergleich einiger Bestimmungsverfahren der Holzfeuchitigkeit. *Holz Roh-Werkst.* 20(12): 461–473.

Langmuir, I. 1918. The adsorption of gases on plane surfaces of glass, mica and platinum. *J. Am. Chem. Soc.* 40: 1361.

Lide, D.R. ed. 1993. *Handbook of Chemistry and Physics.* 74th edition, CRC Press, Boca Raton, FL.

Mantanis, G.I., Young, R.A., and Rowell, R.M. 1994a. Swelling of wood: Part 1. Swelling in water. *Wood Sci. Technol.* 28: 119–134.

Mantanis, G.I., Young, R.A., and Rowell, R.M. 1994b. Swelling of wood: Part 2. Swelling in organic liquids. *Holzforschung* 48: 480–490.

Mantanis, G.I., Young, R.A., and Rowell, R.M. 1995a. Swelling of wood: Part 3. Effect of temperature and extractives on rate and maximum swelling. *Holzforschung* 49: 239–248.

Mantanis, G.I., Young, R.A., and Rowell, R.M. 1995b. Swelling of wood: Part 4. A statistical model for prediction of maximum swelling of wood in organic liquids. *Wood and Fiber Sci.* 27(1): 22–24.

Miller, E.R. and Boxall, J. 1984. The effectiveness of end-grain sealers in improving paint performance on softwood joinery. *Holz als Roh und Werkstoff* 42(1): 27–34.

Nayer, A.N. 1948. Swelling of wood in various organic liquids. Ph.D. thesis, University of Minnesota, Mineapolis, MN.

Okoh, K.A.I. and Skaar, C. 1980. Moisture sorption isotherms of wood and inner bark of ten southern U.S. hardwoods. *Wood and Fiber* 12(2): 98–111.

Robertson, A.A. 1964. Cellulose–liquid interactions. *Pulp Paper Mag. Canada* 65: T171–T178.

Rowell, R.M. ed. 1984. Penetration and reactivity of wood cell wall components. In *American Chemical Society Advances in Chemistry Series No. 207*, Washington, DC, Chapter 4, 175–210.

Rowell, R.M. and Banks, W.B. 1985. Water repellency and dimensional stability of wood. USDA Forest Service General Technical Report FPL 50. Forest Products Laboratory, Madison, WI, 24pp.

Rowell, R.M. and Barbour, J, eds. 1990. *Archaeological Wood: Properties, Chemistry, and Preservation.* American Chemical Society Advances in Chemistry Series 225, Washington, DC, 472pp.

Rowell, R.M. and Ellis. W.D. 1978. Determination of dimensional stabilization of wood using the water-soak method. *Wood and Fiber* 10(2): 104–111.

Rowell, R.M. and Youngs, R.L. 1981. Dimensional stabilization of wood in use. USDA Forest Serv. Res. Note. FPL-0243, Forest Products Laboratory, Madison, WI, 8pp.

Sasaki, T., Norimoto, M., Yamada, T., and Rowell, R.M. 1988. Effect of moisture on the acoustical properties of wood. *J. Japan Wood Res. Soc.* 34(10): 794–803.

Skaar, C. 1984. Wood-water relationships. In: R.M. Rowell, ed., *The Chemistry of Solid Wood*, Advances in Chemistry Series, American Chemical Society, Washington, DC, 207, 127–174.

Spalt, H.A. 1958. Fundamentals of water vapor sorption by wood. *Forest Prod. J.* 8(10): 288–295.

Stamm, A.J. 1935. Shrinking and swelling of wood. *Ind. Eng. Chem.* 27: 401–406.

Stamm, A.J. 1964. *Wood and Cellulose Science*, The Ronald Press Company, New York, NY, 549pp.

Stamm, A.J. and Loughborough, W.K. 1942. Variation in shrinking and swelling of wood. *Trans. Amer. Soc. Mech. Eng.* 64: 379–386.

Stamm, A.J. and Tarkow, H. 1950. Penetrataion of cellulose fibers. *J. Phys. Colloid Chem.* 54: 745–753.

Sumi, Y. Hale, R.D., Meyer, J.A., Leopold, B., and Ranby, B.G. 1964. Accessibility of wood and wood carbohydrates measured with tritiated water. *TAPPI* 47(10): 621–624.

Tarkow, H. and Turner, H.D. 1958. The swelling pressure of wood. *Forest Prod. J.* 8(7): 193–197.

Tiemann, H.D. 1944. *Wood Technology.* Second Edition, Pitman Publishing Company, New York City, NY.

West, H. 1988. Kinetics and meachanism of wood-isocyanate reactions. Ph. D. Thesis. University of North Wales, Bangor, UK.

Winandy, J.E. and Rowell, R.M. 1984. In: R.M. Rowell, ed., *Chemistry of Solid Wood*, American Chemical Society Advances in Chemistry Series No. 207, Washington, DC, pp. 211–255.

Wood Handbook—Wood as an engineering material. 1999. Gen. Tech. Rep. FPL-GTE-113, Madison, WI: U.S. Department of Agriculture, Forest Service, Forest Products Laboratory, 463pp.

Yano, H., Norimoto, M., and Rowell, R.M. 1993. Stabilization of acoustical properties of wooden musical instruments by acetylation. *Wood and Fiber Sci.* 25(4): 395–403.

5 Biological Properties of Wood

Rebecca E. Ibach

CONTENTS

There are numerous biological degradations that wood is exposed to in various environments. Biological damage occurs when a log, sawn product, or final product is not stored, handled, or designed properly. Biological organisms such as bacteria, mold, stain, decay fungi, insects, and marine borers depend heavily on temperature and moisture conditions to grow. Figure 5.1 gives the climate index for decay hazard for the United States of America. The higher the number means a greater decay hazard. The southeastern and northwest coasts have the greatest potential, and the southwest has the lowest potential for decay. This chapter will first focus on the biological organisms and their mechanism of degradation, and then prevention measures. If degradation cannot be controlled by design or exposure conditions, then protection with preservatives is warranted.

5.1 BIOLOGICAL DEGRADATIONS

5.1.1 BACTERIA

Bacteria, the early colonizers of wood, are single-celled organisms that can slowly degrade wood that is saturated with water over a long period of time. They are found in wood submerged in seawater and

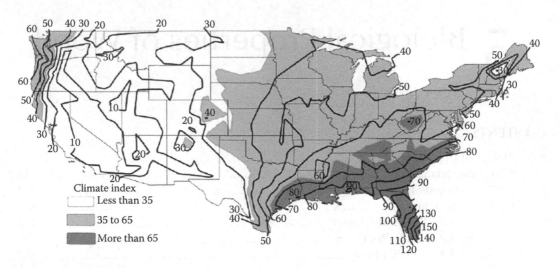

FIGURE 5.1 Climate index for decay potential for wood in service. Higher numbers (darker areas) have greater decay hazard.

freshwater, aboveground exposure, and in-ground soil contact. Logs held under water for months may have a sour smell attributed to bacteria. Bacteria usually have little effect on the properties of wood except over a long time period. Some bacteria can make the wood more absorptive which can make it more susceptible to decay. When dried, the degraded area develops a cross checking on the tangential face. The sapwood is more susceptible than the heartwood and the earlywood more than the latewood.

5.1.2 MOLD AND STAIN

Mold and stain fungi cause damage to the surface of wood, and only differ on their depth of penetration and discoloration. Both grow mainly on sapwood and are of various colors. Molds are usually fuzzy or powdery growth on the surface of wood and range in color from different shades of green, to black or light colors. On softwoods, the fungal hyphae penetrate into the wood, but it can usually be brushed or planed off. On the other hand, on large pored hardwoods, staining can penetrate too deeply to be removed.

The main types of fungus stains are called sapstain or blue stain. They penetrate deeply into the wood and cannot be removed by planing. They usually cause blue, black, or brown darkening of the wood, but some can also produce red, purple, or yellow colors. Figure 5.2 shows the discoloration on a cross section of wood that appears as pie-shaped wedges that are oriented radially.

The strength of the wood is usually not altered by molds and stains (except for toughness or shock resistance), but the absorptivity can be increased making it more susceptible to moisture and then decay fungi. Given moist and warm conditions, mold and stain fungi can establish on sapwood logs shortly after they are cut. To control mold and stain, the wood should be dried to less than 20% moisture content or treated with a fungicide. Wood logs can also be sprayed with water to increase the moisture content to protect wood against fungal stain, as well as decay.

5.1.3 DECAY FUNGI

Decay fungi are single-celled or multicellular filamentous organisms that use wood as food. Figure 5.3 shows the decay cycle of wood. The fungal spores spread by wind, insects, or animals. They germinate on moist, susceptible wood, and the hyphae spread throughout the wood. These hyphae secrete enzymes that attack the cells and cause wood to deteriorate. After serious decay, a new

FIGURE 5.2   Radial penetration of sapstain fungi in a cross section of pine.

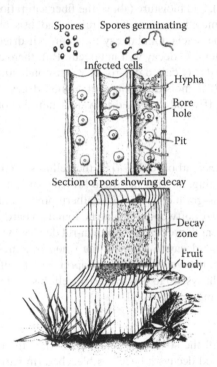

FIGURE 5.3   The wood decay cycle.

fruiting body may form. Brown-, white-, and soft-rot fungi all appear to have enzymatic systems that demethoxylate lignin, produce endocellulases, and use single-electron oxidation system to modify lignin, with some fungi (Eaton and Hale 1993).

In the early or incipient stage of wood decay, serious strength losses can occur before it is even detected (see Chapter 10). Toughness or impact bending is most sensitive to decay. With incipient decay the wood may become discolored on unseasoned wood, but it is harder to detect on dry wood. The advanced stages of wood decay are easier to detect. Decayed wet wood will break across the grain, while sound wood will splinter.

FIGURE 5.4   Brown-rot decay of southern pine wood.

Decay fungi need food (hemicellulose, cellulose, and lignin), oxygen (air), the right temperature (10–35°C; optimum 24–32°C), and moisture (above the fiber saturation point; about 30% moisture content) to grow. Free water must be present (from rain, condensation, or wet ground contact) for the fiber saturation point to be reached and decay to occur. Air-dried wood will usually have no more than 20% moisture content, so decay will not occur. But, there are a few fungi that transport water to dry wood and cause decay called dry-rot or water-conducting fungi. When free water is added to wood to attain 25–30% moisture content or higher, decay will occur. Yet, wood can be too wet or too dry for decay. If wood is soaked in water, there is not enough air for the fungi to develop.

5.1.3.1 Brown-Rot Fungi

Brown-rot fungi decompose the carbohydrates (i.e., the cellulose and hemicelluloses) from wood, which leaves the lignin remaining, making the wood browner in color, hence the name. Figure 5.4 shows the dark color and cross-grain checking of southern pine wood caused by brown-rot decay. Brown-rot fungi mainly colonize softwoods, but can be found on hardwoods as well. Because of the attack on the cellulose, the strength properties of brown-rot decayed wood decrease quickly, even in the early stages. When extreme decay is attained, the wood becomes a very dark, charred color. After the cross-grain cracking, the wood shrinks, collapses, and finally crumbles. Brown-rot fungi first use a low molecular weight system to depolymerize cellulose within the cell wall, and then use endocellulases to further decompose the wood.

5.1.3.2 White-Rot Fungi

White-rot fungi decompose all the structural components (i.e., the cellulose, hemicellulose, and lignin) from wood. As the wood decays it becomes bleached (in part from the lignin removal) or "white" with black zone lines. White-rot fungi occur mainly on hardwoods, but can be found on softwoods as well. The degraded wood does not crack across the grain until it is severely degraded. It keeps its outward dimensions, but feels spongy. The strength properties decrease gradually as decay progresses, except toughness. White-rot fungi have a complete cellulase complex and also the ability to degrade lignin.

5.1.3.3 Soft-Rot Fungi

Soft-rot fungi are related to molds, and occur usually in wood that is constantly wet, but can also appear on surfaces that encounter wet-dry cycling. The decayed wood typically is shallow in growth and "soft" when wet, but the undecayed wood underneath is still firm. Upon drying, the decayed surface is fissured. Figure 5.5 shows surface checking of soft-rotted wood when dry. The wood

FIGURE 5.5   Soft-rot decay of a treated pine pole.

becomes darker (dull-brown to blue-gray) when decayed by soft-rot fungi. Soft-rot fungi first have a system to free the lignin in the wood to then allow the cellulases access to the substrate.

5.1.4   INSECTS

Insects are another biological cause of wood deterioration. Both the immature insect and the adult form may cause wood damage, and they are often not present when the wood is inspected. Therefore, identification is based on the description of wood damage as described in Table 5.1. Figure 5.6 shows pictures of four types of insect damage caused by: (a) termites, (b) powder-post beetles, (c) carpenter ants, and (d) beetles.

5.1.4.1   Termites

Termites are the size of ants and live in colonies. Figure 5.7 is a map of the United States showing the northern limit, (a) of the subterranean termites that live in the ground, and (b) drywood termites or nonsubterranean which live in wood.

5.1.4.1.1   Subterranean Termites

The native subterranean termites live in colonies in the ground; have three stages of metamorphosis (egg, nymph, and adult); and have three different castes (reproductives, workers, and soldiers). They can have winged and wingless adults living in one colony at the same time. Two reproductives (swarmers) are needed to start a colony. Figure 5.8 shows the difference between a winged termite (a) and a winged ant (b). The termite has longer wings and no waist indentation. They are light tan to black, with four wings, three pairs of legs, one pair of antennae, a pair of large eyes, and about 8–13 mm long. Thousands of the swarmers are released from a colony during the daylight hours in the spring or early summer. They fly a short way and then lose their wings. Females attract the males, they find a nesting site, and eggs are laid within several weeks. The worker members are the ones that cause the destruction of the wood.

Moisture is critical for the termites, either from their nest in the soil, or the wood they are feeding on. They will form shelter tubes made of particles of soil, wood, and fecal material. These shelter tubes protect the termites and allow them to go from their nest in the soil to the wood above ground. Termites prefer eating the softer earlywood than the harder latewood.

To protect a house from termites, the soil should be treated with an insecticide, as well as the use of good design and construction practices, such as building the foundation with concrete or

TABLE 5.1
Description of Wood Damage Caused by Insects

Type of Damage	Description	Causal Agent	Damage Begins	Damage Ends
Pin holes	0.25–6.4 mm (1/100–1/4 in.) in diameter, usually circular			
	Tunnels open:			
	Holes 0.5–3 mm (1/50–1/8 in.) in diameter; usually centered in dark streak or ring in surrounding wood	Ambrosia beetles	In living trees and unseasoned logs and lumber	During seasoning
	Holes variable sizes; surrounding wood rarely dark stained; tunnels lined with wood-colored substance	Timber worms	In living trees and unseasoned logs and lumber	Before seasoning
	Tunnels packed with usually fine sawdust:			
	Exit holes 0.8–1.6 mm (1/32–1/16 in.) in diameter; in sapwood of large-pored hardwoods; loose floury sawdust in tunnels	Lyctid powder-post beetles	During or after seasoning	Reinfestation continues until sapwood destroyed
	Exit holes 1.6–3 mm (1/16–1/8 in.) in diameter; primarily in sapwood, rarely in heartwood; tunnels loosely packed with fine sawdust and elongate pellets	Anobiid powder-post beetles	Usually after wood in use (in buildings)	Reinfestation continues; progress of damage very slow
	Exit holes 2.5–7 mm (3/32–9/32 in.) in diameter; primarily sapwood of hardwoods, minor in softwoods; sawdust in tunnels fine to coarse and tightly packed	Bostrichid powder-post beetles	Before seasoning or if wood is rewetted	During seasoning or redrying
	Exit holes 1.6–2 mm (1/16–1/12 in.) in diameter; in slightly damp or decayed wood; very fine sawdust or pellets tightly packed in tunnels	Wood-boring weevils	In slightly damp wood in use	Reinfestation continues while wood is damp
Grub holes	3–13 mm (1/8–1/2 in.) in diameter; circular or oval			
	Exit holes 3–13 mm (1/8–1/2 in.) in diameter; circular; mostly in sapwood; tunnels with coarse to fibrous sawdust or it may be absent	Roundheaded borers (beetles)	In living trees and unseasoned logs and lumber	When adults emerge from seasoned wood or when wood is dried
	Exit holes 3–13 mm (1/8–1/2 in.) in diameter; mostly oval; in sapwood and heartwood; sawdust tightly packed in tunnels	Flatheaded borers (beetles)	In living trees and unseasoned logs and lumber	When adults emerge from seasoned wood or when wood is dried
	Exit holes ~6 mm (~1/4 in.) in diameter; circular; in sapwood of softwoods, primarily pine; tunnels packed with very fine sawdust	Old house borers (a roundheaded borer)	During or after seasoning	Reinfestation continues in seasoned wood in use

Sign	Description	Agent	Where found	Remarks
	Exit holes perfectly circular, 4–6 mm (1/6–1/4 in.) in diameter; primarily in softwoods; tunnels tightly packed with coarse sawdust, often in decay softened wood	Woodwasps	In dying trees or fresh logs	When adults emerge from seasoned wood, usually in use, or when kiln dried
	Nest entry hole and tunnel perfectly circular ~13 mm (~1/2 in.) in diameter; in soft softwoods in structures	Carpenter bees	In structural timbers, siding	Nesting reoccurs annually in spring at same and nearby locations
Network of galleries	Systems of interconnected tunnels and chambers	Social insects with colonies		
	Walls look polished; spaces completely clean of debris	Carpenter ants	Usually in damp partly decayed, or soft-textured wood in use	Colony persists unless prolonged drying of wood occurs
	Walls usually speckled with mud spots; some chambers may be filled with "clay"	Subterranean termites	In wood structures	Colony persists
	Chambers contain pellets; areas may be walled-off by dark membrane	Dry-wood termites (occasionally damp wood termites)	In wood structures	Colony persists
Pitch pocket	Openings between growth rings containing pitch	Various insects	In living trees	In tree
Black check	Small packets in outer layer of wood	Grubs of various insects	In living trees	In tree
Pith fleck	Narrow, brownish streaks	Fly maggots or adult weevils	In living trees	In tree
Gum spot	Small patches or streaks of gum-like substances	Grubs of various insects	In living trees	In tree
Ring distortion	Double growth rings or incomplete annual layers of growth	Larvae of defoliating insect or flatheaded cambium borers	In living trees	In tree
	Stained area more than 25.4 mm (1 in.) long introduced by insects in trees or recently felled logs	Staining fungi	With insect wounds	With seasoning

FIGURE 5.6   Types of insect damage caused by: (a) termites, (b) powderpost beetles, (c) carpenter ants, and (d) beetles.

FIGURE 5.7   Map of termite location in the United States. (a) Subterranean northern limit and (b) drywood termites northern limit.

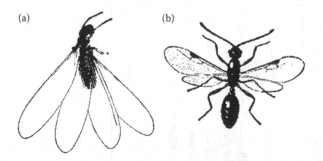

FIGURE 5.8   A winged termite with long wings (a) and a winged ant with a waist indentation (b).

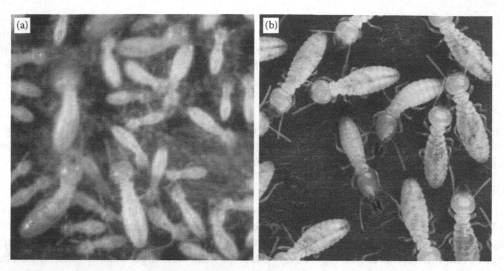

FIGURE 5.9 Subterranean termites (a) and Formosan termites (b).

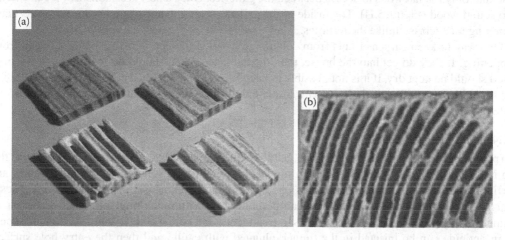

FIGURE 5.10 Subterranean termite attack of wood: (a) "grazing" to final degradation and (b) removal of predominantly spring wood.

pressure-treated wood. If termites get into a building, then a termite control specialist from a national pest control operator association should be contacted.

Figure 5.9 shows subterranean and Formosan termites and Figure 5.10 shows subterranean termite attack from "grazing" to final degradation. Note that this termite attacks the softer spring wood rather than the harder latewood.

5.1.4.1.2 Formosan Subterranean Termites

The Formosan subterranean termite is originally from the Far East. It moved to Hawaii, other Pacific islands, California, Texas, and the southeastern United States. The Formosan termite multiplies and causes damage quicker than the native subterranean species. It nests in wood that is wet from prolonged periods and is good at starting aboveground colonies. Infestation control measures are the same as for the native species, but treatment should be performed within a few months.

5.1.4.1.3 Nonsubterranean (Drywood) Termites

Nonsubterranean termites are found in the southern edge of the continental United States from California to Virginia, the West Indies, and Hawaii. Drywood termites do not multiply as quickly as the subterranean termites, but they can live in drywood without outside moisture or ground contact. Infestations can enter a house in wood products such as furniture. Prevention includes examining all wood and cellulose-based materials before bringing inside, removing woody debris from the outside, and using preservative-treated lumber. Infestations should be treated or fumigated by a professional licensed fumigator. Call the state pest control association.

5.1.4.1.4 Dampwood Termites
Dampwood termites colonize in damp and decaying wood and do not need soil to live, if the wood is wet enough. They are most prevalent on the Pacific Coast. Keeping wood dry is the best protection for preventing colonization and damage by dampwood termites.

5.1.4.2   Carpenter Ants
Carpenter ants use wood for shelter instead of food. They prefer wood that is soft or decayed. They can be black or brown and live in colonies. There are several casts: winged and unwinged queens, winged males, and different sizes of unwinged workers. Carpenter ants have a narrow waist and wings of two different sizes (see Figure 5.8). The front wings are larger than the hind ones. They create galleries along the grain of the wood and around annual rings. They attack the earlywood first, and only the latewood to access between the galleries. Once a nest is established, it can extend into sound wood (Figure 5.11). The inside of the gallery is smooth and clean because the ants keep removing any debris, unlike the termites.

One way to keep carpenter ants from colonizing wood is to keep moisture out and decay from happening. If they do get into the house, then the damaged wood should be removed and the new wood should be kept dry. If it is not possible to keep the wood dry, then a preservative-treated lumber should be used. To treat indoors with insecticides, the state pest control association should be called.

5.1.4.3   Carpenter Bees
Carpenter bees are like large bumblebees, but they differ in that their abdomens shine because the top is hairless. The females make 13 mm tunnels into unfinished softwood to nest. The holes are partitioned into cells, and each cell holds one egg, pollen, and nectar. Carpenter bees reuse nests year after year, therefore some tunnels can be quite long with many branches. They can nest in stained, thinly painted, light preservative salt treatments, and bare wood. To control carpenter bees, an insecticide can be injected in the tunnel, plugged with caulk, and then the entry hole surface treated so that the bees do not use it again the next year. A thicker paint film, pressure preservative treatments, screens, and tight-fitting doors can help prevent nesting damage.

5.1.4.4   Beetles
Table 5.1 describes the types of wood damage that results from various beetles.

5.1.4.4.1 Lyctid Powderpost Beetles
Lyctid beetles cause significant damage to dry hardwood lumber, especially with large pores such as oak, hickory, and ash. They are commonly called powderpost beetles because they make a fine,

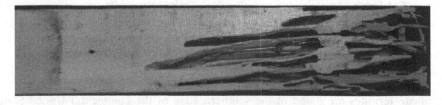

FIGURE 5.11 Carpenter ant damage extending into sound wood.

powdery sawdust during infestation. Activity and damage is greatest when the moisture content of wood is between 10% and 20%, but activity can occur between 8% and 32% moisture content. Infestation can be detected after the first generation of winged adult beetles emerges from the wood producing small holes (0.8–1.6 mm diameter), and a fine wood powder falls out.

5.1.4.4.2 Anobiid Powderpost Beetles

Anobiid beetles are found on older and recently seasoned hardwoods or softwoods throughout the United States of America. They prefer the sapwood that is closest to the bark and their exit holes are 1.6–3 mm in diameter. Their life cycle is 2–3 years and they need about 15% moisture content. If the infestation is old, then there may be very small round (0.8 mm) emergence holes from parasitic wasps larvae that feed on the beetle larvae.

There are several approaches to try to control powderpost beetles. One is to control the environmental conditions by lowering the moisture content of the wood through ventilation, insulation and vapor barriers, as well as good building design. Another is to use chemical treatment by brushing or spraying the wood with insecticides, using boron diffusion treatments, or fumigating if infestation is extensive. Using pressure-treated wood can prevent beetle attack. Another approach is to just eliminate or reduce the beetle population.

5.1.4.4.3 Flatheaded Borers

Flatheaded borers are metallic-colored beetles that vary in size, but a hammer-headed shape produced by an enlarged, flattened body region behind the head characterizes the larvae. The adult flatheaded borer emerges causing 3–13 mm oval or elliptical exit holes in sapwood and heartwood of living trees and unseasoned logs and lumber. Powdery, pale-colored sawdust is found tightly packed in oval to flattened tunnels or galleries in softwoods and hardwood. The adult females lay eggs singly or in groups on the bark or in crevices in the bark or wood. The larvae or young borers mine the inner bark or wood. Since most infestations occur in trunks of weakened trees or logs, the best method of control is to spot treat the local infestations which can be done by applying insecticides to the surface of the wood. This may prevent reinfestation, or kill the larvae that feed close to the surface.

5.1.4.4.4 Cerambycids

The long-horned beetles and old house borers are collectively called the Cerambycids, or the round-headed beetles.

5.1.4.4.4.1 Long-Horned Beetles   The long-horned beetle or roundheaded beetle is the common name of the Asian Cerambycid Beetle, *Anoplophora glabripennis* that is indigenous to southern China, Korea, Japan, and the Isle of Hainan. It is extremely destructive to hardwood tree species and there is no known natural predator in the United States. It attacks not just stressed or aging trees, but healthy trees of any age, and it produces new adults each year, instead of every 2–4 years like other longhorn beetles. The beetle bores into the heartwood of a host tree, eventually killing the tree.

The beetle is believed to have hitchhiked into the United States in wooden crating of a cargo ship in the early 1990s. It was discovered in the United States in August 1996 in Brooklyn, New York. Within weeks another infestation was found on Long Island, in Amityville, New York. Two years later in July of 1998 the beetle was found in Chicago, Illinois. It attacks many healthy hardwood trees, including maple, horsechestnut, birch, poplar, willow, and elm.

The adult beetles have large-bodies; are black with white spots; and have very long black and white antenna. They make large circular holes (3–13 mm diameter) upon emergence and can occur anywhere on the tree, including branches, trunks, and exposed roots. Oval to round, darkened wounds in the bark may also be observed, and these are oviposition sites where adult females chew out a place to lay their eggs. The larvae chew banana-shaped tunnels or galleries into the wood, causing heavy sap flow from wounds and sawdust accumulation at tree bases. These galleries interrupt the flow of water from the roots to the leaves. They feed on, and over-winter in, the interiors of trees. Quarantine is usually imposed on firewood and nursery stock in a known infected area and all infested trees are immediately destroyed.

5.1.4.4.4.2 Old-House Borers

Another roundheaded borer is called the old-house borer. The adult is a large (18–25 mm), black to dark brown, elongated beetle which burrows in structural wood, old and new, seasoned and unseasoned, softwood lumber, but not hardwoods. It is capable of reinfesting wood in use and is found along the Atlantic seaboard.

The adults lay their eggs in the cracks and crevices of wood and they hatch in about 2 weeks. The larvae can live in seasoned softwood for several years. They feed little during the winter months of December through February, but when the larvae are full grown, which usually takes about 5 years, they emerge through oval holes (6–9 mm) in the surface of the wood. Moisture content of 15–25% encourages growth. Emergence happens during June and July. During the first few years of feeding, the larvae cannot be heard, but when they are about 4 years old chewing sounds can be heard in wood during the spring and summer months. Damage depends on the number of larvae feeding, the extent of the infestation, how many years, and whether there has been a reinfestation. To control old-house borers insecticides can be applied to the surface of wood. When there is an extensive and active infestation of the old-house borer, then fumigation may be the best control method. To prevent reinfestation, small infestations can be controlled by applying insecticides to the surface of the wood, which kills the larvae that may feed close to the surface, and contacts the chemical just below the surface.

5.1.5 MARINE BORERS

Marine-boring organisms in salt or brackish waters can cause extensive damage to wood. Attack in the United States is significant along the Pacific, Gulf, and South Atlantic Coasts, and slower along the New England Coast because of cold-water temperatures. The marine borers that cause the most damage in the United States are shipworms, Pholads, crustaceans, and pillbugs (Figure 5.12).

5.1.5.1 Shipworms

Shipworms are worm-like mollusks that cause great damage to wooden boats, piers, and structures. They belong in the family Teredinidae and the genera *Teredo* or *Bankia*. They are found in salt water along the United States Coastal waters, but some can adapt to less saline conditions and live in many of the estuaries. The young larvae swim to wood and bury themselves using a pair of boring shells on their head. They have a tail that has two siphons: one to draw in water containing microscopic organisms for food and oxygen for respiration; and the second siphon to expel waste and for reproduction. The larvae eat the wood and organisms from the ocean, and grow worm-like bodies, but they never leave the wood. The shipworms grow in length and diameter, but their entrance holes are only the size of the young larvae (1.6 mm, see Figure 5.13). The inside of the wood becomes honeycombed and severely degraded. Adults from the genus *Teredo* grow to be 0.3–0.6 m in length and 13 mm in diameter, while those of the genus *Bankia* grown to be 1.5–1.8 m in length and 22 mm in diameter. To protect wood from shipworms, a marine grade preservative treatment is used such as creosote (400 kg/m^3), chromated copper arsenate (CCA, 40 kg/m^3) or ammoniacal copper citrate (CC, 40 kg/m^3).

5.1.5.2 Pholads

Pholads are also wood-boring mollusks, but are different in that they resemble clams, that is, encased in a double shell (Figure 5.14). They belong in the family Pholadidae with two familiar species the *Martesia* and the *Xylophaga*. They enter the wood when they are very young and grow inside the wood, similar to the shipworms. Their entrance holes are about 6 mm in diameter, and most of the damage to the wood is close to the surface. Pholads grow no bigger than 64 mm long

FIGURE 5.12   Damaged to wood due to marine borers.

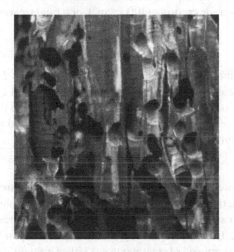

FIGURE 5.13   Internal damage done by shipworms on wood.

FIGURE 5.14   Pholad wood-borers (a) and one boring in wood (b).

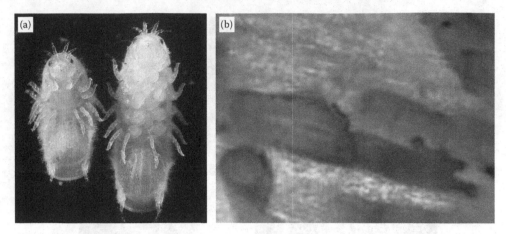

FIGURE 5.15   *Limnoria* wood borer (a) and boring in wood (b).

and 25 mm in diameter, but they can cause extensive damage to wood. They are found in Hawaii, San Diego, California, the Gulf Coast, and from South Carolina down southward. To protect wood from Pholads, either marine-grade creosote (400 kg/m³), or a dual treatment of CCA, and then creosote are effective.

5.1.5.3   Crustaceans

The crustaceans include gribbles, in the family Limnoriidae, genus *Limnoria*, and the pillbugs, from the family Sphaeromatidae, genus *Spaeroma*. They are related to lobsters and shrimp, and differ from the other marine borers in that they are not imprisoned, so they can move from place to place. The boreholes made are shallow, therefore the borers, combined with water erosion, degrade the surface of the wood (Figure 5.15).

5.1.5.3.1   Gribbles

Gribbles or *Limnoria* are quite small (3–4 mm), and their boreholes are usually only about 13 mm deep, but with water erosion, the borers continually bore in deeper. They prefer earlywood, and the attack is usually located between half tide and low tide levels that result in an hourglass shape. Protection with preservative treatment against gribbles depends upon where and what species is present in the water. Two recommended treatments are either a dual treatment of first CCA (16–24 kg/m³) and then marine grade creosote (320 kg/m³), or just using a higher concentration of just CCA (40 kg/m³) or just marine-grade creosote (320–400 kg/m³). To get more information, check with current American Wood Preservers Association (AWPA) Standards (AWPA 2011).

5.1.5.3.2   Pillbugs

Pillbugs or *Spaeroma* are longer (13 mm long) and wider (6 mm wide) than *Limnoria*, and look like a pillbug that lives in damp places. They use the wood for shelter and prefer softer woods. *Spaeroma* are found along the south Atlantic and Gulf Coasts, and from San Francisco southward on the West Coast. It is common to find them in Florida estuaries. Dual treatment with CCA and then creosote is the best protection since they are tolerant to CCA, and with time tolerant to creosote.

5.2   PREVENTION OR PROTECTION OF WOOD

Some wood species have natural decay resistance of the heartwood. It can vary, but there are groupings from very resistant to nonresistant (Table 5.2). When extra protection is needed to protect wood from biological degradation chemical preservatives are applied to the wood either by nonpressure or pressure treatment (Eaton and Hale 1993, Schultz et al. 2008). Penetration and retention of a

TABLE 5.2
Heartwood Decay Resistance of Some Domestic and Imported Woods[a]

Very Resistant	Resistant	Moderately Resistant	Slightly or Nonresistant
		Domestic	
Black locust	Baldcypress, old growth	Baldcypress, young growth	Alder, red
Mulberry, red	Catalpa	Cherry, black	Ashes
Osage-orange	Cedar	Douglas-fir	Aspens
Yew, Pacific	Atlantic white	Honey locust	Beech
	Eastern redcedar	Larch, western	Birches
	Incense	Pine, eastern white, old growth	Buckeye
	Northern white	Pine, longleaf, old growth	Butternut
	Port-orford	Pine, slash, old growth	Cottonwood
	Western redcedar	Redwood, young growth	Elms
	Yellow	Tamarack	Basswood
	Chestnut		Firs, true
	Cypress, Arizona		Hackberry
	Junipers		Hemlocks
	Mesquite		Hickories
	Oaks, white[b]		Magnolia
	Redwood, old growth		Maples
	Sassafras		Pines (other than those listed)[b]
	Walnut, black		Spruces
			Sweetgum
			Sycamore
			Tanoak
			Willows
			Yellow-poplar
		Imported	
Angelique	Aftotmosia (Kokrodua)	Andiroba	Balsa
Azobe	Apamate (Roble)	Avodire	Banak
Balata	Balau[b]	Benge	Cativo
Goncalo alves	Courbaril	Dubinga	Ceiba
Greenheart	Determa	Ehie	Hura
Ipe (lapacho)	Iroko	Ekop	Jelutong
Jarrah	Kapur	Keruing[b]	Limba
Lignumvitae	Karri	Mahogany, African	Meranti, light red[b]
Purpleheart	Kempas	Meranti. dark red[b]	Meranti, yellow[b]
Teak, old	Mahogany, American	Mersawa[b]	Meranti, white[b]
growth	Manni	Sapele	Obeche
	Spanish-cedar	Teak, young growth	Okoume
	Sucupira	Tornillo	Parana pine
	Wallaba		Ramin
			Sande
			Sepitir
			Seraya, white

[a] Decay resistance may be less for members placed in contact with the ground and/or used in warm, humid climates. Substantial variability in decay resistance is encountered with most species, and limited durability data were available for some species listed. Use caution when using naturally durable woods in structurally critical or ground-contact applications.

[b] More than one species included, some of which may vary in resistance from that indicated.

chemical will depend upon wood species and the amount of heartwood (more difficult to treat) or sapwood (easier to treat.) The objective of adding wood preservatives is to obtain long-term effectiveness for the wood product, thus sequestering carbon.

Starting January 2004, the U.S. Environmental Protection Agency (EPA) no longer allows the most widely used wood preservative, CCA, for products for any residential use (i.e., play-structures, decks, picnic tables, landscaping timbers, residential fencing, patios, walkways, and boardwalks). However, it has not concluded that arsenic containing CCA-treated wood poses unreasonable risks to the public from the wood being used around or near their homes (EPA 2002). Alternative preservatives such as ammoniacal copper quat (ACQ) and copper azole (CBA) have replaced CCA for residential use (EPA 2002; PMRA 2002). Looking beyond these replacements for CCA may be wood protection systems not based on toxicity, but rather nontoxic chemical modifications to prevent biological degradation. Chemical modification alters the chemical structure of the wood components thereby reducing the biodegradability of wood, as well as increasing its dimensional stability when in contact with moisture (Rowell 1991) (see Chapter 14).

5.2.1 Wood Preservation

Wood preservatives work by being toxic to the biological organisms that attack wood. The active ingredients in wood preservative formulations are many and varied and each has its own mode of action, some of which are still unknown or unreported. In general mechanisms of toxicity involve denaturation of proteins, inactivation of enzymes, cell membrane disruption causing an increase in cell permeability, and inhibition of protein synthesis.

The degree of protection of a particular preservative and treatment process depends on four basic requirements: (1) toxicity, (2) permanence, (3) retention, and (4) depth of penetration into the wood. Toxicity refers to how effective the chemical is against biological organisms such as decay fungi, insects, and marine borers. Permanence refers to the resistance of the preservative to leaching, volatilization, and breakdown. Retention specifies the amount of preservative that must be impregnated into a specific volume of wood to meet standards, and ensure that the product will be effective against numerous biological agents.

Wood preservatives can be divided into two general classes: (1) oil-type, such as creosote and petroleum solutions of pentachlorophenol; and (2) waterborne salts that are applied as water solutions, such as CCA, ACQ, and CBA. There are many different chemicals in each class and the effectiveness of each preservative can vary greatly depending upon its chemical composition, retention, depth of penetration, and ultimately the exposure conditions of the final product. The degree of protection needed will depend upon geographic location and potential exposures of the wood, expected service life, structural and nonstructural applications, and replacement costs. Wood preservatives should always be used when exposed to ground (soil) contact and marine (salt-water) exposure.

Oilborne preservatives such as creosote and solutions with heavy, less volatile petroleum oils often help to protect wood from weathering, but may adversely influence its cleanliness, odor, color, paintability, and fire performance. Waterborne preservative are often used when cleanliness and paintability of the treated wood are required. In seawater exposure, a dual treatment (waterborne copper-containing salt preservatives followed by creosote) is most effective against all types of marine borers.

Evaluation for efficacy of preservative-treated wood is first performed on small specimens in the laboratory, and then larger specimens with field exposure (ASTM 2010, AWPA 2011) (Figure 5.16). The USDA Forest Service FPL has had in-ground stake test studies on southern pine sapwood ongoing since 1938 in Saucier, Miss., and Madison, Wis (Figure 5.17, Crawford et al. 2002). Table 5.3 shows results of the Forest Products Laboratory studies on 5- by 10- by 46 cm (2- by 4- by 18 in) southern pine sapwood stakes, pressure-treated with commonly used wood preservatives, installed at Harrison Experimental Forest, Mississippi (Figure 5.18). A comparison of preservative-treated

FIGURE 5.16   ASTM soil block test.

FIGURE 5.17   Field test of wood durability.

small wood panels exposed to a marine environment in Key West, Fl. has been evaluated (Figure 5.19) (Johnson and Gutzmer 1990). Outdoor evaluations such as these compare various preservatives and retention levels under each exposure condition at each individual site. These preservatives and treatments include creosotes, waterborne preservatives, dual treatments, chemical modification of wood, and various chemically modified polymers.

Another laboratory test for wood durability is the use of a fungal cellar (Figure 5.20). Small test specimens are placed half their length in unsterile soil of different moisture contents depending on the fungi of interest (see e.g., Stephan et al. 1998). Samples are removed at various times and inspected for fungal attack. Visual damage and weight loss can be determined to give an indication of damage. It is also possible to determine loss of strength which is known to give a faster indicator of fungal attack as compared to weight loss.

The newest laboratory test for fungal decay is the use of a terrestrial microcosm (Nilsson and Björdal 2007). Very small test specimens are placed in test and inspected at different times. Decay is very rapid making this test very inexpensive, fast, and easy (Figure 5.21).

Exposure conditions and length of product lifetime need to be considered when choosing a particular preservative treatment, process, and wood species (Cassens et al. 1995). The AWPA

TABLE 5.3
Results of the Forest Products Laboratory Studies on 5- by 10- by 46 cm (2- by 4- by 18 in)
Southern Pine Sapwood Stakes, Pressure-Treated with Commonly Used Wood
Preservatives, Installed at Harrison Experimental Forest, Mississippi

Preservative	Average Retention Kg/m³ (lb/ft³)	Average Life or Condition at Last Inspection
CCA-type III (Type C)	6.41 (0.40)	No failures after 25 years
Coal-tar creosote	160.2 (10.0)	90% failed after 55 years
Copper naphthenate (0.86% copper in No. 2 fuel oil)	1.31 (0.082)	29.6 years
Oxine copper (copper-8-quinolinolate) (in heavy petroleum)	1.99 (0.124)	No failures after 45 years
No preservative treatment		1.8–3.6 years

Source: Adapted from Crawford, D. M., Woodward, B. M., and Hatfield, C. A. 2002. *Comparison of Wood Preservatives in Stake Tests*. 2000. Progress Report. Res. Note FPL-RN-02. US Department of Agriculture, Forest Service, Forest Products Laboratory, Madison, WI.

FIGURE 5.18   Inspection of a pulled wood stake from a field test.

FIGURE 5.19   Small specimen panels in test in the ocean (a) and inspected (b).

FIGURE 5.20 Federal Institute for Materials Research and Testing Fungus Cellar. Tests of wood preservatives for wood to be used in ground contact. (www.bam.de)

FIGURE 5.21 Terrestrial microcosm.

recently developed the use category system (UCS) standards as a guide to selecting a preservative and loading. The categories are based on the end use and severity of the deterioration hazard (Table 5.4). The categories range from UC1 (interior construction, above ground, dry) to UC5 (seawater and marine borer exposure).

Once the appropriate tests of a durable wood product have been completed, the results are compiled and presented to one of two organizations that reviews and lists durable wood products. Traditionally, durable wood products have been reviewed by AWPA subcommittees, which are composed of representatives from industry, academia and government agencies who have familiarity with conducting and interpreting durability evaluations. More recently the International Code Council–Evaluation Service (ICC–ES) has evolved as an additional route for gaining building code acceptance of new types of pressure-treated wood. The ICC–ES does not standardize preservatives.

TABLE 5.4
Service Conditions for the AWPA Use Category System

Use Category	Service Conditions	Use Environment	Common Agents of Deterioration	Typical Applications
UC1	Interior construction Above ground Dry	Continuously protected from weather or other sources of moisture	Insects only	Interior construction and furnishings
UC2	Interior construction Above ground Damp	Protected from weather, but may be subject to sources of moisture	Decay fungi and insects	Interior construction
UC3A	Exterior construction Above ground Coated and rapid water runoff	Exposed to all weather cycles, not exposed to prolonged wetting	Decay fungi and insects	Coated millwork, siding, and trim
UC3B	Ground contact or fresh water Non-critical components	Exposed to all weather cycles, normal exposure conditions	Decay fungi and insects	Fence, deck, and guardrail posts, crossties, and utility poles (low decay areas)
UC4A	Ground contact or fresh water Non-critical components	Exposed to all weather cycles, normal exposure conditions	Decay fungi and insects	Fence, deck, and guardrail posts, crossties, and utility poles (low decay areas)
UC4B	Ground contact or fresh water Critical components or difficult replacement	Exposed to all weather cycles, high decay potential includes salt-water splash	Decay fungi and insects with increased potential for biodeterioration	Permanent wood foundations, building poles, horticultural posts, crossties, and utility poles (high decay areas)
UC4C	Ground contact or fresh water Critical structural components	Exposed to all weather cycles, severe environments, extreme decay potential	Decay fungi and insects with extreme potential for biodeterioration	Land and fresh-water piling, foundation piling, crossties and utility poles (severe decay areas)
UC5A	Salt or brackish water and adjacent mud zone Northern waters	Continuous marine exposure (salt water)	Salt-water organisms	Piling, bulkheads, bracing
UC5B	Salt or brackish water and adjacent mud zone NJ to GA, South of San Francisco	Continuous marine exposure (salt water)	Salt-water organisms, including creosote-tolerant *Limnoria tripunctata*	Piling, bulkheads, bracing
UC5C	Salt or brackish water and adjacent mud zone South of GA, Gulf Coast, Hawaii, and Puerto Rico	Continuous marine exposure (salt water)	Salt-water organisms, including Martesia, Sphaeroma	Piling, bulkheads, bracing

Instead, it issues Evaluation Reports that provide evidence that a building product complies with the building codes. The tests required by ICC–ES are typically those developed by AWPA. It is important to note that separate toxicity evaluations by appropriate regulatory agencies (e.g., the U.S. Environmental Protection Agency) are mandatory for any durable wood product that incorporates preservative pesticides. For various wood products, preservatives, and their required retention levels see the current AWPA Book of Standards (AWPA 2011).

5.2.2 TIMBER PREPARATION AND CONDITIONING

Preparing the timber for treatment involves carefully peeling the round or slabbed products to enable the wood to dry quickly enough to avoid decay and insect damage and to allow the preservative to penetrate satisfactorily. Drying the wood before treatment is necessary to prevent decay and stain and to obtain preservative penetration, but when treating with waterborne preservatives by certain diffusion methods, high moisture content levels may be permitted. Drying the wood before treatment opens up the checks before the preservative is applied, thus increasing penetration and reducing the risk of checks opening up after treatment and exposing unpenetrated wood.

Treating plants that use pressure processes can condition green material by means other than air and kiln drying, thus avoiding a long delay and possible deterioration. When green wood is to be treated under pressure, one of several methods for conditioning may be selected. The steaming-and-vacuum process is used mainly for southern pines, and the Boulton (or boiling-under-vacuum) process is used for Douglas-fir and sometimes hardwoods.

Heartwood of some softwood and hardwood species can be difficult to treat (see Table 5.5) (Mac Lean 1952). Wood that is resistant to penetration by preservatives, such as Douglas-fir, western hemlock, western larch, and heartwood, may be incised before treatment to permit deeper and more uniform penetration. Incision involves passing the lumber or timbers through rollers that are equipped with teeth that sink into the wood to a predetermined depth, usually 13–19 mm (1/2–3/4 in.). The incisions open cell lumens along the grain that improves penetration, but can result in significant strength reduction. As much cutting and hole boring of the wood product as is possible should be done before the preservative treatment, otherwise untreated interiors will allow ready access of decay fungi or insects.

5.2.3 TREATMENT PROCESSES

There are two general types of wood-preserving methods: (1) pressure processes and (2) nonpressure processes. During pressure processes wood is impregnated in a closed vessel under pressure above atmospheric. In commercial practice wood is put on cars or trams and run into a long steel cylinder, which is then closed and filled with preservative. Pressure forces are then applied until the desired amount of preservative has been absorbed into the wood.

5.2.3.1 Pressure Processes

Three pressure processes are commonly used: full-cell, modified full-cell, and empty-cell. The full-cell process is used when the retention of a maximum quantity of preservative is desired. The steps include: (1) the wood is sealed in a treating cylinder and a vacuum is applied for a half-hour or more to remove air from the cylinder and wood; (2) the preservative (at ambient or elevated temperature) is admitted to the cylinder without breaking the vacuum; (3) pressure is applied until the required retention; (4) the preservative is withdrawn from the cylinder; and (5) a short final vacuum may be applied to free the wood from dripping preservative. The modified full-cell process is basically the same as the full-cell process except for the amount of initial vacuum and the occasional use of an extended final vacuum.

The goal of the empty-cell process is to obtain deep penetration with relatively low net retention of preservative. Two empty-cell processes (the Rueping and the Lowry) use the expansive force of

TABLE 5.5

Penetration of the Heartwood of Various Softwood and Hardwood Species[a]

Ease of Treatment	Softwoods	Hardwoods
Least difficult	Bristlecone pine (*Pinus aristata*)	American basswood (*Tilia americana*)
	Pinyon (*Pinus edulis*)	Beech (white heartwood) (*Fagus*
	Pondersosa pine (*Pinus pondersosa*)	*grandifolia*)
	Redwood (*Sequoia sempervirens*)	Black tupelo (blackgum) (*Nyssa*
		sylvatica)
		Green ash (*Fraxinus pennsylvanica*
		var. *lanceolata*)
		Pin cherry (*Prunus pensylvanica*)
		River birch (*Betula nigra*)
		Red oaks (*Quercus* spp.)
		Slippery elm (*Ulmus fulva*)
		Sweet birch (*Betula lenta*)
		Water tupelo (*Nyssa aquatica*)
		White ash (*Fraxinus Americana*)
Moderately difficult	Baldcypress (*Taxodium distichum*)	Black willow (*Salix nigra*)
	California red fir (*Abies magnifica*)	Chestnut oak (*Quercus montana*)
	Douglas-fir (coast) (*Pseudotsuga*	Cottonwood (*Populus* sp.)
	taxifolia)	Bigtooth aspen (*Populus*
	Eastern white pine (*Pinus strobes*)	*grandidentata*)
	Jack pine (*Pinus banksiana*)	Mockernut hickory (*Carya tomentosa*)
	Loblolly pine (*Pinus taeda*)	Silver maple (*Acer saccharinum*)
	Longleaf pine (*Pinus palustris*)	Sugar maple (*Acer saccharum*)
	Red pine (*Pinus resinosa*)	Yellow birch (*Betula lutea*)
	Shortleaf pine (*Pinus echinata*)	
	Sugar pine (*Pinus lambertiana*)	
	Western hemlock (*Tsuga*	
	heterophylla)	
Difficult	Eastern hemlock (*Tsuga canadensis*)	American sycamore (*Platanus*
	Engelmann spruce (*Picea*	*occidentalis*)
	engelmanni)	Hackberry (*Celtis occidentalis*)
	Grand fir (*Abies grandis*)	Rock elm (*Ulmusthom oasi*)
	Lodgepole pine (*Pinus contorta* var.	Yellow-poplar (*Liriodendron*
	latifolia)	*tulipifera*)
	Noble fir (*Abies procera*)	
	Sitka spruce (*Picea sitchensis*)	
	Western larch (*Larix occidentalis*)	
	White fir (*Abies concolor*)	
	White spruce (*Picea glauca*)	
Very difficult	Alpine fir (*Abies lasiocarpa*)	American beech (red heartwood)
	Corkbark fir (*A. lasiocarpa var.*	(*Fagus grandifolia*)
	arizonica)	American chestnut (*Castanea dentate*)
	Douglas fir (Rocky Mountain)	Black locust (*Robinia pseudoacacia*)
	(*Pseudotsuga taxifolia*)	Blackjack oak (*Quercus marilandica*)
	Northern white cedar (*Thuja*	Sweetgum (redgum) (*Liquidambar*
	occidentalis)	*styraciflua*)
	Tamarack (*Larix laricina*)	White oaks (*Quercus* spp.)
	Western redcedar (*Thaja plicata*)	

[a] As covered in Mac Lean, J. D. 1952. *Preservation of Wood by Pressure Methods.* U.S. Department of Agriculture, Forest Service Washington, DC, 160.

compressed air to drive out part of the preservative absorbed during the pressure period. The Rueping empty-cell process is often called the empty-cell process with initial air. Air pressure is forced into the treating cylinder, which contains the wood, and then the preservative is forced into the cylinder. The air escapes into an equalizing or Rueping tank. The treating pressure is increased and maintained until desired retention is attained. The preservative is drained and a final vacuum is applied to remove surplus preservative. The Lowry process is the same as the Rueping except that there is no initial air pressure or vacuum applied. Hence, it is often called the empty-cell process without initial air pressure.

5.2.3.2 Nonpressure Processes

There are numerous nonpressure processes and they differ widely in their penetration and retention of a preservative. Nonpressure methods consist of (1) surface applications of preservative by brushing or brief dipping, (2) cold soaking in preservative oils or steeping in solutions of waterborne preservative, (3) diffusion processes with waterborne preservatives, (4) vacuum treatment, and (5) various other miscellaneous processes.

5.2.4 PURCHASING AND HANDLING OF TREATED WOOD

The EPA regulates pesticides, and wood preservatives are one type of pesticide. Preservatives that are not restricted by EPA are available to the general consumer for nonpressure treatments, while the sale of others is restricted only to certified pesticide applicators. These preservatives can be used only in certain applications and are referred to as "restricted use." "Restricted-use" refers to the chemical preservative and not to the treated-wood product. The general consumer may buy and use wood products treated with restricted-use pesticides; EPA does not consider treated wood a toxic substance nor is it regulated as a pesticide.

"Consumer Safety Information Sheets" (EPA-approved) are available from retailers of treated-wood products. The sheets provide users with information about the preservative and the use and disposal of treated-wood products. There are consumer information sheets for three major groups of wood preservatives (see Table 5.6): (1) creosote pressure-treated wood, (2) pentachlorophenol pressure-treated wood, and (3) inorganic arsenical pressure-treated wood.

There are two important factors to consider depending upon the intended end use of preservative-treated wood: (1) the grade or appearance of the lumber, and (2) the quality of the preservative treatment in the lumber. The U.S. Department of Commerce American Lumber Standard Committee (ALSC) accredits third party inspection agencies for treated wood products. A list of accredited agencies can be found on the ALSC website at www.alsc.org. The treated wood should be marked with a brand, ink stamp, or end tag. These marks indicate that the producer of the treated-wood product subscribes to an independent inspection agency. The stamp or end tag contains the type of preservative or active ingredient, the retention level, and the intended use category or exposure conditions. Retention levels are usually provided in pounds of preservatives per cubic foot of wood and are specific to the type of preservative, wood species, and intended exposure conditions. Be aware that suppliers often sell the same type of treated wood by different trade names. Depending upon your intended use and location, there will be different types of treated wood available for residential use. Also, be aware that some manufacturers add colorants (such as brown) or water repellents (clear) into some of their preservative treatments. When purchasing treated wood, ask the suppliers for more information to determine what preservative and additives were used, as well as any handling precautions.

Note that mention of a chemical in this article does not constitute a recommendation; only those chemicals registered by the EPA may be recommended. Registration of preservatives is under constant review by EPA and the U.S. Department of Agriculture. Use only preservatives that bear an EPA registration number and carry directions for home and farm use. Preservatives, such as creosote and pentachlorophenol, should not be applied to the interior of dwellings that are occupied by

TABLE 5.6
EPA-Approved Consumer Information Sheets for Three Major Groups of Preservative Pressure-Treated Wood

Preservative Treatment	Inorganic Arsenicals	Pentachlorophenol	Creosote
Consumer information	• This wood has been preserved by pressure-treatment with an EPA-registered pesticide containing inorganic arsenic to protect it from insect attack and decay. Wood treated with inorganic arsenic should be used only where such protection is important. • Inorganic arsenic penetrates deeply into and remains in the pressure-treated wood for a long time. However, some chemical may migrate from treated wood into surrounding soil over time and may also be dislodged from the wood surface upon contact with skin. Exposure to inorganic arsenic may present certain hazards. Therefore, the following precautions should be taken both when handling the treated wood and in determining where to use or dispose of the treated wood.	• This wood has been preserved by pressure-treatment with an EPA-registered pesticide containing pentachlorophenol to protect it from insect attack and decay. Wood treated with pentachlorophenol should be used only where such protection is important. • Pentachlorophenol penetrates deeply into and remains in the pressure-treated wood for a long time. Exposure to pentachlorophenol may present certain hazards. Therefore, the following precautions should be taken both when handling the treated wood and in determining where to use and dispose of the treated wood.	• This wood has been preserved by pressure treatment with an EPA-registered pesticide containing creosote to protect it from insect attack and decay. Wood treated with creosote should be used only where such protection is important. • Creosote penetrates deeply into and remains in the pressure-treated wood for a long time. Exposure to creosote may present certain hazards. Therefore, the following precautions should be taken both when handling the treated wood and in determining where to use the treated wood.
Handling precautions	• Dispose of treated wood by ordinary trash collection or burial. Treated wood should not be burned in open fires or in stoves, fireplaces, or residential boilers because toxic chemicals may be produced as part of the smoke and ashes. Treated wood from commercial or industrial use (e.g., construction sites) may be burned only in commercial or industrial incinerators or boilers in accordance with state and Federal regulations.	• Dispose of treated wood by ordinary trash collection or burial. Treated wood should not be burned in open fires or in stoves, fireplaces, or residential boilers because toxic chemicals may be produced as part of the smoke and ashes. Treated wood from commercial or industrial use (e.g., construction sites) may be burned only in commercial or industrial incinerators or boilers rated at 20 million BTU/hour or greater heat input or its equivalent in accordance with state and Federal regulations.	• Dispose of treated wood by ordinary trash collection or burial. Treated wood should not be burned in open fires or in stoves, fireplaces, or residential boilers, because toxic chemicals may be produced as part of the smoke and ashes. Treated wood from commercial or industrial use (e.g., construction sites) may be burned only in commercial or industrial incinerators or boilers in accordance with state and Federal regulations.

continued

	• Avoid frequent or prolonged inhalation of sawdust from treated wood. When sawing and machining treated wood, wear a dust mask. Whenever possible, these operations should be performed outdoors to avoid indoor accumulations of airborne sawdust from treated wood. • When power-sawing and machining, wear goggles to protect eyes from flying particles. • Wear gloves when working with the wood. After working with the wood, and before eating, drinking, toileting, and use of tobacco products, wash exposed areas thoroughly. • Because preservatives or sawdust may accumulate on clothes, they should be laundered before reuse. Wash work clothes separately from other household clothing.	• Avoid frequent or prolonged inhalation of sawdust from treated wood. When sawing and machining treated wood, wear a dust mask. Whenever possible, these operations should be performed outdoors to avoid indoor accumulations of airborne sawdust from treated wood. • Avoid frequent or prolonged skin contact with pentachlorophenol-treated wood. When handling the treated wood, wear long-sleeved shirts and long pants and use gloves impervious to the chemicals (e.g., gloves that are vinyl-coated). • When power-sawing and machining, wear goggles to protect eyes from flying particles. • After working with the wood, and before eating, drinking, and use of tobacco products, wash exposed areas thoroughly. • If oily preservatives or sawdust accumulate on clothes, launder before reuse. Wash work clothes separately from other household clothing.	• Avoid frequent or prolonged inhalations of sawdust from treated wood. When sawing and machining treated wood, wear a dust mask. Whenever possible these operations should be performed outdoors to avoid indoor accumulations of airborne sawdust from treated wood. • Avoid frequent or prolonged skin contact with creosote-treated wood; when handling the treated wood, wear long-sleeved shirts and long pants and use gloves impervious to the chemicals (e.g., gloves that are vinyl-coated). • When power-sawing and machining, wear goggles to protect eyes from flying particles. • After working with the wood and before eating, drinking, and use of tobacco products, wash exposed areas thoroughly. • If oily preservative or sawdust accumulate on clothes, launder before reuse. Wash work clothes separately from other household clothing.
Use site precautions	• All sawdust and construction debris should be cleaned up and disposed of after construction. • Do not use treated wood under circumstances where the preservative may become a component of food or animal feed. Examples of such sites would be use of mulch from recycled arsenic-treated wood, cutting boards, counter tops, animal bedding, and structures or containers for storing animal feed or human food. • Only treated wood that is visibly clean and free of surface residue should be used for patios, decks, and walkways.	• Logs treated with pentachlorophenol should not be used for log homes. Wood treated with pentachlorophenol should not be used where it will be in frequent or prolonged contact with bare skin (e.g., chairs and other outdoor furniture), unless an effective sealer has been applied. • Pentachlorophenol-treated wood should not be used in residential, industrial, or commercial interiors except for laminated beams or building components which are in ground contact and are subject to decay or insect infestation and where two coats of an appropriate sealer are applied. Sealers may be applied at the installation site. Urethane, shellac, latex epoxy enamel, and varnish are acceptable sealers for pentachlorophenol-treated wood.	• Wood treated with creosote should not be used where it will be in frequent or prolonged contact with bare skin (e.g., chairs and other outdoor furniture) unless an effective sealer has been applied. • Creosote-treated wood should not be used in residential interiors. Creosote-treated wood in interiors of industrial buildings should be used only for industrial building components which are in ground contact and are subject to decay or insect infestation and wood-block flooring. For such uses, two coats of an appropriate sealer must be applied. Sealers may be applied at the installation site.

TABLE 5.6 (continued)
EPA-Approved Consumer Information Sheets for Three Major Groups of Preservative Pressure-treated Wood

Preservative Treatment	Inorganic Arsenicals	Pentachlorophenol	Creosote
Use site precautions	• Do not use treated wood for construction of those portions of beehives which may come into contact with honey.	• Wood treated with pentachlorophenol should not be used in the interiors of farm buildings where there may be direct contact with domestic animals or livestock which may crib (bite) or lick the wood.	• Wood treated with creosote should not be used in the interiors of farm buildings where there may be direct contact with domestic animals or livestock which may crib (bite) or lick the wood.
	• Treated wood should not be used where it may come into direct or indirect contact with drinking water, except for uses involving incidental contact such as docks and bridges.	• In interiors of farm buildings where domestic animals or livestock are unlikely to crib (bite) or lick the wood, pentachlorophenol-treated wood may be used for building components which are in ground contact and are subject to decay or insect infestation and where two coats of an appropriate sealer are applied. Sealers may be applied at the installation site.	• In interiors of farm buildings where domestic animals or livestock are unlikely to crib (bite) or lick the wood, creosote-treated wood may be used for building components which are in ground contact and are subject to decay or insect infestation if two coats of an effective sealer are applied. Sealers may be applied at the installation site. Coal tar pitch and coal tar pitch emulsion are effective sealers for creosote-treated wood-block flooring. Urethane, epoxy, and shellac are acceptable sealers for all creosote-treated wood.
		• Do not use pentachlorophenol-treated wood for farrowing or brooding facilities.	• Do not use creosote-treated wood for farrowing or brooding facilities.
		• Do not use treated wood under circumstances where the preservative may become a component of food or animal feed. Examples of such sites would be structures or containers for storing silage or food.	• Do not use treated wood under circumstances where the preservative may become a component of food or animal feed. Examples of such use would be structures or containers for storing silage or food.
		• Do not use treated wood for cutting-boards or countertops.	• Do not use treated wood for cutting-boards or countertops.
		• Only treated wood that is visibly clean and free of surface residue should be used for patios, decks, and walkways.	

- Do not use treated wood for construction of those portions of beehives which may come into contact with the honey.
- Pentachlorophenol-treated wood should not be used where it may come into direct or indirect contact with public drinking water, except for uses involving incidental contact such as docks and bridges.
- Do not use pentachlorophenol-treated wood where it may come into direct or indirect contact with drinking water for domestic animals or livestock, except for uses involving incidental contact such as docks and bridges.

- Only treated wood that is visibly clean and free of surface residues should be used for patios, decks, and walkways.
- Do not use treated wood for construction of those portions of beehives which may come into contact with the honey.
- Creosote-treated wood should not be used where it may come into direct or indirect contact with public drinking water, except for uses involving incidental contact such as docks and bridges.
- Do not use creosote-treated wood where it may come into direct or indirect contact with drinking water for domestic animals or livestock, except for uses involving incidental contact such as docks and bridges.

humans. Because all preservatives are under constant review by EPA, a responsible State or Federal agency should be consulted as to the current status of any preservative.

REFERENCES

ASTM. 2010. *Annual Book of ASTM Standards*. American Society for Testing and Materials. West Conshohocken, PA.

AWPA. 2011. *AWPA 2011 Book of Standards*. American Wood-Preservers' Association, Birmingham, Alabama.

Cassens, D. L., Johnson, B. R., Feist, W. C., and De Groot, R. C. 1995. *Selection and Use of Preservative-treated Wood*. Forest Products Society, Madison, WI.

Crawford, D. M., Woodward, B. M., and Hatfield, C. A. 2002. *Comparison of Wood Preservatives in Stake Tests*. 2000. Progress Report. Res. Note FPL-RN-02. US Department of Agriculture, Forest Service, Forest Products Laboratory, Madison, WI.

Eaton, R. A. and Hale, M. D. 1993. *Wood: Decay, Pests and Protection*. Chapman & Hall, New York, NY.

EPA. 2002. Whitman announces transition from consumer use of treated wood containing arsenic, U.S. Environmental Protection Agency.

FPL. 2010. *Wood Handbook: Wood as an Engineering Material*. United States Department of Agriculture, Forest Service, Forest Products Laboratory, Madison, WI, General Technical Report-190.

ICC-ES. Evaluation Reports, Section 06070-wood treatment. Whittler, CA: ICC Evaluation Service, Inc. www. icc-es.org.

Johnson, B. R. and Gutzmer, D. I. 1990. *Comparison of Preservative Treatments in Marine Exposure of Small Wood Panels*. USDA Forest Service, Forest Products Laboratory, Madison, WI: 28.

Mac Lean, J. D. 1952. *Preservation of Wood by Pressure Methods*. U.S. Department of Agriculture, Forest Service Washington, DC, 160.

Nilsson T. and Björdal, C. 2007. Personal Communication, Stockholm, Sweden.

PMRA. 2002. *Chromated Copper Arsenate (CCA)*. Canadian Pest Management Regulatory Agency. Ottawa, Ontario, Canada.

Rowell, R. M. 1991. Chemical modification of wood. In: Hon D. N.-S. and Shiraishi, N., *Handbook on Wood and Cellulosic Materials*. Marcel Dekker, Inc. New York, NY, pp. 703–756.

Schultz, T. P., Militz, H, Freeman, M. H., Goodell, B., and Nicholas, D. D. 2008. *Development of Commercial Wood Preservatives*. ACS Symposium Series 982. Washington, DC: American Chemical Society. 655pp.

Stephan, I., Grinda, M., and Rudolph, D. 1998. Comparison of different methods for assessing the performance of preservatives in the BAM fungus cellar test. The International Research Group of Wood Preservation, IRG/WP 98-20149, Stockholm, Sweden.

6 Thermal Properties, Combustion, and Fire Retardancy of Wood

Roger M. Rowell and Mark A. Dietenberger

CONTENTS

One of the greatest assets of cellulosic resources is their compatibility with nature, including their combustibility and degradability which allow for constant turnover and regeneration of these natural resources. A fundamental understanding of these properties and possible methods for controlling them is essential for protection and better utilization of these resources.

Combustion of wood involves a complex series of physical transformations and chemical reactions that are further complicated by the heterogeneity of the substrate. Wood, and cellulosic materials do not burn directly but under the influence of sufficiently strong heat sources they decompose to a mixture of volatiles, tarry compounds, and highly reactive carbonaceous char (Shafizadeh

1984) (Figure 6.1). Gas phase oxidation of the combustible volatiles and tarry products produces flaming combustion. Solid-phase oxidation of the remaining char produces glowing or smoldering combustion, depending on the rate of oxidation.

Lignocellulosic materials decompose on heating and when exposed to an ignition source by two different mechanisms. The first, which is dominated at temperatures below 300°C, involves the degradation of the polymers by breaking internal chemical bonds, dehydration (elimination of water), formation of free radicals, carbonyl, carboxyl and hydroperoxide groups, formation of carbon monoxide (CO) and carbon dioxide (CO_2), and finally, the formation of reactive carbonaceous char. Oxidation of the reactive char results in smoldering or glowing combustion and further oxidation of the combustible volatile gasses gives rise to flaming combustion (Antal 1985, Bridgewater 1999, Czernik et al. 1999, Shafizadeh 1984).

The second mechanism, which takes over at temperatures above 300°C, involves the cleavage of secondary bonds, formation of intermediate products such as anhydromonosaccharides which is converted into low molecular weight products, oligosaccharides and polysaccharides which leads to

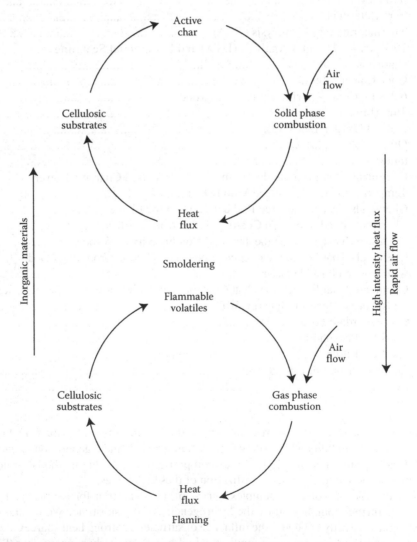

FIGURE 6.1 Graphic presentation of the flaming and smoldering combustion showing the respective roles of combustible volatiles and active char produced by pyrolysis under heat flux at different conditions.

carbonized products (Kawamoto et al. 2003) and high volatile production rates at a high enough heating value to be easily ignitable with air.

6.1 PYROLYSIS AND COMBUSTION

The simplest method of evaluating the thermal property of degradation for wood is by thermogravimetric analysis (TGA). Figure 6.2 shows a schematic of the equipment used for this analysis. A sample is placed in a metal pan in a furnace tube, nitrogen gas is passed through the system, and the furnace tube is slowly heated at a constant rate. The percent weight loss is measured as function of temperature. The temperature is usually raised to 500–600°C in nitrogen at a few °C per minute. The temperature can then be lowered to around 300°C, oxygen is introduced into the system, and the temperature increased again. This second scan will show the combustion of the char in oxygen.

A simple thermogram run in nitrogen is shown in Figure 6.3 for pine (Shafizadeh 1984). As wood is heated from room temperature to 100°C, very little chemical reactions take place. At approximately 100°C, any moisture in the wood is vaporized out. As the wood increases in temperature, very little degradation occur until about 200°C when chemical bonds start to break via dehydration and, possibly, free radical mechanisms to eliminate water and producing volatile gasses. In the absence of, or in limited amounts of oxygen, this thermal degradation process is called pyrolysis. The volatile gasses produced diffuse out of the wood into the surrounding atmosphere. Figure 6.4 shows the first derivative of the TGA curve for pine.

Whole wood starts to thermally degrade at about 250°C according to this TGA (see Figure 6.5). With more sensitive TGA coupled with evolved gas analysis, one can detect volatile organic compounds (VOC) beginning around 130°C, the terpene-related extractives starting around 200°C (McGraw and others 1999), and the water vapor and CO_2 from dehydration and decarboxylation

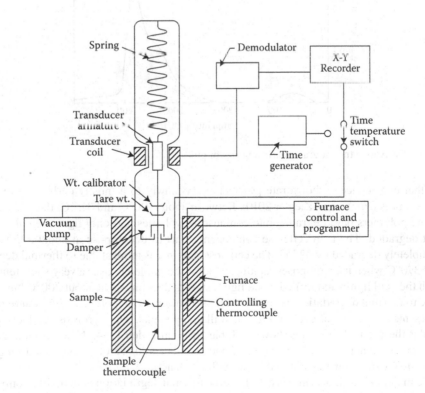

FIGURE 6.2 Schematic diagram of a simple thermogravimetric analysis system.

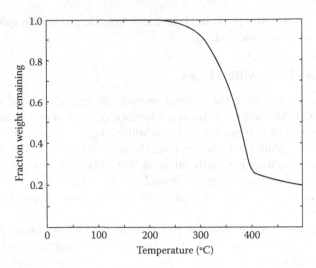

FIGURE 6.3   Thermogravimetric analysis of pine.

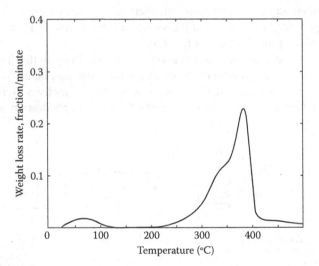

FIGURE 6.4   Derivative thermogravimetric analysis of pine.

processes that can reduce carbohydrate polymers very slowly to a mostly carbon residue within these temperatures (Funke and Ziegler 2010). Between about 300°C and 375°C, the majority of the carbohydrate polymers have degraded into combustible volatiles and only lignin remains as being largely not degraded. The hemicellulose components start to decompose at about 225°C and are almost completely degraded by 325°C. The cellulose polymer is more stable to thermal degradation until about 370°C when it decomposes rapidly and almost completely over a very short temperature range. Both the acid lignin and milled wood lignin start to decompose at about 200°C but are much more stable to thermal degradation as compared to the carbohydrate polymers. The curve for whole wood represents the results of each of the cell wall components. The pyrolysis products given off when wood is thermally degraded is shown in Table 6.1 (Shafizadeh 1984). Note the tar at 28% mass content has many higher molecular weight compounds that can be resolved with modern gas chromatography–mass spectrometry (GC–MS) (www.TarWeb.net).

Since the major cell wall polymer is cellulose, the thermal degradation of cellulose dominates the chemistry of pyrolysis (Shafizadeh and Fu 1973). The decomposition of cellulose leads mainly to

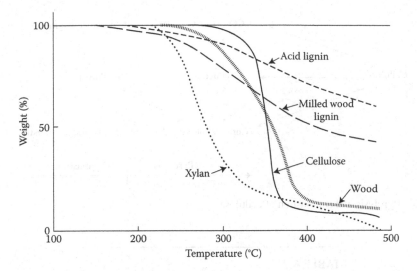

FIGURE 6.5 Thermogravimetric analysis of cottonwood and its cell wall components.

TABLE 6.1
Pyrolysis Products of Wood

Product	Percent in Mixture
Acetaldehyde	2.3
Furan	1.6
Acetone propionaldehyde	1.5
Propenal	3.2
Methanol	2.1
2,3-Butanedione	2.0
1-Hydroxy-2-propanone	2.1
Glyoxal	2.2
Acetic acid	6.7
5-Methyl-2-furaldehyde	0.7
Formic acid	0.9
2-Furfuryl alcohol	0.5
Carbon dioxide	12.0
Water	18.0
Char	15.0
Tar (at 600°C)	28.0

volatile gasses while lignin decomposition leads mainly to tars and char. Figure 6.6 shows the pyrolysis and combustion of cellulose (Shafizadeh 1984). In the early stages of cellulose degradation (below 300°C), the molecular weight is reduced by depolymerization caused by dehydration reactions. The main products are CO, CO_2 produced by decarboxylation and decarbonylation, water, and char residues. In the presence of oxygen, the char residues undergo glowing ignition. The formation of CO and CO_2 are much faster in oxygen than in nitrogen and this rate accelerates as the temperature increases. Table 6.2 shows the rate constants for the depolymerization of cellulose in air and nitrogen (Shafizadeh 1984) for very slow pyrolysis at the lowest temperatures using the simplest mechanistic kinetics equation. These rate constants and mechanistic kinetics equations are

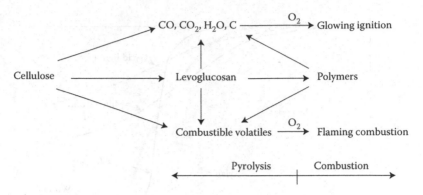

FIGURE 6.6 Pyrolysis and combustion of cellulose.

TABLE 6.2
Rate Constants for the Depolymerization of Cellulose in Air and Nitrogen

Temperature	Condition	$K_o \times 10^7$ (mol/162 g min)[a]
150	N_2	1.1
	Air	6.0
160	N_2	2.8
	Air	8.1
170	N_2	4.4
	Air	15.0
180	N_2	9.8
	Air	29.8
190	N_2	17.0
	Air	48.9

[a] One mol of glucose.

---

continually being updated, particularly due to the fairly recent surge in biomass energy research using slow and fast pyrolysis.

Cellulose while degrading also produces combustible volatiles such as acetaldehyde, propenal, methanol, butanedione, and acetic acid. When the combustible volatiles mix with oxygen and are then heated to the ignition temperature, exothermic combustion occurs. The heat from these reactions in the vapor phase transfer heat back into the wood increasing the rate of pyrolysis in the solid phase via increasing the solid temperature to a higher level. When the burning mixture accumulates sufficient heat, it emits radiation in the visible spectrum. This phenomenon is known as flaming combustion and occurs in the vapor phase. At 300°C, the cellulose molecule is highly flexible and undergoes depolymerization by transglycosylation to primary products such as anhydro-monosaccharides including levoglucosan (1,6-anhydro-β-D-glucopyranose) and 1,6-anhydro-β-D-glucofuranose which is converted into low molecular weight products, randomly linked oligosaccharides and polysaccharides which leads to carbonized products (Figure 6.7).

Figure 6.8 shows the formation of levoglucosan from cellulose. A single unit of cellulose, glucose, undergoes dehydration to form levoglucosan, levoglucosenone, and 1,4:3,6-dianhydro-α-D-glucopyranose. Other products are also known to form such as the 1,2-anhydride and the 1,4-anhydride, 3-deoxy-D-erythrohexosulose, 5-hydroxymethyl-2-furaldehyde, 2-furaldehyde (furfural),

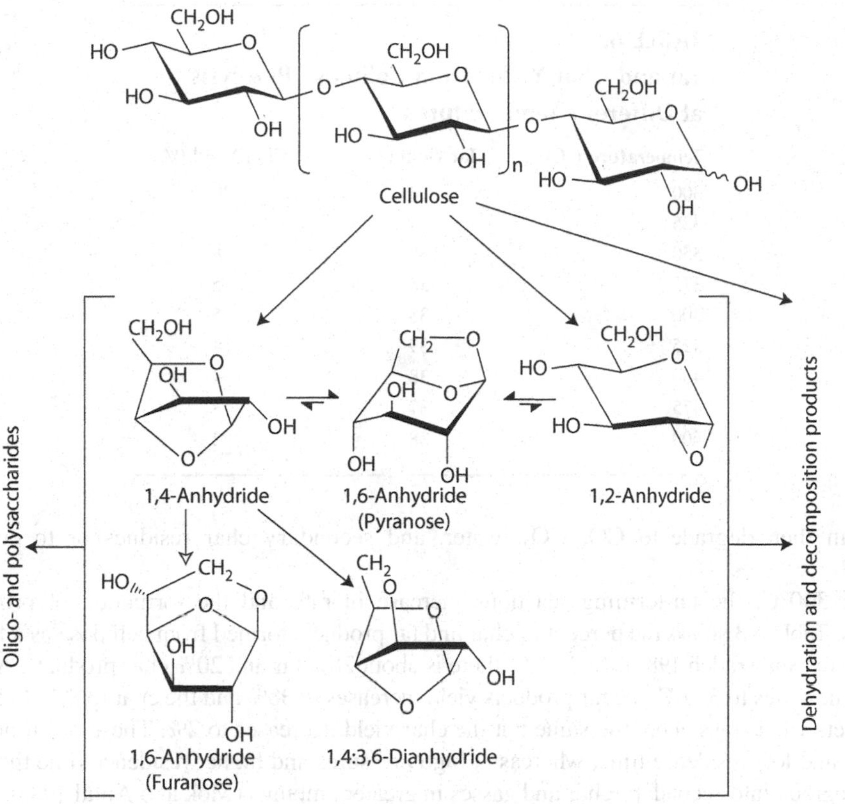

FIGURE 6.7 Pyrolysis of cellulose to anhydro sugars.

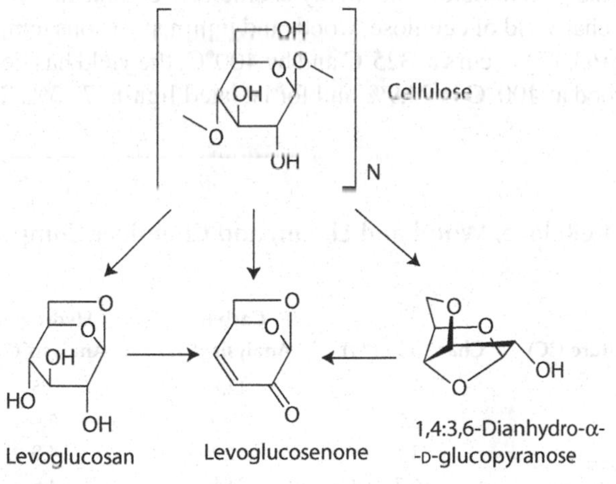

FIGURE 6.8 Formation of levoglucosan and other monomers from cellulose.

other furan derivatives, 1,5-anhydro-4-deoxy-D-hex-1-ene-3-ulose, and other pyran derivatives (Shafizadeh 1984). These dehydration derivatives are important in the intermediate formation of char compounds.

The intermolecular and intramolecular transglycosylation reactions are accompanied by dehydration followed by fission or fragmentation of the sugar units and disproportionation reactions in the gas phase (Shafizadeh 1982). The anhydromonosaccharides can recombine forming polymers

TABLE 6.3
Tar and Char Yields from Cellulose Pyrolysis
at Different Temperatures

Temperature (°C)	Tar Yield (%)	Char Yield (%)
300	28	20
325	37	10
350	38	8
375	38	5
400	38	5
425	39	4
450	38	4
475	37	3
500	38	2

which can then degrade to CO, CO_2, water, and secondary char residues or to combustible volatiles.

Above 300°C, the tar-forming reactions increase in rate and the formation of primary char decreases. Table 6.3 shows the percent of char and tar products formed from cellulose as a function of temperature (Shafizadeh 1984). At 300°C, there is about 28% tar and 20% char products. As the temperature increases to 350°C, the tar products yield increases to 38% and the char to 8%. At 500°C, the tar products yield stays about the same but the char yield decreases to 2%. These occur at the lower pressures and low residence time, whereas at high pressures and higher residence time the cellulosic tar will degrade into secondary char and gasses in greater amounts (Mok and Antal 1983).

The tar products include anhydro sugar derivatives that can hydrolyze to reducing sugars. The evaporation of levoglucosan and other volatile pyrolysis products is highly endothermic. These reactions absorb heat from the system before the highly exothermic combustion reactions take place.

Table 6.4 shows the char yield of cellulose, wood, and lignin at various temperatures. The highest char yield for cellulose (63.3%) occurs at 325°C and by 400°C, the yield has decreased to 16.7%. The char yield for whole wood at 400°C is 24.9% and for isolated lignin, 73.3%. The carbon, hydrogen,

TABLE 6.4
Char Yield from Cellulose, Wood and Lignin, and Chemical Composition
of the Char[a]

Material	Temperature (°C)	Char Yield (%)	Carbon Analysis (%)	Hydrogen Analysis (%)	Oxygen Analysis (%)
Cellulose	Control	—	42.8	6.5	50.7
	325	63.3	47.9	6.0	46.1
	350	33.1	61.3	4.8	33.9
	400	16.7	73.5	4.6	21.9
	450	10.5	78.8	4.3	16.9
	500	8.7	80.4	3.6	16.1
Wood	Control	—	46.4	6.4	47.2
	400	24.9	73.2	4.6	22.2
Lignin	Control	—	64.4	5.6	24.8
	400	73.3	72.7	5.0	22.3

[a] Isothermal pyrolysis, 5 min at temperature.

and oxygen analysis for the various char yields shows the highest carbon content char occurs at 500°C but the char yield is only 8.7%.

The table shows that the lignin component gives the highest char yield (Sharizadeh 1983). Lignin mainly contributes to primary char formation and cellulose and the hemicelluloses mainly to volatile pyrolysis products that are responsible for flaming combustion.

The intensity of combustion can be expressed as:

$$I_R = \frac{-\Delta H \; dm}{dt}$$

where

I_R = reaction density (rate of heat release)
$-\Delta H$ = heat of combustion
dm/dt = rate mass of fuel loss

Table 6.5 shows the heat of combustion for cellulose, whole wood, bark, and lignin as determined from oxygen bomb calorimetry and their pyrolysis yields from a TGA. The highest levels of volatiles are produced by cellulose which has the lowest heat of combustion and the lowest percent char formation. The next level of volatiles is produced by whole wood with slightly higher heat of combustion and char yield as compared to cellulose. Bark has a higher heat of combustion as compared to wood, a char yield of 47% and 52% volatiles. Lignin has the highest heat of combustion and also the highest char yield and the lowest percent of volatiles. We note that the heat of combustion value at oxygen bomb calorimetry conditions for the virgin material are generally less than the heat of combustion value for the developed char, and also higher than the heat of combustion of the developed volatiles for the corresponding material (Dietenberger 2002). Therefore, assigning heat of combustion values to the fuel components during flaming or glowing combustion is not straightforward and requires a very detailed mechanistic kinetics theory.

The high rate of heat released upon flaming combustion provides sufficient energy to pyrolyze the remaining solid wood and thus propagate the fire. For wood, external heat sources in addition to flaming combustion or enhanced flaming with enriched oxygen is usually needed to sustain creeping flame spread (White and Dietenberger 1999, White 1979). Oxidation of the residual char after flaming combustion results in glowing combustion. If the intensity of the heat and the concentration of combustible volatiles fall below the minimum level for flaming combustion, gradual oxidation of the reactive char initiates smoldering combustion. The smoldering combustion process releases nonignitable or unoxidized volatile products and usually occurs in low density woods.

TABLE 6.5
Heat of Combustion of Wood, Cellulose and Lignin, Char and Combustible Volatile Yields

Virgin Fuel	Heat of Combustion ΔH 25°C (cal/g)	Char Yield (%)[a]	Combustible Volatiles (%)[a]
Cellulose	−4143	14.9	85.1
Wood (Poplar)	−4618	21.7	78.3
Bark (Douglas fir)	−5708	47.1	52.9
Lignin (Douglas fir)	−6371	59.0	41.0

[a] Heating rate 200°C/min to 400°C and held for 10 min.

6.2   FIRE RETARDANCY

Wood has been used for many applications because it has poor thermal conductivity properties. In a fire, untreated wood forms a char layer and this layer is an insulation barrier protecting the wood below the burning layer (self-insulating). The comparative fire resistance of wood and metal was never more graphically shown than the pictures taken after many hours of burning in the 1953 fire at a casein plant in Frankfort, New York (Figure 6.9). The steel girders softened and failed at high temperature and fell across the 12 by 16 inch laminated wood beams that were charred but still strong enough to hold the steel girders. This charring protecting property of structural wood has found many applications, such as railroad wooden bridges.

To reduce the flammability of wood, fire retardants have been developed. The use of fire retardants for wood can be documented back to the first century A.D. when the Romans treated their ships with alum and vinegar for protection against fire (LeVan 1984). Later, Gay–Lussac used ammonium phosphates and borax to treat cellulosic textiles. The U.S. Navy specified the use of fire retardants for their ships starting in 1895 and the City of New York required the use of fire retardants in building over 12 stories high starting in 1899 (Eickner 1966).

Standards that pertain to fire safety in structures are specified in building codes and fire codes (White and Dietenberger 1999). Building codes include area and height of the rooms, firestops, doors and other exits, automatic sprinklers, fire detectors, and type of construction. Fire codes include materials combustibility, flame spread, and fire endurance that imply large-scale test performances to obtain the required ratings. However, when developing improved fire-retardant treatments (FRT) it is advantageous to use laboratory scale tests that give an indication of improved fire retardancy of wood-based products. The promising FRTs that pass these initial tests can then be applied to the relatively more expensive large-scale tests to achieve the required rating.

6.3   TESTING FIRE RETARDANTS

6.3.1   THERMOGRAVIMETRIC ANALYSIS (TGA)

There are several ways to test the efficiency of a fire retardant. The most common is to run TGA analysis as described earlier. TGA involves weighing a finely ground sample and exposing it to a heated chamber in the presence of nitrogen. The sample is suspended on a sensitive balance that measures the weight loss of the sample as the system is heated. Nitrogen or another gas flows around the sample to remove the pyrolysis or combustion products. Weight loss is recorded as a function of time and temperature. In isothermal TGA, the change in weight of the sample is recorded as a function of time at a constant temperature. With the use of a derivative computer, the rate of weight loss as a function of time and temperature can also be measured. This is referred to as derivative thermogravimetry (Slade and Jenkins 1966) with results shown in Figure 6.4 for pine.

FIGURE 6.9   Wood beam survives fire in casein plant in Frankfort, NY.

6.3.2 DIFFERENTIAL THERMAL ANALYSIS (DTA) AND DIFFERENTIAL SCANNING CALORIMETER (DSC)

DTA measures the amount of heat liberated or absorbed by a wood sample as it moves from one physical transitions state to another (i.e., melting, vaporization) or when it undergoes any chemical reaction. This heat is determined by measuring the temperature differences between the sample and an inert reference. DTA can be used to measure heat capacity, to provide kinetic data, and to give information on transition temperatures. The test device consists of sample and reference pans exposed to the same heat source. The temperature is measured using thermocouples embedded in the sample and the reference pan. The temperature difference between the sample and reference is recorded against time as the temperature is increased at a linear rate. For calorimetry, the equipment is calibrated against known standards at several temperatures (Slade and Jenkins 1966).

DSC is similar to DTA except the actual differential heat flow is measured when the sample and reference temperature are equal. In DSC, both the sample and reference are heated by separate heaters. If a temperature difference develops between the sample and reference because of exothermic or endothermic reactions in the sample, the power input is adjusted to remove this difference. Thus, the temperature of the sample holder is always kept at the same as the reference.

6.3.3 CONE CALORIMETER

A cone calorimeter (Figure 6.10) is a modern instrument used to study the fire behavior of small samples of various materials in condensed phase. It is widely used in the field of Fire Safety Engineering. It gathers data regarding the ignition time, mass loss rate (MLR), combustion products, heat release rate (HRR), and other parameters as a function of time associated with the finitely thick material's burning properties. A sample is mounted on a load cell that will measure mass loss during a test. The fuel sample can be exposed to different heat fluxes from the cone heater over its 10 by 10 cm surface, typically set at 50 kW/m^2 to simulate the ignition burner heat fluxes in large-scale regulatory tests. The principle for the measurement of the heat release rate is based on the Huggett's

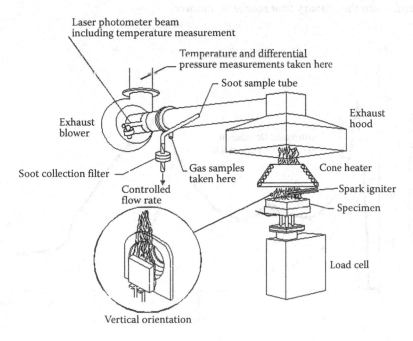

FIGURE 6.10 Schematic diagram of a cone calorimeter.

principle that the net heat of combustion of any organic material is directly related to the amount of oxygen required for combustion (Huggett 1980), which was strongly reaffirmed for wood volatiles and char (Dietenberger 2002) . The cone calorimeter produces large amounts of data including curves of heat release, smoke, and mass loss, CO, CO_2, and other gas yields. These data are automatically collected by the software used with the calorimeters. The ASTM International Standard is ASTM E1354-11a, "Standard Test Method for Heat and Visible Smoke Release Rates for Materials and Products using an Oxygen Consumption Calorimeter" (www.astm.org/Standards/E1354.htm). If only the mass loss measurement under fire conditions is needed to compare fire retardancy of materials, one can use the lower cost mass loss calorimeter, ASTM E2102 or the fire tube, ASTM E69.

6.3.3.1 Cone Calorimeter Tests on Wood

Three distinct uses of the cone calorimeter data, particularly for fire retardancy, have been developed to (1) compare the fire response of materials to assess their fire performance, for materials development, or pyrolysis and burning model developments, (2) derive the material parameters needed as input into mathematical models for the full-/scale room room/corner test assessment, and (3) determine for regulatory purposes the characteristic parameters such as peak HRR or total heat evolved (Schartel and Hull 2007). In the first main use, FRT results in delayed ignition, reduced heat release rate, reduced smoke production rate, and slower spread of flames. HRRs are markedly reduced by FRT (Figure 6.11). Note that the reduction in HRR (and MLR) after the initial peak HRR (PHRR) shortly after ignition is due to the insulating char development in conjunction with thermal wave traveling through the thick material resulting in the pyrolysis front slowing down, while the second peak HRR is due to combined effects of thermal wave termination at the insulated back surface and appearance of glowing combustion (Hagge et al. 2004). The transition from flaming to glowing is evidenced by an almost doubling of the effective heat of combustion (EHC = HRR/MLR) as the combustion transitions from volatiles to char oxidations. Flame-retardant treatment of wood generally improves the fire performance by reducing the amount of flammable volatiles released during fire exposure or by reducing the effective heat of combustion, or both. Both results have the effect of reducing the HRR, particularly during the initial stages of fire, and thus consequently reducing the rate of flame spread over the surface (Dietenberger and White 2001). The wood fire may then self-extinguish when the primary heat source is removed.

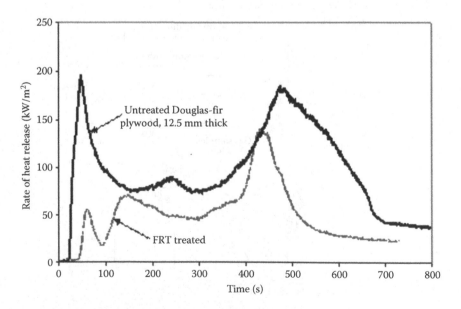

FIGURE 6.11   Heat release curves for untreated and FRT-treated Douglas-fir plywood, 12.5 mm thick.

TABLE 6.6
Comparison of Room/Corner Flashover Times and the Cone's Global Parameters

Material	TFO (s) at 20 kW/m²	TFO (s) predict	Thick (mm)	TIG (s)	PHRR (kW/m²)	THR (MJ/m²)	peakSEA (m²/s)
Wafer board	141	143	13	21.4	208	101	145
OSB	186	211	11.5	23.3	204	91	119
Hardboard	222	299	7	27.6	189	93	119
Particleboard	237	183	12.8	31.4	211	102	136
Southern pine board	240	241	19	21.4	181	143	323
Southern pine plywood	321	274	11	19.1	197	65	126
Douglas-fir plywood ASTM	474	513	11.8	17.6	173	67	81
Redwood lumber	498	465	19	18.9	171	95	200
Douglas-fir plywood	531	528	12	16.6	168	74	97
White spruce lumber	594	559	17.3	17.0	165	89	122
FRT rigid PU foam	621	631	23	2.69	89.9	4.0	1061
FRT Douglas-fir plywood	849	869	12.5	34.2	72	34	10
FRT Southern pine plywood	882[a]	860	11.3	62.5	88	39	75
Type-X gypsum board	NFO	NFO	16	48	118	2.79	6

[a] 300 s added to measured TFO for burner profile glitch.

The global material parameters of the cone calorimeter data that are of significance to flame spread modeling via theoretical analytical solutions of fire growth are time to piloted ignition (TIG), peak HRR (PHRR), total HRR (THRR), and smoke extinction area (SEA), obtained with a horizontal sample exposed to irradiance of 50 kW/m² (Dietenberger and White 2001). These are listed in Table 6.6 for various panel products that span the wide range of flammability ratings and are arranged from the most flammable (waferboard) to the least flammable material (Type-X gypsum board) according to ASTM E84 results (discussed in next section). The flammability ratings for the full-scale room/corner tests (ISO 9705) with the 100/300 kW pilot burner in the corner of wall-only mounted panels is based on time-to-flashover (TFO) measured in various ways, the most objective measure is the occurrence of 20 kW/m² on the floor center. None of the cone's global parameters by themselves give any indication of the room's flammability ratings, not until we utilize the complex exponential root terms of the theoretical upward fire spread model that defines the flame spreading into accelerating, neutral, or damping fire spread modes in a simple empirical relationship (Dietenberger and White 2001) for various full-scale tests. It was interesting that the similar analysis can be applied to plastics with and without FRT, which is helpful in evaluating the wood–plastic composites also.

6.3.4 TUNNEL FLAME-SPREAD TESTS

Building standards designed to control fire growth often require certain flame-spread ratings for various parts of a building. For code regulations in North America, flame-spread ratings are determined by a 25 foot long tunnel test which is an approved standard test method (ASTM E 84, www.astm.org/Standards/E84.htm). A specimen is exposed to an ignition source, and the rate at which the flames travel to the end of the specimen is measured. In the past, red oak flooring was used as a standard and given a rating of 100. In the building codes, the classes for flame spread index are A (FSI of 0–25), B (FSI of 26–75), and C (FSI of 76–200). Generally, codes specify FSI for interior finish based on building occupancy, location within the building, and availability of automatic sprinkler protection. The more restrictive classes, Classes A and B, are generally prescribed for stairways and corridors

TABLE 6.7
Flame Spread Index for Different Woods
Using the 25 foot Tunnel Test

Species	Flame Spread Index
Douglas-fir	70–100
Western hemlock	60–75
Lodgepole pine	93
Western red cedar	70
Redwood	70
Sitka spruce	74–100
Yellow birch	105–110
Cottonwood	115
Maple	104
Red oak	100
Walnut	130–140
Yellow poplar	170–185

that provide access to exits. In general, the more flammable classification (Class C) is permitted for the interior finish of other areas of the building that are not considered exit ways or where the area in question is protected by automatic sprinklers. In other areas, there are no flammability restrictions on the interior finish and unclassified materials (i.e., more than 200 FSI) can be used.

Two-foot tunnel test (ASTM D3806-98(2011), www.astm.org/Standards/D3806.htm) and 8-foot tunnel test (Peters and Eickner 1962) have also been used for research applications. All tunnel tests measure the surface flame spread of the wood although each differs in the method of the exposure and type of flame spreading. The severity of the exposure and the time a specimen is exposed to the ignition source are the main differences between the various tunnel test methods. The 25 foot tunnel test is the most severe exposure where the specimen is exposed for 10 min to a high heat flux burner in an assisted air flow flame spread configuration. An extended test of 30 min is a requirement for FRT products (ASTM E2768-11, www.astm.org/Standards/E2768.htm). Because the 25 foot tunnel test is the most severe exposure, it is used as the standard for building materials. The 2 foot tunnel test is the least severe but because small specimens can be used, it is a tool for development work on fire retardants. Table 6.7 shows average values for the flame spread index of several wood species (White and Dietenberger 1999). Extensive listing of wood materials is at American Wood Council website (http://awc.org/publications/DCA/DCA1/DCA1.pdf).

6.3.5 CRITICAL OXYGEN INDEX TEST

The oxygen index test (ASTM D2863-10, www.astm.org/Standards/D2863.htm) measures the minimum concentration of oxygen in an oxygen–nitrogen mixture that will just support downward creeping flame spread of a test specimen. Highly flammable materials have low oxygen index, less flammable materials have high values (White 1979). One advantage of this test is that very small specimens can be used and it can be used to study the retardant mechanism in the dynamic flaming environment that cannot be done with TGA, DTA, or DSC because they only measure degradation and combustion properties in a quasi-steady environment far removed from real-world fires.

Table 6.8 shows the effect of inorganic additives on oxygen index and the yield of levoglucosan (Fung et al. 1972). Phosphoric acid is the most effective treatment of wood to increase the oxygen index and decrease the formation of levoglucosan. Ammonium dihydrogen orthophosphate, zinc chloride, and sodium borate are also very effective in reducing the yield of levoglucosan.

TABLE 6.8
Effect of Inorganic Additives on Oxygen Index and Levoglucosan Yield

Chemical	Oxygen Index (%)	Levoglucosan Yield (%)
Untreated	17.3	10.1
Potassium dihydrogen phosphate	18.5	0.9
Potassium hydrogen phosphate	18.6	0.2
Sodium borate	19.3	<0.3
Zinc chloride	19.6	0.3
Ammonium dihydrogen orthophosphate	19.6	0.8
Phosphoric acid	20.5	<0.1

Source: Adapted from Fung, D.P.C., Tsuchiya, Y., and Sumi, K. 1972. *Wood Sci.* 5(1): 38–43.

6.3.6 OTHER TESTS

Other tests that can be run on wood and fire retardant-treated wood determine smoke production, toxicity of smoke, and rate of heat release. The production of smoke can be a critical problem with some types of fire retardants. The 25 foot tunnel test uses a photoelectric cell to measure the density of smoke evolved. The effect of fire retardants on smoke production varies depending on the chemical. Chemicals such as zinc chloride and ammonium phosphate generate much larger amounts of smoke as compared to borates. The commercial FRT applied to plywood greatly reduces the smoke production in comparison to untreated plywood in cone calorimeter tests (Table 6.6). The toxicity of the smoke is also a critical consideration for fire retardant-treated wood. A large percentage of fire victims are not touched by flames but are overcome as a result of exposure to toxic smoke (Kaplan et al. 1982). The heat of combustion of wood varies somewhat depending on the species, resin content, moisture content, and other factors. The contribution to fire exposure depends on these factors along with the fire exposure and degree of combustion. Although the overall heat of combustion of wood does not change, fire retardants reduce the rate of heat release by reducing volatization rate and volatile heat of combustion as well as reduce oxidation rates of the chars to dramatically reduce their contribution to unsafe or unwanted fires.

6.4 FIRE RETARDANTS

Fire retardant treatments for wood can be classified into one of six classes: (1) chemicals that promote the formation of increased char at a lower temperature than untreated wood degrades, (2) chemicals which act as free radical traps in the flame, (3) chemicals used to form a coating on the wood surface, (4) chemicals that increase the thermal conductivity of wood, (5) chemicals that dilute the combustible gasses coming from the wood with noncombustible gasses, and (6) chemicals that reduce the heat content of the volatile gasses (LeVan 1984). In most cases, a given fire retardant operates by several of these mechanisms and much research has been done to determine the magnitude and role of each of these mechanisms in fire retardancy.

6.4.1 CHEMICALS THAT PROMOTE THE FORMATION OF INCREASED CHAR AT A LOWER TEMPERATURE THAN UNTREATED WOOD DEGRADATION

Most of the evidence relating to the mechanism of fire retardancy in the burning of wood indicates that retardants alter fuel production by increasing the amount of char and reducing the amount of volatile, combustible vapors, and decrease the temperature where pyrolysis begins.

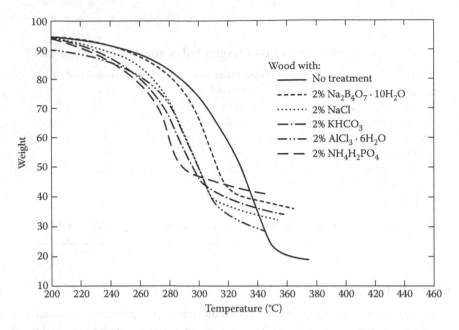

FIGURE 6.12 Thermogravimetric analysis of wood treated with various inorganic additives. (Adapted from Tang, W.K. and Eickner, H.W. 1968. *Effect of Inorganic Salts on Pyrolysis of Wood, Cellulose, and Lignin Determined by Differential Thermal Analysis.* USDA Forest Service Research Paper FPL 82, January.)

Figure 6.12 shows a TGA of untreated wood along with wood that has been treated with several inorganic fire retardants (Tang and Eickner 1968). Fire retardant chemicals such as ammonium dihydrogen orthophosphate greatly increase rate of hydrolysis of glycosidic bonds, increase the amount of residual char and lower the initial temperature of thermal decomposition. The amount of noncondensable gasses increases at the expense of the flammable tar fraction. The chemical mechanism for the reduction of the combustible volatiles involves the ability of the fire retardant to inhibit the formation of levoglucosan (see the earlier discussion) and to catalyze dehydration of the cellulose to more char and fewer volatiles but also enhance the condensation of the char to form cross-linked and thermally stable polycyclic aromatic structures (Shafizadeh 1984). Nanassy (1978) showed that Douglas-fir treated with either sodium chloride and ammonium dihydrogen orthophosphate increased both the char yield and the aromatic carbon content of the char (Table 6.9).

Figure 6.13 shows the TGA of untreated cellulose along with cellulose that has been treated with several inorganic fire retardants (Tang and Eickner 1968). The thermal decomposition pattern for

TABLE 6.9
Effect of Inorganic Additives on Char Yield and Aromatic Carbon Content of the Char

Chemical	Char Yield (wt%)	Aromatic Carbon in Char (wt%)
Untreated	15.3	13.7
Sodium chloride	17.5	16.8
Ammonium dihydrogen orthophosphate	28.9	18.6

Source: Adapted from Shafizadeh, F. 1984. *Advances in Chemistry Series*, Number 207, American Chemical Society, Washington, DC. Chapter 13: pp. 489–529; Nanassy, A.J. 1978. *Journal Wood Sci.* 11(2): 111–117.

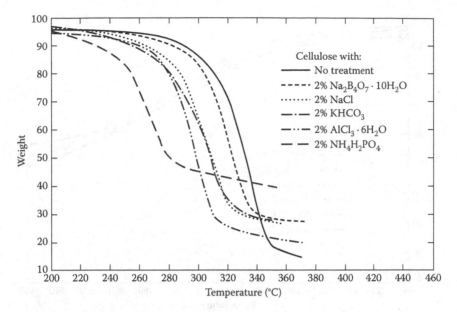

FIGURE 6.13 Thermogravimetric analysis of cellulose treated with various inorganic additives. (Adapted from Tang, W.K. and Eickner, H.W. 1968. *Effect of Inorganic Salts on Pyrolysis of Wood, Cellulose, and Lignin Determined by Differential Thermal Analysis.* USDA Forest Service Research Paper FPL 82, January.)

cellulose is similar to whole wood except pure cellulose is more thermally stable with the hemicelluloses removed.

All inorganic fire retardants reduce the amount of levoglucosan regardless of the relative effectiveness of the fire retardant. This includes the effect of acidic, neutral, and basic additives on the levoglusosan yield. The acid treatment has the most pronounced effect on the reduction in the formation of levoglucosan.

During the heating of cellulose treated with borax or ammonium dihydrogen orthophosphate, the degree of polymerization (DP) of the cellulose decreased (Fung et al. 1972). The DP decreased from 1110 to 650 after only 2 min of heating at 150°C with wood treated with ammonium dihydrogen orthophosphate. The DP dropped from 1300 to 700 after 1 h of heating borax treated wood at 150°C. Both of these chemical treatments suppressed the formation of levoglucosan (see Table 6.8).

Figure 6.14 shows the TGA of untreated lignin along with lignin that has been treated with several inorganic fire retardants (Tang and Eickner 1968). Treating lignin with the fire retardants has very little effect on the thermal decomposition of lignin.

6.4.2 CHEMICALS THAT ACT AS FREE RADICAL TRAPS IN THE FLAME

Certain fire retardants affect vapor-phase reactions by inhibiting the chain reactions as shown below. Halogens such as bromine and chlorine are good free radical inhibitors and have been studied extensively in the plastics industry. Generally, high concentrations of halogen are required (15–30% by weight) to attain a practical degree of fire retardancy. The efficiency of the halogen decreases in the order Br > Cl > F. A mechanism for the inhibition of the chain-branching reactions using HBr as the halogen is:

$$H^{\cdot} + HBr \rightarrow H_2 + Br^{\cdot}$$

$$OH^{\cdot} + HBr \rightarrow H_2O + Br^{\cdot}$$

The hydrogen halide consumed in these reactions is regenerated to continue the inhibition.

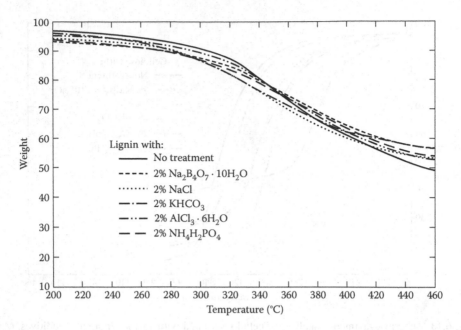

FIGURE 6.14   Thermogravimetric analysis of lignin treated with various inorganic additives. (Adapted from Tang, W.K. and Eickner, H.W. 1968. *Effect of Inorganic Salts on Pyrolysis of Wood, Cellulose, and Lignin Determined by Differential Thermal Analysis.* USDA Forest Service Research Paper FPL 82, January.)

An alternate mechanism has been suggested for halogen inhibition that involves recombination of oxygen atoms (Creitz 1970):

$$O^{\cdot} + Br_2 \rightarrow BrO^{\cdot} + Br^{\cdot}$$

$$O^{\cdot} + OBr^{\cdot} \rightarrow Br^{\cdot} + O_2$$

Thus, the inhibitive effect results from the removal of active oxygen atoms from the vapor phase. Additional inhibition can result from removal of OH radicals in the chain-branching reactions:

$$BrO^{\cdot} + {}^{\cdot}OH \rightarrow HBr + O_2$$

$$BrO^{\cdot} + {}^{\cdot}OH \rightarrow Br^{\cdot} + HO_2$$

Some phosphorus compounds have also been found to inhibit flaming combustion by this mechanism.

6.4.3   CHEMICALS USED TO FORM A COATING ON THE WOOD SURFACE

A physical barrier can retard both smoldering combustion and flaming combustion by preventing the flammable products from escaping and by preventing oxygen from reaching the substrate. These barriers also insulate the combustible substrate from high temperatures. Common barriers include sodium silicates and coatings that intumesce (release a gas at a certain temperature that is trapped in the polymer coating the surface). Intumescent systems swell and char on exposure to fire to form carbonaceous foam and consist of several components. These compounds include a char-producing compound, a blowing agent, a Lewis-acid dehydrating agent, and other chemical components.

In the intumescent systems, the char-producing compound, such as a polyol, will normally burn to produce CO_2 and water vapor and leave flammable tars as residues. However, the compound can esterify when it reacts with certain inorganic acids, usually phosphoric acid. The acid acts as a dehydrating agent and leads to increased yield of char and reduced volatiles. Such char is produced at a lower temperature than the charring temperature of the wood. Blowing agents decompose at determined temperatures and release gases that expand the char. Common blowing agents include dicyandiamide, melamine, urea, guanidine, and they are selected on the basis of their decomposition temperatures. Many blowing agents also act as a dehydrating agent. Other chemicals can be added to the formulation to increase the toughness of the surface foam.

6.4.4 CHEMICALS THAT INCREASE THE THERMAL CONDUCTIVITY OF WOOD

A metal alloy, with a melting point of 105°C, can be used to treat wood. Upon heating, the temperature rise in the metal alloy-treated wood is slower than nontreated wood until the melt temperature is reached (Browne 1958). Above the melt temperature of the metal alloy, the rise in temperature is the same for treated and nontreated wood. Or one could use wood adhesives to fill in the voids of the wood to increase the density for higher thermal conductivity as well as provide blockage to volatile emissions from the subsurface pyrolysis zones.

Another thermal theory suggests that fire retardants cause chemical and physical changes so that heat is absorbed by the chemical to prevent the wood surface from igniting. This theory is based on chemicals that contain a lot of water of crystallization. Water will absorb latent heat of vaporization from the pyrolysis reactions until all the water is vaporized. This serves to remove heat from the pyrolysis zone thereby slowing down the pyrolysis reactions. Once the water is removed, the wood undergoes pyrolysis independent of the past moisture content of the wood.

6.4.5 CHEMICALS THAT DILUTE THE COMBUSTIBLE GASES COMING FROM THE WOOD WITH NONCOMBUSTIBLE GASES

Chemicals such as dicyandiamide and urea release large amounts of noncombustible gasses at temperatures below the temperature at which the major pyrolysis chemistries start. Chemicals such as borax release large amounts of water vapor. Any reduction in the percentage of flammable gasses would be beneficial because it increases the volume of combustible volatiles needed for ignition. Also the movement of gasses away from the wood may dilute the amount of oxygen near the boundary layer between the wood and the vapor-phase reaction that leads to flame extinction. The large emission of water vapor from the dehydration of gypsum boards at temperatures below the wood's degradation temperatures can be a big help for the veneer like panels installed over the gypsum board, or even with a wood/gypsum composite.

6.4.6 CHEMICALS THAT REDUCE THE HEAT CONTENT OF VOLATILE GASES

As seen in Section 6.4.1 before, the addition of inorganic additives lower the temperature at which active pyrolysis begins and this resulting decomposition leads to increased amounts of char and reduced amounts of volatiles. This increased amount of char and reduced volatiles is due to the increased dehydration reactions, mainly the cellulose component of wood. However, other competing reactions also occur such as decarbonylation, decomposition of simpler compounds, and condensation reactions. All these reactions compete with each other. As a result, shifts favoring one reaction over another also change the overall heat of reaction. Differential thermal analysis is used to determine these changes in heats of reactions and can help gain an understanding about these competing reactions. Or one could test these treated materials in the cone calorimeter that provide information on the reduction of heat content as well as provide other properties needed to relate to flammability on a large scale (Schartel and Hull 2007, Dietenberger and White 2001).

DTA of wood in helium shows two endothermic reactions followed by a smaller exothermic one. The first endothermic reaction, which peaks around 125°C, is a result of evaporation of water and desorption of gases; the second, peaking between 200°C and 325°C, indicates depolymerization and volatilization. At around 375°C these endothermic reactions are replaced with a small exothermic peak. When the wood sample is run in oxygen, these endothermic peaks are replaced by strong exothermic reactions. The first exotherm, around 310°C for wood and 335°C for cellulose are attributed to the flaming of volatile products; the second exotherm, at 440°C for wood and 445°C for cellulose and 445°C for lignin are attributed to glowing combustion of the residual char (Tang 1964).

DTA of inorganic fire retardant-treated wood in oxygen shifts the peak position temperatures and/or the amount of heat released. Sodium tetraborate, for example, reduces the volatile products exotherm considerably, increase the glowing exotherm and shows a second glowing peak around 510°C. Sodium chloride also reduces the first exotherm, increases the size of the second, but does not produce a second glowing exotherm as did the sodium tetraborate. Wood treated with ammonium phosphate is the most effective in reducing the amount of volatile products and also reduces the temperature where these products are formed. Ammonium phosphate almost eliminates the glowing exotherm (Tang and Eickner 1968).

Fire retardants treatments of this type reduce the average heat of combustion for the volatile pyrolysis products released at the early stage of pyrolysis below the value associated with untreated wood at comparable stages of volatilization. At 40% volatilization, untreated wood has as a 29% release of accumulated volatile products' heat of combustion; treated wood has only released 10–19% of this total heat. Of all the chemicals tested, only sodium chloride, which is known to be an ineffective fire retardant does not reduce the heat content (Tang and Eickner 1968).

6.4.7 Phosphorus–Nitrogen Synergism Theories

One role phosphoric acid and phosphate compounds play in the fire retardancy of wood is to catalyze the dehydration reaction to produce more char. This reaction pathway is just one of several that are taking place all at once including decarboxylation, condensation, and decomposition. The effectiveness of fire retardants containing both phosphorus and nitrogen is greater than the effectiveness of each of them by themselves.

The interaction of phosphorus and nitrogen compounds produces a more effective catalyst for the dehydration because the combination leads to further increases in the char formation and greater phosphorus retention in the char (Hendrix and Drake 1972). This may be the result of increased cross-linking of the cellulose during pyrolysis through ester formation with the dehydrating agents. The presence of amino groups results in retention of the phosphorus as a nonvolatile amino salt, in contrast to some phosphorus compounds that may decompose thermally and be released into the volatile phase. It is also possible that the nitrogen compounds promote polycondensation of phosphoric acid to polyphosphoric acid. Polyphosphoric acid may also serve as a thermal and oxygen barrier because it forms a viscous fluid coating.

6.4.8 Fire-Retardant Formulations

Many chemicals have been evaluated for their effectiveness as fire retardants. The major fire retardants used today include chemicals containing phosphorus, nitrogen, boron, and a few others. Many fire retardant formulations are water leachable and corrosive so research continues to find more leach resistant and less corrosive formulations.

6.4.8.1 Phosphorus

Chemicals containing phosphorus are one of the oldest classes of fire retardants. Monoammonium and diammonium phosphates are used with nitrogen compounds since the synergistic effect allows for less chemical to be used (Hendrix and Drake 1972, Langley et al. 1980, Kaur et al. 1986). Organophosphorus and polyphosphate compounds are also used as fire retardants. Ammonium

TABLE 6.10
Effects of Inorganic Additives on Thermogravimetric Analysis

Additive	Percent Weight Loss at 500°C
Phosphoric acid	61
Ammonium dihydrogen orthophosphate	66
Zinc chloride	74
Sodium hydroxide	79
Boric acid	81
Sodium chloride	82
Tin chloride	84
Diammonium sulfate	86
Sodium tetraborate decahydrate	89
Sodium phosphate	91
Ammonium chloride	93
Untreated wood	93

polyphosphate at loading levels of 96 kg/m^3 gives a flame-spread index of 15 according to ASTM E84 (Holmes 1977). This treatment generates a low smoke yield but it is corrosive to aluminum and mild steel. Other formulations containing phosphorus are mixture of (1) guanyl urea phosphate (GUP) and boric acid (BA), and (2) phosphoric acid, boric acid, and ammonia. Table 6.10 shows the effectiveness of some of the fire retardants in terms of the percent of weight loss at 500°C. The most effective chemical is phosphoric acid with a weight loss of 61% as compared to 93% weight loss for untreated wood.

6.4.8.2 Boron
Borax (sodium tetraborate decahydrate) and boric acid are the most often used fire retardants. The borates have low melting pints and form glassy films on exposure to high temperatures. Borax inhibits surface flame spread but also promotes smoldering and glowing. Boric acid reduces smoldering and glowing combustion but has little effect on flame spread. Because of this, borax and boric acid are usually used together. The alkaline borates also result in less strength loss in the treated wood and is less corrosive and hydroscopic (Middleton et al. 1965). Boron compounds are also combined with other chemicals such as phosphorus and amine compounds to increase their effectiveness. Table 6.10 shows that wood treated with boric acid shows a weight loss of 81% and borax 89% weight loss at 500°C which is not as effective as phosphorus compounds. Indeed, the combination of GUP and BA was particularly studied because they are considered environmentally acceptable, relatively low in toxicity, relatively noncorrosive, relatively nonhygroscopic, and can be stored relatively long times (Gao et al. 2006). They found a strong synergistic effect between GUP and BA to significantly reduce both HRR and THR of treated wood tested in the cone calorimeter.

6.4.9 Leach-Resistant Fire-Retardants

The most widely studied leach-resistant fire-retardant system is based on amino-resins (Goldstein and Dreher 1964). Basically, the resin system consists of a combination of a nitrogen source (i.e., urea, melamine, guanidine, or dicyandiamide) with formaldehyde to produce a methylolated amine. The product is then reacted with a phosphorus compound such as phosphoric acid. Other formulations can include mixtures of dicyandiamide, melamine, formaldehyde and phosphoric acid or dicyandiamide, urea, formaldehyde, phosphoric acid, formic acid, and sodium hydroxide. Leach resistance is attributed to polymerization of the components within the wood (Goldstein and Dreher 1961). Another formulation uses a urea and melamine amino-resin (Juneja and Fung 1974).

The stability of these resins is controlled by the rate of methylolation of the urea, melamine, and dicyandiamide. The optimum mole ratio for stability of these solutions is 1:3:12:4 for urea or melamine, dicyandiamide, formaldehyde, and orthophosphoric acid. Lee et al. (2004) bonded phosphoramides to wood by reacting phosphorus pentoxide with amines *in situ*. Leach resistance was greatly improved and the mechanism of effectiveness was said to be due to an increase in the dehydration mechanism.

Wood has been reacted with fire-retardant chemicals such as phosphorus pentoxideamine complexes (Lee et al. 2004) or glucose diammonium phosphate (Chen 2002) that results in treatments that are leach resistant (Rowell et al. 1984) (see Chapter 14, 5.10). In another approach, urea and DAP was delivered to the fibers via sol-gel technology that was synergistic in effectively reducing cotton clothing flammability and reducing leaching losses of the FRTs (Chapple and Barkhuysen 2005).

REFERENCES

Antal, M.J. Jr. 1985. Biomass pyrolysis: A review of the literature. Part 2: Lignocellulose pyrolysis. *Adv. Solar Energy* 2: 175–256.

Bridgewater, A.V. 1999. Principles and practice of biomass fast pyrolysis processes for liquids. *J Anal. Appl. Pyrolysi*s 51: 3–22.

Browne, F.L. 1958. Theories of the combustion of wood and its control. Forest Service Report No. 2136, Forest Products Laboratory, Madison, WI.

Chapple, S.A. and Barkhuysen, F.A. 2005. Sol-gel flame retardant treatment of cotton, *Proceedings of the 13th National Textile Centre Forum/84th Textile Institute Annual World Conference*, Raleigh, North Carolina.

Chen, G.C. 2002. Treatment of wood with glucose-diammonium phosphate for fire and decay protection. In: *Proceedings of the 6th Pacific Rim Bio-Based Composite Symposium*, Volume 2, 616–622.

Creitz, E.C. 1970. Literature survey of the chemistry of flame inhibition. *J. Res. Natl. Bur. Stand.*, Section A, 74(4): 521–30.

Czernik, S., Maggi, R., and Peacocke, G.V.C. 1999. A review of physical and chemical methods of upgrading biomass-derived fast pyrolysis liquids. In: Overend, R.P. and Cornet, E. (eds), *Biomass: Proceedings of the 4ᵗʰ Biomass Conference of the Americas*. Volume 2. Pergamon, New York, NY, pp. 1235–1240.

Dietenberger, M. 2002. Update for combustion properties of wood components. *Fire and Materials J.*, 26: 255–267.

Dietenberger, M.A. and White, R.H. 2001. Reaction-to-fire testing and modeling for wood products. In: Proceedings *12th Annual BCC Conference on Flame Retardancy*, Stanford, CT, May 21–23, 2001. Norwalk, CT: Business Communications Co. Inc. pp. 54–69.

Eickner, H.W. 1966. Fire-retardant treated wood. *J. Mater.* 1(3): 625–44.

Fung, D.P.C., Tsuchiya, Y., and Sumi, K. 1972. Thermal degradation of cellulose and levoglucosan. Effect of inorganic salts. *Wood Sci.* 5(1): 38–43.

Funke, A. and Ziegler, F. 2010. Hydrothermal carbonization of biomass: A summary and discussion of chemical mechanisms for process engineering. *Biofuels, Bioprod. Bioref.* 4: 160–177.

Gao, M., Niu, J., and Yang, R. 2006. Synergism of GUP and boric acid characterized by cone calorimetry and thermogravimetry. *J. of Fire Sciences* 24: 499–511.

Goldstein, I.S. and Dreher, W.A. 1961. A non-hygroscopic fire retardant treatment for wood. *Forest Prod. J.* 11(5): 235–237.

Goldstein, I.S. and Dreher, W.A. 1964. Method of imparting fire retardance to wood and the resulting product. U.S. Patent 3,159,503.

Hagge, M.J., Bryden, K.M., and Dietenberger, M.A., 2004. Effect of backing board materials on wood combustion performance. *Wood & Fire Safety: Proceedings, 5th Intl Scientific Conf*, April 18–22.

Hendrix, J.S. and Drake, G.L. Jr. 1972. Pyrolysis and combustion of cellulose. III. mechanistic basis for synergism involving organic phosphates and nitrogenous bases. *J. Applied Polymer Sci.* 16: 257–274.

Holmes, C.A. 1977. Wood technology: Chemical aspects. In: Goldstein, I.S. (ed), *Am. Chemical Soc. Symposium Series 43*, Washington, DC, pp. 82–106.

Huggett, C. 1980. Estimation of rate of heat release by means of oxygen consumption measurements. *Fire and Materials* 4: 61–65.

Juneja, S.C. and Fung, D.P.C. 1974. Stability of amono resin fire retardants. *Wood Sci.* 7(2): 160–163.

Kaplan, H.L., Grand, A.F., and Hartzell, G.E. 1982. *A Critical Review of the State-of-the-Art of Combustion Toxicology*. Southwest Research Institute, San Antonio, TX.

Kaur, B., Gur, I.S., and Bhatnagar, H.L. 1986. Studies on thermal degradation of cellulose and cellulose phosphoramides. *J. Applied Polymer Sci.* 31: 667–683.

Kawamoto, H., Murayama, M., and Saka, S. 2003. Pyrolysis behavior of levoglucosan as an intermediate in cellulose pyrolysis: Polymerization into polysaccharide as a key reaction to carbonized product formation. *J. Japanese Wood Soc.* 49: 469–473.

Langley, J.T., Drews, M.J., and Barkeer, R.H. 1980. Pyrolysis and combustion of cellulose. VII. Thermal analysis of the phosphorylation of cellulose and model carbohydrates during pyrolysis in the presence of aromatic phosphates and phosphoramides. *J. Applied Polymer Sci.* 25: 243–262.

Lee, H.L., Chen, G.C., and Rowell, R.M. 2004. Thermal properties of wood reacted with a phosphorus pentoxide-amine system. *J. Applied Polymer Sci.* 91(4): 2465–2481.

LeVan, S.L. 1984. Chemistry of fire retardancy. The chemistry of solid wood. In: R.M. Rowell (ed.), *Advances in Chemistry Series*, Number 207, American Chemical Society, Washington, DC, Chapter 14, pp. 531–574.

McGraw, G., Hemingway, R., Ingram, L., Canady, C., and Mcgraw, W. 1999. Thermal degradation of terpenes: Camphene, A-carene, limonene, and a-terpinene. *Environ. Sci. Technol.* 33: 4029–4033.

Middleton, J.C., Draganov, S.M., and Winters, F.T. Jr. 1965. Evaluation of borates and other inorganic salts as fire retardants for wood products. *For. Prod. J.* 15(12): 463–467.

Mok, W.S. and Antal, M.J. 1983. Effects of pressure on biomass pyrolysis. II. Heats of reaction of cellulose pyrolysis. *Thermochimica Acta.* 68: 165–186.

Nanassy, A.J. 1978. Treatment of Douglas-fir with fire retardant chemicals. *Journal Wood Sci.* 11(2): 111–117.

Peters, C.C. and Eickner, H.W. 1962. Surface flammability as determined by the FPL 8-foot tunnel. Forest Service Report No. 2257, Forest Products Laboratory, Madison, WI.

Rowell, R.M., Susott, R.A., DeGroot, W.F., and Shafizadeh, F. 1984. Bonding fire retardants to wood. Part 1: Thermal behavior of chemical bonding agents. *Wood and Fiber Sci.* 16(2): 214–223.

Schartel, B. and Hull, T.R. 2007. Development of fire-retarded materials—Interpretation of cone calorimeter data. *Fire and Materials* 31: 327–354.

Shafizadeh, F. 1982. Introduction to pyrolysis of biomass. *J. Anal Appl. Pyrolysis* 3: 283–305.

Shafizadeh, F. 1984. The chemistry of pyrolysis and combustion. The chemistry of solid wood. In: Rowell, R.M., (ed), *Advances in Chemistry Series*, Number 207, American Chemical Society, Washington, DC. Chapter 13: pp. 489–529.

Shafizadeh, F. and Fu, Y.L. 1973. Pyrolysis of cellulose. *Carbohydrate Res.* 29: 113–122.

Slade, P.E. Jr. and Jenkins, L.T. 1966. *Techniques and Methods of Polymer Evaluation*, Marcel Dekker, Inc., New York, NY.

Tang, W.K. 1964. *Effect of Inorganic Salts on the Pyrolysis, Ignition, and Combustion of Wood, Cellulose, and Lignin.* Ph.D. Thesis, Chemical Engineering, University of Wisconsin–Madison, 275pp.

Tang, W.K. and Eickner, H.W. 1968. *Effect of Inorganic Salts on Pyrolysis of Wood, Cellulose, and Lignin Determined by Differential Thermal Analysis.* USDA Forest Service Research Paper FPL 82, January.

White, R.H. 1979. Oxygen index evaluation of fire-retardant-treated wood. *Wood Sci.* 12(2): 113–121.

White, R.H. and Dietenberger, M.A. 1999. In: *Wood Handbook—Wood as an Engineering Material.* Fire Safety. Chapter 17, Gen. Tech. Rep. FPL-GTE-113, Madison, WI: U.S. Department of Agriculture, Forest Service, Forest Products Laboratory, 17.1–16.

7 Weathering of Wood and Wood Composites

Philip D. Evans

CONTENTS

7.1 INTRODUCTION

All common building materials (concrete, metals, and wood) are susceptible to environmental degradation when they are used outdoors. The degradation that occurs when wood is used outdoors and
above ground is termed weathering (Feist 1990a). Weathered wood is gray and its surface is often
cracked and rough (Figure 7.1). However, under the weathered surface layer, which is no more than
a few millimeters deep, the wood is essentially sound. The surface of weathered wood is often colonized by microorganisms and lichens, but conditions at exposed wood surfaces generally do not
favor decay by basidiomycete fungi, which can extend deeply into wood and significantly reduce the
strength of structural timber (Feist 1990a). Hence, weathering has little effect on the mechanical
properties of structural timber and, accordingly, there are many examples of wooden buildings such
as the stave (pole) churches in Norway that are still structurally sound despite having been exposed
to the weather for over 1000 years (Borgin 1970).

FIGURE 7.1   Appearance of weathered wood (clockwise from top left); (1) end grain of coniferous log used
in the construction of a log cabin showing severe checking and erosion of wood (top left); (2) longitudinal
surface of a log from the same building showing gray coloration, checking and colonization of wood by
lichens (top right); (3) Artists rendition of the effects of 100 years of natural weathering on the appearance of
round and square timbers. The cutaway shows that wood below the unweathered surface is largely unchanged.
Note that the top of the posts are severely checked (bottom center). (Adapted from Feist, W.C. 1982. In: Meyer,
R.W. and Kellogg, R.M. (Eds.), *Structural Use of Wood in Adverse Environments*. Van Nostrand Reinhold Co.,
New York.)

Weathered wood is mentioned in the Bible, as pointed out by Feist and Hon (1984), and they also mention the early research on the weathering of wood by Berzelius (1827), Wiesner (1864), and Schramm (1906a,b). Baker (1900) also remarked on the weathering of wood in his descriptions of two species of Australian acacias. Wiesner (1864) reported that "the intercellular substance of wood had been lost because of weathering" and concluded that "the remaining grey layer consists of cells that, leached by atmospheric precipitation, have been robbed entirely or in large part of their infiltrated products so much that the remaining membranes consist of chemically pure or nearly chemically pure cellulose." In related research Wiesner observed that the yellowing of paper was caused by sunlight, and he correctly ascribed this effect to the photooxidation of "ligneous substances," noting that yellowing of paper did not occur when "through some chemical (pulping) process the lignine that forms the essential part of the wood is removed" (Anon 1887). The yellowing of paper was also examined by Richter (1935). He also carried out a preliminary experiment that examined chemical changes in spruce, birch, and poplar wood exposed to sunlight for 100 h. He found evidence of photooxidation of wood carbohydrates in all species and delignification of birch and poplar. Despite these insights into the chemistry of photo degradation, research in the early part of the twentieth century on the weathering of wood mainly focused on the use of paints and preservatives to restrict the adverse effects of weathering on wood's appearance (Browne 1927, Curtin et al. 1927, Gillander et al. 1934, Salzberg et al. 1931). These studies and others emphasized the role that water plays in the creation of defects such as checking, loose grain, cupping, and warping. For example, Pilson et al. (1931) in a thesis on the properties and uses of western larch (*Larix occidentalis* Nutt) compared the weathering resistance of western larch with other species in terms of the checking that developed when the woods were exposed to natural weathering. They wrote that "weathering of western larch takes place in a comparatively short time." "Unpainted western larch showed some checking after 3 weeks exposure." "At the end of 4 years the weathering in the western larch panels was more pronounced than in Douglas fir (*Pseudotsuga menziesii* (Mirb.) Franco) or southern yellow pine (*Pinus* spp.) and considerably more pronounced than in the cedars and southern cypress." The weathering of wood was attributed by Browne (1927) to "the disintegrating effect of internal stresses set up in the wood as the result of fluctuating moisture content of those portions exposed directly to the weather." He also acknowledged that "abrasion and photochemical oxidation of the cellulose doubtless play parts in the weathering" (of wood). The importance of sunlight in the weathering of wood was apparent from Jemison's study of the loss of weight of small wooden dowels exposed to the weather in full sunlight or under moderate or dense shade (Jemison 1937). His results clearly demonstrated a decrease in weight losses due to weathering as dowels were shaded from full sunlight (Jemison 1937). The role that light plays in the weathering of wood has been the focus of much of the subsequent research on the weathering of wood and is described below. Research to protect wood from weathering initially focused on the use of coatings and wood preservatives, as mentioned above, but soon expanded to include wood modification technologies to protect both wood and wood composites from moisture and weathering (Stamm and Seborg 1936, 1939). The techniques used to protect wood and wood composites from weathering are also described in this review. There have been several excellent reviews on the weathering and protection of wood including those by Kalnins (1966), Feist (1982, 1990a), Feist and Hon (1984), and Williams (2005). This chapter updates these reviews in places and emphasizes some of the more recent research on the weathering and protection of wood and particularly wood composites.

7.2 ENVIRONMENTAL FACTORS AFFECTING THE WEATHERING OF WOOD

The main environmental factors involved in the weathering of wood are solar electromagnetic radiation (ultraviolet [UV] and visible light), molecular oxygen (O_2), water, heat, atmospheric pollutants, wind-blown particulate matter and certain specialized microorganisms (Feist 1990a).

7.2.1 Solar Radiation

The maximum amount of solar radiation available on the Earth's surface on a clear day is normally 1000 W/m^2. Such radiation consists of ~5% UV (286–380 nm), 45% visible (380–780 nm), and 50% infrared (780–3000 nm). Solar radiation is the main environmental factor responsible for the surface weathering of wood. UV light particularly UVB (280–320 nm) is more energetic than visible light and capable of cleaving the carbon–carbon, carbon–oxygen, and carbon–hydrogen bonds that connect the polymeric components of wood; that is, cellulose, hemicelluloses, and lignin (see Equation 7.1). Nevertheless, wood exposed to visible light degrades at about half the rate of material exposed to the full solar spectrum (Derbyshire and Miller 1981). The component of visible light that has sufficient energy to photodegrade lignin is violet light (Kataoka et al. 2007). Light with wavelengths greater than 400 nm has insufficient energy to significantly degrade lignin although wood's low-molecular weight extractives can be degraded by visible light with longer wavelengths (Kataoka et al. 2007).

$$E = \frac{hc}{\lambda} \tag{7.1}$$

where E = energy of a photon; h = Planck's constant; c = speed of light; λ = wavelength.

The surface photodegradation of wood is more rapid when levels of solar radiation are elevated. Table 7.1 shows the seasonal variation in weight losses of thin radiata pine (*Pinus radiata* D. Don) wood veneers inclined at three angles (0° [horizontal], 45°, and 90° [vertical]) and exposed to the weather in Canberra, Australia (Evans 1996). Weight losses were greatest in the summer (December to February in the Southern Hemisphere) when levels of solar radiation were high (Evans 1996). Degradation was also highest for samples exposed horizontally to the weather (Evans 1989a, 1996). In accord with these observations, Williams et al. (2001a) found that the erosion of wood during natural weathering was generally greatest for samples exposed horizontally. These effects are related to the fact that low angles of exposure maximize the levels of UV radiation at exposed surfaces (Qayyum and Davis 1984). Increased levels of solar radiation at high altitudes are also thought to be responsible for more pronounced bleaching of wood exposed in mountainous regions compared to wood exposed at sea level (Arndt and Willeitner 1969).

Losses in tensile strength of thin wood veneers exposed outdoors can be related to the level of photochemically active radiation received by the veneers using Equation 7.2.

$$S = S_0 \exp. \, (-kD) \tag{7.2}$$

where S is the tensile strength of veneers after exposure to a radiation dose D, and S_0 the initial strength of the veneers prior to weathering. The constant k in the exponent is the degradation rate (Derbyshire and Miller 1981).

This equation has been used to derive photodegradation rates for wood exposed to natural or artificial weathering, and different wood species exposed to natural weathering (Derbyshire et al. 1995, 1996). These studies found small differences in the rate of photodegradation of six softwood species, and concluded that artificial weathering is a valid alternative to natural weathering for studies of the effect of temperature and wood moisture content on photodegradation rate (Derbyshire et al. 1995, 1996).

7.2.2 Oxygen

Oxygen is an abundant atmospheric gas that plays "an important role in many photochemical processes" (Feist and Hon 1984). Oxygen is involved in the free radical reactions that cause the

TABLE 7.1

Effect of Season and Angle of Exposure on the Weight Losses of Thin Wood Veneers Exposed to Natural Weathering in Canberra, Australia

Angle (deg.)	% Weight Losses of Veneers during Different Exposure Periods						
	April/May	June/Jul	Aug/Sept	Oct/Nov	Dec/Jan	Feb/Mar	Average
0 (Horiz.)	44.2	27.9	41.4	40.9	51.1	42.0	41.2
45	40.9	31.5	38.7	37.8	47.6	39.7	39.4
90 (Vert.)	24.5	19.0	25.4	23.3	23.8	24.2	23.4
Average	36.5	26.1	35.1	34.0	40.1	35.3	

photodegradation of wood. Wengert (1966a) observed that exposure of redwood (*Sequoia semper-virens* (D. Don.) End.) and birch (*Betula* spp.) to UV radiation in air, oxygen, nitrogen, or argon atmospheres caused the woods to darken during the first several hours of exposure. Thereafter, samples exposed in air and oxygen atmospheres became lighter whereas samples exposed in argon and nitrogen atmospheres continued to darken. Wengert (1966a) concluded that the lightening of redwood and birch exposed to UV radiation in air or an oxygen atmosphere was due to photooxidation. Dirckx et al. (1992) observed that the photo-yellowing of grand fir wood (*Abies grandis* Lindl.) exposed to UV light for prolonged periods of time (1–25 h) was greater in an oxygen than in air or nitrogen atmospheres, although initially (0–15 min) "the yellowing rate seemed to be independent of the gaseous atmosphere" Paulsson et al. (2001) found that the "degree of photo-yellowing of untreated chemi-thermomechanical pulp decreased when the air in the surrounding atmosphere was replaced with oxygen-free argon." However, the decrease was only moderate and they suggested that "only a trace amount of oxygen (strongly adsorbed to the fiber material) is necessary to cause discoloration."

7.2.3 WATER

Water plays an important role in the weathering of wood. Dimensional change as a result of the wetting and drying of wood generates surface tensile stresses that cause checking and warping of timber (Schniewind 1963). Wood composites are particularly susceptible to dimensional change caused by water. Water generates capillary forces that cause the collapse of photodegraded earlywood cells, according to Yata and Tamura (1995). Water leaches degraded lignin fragments and hemicellulose-derived sugars from photodegraded wood surfaces, and hence under natural weathering conditions losses in weight of wood veneers are correlated with rainfall events (Evans et al. 1993, Figure 7.2). The leaching of colored unsaturated (quinoid) lignin breakdown products from weathered wood surfaces by rain explains why weathered wood, which yellows initially, eventually becomes gray and is mainly composed of cellulose. In the absence of such leaching, photodegraded lignin fragments accumulate within the wood and the wood becomes darker (Frey Wyssling 1950, Kleinert 1970). For example, Frey Wyssling (1950) noted that weathered wood in chalets in Switzerland where precipitation falls as snow is black, because photodegraded lignin is not leached from the wood and is transformed into dark humic substances.

Water can hydrolyze the noncellulosic components of wood and prolonged exposure of wood to water at mild temperatures (50–65°C) *in vitro* results in degradation of hemicelluloses and lignin with little degradation of cellulose (Evans and Banks 1988, 1990). Moisture in the presence of superficial solar heating (see below) may thus degrade the noncellulosic components of wood although measurements in the field to confirm this suggestion are lacking.

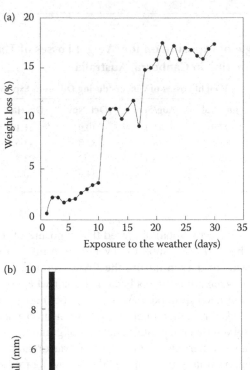

FIGURE 7.2   Loss in weight of thin wood veneers exposed to the weather for 30 days in Canberra, Australia (a), and rainfall during the same period (b); note the correlation between weight loss of veneers and rainfall events.

7.2.4   HEAT

Wood surfaces exposed outdoors in the continental United States can reach maximum temperatures of 50–60°C, depending on the reflectivity of the wood, incident solar radiation and air temperature (Wengert 1966b). These temperatures are not close to the glass transition temperature of lignin (130–150°C) or those at which significant structural degradation of wood's chemical components occurs (200°C). Therefore, it is unlikely that heat directly causes surface degradation of wood in temperate climates, but rather accelerates photooxidation (Futó 1976a,b), and hydrolysis of wood caused by light and water, respectively. Evidence to support this suggestion is the finding that loss in tensile strength of thin softwood veneers subjected to artificial weathering is temperature dependent, and increases with rising temperature within the range 25–55°C (Derbyshire et al. 1997). Furthermore, Mitsui and Tsuchikawa (2005) found that there was a positive relationship between temperature (−40–50°C) and color changes/lignin degradation of Japanese cypress (*Chamaecyparis obtusa* Sieb. et Zucc.) and Japanese beech (*Fagus crenata* Blume) exposed to UV radiation.

Heat may also accelerate the surface drying of wood thereby generating stresses that cause checking of wood. Exposure of wood to low temperatures and repeated freezing and thawing, may also cause checking of wood (Borgin 1970).

7.2.5 ENVIRONMENTAL POLLUTANTS AND PARTICULATES

Observations in the field and laboratory experiments have all tended to suggest that the weathering of wood is more rapid in polluted than in unpolluted atmospheres. The main pollutants in the atmosphere are dust and smoke particles, and volatile pollutants including sulfur compounds, ammonia, nitrogen oxides, carbon monoxide, and saturated/unsaturated aliphatic and aromatic hydrocarbons and their derivatives. Nitric dioxide is an initiator of photooxidation and sulfur dioxide in the atmosphere may be converted to sulfuric acid, which can degrade wood surfaces (Spedding 1970). Elevated levels of airborne sulfur dioxide arising from the burning of coal were thought to be responsible, in part, for significant losses in tensile strength of Norway spruce (*Picea abies* (L.) Karsten) wood veneers exposed to the weather during the winter in Poland (Raczkowski 1980). Presoaking wood samples in dilute (pH 3.0–4.0) sulfuric or nitric acid increases the degradation of wood during artificial weathering (Park et al. 1996, Williams 1987). The degradation of wood during accelerated weathering can also be accelerated by iron. For example, Helinska-Raczkowska and Raczkowski (1978) examined the creep in bending and creep recovery of Scots pine (*Pinus sylvestris* L.) samples subjected to accelerated weathering in contact with rusting iron. They noted that deformation and permanent strain after removal of load were greater in the presence of rusting iron than in control samples. These effects were thought to be due to the catalytic degradation of cellulose in wood by rusting iron.

Degradation of wood in cold climates can occur due to abrasion caused by wind-blown particles of ice. For example, the iconic Australian explorer Mawson (Mawson 1915) wrote in 1915 following his pioneering expedition to the Antarctic: "the abrasion-effects produced by the impact of the snow particles were astonishing ... A deal (*Pinus* spp.) box, facing the wind, lost all its painted bands and in a fortnight was handsomely marked; the hard knotty fibers (latewood) being only slightly attacked, whilst the softer, pithy laminae (earlywood) were corroded to a depth of one-eighth of an inch (approximately 3 mm)." Windblown sand and salt can also cause similar, if less spectacular, abrasion of wood exposed outdoors (Feist 1990a). The growth of salt crystals in wood cell walls exposed to salt water can contribute to micro-checking and separation of tracheids in southern pine (*Pinus* spp.). This effect accounts for the fuzzy wood surface that sometimes "develops on wharves, piling and buildings exposed to the ocean" (Johnson et al. 1992).

7.2.6 ORGANISMS

Mobius (1924) established a link between the gray color of weathered wood and its superficial colonization by fungi. Since then it has been found that a wide range of organisms are able to colonize wood surfaces exposed outdoors creating a biofilm, which can include fungi, bacteria, and algae (Dickinson 1971, Duncan 1963, Sailer et al. 2010). The fungi colonizing weathered wood surfaces mainly belong to the subphylum ascomycotina. The fungus, *Aureobasidium pullulans* (de Bary) Arnaud, in particular, is frequently isolated from weathered wood in both temperate and tropical climates. *A. pullulans* is capable of withstanding temperatures of 80°C, growing over a pH range of 1.9–10.1, surviving for long periods without moisture and metabolizing lignin breakdown products (Cooke and Matsuura 1963, Duncan 1963, Schoeman and Dickinson 1996, 1997). Hence, *A. pullulans* is particularly well suited to the micro-environment of weathered wood and also painted wood surfaces (Bardage and Bjurman 1998, Cronin et al. 2000, Seifert 1964, Schmidt and French 1976, Schoeman and Dickinson 1997). Fungi colonizing weathered wood have been identified by several researchers. Sell and Wälchli (1969) observed *A. pullulans, Macrosporium* spp., *Tetracoccosporium* spp., *Cladosporium* spp., and *Sclerophoma* spp., growing on wooden avalanche barriers exposed to the weather in Switzerland. Subsequently, Dickinson (1971) isolated a range of fungi including *A. pullulans, Cladosporium* spp., *Alternaria* spp., *Stemphylium* spp., and *Torula* spp. from Scots pine and western red cedar (*Thuja plicata* Donn ex D. Don) wood exposed to the weather in Northern Europe (UK and Sweden). More recently Lim et al. (2005, 2007) isolated a wide range of basidio-mycetes and ascomycetes from weathered western red cedar decks and fences in Vancouver, Canada. Hansen (2008) summarized the results of studies of the mold fungi colonizing weathered wood exposed in different climates. His results indicate that *Cladosporium* spp. are ubiquitous on weathered wood surfaces, and they also point to the frequent colonization of wood surfaces exposed in tropical climates by fungi from the genus *Phoma* (Table 7.2). Even in the extreme climate of Antarctica, fungi are able to colonize weathered wood. For example, Held et al. (2006) isolated *Candophora* spp. plus, *Cladosporium* spp., *Hormonena dematiodes, Lecythophora hoffmanii*, and *Penicilium* spp. from a more than 40-year-old wooden structure located in New Harbor, Antarctica (Held et al. 2006).

Many of the fungi found on weathered wood surfaces including *Alternaria* spp., *Aspergillus* spp., *Coniochaeta lignaria, Lecythophora hoffmanii, Penicillium* spp., *Phialophora* spp., *Phoma* spp., *Trichoderma* spp. are capable of enzymatically degrading cellulose and hemicellulose and, in the-ory, causing soft rot of wood (Bugos et al. 1988, Lim et al. 2007, Savory 1954, Smith and Swann 1976). *A. pullulans* is able to degrade cellulose and hemicelluloses in wood to a limited extent, and Seifert (1964) characterized it as a "soft-rot of low ability." However, soft rot is not common at weathered wood surfaces possibly because the intermittent wetting and drying of exposed surfaces does not favor microbial degradation. In humid tropical climates, where conditions at wood surfaces are much more favorable for microbial activity, soft rot of weathered wood surfaces may be more common. In support of such a suggestion Sudiyani et al. (2002) noted that in the tropical climate of Indonesia weathering of wood may encourage colonization by decay fungi, particularly in wood species that lack natural durability.

Wood surfaces exposed under shade in wet climates can be colonized by algae. For example the algal species *Chlorococcum* spp. and *Amphora* spp. are able to grow on wood surfaces, and also beneath coatings (De Souza and Gaylarde 2002). The moisture dynamics at weathered wood sur-faces generally do not favor algal growth, but algae can survive wetting and drying at wood surfaces by forming symbiotic relationships with fungi to form lichens. One in five fungi can form such a relationship (lichenization) and approximately 46% of these fungi belong to the ascomycota (Hawksworth and Hill 1984). *A. pullulans* is capable of forming symbiotic lichenized associations

TABLE 7.2
Fungal Species Colonizing Weathered Wood Surfaces in Temperate and Tropical Locations

Species	Location				
	Germany	N.W. USA	Malaysia	Thailand	Brazil
Alternaria spp.	7	3	15	4	7
Aureobasidium spp.	-	44		7	6
Cladosporium spp.	24	25	8	11	6
Curvularia spp.	-	-	-	-	7
Fusarium spp.	-	-	-	-	6
Nigrospora spp.	-	5		-	-
Penicillium spp.	2	-		-	-
Phoma spp.	-	-	18	59	13
Stemphyllium spp.	2	-	-	-	-
Others	65	23	52	19	55

with the green algae *Lecidea granulose* (Hoffm. Ach) (Schmidt and French 1976), and lichens are frequently observed colonizing weathered wood surfaces particularly in unpolluted atmospheres (Figure 7.1). The role that lichens play in the weathering of wood has received little attention, however, it is possible that they alter the micro-environment of weathered wood through the production of lichen acids such as usnic, atranorin and evernic acid, which are known to show antimicrobial and antioxidant activity (Aalto-Korte et al. 2005). The colonization of weathered western red cedar roofing shakes by lichens may retard the drying of the shakes and possibly accelerate decay, according to Smith and Swann (1976). Contact dermatitis of forestry workers (Woodcutters disease) has been linked to exposure to lichen acids through handling weathered firewood colonized by lichens (Aalto-Korte et al. 2005).

Bacteria have been observed colonizing weathered wood surfaces (Evans 1989b, Paajanen 1994). They are very well adapted to the micro-environment of weathered wood surfaces. There is little information on species of bacteria that colonize weathered wood and their role (if any) in the weathering process.

A number of insect species, particularly paper wasps (Polistinae), yellow jackets and hornets (Vespinae) mine-weathered wood surfaces to obtain delignified wood fragments (Figure 7.3). The weathered wood is intensively chewed and mixed with the insect's saliva to form a water-repellent fibrous, paper-like, composite that the insects use to construct their elaborate nests. These insects show a preference for weathered wood and rarely utilize fresh wood (Schmolz et al. 2000).

7.2.7 INTERACTIVE EFFECTS

The individual factors involved in the weathering of wood rarely act in isolation, but tend to operate together and sometimes synergistically (Groves and Banana 1986). The most obvious example of such synergy is the role that heat plays in accelerating photodegradation (Derbyshire et al. 1997, Futó 1976a,b, Mitsui and Tsuchikawa 2005) and hydrolysis (Evans and Banks 1988, 1990). Photodegradation weakens the wood surface and degrades its microstructure, which makes wood more susceptible to checking (Evans et al. 2008). Changes in the surface chemistry of wood influences subsequent colonization of wood by microorganisms (Schoeman and Dickinson 1997). The photo-depolymerization of lignin and holocellulose and the production of low-molecular weight phenolic compounds and sugars, respectively, provides a nutritional source for microorganisms and explain, in part, why *A. pullulans* is frequently isolated from weathered wood (Schoeman and Dickinson 1997). Exposure of wood to artificial UV radiation also encourages colonization and

FIGURE 7.3 A wasp (*Vespula vulgaris* L.) removing weathered wood from the surface of weathered pine; note the lighter areas where wasps have removed weathered wood.

degradation of wood by mold fungi (Desai and Shields 1971). UV light has also been shown to enhance the rate of discoloration of wood caused by sulfur dioxide and nitrogen dioxide (Hon and Feist 1993). As mentioned above, water leaches lignin photodegradation products from wood surfaces, which reduces the nutrients available for organisms such as *A. pullulans*. Hence, fully exposed wood surfaces may show less colonization by *A. pullulans* than surfaces, such as those beneath clear finishes, which retain lignin photodegradation products. Photodegradation of wood is greater in the presence of moisture than under dry conditions (Anderson et al. 1991a,b, Arnold et al. 1991, Horn et al. 1994, Turkulin et al. 2004), possibly because water molecules swell wood thereby opening up inaccessible regions of the cell wall to photodegradation (Feist and Hon 1984). The full extent of the interaction of the factors involved in the weathering of wood has yet to be elucidated.

7.3   MECHANISMS INVOLVED IN WEATHERING

The range of environmental factors involved in the weathering of wood suggests that it may be difficult to define a single mechanism responsible for the weathering of wood. The superficial nature of weathering, however, is one of its defining features, and this can be linked to the limited penetration of light into wood (Kataoka et al. 2004). The other factors involved in the weathering of wood (water, heat, microorganisms, etc.), have the ability, in principle, to degrade wood to a much greater depth, yet beyond the zone affected by light, the chemical and physical properties of weathered wood are largely unchanged. Clearly, light plays the major role in the weathering of wood. An exception to the rule that weathering is only superficial in nature is the surface checking caused by moisture-induced dimensional changes, which may extend several millimeters into wood (Evans et al. 2000b). Knowledge of both of the mechanisms responsible for the photodegradation and checking of wood is essential in order to understand the weathering of wood and develop more effective protection systems.

7.3.1   PHOTODEGRADATION OF WOOD

In order to act upon wood, solar radiation must be absorbed by one of wood's chemical constituents, cellulose, hemicelluloses, and lignin or low-molecular weight extractives. Experimentation has shown that lignin strongly absorbs UV light with a distinct maximum at 280 nm, and decreasing absorption extending beyond 380 nm into the visible region of the spectrum (Kalnins 1966). The entities in lignin that absorb UV light are double bonds, phenolic and carbonyl groups, quinones, quinonemethides, and biphenyls (Hon and Glasser 1979). Cellulose and hemicelluloses absorb UV light between 200 and 300 nm (Bos 1972), and show some absorption of visible light, but the absorption of light by lignin is thought to be the first step leading to the photodegradation of wood. In fact lignin is so effective at absorbing light that very little (<10%) photoactive light (<420 μm) penetrates wood beyond a depth of 220 μm (Kataoka et al. 2004). The penetration of light into wood is inversely proportional to the logarithm of wood density, but it is also affected by the chemical composition of the wood (Kataoka et al. 2005). Species containing high concentrations of phenolic extractives, which absorb UV light, show less penetration of light than predicted by their density (Kataoka et al. 2005). The accumulation of lignin photodegradation products at wood surfaces also restricts the penetration of light into wood (Kataoka et al. 2005).

Energy absorbed from solar radiation can be dissipated in wood (and polymers in general) through the cleavage of molecular bonds resulting in the formation of a free radical, a molecular species that is highly reactive because it has an unpaired valence electron. Aromatic radicals can be readily detected using electron spin resonance spectroscopy (ESR) when wood is irradiated with UV light, indicating the photolysis of lignin (Hon et al. 1980, Hon and Ifju 1978, Schmid et al. 2000). Hon and coworkers used ESR to examine free radical formation in wood and model wood compounds following irradiation with UV light (Hon 1983, Hon et al. 1980). In summary, they found that sunlight has sufficient energy to dissociate bonds in those elements of lignin with carbonyl, biphenyl, or ring-conjugated double bonds. In contrast, "lignin structural units having a saturated propane side chain such as such as guaiacylgycerlguaiacyl ether, phenylcoumaran and pinoresinol do not absorb light (305–420 nm) and are less likely to dissociate and form radicals" (Lin 1982, Lin and Kringstad 1970, Williams 2005). The formation of free radicals is influenced by temperature, atmospheric gases (as described above), and increases with wood's moisture content from 0% to 6% and then decreases at higher moisture contents (Feist and Hon 1984). Free radical formation in wood exposed to UV light is also influenced by the chemical composition of irradiated surfaces and in particular the concentration of lignin. Hence, free radical formation also varies with wood species and surface type (radial v. tangential surfaces) (Feist and Hon 1984).

Generalized schemes for the photodegradation of polymers describe the reaction of free radicals with atmospheric oxygen to form peroxy radicals. Such radicals typically attack polymer chains via hydrogen abstraction forming a hydroperoxide and another free radical. Hydroperoxides are susceptible to UV radiation and undergo photolysis to form additional free radicals. Evidence for the photooxidation of wood and the participation of oxygen in photodegradation is the finding that free radicals formed in wood subjected to irradiation are stable in a vacuum or inert atmosphere, but readily disappear in the presence of oxygen (Feist and Hon 1984). The literature indicates that free radical formation and the photodegradation of wood fits some, but not all of the aforementioned scheme for the photodegradation of synthetic polymers. The photodegradation of wood is undoubtedly more complicated than that of synthetic homopolymers because it consists of a blend of polymers (lignin, cellulose, and hemicellulose) and low-molecular weight extractives that differ in their susceptibility to solar radiation. Furthermore, it is clear that the precise mechanisms and reaction pathways involved in the photodegradation of each of these components have not been fully elucidated. However, the key step involved in the photodegradation of wood is the photooxidation and fragmentation of lignin resulting in the formation of different aromatic radicals (Figure 7.4).

The photooxidation of lignin involves four pathways, as summarized by Schaller and Rogez (2007): (1) direct absorption of UV light by conjugated phenolic groups to form phenoxyl free radicals; (2) abstraction of phenolic hydrogen as a result of aromatic carbonyl triplet excitation to produce a ketyl and phenoxyl free radical (Kringstad and Lin 1970); (3) cleavage of nonphenolic phenacyl-α-O-arylethers to form phenacyl phenoxyl free radical pairs (Gierer and Lin 1972); and (4) abstraction of benzylic hydrogen of the α-guaiacylglycerol-β-arylether group to form ketyl free radicals which then undergoes cleavage of the β-O-4 arylether bond to produce an enol and phenoxyl

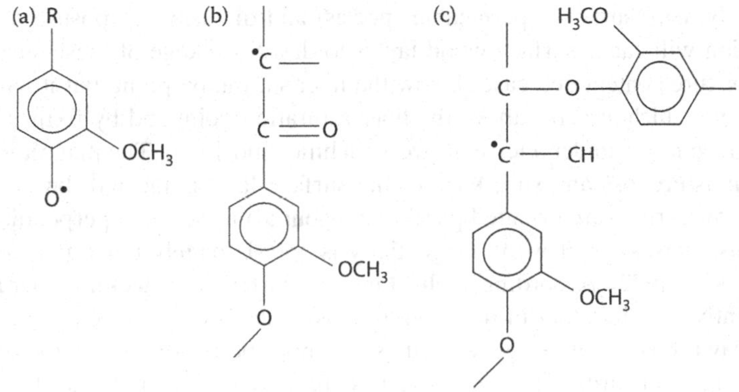

FIGURE 7.4 Structure of different lignin-derived radicals: (a) gaiacoxyl radical; (b) phenacyl radical; (c) cetyl radical. (Adapted from George, B. et al. 2005. *Polymer Degradation & Stability*. 88(2): 268–274.)

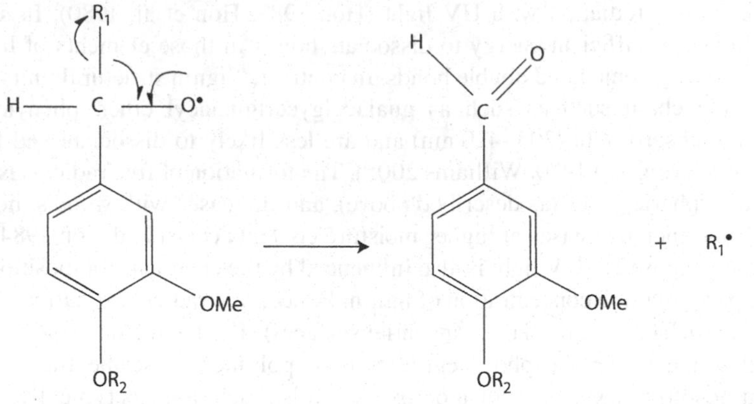

FIGURE 7.5 Fragmentation mechanism for lignin during the photodegradation of wood.

free radical (Scaiano et al. 1991, Schmidt and Heitner 1993). The enol then tautomerizes to a ketone. Alkoxyl and peroxyl free radicals produced from the reaction of oxygen and "lignin free radicals" react with phenoxyl free radical to produce colored chromophores, for example, quinoides, aromatic ketones, aldehydes, and acids as photodegradation products (Leary 1994). The formation of these chromophores explains why wood initially yellows when exposed to light (Feist and Hon 1984, Kishino and Nakano 2004b).

Vanillin has been consistently identified as one of the end products of the photodegradation of lignin (Sandermann and Schlumbom 1962). A scheme for the photodegradation of lignin that is consistent with the formation of vanillin is shown in Figure 7.5. This scheme involves cleavage of the α-β bond in a typical aryl-ether lignin subunit. Fragmentation of the same bond is also important in lignin breakdown during chemical oxidation (Schultz and Templeton 1986) and fungal decay (Schoemaker et al. 1985).

7.3.2 CHECKING OF WOOD

Checks represent the macroscopic response of wood to surface stresses generated by anisotropic shrinkage of wood (Schniewind 1963). An alternative response to such stresses is for wood to warp. Both forms of degrade occur when wood is exposed to the weather, although warping may be restrained for a period of time by fixings (nails or screws).

When wood surfaces are exposed outdoors, water absorbed by wood surfaces will penetrate subsurface layers by capillarity (in permeable species) and diffusion. Exposure of wood to solar (infra-red) radiation will cause surface wood layers to dry. Shrinkage of these surface layers will occur as their moisture content decreases below the fiber saturation point, but it will be restrained by subsurface layers which are still above the fiber saturation point and by fixings. Such restraint will place the surface layers under a tension stress (Schniewind 1963). The magnitude of this stress will depend on moisture loss and shrinkage of the surface layers, and will be greatest for wood surfaces whose growth rings are oriented parallel (tangential) rather than perpendicular (radial) to the surface because shrinkage of wood tangentially is approximately twice that occurring in the radial direction (Skaar 1972). Accordingly, checking of quarter-sawn decking boards exposed outdoors is significantly less than that in flat sawn boards (Yata 2001). Less checking has also been observed in flat sawn boards whose growth rings are more perpendicular to the exposed surface (i.e., pith side up) than in boards whose growth rings are more parallel to the surface (bark side up) (Urban and Evans 2005). However, flat sawn boards used for decking, benches and outdoor table tops are usually laid pith side down to reduce splintering caused by raised or loose grain (Pilson et al. 1931). Surface tensile strains that develop when wood surfaces dry clearly play an important part in the initiation and development of checks in wood exposed outdoors (Schniewind 1963). Strains also develop when wood absorbs moisture. These strains are compressive at the surface of decking boards, but high tensile strains develop in the rays and in the inner core of boards (Ribarits and Evans 2010). The high tensile strains in rays probably explain why checks propagate radially along the rays in flat-sawn boards. The high tensile strains that develop in the core of boards may explain the development of internal checks in decking boards, and how large cracks can develop in decking boards as a result of surface and internal checks coalescing (Zahora 2000). Previous explanations of the checking of decking boards exposed outdoors have largely overlooked the importance of the tensile stresses that develop in rays and in the core of boards when they become wet (Ribarits and Evans 2010). Its appears likely that such strains play a very important role in the development of checks, which would explain why water repellents are so effective at preventing the checking of wood exposed outdoors (Evans et al. 2003, 2009, Zahora 2000).

Checking of wood during kiln drying occurs when stresses cause surface tensile strains of 0.2. Checking at weathered wood surfaces may require lower stresses and strains due the weakening of wood surfaces caused by photodegradation, and the creation of voids or microchecks in the wood, which act as focal points for the development of visible checks (Evans 1989b, Evans et al. 2008).

7.4 EFFECTS OF WEATHERING

The effects of weathering can be observed at different length scales, molecular, microscopic, and macroscopic. Changes at the molecular level occur rapidly, and our understanding of the rate of change occurring at wood surfaces exposed outdoors has advanced considerably in recent years due to the widespread availability of spectroscopic and other techniques that can probe the chemical composition of surfaces in great detail (e.g., Fourier transform infrared spectroscopy (FTIR), x-ray photoelectron spectroscopy (XPS), nuclear magnetic resonance (NMR) spectroscopy and, more recently, atomic force microscopy (Meincken and Evans 2009, 2010).

7.4.1 CHANGES AT THE MOLECULAR LEVEL

All of the macromolecular components of wood are degraded during weathering. Over 145 years ago Wiesner (1864) reported that weathered wood surfaces (the gray layer at the surface) consisted of a layer of pure or nearly pure cellulose with a very low lignin content (Feist 1990a, Feist and Hon 1984). Studies nearly a century later confirmed this finding and have established that the photodegradation of lignin at wood surfaces exposed to UV radiation and/or natural weathering occurs very rapidly. Infra-red spectroscopy of wood exposed to natural weathering in North America revealed that wood surfaces were completely delignified after 30 days exposure (Feist and Hon 1984). Delignification of wood surfaces exposed outdoors to natural weathering can occur much faster. Evans et al. (1993, 1996) used FTIR internal reflectance spectroscopy to examine the chemical changes at the surface of radiata pine veneers exposed outdoors in the summer in Canberra, Australia for short periods of time. Their spectra showed a remarkably rapid decrease in the peak at 1505 cm^{-1}, which corresponds to aromatic C=C bond stretching in lignin (Evans et al. 1996). Spectra suggested perceptible surface (1–2 μm) delignification after only 4 h exposure, substantial delignification after 3 days and almost complete surface delignification after 6 days (Figure 7.6). Changes in peaks at 1601 cm^{-1} (C=C bond stretching), 1263 cm^{-1} (C–O bond stretching vibration in lignin and hemicelluloses), and 870 cm^{-1} (CH out-of-plane bending vibration in lignin) also suggested substantial and rapid degradation of lignin (Evans et al. 1993, 1996). These findings were subsequently confirmed for rubberwood (*Hevea brasiliensis* Müll. Arg) exposed to natural weathering in the spring in Bangalore, India (Pandey and Pitman 2002).

XPS of weathered and UV-irradiated wood surfaces have shown increases in the intensities of carbon–oxygen and oxygen–carbon–oxygen bond signals, and decreases in the intensities of carbon–carbon and carbon–hydrogen bond signals, suggestive of degradation of lignin and enrichment

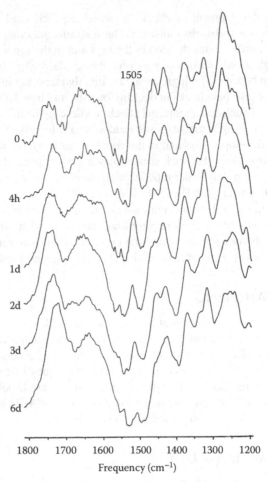

FIGURE 7.6 Fourier transform infrared spectra of radiata pine veneers weathered for short periods of time in Canberra, Australia.

of the surface with cellulose (Hon 1984). Degradation of lignin at wood surfaces is accompanied by reduction in methoxyl content and an increase in surface acidity and carbonyl content (Leary 1967, 1968). These changes occur as a result of the depletion of lignin at the wood surface and the accumulation of lignin degradation products in the wood including vanillin, syringaldehyde, and organic (carboxylic) acids. Formaldehyde, carbon monoxide, carbon dioxide, hydrogen, water, and methanol have also been identified as degradation products of wood during weathering (Kalnins 1966).

Hemicelluloses are degraded during weathering and are leached from weathered wood surfaces (Kalnins 1966). Chromatography of leachates from weathered softwood surfaces showed a high proportion of mannose and xylose suggestive of degradation of the hemicelluloses galactogluco-mannan and arabinoglucuronoxylan, respectively (Evans et al. 1992). FTIR spectroscopy of weathered softwood surfaces has shown reductions in the peak at 1728 cm^{-1} (C=O stretching vibration in acetyl and carboxyl in hemicelluloses) and 809 cm^{-1} (mainly vibration of mannan in hemicelluloses and CH out-of-plane bending vibration in lignin) confirming that hemicelluloses in wood are degraded during weathering (Evans et al. 1992).

Weathered wood surfaces are rich in cellulose, as mentioned above, and this led to suggestions that cellulose was resistant to the effects of weathering. This is clearly not the case as cellulose at wood surfaces exposed under natural exposure conditions is rapidly depolymerized (Derbyshire and

Miller 1981, Evans et al. 1996). This has been confirmed by both viscometry and high-performance size-exclusion chromatography of solutions of holocellulose from weathered wood (Derbyshire and Miller 1981, Evans et al. 1996, 2000a). Reductions in the degree of polymerization of cellulose account for the rapid and large losses in tensile strength of thin (80 μm) wood veneers exposed to natural or artificial accelerated weathering (Derbyshire and Miller 1981, Derbyshire et al. 1995, 1996, 1997).

Extractives are degraded by UV and short wavelength visible light, which accounts for why many distinctly colored woods fade over time, even when exposed indoors (Kitamura et al. 1989). Woods with high extractive contents show greater change in color when exposed to UV light and leaching with water than woods with low extractive contents (Barreto and Pastore 2009). There is a divergence of opinion about how extractives influence the photodegradation of wood's structural components (cellulose, hemicelluloses, and lignin). Nzokou and Kamdem (2006) concluded from their work on the influence of wood extractives on the photo-discoloration of wood during artificial weathering that "extractives in wood act as anti-oxidants and are able to provide some protection to wood surfaces against weather." In contrast, Pandey (2005) suggested that "presence of extractives increases rate of photo-discoloration and result in an apparent increase in rate of delignification of wood surfaces in the initial period of exposure to UV light." Sharratt et al. (2009) found that early color changes in Scots pine exposed to UV light were "unaffected by the removal of extractives from the wood and independent of temperature."

7.4.2 Microscopic Changes

Changes in the microscopic structure of wood during weathering were first noted by Wiesner (1864). He found that intercellular material was lost in ray cells and vessels first, and only later in the other wood cells. Structural changes included spirally oriented cracks in cell walls and in pits. Since these early observations there have been several studies that have focused on the structural changes that occur at wood surfaces exposed to natural weathering or to UV radiation/artificial accelerated weathering. These studies report that microscopic changes are pronounced in regions of cell walls such as the middle lamella that contain high concentrations of lignin, or are perforated by pits. Thin walled cells are very susceptible to degradation, particularly where they are found adjacent to thicker-walled cells.

Erosion of the middle lamella occurs rapidly during exterior exposure and can be clearly seen in transverse surfaces of radiata pine exposed outdoors for 5 days (Evans et al. 1993). The warty layer that covers the S₃ layer of the secondary wall in softwood tracheids is rich in polyphenolic materials and is rapidly eroded during exposure of wood to UV light (Kuo and Hu 1991) or natural weathering (Kim et al. 2008). Erosion of the middle lamella is more pronounced in the cell corners where the concentration of lignin is highest (Evans 1989b, Kim et al. 2008), between radial rather than tangential walls, and between latewood tracheids (Evans 1989b). Longer periods of exposure lead to complete erosion of the middle lamella and partial separation of individual tracheids along their length (Borgin 1971, Evans 1989b) (Figure 7.7). This exposes the primary and secondary wall layers to the weather and, as a result, they delaminate and cells become progressively thinner and more distorted (Evans 1989b).

Degradation of the middle lamella can also be seen in longitudinal surfaces where double cell walls are exposed to the weather (Miniutti 1964, 1967). In this case erosion of the middle lamella again causes cells to separate, but whole tracheids are not always detached from the surface, because microchecks that develop in pits cause tracheids or fibers to fragment.

Micro-checking of the cell wall in softwoods develops in bordered pits (Evans 1989b) and in cell walls themselves (Paajanen 1994). Microchecking of bordered pits starts as two small notches in the aperture (Figure 7.8). These notches are orientated at an angle to the long axis of the tracheid. Subsequently these notches propagate on either side of the pit aperture creating a diamond-shaped micro-check (Evans 1989b) (Figure 7.8). The enlargement of these micro-checks in areas of the cell

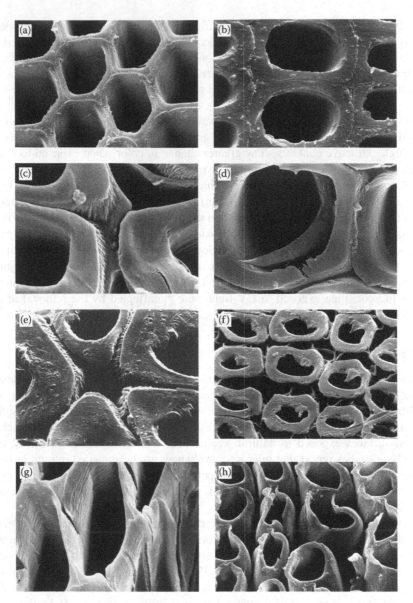

FIGURE 7.7 Microstructural changes at transverse surfaces of radiata pine during natural weathering in Canberra, Australia; (a) unweathered earlywood tracheids, ×1100; (b) unweathered latewood tracheids, ×1580; (c) latewood weathered for 30 days showing erosion of the middle lamella and delamination of the S_1 layer of the secondary wall, ×4200; (d) latewood weathered for 30 days showing delamination of the S_3 layer of the secondary wall, ×2030; (e) latewood weathered for 40 days showing erosion of the middle lamella in a cell corner and between tracheids, ×3500; (f) latewood weathered for 40 days showing erosion of middle lamellae, delamination of S_1 layer of secondary wall and partial separation of tracheids, ×1110; (g) earlywood weathered for 200 days showing erosion and thinning of secondary walls, ×1370; (h) latewood weathered for 200 days showing erosion and distortion of tracheids, ×980.

wall where several pits are found close together creates thin bridges of cell wall material between the microchecks (Evans 1989b). Micro-checking within these bridges creates small cell wall fragments that are easily detached from the cell wall (Figure 7.8). Micro-checking of half bordered pits in rays also occurs during weathering and creates fragile areas of cell wall that are easily eroded from the exposed surface (Chang et al. 1982, Evans 1989b, Miniutti 1967).

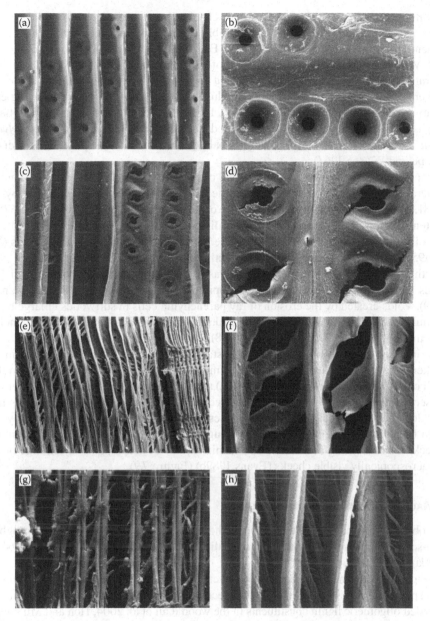

FIGURE 7.8 Microchecking of radiata pine tracheids at radial longitudinal surfaces during natural weathering; (a) bordered pits aligned in single files (uniseriate) in unexposed earlywood, ×420; (b) bordered pits with circular apertures in unexposed earlywood, ×1290; (c) progressive microchecking of bordered pits in earlywood tracheids following 30 days exposure to natural weathering, ×350; (d) microchecking of bordered pits in earlywood tracheids following 30 days exposure to natural weathering, note colonization of pit borders by bacteria (upper left), ×1100; (e) loss of integrity of wood surface as a result of microchecking of bordered pits in tracheids and half bordered pits in rays (center right) following 50 days exposure to natural weathering, ×125; (f) microchecking of the bridge of material between two bordered pit microchecks following 100 days exposure to natural weathering, ×1020; (g) voids formed at a wood surface as a result of microchecking and loss of cell wall material following 200 days exposure to the weather, ×370; (h) microchecking of wood cell wall in the absence of bordered pits in wood exposed to the weather for 100 days, ×1100.

The aforementioned description of the micro-checking of bordered pits in radial cell walls of softwoods is a generalized scheme and considerable variation exists in the form that the checking takes, depending on wood species and exposure. For example, in Californian redwood exposed to artificial UV light, micro-checking and erosion of adjacent, biseriate pits created large voids that were separated by cell wall material, crassulae, that appeared to be resistant to degradation (Miniutti 1967). Where pitting was uniseriate the pit annulus was eventually lost with increasing exposure, again creating voids in cell walls (Miniutti 1967). It has also been reported that micro-checking of bordered pits did not develop in a range of hardwoods including European beech (*Fagus sylvatica* L.), opepe (*Nauclea diderrichii* Merrill) and English oak (*Quercus robur* L.) exposed to artificial weathering (Coupe and Watson 1967).

Microchecks can also develop in the absence of pits in wood cell walls, as mentioned above. These generally follow the microfibril angle of the S_2 layer of the secondary wall (Derbyshire and Miller 1981, Paajanen 1994) or the direction of orientation of tracheids (Miniutti 1967). Larger checks often develop at the interfaces between different cell types, for example, at growth ring boundaries in softwoods (Evans 1989b), and at the interfaces between rays and tracheids (Yata and Tamura 1995). In each case the check develops within the thinner walled tissues (Figure 7.9). Thin-walled epithelial cells in resin canals are also easily degraded during weathering which creates small voids in transverse surfaces and microscopic checks in transverse surfaces (Evans 1989b) (Figure 7.9). Voids created by the erosion of ray parenchyma cells in softwoods enlarge beyond the original boundaries of the ray as a result of degradation and separation of the middle lamella above and below the ray (Evans 1989b, Evans et al. 2008).

Multiseriate rays in hardwoods are more resistant to weathering than the surrounding ground tissue. Baker (1900) noted that weathering acted more on the wood between the large multiseriate rays in bull oak (*Casuarina luehmannii* R. T. Bak.) than on the rays themselves, and Kučera and Sell (1987) observed that large multiseriate rays in beech remained intact during weathering and could be readily detached from weathered wood surfaces. The extractives within softwood rays are also reported to be resistant to degradation, even though the rays themselves are not (Miniutti 1967). Micro-checks at growth ring boundaries and within rays and resin canals may enlarge or coalesce to form macroscopically visible checks (Evans 1989b) (Figure 7.9).

7.4.3 MACROSCOPIC AND COLOR CHANGES

The most obvious macroscopic features of weathered wood are its gray color and rough surface texture caused by the presence of checks and the differential swelling and erosion of woody tissues (Feist 1990a).

Wood exposed to the weather changes color very rapidly. Light colored woods, including most coniferous species, darken in color initially and become yellow or brown due to the accumulation of photodegraded oligomeric lignin constituents in the wood (Cui et al. 2004, Hon and Minemura 1991, Kishino and Nakano 2004b, Tolvaj and Faix 1995). Dark-colored woods that are rich in phenolic extractives may fade initially before becoming yellow or brown (Hon and Minemura 1991). However, there are differences in the rate at which different wood species yellow (Cui et al. 2004, Hon and Feist 1986, Nzokou and Kamdem 2002). Irrespective of these initial color changes, wood exposed outdoors for 6–12 months (depending on climatic conditions) becomes gray as extractives and photodegraded lignin fragments are leached from the wood resulting in surface layers that are rich in cellulose (Arndt and Willeitner 1969). Subsequently, the wood may acquire an unattractive dark gray, blotchy, appearance due to the presence of fungal hyphae, spores and pigments within weathered wood surface layers (Feist 1990a). Wood exposed outdoors in coastal (exposed to salt) or very dry environments where microbial activity is restricted, however, is often an attractive silvery-gray color (Feist 1990a).

The lustre of wood may decrease during weathering as the surface becomes rougher and the scattering of light becomes more diffuse (Hon and Minemura 1991). These increases in surface roughness are due to preferential erosion of low-density earlywood (Figure 7.10).

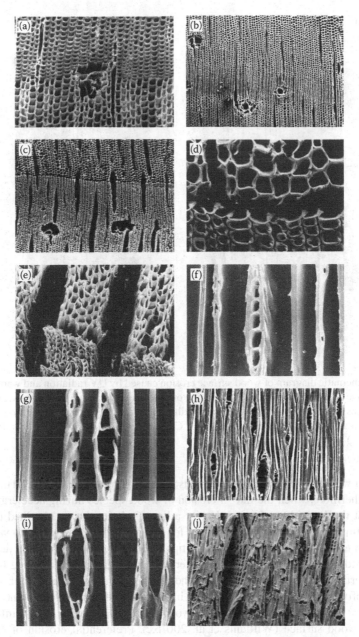

FIGURE 7.9 Microchecking of radiata pine originating in resin canals, rays and at growth ring boundaries at transverse surfaces (a–e) and tangential longitudinal surfaces (f–j) during natural weathering; (a) growth ring boundary in unexposed wood showing earlywood (bottom) and latewood (top), note resin canal in center, ×118; (b) formation of small checks as a result of degradation and separation of rays following 20 days exposure to natural weathering, ×41; (c) enlargement of ray checks and formation of voids and checks within resin canals and at a growth ring boundary, respectively, following 100 days exposure to natural weathering, ×35; (d) failure of early-wood tracheids at the interface of a check formed at a growth ring boundary following 100 days exposure to natural weathering, ×360; (e) large checks formed within rays and at the growth ring boundary following 200 days exposure to natural weathering, ×153; (f) ray in unexposed wood, ×640; (g) degradation of thin walled parenchyma cells in a ray following 30 days exposure to natural weathering, ×770; (h) voids created by erosion of thin walled parenchyma cells in rays and separation of middle lamella above and below rays following 50 days exposure to natural weathering, ×122; (i) large void created by erosion of thin-walled parenchyma cells in a ray following 50 days exposure to natural weathering, ×122; (j) enlargement of ray checks following 2 years exposure to natural weathering, ×112.

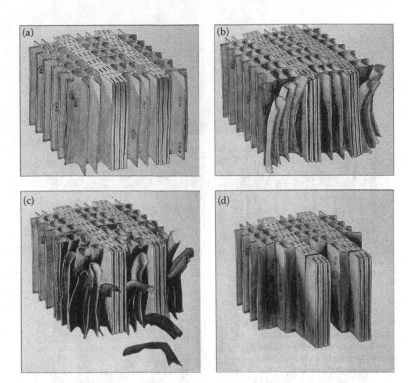

FIGURE 7.10 Schematic diagram of wood surface erosion caused by UV radiation and water: (a) unexposed wood; (b) early stage of weathering showing loosening of fibres; (c) early stage of fiber loss; and (d) later stage showing greater loss of low-density earlywood leading to a rough corrugated surface. (Adapted from McDonald, K. A. et al. 1996. *Wood Decks. Materials, Construction, and Finishing.* Forest Products Society, Madison, WI, 93pp.)

The erosion of wood during exterior exposure is inversely proportional to wood density (Sell and Feist 1986). Furthermore, when the rate of erosion of wood species during accelerated weathering is plotted against the logarithm of wood density a linear relationship is obtained (Kataoka et al. 2005). This relationship is similar to that found for the penetration of light into wood and wood density (Kataoka et al. 2004). Increased penetration of light and depth of photodegradation may thus explain why low-density species such as western red cedar erode at a rate of 12 mm per century when exposed vertically facing south in the northern hemisphere, whereas comparable figures for higher density softwoods such as Douglas fir and high density hardwoods are 6 and 3 mm per century, respectively (Feist 1990a). The erosion of wood during weathering is also influenced by the topology of the wood surface (Williams et al. 2001b,c). Preferential erosion of earlywood may expose a greater surface area of denser latewood to light resulting in higher than expected erosion of such tissue (Williams et al. 2001b).

Weathered wood often contains checks which can penetrate wood to a depth of several millimeters and extend longitudinally for hundreds of millimeters (Evans et al. 2003). A check is defined as a longitudinal separation that extends into wood to a depth that does not exceed 75% of the thickness of the board (Flæte et al. 2000). In contrast, a crack can penetrate completely through a board and lead to its separation into two pieces. Both checks and cracks are usually much longer than they are wide (Evans et al. 2003).

The macroscopically visible checking that develops in wood exposed outdoors has its origins in the micro-checking described above that causes separation of different woody tissues (Evans 1989b). Hence, the large checks that cause growth rings to separate on end grain surfaces of softwoods (Figure 7.1) result from micro-checking at the interface between latewood and earlywood

(Evans 1989b, Evans et al. 2008). Checking at tangential softwood surfaces exposed to the weather is more numerous, and the checks are larger than those found on similarly exposed radial surfaces (Yata 2001, Sandberg 1999). In the former, checks develop as a result of degradation and separation of rays particularly in latewood whereas at radial surfaces checks often develop at the interface between latewood and earlywood. Checks often develop within the resin ducts that are found in *Pinus* species (Evans 1989b). In hardwoods checks develop within rays and in vessels. There is considerable variation between species in the pattern of checking. For example, aspen (*Populus tremula* L.) develops a high number of relatively short checks when exposed outdoors, whereas spruce (*Picea* spp.) develops fewer, longer, checks (Flæte et al. 2000). Western red cedar boards develop fewer, smaller, checks than similarly exposed lodgepole pine (*Pinus contorta* Dougl. var. *latifolia* Wats.) and Douglas fir boards.

7.4.4 Changes in Wood Properties and Ease of Processing

Many surface properties of wood in addition to color are affected by weathering. The surface wettability of wood increases during weathering due to loss of extractives and lignin from exposed surfaces (Williams 2005). For example, Kalnins and Feist (1993) used contact angle measurements to evaluate changes in the wettability of western red cedar exposed to 4 weeks of natural weathering. They observed that the contact angle of water droplets with wood surfaces was 77° on unweathered wood and dropped to 55° on weathered wood. Jirous-Rajkovic et al. (2007) also noted that the wettability of fir and pine wood increased during natural weathering. In accord with these observations, Kang et al. (2002), Kishino and Nakano (2004a), and Zhang et al. (2009) observed that the wettability of wood increased during artificial weathering. Kishino and Nakano (2004a) explained their observations that artificial weathering increased the wettability of eight tropical wood species "in terms of the increase in hydroxyl groups originating from both the exposed cellulose and adsorbed water." They noted that there was variation in the wettability of the different wood species after artificial weathering. Such variation was explained in terms of differences in microchecking of the various wood species (Kishino and Nakano 2004a). The pH of redwood and Engelmann spruce (*Picea engelmannii* (Parry) Engelm.) exposed to UV light for 128 h decreased, whereas pH increased when the same wood species were exposed to UV light and water spray for the same period of time (Webb and Sullivan 1964). Exposure to UV radiation lowered the affinity of Scots pine for moisture (Sharratt et al. 2010). This effect was thought to be due to cross-polymerization of lignin, blocking water sorption sites.

Processing technologies and applications of wood that depend on its surface properties are severely affected by weathering. Notable in this regard is the painting and finishing of wood. Weathering of wood for 2–4 weeks prior to painting has been shown to significantly reduce the adhesion and performance of finishes applied to wood (Ashton 1967, Boxall 1977, Desai 1967, Evans et al. 1996, Jirous-Rajkovic et al. 2007, Underhaug et al. 1983, Williams and Feist 1993, 1994, 2001, Williams et al. 1987, 1999). Evans et al. (1996) found that the adhesion of exterior acrylic primers on radiata pine was reduced if the wood was exposed to the weather for only 5–10 days prior to painting. They also found that "primer adhesion was lower on weathered radial surfaces than on similarly exposed tangential surfaces."

There are reports of imperfect hardening of cement in contact with weathered plywood shuttering (form-ply) (Yoshimoto et al. 1967). It was suggested that the presence of sugars, mainly arabinose and polysaccharides produced by the photodegradation of hollocellulose, at the surface of the weathered plywood interfered with the curing of the cement (Yoshimoto et al. 1967).

Weathering also reduces the natural durability of western red cedar roofing shakes by leaching fungitoxic thujaplicins from wood (Schmidt and French 1976). Because weathering is confined to wood surface layers, the mechanical properties of wood, assuming decay to be absent, are largely unaffected by prolonged exposure to the weather. In contrast, the mechanical properties of wood composites, which depend in part on the strength of wood-adhesive bonds, can be significantly

reduced by moisture-induced dimensional changes when they are used outdoors. Despite the deleterious effects of weathering on the uses of wood, in certain applications weathered wood is
preferred to fresh wood. A good example of this is the use of weatherboards for the construction
of "New England" type barns, where the wood may be treated prior to building construction in
order to give the building an aged appearance in keeping with its rural surroundings (Anon 1976).
The treatments involve physical distressing (rough sawing, sand-blasting, wire-brushing, and planing with notched knives) to give the surface the texture of weathered wood (Anon 1976, Cassens
and Feist 1991), followed by chemical treatment to impart a gray color to the wood. Various chemicals have been used for the latter purpose including bleaches, or tannins in combination with ferrous ammonium sulfate (Cassens and Feist 1991). Alternatively, a gray stain can be applied to
wood which is "durable enough to allow a soft transition between its own degradation and the
eventual development of the gray color produced by natural weathering when the wood is exposed
outdoors" (Podgorski et al. 2009).

7.4.5 DEPTH OF DEGRADATION

It has been known for a long time that weathering only degrades the surface of wood. For example,
the *Annals of Horticulture* in 1850 mentioned that when old wooden planks have their weathered
surface planed off by the adze they are "then undistinguishable from new wood" (Anon 1850). The
popular prose writer of the 1930s Julia Peterkin described the old wooden African-American plantation houses as "weathered by long years of rain and wind and sunshine into a soft gray, but underneath this gentle color their yellow wood stands as solid and steadfast as it was a hundred years ago"
(Peterkin 1934). Changes in weathered wood are confined to surface layers because light does not
penetrate deeply into the wood, as mentioned above (Browne and Simonson 1957, Hon and Ifju
1978). Degradation of lignin has been detected at depths of approximately 400–700 µm in low-
density wood species exposed to artificial UV light (Bamber and Summerville 1981, Kataoka and
Kiguchi 2001, Kataoka et al. 2004, 2005, 2007, Yata and Tamura 1995). The depth to which photodegradation extends into wood depends on the spectral characteristics of the light source, duration
of exposure, density, and orientation of the wood elements and their chemical composition (Kataoka
and Kiguchi 2001, Kataoka et al. 2004, 2005, 2007). Kataoka et al. (2007) found that there was a
positive correlation between the penetration of light into sugi earlywood and the wavelength of the
incident radiation within the range 246–496 nm. The depth of photodegradation also increased with
wavelength up to and including the violet region (403 nm) of the visible spectrum. Blue light (434–
496 nm) penetrated wood to a greater extent than violet light and was capable of bleaching the
wood, but it did not significantly modify lignin, and hence it was not responsible for subsurface
photodegradation of wood. Kataoka et al. (2007) concluded that violet light is the component of the
visible spectrum that extends photodegradation into wood beyond the zone affected by UV radiation. Browne and Simonson (1957) indicated that the depth of weathering degradation sometimes
extends more than 2540 µm in "well-weathered wood." This observation may be due to redistribution of water soluble lignin degradation products from the surface to subsurface layers. It has also
been suggested that radiation can penetrate more deeply through the open pores of transverse surfaces resulting in more complete degradation of such surfaces compared to radial surfaces (Futó
1974).

7.5 PROTECTION

The ability of protection systems to restrict the weathering of wood surfaces depends on the properties of the wood substrate, particularly its density and shrinkage characteristics, and also on the
effectiveness of treatments at preventing the photodegradation of wood and development of unbal-
anced surface stresses that cause checking. The following sections examine the photoprotection of
wood and the ability of current treatments to restrict the weathering of wood. Excellent reviews are

available on treatments to protect wood from water and the reader is referred to these publications for more detailed information on this subject (Rowell and Banks 1985, Stamm 1964).

7.5.1 SURFACE COATINGS

The most common method of protecting wood from weathering and photodegradation is through the use of paints, varnishes, stains, or water repellents (Feist 1990a). Wood coatings are generally classed as either film forming or penetrating, and the latter can be used as a pretreatment or as a final finish (Feist 1990a). Film-forming finishes such as paints and "solid-body stains" contain pigments that screen wood from solar radiation and, because they form a barrier over the wood surface, they also prevent surface wetting and erosion (Feist 1990a). Feist points out that "a correctly applied and maintained paint system including a primer and at least two top-coats can greatly reduce the deleterious effects of weathering on wood" (Feist 1990a). However, a major problem with *clear* film forming finishes on wood is their loss of adhesion during weathering, which is due to photodegradation of the underlying wood (Macleod et al. 1995, Singh and Dawson 2003). Paints are sometimes less effective than penetrating finishes at controlling decay and dimensional movement and therefore they often perform better on wood that has been pretreated with a water-repellent preservative (Feist 1990a). The high maintenance requirements of paints when they fail and their tendency (when used on nondurable timber) to trap water and encourage decay has led to the increased use of penetrating water-repellent stains as a way of protecting wood used outdoors (Feist 1990a).

Penetrating finishes typically contain a hydrophobe such as wax, as well as an oil- or resin-based binder which penetrates the wood and cures. Unlike paints, which impose a physical barrier, water repellents rely on the formation of a hydrophobic coating which raises the contact angle of the treated wood and applied water droplets to over 90°, preventing water from being taken up by the surface or subsurface capillaries (Borgin 1965). Water repellents reduce moisture absorption so they impart a certain degree of dimensional stability to wood and restrict leaching of photodegraded fibers from wood surfaces (Feist 1988a). Biocides are often added to water repellents to create water-repellent preservatives, which are capable of retarding the growth of microorganisms on finished wood surfaces (Feist and Mraz 1980). Effective penetration of end grain by water repellents is essential to obtain good performance from finished joinery (Voulgaridis and Banks 1981, 1983). In practice, however, the hydrophobic system eventually breaks down due to the presence of impurities and imperfections in the external and internal coatings.

Penetrating stains are water-repellent preservatives that contain a variety of additives to reduce the weathering of wood, including pigments and UV stabilizers to screen wood from solar radiation. Penetrating stains tend to fail on wood during exterior exposure through cracking of the wood substrate and erosion of pigments from wood surfaces (Kiguchi et al. 1997a). This leads to discoloration of the finish through loss of pigmentation and accumulation of atmospheric pollutants (Kiguchi et al. 1997a). Semitransparent stains contain pigments that partially obscure the wood surface and hence they reduce the amount of light reaching the wood. At a pigment concentration of 8.4%, stains reduce erosion of wood during accelerated weathering by 65%. The use of water repellents in the formulation provides added protection, but photodegradation cannot be prevented completely (Feist 1988a). Stains applied as successive coats may form high-build films similar to paints and may be semitransparent to opaque, thus obscuring the wood surface. Their behavior and mechanism of failure then becomes similar to that of paints (Hilditch and Crookes 1981). Stains can provide protection against weathering for 2–6 years depending on wood species and surface texture, type and quantity of stain applied to the wood and degree of exposure to the weather. However, much earlier failure of stains occurs on knots and compression wood than on "clear" defect-free wood.

Paints, and to a lesser extent, stains, modify the appearance of wood. For end uses where it is important to retain the woods natural color, the wood can be finished with a clear coating. Clear film-forming finishes, although they often contain UV stabilizers and a biocide, are limited in their ability to protect wood from weathering because they transmit visible light and some UV light, which can degrade the underlying wood surface (Macleod et al. 1995, Singh and Dawson 2003). Hence, they perform poorly on wood used outdoors and invariably fail by peeling and cracking within 1–3 years of application. One means of increasing the performance of clear finishes on wood is to photostabilize the underlying wood surface prior to application of the clear finish using UV absorbers, hindered amine light stabilizers or chromic acid and related inorganic compounds.

7.5.2 Photoprotective Additives

Photoprotection of wood can be achieved through the use of additives that reflect or harmlessly absorb the light responsible for photodegradation or terminate the free radicals that degrade wood's constituents. Such additives are often added to water repellents, stains, and film-forming finishes to restrict the photodegradation of the finish and the underlying wood.

7.5.2.1 Inorganic Particles

Inorganic particles can block light from reaching wood substrates and protect wood from photodegradation. Hence, inorganic metal oxide (Fe, Ti) particles are commonly added to opaque exterior finishes (paints and heavy bodied stains). Smaller inorganic particles are also added to wood finishes, especially when the coating needs to be more transparent. The ability of small particles to scatter light is inversely proportional to the fourth power of the wavelength (Rayleigh 1871 cited by Blackburn et al. 1991). In other words, the shorter the wavelength, the greater the scattering by smaller particles. Small particles below a certain size are thus able to scatter UV light while having little effect on the visible component of the spectrum. This property of small particles and also their ability to absorb UV light underpins the use of minute particles of titanium dioxide and iron and zinc oxides as "transparent" photoprotective agents for coatings applied to wood. Transparent synthetic iron oxides are often used in stains and are available as red, orange, or yellow crystals (Sharrock 1990). They differ from iron oxide pigments as they are smaller (0.01–0.15 μm) and transmit light in the visible spectrum while screening UV radiation. Such screening of UV radiation can be achieved at a concentration of 2 g/m^2 or 2% for a coating applied at 100 g/m^2 (Sharrock 1990). Furthermore, transparent iron oxides are reported to retain their ability to protect wood from photodegradation over longer periods of exterior exposure compared to organic UV absorbers (Sharrock 1990). One obvious disadvantage of iron oxides in applications where it is essential to retain the natural appearance of wood is their yellow color. Furthermore, Aloui et al. (2007) reported that "transparent" iron oxide caused discoloration of clear-coated wood during natural weathering. Hence, there is also interest in using micronized titanium dioxide and zinc oxides as photoprotection agents for wood (Schulte 2001). Titanium dioxide is a highly effective white pigment that is commonly added to opaque paint. Micronized titanium dioxide does not scatter visible light, but has the capacity to protect wood from UV light when it is in the rutile form and treatments are applied to reduce its photocatalytic effect (Schulte 2001). Micronized titanium dioxide has been shown to provide good protection to wood surfaces when used on its own at application rates of 0.5–1.0% (Blackburn et al. 1991) or in combination with iron oxide. There has also been interest in using micronized zinc and cerium oxides to protect wood from photodegradation (George et al. 2005, Liu et al. 2010). The advantages of zinc oxide compared to titanium dioxide for this application is that zinc oxide is more protective against UV radiation and is less white at a given concentration (Pinnell et al. 2000). The transparency of inorganic particles can be increased by reducing their particle size and this has led to interest in using metallic oxides with very small average particle size (nanoparticles) to improve the weathering resistance of clear coatings on wood. Recent research has demonstrated that metallic oxide nanoparticles can increase the durability of clear coatings on wood exposed to accelerated weathering. For example, Allen et al. (2002) found that the rutile form of titanium dioxide nanoparticles was an effective stabilizer for clear coatings on pine wood with performance greater than or equal to organic UV absorbers and hindered amine light stabilizers.

Particles 70 nm in size were more effective than 90 nm particles. Weichelt et al. (2010) found that spruce wood samples finished with clear coatings containing zinc oxide nanoparticles were less discolored after accelerated weathering than samples finished with an unmodified coating. The effectiveness of the modified coating at reducing discoloration was further enhanced if the wood was pretreated with an organic "lignin stabilizer." Similarly, Cristea et al. (2010) showed that zinc oxide and titanium dioxide nanoparticles improved the durability of clear coatings on black spruce wood (*Picea mariana* (Mill.) Britten, Sterns, and Poggenberg) exposed to accelerated weathering. Zinc oxide "nanofilms" can also photostabilize wood surfaces (Yu et al. 2010), but cerium oxide nanoparticles were less effective at restricting weight and tensile strength losses of wood veneers exposed to natural weathering than conventional photostabilizers (UV absorbers, hindered amine light stabilizers, and micronized iron oxide) (Liu et al. 2010).

7.5.2.2 UV Absorbers

UV absorbers absorb incident UV radiation, undergo tautomeric conversion and dissipate the energy as nonradiative heat. A wide range of UV absorbers are available commercially including benzophenone, benzotriazole, triazine UV absorbers, cinnamic derivatives, and oxalanilides (Hayoz et al. 2003) (Figure 7.11). Within these basic types there is further variation in molecular weight and type of substituent groups. The absorption characteristics and light fastness of the UV absorber are determined by the chromophoric group whereas the volatility, polarity and resistance of the UV absorber to leaching are controlled by its molecular weight and the nature of substituent groups (Rabek 1990).

UV absorbers are used to improve the resistance of plastics and coatings to photodegradation (Davis and Sims 1983, Rabek 1990). They are also used to restrict the photodegradation of wood finishes and the underlying wood. An early study found that treatment of Douglas fir veneer with a dibenzoylresorcinol UV absorber reduced the production of gaseous and volatile photodegradation products produced during irradiation of the veneer with artificial UV light (Kalnins 1966). Pretreatment of wood surfaces with the benzotriazole UV absorber Tinuvin 1130 was able to restrict color change of Taiwania (*Taiwania cryptomeriodes* Hay.) heartwood exposed to fluorescent UV light (Chang et al. 1998). Tinuvin 1130 also prevented clear coats on grand fir, tauari (*Couratari* spp.) and European oak (*Quercus petraea* (Mattuschka) Liebl. and *Quercus robur* L.) from cracking when coated specimens were exposed to 840 h of accelerated weathering (Aloui et al. 2007). In contrast, inorganic photostabilizers were much less effective at preventing coatings from cracking during accelerated weathering (Aloui et al. 2007).

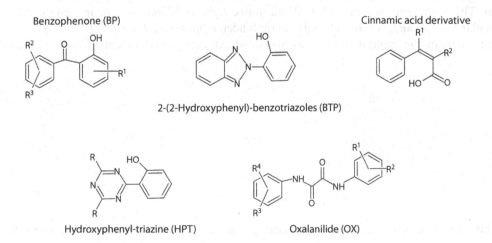

FIGURE 7.11 Chemical structures of UV absorbers used to photostabilize polymers and wood.

The sensitivity of lignin in wood to degradation by UV and visible light led to the development and use of a new substituted tris-resorcinol triazine derivative (1-isoctyloxycarbonyl ethylated 2,4,6 tris(2,4-hydroxyphenyl)-1,3,5 triazine derivatives) for clear coatings applied to wood (Schaller and Rogez 2007). This UV absorber has "improved photopermanence, highest extinction and broadest-absorption of long-wavelength UV radiation (i.e., red shifted)." The effectiveness of this triazine UVA and other benzotriazole UV absorbers at restricting the photodegradation of wood can be enhanced by using them in combination with chemicals that terminate the free radicals which cause photooxidation of wood (Hayoz et al. 2003).

7.5.2.3 Radical Scavengers

A range of compounds including hindered amine light stabilizers, phenolic antioxidants, heavy hydrocarbons, and glycols are able to scavenge or terminate free radicals, and some of these have been employed in polymer and wood photoprotection systems.

Hindered amine light stabilizers are thermally and light stable derivatives of 2,2,6,6-tetramethyl piperidine that provide photoprotection to wood when they are used in combination with UV absorbers (Figure 7.12) (Hayoz et al. 2003). They are also able to protect polymers from photodeg-radation at relatively low concentrations (Rabek 1990). The mechanisms by which hindered amine light stabilizers photostabilize wood have not been studied in any detail, however, much more is known about how they photostabilize polymers such as polyethylene and polypropylene. In general terms hindered amine light stabilizers photostabilize polymers by terminating the free radicals involved in photooxidation (Gugumus 1993). Model studies of the interaction of hindered amine light stabilizers with simple hydroperoxides, peroxy, and acylperoxy radicals have yielded important insights into their modes of action (Klemchuk and Gande 1988). These studies have shown that hindered amine light stabilizers trap acylperoxy radicals converting them into carboxylic acids and in the process are converted into nitroxyl radicals (Klemchuk and Gande 1988). These nitroxyl radicals are extremely efficient at scavenging a range of free radicals including alkyl, alkoxy, and alkyl-peroxy radicals (Brede and Goettinger 1998). Nitroxyls are rapidly regenerated from N-alkyloxy hindered amines by reacting with acylperoxy radicals (Klemchuk and Gande 1988). Quenching of excited polymer oxygen complexes (Gugumus 1993), and chelation of transition metal impurities (Fairgrieve and MacCallum 1984) have also been proposed as means by which hindered amine light stabilizers protect polymers from photooxidation.

Hindered amine light stabilizers are available that vary in molecular weight and contain different substituent groups (R1 and R2) (Figure 7.12). The radical scavenging effectiveness and basicity of the molecule are determined by the substituent group at R1 whereas secondary properties including solubility, thermal stability, and migration resistance are influenced by the group at R2 (Rabek 1990). The migration resistance of hindered amine light stabilizers is also influenced by their molecular weight, and, accordingly, oligomeric hindered amine light stabilizers have been developed for demanding applications such as fibers and extracting environments (Malik et al. 1998).

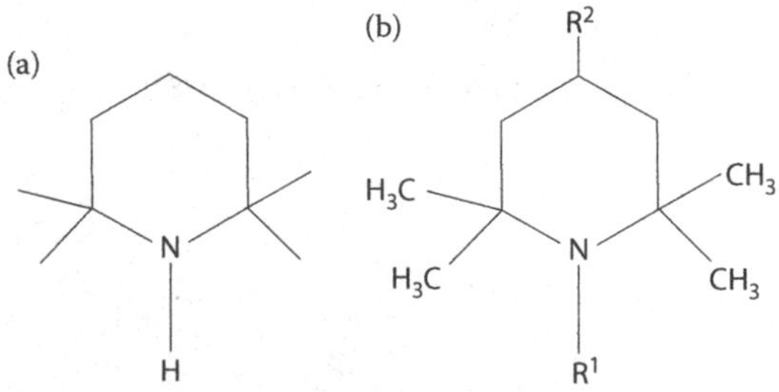

FIGURE 7.12   Structure of 2,2,6,6-tetramethyl piperidine (a) and a hindered amine light stabilizer (HALS) (b).

Polymer graftable hindered amine light stabilizers have also been developed that are highly resistant to leaching and volatilization (Malik et al. 1998).

Pretreatment of wood surfaces with the hindered amine light stabilizer Tinuvin 292 restricted color changes of Taiwania heartwood exposed to fluorescent UV light, and a synergistic effect was observed between Tinuvin 292 and the UV absorber Tinuvin 1130 on heartwood color stability (Chang et al. 1998). George et al. (2005) suggested that impregnation of wood surfaces with HALS prior to the application of a clear coat containing a UVA could be a useful strategy for protecting wood from photodegradation and enhancing the performance of clear coats used outdoors. Rogez and coworkers described the synergistic effects of combinations of UV absorbers and hindered amine light stabilizers on the performance of clear coatings on wood, and the progress made to develop more effective combinations of UV absorbers and hindered amine light stabilizers for the photoprotection of wood and clear finishes (Hayoz et al. 2003, Rogez 2002, Schaller and Rogez 2007). They described a state of the art system for photostabilizing wood beneath clear coatings. This system consisted of a clear coating containing a red shifted triazine UV absorber (1–1.5%) and the HALS (decanedioic acid,*bis*(2,2,6,6-tetramethyl-1-(octyloxy)-4-piperidinyl) ester)) (0.5%) applied to wood that had been pretreated with a HALS (Lignostab, 4-hydroxy-2,2,6,6-tetramethylpiperidinoxyl, 2%) specifically designed to trap the radicals formed as a result of the degradation of lignin by visible light. This system was able to successfully prevent the degradation of clear-coated radiata pine panels exposed to the weather in Switzerland for 18 months (Schaller and Rogez 2007). Exposed wood panels lost very little of their original color and no coating defects were present after outdoor exposure (Schaller and Rogez 2007). Forsthuber and Grüll (2010) further examined the effectiveness of the photoprotective system described by Schaller and Rogez (2007), but they extended the concept by using inorganic photostabilizers as well as UVA/HALS combinations in clear coatings. They concluded that "careful selection of light stabilizers, such as TiO_2 or UVA in combination with HALS in the topcoat, as well as a lignin stabilizer in an aqueous primer can significantly improve the color retention of wood" (exposed to UV radiation). Hindered amine light stabilizers have also been shown to be useful inhibitors of iron oxidation and water degradation of wood (Hussey and Nicholas 1985).

Hindered phenolic antioxidants are traditionally used to terminate free radicals originating from heat-induced degradation of coatings (Rabek 1990). They have similar functions to hindered amine light stabilizers, so they can be used to retard the degradative effects of UV light on wood (Rabek 1990). Unlike hindered amine light stabilizers, however, they are nonregenerative and decrease in concentration during the photostabilization process (Figure 7.13). There are no published accounts of the use of phenolic antioxidants to photostabilize wood, but a copper II complex of the antioxidant curcumin restricted color changes of spruce exposed to UV light. However, more complete photostabilization was achieved if the antioxidant was used in combination with a benzotriazole UV absorber (Sundaryono et al. 2003).

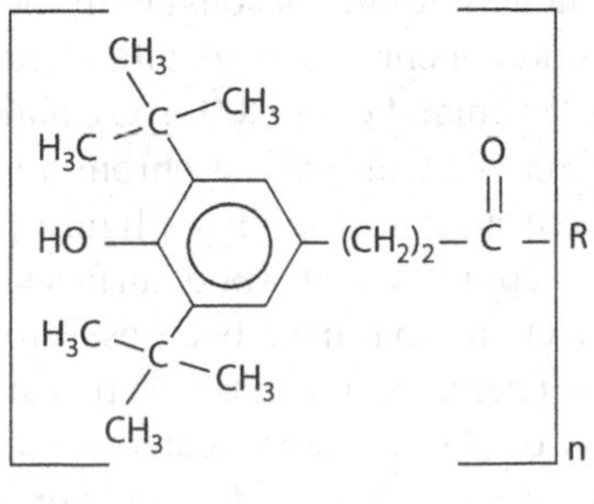

FIGURE 7.13   Structure of a phenolic antioxidant (Irganox series).

Ohkoshi (2002) found that impregnation of wood with polyethylene glycol decreased the generation of carbonyl groups and degradation of lignin in wood exposed to UV light. Salaita et al. (2008) recently confirmed that polyethylene glycol can photostabilize wood. However, polyethylene glycol is degraded by UV light and hence Ohkoshi (2002) suggested that this would limit the long-term ability of poly-ethylene glycol to photostabilize wood. Nevertheless, pretreatment of Douglas-fir plywood with a 10% solution of polyethylene glycol greatly reduced film failure of clear finishes on panels exposed outdoors for 2 years (Kiguchi et al. 1997b).

7.5.3 REACTIVE METAL COMPOUNDS

Modification of lignin to make it less susceptible to photodegradation is an obvious route to increasing the photostability and weathering resistance of wood. Treatments that fall into this category include reactive transition metal compounds, most notably chromic acid, but also copper and cobalt chromates, ferric chloride and nitrate, and various manganese, titanium, and zirconium compounds (Black and Mraz 1974, Chang et al. 1982, Feist 1979, Kubel and Pizzi 1981, Schmalzl and Evans 2003, Upreti and Pandey 2005). Chromic acid is remarkably effective at photostabilizing wood, and this led to a great deal of interest in the mechanisms responsible for its effectiveness.

The formation of chromium complexes with lignin was suggested as the likely chemical basis for the ability of chromium trioxide to photostabilize wood (Black and Mraz 1974, Hon and Chang 1985, Williams and Feist 1984). FTIR spectroscopy and ESCA of chromium trioxide modified wood surfaces has confirmed that oxidation and modification of lignin occurs (de Lange et al. 1992, Evans et al. 1992, Kiguchi 1992, Michell 1993, Pandey and Khali 1998, Williams and Feist 1984). On chromium trioxide modified wood surfaces, absorptions due to aromatic C=C vibrations and C-O vibrations of lignin were substantially reduced compared with those of the parent wood (Evans et al. 1992). The spectrum of chromium trioxide treated wood veneers was virtually unchanged after 35 days of natural weathering (Evans et al. 1992) providing clear evidence that chromium–lignin complexes formed at wood surfaces treated with chromium trioxide are photostable.

The reaction of model lignin compounds with chromium trioxide and ferric salts has been studied as a means of better understanding the chemical basis for the stabilization of wood surfaces with chromium trioxide (Hon and Chang 1985, Jorge et al. 1999, Michell 1993, Pizzi 1980, Schmalzl et al. 1995, 2003). These studies have yielded insights into how chromic acid modifies lignin to make it photostable. Pizzi used a dilute solution of guaiacol as a lignin model to study the reaction of chromium (VI) with wood (Pizzi 1980). He suggested that chromium (VI) formed an insoluble complex with lignin (Pizzi 1980). However, magnetic susceptibility measurements of adducts formed by reacting chromium trioxide with lignin and model lignin compounds demonstrated that chromium in adducts was mainly in the trivalent state (Wright and Banks 1989). A reexamination of the reaction of guaiacol with excess aqueous chromic acid (Schmalzl et al. 1995) found that guaiacol oligomers were bound or cross-linked by hydroxylated trivalent (presumably octahedral) chromium species to form complicated three-dimensional high-molecular weight polymers. Further studies have confirmed that the reaction of chromic acid with guaiacol results in the formation of an amorphous chromium III complex (Schmalzl et al. 2003). Oxidation of 2,6-dimethoxyphenol with chromic acid resulted in the formation of an amorphous chromium III coerulignone (quinone) complex. It is therefore plausible that chromic acid oxidizes lignin phenols in wood resulting in the formation of chromium III quinone complexes at wood surfaces. Quinone and its derivatives are efficient antioxidants and UV protectants that have been used to photostabilize polymers (Sabaa et al. 1988, Yassin and Sabaa 1982). Therefore, it is possible that stable metal–lignin–quinone complexes account for the photostability of chromic acid treated wood surfaces.

Pretreatment of wood surfaces with chromic acid significantly improves the photostability of wood and, predictably, when used as a wood surface pretreatment increases the service life of natural finishes used outdoors (Williams and Feist 1988). Chromic acid is very effective at restricting

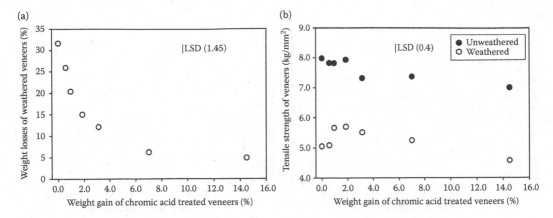

FIGURE 7.14 Photostability of radiata pine veneers modified with chromic acid to different weight gains: (a) percentage weight losses of veneers exposed to natural weathering for 35 days in Canberra, Australia; (b) zero-span tensile strength of veneers modified with chromic acid before (filled circles) and after (open circles) natural weathering for 35 days in Canberra.

weight losses of wood veneers exposed to natural weathering (Figure 7.14a), which is a good indicator of its ability to photostabilize lignin (Evans and Schmalzl 1989). Chromic acid is less effective at reducing losses in tensile strength of wood veneers during natural weathering (Figure 7.14b), but it can reduce the checking of simple and bordered pits in wood exposed to UV light and natural weathering, and also restrict the surface erosion of wood during artificial accelerated weathering (Black and Mraz 1974, Evans et al. 1994).

Chromic acid treatment also imparts some dimensional stability to wood, and reduces its hygroscopicity (Williams and Feist 1985, 1988). Pretreatment of wood with hexavalent chromium compounds results in a green coloration, which is undesirable when the aim of finishing is to preserve the natural appearance of wood. Hexavalent chromium compounds are also carcinogenic (Ruetze et al. 1994), a fact that has prevented their commercial use as photostabilizing compounds for wood (except in Japan where pretreatment of wooden doors with chromic acid was used in the 1980s to enhance the weathering resistance of acrylic-urethane finishes) (Ohtani 1987). Trivalent chromium compounds are less toxic than chromic acid, but they are also less effective at protecting wood from photodegradation (Williams and Feist 1988). Similarly, other transition metal compounds are also less effective than chromic acid at photostabilizing wood. For example, ferric chloride and ferric nitrate were unable to significantly reduce losses in weight and tensile strength of radiata pine veneers exposed to natural weathering (Evans and Schmalzl 1989), even though earlier research had shown that ferric chloride could reduce the microchecking of bordered pits in wood exposed to artificial accelerated weathering (Chang et al. 1982). Treatment of wood surfaces with oxidative manganese compounds, such as manganese (III) acetate dihydrate or potassium permanganate is able to protect wood from photodegradation (Figure 7.15). However, these compounds are less effective than chromic acid and they also discolor the treated wood (Schmalzl and Evans 2003).

Titanates and zirconates are colourless and there has been some interest in their use as photoprotective agents for wood (Schmalzl and Evans 2003). Tetrabutyl, tetraisopropyl, and ethylhexyl titanate enhanced the tensile strength of wood both before and after weathering, possibly due to the formation of complexes between the titanates and cellulose in wood. Figure 7.15b shows the positive effect of ethylhexyl titanate (ET) on the tensile strength of radiata pine veneers both before and after natural weathering. Ethylhexyl titanate and other titanates, however, were unable to restrict veneer weight losses during natural weathering, indicating that they did not form stable complexes with lignin (Figure 7.15a). Zirconates such as tetrapropyl zirconate (TpZ in Figure 7.15) and

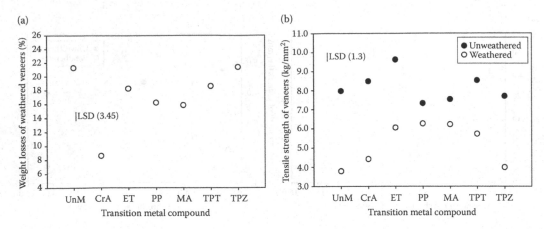

FIGURE 7.15 Photostability of radiata pine veneers modified with different transition metal compounds: (a) percentage weight losses of veneers exposed to natural weathering for 35 days in Canberra, Australia; (b) zero-span tensile strength of modified veneers before (filled circles) and after (open circles) natural weathering for 35 days in Canberra. Each treatment used 0.64 M solutions of the following compounds, CrA (chromic acid in water), ET (ethylhexyl titanate in white spirit), PP (potassium permanganate in water), MA (manganese III acetate dihydrate in methanol), TpT (tetraisopropyl titanate in white spirit), TpZ (tetrapropyl zirconate in white spirit). UnM = unmodified veneers (control).

tetrabutyl zirconate were even less effective at protecting wood from weathering than titanates (Schmalzl and Evans 2003).

7.5.4 WOOD PRESERVATIVES

Inorganic compounds used as components of some wood preservatives can also provide photoprotection to wood. Feist and Williams examined the ability of chromated-copper-arsenate (CCA) to reduce the weathering degradation of unfinished wood and improve the durability of semi-transparent and solid-color stains applied to wood (Feist and Williams 1991). CCA provided long-term protection against weathering-induced erosion and greatly extended the lifetime and durability of a partially UV-transparent stain (Feist and Williams 1991). The copper component of ammonia-cal copper quat (ACQ) and copper azole wood preservatives also enhances the resistance of wood to weathering. Experimental studies have shown that both ACQ and copper azole restrict the rate of delignification of wood during exterior exposure (Cornfield et al. 1994, Liu et al. 1994). Copper-ethanolamine also retarded the surface degradation of lignin in southern pine exposed to artificial accelerated weathering (Zhang et al. 2009). In contrast, metal free organic wood preservatives provide little photoprotection to wood unless pigments or photostabilizers are added to the formulation (Schauwecker et al. 2009). Organic wood preservatives containing alkylammonium compounds may even accelerate photodegradation (Jin et al. 1991, Zhang 2003) because alkylammonium compounds can catalyze the delignification of wood (Tanczos and Schmidt 2002).

Wood preservatives commonly contain hydrophobic additives such as waxes or oils to restrict the cupping and checking that occurs when treated wood is exposed to natural weathering. Both wax and oil emulsion additives are effective at increasing the water repellency and dimensional stability of treated wood (Fowlie et al. 1990, Greaves 1990, 1992, Zahora and Rector 1990, Zahora 1991, 2000, Jin et al. 1992, Cui and Zahora 2000). Wax emulsion additives also reduce the cupping and checking that develops when treated wood is exposed to artificial accelerated or natural weathering (Christy et al. 2005, Evans et al. 2003, Fowlie et al. 1990, Zahora 1992, 2000) and oil emulsion additives restrict the checking of treated wood exposed to natural weathering (Evans et al. 2009).

7.5.5 WOOD MODIFICATION

7.5.5.1 Chemical Modification

Most chemical modification systems have been investigated and, in some cases, commercialized because they have the potential to improve the dimensional stability and decay resistance of wood (Hill 2006). Nevertheless, the ability of such systems to protect wood from weathering and improve the performance of finishes on wood has often been examined. Chemical modification of wood by methylation, acetylation, or alkylation improved the color stability of wood during weathering because the "blocking" of phenolic hydroxyl groups retarded the formation of quinones (Kalnins 1984). Accordingly, acetylation of wood to weight gains of 10–20% reduces the photo-yellowing (Plackett et al. 1992, Tarkow et al. 1946), checking (Dunningham et al. 1992) and erosion of wood exposed to natural or artificial weathering (Feist et al. 1991). The ability of acetylation to reduce checking and erosion of wood during weathering may be explained by the increased dimensional stability and hydrophobicity of the modified wood. Acetylation of Scots pine veneers to low weight gains of 5–10%, however, reduced the photostability of the modified veneers (Figure 7.16a). When wood is acetylated to low weight gains with acetic anhydride and using pyridine as a solvent/catalyst, preferential substitution of lignin phenolic hydroxyl groups occurs. This may prevent the termination of free radicals via the formation of quinones, thus increasing photodegradation (Evans et al. 2000a). When wood is acetylated to a higher weight gain of 20% the photostability of acetylated wood is increased (Figure 7.16). At higher weight gains substitution of cellulosic hydroxyl groups occurs which may provide protection to wood, in accord with observations that cellulose derivatives are generally less susceptible to photodegradation than unmodified cellulose (Usmanov 1978). Thus, acetylation can protect cellulose to some degree (Feist et al. 1991), but it is unable to photostabilize lignin (Kalnins 1984). The photoprotective effects of acetylation, however, are reduced with prolonged exposure of the modified wood to the weather because deacetylation of wood surfaces occurs (Evans et al. 2000a). The presence of acetic acid in acetylated wood may also influence the weathering of the modified wood in accord with observations that dilute acids increase the degradation of wood during artificial accelerated weathering (Williams 1987). Further research is needed, however, to confirm this hypothesis. Nevertheless, acetylation generally has positive effects on the performance of coatings (Beckers et al. 1998, Nienhuis et al. 2003). Acetylation was able to significantly increase the service life of an opaque acrylic finish on Scots pine, but it had no such effect on the performance of an alkyd topcoat, possibly because residual acetic acid in the acetylated wood increased ageing of the alkyd paint (Brelid and Westin 2007).

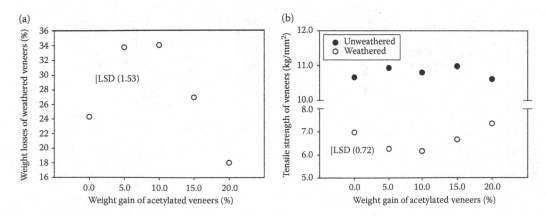

FIGURE 7.16 Photostability of Scots pine veneers modified with acetic anhydride in pyridine to different weight gains: (a) percentage weight losses of modified and unmodified veneers exposed to natural weathering for 35 days in Canberra, Australia; (b) zero-span tensile strength of acetylated veneers before (filled circles) and after (open circles) natural weathering for 35 days in Canberra.

The ability of many other chemical modification systems to photostabilize wood, and improve the performance of finishes has also been examined. Hill et al. (2001) found that modification of wood with methacrylic anhydride followed by grafting with styrene was ineffective at preventing the photodegradation of Corsican pine (*Pinus nigra* J.F. Arnold) and Scots pine wood veneers. Esterification with dicarboxylic acid anhydrides was also unable to significantly reduce losses in weight and tensile strength of modified wood veneers exposed to accelerated weathering (Figure 7.17), suggesting that the treatments were unable to photostabilize lignin or cellulose (Evans 1998). However, a more recent study found that esterification using succinic or proprionic anhydride was able to enhance the weathering resistance of two acacia species exposed to natural weathering for 1 year (Bhat et al. 2010).

Esterification of wood with a mixture of phthalic anhydride and the epoxide epichlorohydrin was also able to enhance the weathering resistance of Japanese cypress exposed to accelerated weathering (Murakami and Matsuda 1990). Gowdra et al. (2006) found that esterification of wood with octanoyl chloride restricted photo-yellowing, but did not protect lignin in wood from photodegradation. Other studies have also shown that various esterification reactions can prevent photo-yellowing of wood. For example, Chang and Chang (2001a) showed that butyrylation with butyric anhydride inhibited the photo-discoloration of wood. Esterification of wood surfaces with acetic or succinic anhydride also reduced the surface yellowing of UV-irradiated Chinese fir (*Cunninghamia lanceolata* (Lamb.) Hook. F.) wood, whereas treatment of wood with phthalic anhydride accelerated photo-discoloration (Chang and Chang 2001b). Chang and Chang (2006) also showed that etherification with isopropyl glycidyl ether reduced the photo-yellowing of wood exposed to UV radiation. Modification of wood with butylene and propylene oxide, however, has little positive effect on the weight and tensile strength losses of Scots pine veneers exposed to UV radiation and leaching with hot water (Figure 7.18).

In accord with these observations, Rowell et al. (1981) and Feist and Rowell (1982) found that modification with butylene oxide, methyl isocyanate, or butyl isocyanate at weight gains of 25% or more was unable to photostabilize wood. Pandey et al. (2010) found that modification of wood with propylene oxide or butylene oxide provided limited or no photostability to wood. Similarly, various etherification reactions including cyanoethyolation, benzylation or allylation did not significantly improve the UV resistance of wood, but the last two treatments improved the performance of clear coatings applied to modified wood (Kiguchi 1990). In contrast, chemical modification of wood with a maleic acid–glycerol mixture enhanced the resistance of wood to

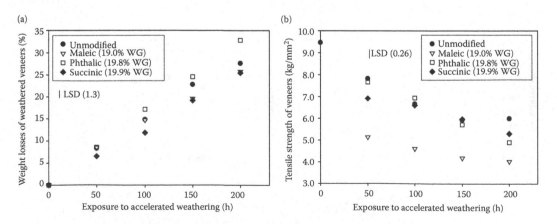

FIGURE 7.17 Photostability of Scots pine modified to ~20% weight gain with different carboxylic acid anhydrides: (a) percentage weight losses of veneers exposed to artificial accelerated weathering in a xenon-arc weatherometer (continuous exposure to UV radiation and 4 h water spray every 24 h); (b) zero-span tensile strength of modified veneers after accelerated weathering.

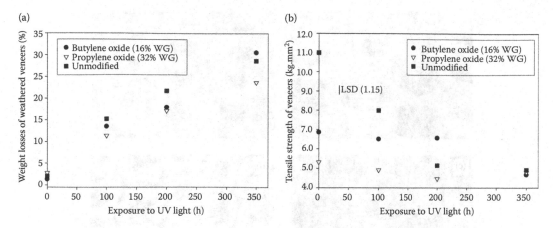

FIGURE 7.18 Photostability of Scots pine veneers modified to 16% weight gain with butylene oxide or 32% weight gain with propylene oxide: (a) percentage weight losses of veneers exposed to UV radiation in a xenon-arc weatherometer and then leached in hot (80°C) water for 3 h; (b) zero-span tensile strength of modified veneers and unmodified controls after exposure to UV radiation and leaching in hot water.

weathering (Fujimoto 1992). Similarly, benzoylation with benzoyl chloride (to high weight gains) was also effective at photostabilizing wood (Evans et al. 2002, Pandey and Chandrashekar 2006). Figure 7.19 shows the effect of benzoylation on the weight losses and tensile strength of Scots pine veneers exposed to natural weathering. Benzoylation was very effective at reducing weight losses of veneers indicating that (like chromic acid) it is able to photostabilize lignin (Figure 7.19a).

Scanning electron photomicrographs of benzoylated wood exposed to natural weathering clearly show that benzoylation, unlike acetylation, is able to restrict the degradation of the lignin-rich middle lamella (Figure 7.20). It was suggested that the ability of benzoylation to photostabilize lignin in wood was due to the absorption of UV light or termination of free radicals by benzoyl groups (Evans et al. 2002). The ability of benzoylation to photostabilize lignin has led to interest in finding out whether other modification systems that deposit aromatic groups in cell walls can photostabilize wood. For example, Jebrane et al. (2009) showed that esterification with the aromatic vinyl ester,

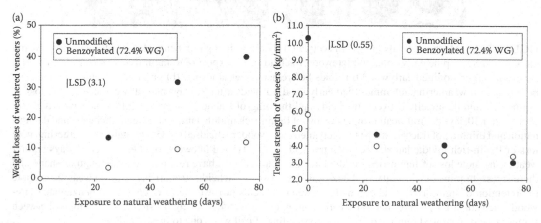

FIGURE 7.19 Photostability of Scots pine veneers modified to 72.4% weight gain with benzoyl chloride: (a) percentage weight losses of benzoylated and unmodified veneers after exposure to natural weathering in Canberra, Australia; (b) zero-span tensile strength of benzoylated veneers and unmodified controls after exposure to natural weathering in Canberra.

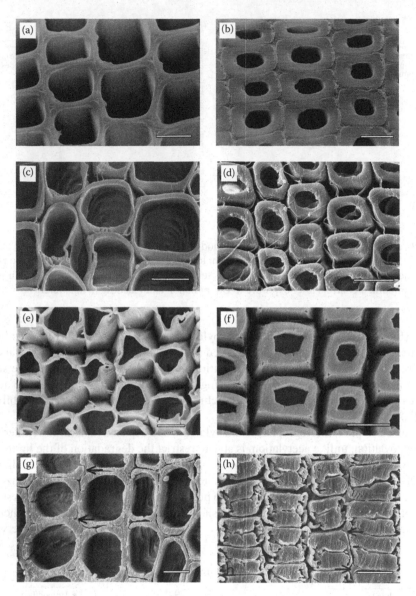

FIGURE 7.20 Morphological changes in acetylated (17.5% weight gain), benzoylated (34% weight gain) and unmodified Scots pine earlywood and latewood tracheids after exposure to natural weathering in Canberra, Australia: (a) unmodified earlywood tracheids before natural weathering; (b) unmodified latewood tracheids before natural weathering; (c) unmodified earlywood tracheids after 30 days natural weathering, note loss of lignin-rich middle lamella between tracheids and thinning of tracheid walls; (d) unmodified latewood tracheids after 30 days natural weathering, note loss of lignin-rich middle lamella between tracheids, and delamination and thinning of tracheid walls; (e) acetylated earlywood tracheids after 35 days natural weathering, note loss of lignin-rich middle lamella between tracheids; (f) acetylated latewood tracheids after 35 days natural weathering, note loss of lignin-rich middle lamellae, but tracheids have retained their rectangular shape and there is less thinning of cell walls; (g) benzoylated earlywood tracheids after 30 days natural weathering, note the retention of lignin-rich middle lamella between tracheids (arrowed left of center); (h) benzoylated latewood tracheids after 30 days natural weathering, note the retention of lignin-rich middle lamella between some tracheids (arrowed right) and pronounced swelling of cell walls due to benzoylation.

vinyl benzoate, was effective at photostabilizing lignin in wood, whereas modification of wood with two other aromatic vinyl esters, vinyl cinnamate and vinyl-4-T-butylbenzoate, was less effective.

7.5.5.2 Thermal Modification

Thermal modification improves the dimensional stability and decay resistance of wood to levels where it can be used outdoors for cladding and decking (Militz 2008). In these applications thermally modified wood is exposed to the weather and it is often left uncoated (Ala-Viikari and Mayes 2009). Hence, there has been interest in understanding the effects of thermal modification on the photostability and weathering resistance of wood. The short-term color stability of thermally modified wood exposed to artificial UV radiation was reported to be better than that of untreated wood (Ayadi et al. 2003, Shi and Jiang 2011). Thermally modified beech wood exposed to natural weathering was also less susceptible to discoloration and colonization by *A. pullulans* and other staining fungi than unmodified wood (Feist and Sell 1987). Thermal modification reduced the erosion of beech wood during artificial accelerated weathering (Feist and Sell 1987). Thermal modification of Norway spruce samples at 185°C and 1000 kPa for 3 h also reduced the erosion of samples during accelerated weathering whereas modification at 175°C and 1000 kPa for 2 h had the opposite effect (Feist and Sell 1987).

Lignin at the surface of thermally modified wood is reported to be slightly less susceptible to photodegradation than lignin in unmodified wood, possibly due to increased condensation of lignin induced by thermal treatment (Nuopponen et al. 2004). Feist and Sell (1987) suggested that the low equilibrium moisture content of thermally modified wood might reduce its susceptibility to photodegradation. The increased (short-term) weathering resistance of thermally modified wood might also be due, in part, to its increased water repellency, which would restrict the leaching of photodegraded lignin and hemicelluloses from exposed surfaces. Nevertheless, the brown color of thermally modified wood fades quite quickly when the wood is exposed outdoors and it eventually becomes gray (Jämsä et al. 2000), indicating photodegradation and loss of lignin at exposed wood surfaces. In accord with these observations it has been noted that thermally modified wood used as cladding on buildings weathers from a brown to a gray color and also develops surface checks (Ala-Viikari and Mayes 2009). Weathering also causes significant erosion of tracheids from the surface of thermally modified wood (Ala-Viikari and Mayes 2009). Such erosion is caused in-part by wasps (*V. vulgaris*) mining weathered tracheids, which they use to construct their paper-like nests (Ala-Viikari and Mayes 2009).

There are discrepancies in the literature about whether thermal modification reduces the susceptibility of wood to check (crack) when it is exposed outdoors. Vernois (2001) stated that "cracking due to dimensional motion is reduced in comparison with natural wood," and Dubey et al. (2010) found no surface checks in oil-heat-treated radiata pine exposed to accelerated weathering for 2100 h. In contrast, Jämsä et al. (2000) found that thermally modified pine and spruce checked to the same extent as unmodified wood when they were exposed outdoors, and the application of unpigmented or low-build stains and oils did not reduce the severity of checking of the modified wood (Jämsä et al. 2000). Feist and Sell (1987) found that there was much more grain-raising and cracking of thermally modified spruce exposed to 14 months natural weathering compared to similarly exposed, unmodified controls. Furthermore, the surfaces of thermally modified spruce samples were much rougher than the unmodified weathered samples. In contrast, the surfaces of thermally modified beech samples were smoother than unmodified samples after the samples were exposed outdoors for 14 months, and there was little difference in cracking of the modified and unmodified samples. Feist and Sell (1987) also found that semi-transparent and film-forming stains performed slightly better on thermally modified beech, but their performance was slightly worse on thermally modified spruce compared to their performance on unmodified spruce. Nevertheless, Jämsä et al. (2000) concluded that thermal modification does not adversely affect the performance of coatings on wood and no alterations in coating protocols are required when finishing thermally modified wood.

7.5.5.3   Impregnation Modification

Impregnating wood with polymerizable monomers improves wood's resistance to weathering (Desai and Juneja 1972, Feist and Rowell, 1982, Feist et al. 1991, Raczkowski 1982, Rowell et al. 1981). For example, impregnation with and polymerization of styrene significantly increased the resistance of beech wood to accelerated weathering in contact with rusting iron (Raczkowski 1982). Impregnation of wood with methyl methacrylate monomer and subsequent polymerization of the monomer in wood cell lumens reduced the rate of moisture sorption and erosion of the modified wood during accelerated weathering (Rowell et al. 1981). The combination of butylene oxide, methyl- or butyl-isocyanate modification of wood followed by methyl methacrylate monomer treatment further enhanced the resistance of wood to UV light and accelerated weathering (Feist and Rowell 1982, Rowell et al. 1981), despite findings that modification of wood with butylene oxide, methyl- or butyl-isocyanate on their own was ineffective at enhancing the photostability of wood (Feist and Rowell 1982, Rowell et al. 1981). The combination of acetylation followed by methyl methacrylate monomer treatment also increased the weathering resistance of wood (Feist et al. 1991). Impregnation of wood and wood composites with low-molecular weight phenol formaldehyde resin improves their resistance to weathering (Lloyd and Stamm 1958, Sudiyani et al. 2001, V'Icheva 1986). More recent research on the use of polymer treatments to restrict the photodegradation of wood has focused on the use of water-soluble thermosetting amino resins, textile cross-linking agents, silicone compounds and waxes. Impregnation of wood with melamine formaldehyde resin protected wood from photodegradation (Pittman et al. 1992, Rapp and Peek 1999) and reduced the discoloration and cracking of wood exposed outdoors (Hansmann et al. 2006). Impregnation of wood with furan polymers formed from the condensation of furfuryl alcohol, however, was ineffective at restricting the discoloration and delignification of wood exposed to artificial accelerated weathering (Temiz et al. 2007). The textile cross-linking resin 1,3-dimethylol-4,5-dihydroxyethyleneurea (DMDHEU) can provide modest protection to wood exposed to artificial accelerated weathering (Figure 7.21), but the resin was ineffective at preventing delignification caused by UV light (Xie et al. 2005). In contrast, impregnation of wood with high loadings of certain waxes, for example montan wax, restricted the rate of delignification of wood exposed to artificial accelerated weathering (Lesar et al. 2011).

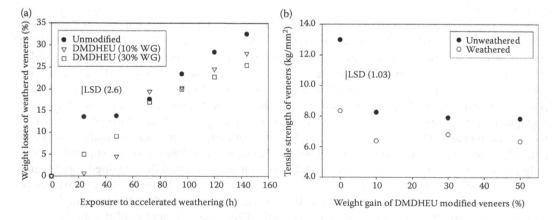

FIGURE 7.21   Photostability of Scots pine veneers modified to 10 or 30% weight gain with 1,3-dimethylol-4,5-dihydroxyethyleneurea (DMDHEU): (a) percentage weight losses of modified and unmodified veneers exposed for up to 48 artificial accelerated weathering cycles in a QUV weatherometer (1 cycle = 2.5 h UV radiation and 0.5 h cold water spray); (b) zero-span tensile strength of modified and unmodified veneers after 144 h of accelerated weathering (48 cycles).

A range of silicon compounds have also been tested to determine if they can increase the weathering resistance of wood. These compounds react in "a sol-gel" process to create a three-dimensional xerogel within wood, modifying substrate properties (Mahltig et al. 2008). Changes to substrate properties depend on the choice of silicon compound and whether additives are combined with the inorganic nano-sol (Mahltig et al. 2008). Treatment of wood with silicon nanosols derived from colloidal silica solutions reduced photo-discoloration of wood during artificial weathering (Temiz et al. 2006). Hydrophobic nano-sols derived from monomeric or oligomeric silanes also reduced the discoloration of wood exposed to artificial weathering, and discoloration could be further reduced by adding UV stabilizers to treatment solutions (Donath et al. 2007). Hydrophobic nano-sols also reduced the water-uptake of wood during natural weathering, but they were unable to prevent checking of the wood (Donath et al. 2007).

7.5.5.4 Grafting of UV Stabilizers and Polymerizable UV Absorbers

The permanence and effectiveness of UV absorbers at preventing the photodegradation of polymers can be improved by chemically bonding the UV absorber to the polymer, and this approach has also been used to protect wood from photodegradation. The majority of published research in this area has employed either benzophenone or triazine UV absorbers containing an epoxy group. These UV absorbers can be permanently bonded to wood at high temperatures in the presence of an amine catalyst. Grafting of epoxy-functionalized UV absorbers reduced the erosion of unfinished wood during artificial accelerated weathering, and the weight and tensile strength losses of wood veneers during natural weathering (Williams 1983, Kiguchi and Evans 1998, Kiguchi et al. 2001). Grafting of 2-hydroxy-4(2,3-epoxypropoxy)-benzophenone (HEPBP) to wood was effective at reducing both weight and tensile strength losses of Scots pine veneers exposed to natural weathering (Figure 7.22). Grafting of HEPBP to wood was as effective as chromium trioxide at restricting mass losses of thin wood veneers exposed to natural weathering and was better than chromium trioxide at reducing tensile strength losses of veneers (Kiguchi and Evans 1998). XPS indicated that the photoprotective effect of grafting on weight losses of veneers was due to the protection of lignin at exposed surfaces (Kiguchi and Evans 1998). Grafting of epoxy-functionalized UV absorbers to wood also improved the performance of clear coatings on modified veneer surfaces (Kiguchi and Evans 1998, Kiguchi et al. 2001).

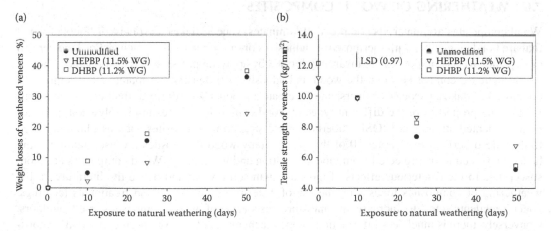

FIGURE 7.22 Photostability of Scots pine veneers grafted with 2-hydroxy-4(2,3-epoxypropoxy)-benzophenone (HEPBP) or impregnated with 2,4-dihydroxybenzophenone (DHBP) to weight gains of ~11%: (a) percentage weight losses of modified and unmodified veneers exposed to natural weathering in Canberra, Australia; (b) Zero-span tensile strength of modified and unmodified veneers after exposure to natural weathering in Canberra.

Another route to grafting UV absorbers to wood is to employ UV absorbers containing an isocyanate group. Grelier et al. (1997) reported that the photo-yellowing of wood could be restricted by grafting an isocyanate-modified benzotriazole UV absorber to wood, and pretreating the surface with radical scavengers such as polyethylene glycol or hindered amine light stabilizers. Grelier et al. (2007) also grafted a polystyrene-maleic anhydride copolymer containing polyethylene glycol chains and a hydroxyphenylbenzotriazole UV absorber (both with acyl azide functionality) to wood. This grafting treatment improved the color stability of wood exposed to UV light and the adhesion of clear varnishes on grafted wood exposed to artificial accelerated weathering, but it was less effective at preventing the discoloration of finished wood.

An alternative approach to grafting, which also increases the permanence and effectiveness of UV absorbers in polymers, is to increase their molecular weight. This can be achieved by reacting the parent UV absorber with other compounds (Bailey and Vogl 1976). For example, reaction of HEPBP with dicarboxylic acid anhydrides can create higher molecular weight (MW) polyester-type UV absorbers without destroying the UV-absorbing 2-hydroxybenzophenone unit (MW <1600) (Luston, et al. 1973). Such higher molecular weight UV absorbers are effective at protecting plastics and wood from photo-degradation (Luston et al. 1973, Evans and Chowdhury 2010). Polyester-type UV absorbers created from the reaction of HEPBP with phthalic anhydride were more effective at restricting the weight and tensile strength of thin wood veneers exposed to natural weathering than UV absorbers created from the reaction of HEPBP and maleic or succinic anhydride (Evans and Chowdhury 2010). Sundell and Sundholm (2004) reacted benzotriazole UV absorbers and a hindered amine light stabilizer with a polyester binder. One of the benzotriazole UV absorbers was able to bind to the polyester. This modified polyester was the most effective of the polyesters at restricting color changes of Scots pine exposed to accelerated weathering, but the change in color was similar to that of untreated samples treated with unmodified polyester. Samples treated with polyester that was not bound to either the benzotriazole UV absorber or HALS showed greater color changes than the control (Sundell and Sundholm 2004). In contrast, Hon et al. (1985) found much greater improvements in the photostability of wood coated with a copolymer of 2-hydroxy-4(3-methacryloxy-2-hydroxy-propoxy)benzophenone (poly(HMHBP)). The combination of poly(HMHBP) and another photostabilizer, the antioxidant butylated hydroxytoluene, provided "enhanced protection against photodegradation" (Hon et al. 1985).

7.6   WEATHERING OF WOOD COMPOSITES

Wood composites are materials composed of laminates, veneer, flakes, particles, or fibers of wood that are bonded together with thermosetting adhesives or inorganic binders such as Portland cement or gypsum. Wood composites are invariably formed by applying pressure to the wood and adhesive to: (1) increase contact between the wood and adhesive; (2) densify the material for composites composed of flakes, particles or fibers; and (3) create a product with uniform thickness. The manufacture and properties of the different types of wood composites, for example, plywood, particleboard, oriented strandboard (OSB), fiberboard and structural composites such as glue-laminated timber are described in Chapter 10 of this book. Many wood composites are used outdoors and hence they need to be protected from biodeterioration and weathering. Wood composites are more susceptible to the deleterious effects of moisture than solid wood and hence the literature on the weathering of composites focuses on the use of water-proof adhesives and treatments (coatings, water repellents, etc.) to increase the moisture resistance of wood composites used outdoors. Conversely, there is much less information on the mechanisms of the weathering of wood composites compared to the information available on this subject for solid wood.

7.6.1   PLYWOOD

Plywood is used outdoors as sheathing on buildings and boats, backing boards for signs, soffits, fences, fruit bins, and buildings to house farm animals, such as portable pig houses. In these applications plywood is exposed to the weather and its surface erodes, roughens and becomes gray like solid wood (Feist 1990a). The surface of plywood exposed outdoors also checks. Surface checking of plywood can be more pronounced than that of solid wood, because lathe checks formed when plywood veneer is rotary peeled become larger when plywood is exposed to the weather. The veneer sheets in plywood may also delaminate and, as a result, plywood can lose strength when it is exposed to the weather (unlike solid timber). For this reason, the veneer sheets in exterior grade plywood must be bonded together with a waterproof adhesive that can resist the stresses generated by the swelling and shrinkage of the outer veneer sheet that is exposed to the weather. Several studies have compared the resistance to weathering of plywood bonded with different adhesives (Knight 1947, Knight and Newall 1954, Tanaka 1964, Rice 1965, Neusser and Schall 1972, Gillespie and River 1976, Yoshida 1986, McLaughlan 1991, Damodaran and Padmanabhan 1992). Early studies showed that protein-based (blood–albumen) and urea formaldehyde adhesives were unsuitable for plywood used outdoors, whereas plywood's bonded with phenolic and resorcinol adhesives performed well when they were exposed to the weather for seven and a half years (Knight 1947, Knight and Newall 1954). Later studies have confirmed the suitability of phenolic adhesives for exterior plywood (Tanaka 1964, Neusser and Schall 1972, Gillespie and River 1976). The weathering resistance of plywood bonded with other thermosetting adhesives, for example, melamine formaldehyde, cross-linked polyvinyl acetate and isocyanate adhesives have also been examined (Neusser and Schall 1972, Yoshida 1986, McLaughlan 1991). Neusser and Schall (1972) examined the moisture resistance of European beech plywood bonded with melamine–urea formaldehyde adhesives. They concluded that "the resistance to weathering of a good melamine resin is not equal to that of a good phenolic resin" (Neusser and Schall 1972). In accord with this conclusion, McLaughlan (1991) observed that uncoated radiata pine plywood bonded with melamine urea formaldehyde resin and exposed to the weather easily delaminated when it was tested in shear. The durability of polymer-isocyanate and melamine phenol formaldehyde adhesive bonds in plywood exposed to accelerated ageing or 36 months of natural weathering was examined by Yoshida (1986). His accelerated ageing tests indicated that adhesive bonds in plywood bonded with polymer-isocyanate were as durable as those in plywood bonded with melamine phenol formaldehyde adhesive. However, the quality of polymer-isocyanate adhesive bonds rapidly deteriorated after 6–12 months natural weathering and most bond quality was lost after 36 months exposure (Yoshida 1986). The properties of individual adhesives also influence the weather resistance of plywood. For example, Tanaka (1964) found that the glue-line strength of plywood bonded with powdered urea formaldehyde adhesive and exposed to 22 months outdoor weathering was superior to that of similarly exposed plywood that had been bonded with conventional (liquid) urea formaldehyde adhesive. Rice (1965) found that the durability of bonds in plywood exposed to accelerated weathering was better in boards with thicker glue-lines and reduced penetration of the adhesive into voids in the surface of veneer. The durability of bonds in plywood is also influenced by wood species. Krahmer et al. (1992) used shear tests to compare the bond durability (% wood failure) of south East Asian plywood exposed to 10 years of natural weathering. Bond durability was better in red meranti (*Shorea* spp.) and two species of keruing (*Dipterocarpus crinitus* and *D. kerrii*) (wood failure >85%) than in red balau (Shorea spp.), one keruing (*Dipterocarpus* spp.) and two kapurs (*D. aromatica* and *D. oblongifolia*). The durability of bonds in plywood made from southeast asian and Japanese wood species and exposed to natural weathering for 1, 3, and 7 years was assessed by Yoshida and Taguchi (1978a, b). Plywood made from kapur showed the greatest reductions in shear strength during weathering whereas the greatest reductions in impact bending strength during weathering occurred in basswood and birch. Shear strength tests were used by Koch (1967) to evaluate bond durability of southern pine plywood exposed to natural weathering for 6 and 12 months. He found that heavier woods showed more delamination than lighter woods, especially if peeled loose, but growth rate was not correlated with delamination.

Wood species also affects the surface checking of plywood exposed to natural weathering. For example, Biblis (2000) compared the effects of weathering on the checking and surface quality of plywood made from Douglas-fir, redwood, bald cypress (*Taxodium distichum* (L.) Rich.), western red cedar, southern yellow pine (*Pinus taeda* L.) and lauan (*Shorea* spp.). Southern yellow pine and lauan both developed severe surface checking whereas checking in the other 4 species was less pronounced. Biblis (2000) ranked Douglas fir plywood most highly out of all of the species in terms of its color, surface quality, and the retained mechanical properties after 6 years outdoor exposure. Knight (1947) compared the surface checking of plywood made from several different softwood and hardwood species. He was able to classify the species into three groups based on the amount of checking in face plies exposed to weathering. Knight and Doman (1962) also examined the influence of surface veneer thickness and the geometry of the board on surface checking of beech plywood exposed indoors or to natural weathering for 5 years. They concluded that surface checking "depends entirely on the thickness of the face veneers and is independent of the total thickness of the board and of the number and gauge of the veneers forming the core." They noted that plywood with the thickest face veneer (3.175 mm) checked when exposed indoors and outdoors (more severely for the latter), whereas plywood with thinner face veneers only checked when exposed outdoors. Thin veneers, in general, "are less likely to develop face checks than thicker veneers of the same species under the same conditions" (Anon 1964). In accord with these observations Yagishita (1958) found that plywood made from thinner veneers performed better during accelerated weathering than plywood made from thicker veneers.

Checking of plywood develops more rapidly in panels when the "loose side" of the veneer, which contains lathe checks, is uppermost (Anon 1964); however, Elmendorf and Vaughan (1960) observed that checking of Douglas fir plywood containing stress-relieving grooves and exposed to changes in atmospheric humidity was much more conspicuous in panels with the tight side out. Stress relieving grooves cut into the face veneer of plywood prevents checking of plywood exposed to natural weathering (Elmendorf and Vaughan 1960), but more common methods of reducing the surface checking of plywood involve application of coatings or moisture-resistant overlays to face veneers or impregnating plywood veneers with resins. All of these techniques can reduce the severity of checking of plywood exposed to the weather because they reduce cyclic variation in moisture content and therefore the stresses that cause face veneers to check.

Opaque coating systems can reduce the severity of checking of plywood exposed to the weather. For example, Feist (1988b) found that a water-repellent wood preservative in combination with an alkyd primer and acrylic top-coat reduced the face checking of southern pine plywood siding exposed to the weather. Conversely, transparent finishes and semi-transparent and solid color stains failed to prevent checking, although they protected the surface from photodegradation to various degrees. Feist (1988b) concluded that rough-sawn southern pine plywood has better finishing and weathering characteristics than solid southern pine. A subsequent study by Feist (1990b) confirmed that water-repellent preservative pretreatment improved paint performance on southern pine plywood. Pretreatment with chromium compounds improved the performance of transparent and semi-transparent coatings on Douglas fir plywood exposed to the weather for 30 months (Williams and Feist 1988). Coatings can also reduce the delamination and losses in glue-bond strength that occur when plywood is exposed to the weather, particularly if there is regular maintenance of the coating (Damodaran and Naidu 1995, Damodaran and Padmanabhan 1992, Hunt and Matteson 1976, Kaneda and Maku 1973a,b, McLaughlan 1991). For example, Hunt and Matteson (1976) found no losses in strength of exterior Douglas fir plywood roof panels taken from pig houses after 21–29 years in service (and repainted every 3 years). However, weather resistant opaque film-forming finishes are reported to increase the incidence of decay in plywood made from nondurable hardwood species and exposed to the weather for prolonged (12 years) periods of time (Williams and Feist 2004).

An effective approach to protecting plywood from weathering is to bond resin-impregnated paper or polymer overlays to one or both of the face veneers. Knight and Doman (1967) bonded paper overlays impregnated with phenolic or cresylic resin to African mahogany (*Khaya* spp.) plywood and exposed the over-laid plywood to natural weathering. Some of the overlays "proved extremely effective in reducing color change and surface checking (of the plywood) although none of the overlays had completely retained their color and gloss after 9 years exposure." Acrylic polymer overlays were able to significantly reduce weathering failure of southern pine plywood exposed to natural weathering in Mississippi for 2 years, and the same study noted that thick kraft paper overlays performed well especially when they contained a durable resin (Barnes 1976). Overlays enhance the performance of paints applied to plywood (Knight and Doman 1967, Selbo 1969). For example, Black et al. (1976) observed that overlaid and painted softwood plywood was free of checks after 10 years exposure in Wisconsin or Mississippi. The impregnation of plywood with phenol-resin (Lloyd and Stamm 1958) or polyethylene glycol (Sakuno and Goto 1975) is also effective at reducing the surface checking of plywood exposed to natural weathering.

7.6.2 PARTICLEBOARD AND ORIENTED STRANDBOARD

Particleboard and oriented strandboard (OSB) show pronounced thickness swelling when exposed to water and hence they are unsuitable for outdoors use unless they contain a moisture resistant glue, hydrophobe, and have a barrier treatment (coating system or overlay) to prevent the ingress of moisture into the composites. Particleboard and OSB exposed outdoors also require preservative treatment if they are made from nondurable species. Hence, particleboard is mainly used for furniture, floor underlayment, and decking for mobile homes where it is not exposed to the weather. OSB is also used for these applications, but a far more important use of OSB is as a sheathing material for residential buildings where it can be exposed to the weather for short periods of time until the composite is covered with a moisture barrier or the building envelope is closed. OSB and to a lesser extent particleboard have been used for the manufacture of siding (cladding) on buildings where there is direct exposure of the composites to the weather. For example, Haygreen and Bowyer (1982) show photographs of two different siding products made from particleboard, one of which was used as a covering for exterior walls of buildings in the tropical climate of Indonesia. OSB is also used in North America for the manufacture of exterior siding on buildings. These uses of particleboard and OSB for exterior siding have been underpinned by a body of research on the weathering of the two different types of composites. This research essentially mirrors the research carried out on the weathering of plywood and focuses on evaluating the weather resistance of boards made with different adhesives and/or modified to make them more resistant to moisture and weathering. There has been less research on the effects of weathering on the surface properties of particleboard and OSB, although both composites swell irreversibly and lose strength when exposed outdoors, and the exposed surface roughens and becomes gray. The effects of weathering on particleboard are most severe at the exposed surface (like solid wood) (Clad and Schmidt-Hellerau 1965), but unlike solid wood, the strength losses that can occur when particleboard and the related composite waferboard are exposed to weathering are significant. For example, Alexopoulos (1992) observed that long-term exposure of structural-size aspen (*Populus* spp.) waferboard to natural weathering reduced bending strength by as much as 39%, stiffness by 52% and nail-holding properties by 60%. Poplar particleboard lost more than 50% of its bending properties after 27 months exposure to natural weathering at a site in Bulgaria with high rainfall (Iosifov and V'Icheva 1986).

Research on the weathering of particleboard has clearly shown that boards made with phenol formaldehyde adhesive are less susceptible to weathering than boards made using melamine–urea formaldehyde and urea formaldehyde adhesives. For example, Beech et al. (1974) concluded that "UF boards were unsuitable for external uses; laboratory-made MF/UF boards were superior to UF boards but inferior to the commercial MF/UF board studied and to all PF boards." Similarly, Barnes and Lyon (1978a,b) found that weathering reduced bending properties of particleboard more for UF than for PF bonded boards, and Gressel (1969) concluded from measurements of the strength and swelling of particleboard exposed to the weather for 4 years that PF-bonded boards were superior to both UF and melamine-fortified UF boards. Neusser and Zentner (1970) found that particleboards bonded with PF resin, fortified urea–melamine resin, and sulphite waste liquor gave good performance, (especially when treated with preservatives) when exposed outdoors for 5 years. Sell et al. (1979) also exposed particleboards bonded with different adhesives (coated and uncoated) to natural weathering for 4 years. They concluded from measurements of the thickness swelling and shear strength of boards that "PF-bonded boards were more resistant to weathering than UF-bonded boards, and their performance was improved by the surface coatings." Coatings are able to improve the weathering resistance of particleboard, but particular attention has to be paid to making sure that the edges and joints of boards are well sealed (Sell 1973, Meierhofer and Sell 1983, 1984). Furthermore, coatings do not act as wood preservatives and decay of coated particleboards can occur when they are exposed to the weather (Carll and Feist 1987, 1989). The effectiveness of coatings at protecting particleboard from weathering depends on pretreatment, coating type and the number of coats of paint or stain. Paints were superior to stains at preventing the cracking and delamination of wafers at the surface of waferboard exposed to the weather (Carll and Feist 1987), and Chen (1978) found that an opaque white paint was better at protecting particleboard than clear varnish. Water repellents are also effective at preventing the swelling of particleboard exposed to the weather. For example, Sell (1973) found that a water repellent containing paraffin wax protected the surfaces and edges of particleboard exposed for 20 months to open air weathering better than impermeable and permeable film-forming coatings. He later concluded with coworker Meierhofer that "an effective surface treatment for particleboard exposed to the weather should have a high degree of water repellency and also sealing ability" (Meierhofer and Sell 1977). However, in a follow-up study they found that a coating of solvent-based plaster was better at protecting melamine–urea–phenol formaldehyde bonded particleboard from the effects of weathering than coating systems consisting of a water-repellent primer and top coats of alkyd resin enamel paint (Meierhofer and Sell 1983).

The weather resistance of particleboard is also influenced by their resin content and the wood species that they are made from. For example, Stofko (1964) exposed particleboard made from beech and bonded with different levels of a PF adhesive (8%, 10%, 12%, 14%, or 16%) to artificial or natural weathering. He found that boards with the lowest resin content of 8% were "significantly inferior to all others." Particleboards made from the hardwood beech and bonded with a modified melamine formaldehyde adhesive did not perform as well as boards made from softwoods in open air weathering tests, with the exception of lifting of wood particles from the surface of boards (Clad and Schmidt-Hellerau 1976). In accord with the latter finding, Grigorion (1981) also concluded that the "effect of climate on the surface roughness of particleboard was generally greater for boards with surface layers of softwoods than for those of hardwoods." The proportions of softwood and hardwood in particleboard also influence the resistance of boards to natural weathering. For example, Béldi and Balint (1982) found that softwood particleboard disintegrated to a greater extent after 36 h of artificial weathering than boards made from a mixture of softwood and the hardwood, robinia (*Robinia pseudoacacia* L.).

Various treatments, apart from coatings, have been tested to increase the weather resistance of particleboard. A steam posttreatment was mildly successful at reducing the thickness swelling (springback) of particleboard exposed to the weather for 2 years (Geimer et al. 1973). Treatments that are more successful at increasing the weather resistance of particleboard are moisture resistant overlays (Barnes 1976), chemical modification (acetylation or PF resin treatment) of wood flakes, or polymer impregnation of finished boards (Schaudy and Proksch 1980, V'Icheva 1986). Overlays have also been used to increase the weather and moisture resistance of OSB. For example, Biblis (1990) found that "when OSB panels were overlaid with PF resin impregnated paper, primed, painted and edge sealed, thickness swelling and checking at the edges of boards was effectively controlled during 21 days of continuous exposure to water-spray." Weather-resistant OSB siding is currently manufactured in the United States. This product is made from a moisture resistant, zinc-borate-treated OSB, which is protected from the weather by a primed moisture resistant overlay. Top-coats of paint provide further protection against the deleterious effects of the weather.

Heat treatment and impregnation with PF resin can also increase the weather resistance of OSB (V'Icheva 1986, Del Menezzi et al. 2008), but these treatments are not being used commercially to produce weather resistant OSB. However, a range of treatments are being used commercially to increase the short-term resistance of OSB sheathing to the weather. Such treatments protect OSB from rain water during the construction of buildings. The need to protect OSB from exposure to rain was pointed out by Winterowd et al. (2004) in their patent on an edge sealant formulation for wood-based panels. They stated that "an uncovered subfloor can accumulate as much as two inches (50.8 mm) of water during a rainstorm." They pointed out that "the accumulated water will be left to absorb into the sub-floor panels for several days during the home-building process" (Winterowd et al. 2004). Such absorption of water causes thickness swelling of panels (Brochmann et al. 2004, Gu et al. 2005), which can be 0.5–3.8 mm greater at the edges of the panel (Winterowd et al. 2004). This differential edge swelling or edge-flare is a particular problem when boards are used as an underlay for floor coverings because the edges protrude above the surface of the floor and costly on-site sanding is sometimes required to level the panels before the application of floor coverings (Taylor and Wang 2010). In response to this problem, a number of different North American companies have developed enhanced OSB subfloor products that are less susceptible to edge swelling (Taylor and Wang 2010). These products usually contain higher loadings of phenol formaldehyde or isocyanate adhesive, and wax to increase their dimensional stability, and all of them employ an edge seal to reduce the ingress of moisture into the edges of boards (Kornicer and Palardy 2003, Winterowd et al. 2004). The ingress of moisture into boards and edge swelling can also be reduced by cutting drainage channels (notches) into the tongue of boards, reducing board density, particularly in surface layers, improving the alignment of strands or by applying an edge taper to boards.

7.6.3 FIBERBOARDS

Fiberboards vary greatly in their density, properties, and applications (Suchsland and Woodson 1991). Medium density fiber board is mainly used for the manufacture of furniture, cabinets, baseboards (trim), and toys, whereas hardboard (Masonite) has been used for a wide range of products, including house siding, that are exposed to weathering. The long-term satisfactory performance of hardboard siding, as pointed out by Suchsland and Woodson (1991), requires moisture control before and after application, and proper installation techniques. The processes used to improve the moisture resistance of hardboard siding are described in detail by Suchsland and Woodson (1991), and involve the addition of wax, oil, thermoplastic resin, and biocides to boards when they are being made. After the manufacture of boards they are exposed to dry heat (~150°C) and steamed to reduce water absorption and permanent thickness swelling when boards are exposed to moisture or the weather. The moisture resistance of hardboard siding is further improved by factory application of overlays (vinyl or resin impregnated overlays) and coatings. Finishing of hardboard siding involves automated application of a primer and a series of top coats (Suchsland and Woodson 1991).

The performance of commercially produced hardboard siding products exposed to the weather for 38 months was examined by Carll and TenWolde (2004). The thickness swelling of the different commercially produced boards varied significantly (Carll and TenWolde 2004). Fraipont (1973) also observed significant variation in the performance of 24 different commercial hardboards exposed to the weather for 1 year. For example, losses in internal bond strength of the hardboards varied from 15% to 85%, and losses in bending strength varied from 15% to 50% (Fraipont 1973). Boards that performed badly during exterior exposure cracked along their edges and showed excessive thickness swelling. The performance of hardboard exposed outdoors can be improved by applying paint to the "drip edge" (edge exposed to water) (Carll and TenWolde 2004). Alston (1988) examined the factors affecting drip edge swell of hardboard exposed outdoors. He showed that extending the steaming period during the manufacture of boards was most effective in reducing drip edge swell, and the wax content of boards also influenced swelling (Alston 1988). The impregnation of hardboard with asphalt, phenolic resin or mixtures of resins and monomers has been used to improve the weather resistance of hardboard (Virtala and Oksanen 1948, McNamara and Shaw 1972, Schaudy and Proksch 1980). For example, McNamara and Shaw (1972) found that impregnation with phenol-formaldehyde resin improved the weather resistance of hardboard, and Saotome et al. (2009) suggested that high-density fiberboard impregnated with phenol resin was "expected to withstand actual exterior use." However, Schaudy and Proksch (1980) found that hardboard impregnated with mixtures of resins and polymers (followed by radiation curing) only had moderate weather resistance, although the treatments reduced the moisture-induced swelling of boards.

7.6.4 WOOD CEMENT COMPOSITES

Wood cement composites are made from wood strands (wood-wool), particles or fibers, bonded together using an inorganic binder, usually Portland cement. Wood cement composites composed of wood-wool and Portland cement were first developed in the 1920s. Wood-wool cement boards have mainly been used as sound barriers and to provide thermal and acoustic insulation in buildings. In these applications the boards have no structural function and therefore strength changes due to weathering are of little consequence (Lempfer and Sattler 1989). Hence, there are few studies of the weathering of wood-wool cement boards. Nevertheless, wood-wool cement boards are reported to have greater resistance to weathering than resin-bonded wood composites (Pinion 1975). The weathering resistance of two other types of wood-cement composites, cement-bonded particleboard and cement bonded fiberboards is more important because both of these products are used in applications such as cladding and soffits where they are subjected to partial loads (Lempfer and Sattler 1989).

Early reports on the weather resistance of cement-bonded particleboard describe their weathering resistance as excellent (Deppe 1973), and "superior to that of light weight wood wool/cement boards" (Pampel and Schwarz 1979). Herzig et al. (1980) concluded that one of the most outstanding properties of cement bonded particleboard was their durability and weather resistance. In accord with this conclusion, Sergeev et al. (1995) found that cement bonded particleboard exposed outdoors in Russia for over 12 years retained a high average bending strength. Sekino and Suzuki (2002) commented that cement bonded particleboard exposed outdoors for 10 years in Morioka in Japan "showed excellent dimensional stability and no distinct reduction in bending strength, stiffness or internal bond strength." Dinwoodie and Paxton (1989) also found that the stiffness and strength of cement-bonded particleboards exposed to the weather for 10 years was largely unchanged because losses in strength due to weathering were offset by strength gains due to age-related hardening (carbonation) of cement. However, Dinwoodie and Paxton (1989) noted that there was a marked loss of internal bond strength after 5 years exposure in one of the boards that they tested, and the surface of all boards became rougher and colonized by algae and lichens after more than 7 years exposure. Deppe and Schmidt (1986) found that the bending strength of uncoated cement bonded particleboard was 40% to 50% of their initial values after 12 years outdoor weathering. However, boards coated with a polyurethane-alkyd resin retained almost all of their strength during weathering (Deppe and Schmidt 1986). Other types of coatings were much less successful at preventing losses in bending strength (Deppe and Schmidt 1986). Cabangon et al. (2003) found significant variation in the ability of different coatings to restrict the creep of high-density wood-wool cement boards, which are used as a substitute for cement bonded particleboards in the Philippines. Film forming coatings with a high moisture excluding efficiency were the most effective ones at reducing creep of boards, and also the losses of bending strength and stiffness that occurred when boards were exposed to cyclic variation in relative humidity (Cabangon et al. 2003). In accord with these findings, Ahmad Shakri and Rahim (1989) noted that film-forming finishes were effective in retarding moisture absorption into cement-bonded particleboard and provided better protection against weathering than penetrating type finishes.

Wood fiber cement boards are used for external siding of houses where they compete with solid wood, plastic and overlaid wood composite siding. Wood fiber cement boards are manufactured from sand (56%), Portland cement (30%), unbleached or bleached softwood kraft wood pulp fibers (7–8.5%), and aluminium trihydrate (3–4%). The manufactured boards consist of 3–6 thin fiber-cement lamellae in which the fibers are orientated in opposite directions. The fibers provide reinforcement to the composite, a role that asbestos fibers played in an earlier generation of composite panels.

Wood fiber cement composites exposed to the weather become stronger at first, but they lose their ductility (Akers et al. 1989, Akers and Studinka 1989). Then the strength properties of the composite remain largely unchanged for 3–5 years (Sharman 1983, West and Majumdar 1991). Serious losses in strength of wood fiber cement composites only occur in uncoated composites exposed to severe conditions (tropical climates and accelerated weathering, for example) (Cooke 2000, Guntekin and Sahin 2009).

Strength losses of wood fiber cement composites exposed to the weather occur due to moisture-induced swelling of wood cement fiber boards, the formation of micro-cracks that weakens interfacial bonding between fibers and the cement matrix, and reduction in the strength of fibers in the alkaline environment of cement (Akers et al. 1989, Lempfer and Sattler 1989, Tait and Akers 1989). Moisture movement in wood fiber cement composites is increased by carbonation (Sharman and Vautier 1986). This synergistic effect of carbonation on the ingress of moisture into wood fiber cement composites can be minimized by applying a coating to the surface of the composite (Cooke 2000). A well maintained coating can also reduce freeze–thaw damage to wood fiber cement composites (Cooke 2000). Strength losses of fibers in the alkaline environment of cement can be achieved by changing the composition of the cementatious binder (Lempfer and Sattler 1989).

The durability of wood fiber cement composite siding also depends on correct installation of panels and, most importantly, the application and maintenance of coatings. For example, siding should not be in direct contact with the ground or water, and the edges of panels should be sealed with a coating. Wood fiber cement sheeting when properly installed, finished and maintained has performed well after 18 years exposure, and Cooke (2000) suggested that it was not unreasonable to expect 50 years service "providing that it is selected, installed and maintained in a manner appropriate for its anticipated exposure."

7.6.5 GLULAM AND LAMINATED VENEER LUMBER

Glue-laminated timber (glulam) and laminated veneer lumber (LVL) are structural wood composites that are used outdoors as supporting elements in bridges and buildings. There have been several studies of the weathering of glulam and LVL, but there is little information on the weathering of other structural wood composites such as parallam, oriented strand lumber, laminated strand lumber and scrimber. The focus of the studies on the weathering of glulam and LVL is on the durability of the adhesives in the composites and the use of preservatives to restrict weathering.

Selbo examined the effects of adhesive type, preservative treatment and wood species on the resistance of glulam to long-term natural weathering (Selbo 1952, 1964, 1965, 1967, Selbo et al. 1965). He observed that glulam treated with oil-borne wood preservatives such as creosote, pentachlorphenol and copper napthenate showed little delamination or checking when exposed to the weather for prolonged periods of time (10 or 20 years) (Selbo 1964, Selbo et al. 1965). Creosote was the most effective oil-borne preservative at restricting end-checking of glulam beams exposed to the weather (Selbo 1952). However, in one study Selbo noted that glulam made from maple treated with oil-borne preservatives delaminated badly after 6 years outdoor exposure, whereas similarly treated and exposed glulam made from red oak, southern yellow pine and Douglas fir showed little delamination (Selbo 1952). He concluded that glulam treated with oil-borne preservatives is capable of giving good long-term performance when exposed outdoors (Selbo 1967). In contrast, glulam treated with water-borne preservatives, or glulam made from laminates treated with aqueous preservatives, and exposed to natural weathering for 10 or 20 years showed "considerable checking and joint separation" (Selbo 1964, Selbo et al. 1965). Meierhofer (1986, 1988), however, demonstrated that waterborne and oil preservatives had a positive effect on moisture change, linear expansion as well as the checking and delamination of glulam exposed outdoors. It is also possible to protect glulam to some degree against surface and end-checking by using coatings (Oviatt 1975). Nevertheless, due to the slow diffusion of moisture through the coating, swelling of the wood and thus cracks at the coated surface may still occur. To prevent this phenomenon from occurring the surface coating needs to be frequently renewed. It has been found that diffusion resistance increases linearly with increasing coating thickness. Thus thicker coatings give better long term protection (Sell 1983).

Chemical and impregnation modification has great potential to improve the weather resistance of glulam, and acetylation was recently used to improve the weather resistance and durability of a very large glulam bridge in the Netherlands (Tjeerdsma and Bongers 2009). Good weather resistance and durability of glulam can also be achieved by using certain naturally durable wood species. For example, Roos et al. (2009) mentioned that Alaska yellow-cedar is being used as lamstock to manufacture beams for exterior weather-exposed conditions.

The adhesive used in glulam exposed outdoors needs to be resistant to moisture, like those used in exterior grade plywood. Glulam bonded with resorcinol, phenol-resorcinol or melamine formaldehyde adhesives gives good long-term performance when exposed outdoors (Selbo 1964, Selbo et al. 1965). For example, Douglas-fir beams bonded with melamine formaldehyde adhesive and exposed to the weather for 20 years showed practically no delamination (Selbo 1965). However, melamine formaldehyde gluelines were more susceptible to degradation by sea water than those bonded with phenol-resorcinol (Selbo 1965). The fortification of a phenol resin with 16% or more of resorcinol enhanced the durability and lowered the delamination of CCA treated radiata pine glulam during a cyclic delamination test (Lisperguer and Becker 2005).

There is less information on the resistance of LVL to weathering than is available for glulam. LVL exposed to accelerated weathering was reported to be more stable than glulam "as indicated by its low incidence of delamination and high retention of glueline-shear and tensile strengths," according to Laufenberg (1982). Nevertheless, Bodig and Fyie (1986) observed that the rate of degradation of LVL during artificial weathering "appeared to be higher than that of solid Douglas fir lumber." A natural weathering trial in Japan of untreated and uncoated LVL made from six different wood species (Douglas fir, Siberian larch [*Larix sibirica* Ledeb], Japanese larch [*Larix leptolepis* Gord], western hemlock [*Tsuga heterophylla* Sarg], grand fir [*Abies grandis* Lindl], radiata pine, and meranti [*Shorea* spp.]) showed that on average the LVL retained 78% and 77% of their stiffness and compressive strength, respectively, after 6 years outdoor exposure (Hayashi et al. 2002). There was no significant delamination of the glulam, but slight longitudinal checks developed after a few months exposure as lathe checks opened up. The color of the LVL faded after a few months and "after 2 years algae grew on the surface of some of the LVL specimens, and they turned green." There was no softening of the surface of weathered LVL specimens except for a few grand fir specimens that were heavily decayed by brown rot fungi (Hayashi et al. 2002).

A range of textile cross-linking agents, including tetraoxane, gluteraldehyde, glyoxal and the *N*-methylol compound 1,3-dimethylol-4,5-dihydroxyethyleneurea (DMDHEU), have been tested to determine whether they can improve the weathering resistance of laminated veneer lumber manufactured from radiata pine (Yusuf et al. 1995). DMDHEU was the most effective compound at restricting photo-discoloration and checking of the treated composites.

7.7 WEATHERING OF WOOD PLASTIC COMPOSITES

Wood plastic composites are a class of materials composed of plastics such as high-density polyethylene or polypropylene and wood flour (as a filler). Wood plastic composites are commonly used outdoors for deck boards and deck components (posts, rails, and caps). Wood plastic composites have captured a significant share of the decking market in the United States, at the expense of preservative treated wood, because they require less maintenance than wood. The plastics in wood plastic composites are all susceptible to photodegradation in their native state, but they can be photostabilized by carbon black or other UV stabilizers (mainly UV absorbers and hindered amine light stabilizers) (Davis and Sims 1986). Wood plastic composites, however, are more susceptible to surface photooxidation than the parent plastic (Stark and Matuana 2004a,b). Furthermore, photodegradation is more severe when the proportion of wood in the composite is higher or when the composites are exposed to both light and water together rather than to light on its own (Matuana and Kamdem 2002, Fabiyi et al. 2008, La Mantia and Morreale 2008, Muasher and Sain 2006, Stark 2006, Stark and Matuana 2004a,b). For example, Matuana and Kamdem (2002) found that a wood-polyvinyl chloride composite degraded faster during accelerated weathering than PVC. Stark and Matuana (2004b) found that a wood plastic composite composed of high-density polyethylene and 50% wood flour was more susceptible to surface photooxidation than the parent polymer, and the same group also observed that wood plastic composites with "more wood component at the surface (as a result of planing the surface of the composite) experienced a larger percentage loss in flexural properties after 3000 hours of accelerated weathering than the unplaned control" (Stark et al. 2004a). In contrast to these observations that wood plastic composites are more susceptible to weathering than the parent polymer, Lundin et al. (2002) found that a wood/PVC composite showed no loss in tensile properties when exposed to 400 h of accelerated weathering, whereas similarly exposed samples of PVC lost almost 50% of their strength.

Weathering causes the color of wood plastic composites to fade. Such losses of color have been attributed to degradation of coloring matter by free radicals generated from the photooxidation of lignin (Muasher and Sain 2006, Fabiyi et al. 2009). Degradation of lignin according to Andrady et al. (2011) also "yields low molecular weight materials that support the growth of fungi and therefore helps to initiate biodegradation of wood plastic composites."

Wood plastic composites are also susceptible to surface degradation by moisture. Moisture-induced swelling of photodegraded wood particles at the surface of wood composites causes minute checks to develop at the interface between wood particles and plastic (Adhikary et al. 2010, Andrady et al. 2011). The development of these small checks and also loss of wood particles from exposed surfaces increases the surface roughness of wood plastic composites exposed to the weather (Mehta et al. 2006). Wood plastic composites also lose strength when they are exposed to accelerated weathering (Stark et al. 2004a, Mehta et al. 2006, Stark 2006), or moisture and freezing and thawing cycles (Adhikary et al. 2010). All these undesirable effects of weathering on the properties of wood plastic composites have generated a great deal of interest in methods of photostabilizing wood plastic composites and a variety of approaches have been tested. These approaches include: (1) the use of additives to restrict photooxidation; (2) changing the components (wood or plastic) used to make wood plastic composites; (3) the use of coatings or a capping layer of polymer on the surface of the wood plastic composite. The use of additives such as UV absorbers, hindered amine light stabilizers and pigments is an obvious approach to protecting wood plastic composites from the undesirable effects of weathering since such additives can restrict the photooxidation of both plastics and wood. The addition of a benzotriazole UV absorber and a zinc-ferrite pigment reduced undesirable changes in crystallininty, stiffness and cracking of a high-density wood–flour–polyethylene composite exposed to accelerated weathering (Stark and Matuana 2003, Stark et al. 2004b). These studies concluded that a UV absorber and pigment were more effective at restricting the photodegradation of the wood-polymer composites than hindered amine light stabilizers (Stark and Matuana 2003, Stark et al. 2004b). A later study by Stark and Matuana (2006) concluded that UV absorbers and pigments provide protection to wood plastic composites from the deleterious effects of weathering with the amount of protection afforded to the composite dependent upon concentration of photostabilizer and exposure variables. There are also studies that show that wood plastic composites can be photostabilized using hindered amine light stabilizers (HALS). For example, Muasher and Sain (2006) compared the ability of different photostabilizers to restrict changes in color of wood plastic composites exposed to natural weathering for 2000 h. They concluded that high-molecular weight diester HALS were the most effective stabilizer at restricting the yellowing and long-term fading of wood plastic composites. The positive effect of the diester HALS on color changes of the composite was enhanced (synergistically) by combining it with a benzotriazole UV absorber (Muasher and Sain 2006). The positive effect of HALS at restricting color changes of wood composites exposed to weathering was recently confirmed by Bouza et al. (2011). They observed that HALS restricted the surface cracking of a wood-flour polypropylene composite exposed to accelerated weathering, and concluded that the combination of HALS and UV absorbers was effective at preventing the UV-induced degradation of the composite.

An alternative route to protecting polymers from photodegradation is to use pigments rather than organic UV stabilizers (Davis and Sims 1986). Hence, there has been interest in using pigments to photostabilize wood plastic composites. For example, an early study by Matuana et al. (2001) found that the addition of a photoactive pigment (rutile titanium dioxide) to a polyvinyl chloride wood fiber composite enhanced the UV stability of the composite. More recently Du et al. (2010) compared the effect of different colored pigments on surface degradation of a wood plastic composite exposed to accelerated weathering. They found that the addition of pigment to the composite "results in less weather-related damage," and concluded that carbon black was the most effective inorganic pigment at restricting the color changes when composites were exposed to accelerated weathering. Wang et al. (2011) found that inorganic pigments were more effective at restricting photodegradation of wood plastic composites than organic dyes.

Wood plastic composites are more susceptible to photodegradation than the parent plastic because lignin in wood is a photosensitizer. Accordingly, modification of wood to make it less susceptible to photodegradation is another route to improving the resistance of wood plastic composites to weathering. There have been several studies that have taken this approach. For example, Kiguchi et al. (2000) found that grafting of the epoxy functionalized benzophenone-type UV absorber (2-hydroxy-4-(2,3-epoxypropoxy)-benzophenone) to wood fibers used in wood plastic composites improved the color stability of the composite during accelerated weathering. The color stability of wood plastic composites exposed to accelerated weathering can also be improved by pretreating wood flour used to make the composites with a pigmented oil-based stain, but washing the flour with water to remove photosensitive extractives or pretreating the wood flour with water-based dye were ineffective at "improving the color stability of the composite during weathering" (Stark and Mueller 2008). Fabiyi et al. (2009) confirmed that removal of extractives has no positive effect on the color stability of wood plastic composites exposed to accelerated and natural weathering, but they showed that removal of lignin from the wood used to make the composite has a positive effect on the color stability of the composite. In contrast, Beg and Pickering (2008) found that there was no difference in the losses of tensile strength and stiffness of wood fiber-reinforced polypropylene composites containing either bleached or unbleached kraft wood fiber when the composites were exposed to accelerated weathering. However, the type of wood that is added to wood plastic composites does influence the color stability of the composites when they are exposed to accelerated weathering. For example, Fabiyi and McDonald (2010) compared the weathering resistance of wood plastic composites manufactured from high-density polyethylene (HDPE) and a range of different North American softwood and hardwood species. They found that composites containing hybrid poplar (*Populus* spp.) or ponderosa pine (*Pinus ponderosa* Dougl.) wood flour had better color stability during accelerated weathering than composites containing wood flour obtained from robinia (black locust), Douglas fir, or white oak (*Quercus* spp.) (Fabiyi and Macdonald 2010).

The type of plastic that is used to make wood plastic composites also influences their susceptibility to weathering. Fabiyi et al. (2008) compared the susceptibility of wood plastic composites made of high-density polyethylene or polypropylene to surface degradation caused by natural or artificial weathering. They observed that there was less fading and loss of wood from composites made from high-density polyethylene compared to those made from polypropylene. Wood plastic composites made from recycled plastic are also reported to weather differently to composites made from fresh plastic. For example, Adhikary et al. (2010) examined the effects of wetting and freeze–thaw cycling on the physical and mechanical properties of wood plastic composites made from fresh or recycled high-density polyethylene or polypropylene. They found that composites made from recycled plastics lost crystallinity as a result of exposure to wetting and freeze–thaw cycling, whereas the crystallinity of the composites made from fresh plastic increased during exposure (Adhikary et al. 2010).

Most recently, there has been interest in using light-stabilized clear coatings or a clear layer of plastic to restrict the surface photooxidation of wood plastic composites (Waldron and Moyer 2009, Matuana et al. 2011, Pattamasattayasonthi et al. 2011). Matuana et al. (2011) found that coextruding a clear layer of HDPE over the surface of a wood plastic composite made from HDPE "significantly decreased discoloration during the weathering process," but they noted some failure of the plastic coating during accelerated weathering. An acrylic coating containing cerium dioxide as a UV absorber prevented losses in surface hydrophobicity and mechanical properties of a polyvinylchloride/wood flour composite exposed to UV light and moisture (Pattamasattayasonthi et al. 2011). There was no mention of failure of the coating on the composite and, furthermore, there is little discussion in the literature of the maintenance requirements of coatings on wood plastic composites, which is important since wood plastic composites are promoted as a maintenance free alternative to treated or naturally durable wood.

7.8 FUTURE CONSIDERATIONS

Society would like to have materials and products that have small environmental impacts when they are manufactured and used. These environmental impacts can be quantified using life cycle analysis. Life cycle analyses have shown that products (siding, windows, etc.) made from a renewable material such as wood can have greater environmental impacts than products made from nonrenewable materials (fiber-glass, plastics, etc.), partly because of increased maintenance caused by weather-induced failure of coatings on wood. Weathering-induced damage is responsible for the large-scale replacement of treated wooden decking in North America by wood plastic composite decking. In response to the success of wood plastic composites and also other hybrid materials with good resistance to weathering (wood fiber cement composites, for example), chemical and coating companies are showing greater interest in developing treatments to improve the weather resistance of wood products. Such interest is likely to continue in future, but with greater emphasis on treatments (coatings, chemicals, and additives) that are derived from biomass. The development of such coatings, chemicals, and additives will be increasingly important in the next decade, as will be the evaluation of the weathering resistance of products that employ such protective systems. There will also be pressure in future to use plastics in wood plastic composites that are derived from renewable resources and to incorporate a greater percentage of wood flour or fiber in the composites. Similar pressures are likely to see the development of wood fiber cement composites that employ cements that have greater capacity to sequester CO_2 or generate less CO_2 when they are made. The evaluation of the weathering resistance of this new generation of hybrid composite materials will also become increasingly important in future.

The importance of maintaining fundamental research into the weathering and photoprotection of wood cannot be overstated. There are many areas where our knowledge of the weathering of wood is very superficial. For example, the role of the specialized, deeply pigmented, organisms that thrive at weathered wood surfaces, and the mechanisms responsible for the photodegradation of hemicelluloses and cellulose in wood are not fully understood. Fortunately, the research community now has the tools to probe biological organisms and chemical reactions in exquisite detail, and, in theory, great progress could be made in understanding these phenomena and many of the other complex processes involved in the weathering of wood. Obtaining a deeper understanding of the weathering of wood will be a challenge, but scientists working on such fundamental research are more likely to stumble upon very effective treatments to block the weathering of wood, than scientist involved in directed research using known approaches (as has been demonstrated numerous times in many other fields). Hence, it is important for fundamental research on the weathering of wood continues, and for support to be made available to attract young scientists to work on understanding this important natural phenomenon.

REFERENCES

Aalto-Korte, K., Lauerma, A., and Aalanko, K. 2005. Occupational allergic contact dermatitis from lichens in present-day Finland. *Contact Dermatitis.* 52(1): 36–38.

Adhikary, K.B., Pang, S., and Staiger, M.P. 2010. Effects of the accelerated freeze-thaw cycling on physical and mechanical properties of wood flour-recycled thermoplastic composites. *Polym. Comp.* 31(2): 185–194.

Ahmad Shakri, M.S. and Rahim, S. 1989. Finishing properties of coated cement-bonded particleboard. *J. Tropical Forest Science.* 2(2): 122–128.

Akers, S.A.S., Crawford, D., Schultes, K., and Gerneka, D.A. 1989. Micromechanical studies of fresh and weathered fibre cement composites. Part 1: Dry testing. *Internationa J. Cement Composites and Lightweight Concrete.* 11(2): 117–124.

Akers, S.A.S. and Studinka, J.B. 1989. Aging behavior of cellulose fibre cement composites in natural weathering and accelerated tests. *International J. Cement Composites and Lightweight Concrete.* 11(2): 93–97.

Ala-Viikari, J. and Mayes, D. 2009. New generation ThermoWood® - How to take ThermoWood® to the next level. In: Englund, F., Segerholm, B.K., Hill, C.A.S., and Militz, H. (Eds.). Proceedings of 4th European Conference on Wood Modification, Stockholm, Sweden, pp. 23–29.

Alexopoulos, J. 1992. Accelerated aging and outdoor weathering of aspen waferboard. *Forest Products J.* 42(2): 15–22.

Allen, N.S., Edge, M., Ortega, A., Liauw, C.M., Stratton, J., and McIntyre, R.B. 2002. Behaviour of nanoparticle (ultrafine) titanium dioxide pigments and stabilisers on the photooxidative stability of water based acrylic and isocyanate based acrylic coatings. *Polymer Degradation & Stability.* 78(3): 467–478.

Aloui, F., Ahajji, A., Irmouli, Y., George, B., Charrier, B., and Merlin, A. 2007. Inorganic UV absorbers for the photostabilisation of wood-clearcoating systems: comparison with organic UV absorbers. *Applied Surface Science.* 253(8): 3737–3745.

Alston, M.J. 1988. Weathering of hardboard—Some drip edge swell effects. *Appita J.* 41(2): 124–128.

Anderson, E.L., Pawlak, Z., Owen, N.L., and Feist, W.C. 1991a. Infrared studies of wood weathering. Part I: Softwoods. *Applied Spectroscopy.* 45(4): 641–647.

Anderson, E.L., Pawlak, Z., Owen, N.L., and Feist, W.C. 1991b. Infrared studies of wood weathering. Part II: Hardwoods. *Applied Spectroscopy.* 45(4): 648–652.

Andrady A.L., Hamid H., and Torikai A. 2011. Effects of solar UV and climate change on materials. *Photochemical & Photobiological. Sciences.* 10(2): 292–300.

Anon. 1850. *The Annals of Horticulture and Year Book of Information on Practical Gardening for 1850.* Charles Cox. London.

Anon. 1887. Effect of the electric light upon books. *English Mechanic and World of Science.* 1(165): 483.

Anon. 1964. Manufacture and general characteristics of flat plywood. U.S. For. Serv. Res. Note. FPL-064, USDA For. Serv., For. Prod. Lab., Madison, Wisconsin, 18 pp.

Anon. 1976. New wood with an old look. *Wood & Wood Products.* 81: 33–35.

Arndt, U. and Willeitner, H. 1969. On the resistance behaviour of wood in natural weathering. *Holz als Roh- und Werkstoff.* 27(5):179–188.

Arnold, M., Sell, J., and Feist, W.C. 1991. Wood weathering in fluorescent ultraviolet and xenon arc chambers. *Forest Products J.* 41(2): 40–44.

Ashton, H.E. 1967. Clear finishes for exterior wood field exposure tests. *J. Paint Technology.* 39(507): 212–224.

Ayadi, N., Lejeune, F., Charrier, F., Charrier, B., and Merlin, A. 2003. Colour stability of heat-treated wood during artificial weathering. *Holz als Roh- und Werkstoff.* 61(3): 221–226.

Bailey, D. and Vogl, O. 1976. Polymeric ultraviolet absorbers. *J. Macromolecular Sci. Reviews.* 14(2): 267–293.

Baker, R.T. 1900. On two new species of Casuarina. *Proc. Linn. Soc. of N.S.W.* 24: 605–611.

Bamber, R.K. and Summerville. R. 1981. Microscopic studies of the weathering of radiata pine sapwood. *J. of the Institute of Wood Science.* 9(2): 84–88.

Bardage, S.L. and Bjurman, J. 1998. Isolation of an Aureobasidium pullulans polysaccharide that promotes adhesion of blastospores to water-borne paints. *Canadian J. Microbiology.* 44(10): 954–958.

Barnes, H.M. 1976. Overlays for Southern pine substrates. *Forest Products J.* 26(6): 36–42.

Barnes, H.M. and Lyon, D.E. 1978a. Effect of weathering on the dimensional properties of particleboard decking. *Wood & Fiber.* 10(3): 175–185.

Barnes, H.M. and Lyon, D.E. 1978b. Effect of aging on the mechanical properties of particleboard decking. *Wood & Fiber.* 10(3): 164–174.

Barreto, C.C.K. and Pastore, T.C.M. 2009. Resistance to artificial weathering of four tropical woods: The effect of the extractives. *Ciência Florestal.* 19(1/2): 23–30.

Beckers, E.P.J., de Meijer, M., Militz, H., and Stevens, M. 1998. Performance of finishes on wood that is chemically modified by acetylation. *J. Coatings Technology.* 70(878): 59–67.

Beech, J.C., Hudson, R.W., Laidlaw, R.A., and Pinion, L.C. 1974. Studies on the performance of particle board in exterior situations and the development of laboratory predictive tests. *Current Papers, Building Research Establishment.* 77(74): 16.

Beg, M.D.H. and Pickering, K.L. 2008. Accelerated weathering of unbleached and bleached Kraft wood fibre reinforced polypropylene composites. *Polymer Degradation & Stability.* 93(10): 1939–1946.

Béldi, F. and Balint, J. 1982. On the natural and artificial weathering of particleboards. *Holztechnologie.* 23(2): 107–110.

Berzelius, J.J. 1827. *Lehrbuch der Chemie.* Arnold, Dresden, Germany.

Bhat, I., Khalil, H.P.S.A., Awang, K.B., Bakare, I.O., and Issam, A.M. 2010. Effect of weathering on physical, mechanical and morphological properties of chemically modified wood materials. *Materials & Design.* 31(9): 4363–4368.

Biblis, E.J. 1990. Performance of southern OSB overlaid with resin-impregnated paper. *Forest Products J.* 40(4): 55–62.

Biblis, E.J. 2000. Effect of weathering on surface quality and structural properties of six species of untreated commercial plywood siding after 6 years of exposure in Alabama. *Forest Products J.* 50(5): 47–50.

Black, J.M., Lutz, J.F., and Mraz, E.A. 1976. Performance of softwood plywood during 10 years' exposure to weather. *Forest Products J.* 26(4): 24–27.

Black, J.M. and Mraz, E.A. 1974. Inorganic surface treatments for weather-resistant natural finishes. U.S Forest Service Res. Pap. FPL 232, U.S. Department of Agriculture, Forest Service, Forest Products Laboratory, Madison, WI, 40 pp.

Blackburn, S.R., Meldrum, B.J., and Clayton. J. 1991. The use of fine particle titanium dioxide for UV protection in wood finishes. *Faerg och Lack Scandinavia.* 37(9): 192–196.

Bodig, J. and Fyie, J. 1986. Performance requirements for exterior laminated veneer lumber. *Forest Products J.* 36(2): 49–54.

Borgin, K. 1965. The effect of water repellents on the dimensional stability of wood. *Norsk Skogindustri.* 15(11): 507–521.

Borgin, K. 1970. The use of scanning electron microscope for the study of weathered wood. *Journal of Microscopy.* 92(1):47–55.

Borgin, K. 1971. The mechanism of the breakdown of the structure of wood due to environmental factors. *J. of the Institute of Wood Science.* 5(4): 26–30.

Bos, A. 1972. The UV spectra of cellulose and some model compounds. *J. Applied Polymer Science.* 16(10): 2567–2576.

Bouza, R., Abad, M.J., Barral, L., Lasagabaster, A., and Pardo, S.G. 2011. Efficacy of hindered amines in woodflour-polypropylene composites compatibilized with vinyltrimethoxysilane after accelerated weathering and moisture absorption. *J. Applied Polymer Science.* 120(4): 2017–2026.

Boxall, J. 1977. Painting weathered timber. *Buildg. Res. Est. (UK) Information Sheet.* 20(77): 1–2.

Brede, O. and Goettinger, H.A. 1998. Transformation of sterically hindered amines (HALS) to nitroxyl radicals: What are the actual stabilizers? *Angewandte Makromol. Chem.* 261/262(1): 45–54.

Brelid, P.L. and Westin, M. 2007. Acetylated wood-Results from long-term field tests. In: Hill, C.A.S., Jones, D., Militz, H., and Ormondroyd, G.A. (Eds.). Proceedings of 3rd European Conference on Wood Modification, Cardif, UK, pp. 71–78.

Brochmann J., Edwardson C., and Shmulsky, R. 2004. Influence of resin type and flake thickness on properties of OSB. *Forest Products J.* 54(3): 51–55.

Browne, F.L. 1927. A principle for testing the durability of paints as protective coatings for wood. *Industrial & Engineering Chemistry.* 19(9): 982–985.

Browne, F.L. and Simonson, H.C. 1957. The penetration of light into wood. *Forest Products J.* 7(10): 308–314.

Bugos, R., Sutherland, J., and Adler, J. 1988. Phenolic compound utilization by the soft rot fungus *Lecythophora hoffmannii. Applied & Environmental Microbiology.* 54(7): 1882–1885.

Cabangon, R.J., Cunningham, R.B., and Evans, P.D. 2003. Effects of surface coatings on the mechanical properties of wood-wool cement board. *Surface Coatings Australia.* 40(11): 20–25.

Carll, C.G. and Feist, W.C. 1987. Weathering and decay of finished aspen waferboard. *Forest Products J.* 37(4): 27–30.

Carll, C.G. and Feist, W.C. 1989. Long-term weathering of finished aspen waferboard. *Forest Products J.* 39(10): 25–30.

Carll, C. and TenWolde, A. 2004. Durability of hardboard lap siding—determination of performance criteria. *USDA Forest Service Report*, FPL-RP-622, iii + 29 pp.

Cassens, D.L. and Feist, W.C. 1991. Exterior wood in the south: Selection, application and finishes. Gen. Tech. Rep. FPL-GTR-69. Madison WI: USDA, Forest Service, Forest Products Laboratory, 60 pp.

Chang, S.T. and Chang, H.T. 2001a. Inhibition of the photodiscoloration of wood by butyrylation. *Holzforschung*. 55(3): 255–259.

Chang, S.T. and Chang, H.T. 2001b. Comparison of the photostability of esterified wood. *Polymer Degradation & Stability*. 71(2): 261–266.

Chang, S.T. and Chang, H.T. 2006. Modification of wood with isopropyl glycidyl ether and its effects on decay resistance and light stability. *Bioresource Technology*. 97(1): 1265–1271.

Chang, S.T., Hon, N.S., and Feist, W.C. 1982. Photodegradation and photoprotection of wood surfaces. *Wood & Fiber*. 14(2): 104–117.

Chang, S.T., Wang, S.Y., and Su, Y.C. 1998. Retention of red color in Taiwania *(Taiwania cryptomeriodes* Hay.) heartwood. *Holzforschung*. 52(1): 13–17.

Chen, T.Y. 1978. *Studies on the Weathering of Wood Based Materials and its Control. 1.* Technical Bulletin, Department of Forestry, Chung Hsing University, Taiwan, 160, 472 pp.

Christy, A.G., Senden, T.J., and Evans, P.D. 2005. Automated measurement of checks at wood surfaces. *Measurement*. 37(2): 109–118.

Clad, W. and Schmidt-Hellerau, C. 1965. The durability of urea and phenolic resins in natural weathering of particle boards. *Holzzentralblatt*. 91(44/45Suppl.): 349–352.

Clad, W. and Schmidt-Hellerau, C. 1976. Tests on wood-based materials bonded with melamine resin. 3. The effect of the wood species used on the durability of particle board. *Holz-Zentralblatt*. 102(40): 543.

Cooke, A.M. 2000. Durability of autoclaved cellulose fiber cement composites. In *Proceedings 7th Inorganic-Bonded Wood and Fiber Conference*, pp. 37. available at http://www.fibrecementconsulting.com/publica-tions/990925.DurabilityPaper.pdf (accessed 8th September 2011).

Cooke W.B. and Matsuura, G. 1963. Physiological studies in the black yeasts. *Mycopathologia*. 21(3–4): 225–271.

Cornfield, J.A., Hale, M., and Fellis, G. 1994. A comparison of analytical and visual techniques used for assess-ment of weathering properties of chromium and copper azole treated timber. *International Research Group on Wood Preservation Doc*. IRG/WP/94-20023.

Coupe, C. and Watson, R.W. 1967. Fundamental aspects of weathering. *Record 17th Annual Convention*. British Wood Preservers Association. 2:37–49.

Cristea, M.V., Riedl, B., and Blanchet, P. 2010. Enhancing the performance of exterior waterborne coatings for wood by inorganic nanosized UV absorbers. *Progress in Organic Coatings*. 69(4): 432–441.

Cronin, L.A., Tiffney, W.N. Jr and Eveleigh, D.E. 2000. The graying of cedar shingles in a maritime climate—A fungal basis? *J. Industrial Microbiology & Biotechnology*. 24(5): 319–322.

Cui, W.N., Kamdem, D.P., and Rypstra, T. 2004. Diffuse reflectance infrared Fourier transform spectroscopy (DRIFT) and color changes of artificial weathered wood. *Wood & Fiber Science*. 36(3): 291–301.

Cui, F. and Zahora A. 2000. Effect of a water repellent additive on the performance of ACQ treated decks. *International Research Group on Wood Preservation Doc*. IRG/WP/00-40168.

Curtin, L.P., Kline, B.L., and Thordarson, W. 1927. Experiments in wood preservation. V-Weathering tests on treated wood. *Industrial & Engineering Chemistry*. 19(12): 1340–1343.

Damodaran, K. and Padmanabhan, S. 1992. Studies on weathering behaviour of Gurjan and Hollong plywood. *Wood News*. 2(2/3): 38–42.

Damodaran, K. and Naidu, M.V. 1995. Weathering behaviour of coated plywood. *Wood News*. 5(1): 5–9.

Davis, A. and Sims, D. 1983. *Weathering of Polymers*. Applied Science Publishers, London, 294pp.

de Lange, P.J., de Kreek, A.K., van Linden, A., and Coenjaarts, N.J. 1992. Weathering of wood and protection by chromium studied by XPS. *Surface and Interface Analysis*. 19(1–12): 397–402.

De Souza, A. and Gaylarde C.C. 2002. Biodeterioration of varnished wood with and without biocide: implica-tions for standard test methods. *International Biodeterioration & Biodegradation*. 49(1): 21–25.

Del Menezzi, C.H.S., de Souza, R.Q., Thompson, R.M., Teixeira, D.E., Okino, E.Y.A., and da Costa, A.F. 2008. Properties after weathering and decay resistance of a thermally modified wood structural board. *International Biodeterioration & Biodegradation*. 62(4): 448–454.

Deppe, H.J. 1973. The manufacture and use of particle board with cement binder. *Holz-Zentralblatt*. 99(49/50): 737–739.

Deppe, H.J. and Schmidt, K. 1986. Testing the weathering resistance of wood cement. *Holz als Roh- und Werkstoff.* 44(10): 395–397.

Derbyshire, H. and Miller, E.R. 1981. The photodegradation of wood during solar irradiation. Part 1. Effects on the structural integrity of thin wood strips. *Holz als Roh- und Werkstoff.* 39(8): 341–350.

Derbyshire, H., Miller, E.R., and Turkulin, H. 1995. Investigations into the photodegradation of wood using microtensile testing. Part 1. The application of microtensile testing to measurement of photodegradation rates. *Holz als Roh- und Werkstoff.* 53(5): 339–345.

Derbyshire, H., Miller, E.R., and Turkulin, H. 1996. Investigations into the photodegradation of wood using microtensile testing. Part 2. An investigation of the changes in tensile strength of different softwood species during natural weathering. *Holz als Roh- und Werkstoff.* 54(1): 1–6.

Derbyshire, H., Miller E.R., and Turkulin, H. 1997. Investigations into the photodegradation of wood using microtensile testing. Part 3: The influence of temperature on photodegradation rates. *Holz als Roh- und Werkstoff.* 55(5): 287–291.

Desai, R.L. 1967. Coating adhesion to weathered wood. *Canadian Dept. Fisheries and Forestry. Bi-monthly Res. Notes.* 23: 36–37.

Desai, R.L. and Shields, J.K. 1971. Effect of near ultraviolet light on fungi colonizing hardwood chips. *International Biodeterioration Bulletin.* 7(1): 11–13.

Desai, R.L. and Juneja, S.C. 1972. Weather-O-meter studies on wood-plastic composites. *Forest Products J.* 22(9): 100–103.

Dickinson, D.J. 1971. Disfigurement of decorative timbers by blue stain fungi. *Record of the Annual Convention of the British Wood Preserving Association,* pp. 151–169.

Dinwoodie, J.M. and Paxton, B.H. 1989. A technical assessment of cement-bonded particleboard. In: Moslemi, A.A. (Ed.) *Fiber Particleboards Bonded with Inorganic Binders.* Forest Products, Society, Madison, pp. 115–22.

Dirckx, O., Triboulot-Trouy, M.C., Merlin, M., and Deglise, X. 1992. Modification de la couleur du bois d' *Abies grandis* exposé à la lumière solaire. *Annals of Forest Science.* 49(5): 425–447.

Donath, S., Militz, H., and Mai, C. 2007. Weathering of silane treated wood. *Holz als Roh-und Werkstoff.* 65(1): 35–42.

Du, H., Wang, W., Wang, Q., Zhang, Z., Sui, S., and Zhang, Y. 2010. Effects of pigments on the UV degradation of wood-flour/HDPE composites. *J. Applied Polymer Science.* 118(2): 1068–1076.

Dubey, M.K., Pang, S., and Walker, J. 2010. Color and dimensional stability of oil heat-treated radiata pinewood after accelerated UV weathering. *Forest Products J.* 60(5): 453–459.

Duncan, C. 1963. Role of microorganisms in the weathering of wood and degradation of exterior finishes. *Official Digest Federation Societies Paints Technology.* 35(465): 1003–1012.

Dunningham, E.A., Plackett, D.V., and Singh, A.P. 1992. Weathering of chemically modified wood. Natural weathering of acetylated radiata pine: Preliminary results. *Holz als Roh- und Werkstoff.* 50(11): 429–432.

Elmendorf, A. and Vaughan, T.W. 1960. Means for reducing the checking of Douglas fir plywood. *Forest Products J.* 10(1): 45–47.

Evans, P.D. 1989a. Effect of angle of exposure on the weathering of wood surfaces. *Polymer Degradation & Stability.* 24(1): 81–87.

Evans, P.D. 1989b. Structural changes in *Pinus radiata* during weathering. *J. of the Institute of Wood Science.* 11(5): 172–181.

Evans, P.D. 1996. The influence of season and angle of exposure on the weathering of wood. *Holz als Roh- und Werkstoff.* 54(3): 200.

Evans, P.D. 1998. Weather resistance of wood esterified with dicarboxylic acid anhydrides. *Holz als Roh- und Werkstoff.* 56(5): 294.

Evans, P.D. and Banks, W.B. 1988. Degradation of wood surfaces by water: Changes in mechanical properties of thin wood strips. *Holz als Roh- und Werkstoff.* 46(11): 427–435.

Evans, P.D. and Banks, W.B. 1990. Degradation of wood surfaces by water: Weight losses and changes in ultra-structural and chemical composition. *Holz als Roh- und Werkstoff.* 48(4): 159–163.

Evans, P.D. and Chowdhury, M.J.A. 2010. Photoprotection of wood using polyester-type UV-Absorbers derived from the reaction of 2-hydroxy-4(2,3-epoxypropoxy)-benzophenone with dicarboxylic acid anhydrides. *J. Wood Chemistry & Technology.* 30(2): 1–19.

Evans, P.D., Donnelly C.F., and Cunningham R.B. 2003. Checking of CCA-treated radiata pine decking timber exposed to natural weathering. *Forest Products J.* 53(4): 66–71.

Evans, P.D., Michell, A.J., and Schmalzl, K.J. 1992. Studies of the degradation and protection of wood surfaces. *Wood Science & Technology.* 26(2): 151–163.

Evans, P.D, Owen, N.L., Schmid, S., and Webster, R.D. 2002. Weathering and photostability of benzoylated wood. *Polymer Degradation & Stability.* 76(2): 291–303.

Evans, P.D., Pirie, J.D.R., Cunningham, R.B., Donnelly, C.F., and Schmalzl, K.J. 1994. A quantitative weathering study of wood surfaces modified by chromium VI and Iron III compounds: Part 2. Image analysis of cell wall pit micro-checking. *Holzforschung.* 48(4): 331–336.

Evans, P.D. and Schmalzl, K.J. 1989. A quantitative weathering study of wood surfaces modified by chromium VI and iron III compounds. Part 1. Loss in zero-span tensile strength and weight of thin wood veneers. *Holzforschung.* 43(5): 289–292.

Evans P.D., Schmalzl K.J., and Michell A.F. 1993. Rapid loss of lignin at wood surfaces during natural weathering. In: Kennedy, J.F., Phillips, G.O., and Williams, P.A. (Eds.), *Cellulosics: Pulp, Fibre and Environmental Aspects*, Ellis Horwood Ltd, Chichester, UK, pp. 335–340.

Evans, P.D., Thay, P.D., and Schmalzl, K.J. 1996. Degradation of wood surfaces during natural weathering. Effects on lignin and cellulose and on the adhesion of acrylic latex primers. *Wood Science & Technology.* 30(6): 411–422.

Evans, P.D., Urban, K., and Chowdhury, M.J.A. 2008. Surface checking of wood is increased by photodegradation caused by ultraviolet and visible light. *Wood Science & Technology.* 42(3): 251–265.

Evans, P.D., Wallis, A.F.A., and Owen, N.L. 2000a. Weathering of chemically modified wood surfaces— Natural weathering of Scots pine acetylated to different weight gains. *Wood Science & Technology.* 34(2): 151–165.

Evans, P.D., Wingate-Hill R., and Barry, S.C. 2000b. The effects of different kerfing and center-boring treatments on the checking of ACQ treated pine posts exposed to the weather. *Forest Products J.* 50(2): 59–64.

Evans, P.D., Wingate-Hill, R., and Cunningham, R.B. 2009. Wax and oil emulsion additives: How effective are they at improving the performance of preservative-treated wood. *Forest Products J.* 59(1/2): 66–70.

Fabiyi, J.S. and McDonald, A.G. 2010. Effect of wood species on property and weathering performance of wood plastic composites. *Composites Part A-Applied Science & Manufacturing.* 41(10): 1434–1440.

Fabiyi, J.S., McDonald, A.G., and McIlroy, D. 2009. Wood modification effects on weathering of HDPE-based wood plastic composites. *J. Polymers & the Environment.* 17(1): 34–48.

Fabiyi, J.S., McDonald, A.G., Wolcott, M.P., and Griffiths, P.R. 2008. Wood plastic composites weathering: Visual appearance and chemical changes. *Polymer Degradation & Stability.* 93(8): 1405–1414.

Fairgrieve, S.P. and MacCallum, J.R. 1984. Hindered amine light stabilizers: A proposed photo-stabilization mechanism. *Polymer Degradation & Stability.* 8(2): 107–121.

Feist, W.C. 1979. Protection of wood surfaces with chromium trioxide. USDA Forest Service. *Forest Service Res. Pap.* FPL 339, pp. 11.

Feist, W.C. 1982. Weathering of wood in structural uses. In: Meyer, R.W. and Kellogg, R.M. (Eds.), *Structural Use of Wood in Adverse Environments*. Van Nostrand Reinhold Co., New York.

Feist, W.C. 1988a. Weathering of wood and its control by water-repellent preservatives, in M.P. Hamel (ed.), *Wood Protection Techniques and the Use of Treated Wood in Construction*, Forest Products Research Society, Madison, WI, pp. 82–88.

Feist, W.C. 1988b. Weathering performance of finished southern pine plywood siding. *Forest Products J.* 38(3): 22–28.

Feist, W.C. 1990a. Outdoor wood weathering and protection. In Rowell, R. M. and Barbour, J. R. (Ed), *Archaeological Wood: Properties, Chemistry, and Preservation*. Advances in Chemistry Series 225. *Proceedings of 196th Meeting*, American Chemical Society; 1988 September 25–28, Los Angeles. American Chemical Society, Washington, DC, Ch. 11. pp. 263–298.

Feist, W.C. 1990b. Weathering performance of painted wood pretreated with water-repellent preservatives. *Forest Products J.* 40(7/8): 21–26.

Feist, W.C. and Hon, D.N.S. 1984. Chemistry of weathering and protection. In R. M. Rowell, (Ed.), *The Chemistry of Solid Wood*. Advances in Chemistry Series 207. American Chemical Society, Washington, DC, Ch. 11. pp. 401–451.

Feist, W.C. and Mraz, E.A. 1980. Performance of mildewcides in a semitransparent stain wood finish. *Forest. Products J.* 30(5): 43–46.

Feist, W.C. and Rowell, R.M. 1982. Ultraviolet degradation and accelerated weathering of chemically modified wood. In: Hon, D.N.S., (Ed.), *Graft Copolymerization of Lignocellulosic Fibers*. American Chemical Society Symposium Series 187. American Chemical Society, Washington, DC. Chapter 21, pp. 349–370.

Feist, W.C., Rowell, R.M., and Ellis, W.D. 1991. Moisture sorption and accelerated weathering of acetylated and methacrylated aspen. *Wood & Fiber Science.* 23(1): 128–136.

Feist, W.C. and Sell, J. 1987. Weathering behavior of dimensionally stabilized wood treated by heating under pressure of nitrogen gas. *Wood & Fiber Science.* 19(2): 183–195.

Feist, W.C. and Williams, R.S. 1991. Weathering durability of chromium-treated southern pine. *Forest Products J.* 41(1): 8–14.

Flæte, P.O., Høibø, O.A., Fjærtoft, F., and Nilsen, T.N. 2000. Crack formation in unfinished siding of aspen (*Populus tremula L.*) and Norway spruce (*Picea abies* (L.) Karst.) during accelerated weathering. *Holz als Roh- und Werkstoff.* 58(3): 135–139.

Forsthuber, B. and Grüll, G. 2010. The effects of HALS in the prevention of photo-degradation of acrylic clear topcoats and wooden surfaces. *Polymer Degradation & Stability.* 95(5): 746–755.

Fowlie, D.A., Preston, A.F., and Zahora, A.R. 1990. Additives: An example of their influence on the performance and properties of CCA-treated southern pine. *In: Proceedings Eighty-Sixth Annual Meeting of the American Wood-Preservers' Association.* pp. 11–21.

Frey Wyssling, A. 1950. The discoloration of untreated wood by weathering. *Schweizerische Zeitschrift fur Forstwesen.* 101(6): 278–282.

Fraipont, L. 1973. Changes in some characteristics of hardboards exposed to one year's weathering. *Rapport d'Activite, Station de Technologie Forestiere, Gembloux.* 1974: 175–240.

Fujimoto, H. 1992. Weathering behaviour of chemically modified wood with a maleic acid-glycerol (MG) mixture. In; *Chemical Modification of Lignocellulosic*, Rotorua, New Zealand, Forest Research Institute Bulletin. 176:87–96.

Futó, L.P. 1974. Der photochemische abbau des holzes als präparations-und analysenmethode. *Holz als Roh- und Werkstoff.* 32(1): 303–314.

Futó, L.P. 1976a. Effects of temperature on the photochemical degradation of wood. I. Experimental presentation. *Holz als Roh- und Werkstoff.* 34(1): 31–36.

Futó, L.P. 1976b. Effects of temperature on the photochemical degradation of wood. 2, Scanning electron microscope presentation. *Holz als Roh- und Werkstoff.* 34(2): 49–54.

Geimer, R.L., Heebink, B.G., and Hefty, F.V. 1973. Weathering characteristics of particleboard. *USDA Forest Service Research Paper*, FPL 212, pp. 21.

George, B., Suttie, E., Merlin, A., and Deglise, X. 2005. Photodegradation and photostabilisation of wood— The state of the art. *Polymer Degradation & Stability.* 88(2): 268–274.

Gierer, J. and Lin, S.Y. 1972. Photodegradation of lignin. A contribution to the mechanism of chromophore formation. *Svensk Papperstidn.* 75(7): 233–239.

Gillander, H.E., King, C.G., Rhodes E.O., and Roche, J.N. 1934. The weathering of creosote. *Industrial & Engineering Chemistry.* 26(2): 175–183.

Gillespie, R.H. and River, B.H. 1976. Durability of adhesives in plywood. *Forest Products J.* 26(10): 21–25.

Gowdru, K.P., Pandey, K.K., Ram, R.K.D., and Mahadevan, K.M. 2006. Dimensional stability and photostability of octanoylated wood. *Holzforschung.* 60(5): 539–542.

Greaves, H. (1990). Current trends in the protection of timber. In: *Proceedings of 13th All Australia Timber Congress* pp. 14 pp.

Greaves, H. 1992. Recent developments in wood preservation research in Australia.: *Record of the Annual Convention of the British Wood Preserving and Damp-Proofing Association.* pp. 24–30.

Grelier, S., Castellan, A., Desrousseaux, S., Nourmamode, A., and Podgorski, L. 1997. Attempt to protect wood colour against UV/visible light by using antioxidants bearing isocyanate groups and grafted to the materials with microwave. *Holzforschung.* 51(6): 511–518.

Grelier, S., Castellan, A., and Podgorski, L. 2007. Use of low molecular weight modified polystyrene to prevent photodegradation of clear softwoods for outdoor use. *Polymer Degradation & Stability.* 92(8): 1520–1527.

Gressel, P. 1969. Further investigations on weathered wood particle boards. Comparison between four-year outdoor weathering and three rapid test methods. *Holz als Roh- und Werkstoff.* 27(10): 366–371.

Grigorion, A. 1981. The influence of different wood species on the properties of three-layer particleboards and their surface layers. *Holz als Roh- und Werkstoff.* 39(3): 97–105.

Groves, K.W. and Banana, A.Y. 1986. Weathering characteristics of Australian grown radiata pine. *J. of the Institute of Wood Science.* 10(5): 210–213.

Gu, H., Wang, S., Neimsuwan, T., and Wang, S. 2005. Comparison study of thickness swell performance of commercial oriented strandboard flooring products. *Forest Products J.* 55(12): 239–245.

Gugumus, F. 1993. Current trends in mode of action of hindered amine light stabilizers. *Polymer Degradation & Stability.* 40(2): 167–215.

Guntekin, E. and Sahin, H.T. 2009 Accelerated weathering performance of cement bonded fiberboard. *Scientific Research and Essays.* 4(5): 484–492.

Hansen, K. 2008. Molds and moldicide formulations for exterior paints and coatings. In Schultz, T.P., Militz, H., Freeman, M.H., Goodell, B., Nicholas, D.D. (Eds). *Development of Commercial Wood Preservatives. Efficacy, Environmental, and Health Issues.* ACS Symposium series Vol. 982, pp. 198–213.

Hansmann, C., Deka, M., Wimmer, R., and Gindl, W. 2006. Artificial weathering of wood surfaces modified by melamine formaldehyde resins. *Holz als Roh- und Werkstoff.* 64(3): 198–203.

Hawksworth, D. and Hill, D. 1984. *The Lichen-forming Fungi.* Blackie, New York.

Hayashi, T., Miyatake, A., and Harada, M. 2002. Outdoor exposure tests of structural laminated veneer lumber. I: Evaluation of the physical properties after six years. *J. Wood Science.* 48(1): 69–74.

Haygreen, J.G. and Bowyer, J.L. 1982. *Forest Products and Wood Science. An Introduction.* 1st Ed. The IOWA State University Press. Ames, IA.

Hayoz, P., Peter, W., and Rogez, D. 2003. A new innovative stabilization method for the protection of natural wood. *Progress in Organic Coatings.* 48(2–4): 297–309.

Held, B.W., Jurgens, J.A., Duncan, S.M., Farrell, R.L., and Blanchette, R.A. 2006. Assessment of fungal diversity and deterioration in a wooden structure at New Harbor, Antarctica. *Polar Biology.* 26(6): 526–531.

Helinska-Raczkowska, L., and Raczkowski, J. 1978. Creep in pine wood subjected previously to atmospheric corrosion in contact with rusting iron. *Holzforschung und Holzverwertung.* 30(3): 50–54.

Herzig, E., Meierhofer, U., and Sell, J. 1980. Stand der entwicklung und anwendung zementgebundener holzspanplatten. *Schweizer Ingenieur und Architekt Heft.* 13(80): 4.

Hilditch, E.A. and Crookes, J.V. 1981. Exterior wood stains, varieties, performance and appearance. *Record Annual Convention, British Wood Preserving Association*, pp. 59–66.

Hill, C.A.S. 2006. *Wood Modification. Chemical, Thermal and other Processes.* John Wiley & Sons. Chichester.

Hill, C.A.S., Cetin, N.S., Quinney, R.F., Derbyshire, H., and Ewen, R.J. 2001. An investigation of the potential for chemical modification and subsequent polymeric grafting as a means of protecting wood against photodegradation. *Polymer Degradation & Stability.* 72(1): 133–139.

Hon, D.N.S. 1983. Weathering reactions and protection of wood surfaces. *J. Applied Polymer Science.* 37(1): 845–864.

Hon, D.N.S. 1984. ESCA study of oxidized wood surfaces. *J. Applied Polymer Science.* 29(9): 2777–2784.

Hon, D.N.S. and Chang S.T. 1985. Photoprotection of wood surfaces by wood-ion complexes. *Wood & Fiber Science.* 17(1): 92–100.

Hon, D.N.S., Chang, S.T., and Feist, W.C. 1985. Protection of wood surfaces against photooxidation. *J. Applied Polymer Science.* 30(4): 1429–1448.

Hon, D.N.S. and Feist, W.C. 1986. Weathering characteristics of hardwood surfaces. *Wood Science & Technology.* 20(2): 169–183.

Hon, D.N.S. and Feist, W.C. 1993. Interaction of sulfur dioxide and nitric oxide with photoirradiated wood surfaces. *Wood & Fiber Science.* 25(2): 136–141.

Hon, D.N.S. and Glasser, W. 1979. On possible chromophoric structures in wood and pulps. *Polymer Plastics Technology Eng.* 12(2): 159–179.

Hon, D.N.S. and Ifju, G. 1978. Measuring penetration of light into wood by detection of photo-induced free radicals. *Wood Science.* 11(2): 118–127.

Hon, D.N.S., Ifju, G., and Feist, W.C. 1980. Characteristics of free radicals in wood. *Wood & Fiber.* 12(2): 121–130.

Hon, D.N.S. and Minemura, N. 1991. Color and discoloration. In: Hon, D.N.S., Shiraishi, N. (Eds.) *Wood and Cellulose Chemistry.* Marcel Dekker, New York. 1991, pp. 395–454.

Horn, B.A., Qiu, J., Owen, N.L., and Feist, W.C. 1994. FT-IR studies of weathering effects in western redcedar and southern pine. *Applied Spectroscopy.* 48(6): 662–668.

Hunt, M.O. and Matteson, D.A. Jr. 1976. Structural characteristics of weathered plywood. *Journal of the Structural Division, Proceedings of the American Society of Civil Engineers.* 102(ST4): 759–768.

Hussey, B.E. and Nicholas, D.D. 1985. The effect of light stabilizers on the iron and water degradation of wood. *Proceeding American Wood Preserver's Association.* 81: 169–173.

Iosifov, N. and V'lcheva, L. 1986. Weather resistance of particleboards of poplar wood in natural conditions. *Gorsko Stopanstvo Gorska Promishlenost.* 42(11): 13–15.

Jämsä, S., Ahola, P., and Viitaniemi, P. 2000. Long-term natural weathering of coated ThermoWood. *Pigment & Resin Technology.* 29(2): 68–74.

Jebrane, M., Sèbe, G., Cullis, I., and Evans, P.D. 2009. Photostabilization of wood using aromatic vinyl esters. *Polymer Degradation & Stability.* 94(2): 151–157.

Jemison, G.M. 1937. Loss of weight of wood due to weathering. *J. Forestry.* 35(5): 460–462.

Jin, L., Archer, K., and Preston, A. 1991. Surface characteristics of wood treated with various AACs, ACQ and CCA formulations after weathering. *International Research Group on Wood Preservation Doc.* IRG/WP/2369.

Jin, L., Roberts D.M., and Preston A.F. 1992. Influence of water-borne preservatives on water repellency and the impact of addition of water repellent additives. *International Research Group on Wood Preservation Doc.* IRG/WP/3704–92.

Jirous-Rajkovic, V., Bogner, A., Mihulja, G., and Vrsaljko, D. 2007. Coating adhesion and wettability of aged and preweathered fir wood and pine wood surfaces. *Wood Research.* 52(2): 39–48.

Johnson, B.R., Ibach, R.E., and Baker, A.J. 1992. Effect of salt water evaporation on tracheid separation from wood surfaces. *Forest Products J.* 42(7/8): 57–59.

Jorge, F.S., Santos, T.M., de Jesus, J.P., and Banks, W.B. 1999. Reactions between Cr(VI) and wood and its model compounds. Part 2: Characterisation of the reaction products by elemental analysis, magnetic susceptibility and FTIR. *Wood Science & Technology.* 33(6): 501–517.

Kalnins M.A. 1966. Surface characteristics of wood as they affect durability of finishes. Part II. Photochemical degradation of wood. *U.S Forest Service Res. Pap.* FPL 57, pp. 23–57.

Kalnins, M.A. 1984. Photochemical degradation of acetylated, methylated, phenylhydrazine-modified and ACC-treated wood. *J. Applied Polymer Science.* 29(1): 105–115.

Kalnins, M.A. and Feist, W.C. 1993. Increase in wettability of wood with weathering. *Forest Products J.* 43(2): 55–57.

Kaneda, H. and Maku, T. 1973a. Studies on the weatherability of composite wood. I. A few changes of material property of Lauan plywood by exterior exposure. *Mokuzai Gakkaishi.* 19(4): 157–164.

Kaneda, H. and Maku, T. 1973b. Studies on the weatherability of composite wood. II. An observation of surface deterioration of Lauan plywood by exterior exposure. *Mokuzai Gakkaishi.* 19(5): 215–220.

Kang, H.Y. Park, S.J., and Kim, Y.S. 2002. Moisture sorption and ultrasonic velocity of artificially weathered spruce. *Mokchae Konghak.* 30(1): 18–24.

Kataoka, Y. and Kiguchi, M. 2001. Depth profiling of photo-induced degradation in wood by FTIR microspectroscopy. *J. Wood Science.* 47(4): 325–327.

Kataoka, Y., Kiguchi, M., and Evans, P.D. 2004. Photodegradation depth profile and penetration of light in Japanese cedar earlywood (*Cryptomeria japonica* D Don) exposed to artificial solar radiation. *Surface Coatings International Part B: Coatings Transactions.* 87(3): 187–193.

Kataoka, Y., Kiguchi, M., Fujiwara, T., and Evans, P.D. 2005. The effects of within-species and between-species variation in wood density on the photodegradation depth profiles of sugi (*Cryptomeria japonica*) and hinoki (*Chamaecyparis obtusa*). *J. Wood Science.* 51(5): 531–536.

Kataoka, Y., Kiguchi, M., Williams, R.S., and Evans, P.D. 2007. Violet light causes photodegradation of wood beyond the zone affected by ultraviolet light. *Holzforschung.* 61(1): 23–27.

Kiguchi, M. 1990. Chemical modification of wood surfaces by etherification. II. Weathering ability of hot-melted wood surfaces and manufacture of self hot-melt bonded particleboard. *Mokuzai Gakkaishi.* 36(10): 867–875.

Kiguchi, M. 1992. Photo-deterioration of chemically modified wood surfaces: preliminary study with ESCA. In: *Chemical Modification of Lignocellulosic*, Rotorua, New Zealand, Forest Research Institute Bulletin, 176:77–86.

Kiguchi, M. and Evans, P.D. 1998. Photostabilisation of wood surfaces using a grafted benzophenone UV absorber. *Polymer Degradation & Stability.* 61(1): 33–45.

Kiguchi, M., Evans, P.D., Ekstedt, J., Williams, R.S., and Kataoka, Y. 2001. Improvement of the durability of clear coatings by grafting of UV-absorbers on to wood. *Surface Coating International Part B: Coating Transactions.* 84(B4): 263–270.

Kiguchi, M., Kataoka, Y., Doi, S., Mori, M., Hasegawa, M., Morita, S., Kinjo, M., Kadekaru, Y., and Imamura, Y. 1997b. Improvement of weather resistance of film-forming type clear finishes by pre-treatment with PEG and influence of exposure test sites. *Mokuzai Hozon.* 22(4): 10–17.

Kiguchi, M., Kataoka, Y., Kaneiwa, H., Akita, K., and Evans, P.D. 2000. Photostabilisation of woodfibre-plastic composites by chemical modification of woodfibre. *Proc. 5th Pac. Rim Bio-based Comp. Symp.*, 10–13 Dec., Canberra, Australia. pp. 145–150.

Kiguchi, M., Suzuki, M., Kinoshita, T., and Kawamura, J. 1997a. Evaluation of exterior pigmented stains by a new criterion of refinishing. *Mokuzai Kogyo.* 52: 612–617.

Kim, J.S., Singh, A.P., Wi, S.G., Koch, G., and Kim Y.S. 2008. Ultrastructural characteristics of cell wall disintegration of *Pinus* spp. in the windows of an old Buddhist temple exposed to natural weathering. *International Biodeterioration & Biodegradation.* 61(2): 194–198.

Kishino, M and Nakano, T. 2004a. Artificial weathering of tropical woods. Part 1: Changes in wettability. *Holzforschung.* 58(5): 552–557.

Kishino, M. and Nakano, T. 2004b. Artificial weathering of tropical woods. Part 2. Color change. *Holzforschung.* 58(5): 558–565.

Kitamura, Y., Setoyama, K., and Kurosu, H. 1989. Wavelength dependancy of light-induced discoloration in wood and dyed wood. *In:* J.F. Kennedy, G.O. Phillips, and P.A. Williams (Eds.), *Wood Processing and Utilization,* Ellis Horwood Ltd, Chichester, UK, 51:387–392.

Kleinert, T.N. 1970. Physikalisch-chemische holzveränderungen im freien. *Holzforschung und Holzverwertung.* 22(2): 21–24.

Klemchuk, P.P. and Gande, M. 1988. Stabilization mechanisms of hindered amines. *Polymer Degradation & Stability.* 22(3): 241–274.

Knight, R.A.G. 1947. The surface-checking of plywood in service. *Wood.* 10(12): 285–287.

Knight, R.A.G. and Doman, L.S. 1962. Reducing the surface checking of plywood exposed to weather. *J. of the Institute of Wood Science.* 10: 66–73.

Knight, R.A.G. and Doman, L.S. 1967. Surface protection of plywood by resin-treated papers. *Wood.* 32(1): 31–33.

Knight, R.A.G. and Newall, R.J. 1954. Seven and a half years of durability trials on adhesives in plywood. *Wood.* 19(7): 287–90.

Koch, P. (1967). Minimizing and predicting delamination of Southern [Pine] plywood in exterior exposure. *Forest Products J.* 17(2): 41–47.

Kornicer, D.R. and Palardy, R.D. 2003. Manufacture of multi-layered board with a unique resin system. United States Patent Application 2003/0035921, p. 6.

Krahmer, R.L., Lowell, E.C., Dougal, E.F., and Wellons, J.D. 1992. Durability of southeast-Asian hardwood plywood as shown by accelerated-aging tests and 10-year outdoor exposure. *Forest Products J.* 42(4): 40–44.

Kringstad, K.P. and Lin, S.Y. 1970. Mechanisms in the yellowing of high yield pulps by light: structure and reactivity of free radical intermediates in the photodegradation of lignin. *Tappi.* 53(12): 2296–2301.

Kubel, H. and Pizzi, A. 1981. Protection of wood surfaces with metallic oxides. *J. Wood Chemistry & Technology.* 1(1): 75–92.

Kučera, L.J. and Sell J. 1987. Weathering behavior of beech wood in the ray tissue region. *Holz als Roh- und Werkstoff.* 45(3): 89–93.

Kuo, M. and Hu, N. 1991. Ultrastructural changes of photodegradation of wood surfaces exposed to UV. *Holzforschung.* 45(5): 347–353.

La Mantia, F.P., and Morreale, M. 2008. Accelerated weathering of polypropylene/wood flour composites. *Polymer Degradation & Stability.* 93(7): 1252–1258.

Laufenberg, T. 1982. Exposure effects upon performance of laminated veneer lumber [LVL] and glulam materials. *Forest Products J.* 32(5): 42–48.

Leary, G.J. 1967. The yellowing of wood by light. *Tappi.* 50(1): 17–19.

Leary, G.J. 1968. The yellowing of wood by light: Part II. *Tappi.* 51(6): 257–260.

Leary, G.J. 1994. Recent progress in understanding and inhibiting the light-induced yellowing of mechanical pulps. *J. Pulp & Paper Science.* 20(6): J154–J160.

Lempfer, K. and Sattler, H. 1989. Long-term performance of cement-bonded particleboard and fiberboard. In: Moslemi, A.A. (Ed.), *Fiber and Particleboards Bonded with Inorganic Binders.* Forest Products Society, Madison pp. 125–132.

Lesar, B., Pavlic, M., Petric, M., Skapin, A.S., and Humar, M. 2011. Wax treatment of wood slows photodegradation. *Polymer Degradation & Stability.* 96(7): 1271–1278.

Lim, Y.W., Chedgy, R., Amirthalingam, S., and Breuil, C. 2007. Screening fungi tolerant to western red cedar (*Thuja plicata* Donn.) extractives. Part 2. Development of a feeder strip assay. *Holzforschung.* 61(2): 195–200.

Lim, Y.W., Kim, J.J., Chedgy, R., Morris, P., and Breuil, C. 2005. Fungal diversity from western red cedar fences and their resistance to β-thujaplicin. *Antonie van Leeuwenhoek.* 87(2): 109–117.

Lin, S.Y. 1982. Photochemical reaction mechanisms of lignin. *Forest Products Industries, Dept. of Forestry, National Taiwan Univ.* 1(2): 2–19.

Lin, S.Y. and Kringstad, K.P. 1970. Photosensitive groups in lignin and lignin model compounds. *Tappi.* 53(4): 658–663.

Lisperguer, J.H. and Becker, P.H. 2005. Strength and durability of phenol-resorcinol formaldehyde bonds to CCA-treated radiata pine wood. *Forest Products J.* 55(12): 113–116.

Liu, C., Ahniyaz, A., and Evans, P.D. 2010. Preliminary observations of the photostabilization of wood surfaces with cerium oxide nanoparticles. *International Research Group on Wood Protection Doc.* IRG/WP/10–40504.

Liu, R., Ruddick, J.N.R., and Jin, L. 1994. The influence of copper (II) chemicals on the weathering or treated wood, Part I ACQ treatment of wood on weathering. *International Research Group on Wood Preservation Doc.* IRG/WP/94–30040.

Lloyd, R.A. and Stamm, A.J. 1958. Effect of resin treatment and compression upon the weathering properties of veneer laminates. *Forest Products J.* 8(8): 230–235.

Luston, J., Guniš, J., and Manásek, Z. 1973. Polymeric-UV absorbers of 2-hydroxybenzophenone type 1. Polyesters on the base of 2-hydroxy-4-(2,3-epoxypropoxy)benzophenone. *J. Macromol. Sci.-Chem.* A7(3): 587–599.

Lundin, T., Falk, R.H., and Felton, C. 2002. Accelerated weathering of natural fiber-thermoplastic composites: Effect of ultraviolet exposure on bending strength and stiffness. *Proceedings: International Conference on Woodfiber-Plastic Composites, 6th*, Madison, WI, USA, pp. 87–93.

MacLeod, I.T., Scully, A.D., Ghiggino, K.P., Ritchie, P.J.A., Paravagna, O.M., and Leary, B. 1995. Photodegradation at the wood-clearcoat interface. *Wood Science & Technology.* 29(3): 183–189.

Mahltig, B., Swaboda, C., Roessler, A., and Böttcher, H. 2008. Functionalising wood by nanosol application. *J. Materials Chemistry.* 18(27): 3180–3192.

Malik, J., Ligner, G., and Avar, L. 1998. Polymer bound HALS—expectations and possibilities *Polymer Degradation & Stability.* 60(1): 205–213.

Matuana, L.M. and Kamdem, D.P. 2002. Accelerated ultraviolet weathering of PVC/wood-flour composites. *Polymer Engineering & Science.* 42(8): 1657–1666.

Matuana, L.M., Kamdem, D.P., and Zhang, J. 2001. Photoaging and stabilization of rigid PVC/wood-fiber composites. *J. Applied Polymer Science.* 80(11): 1943–1950.

Matuana, L.M., Jin, S., and Stark, N.M. 2011. Ultraviolet weathering of HDPE/wood-flour composites coextruded with a clear HDPE cap layer. *Polymer Degradation & Stability.* 96(1): 97–106.

Mawson, D. 1915. *The Home of the Blizzard. Being the Story of the Australasian Antarctic Expedition, 1911–1914.* William Heinemann, London.

McDonald, K.A., Falk, R.H., Williams, R.S., and Winandy, J.E. 1996. *Wood Decks. Materials, Construction, and Finishing.* Forest Products Society, Madison, WI, 93pp.

McLaughlan, J.M. 1991. Properties of treated and untreated *Pinus radiata* plywood after 12 years' weathering. *New Zealand J. Forestry Science.* 21(1): 96–110.

McNamara, W.S. and Shaw, M.D. 1972. Vacuum-pressure impregnation of medium-density hardboard with phenolic resin. *Forest Products J.* 22(11): 19–22.

Mehta, G., Mohanty, A.K., Drzal, L.T., Kamdem, D.T., and Misra, M. 2006. Effect of accelerated weathering on biocomposites processed by SMC and compression molding. *J. Polymers & Environment* 14(4): 359–368.

Meierhofer, U.A. 1986. Behavior of weather exposed glue-laminated wood impregnated with waterborne and oil prevervatves. *Holz als Roh-und Werkstoff.* 44(5): 173–177.

Meierhofer, U.A. 1988. Weathering behaviour of preservative-treated glulam after five-year outdoor exposure. *Holz als Roh- und Werkstoff.* 46(2): 53–58.

Meierhofer, U.A. and Sell, J. 1977. Influence of surface treatments on the properties of weathered particleboards. *Forest Products J.* 27(9): 24–27.

Meierhofer, U.A. and Sell, J. 1983. Optimizing the surface treatment of timber construction elements. Part 3: Weathering tests of particleboard specimens. *Holz als Roh- und Werkstoff.* 41(11): 449–454.

Meierhofer, U.A. and Sell, J. 1984. Optimizing the surface treatment of timber constructional elements——Exterior wall cladding using particleboards with various coatings. *Holz als Roh- und Werkstoff.* 42(7): 253–259.

Meierhofer, U.A. and Sell, J. 1983. Behaviour of various full-size coated particleboards under natural weathering tests. *Bericht, Eidgenössische Materialprüfungs- und Versuchsanstalt.* 115/4: 33.

Meincken, M. and Evans, P.D. 2009. Nanoscale characterization of wood photodegradation using atomic force microscopy. *Holz als Roh- und Werkstoff.* 67(2): 229–231.

Meincken, M. and Evans, P.D. 2010. Use of atomic force microscopy to detect wavelength dependent changes in wood veneers and spin-coated lignin and cellulose films exposed to solar radiation. *International Wood Products Journal.* 1(2): 75–80.

Michell, A.J. 1993. FTIR spectroscopic studies of the reactions of wood and of lignin model compounds with inorganic agents. *Wood Science & Technology.* 27(2): 69–80.

Militz, H. 2008. Processes and properties of thermally modified wood manufactured in Europe. In Schultz, T.P., Militz, H., Freeman, M.H., Goodell, B., Nicholas, D.D. (Eds). *Development of Commercial Wood Preservatives. Efficacy, Environmental, and Health Issues*. ACS Symposium Series Vol. 982, pp. 372–388.

Miniutti, V.P. 1964. Preliminary observations. Microscale changes in cell structure at softwood surfaces during weathering. *Forest Products J.* 14(12): 571–576.

Miniutti, V.P. 1967. Microscopic observations of ultraviolet irradiated and weathered softwood surfaces and clear coatings. *U.S. For. Serv. Res. Pap. FPL* 74, pp. 1–32.

Mitsui, K. and Tsuchikawa, S. 2005. Low atmospheric temperature dependence on photodegradation of wood. *J. Photochemistry and Photobiology B: Biology* 81(2): 84–88.

Mobius, M. 1924. The grey discoloration of wood. *Berichte der Deutschen Botanischen Gesellschaft.* 42(1–2): 15–18.

Muasher, M. and Sain, M. 2006. The efficacy of photostabilizers on the color change of wood filled plastic composites. *Polymer Degradation & Stability.* 91(5): 1156–1165.

Murakami, K. and Matsuda, H. 1990. Oligoesterified woods based on anhydride and epoxide VIII. Resistance of oligoesterified woods against weathering and biodeterioration. *Mokuzai Gakkaishi.* 36(7): 538–544.

Neusser, H. and Schall, W. 1972. Studies on some promising possibilities for improvement of plywood. I. Gluing with aminoplasts. *Holzforschung und Holzverwertung.* 24(5): 108–116.

Neusser, H., Krames, U., and Schall, W. 1971. Weathering behaviour of unprotected and surface-faced plywood. *Holzforschung und Holzverwertung.* 23(4): 61–73.

Neusser, H. and Zentner, M. 1970. Weathering experiments with chipboards and fibreboards. *Holzforschung und Holzverwertung.* 22(3): 50–60.

Nienhuis, J.G., van de Velde, B., Cobben, W.N.H., and Beckers, E.P.J. 2003. Exterior durability of coatings on modified wood. In: Van Acker, J., and Hill, C. (Eds.), *First European Conference on Wood Modification*, pp. 203–206.

Nuopponen, M., Wikberg, H., Vuorinen, T., Maunu, S.L., Jämsä, S., and Viitaniemi, P. 2004. Heat-treated softwood exposed to weathering. *J. Applied Polymer Science.* 91(4): 2128–2134.

Nzokou, P. and Kamdem, D.P. 2002. Weathering of two hardwood species: African padauk (*Pterocarpus soyauxii*) and red maple (*Acer rubrum*). *J. Tropical Forest Products.* 8(2): 200–209.

Nzokou, P. and Kamdem, D.P. 2006. The influence of wood extractives on the photodiscoloration of wood surfaces exposed to artificial weathering. *Color Research & Application.* 31(4): 425–434.

Ohkoshi, M. 2002. FTIR-PAS study of light-induced changes in the surface of acetylated or polyethylene glycol-impregnated wood. *J. Wood Science.* 48(5): 394–401.

Ohtani, K. 1987. Chromium trioxide enhances the durability of wooden doors. *Chromium Review.* 8: 4–7.

Oviatt, A.E. 1975. Protecting exposed ends of timber beams in the Puget Sound area. *USDA Forest Service Research Note, Pacific Northwest Forest and Range Experiment Station*, PNW-263, 15 pp.

Paajanen, L.M. 1994. Structural-changes in primed Scots pine and Norway spruce during weathering. *Materials & Structures.* 27(168): 237–244.

Pampel, H. and Schwarz, H.G. 1979. Technology and processing of cement-bonded particleboards. *Holz als Roh- und Werkstoff.* 37(5): 195–202.

Pandey, K.K. 2005. A note on the influence of extractives on the photo-discoloration and photo-degradation of wood. *Polymer Degradation & Stability.* 87(2): 375–379.

Pandey, K.K. Hughes, M., and Vuorinen, T. 2010. Dimensional stability, UV resistance, and static mechanical properties of Scots pine chemically modified with alkylene epoxides. *Bioresources.* 5(2): 598–615.

Pandey, K.K. and Chandrashekar, N. 2006. Photostability of wood surfaces esterified by benzoyl chloride. *J. Applied Polymer Science.* 99(5): 2367–2374.

Pandey, K.K. and Khali, D.P. 1998. Accelerated weathering of wood surfaces modified by chromium trioxide. *Holzforschung.* 52(5): 467–471.

Pandey, K.K. and Pitman, A.J. 2002. Weathering characteristics of modified rubberwood (*Hevea brasiliensis*). *J. Applied Polymer Science.* 85(3): 622–631.

Park, B.S., Furuno, T., and Uehara, T. 1996. Histochemical changes of wood surfaces irradiated with ultraviolet light. *Mokuzai Gakkaishi.* 42(1): 1–9.

Pattamasattayasonthi, N., Chaochanchaikul, K., Rosarpitak, V., and Sombatsompop, N. 2011. Effects of UV weathering and a CeO(2)-based coating layer on the mechanical and structural changes of wood/PVC composites. *J. Vinyl & Additive Technology.* 17(1): 9–16.

Paulsson, M., Lucia, L.A., Ragauskas, A.J., and Li, C. 2001. Photoyellowing of untreated and acetylated aspen chemithermomechanical pulp under argon, ambient, and oxygen atmosphere, *J. Wood Chemistry & Technology.* 21(4): 343–360.

Peterkin, J.A. 1934. *Plantation Christmas*. Houghton Mifflin Company, Boston, USA.

Pilson, R., Johnson, A., and Bradner, M.I. 1931. *Properties of Western Larch and their Relation to uses of the Wood*. B. Eng. (Civil) Thesis, Univ. Wisconsin, USA.

Pinion, L.C. 1975. The properties of wood-wool/cement building slabs. *BRE Information Paper IS 22/75, Princes Risborough Laboratory*, p. 2.

Pinnell, S.R., Fairhurst, D., Gillies, R., Mitchnick, M.A., and Kollias, N. 2000. Microfine zinc oxide is a superior sunscreen ingredient to microfine titanium dioxide. *Dermatologic Surgery*. 26(4): 309–314.

Pittman, C.U., Kim, M.G., Nicholas, D.D., Wang, L.C., Kabir, F.R.A., Schultz, T.P., and Ingram, L.I. 1992. Wood enhancement treatments. 1. Impregnation of southern yellow pine with melamine-formaldehyde and melamine-ammeline-formaldehyde resins. *Holzforschung*. 46(5): 395–401.

Pizzi, A. 1980. Wood waterproofing and lignin crosslinking by means of chromium trioxide/guajacyl units complexes. *J. Applied Polymer Science*. 25(11): 2547–2553.

Plackett, D.V., Dunningham, E.A., and Singh, A.P. 1992. Weathering of chemically modified wood. Accelerated weathering of acetylated radiata pine. *Holz als Roh- und Werkstoff*. 50(4): 135–140.

Podgorski, L., Georges, V., Izaskun Garmendia, I., and Sarachu, B.S. 2009. A fast and economic method to produce grey wooden surfaces for decking and cladding: preliminary results. *International Research Group on Wood Protection Doc*. IRG/WP/09–40474.

Qayyum, M.M. and Davis, A. 1984. Ultraviolet radiation for various angles of exposure at Jeddah and its relation to the weathering of polyacetal. *Polymer Degradation & Stability*. 6(4): 201–209.

Rabek, J.F. 1990. *Photostabilization of Polymers: Principles and Application*. Elsevier, London.

Raczkowski, J. 1980. Seasonal effects on the atmospheric corrosion of spruce micro-sections. *Holz als Roh- und Werkstoff*. 38(6): 231–234.

Raczkowski, J. 1982. Modification of wood with polystyrene improves its resistance against accelerated weathering in contact with rusting iron. In: Meyer, R.W., and Kellogg. R.M. (Ed.), *Structural Uses of Wood in Adverse Environments*, Van Nostrand Reinhold Co., New York, pp. 150–155.

Rapp, A.O. and Peek, R.D. 1999. Melamine resin treated as well as varnish coated and untreated solid wood during two years of natural weathering. *Holz als Roh- und Werkstoff*. 57(5): 331–339.

Ribarits, S. and Evans, P.D. 2010. Finite element modelling of the checking of wood exposed to accelerated weathering. *International Research Group on Wood Protection Doc*. IRG/WP/10–20459.

Rice, J.T. 1965. The effect of urea formaldehyde resin viscosity on plywood wood bond durability. *Forest Products J*. 15(3): 107–12.

Richter, G.A. 1935. Relative permanence of papers exposed to sunlight. II. *Industrial & Engineering Chemistry*. 27(4): 432–439.

Rogez, D. 2002. Color stabilization of wood and durability improvement of wood coatings: A new UV light-protection concept for indoor and outdoor applications. *Paint & Coatings Industry*. 18(3): 56–65.

Roos, J.A., Brackley, A.M., and Sasatani, D. 2009. The U.S. glulam beam and lamstock market and implications for Alaska lumber. Pacific Northwest Research Station, USDA Forest Service, Report PNW-GTR-796, i + 19 pp.

Rowell, R.M. and Banks, W.B. 1985. Water repellency and dimensional stability of wood, Gen. Tech. Rep., PPL-50, US Department of Agriculture, Forest Service, Forest Products Laboratory, Madison, WI. pp. 1–24.

Rowell, R.M., Feist, W.C., and Ellis, W.D. 1981. Weathering of chemically modified southern pine. *Forest Products J*. 12(4): 202–208.

Ruetze, M., Schmitt, U., Noack, D., and Kruse, S. 1994. Investigations on the potential role of chromium-containing wood stains in the development of nasal adenocarcinomas. *Holz als Roh- und Werkstoff*. 52(2): 87–93.

Sabaa, M.W., Madkour, T.M., and Yassin, A.A. 1988. Polymerization products of *p*-benzoquinone as bound antioxidants for SBR. Part II—The antioxidizing efficiency. *Polymer Degradation & Stability*. 22(3): 205–222.

Sailer, M. van Nieuwenhuijzen, E., and Knol. W. 2010. Forming of a functional biofilm on wood surfaces. *Ecological Engineering*. 36(2): 163–167.

Sakuno, T. and Goto, T. 1975. Weather effects on gluability and surface checking of plywood treated with PEG. *Bulletin of the Faculty of Agriculture, Tottori University*. 27: 112–116.

Salaita, G.N., Ma, F.M.S., Parker, T.C., and Hoflund, G.B. 2008. Weathering properties of treated southern yellow pine wood examined by X-ray photoelectron spectroscopy, scanning electron microscopy and physical characterization. *Applied Surface Science*. 254(13): 3925–3934.

Salzberg, H.K., Browne, F.L., and Odell, I.H. 1931. Paint thinners II-Results of accelerated weathering tests of white house paints reduced with different types of thinners. *Industrial & Engineering Chemistry*. 23(11): 1214–1220.

Sandberg, D. 1999. Weathering of radial and tangential wood surfaces of pine and spruce. *Holzforschung.* 53(4): 264–355.

Sandermann, W. and Schlumbom, F. 1962. On the effect of filtered ultraviolet light on wood. Part 1. Photometric and chromatographic investigations on wood powders. *Holz als Roh- und Werkstoff.* 20(7): 245–252.

Saotome, H., Ohmi, M., Tominaga, H., Fukuda, K., Kataoka, Y., Kiguchi, M., Hiramatsu, Y., and Miyatake, A. 2009. Improvement of dimensional stability and weatherability of composite board made from water-vapor-exploded wood elements by liquefied wood resin impregnation. *J. Wood Science.* 55(3): 190–196.

Savory, J. 1954. Breakdown of timber by ascomycetes and fungi imperfecti. *Ann. Applied Biology.* 41(2): 336–347.

Scaiano, J.C., Netto-Ferreira, J.C., and Wintgens, V. 1991. Fragmentation of ketyl radicals derived from a-phenoxyacetophenone: An important mode of decay for lignin-related radicals. *J. Photochemistry & Photobiology. A. Chemistry.* 59(2): 265–268.

Schaller, C. and Rogez, D. 2007. New approaches in wood coating stabilization. *J. Coatings Technology & Research.* 4(4): 401–409.

Schramm, W.H. 1906a. The yellowing of woods. *Jahresber Angew. Bot.* 3: 116–139.

Schramm, W.H. 1906b. The graying of woods. *Jahresber Angew. Bot.* 3: 140–153.

Schaudy, R. and Proksch, E. 1980. Electron radiation curing of wood particleboards and fibre building boards impregnated with synthetic resins. *Holzforschung.* 34(3): 104–109.

Schauwecker, C., Preston, A., and Morrell, J.J. 2009. A new look at the weathering performance of solid wood decking materials. *J. Coatings Technology.* 6(9): 32–38.

Schmalzl, K.J. and Evans, P.D. 2003. Wood surface protection with some titanium, zirconium and manganese compounds. *Polymer Degradation & Stability.* 82(3): 409–419.

Schmalzl, K.J., Forsyth, C.M., and Evans, P.D. 1995. The reaction of guaiacol with iron III and chromium VI compounds as a model for wood surface modification. *Wood Science & Technology.* 29(4): 307–319.

Schmalzl, K.J., Forsyth, C.M., and Evans, P.D. 2003. Evidence for the formation of chromium III diphenoquinone complexes during oxidation of guaiacol and 2,6-dimethoxyphenol with chromic acid. *Polymer Degradation & Stability.* 82(3): 399–407.

Schmid, S., Webster, R.D., and Evans, P.D. 2000. The use of ESR spectroscopy to assess the photostabilising effects of wood preservatives. *International Research Group on Wood Protection Doc.* IRG/WP/00–20186.

Schmidt, E.L. and French, D.W. 1976. *Aureobasidium pullulans* on wood shingles. *Forest Products J.* 26(7): 34–37.

Schmidt, J.A. and Heitner, C. 1993. Light-induced yellowing of mechanical and ultra-high yield pulps. Part 2. radical-induced cleavage of etherified guaiacylglycerol-β-arylether groups is the main degradative pathway. *J. Wood Chemistry & Technology* 13(3): 309–325.

Schmolz, E., Brüders, N., Daum, R., and Lamprecht, I. 2000. Thermoanalytical investigations on paper covers of social wasps. *Thermochimica Acta.* 361(1–2): 121–129.

Schniewind, A.P. 1963. Mechanism of check formation. *Forest Products J.* 13(11): 475–480.

Schoemaker, H.E., Harvey, P.J., Bowen, R.M., and Palmer, J.M. 1985. On the mechanism of enzymatic lignin breakdown. *Febs Lett.* 183(1): 7–12.

Schoeman, M.W. and Dickinson, D.J. 1996. *Aureobasidium pullulans* can utilize simple aromatic compounds as a sole source of carbon in liquid culture *Letters in Applied Microbiology.* 22(2): 129–131.

Schoeman, M. and Dickinson, D. 1997. Growth of *Aureobasidium pullulans* on lignin breakdown products at weathered wood surfaces. *The Mycologist.* 11(4): 168–172.

Schulte, K. 2001. Application of micronized titanium dioxide as inorganic UV-absorber. Presentation to 11th Asia Pacific Coatings Conference, 26–27th June, Bangkok, Thailand. http://www.sachtleben.de/publications/0160e072.pdf (Accessed 8th September 2011).

Schultz, T.P. and Templeton, M.C. 1986. Proposed mechanism for the nitrobenzene oxidation of lignin. *Holzforschung.* 40: 93–97.

Sergeev, Y.V., Linkov, I.M., and Linkov, V.I. 1995. Effect of atmospheric factors on the strength of wood-cement boards. *Derevoobrabatyvayushchaya Promyshlennost.* 3(1): 19–20.

Seifert, K. 1964. Changes in the chemical composition of wood by the blue stain fungus A. pullulans. *Holz als Roh- und Werkstoff.* 22(11): 405–409.

Sekino, N. and Suzuki, S. 2002. Durability of wood-based panels subjected to ten-year outdoor exposure in Japan. In: *Proceedings 6th Pacific Rim Bio-based Composites Symposium*, pp. 323–332.

Selbo, M.L. 1952. Durability of glue joints in preservative-treated wood. *Southern Lumberman.* 185(2321): 203–206.

Selbo, M.L. 1964. Durability of glue joints in preservative-treated wood. Ten-year exposure of laminated beams treated with oilborne and waterborne preservatives. *Forest Products J.* 14(11): 517–520.

Selbo, M.L. 1965. Performance of melamine resin adhesives in various exposures. *Forest Products J.* 15(12): 475–83.

Selbo, M.L. 1967. Long term effect of preservatives on gluelines in laminated beams. *Forest Products J.* 17(5): 23–32.

Selbo, M.L. 1969. Performance of Southern pine plywood during five years' exposure to weather. *Forest Products J.* 19(8): 56–60.

Selbo, M.L., Knauss, A.C., and Worth, H.E. 1965. After two decades of service, glulam timbers show good performance. *Forest Products J.* 15(11): 466–472.

Sell, J. 1973. Surface treatment of particle boards for exterior use. Experiments to elucidate principal requirements. *Holz-Zentralblatt.* 99(147): 2337–2338.

Sell, J. 1983. Rissbildung bei wetterbeanspruchten Brettschichttraegern. *Holz Zentralblatt.* 109: 47.

Sell, J. and Feist, W.C. 1986. Role of density in the erosion of wood during weathering. *Forest Products J.* 36(3): 57–60.

Sell, J., Sommerer, S., and Meierhofer, U. 1979. Investigations on weathered particleboards. 4. Four-year weathering tests on boards with water repellent surface treatments. *Holz als Roh- und Werkstoff.* 37(10): 373–378.

Sell, J. and Wälchli, O. 1969. Changes in the surface texture of weather-exposed wood. *Material und Organismen.* 4(2): 81–87.

Sharman, W.R. 1983. Durability of fibre cement sheet claddings, *New Zealand Concrete Construction.* 27(2): 3–7.

Sharman, W.R. and Vautier, B.P. 1986. Durability studies of wood fibre reinforced cement-sheet. RILEM Symposium FRC 86. Developments in fibre reinforced cement and concrete, Vol. 2. Building Research Association of New Zealand, Porirua City, New Zealand.

Sharratt, V., Hill, C.A.S., and Kint, D.P.R. 2009. A study of early colour change due to simulated accelerated sunlight exposure in Scots pine (*Pinus sylvestris*). *Polymer Degradation & Stability.* 94(9): 1589–1594.

Sharratt, V., Hill, C.A.S., Zaihan, J., and Darwin, K.P.R. 2010. Photodegradation and weathering effects on timber surface moisture profiles as studied using dynamic vapour sorption. *Polymer Degradation & Stability.* 95(12): 2659–2662.

Sharrock, R.F. 1990. A European approach to UV protection with a novel pigment. *J. Coatings Technology.* 62(789): 125–130.

Shi, Q. and Jiang, J. 2011. Color stability of heat-treated Okan sapwood during artificial weathering. In: Zhou, H.Y., Gu, T.L., Yang, D.G., Jiang, Z.Y., and Zeng, J.M. (Eds) *New And Advanced Materials*, Pts 1 and 2, Book Series: Advanced Materials Research. 197–198:13–16.

Singh, A.P. and Dawson, B.S.W. 2003. The mechanism of failure of clear coated wooden boards as revealed by microscopy. *International Association Wood Anatomists J.* 24(1): 1–11.

Skaar, C. 1972. *Water in Wood.* Syracuse University Press, Syracuse, New York.

Smith, R.S. and Swann, G.W. 1976. Colonization and degradation of western red cedar shingles and shakes by fungi. *Material und Organismen.* 3: 253–262.

Spedding, D.J. 1970. Sorption of sulphur dioxide by indoor surfaces. II. Wood. *J. Applied. Chem.* 20(7): 226–228.

Stamm, A.J. 1964. *Wood and Cellulose Science.* The Ronald Press Co. New York.

Stamm, A.J. and Seborg, R.M. 1936. Minimizing wood shrinkage and swelling-treating with synthetic resin-forming materials. *Industrial & Engineering Chemistry.* 28(10): 1164–1169.

Stamm, A.J. and Seborg, R.M. 1939. Resin-treated plywood. *Industrial & Engineering Chemistry.* 31(7): 897–902.

Stark, N.M. 2006. Effect of weathering cycle and manufacturing method on performance of wood flour and high-density polyethylene composites. *J. Applied Polymer Science.* 100(4): 3131–3140.

Stark, N.M. and Matuana, L.M. 2003. Ultraviolet weathering of photostabilized wood flour-filled high-density polyethylene composites. *J. Applied Polymer Science.* 90(10): 2609–2617.

Stark N.M. and Matuana, L.M. 2004a. Surface chemistry and mechanical property changes of wood-flour/high-density-polyethylene composites after accelerated weathering. *J. Applied Polymer Science.* 94(6): 2263–2273.

Stark, N.M. and Matuana, L.M. 2004b. Surface chemistry changes of weathered HDPE/wood-flour composites studied by XPS and FTIR spectroscopy. *Polymer Degradation & Stability.* 86(1): 1–9.

Stark, N.M. and Matuana, L.M. 2006. Influence of photostabilizers on wood flour - HDPE composites exposed to xenon-arc radiation with and without water spray. *Polymer Degradation & Stability.* 91(12): 3048–3056.

Stark, N.M. and Mueller, S.A. 2008. Improving the color stability of wood-plastic composites through fiber pre-treatment. *Wood & Fiber Science.* 40(2): 271–278.

Stark, N.M., Matuana, L.M., and Clemons, C.M. 2004. Effect of processing method on surface and weathering characteristics of wood—flour/HDPE composites. *J. Applied Polymer Science.* 93(3): 1021–1030.

Stofko, J. 1964. On the resistance of particle boards to exterior weather conditions. *Drevarsky Vyskum,* 4: 203–222.

Suchsland, O. and Woodson, G.E. 1991. *Fiberboard Manufacturing Practices in the United States.* Forest Products Research Society, Madison, Wisconsin.

Sudiyani, Y., Horisawa, S., KeLi, C., Doi, S., and Imamura, Y. 2002. Changes in surface properties of tropical wood species exposed to the Indonesian climate in relation to mold colonies. *J. Wood Science.* 48(6): 542–547.

Sudiyani, Y., Ryu, J.Y., Hattori, N., and Imamura, Y. 2001. Phenolic resin treatment of wood for improving weathering properties. In: Imamura, Y. (Ed.), *High-Performance Utilization of Wood for Outdoor Uses,* World Research Institute, Kyoto University, Kyoto, Japan, pp. 85–96.

Sundaryono, A., Nourmamode, A., Gardrat, C., Grelier, S., Despeyroux, O., and Castellan, A. 2003. Attempt to protect wood using copper(II) complex with 1,7-diphenyl-1,6-heptadiene-3,5-dione, a non-phenolic cur-cuminoid. *Holz als Roh- und Werkstoff.* 61(5): 377–381.

Sundell, P. and Sundholm, F. 2004. Polyester binders for wood containing benzotriazole and HALS light stabilizers. *J. Applied Polymer Science.* 92(3): 1413–1421.

Tait, R.B. and Akers, S.A.S. 1989. Micromechanical studies of fresh and weathered fibre cement composites. Part 2: Wet testing. *International J. Cement Composites and Lightweight Concrete.* 11(2): 125–131.

Tanaka, A. 1964. Studies on the powdered dry-bonding process in the manufacture of plywood. I. On the relation between the gluing addition and the strength of adhesion. *Mokuzai Gakkaishi.* 10(4): 131–135.

Tanczos, I. and Schmidt, H. 2002. Quantam Process-New sulfur-free delignification. *J. Wood Chemistry & Technology* 22(4): 219–233.

Tarkow, H., Stamm, A.J., and Erickson, E.C.O. 1946. Acetylated wood. *U.S. Forest Products Laboratory Report.* 1593:1–15.

Taylor, A. and Wang, S. 2010. Properties of 'enhanced' OSB subfloor panels. Wood Products Information, Forest Products Center, University of Tennessee, Knoxville, W176, 3 pp, http://trace.tennessee.edu/utk_agexfores/100 (accessed July 4th 2010).

Temiz, A., Terziev, N., Eikenes, M., and Hafren, J. 2007. Effect of accelerated weathering on surface chemistry of modified wood. *Applied Surface Science.* 253(12): 5355–5362.

Temiz, A., Terziev, N., Jacobsen, B., and Eikenes, M. 2006. Weathering, water absorption, and durability of silicon, acetylated, and heat-treated wood. *J. Applied Polymer Science.* 102(5): 4506–4513.

Tjeerdsma, B.F. and Bongers, F. 2009. The making of a traffic bridge of acetylated radiata pine. Proceedings of 4th European Conference on Wood Modification. Stockholm, Sweden. pp. 15–22.

Tolvaj, L. and Faix, O. 1995. Artificial ageing of wood monitored by DRIFT spectroscopy and CIE L*a*b* color measurements. *Holzforschung.* 49(5): 397–404.

Turkulin, H., Derbyshire, H., and Miller, E.R. 2004. Investigations into the photodegradation of wood using microtensile testing Part 5: The influence of moisture on photodegradation rates. *Holz als Roh- und Werkstoff.* 62(4): 307–312.

Underhaug, Å., Lund, T.J., and Kleive, K. 1983. Wood protection–the interaction between substrate and product and the influence on durability. *J. of the Oil and Colour Chemists' Association.* 66(11): 345–355.

Upreti, N.K. and Pandey, Y.K. 2005. Role of pretreatments in the protection of wood surface and finishes in the weathering of *Pterocarpus marsupium* wood. *J. Tropical Forest Science.* 17(1): 141–150.

Urban, K. and Evans, P.D. 2005. Preliminary observations of the effect of growth ring orientation on the surface checking of flat sawn Southern pine decking. *International Research Group on Wood Protection Doc,* IRG/WP/05-20313.

Usmanov, K.U. 1978. Degradation and stabilisation of cellulose acetates and some polymers based on vinyl fluoride. *Polymer Science Series A (USSR).* 20(8): 1683–1690.

Vernois, M. 2001. Heat treatment of wood in France-state of the art. In: Review on heat treatments of wood. *Proc. Special Seminar of Cost Action E22,* Antibes, France, pp. 6.

Virtala, V. and Oksanen, S. 1948. Experiments in making roofing shingles of hard fibre board. *Valtion Teknillinen Tutkimuslaitoksen tiedotus.* 60: 14.

V'Icheva, L. 1986. Increasing the resistance of particleboards to weather conditions. *Gorsko Stopanstvo Gorska Promishlenost.* 42(8): 20–23.

Voulgaridis, E.V. and Banks, W.B. 1981. Degradation of wood during weathering in relation to water repellent long-term effectiveness. *J. Institute of Wood Science.* 9(2): 72–83.

Voulgaridis, E.V. and Banks, W.B. 1983. Laboratory evaluation of the performance of water repellants applied to long wood specimens. *Holzforschung.* 37(5): 261–266.

Waldron, R. and Moyer, B. 2009. Light-stabilized coatings for preservation of wood-plastic composite decking. *JCT Coatings Technology.* 6(3): 24–31.

Wang, W., Zhang, Z., Wang, Q., and Cui, Y. 2011. Application of organic and inorganic dye to wood plastic composite. *Env Biotech Mat Eng. PTS 1-3 Book Series: Advanced Materials Research.* 183–185: 2293–2297.

Webb, D.A. and Sullivan, J.D. 1964. Surface effect of light and water on wood. *Forest Products J.* 14(11): 531–534.

Weichelt, F., Emmler, R., Flyunt, R., Beyer, E., Buchmeiser, M.R., and Beyer, M. 2010. ZnO-based UV nano-composites for wood coatings in outdoor applications. *Macromolecular Materials & Engineering.* 295(2): 130–136.

Wengert, E.M. 1966a. Effect of atmospheric gases on color changes in wood exposed to ultraviolet light. *J. Paint Technology.* 38(493): 71–76.

Wengert, E.M. 1966b. Parameters for predicting maximum surface temperatures of wood in exterior exposures. *US Forest Service, Research Paper* FPL 62.

West, J.M. and Majumdar, A.J. 1991. Durability of non-asbestos fibre-reinforced cement. *BRE Information Paper IP.* 1/91.

Wiesner, J. 1864. The decomposition of wood in the atmosphere. *Sitzungsber. Akad. Wiss. Wien.* 49: 61–94.

Williams, R.S. 1983. Effect of grafted UV stabilizers on wood surface erosion and clear coating performance. *J. Applied Polymer Science.* 28(6): 2093–2103.

Williams, R.S. 1987. Acid effects on accelerated wood weathering. *Forest Products J.* 37(2): 37–38.

Williams, R.S. 2005. Weathering of wood. In: Rowell, R, (ed) *Handbook of Wood Chemistry and Wood Composites.* CRC Press, Boca Raton, pp. 139–185.

Williams, R.S. and Feist, W.C. 1984. Application of ESCA to evaluate wood and cellulose surfaces modified by aqueous chromium trioxide treatment. *Colloids & Surfaces.* 9(3): 253–271.

Williams, R.S. and Feist, W.C. 1985. Wood modified by inorganic salts: mechanism and properties. I. Weathering rate, water repellency, and dimensional stability of wood modified with chromium (III) nitrate versus chromic acid. *Wood & Fiber Science.* 17(2): 184–198.

Williams, R.S. and Feist, W.C. 1988. Performance of finishes on wood modified with chromium nitrate versus chromic acid. *Forest Products J.* 38(11/12): 32–35.

Williams, R.S. and Feist, W.C. 1993. Durability of paint or solid-color stain applied to preweathered wood. *Forest Products J.* 43(1): 8–14.

Willaims, R.S. and Feist, W.C. 1994. Effects of preweathering, surface roughness, and wood species on the performance of paint and stains. *J. Coatings Technology.* 66(828): 109–121.

Williams, R.S. and Feist, W.C. 2001. Duration of wood preweathering: effect on the service life of subsequently applied paint. *J. Coatings Technology.* 73(930): 65–72.

Williams, R.S. and Feist, W.C. 2004. Durability of yellow-poplar and sweetgum and service life of finishes after long-term exposure. *Forest Products J.* 54(7/8): 96–101.

Williams, R.S., Winandy, J.E., and Feist, W.C. 1987. Paint adhesion to weathered wood. *J. Coatings Technology.* 59(749): 43–49.

Williams, R.S., Knaebe, M.T., Evans, J.W., and Feist, W.C. 2001a. Erosion rates of wood during natural weathering: Part III. Effect of exposure angle on erosion rate. *Wood & Fiber Science.* 33(1): 50–57.

Williams, R.S., Knaebe, M.T., and Feist, W.C. 2001b. Erosion rates of wood during natural weathering: Part II. Earlywood and latewood erosion rates. *Wood & Fiber Science.* 33(1): 43–49.

Williams, R.S., Knaebe, M.T., Sotos, P.G., and Feist, W.C. 2001c. Erosion rates of wood during natural weathering: Part I. Effect of grain angle and surface texture. *Wood & Fiber Science.* 33(1): 31–42.

Williams, R.S., Sotos, P., and Feist, W.C. 1999. Evaluation of several finishes on severely weathered wood. *J. Coatings Technology.* 71(895): 97–102.

Winterowd, J.G., Lewis, C.E., Izan, J.D., and Shantz, R.M. 2004. Edge sealant formulation for wood-based panels. United States Patent Application 2004/0013857.

Wright, J.K. and Banks, W.B. 1989. The valence state of chromium in treated wood studied by magnetic susceptibility. *J. Wood Chemistry & Technology.* 9(4): 569–572.

Xie, Y., Krause, A., Mai, C., Militz, H., Richter, K., Urban, K., and Evans, P.D. 2005. Weathering of wood modified with the N-methylol compound 1,3-dimethylol-4,5-dihydroxyethyleneurea. *Polymer Degradation & Stability.* 89(2): 189–199.

Yagishita, M. 1958. The results of weathering tests on veneer-overlay plywood. *Wood Industry.* 13(3): 17–20.

Yassin, A.A. and Sabaa, M.W. 1982. *p*-Benzoquinone-tin polycondensates as stabilisers for polybutadiene rubber against photo-degradation. *Polymer Degradation & Stability*. 4(4): 313–318.

Yata, S. 2001. Occurrence of drying checks in softwood during outdoor exposure. In: Y. Imamura. (Ed.), *High-Performance Utilization of Wood for Outdoor Uses*, Wood Research Institute, Kyoto University, Kyoto, Japan, pp. 65–70 .

Yata, S. and Tamura, T. 1995. Histological changes of softwood surfaces during outdoor weathering. *Mokuzai Gakkaishi*. 41(11): 1035–1042.

Yoshida, H. 1986. Bond durability of water-based polymer-isocyanate adhesives (API resin) for wood I. Degradation of bond quality of API resin-bonded plywood by accelerated aging treatments and natural weathering. *Mokuzai Gakkaishi*. 32(6): 432–438.

Yoshida, H. and Taguchi, T. 1978a. Shear properties of weathered plywood. *Mokuzai Gakkaishi*. 24(10): 720–725.

Yoshida, H. and Taguchi, T. 1978b. Impact bending and rolling shear properties of weathered plywood. *Mokuzai Gakkaishi*. 24(8): 546–551.

Yoshimoto, T., Minami, K., and Kondo, M. 1967. Photodegradation of wood. II. Imperfect hardening of cement caused with reducing substances produced in wooden moulding board exposed to sunlight. *Mokuzai Gakkaishi*. 13(3): 96–101.

Yu, Y., Jiang, Z., Wang, G., and Song, Y. 2010. Growth of ZnO nanofilms on wood with improved photostability. *Holzforschung*. 64(3): 385–390.

Yusuf, S., Imamura, Y., Takahashi, M., and Minato, K. 1995. Weathering properties of chemically modified wood with some cross-linking agents and its decay resistance after weathering. *Mokuzai Gakkaishi*. 41(8): 785–793.

Zahora, A.R. 1991. Interactions between water-borne preservatives and emulsion additives that influence the water repellency of wood. *International Research Group on Wood Preservation Doc*. IRG/WP/2374.

Zahora, A.R. 1992. A water repellent additive's influence on the field performance of southern yellow pine lumber. *Proceedings Eighty-Eighth Annual Meeting of the American Wood-Preservers' Association*. pp. 148–159.

Zahora, A.R. 2000. Long-term performance of a "wax" type additive for use with water-borne pressure preservative treatments. *International Research Group on Wood Preservation Doc*. IRG/WP/00–40159.

Zahora, A.R. and Rector, C.M. 1990. Water repellent additives for pressure treatments. *Proceedings of the Eleventh Annual Meeting of the Canadian Wood Preservation Association*. pp. 22–41.

Zhang, J., Kamdem, D.P., and Temiz, A. 2009. Weathering of copper-amine treated wood. *Applied Surface Science*. 256(3): 842–846.

Zhang, X. 2003. *Photo-Resistance of Chemically Treated Wood*. M.Sc. thesis, Department of Wood Science, University of British Columbia, Vancouver, BC, Canada.

8 Surface Characterization

Mandla A. Tshabalala, Joseph Jakes, Mark R. VanLandingham, Shaoxia Wang, and Jouko Peltonen

CONTENTS

8.1 OVERVIEW OF SURFACE PROPERTIES

Surface properties of wood play an important role when wood is used or processed into different commodities such as siding, joinery, textiles, paper, sorption media, or wood composites. Thus, for example, the quality and durability of a wood coating are determined by the surface properties of the wood and the coating. The same is true for wood composites where the efficiency of stress transfer from the wood component to the nonwood component is strongly influenced by the surface properties of both components.

Surface properties of wood can be divided into two major groups: physical and chemical properties. Physical properties include morphology, roughness, smoothness, specific surface area, and permeability. Chemical properties include elemental and molecular or functional group composition. Taken together these two major groups of properties determine the thermodynamic characteristics of the wood surface such as surface-free energy and surface acid–base acceptor and donor numbers, and mechanical properties such as hardness, modulus of elasticity, and toughness.

Wood has a cellular structure whose cell walls are composed of three major constituents: cellulose, hemicelluloses, and lignin. In addition to these major constituents the cell wall also contains pectins, extractives, and trace metals. The surface properties of wood are, therefore, determined by the morphology of the cell wall at the surface of a wood element (particle, fiber, flake, or chip), and the distribution of the major and minor constituents in the cell wall. Hence, in order to optimize the quality of interaction of the wood surface with a coating, or with the matrix in a wood composite, the surface properties of both the wood and the coating or the matrix in a composite must be known.

Methods for characterizing surface properties of wood may be divided into four broad categories: microscopic, spectroscopic, thermodynamic, and nanomechanical. Microscopic methods provide information about surface morphology; spectroscopic methods provide information about surface chemistry; thermodynamic methods provide information about the surface energy; and nanomechanical methods provide information about the mechanical strength of wood cell walls.

8.2 MICROSCOPIC METHODS FOR CHARACTERIZING SURFACE PROPERTIES

Many types of microscopic methods are available for characterizing physical properties of surfaces of various materials, but only a few have been particularly useful and productive in characterizing physical properties of wood surfaces. These are confocal laser scanning microscopy (CLSM), scanning electron microscopy (SEM), and atomic force microscopy (AFM).

8.2.1 CONFOCAL LASER SCANNING MICROSCOPY (CLSM)

A typical CLSM instrument consists of a light microscope equipped with scanning mechanisms and motorized focus, a laser source (He/Ne, krypton, or argon), and computer system with software for instrument control, and for three-dimensional (3-D) reconstruction, image processing, and image analysis.

The basic principle of CLSM is illustrated schematically in Figure 8.1. A collimated, polarized laser beam from an aperture is reflected by a beam splitter (dichroic mirror) into the rear of the objective lens and is focused on the specimen. The reflected or emitted, longer-wavelength fluorescent light returning from the specimen passes back through the same lens. The light beam is focused by the beam splitter into a small pinhole, the confocal aperture, to eliminate all the out-of-focus light that comes from regions of the specimen above or below the plane of focus. A photo multiplier tube (PMT) positioned behind the confocal aperture converts the detected in-focus light beam from each specimen point into an analog output signal that is stored in a computer in digital form. A point-by-point digital image is obtained by scanning the beam over an XY plane. The thickness of such an XY image in the Z direction depends on the diameter of the detector pinhole: the more open

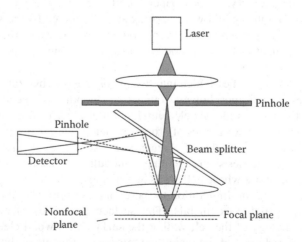

FIGURE 8.1 Principle of CLSM. Out-of-focus light beam from nonfocal plane is excluded from the detector by the confocal detector pinhole. (Courtesy of Lloyd Donaldson, M.Sc. Hons, D. Sc. Cell Wall Biotechnology Center, Forest Research, New Zealand.)

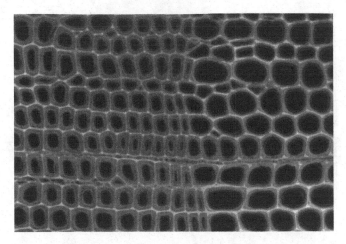

FIGURE 8.2 Transverse view of thick-wall latewood (left) and thin-wall early wood (right) tracheids of *Pinus radiata*. (Micrographs provided by Lloyd Donaldson, M.Sc. Hons, D.Sc. Cell Wall Biotechnology Center, Forest Research, New Zealand.)

the pinhole, the thicker the XY image (Pawley 1990, Boyde 1994, Lichtman 1994, Béland and Mangin 1995, Leica 1999).

Perhaps the greatest advantage of CLSM over other forms of optical microscopy is that it allows 3-D imaging of thick and opaque specimens such as wood surfaces without physically slicing the specimen into sections. Figures 8.2 through 8.4 show 625 μm × 625 μm micrographs of a wood specimen obtained on a Leica TCS/NT confocal microscope. The wood specimen was stained with acriflavin (Donaldson 2003). The micrographs clearly show the anisotropic morphology of wood surfaces. In transverse view (Figure 8.2) thin-walled early wood and thick-walled late wood tracheids are clearly distinguishable. In the radial view (Figure 8.3) radial bordered pits and an uniseriate heterogeneous ray are clearly visible. In the tangential view (Figure 8.4) uniseriate rays, small tangential pits in late wood, and larger tangential pits in early wood are also clearly visible.

CLSM has also been used in the study of resin distribution in medium density fiberboard (Loxton et al. 2003). In another study, CLSM was used to study the effect of simulated acid rain on coatings for exterior wood panels (Lee et al. 2003).

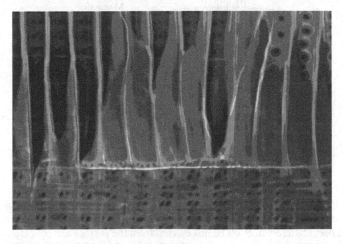

FIGURE 8.3 Radial view of tracheids of *Pinus radiata* showing radial bordered pits (upper left-hand corner) and uniseriate heterogeneous ray (center of micrograph).

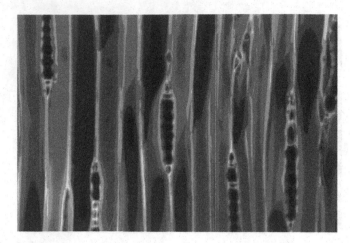

FIGURE 8.4   Tangential view of tracheids of *Pinus radiata* showing fusiform rays, and uniseriate heterogeneous and homogeneous rays.

8.2.2 Scanning Electron Microscopy (SEM)

The basic principle of SEM is illustrated in Figure 8.5. The primary electron beam, which is produced under high vacuum at the top of the microscope by heating a metallic element, is scanned across the surface of a specimen. When the electrons strike the specimen, a variety of signals are generated, and it is the detection of specific signals that produces an image of the surface, or its elemental composition. The three signals, which provide the greatest amount of information in SEM, are the secondary electrons, backscattered electrons, and x-rays.

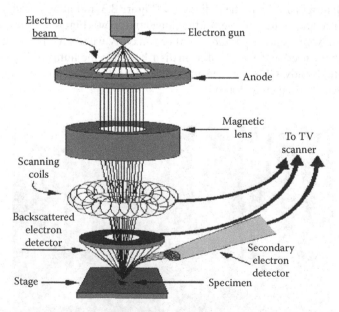

FIGURE 8.5   Principle of SEM. The primary electron beam produced at the top of the microscope by heating of metallic filament is focused and scanned across surface of a specimen by means of electromagnetic coils. Secondary and backscattered electron beams ejected from the surface are collected in detectors, which convert them to a signal that is processed into an image. (Courtesy of Prof. Scott Chumbley, Iowa State University, Materials Science and Engineering Department.)

Secondary electrons are emitted from the atoms occupying the top surface layer, and produce an image of the surface. The contrast in the image is determined by the surface morphology. A high-resolution image can be obtained because of the small diameter of the primary electron beam.

Backscattered electrons are primary beam electrons, which are reflected by the atoms in the surface layers. The contrast in the image produced is determined by the atomic number of the elements in the surface layers. The image will therefore show the distribution of different chemical phases in the specimen surface. Because backscattered electrons are emitted from a depth in the specimen, the resolution in the image is usually not as good as for secondary electrons.

Interaction of the primary beam with atoms in the specimen causes electron shell transitions, which result in the emission of an x-ray. The emitted x-ray has an energy that is characteristic of its parent element. Detection and measurement of x-ray energy permits elemental analysis, and is commonly referred to as energy dispersive x-ray spectroscopy (EDS or EDX or EDXA). EDS can provide rapid qualitative or, with adequate calibration standards, quantitative analysis of elemental composition with a sampling depth of 1–2 μm. X-rays may also be used to form maps or line profiles, showing the elemental distribution in a specimen surface.

One of the latest innovations in SEM is environmental scanning electron microscopy (ESEM). ESEM differs from conventional SEM in two crucial aspects (Danilatos 1993, Donald 2003). First, ESEM allows the introduction of a gaseous environment in the specimen chamber although the electron gun itself is kept under the standard SEM high vacuum. Second, the wood specimens do not need to be coated with a metallic layer, as is the case in conventional SEM.

It should be noted that the gaseous environment in the specimen chamber plays a key role in signal detection (Danilatos 1988, Meredith et al. 1996). As the secondary electrons travel toward the positively charged detector they collide with the gas molecules. Each such collision leads to the generation of an additional daughter electron, so that an amplified cascade of electrons reaches the detector. Along with additional daughter electrons, positive ions are also produced. These positive ions drift down toward the specimen, and hence serve to compensate charge build-up at the surface of nonconductive specimens such as wood. It is for this reason that nonconductive specimens do not need to be coated with a metallic layer to prevent charging artifacts and consequent loss of image quality.

Figures 8.6, 8.7, and 8.8 show ESEM micrographs of specimens prepared from the softwood *Pinus taeda*.

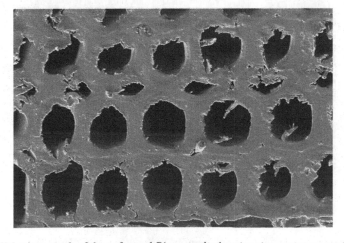

FIGURE 8.6   ESEM micrograph of the softwood *Pinus taeda* showing the transverse surface of thick-walled latewood tracheids. A uniseriate ray is visible above the bottom row of tracheids. (Courtesy T.A. Kuster, USDA Forest Service, Forest Products Laboratory, Madison, WI.)

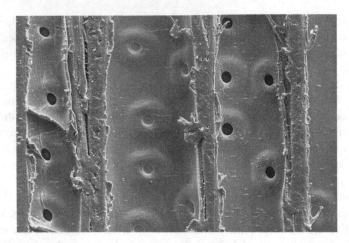

FIGURE 8.7 ESEM micrograph of the softwood *Pinus taeda* showing the radial surface of tracheids. Bordered pits are clearly visible. The warty membrane that lines the lumen is also clearly visible, especially in the left lumen where it has peeled back at some spots. (Courtesy T.A. Kuster, USDA Forest Service, Forest Products Laboratory, Madison, WI.)

8.2.3 ATOMIC FORCE MICROSCOPY (AFM)

The basic principle and application of AFM has been the subject of a number of excellent reviews (Meyer 1992, Frommer 1992, Hoh and Engel 1993, Frisbie et al. 1994, Hanley and Gray 1994, 1995, Louder and Parkinson 1995, McGhie et al. 1995, Rynders et al. 1995, Schaefer et al. 1995, Hansma et al. 1998). In AFM, a probe consisting of a sharp tip (nominal tip radius on the order of 10 nm) located near the end of a microcantilever beam is raster scanned across the surface of a specimen using piezoelectric scanners. Changes in the tip-specimen interaction are often monitored using an optical lever detection system, in which a laser beam is reflected off the cantilever and onto a position-sensitive photodiode. During scanning, a particular operating parameter is maintained at a constant level, and topographic images are generated through a feedback loop between the optical detection system and the piezoelectric scanners.

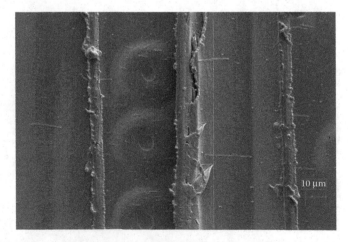

FIGURE 8.8 ESEM micrograph of the softwood *Pinus taeda* showing the tangential surface of thick-walled latewood tracheids. The tangential surface has relatively fewer bordered pits compared to the radial surface. (Courtesy T.A. Kuster, USDA Forest Service, Forest Products Laboratory, Madison, WI.)

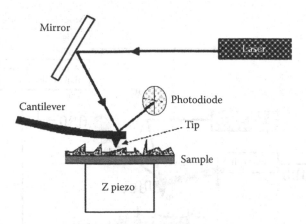

FIGURE 8.9 Schematic diagram of contact mode AFM. The tip makes soft physical contact with the sample. Optical detection of the position of the cantilever leads to a topographic map of the sample surface. (Courtesy of National Physical Laboratory, Teddington, Middlesex, UK, TW11 0LW. © Crown Copyright 2004. Reproduced by permission of the Controller of HMSO.)

Three imaging modes, contact mode, noncontact mode, and intermittent contact or tapping mode can be used to produce a topographic image of the surface. In contact mode (see Figure 8.9), the probe is essentially dragged across the surface of the specimen. During scanning, a constant bend in the cantilever is maintained. A bend in the cantilever corresponds to a displacement of the probe tip, z_t, relative to an undeflected cantilever, and is proportional to the applied normal force, $P = k \cdot z_t$, where k is the cantilever spring constant. As the topography of the surface changes, the z-scanner must move the position of the tip relative to the surface to maintain this constant deflection. Using this feedback mechanism, the topography of the specimen surface is thus mapped during scanning by assuming that the motion of the z-scanner directly corresponds to the surface topography. To minimize the amount of applied force used to scan the sample, low-spring constant ($k \leq 1$ N/m) probes are normally used. However, significant deformation and damage of soft samples (e.g., biological and polymeric materials) often occurs during contact mode imaging in air, because significant force must be applied to overcome the effects of surface roughness or adsorbed moisture as is the case with wood specimens. The combination of a significant normal force, the lateral forces created by the dragging motion of the probe tip across the specimen, and the small contact areas involved result in high contact stresses that can damage either the specimen surface, or the tip, or both. To overcome this limitation, contact mode imaging can be performed under a liquid environment, which essentially eliminates problems due to surface moisture such that much lower contact forces can be used. In fact, the ability to image samples under a liquid environment is often a desirable capability of AFM, but in some cases it might not be practical or feasible. Also, working with liquid cells for many commercial AFM systems can be tricky because of the potential for spills and leaks that can introduce liquid into the scanners.

To mitigate or completely eliminate the damaging forces associated with contact mode, the cantilever can be oscillated near its first bending mode resonance frequency (normally on the order of 100 kHz) as the probe is raster scanned above the specimen surface in either noncontact mode or tapping mode. In noncontact mode (see Figure 8.10), both the tip-specimen separation and the oscillation amplitude are on the order of 1–10 nm, such that the tip oscillates just above the specimen. The resonance frequency and amplitude of the oscillating probe decrease as the specimen surface is approached due to interactions with van der Waals and other long-range forces extending above the surface. These types of forces tend to be quite small relative to the repulsive forces encountered in contact mode. Either a constant amplitude or constant resonance frequency is maintained through a feedback loop with the scanner and, similar to contact mode, the motion of the scanner is used to

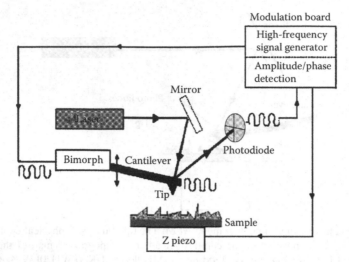

FIGURE 8.10 Schematic diagram of noncontact mode AFM. The cantilever is vibrated near the surface of the sample. Changes in the resonance frequency or vibration amplitude of the cantilever as the tip approaches the surface are used to measure changes in the sample surface topography. (Courtesy of National Physical Laboratory, Teddington, Middlesex, UK, TW11 0LW. © Crown Copyright 2004. Reproduced by permission of the Controller of HMSO.)

generate the topographic image. To reduce the tendency of the tip to be pulled down to the surface by attractive forces, the cantilever spring constant is normally much higher compared to contact mode cantilevers. The combination of weak forces affecting feedback and large spring constants causes the noncontact AFM signal to be small, which can lead to unstable feedback and requires slower scan speeds than either contact mode or tapping mode. Also, the lateral resolution in noncontact mode is limited by the tip-specimen separation and is normally lower than that in either contact mode or tapping mode.

Tapping mode tends to be more applicable to general imaging in air, particularly for soft samples, as the resolution is similar to contact mode while the forces applied to the specimen are lower and less damaging. In fact, the only real disadvantages of tapping mode relative to contact mode are that the scan speeds are slightly slower and the AFM operation is a bit more complex, but these disadvantages tend to be outweighed by the advantages. In tapping mode, the cantilever oscillates close to its first bending mode resonance frequency, as in noncontact mode. However, the oscillation amplitude of the probe tip is much larger than for noncontact mode, often in the range of 20–200 nm, and the tip makes contact with the sample for a short time in each oscillation cycle. As the tip approaches the specimen, the tip-specimen interactions alter the amplitude, resonance frequency, and phase angle of the oscillating cantilever. During scanning, the amplitude at the operating frequency is maintained at a constant level, called the set-point amplitude, by adjusting the relative position of the tip with respect to the specimen. In general, the amplitude of oscillation during scanning should be large enough such that the probe maintains enough energy for the tip to tap through and back out of the surface layer.

One recent development in tapping mode is the use of the changes in phase angle of the cantilever probe to produce a second image, called a phase image or phase contrast image. This image often provides significantly more contrast than the topographic image and has been shown to be sensitive to material surface properties, such as stiffness, viscoelasticity, and chemical composition. In general, changes in phase angle during scanning are related to energy dissipation during tip-sample interaction and can be due to changes in topography, tip-specimen molecular interactions, deformation at the tip-specimen contact, and even experimental conditions. Depending on the operating conditions, different levels of tapping force might be required to produce an accurate and reproducible

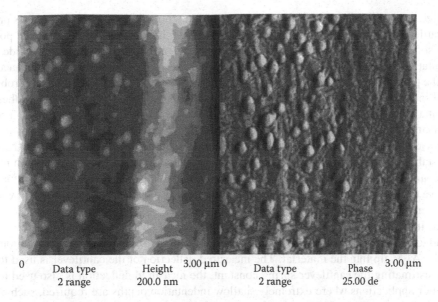

| 0 | Data type
2 range | Height
200.0 nm | 3.00 μm | 0 | Data type
2 range | Phase
25.00 de | 3.00 μm |

FIGURE 8.11 Topographical (height) and phase contrast images of a bordered pit showing the fibrillar morphology of the margo. The phase contrast image clearly shows the network of nanofibrils that constitute the margo and nodules located close to the pores that permeate the margo.

image of a surface. Also, the amount of tapping force used will often affect the phase image, particularly with regard to whether local tip-specimen interactions are attractive or repulsive.

Similar contrast images can be constructed concurrently with the topographic image in contact mode. One example is lateral force or friction force imaging, in which torsional rotation of the probe is detected while the probe is dragged across the surface in a direction perpendicular to the long axis of the cantilever. Friction force imaging with a chemically modified probe (i.e., a probe that has been coated with a monolayer of a specific organic group) is often referred to as chemical force microscopy. Another example of contrast imaging is force modulation, which combines contact mode imaging with a small oscillation of the probe tip at a frequency far below resonance frequency. This oscillating force should deform softer regions more than harder regions of a heterogeneous sample such that contrast between these regions is observed. In practice, however, the difference between the elastic modulus of the different regions usually has to be substantial (e.g., rubber particles in a plastic, carbon fibers in an epoxy) for contrast to be realized. Thus far, these types of AFM contrast images are purely qualitative due to inaccurate or unknown spring constants, unknown contact geometry, and contributions from different types of surface properties.

Figure 8.11 shows topographic (height) and a phase contrast images of a bordered pit on the surface of the cell wall of a softwood specimen. The images were acquired and recorded simultaneously by tapping mode AFM at ambient conditions, using a Nanoscope MultiMode SPM™.

8.2.4 NANOINDENTATION

Surface mechanical properties of the wood cell wall can be determined by nanoindentation. Nanoindentation is a type of hardness test in which a specifically shaped probe is pressed into a material and withdrawn following a prescribed loading profile. Load and displacement are continuously monitored and recorded during the experiment. The resulting load-depth trace can be analyzed to assess mechanical properties including, most often, elastic modulus and hardness. The primary advantage of nanoindentation over other types of mechanical tests is its ability to assess mechanical properties from small volumes of materials. Depths of penetration typically range from 10 nm to 10 μm. The initial motivation for development of nanoindentation in the 1980s was to characterize hard, thin inorganic films. Since its inception, however, nanoindentation has proven to be a versatile technique amendable to nearly any material, including ceramics, metals, polymers, and both soft and hard biological tissues. Basic requirements for a nanoindenter include precise instrumentation to separately measure load and displacement, an actuation mechanism to accurately perform the prescribed loading profile, and translational stages to precisely position the probe to test regions of interest. Numerous commercial instruments are available, and each achieves these basic instrumentation requirements in different ways. Two examples are given in Figure 8.12. These and other instruments are also described in greater detail in Fischer-Cripps (Fischer-Cripps 2004). Typical noise floors in commercial nanoindenters are sub-nm in displacement and sub-μN in force. Thermal stability and vibration isolation are critical to achieving quality nanoindentation measurements, so nanoindenters are usually housed in their own, specially designed enclosures. For materials sensitive to changes in moisture content, relative humidity also needs to be controlled within the nanoindenter enclosure.

Surface indentation can also be performed with an AFM, which is more sensitive at shallow depths and lower loads than a nanoindenter. During AFM-based indentation, piezo actuation is used to press the AFM tip into the material. The measured deflection of the cantilever is used to assess depth. By estimating the cantilever spring constant, the measured deflection is also used to assess the load. For applications where extremely shallow indentation depths are required, such as in the characterization of individual cellulose nanocrystals (Lahiji et al. 2010), AFM-based indentation is more suitable than conventional nanoindentation. However, uncertainties in probe shape and cantilever spring constant can produce large errors in AFM-based indentation measurements. In addition, the cantilever beam can impart lateral forces to the surface in addition to the usual normal forces, a factor that is not accounted for in usual nanoindentation analysis.

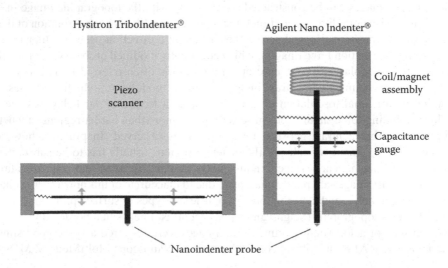

FIGURE 8.12 Two examples of the basic instrumentation used in commercial nanoindenters. The Hysitron Inc. (Minneapolis, MN, USA) TriboIndenter® uses a three plate capacitive force/displacement transducer. The probe is attached to the center plate which is suspended between the two outer plates by leaf springs. To measure displacement, an AC signal 180° out of phase is applied to the two outer plates creating an electric field potential between the two outer plates that is zero in the center. The voltage measured from the middle plate is calibrated to measure displacement. Force is applied to the center plate by applying a large DC offset to the bottom or top plate, which creates an electrostatic attraction with the center plate that can be calibrated as a force. The transducer is attached to a piezo scanner, which allows the nanoindenter probe to also be used as an imaging probe, similar to contact mode AFM. The Agilent (Santa Clara, CA, USA) Nano Indenter® uses an electromagnetic force transducer and capacitance gauge to measure displacement. The probe assembly is also supported by leaf springs.

The size-scale of nanoindentation makes it an ideal tool for many areas of wood science research, including studying both unmodified and modified wood cell walls, adhesives, and coatings. Wimmer and coworkers were the first to use nanoindentation to assess the mechanical properties of S_2 cell wall laminae and compound corner middle lamellae in wood (Wimmer and Lucas 1997, Wimmer et al. 1997). They assessed the properties in the longitudinal direction, similar to those shown in Figure 8.13a for loblolly pine (*Pinus taeda*). The indents were made using a Berkovich probe, a three-sided pyramid commonly used in nanoindentation. Other types of probes, such as cube corner (a three-sided pyramid that is sharper than the Berkovich), conical, spherical, and flat punches, can also be used (Fischer-Cripps 2004). The load-depth traces for the same indents are also shown in Figure 8.13b. They are comprised of three segments: a loading segment, a hold segment at maximum load, and an unloading segment. Accurate depth measurements require accounting for the machine compliance of the nanoindenter. The machine compliance arises from deformation in the instrument itself, such as compression of the shaft connecting the probe to the load actuator. The machine compliance can easily be measured by performing a series of indents with varying loads in a standard material like fused silica and employing the SYS correlation (Stone et al. 1991, Jakes et al. 2008a). Commercial instruments already have built into them the calibration routines that rely on a version of this correlation and can be set up to automatically remove displacement arising from the machine compliance from the depth measurements.

The Oliver–Pharr method (Oliver and Pharr 1992, 2004) is the most common nanoindentation analysis used to calculate Meyer hardness (i.e., load divided by projected area) and elastic modulus. In this method, contact area of the indent is estimated using the contact depth calculated directly from the load-depth trace and a predetermined area function for the probe used in the experiment (Oliver and Pharr 1992, 2004). Therefore, unlike traditional hardness tests, there is no need to image the residual indent impressions to determine the contact area. If necessary, high-resolution images of residual indent impressions, usually obtained from AFM or SEM, can be used to manually measure the contact area. Meyer hardness is a metric of deformation resistance and is usually defined as the maximum load achieved just prior to unloading divided by the contact area of the indent. For highly elastic materials, such as rubber, the Meyer hardness is proportional to the Young's modulus. When yielding takes place beneath the indenter the hardness becomes proportional to flow stress (yield stress at some particular strain, 7–8% for a Berkovich indenter). The proportionality "constant" between hardness and flow stress depends on the hardness/modulus ratio of the material

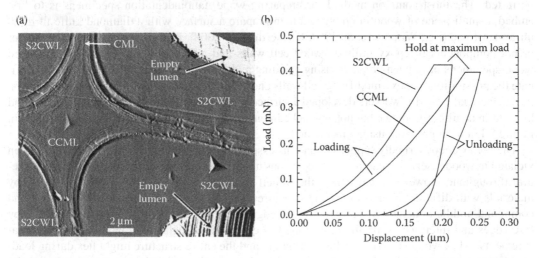

FIGURE 8.13 AFM images (a) of nanoindents placed in the compound corner middle lamella (CCML) and S2 cell wall lamina (S2CWL). The CCML is formed at the junction of three or more tracheids and is part of the compound middle lamella (CML). Load-depth traces (b) of these nanoindents are also shown.

tested (Johnson 1985, Jakes et al. 2011b). Values of 350 and 400 MPa were calculated using the Oliver–Pharr method for the compound corner middle lamella and S_2 cell wall lamina, respectively, in Figure 8.13. Elastic modulus is calculated from the initial slope of the unloading segment in a load-depth trace and the contact area. Note that both the indents in Figure 8.13 are approximately the same size, but the initial slope of the unloading segment in the load-depth trace for the compound corner middle lamella is noticeably lower than for the S_2 cell wall lamina. This indicates the elastic modulus is lower in the compound corner middle lamella and indeed, for these particular indents elastic moduli of 5 GPa and 18 GPa were calculated using the Oliver–Pharr method for the compound corner middle lamella and S2 cell wall lamina, respectively.

Ever since the pioneering work of Wimmer et al. (Wimmer and Lucas 1997, Wimmer et al. 1997), nanoindentation has been used in many areas of wood science research including mechanical properties of different types of wood cells (Gindl et al. 2002, 2004a, Tze et al. 2007); unmodified wood cell walls from different tree species (Wu et al. 2009, Gershon et al. 2010); wood adhesives and wood-adhesive bond lines (Gindl et al. 2004b,c, Konnerth and Gindl 2006, 2008, Konnerth et al. 2006, 2007, Follrich et al. 2010, Stöckel et al. 2010a,b); cell walls conditioned at different values of relative humidity (Yu et al. 2010); thermally modified cell walls (Stanzl-Tschegg et al. 2009); melamine-modified cell walls (Gindl et al. 2002); pyrolyzed cell walls (Zickler et al. 2006, Brandt et al. 2010); cell walls damaged in the process of making wood composites (Xing et al. 2009, Gacitua et al. 2010); biologically degraded cell walls (Konnerth et al. 2010b, Lehringer et al. 2011); and bleached pulp fiber cell walls (Adusumalli et al. 2010a). All these studies employed the standard Oliver–Pharr method. Even though the Oliver–Pharr method has provided wood researchers with useful results, the validity and significance of the nanoindentation results has been questioned (Gindl and Schoberl 2004, Jakes et al. 2008a, Jäger et al. 2011). Of particular concern are the violations of the basic assumptions of the Oliver–Pharr method. The Oliver–Pharr method assumes the material tested is homogeneous, is rigidly supported in the nanoindenter, and is isotropic. Experiments performed in wood cell walls violate these assumptions. The Oliver–Pharr method also does not address issues associated with testing materials with time-dependent mechanical properties, which wood cell walls possess. Much recent work has focused on overcoming these shortfalls of the standard Oliver–Pharr method to better understand the meaning of nanoindentation results in wood science research. In addition, improved methods for sample preparation have been developed.

Surfaces prepared for nanoindentation must be smooth on the nanometer-scale. Also, the technique used to prepare the surface must not affect the mechanical properties of the material that is to be tested. The most common method for preparing wood nanoindentation specimens is to first embed a small piece of wood in epoxy and then prepare a surface with a diamond knife fit in an ultramicrotome (e.g., Wimmer et al. 1997, Konnerth et al. 2008). However, it is uncertain whether or not components of epoxy infiltrate wood cell walls and affect their properties. Unembedded wood specimens have been prepared using grinding and polishing (Zickler et al. 2006). To eliminate the possibility of epoxy-modifying cell walls and to limit the amount of mechanical damage on the surface, Jakes and coworkers developed a surface preparation method that also uses a diamond knife fit in an ultramicrotome but not any embedment (Jakes et al. 2008a,b, 2009a). The surfaces in Figure 8.13a were prepared using this method.

The homogeneous, rigidly supported material assumptions of the Oliver–Pharr method are often violated in wood science research. A homogeneous material has a consistent composition and structure throughout. However, for instance, the S_2 cell wall lamina (Figure 8.13a), is surrounded by materials with different compositions and structures, such as the S_1 and S_3 cell wall laminae and compound middle lamellae. Also, there are free edges because of the empty lumina. These nearby free edges and interfaces with different properties can produce artifacts in nanoindentation measurements. Also, wood is an open cellular structure and the entire structure might flex during loading, which violates the rigid-support assumption. Jakes and coworkers recently developed the structural compliance method to better understand the effects of edges and specimen-scale flexing in nanoindentation measurements (Jakes et al. 2008a, 2009a). They found that the effect of a nearby edge or specimen-scale flexing is to introduce a structural compliance into the measurement, which behaves similar to the machine compliance. Therefore, a method similar to the one used to assess the machine compliance, which utilizes the SYS correlation, can be used to assess the structural compliance so its affects can be removed from the measurement. Jakes and Stone also developed expressions predicting the effect of a free edge on hardness and elastic modulus measurements made using the Oliver–Pharr method (Jakes and Stone 2011), which allows researchers to both anticipate the effects of edges and design experiments that minimize edge effects.

The mechanical properties of an isotropic material are the same in all directions. Materials like the compound corner middle lamella and wood adhesives can likely be considered isotropic. However, properties of S_2 cell wall laminae cannot be considered isotropic because mechanical properties, especially elastic modulus, are influenced by the orientation of the stiff cellulose microfibrils. The microfibrils are wound helically around a wood cell and oriented more closely with the longitudinal axis than the transverse axis in a typical wood cell. As a consequence, the longitudinal elastic modulus is higher than the transverse elastic modulus. Because stress fields beneath an indenter are multiaxial, the elastic modulus assessed by nanoindentation will be influenced by the different elastic moduli and fall between the actual longitudinal and transverse elastic moduli. Using previously derived nanoindentation solutions (Vlassak et al. 2003), Jäger and coworkers (Jäger et al. 2011) developed the relationship between nanoindentation elastic modulus and S_2 cell wall laminae elastic constants assuming transverse isotropy. Using this relationship and experimental nanoindentation elastic moduli from S_2 cell wall laminae tested at different microfibril angles, Konnerth and coworkers (Konnerth et al. 2010a) calculated a longitudinal elastic modulus of 26 GPa and a transverse elastic modulus of 5 GPa for the S_2 cell wall lamina. For comparison, the nanoindentation elastic modulus when the cellulose microfibrils were oriented parallel to the indentation direction was about 18 GPa.

Most materials of interest to wood science, including wood cell walls, adhesives, composite matrices, and coatings, possess time-dependent mechanical properties. A better understanding of these time-dependent material properties would have many benefits in wood science research, such as an improved understanding of the time-dependent properties of bulk wood and wood-based composites. Also, time-dependent properties provide clues into the physical mechanisms that control properties. Not surprisingly, the assessed hardness and elastic modulus measured from nanoindentation depend on the times used in the loading profile (Tang and Ngan 2003, Fujisawa and Swain 2008). For instance, a longer hold segment allows the indenter to penetrate deeper into the material, resulting in a lower Meyer hardness. Therefore, reported property values are not unique and depend on the loading profile used. Numerous approaches have been proposed to better understand and assess time-dependent properties using nanoindentation (e.g., Loubet et al. 2000, Asif et al. 2001, Cheng and Cheng 2004, Vanlandingham et al. 2005, Herbert et al. 2009, Jakes et al. 2011a,b, Oyen 2011). In some approaches, dynamic experiments are performed by superimposing a small sinusoidal signal on the static load, allowing viscoelastic properties such as storage and loss moduli to be assessed over a range of frequencies (Loubet et al. 2000, Asif et al. 2001, Herbert et al. 2009). A common practice in interpreting time-dependent nanoindentation data is to assume a particular material constitutive model and solve the associated contact problem for a given probe geometry (Cheng and Cheng 2004, Vanlandingham et al. 2005, Oyen 2011). One must apply this approach with caution, however, because the results are at best as good as the assumed constitutive model, and many varieties of constitutive model can be used to fit experimental data. A more recent approach to assessing time-dependent properties is the application of broadband nanoindentation (Puthoff et al. 2009, Jakes et al. 2011a,b), in which viscoplastic and viscoelastic properties are independently assessed over a wide range of time-scale (approximately 4–5 orders of magnitude) without need to rely on a specific constitutive model to interpret the results. Broadband nanoindentation creep, which is a viscoplastic measurement, has already been used to better understand the effects on wood cell walls of ethylene glycol plasticization (Jakes et al. 2008b) and polymeric diphenyl diisocyanate adhesive infiltration (Jakes et al. 2009b).

Other modes of nanoindentation testing are also available (Fischer-Cripps 2004). Many commercial instruments are capable of applying a lateral force and measuring lateral displacement, which facilitates scratch testing. Scratch testing is particular useful for understanding scratch resistance in protective coating and investigating substrate-coating interface failures. Nanoindentation probes can also be used as scanning probes to create images, similar to the operation of an AFM. During imaging, a small force, typically 1 μN or less, is used to maintain contact with the probe. By applying a small sinusoidal load signal during scanning, elastic modulus maps can be created (Asif et al. 2001). By changing the force applied during probe scanning or changing the number of times an area is scanned, wear testing can be performed. Fracture events during nanoindentation can be investigated by analyzing the associated discontinuities in displacements in load-depth traces, or in some specialized instruments using acoustic emissions. Nanoindenters are also available to perform impact testing. Variable temperature nanoindentation is becoming increasingly common and has been used to study the effects of temperature on wood adhesives (Konnerth and Gindl 2008). Instrumentation is available to combine nanoindentation with SEM, which allows researchers to observe deformation mechanisms as they occur. This technique has been used to study *in situ* deformation of S_2 cell wall laminae (Adusumalli et al. 2010b) and single wood pulp fibers in the transverse direction (Adusumalli et al. 2010c). Instrumentation is also available to perform indentation inside a transmission electron microscope (TEM), which facilitates observation of atomic-scale deformations. However, so far at least, *in situ* TEM nanoindentation has primarily been used in metallurgical research (Minor et al. 2001, De Hosson et al. 2006).

The future of nanoindentation in wood science research holds numerous opportunities. In areas already benefiting from nanoindentation, like understanding effects of wood modifications on cell wall mechanical properties, nanoindentation will continue to aid researchers. In addition to assessing changes in mechanical properties, nanoindentation is also capable of becoming a tool to better understand the mechanisms controlling the properties. By combining techniques to assess properties over a wide range of time-scale, particularly broadband nanoindentation (Jakes et al. 2011a,b), with variable temperature or variable relative humidity capabilities, nanoindentation is transformed into a localized tool for mechanical spectroscopy. The kinetic signatures arising from mechanical spectroscopy results can be analyzed to gain insight into the physical mechanisms controlling the properties. *In situ* TEM indentation presents researchers with the exciting possibility to actually observe these mechanisms while they are occurring. Another exciting possibility is combining indentation with spectroscopic techniques, which gives the opportunity to observe changes in polymer chemistry or orientation during deformation beneath an indenter.

8.3 SPECTROSCOPIC METHODS FOR CHARACTERIZING SURFACE PROPERTIES

Although a broad array of spectroscopic methods for surface analysis of materials exists, only a handful of these have been applied in the characterization of wood surfaces. The focus of this section will be on those methods that the author deemed to be of greatest interest.

Spectroscopic methods may be divided into three classes: molecular, electronic, and mass spectroscopies. Molecular spectroscopy includes Fourier transform infrared (FTIR), Fourier transform attenuated total reflectance infrared (FT–ATR), and Fourier transform Raman (FT-Raman) methods. Electronic spectroscopy includes x-ray photoelectron (XPS), and energy dispersive x-ray spectroscopy (EDXA, EDX, or EDS). Mass spectroscopy includes secondary ion mass spectroscopy (SIMS) and static secondary ion mass spectroscopy (SSIMS).

8.3.1 MOLECULAR SPECTROSCOPY

Molecular spectroscopic methods can provide information about the chemical composition of a surface. However, successful characterization of a wood surface by any of the molecular spectroscopic techniques depends upon judicious choice of sampling techniques and critical interpretation of the data. Detailed descriptions of the fundamentals and application of molecular spectroscopy have been given elsewhere (Mirabella 1998, Stuart 2004). As significant advances in instrumentation are realized, new applications of infrared spectroscopy in surface characterization of wood have emerged.

FTIR and FT–ATR spectra of wood samples have been reported in the literature (Deshmukh and Aydil 1996, Tshabalala et al. 2003). Assignment of infrared absorption bands measured by different FTIR techniques is given in Table 8.1. Figure 8.14 shows examples of infrared spectra of the surface of a softwood wafer obtained by FTIR–ATR.

Infrared analysis of wood surfaces is generally not considered to be sufficiently surface sensitive because the sampling depth of infrared radiation is of the order of 100 μm. Consequently changes in infrared spectral features due to changes in surface chemistry are often masked by the spectral features of the underlying bulk chemistry of the wood specimen. In the example shown in Figure 8.14b, the peaks at 2916, 2847, 1269, 894, and 769 cm^{-1} were attributed to the alkoxysilane thin film that was deposited on the wood wafer (Tshabalala et al. 2003). The peaks at 2916 and 2847 cm^{-1} were assigned to the asymmetric and symmetric C–H stretch modes of the –CH$_3$ and –CH$_2$ groups (Sahli et al. 1994, Deshmukh and Aydil 1996, Conley 1996); the peak at 1269 cm^{-1} was assigned to the Si–CH$_3$ bending mode; and the peaks at 894 and 769 cm^{-1} were assigned to Si–C, Si–O, and

---

TABLE 8.1
Assignment of IR Bands of Solid Wood or Wood Particles Measured by Diffuse Reflectance FTIR

Band Position (cm^{-1})	Assignment
3380 (3418)[a]	O–H stretch vibration (bonded)
2928 (2928)	C–H stretch vibration
1736 (1745)	C=O stretch vibration (unconjugated)
1655 (1660)	H–O–H deformation vibration of adsorbed water and conjugated C=O stretch vibration
1595 (1596)	Aromatic skeletal and C=O stretch vibration
1504 (1507)	Aromatic skeletal vibration
1459 (1465)	C–H deformation (asymmetric) and aromatic vibration in lignin
1423 (1427)	C–H deformation (asymmetric)
1372 (1374)	C–H deformation (symmetric)
1326 (1329)	–CH$_2$ wagging vibration in cellulose
1267 (1270)	C–O stretch vibration in lignin, acetyl, and carboxylic vibration in xylan
1236 (1242)	C–O stretch vibration in lignin, acetyl, and carboxylic vibration in xylan
1158 (1165)	C–O–C asymmetric stretch vibration in cellulose and hemicellulose
1113 (1128[b])	O–H association band in cellulose and hemicellulose
1055[b] (1083)	C–O stretch in cellulose and hemicellulose
1042 (1036)	C–O stretch
1000 (1003)	C–O stretch in cellulose and hemicellulose
897 (899)	C1 group frequency in cellulose and hemicellulose
667 (670)	C–O–H out-of-plane bending mode in cellulose

Source: Adapted from Pandey, K.K. and Theagarajan, K.S. 1997. *Holz als Roh- und Werkstoff* 55:383–390.

[a] Quantities given in parenthesis correspond to band positions obtained from solid wood chips or wood particles undiluted with KBr.

[b] Highest intensity band.

(a)   (b)

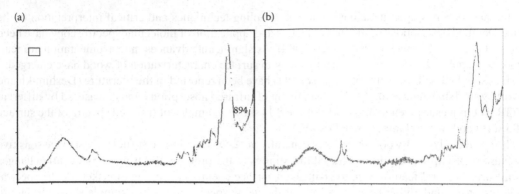

FIGURE 8.14 FTIR–ATR spectra of a wood wafer before (a) and after (b) coating with a thin film of alkoxysilanes.

Si–O–CH$_3$ groups (Selamoglu et al. 1989, Fracassi et al. 1992, Libermann and Lichtenberg 1994, Conley 1996).

Raman and infrared spectra give basically the same kind of molecular information, and both methods can be used to supplement or complement each other. Although Raman spectra are theoretically simpler than the corresponding infrared spectra, primarily because overlapping bands are much less common in Raman spectroscopy, its use in the study of wood surfaces is not widespread. The primary reason for this lack of popularity has been the difficulty of obtaining sufficiently intense Raman bands, which are not completely masked by the fluorescence bands emitted by the wood components when irradiated with visible light. However, with the advent of FT-Raman and introduction of the red laser source (excitation wavelength, 1064 nm), this difficulty has been largely overcome, and application of FT-Raman to the study of wood surfaces is likely to increase.

FT-Raman spectra of wood and wood components have been reported in the literature (Agarwal and Ralph 1997, Agarwal 1999). Table 8.2 gives a summary of band assignment in the FT-Raman spectra of wood components.

8.3.2 Electron Spectroscopy

Electron spectroscopy is concerned with the energy analysis of low energy electrons (generally in the range of 20–2000 eV) liberated from the surface of a specimen when it is irradiated with soft x-rays or bombarded with an electron beam (Watts 1990). Two methods of electron spectroscopy have been developed; x-ray photoelectron spectroscopy (XPS), also known as electron spectroscopy for chemical analysis (ESCA); and Auger electron spectroscopy (AES). Since AES has not been applied to the study of wood surfaces, the rest of this discussion will focus on XPS, which has been widely used for the characterization of the elemental composition of wood surfaces.

In XPS, the surface of a sample maintained in a high vacuum (10^{-6}–10^{-11} torr) is irradiated with x-rays, usually from Mg or Al anodes, which provide photons of 1253.6 eV and 1486.6 eV, respectively. Electrons are ejected from the core levels of atoms in the surface, and their characteristic binding energies are determined from the kinetic energy of the ejected electrons and the energy of the incident x-ray beam.

The kinetic energy (E_K) of the electron is the experimental quantity measured by the spectrometer. The binding energy of the electron (E_B) is related to the experimentally measured kinetic energy by the following relationship:

$$E_B = h\nu - E_K - W \tag{8.1}$$

where $h\nu$ is the incident photon energy and W is the spectrometer work function.

TABLE 8.2

Raman Band Assignment in the Spectra of Softwood Cellulose and Lignin

Band Position (cm⁻¹)	Assignment	Attributed to
3065 mᵃ	Aromatic stretch	Lignin
3007 sh	C–H stretch in OCH_3, asymmetric	Lignin
2938 m	C–H stretch in OCH_3, asymmetric	Lignin
2895 vs	C–H and CH_2 stretch	Cellulose
2886 sh	C–H stretch in R_3C–H	Lignin
2848 sh	C–H and CH_2 stretch	Cellulose
2843 m	C–H stretch in OCH_3, symmetric	Lignin
1658 s	Ring conj. C=C stretch of coniferyl alcohol; C=O stretch of coniferylaldehyde	Lignin
1620 sh	Ring conjugated C=C stretch of coniferylaldehyde	Lignin
1602 vs	Aryl ring stretch, symmetric	Lignin
1508 vw	Aryl ring stretch, asymmetric	Lignin
1456 m	H–C–H and H–O–C bending	Cellulose
1454 m	O–CH_3 deformation; CH_2 scissoring; guaiacyl ring vibration	Lignin
1428 w	O–CH_3 deformation; CH_2 scissoring; guaiacyl ring vibration	Lignin
1393 sh	Phenolic O–H bend	Lignin
1377 m	H–C–C, H–C–O, and H–O–C bending	Cellulose
1363 sh	C-H bend in R_3C–H	Lignin
1333 m	Aliphatic O–H bend	Lignin
1298 sh	H–C–C and H–C–O bending	Cellulose
1297 sh	Aryl-O of aryl-OH and aryl-O-CH_3; C=C stretch of coniferyl alcohol	Lignin
1271 m	Aryl-O of aryl-OH and aryl-O-CH_3; guaiacyl ring (with C=O group)	Lignin
1216 vw	Aryl-O of aryl-OH and aryl-O-CH_3; guaiacyl ring (with C=O group)	Lignin
1191 w	A phenol mode	Lignin
1149 sh	C–C and C–O stretch plus H–C–C and H–C–O bending	Cellulose
1134 m	A mode of coniferaldehyde	Lignin
1123 s	C–C and C–O stretch	Cellulose
1102 w	Out of phase C–C–O stretch of phenol	Lignin
1095 s	C–C and C–O stretch	Cellulose
1073 sh	C–C and C–O stretch	Cellulose
1063 sh	C–C and C–O stretch	Cellulose
1037 sh	C–C and C–O stretch	Cellulose
1033 w	C–O of aryl-O-CH_3 and aryl-OH	Lignin
1000 vw	C–C and C–O stretch	Cellulose
971 vw	C–C and C–O stretch	Cellulose
969 vw	–C–C–H and –HC = CH– deformation	Lignin
926 vw	–C–C–H wag	Lignin
900 vw	Skeletal deformation of aromatic rings, substituent groups, and side chains	Lignin
899 m	H–C–C and H–C–O bending at C6	Cellulose
787 w	Skeletal deformation of aromatic rings, substituent groups, and side chains	Lignin
731 w	Skeletal deformation of aromatic rings, substituent groups, and side chains	Lignin

continued

TABLE 8.2   (continued)
Raman Band Assignment in the Spectra of Softwood Cellulose and Lignin

Band Position (cm⁻¹)	Assignment	Attributed to
634 vw	Skeletal deformation of aromatic rings, substituent groups, and side chains	Lignin
591 vw	Skeletal deformation of aromatic rings, substituent groups, and side chains	Lignin
555 vw	Skeletal deformation of aromatic rings, substituent groups, and side chains	lignin
537 vw	Skeletal deformation of aromatic rings, substituent groups, and side chains	Lignin
520 m	Some heavy atom stretch	Cellulose
491 vw	Skeletal deformation of aromatic rings, substituent groups, and side chains	Lignin
463 vw	Skeletal deformation of aromatic rings, substituent groups, and side chains	Lignin
458 m	Some heavy atom stretch	Cellulose
435 m	Some heavy atom stretch	Cellulose
384 w	Skeletal deformation of aromatic rings, substituent groups, and side chains	Lignin
380 m	Some heavy atom stretch	Cellulose
357 w	Skeletal deformation of aromatic rings, substituent groups, and side chains	Lignin
351 w	Some heavy atom stretch	Cellulose

Source: Adapted from Agarwal, U.P. 1999. In: *Advances in Lignocellulosic Characterization.* Dimitris S. Argyropoulos (ed.), TAPPI Press, Atlanta, Georgia.

[a] *Note*: vs = very strong; s = strong; m = medium; w = weak; vw = very weak; sh = shoulder. Band intensities are relative to other peaks in the spectrum.

The kinetic energy of the electrons ejected from the surface of a specimen is determined by means of an analyzer, which gives a photoelectron spectrum. The photoelectron spectrum gives an accurate representation of the electronic structure of an element, as all electrons with a binding energy less than the incident photon energy will be featured in the spectrum.

The characteristic binding energy of a given atom is also influenced by its chemical environment. This dependence allows the assignment of a given atom to a particular chemical functional group on the surface. The sampling depth of XPS is about 5–10 nm.

Examples of survey and high-resolution XPS spectra of a wood specimen are shown in Figures 8.15 and 8.16 respectively. The survey spectrum gives the surface elemental composition of the wood specimen. As expected the spectrum is dominated by carbon and oxygen peaks, which are the elements that make up the constituents of wood. The high-resolution spectrum of the carbon peak shows the presence of different chemical states or classes of carbon on the wood surface.

In early pioneering studies on the surface analysis of paper and wood fibers by XPS, C1s, and O1s spectra were obtained for samples of Whatman™ filter paper, bleached kraft and bleached sulfite paper, spruce dioxane lignin, stone groundwood pulp, refiner mechanical pulp and thermomechanical pulp.

The C1s peak was observed to consist of four main components, which were ascribed to four classes of carbon atoms present in wood components. Class-I carbon atoms are those bonded to carbon or hydrogen; Class-II carbon atoms are bonded to a single noncarbonyl oxygen atom; Class-III carbon atoms are bonded to two noncarbonyl atoms or to a single carbonyl oxygen atom;

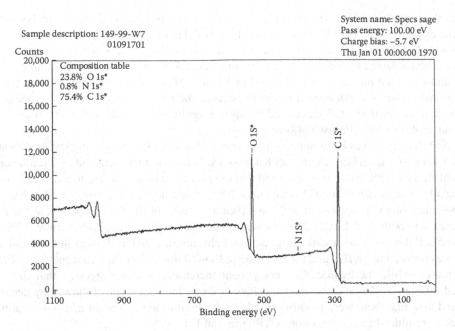

FIGURE 8.15 XPS survey spectrum of a wood specimen of *Pinus taeda* shows the elemental composition of the surface.

and Class-IV carbon atoms are ascribed to the carboxyl carbon. The four components were designated C1, C2, C3, and C4 respectively. The change in the relative magnitude of these components as a function of the oxygen ratio was taken to suggest that both lignin and extractives contribute to the change in surface composition on going from pure cellulose to mechanical wood pulps (Dorris and Gray 1978a,b, Gray 1978).

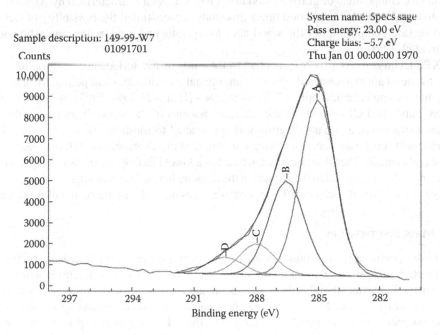

FIGURE 8.16 High-resolution XPS spectrum of the $C1_{s*}$ carbon showing the existence of different states or classes of carbon atoms on the surface of the wood specimen (A = C1; B = C2; C= C3; and D = C4).

In another pioneering study on the applicability of XPS to the chemical surface analysis of wood fibers, the oxygen-to-carbon (O/C) ratio of wood and wood fibers prepared by different methods was determined. The deviation of the observed O/C ratios from theoretical values was used to provide qualitative characterization of the surface chemical composition. For example, samples of unextracted and extracted pine chips showed O/C ratios of 0.26 and 0.42, respectively. This was interpreted as indicative of surfaces rich in lignin because the observed ratios were very close to the theoretical values of 0.33–0.36 calculated for spruce lignin and considerably lower than the theoretical value of 0.83 for cellulose (Mjörberg 1981).

The XPS technique has been applied in a number of studies for monitoring the modification of wood surfaces. Wood surfaces treated with aqueous solutions of nitric acid and of sodium periodate were analyzed by XPS, and it was observed that the periodate treatment led to a dramatic increase in the relative magnitude of the C2 component. The nitric acid treatment on the other hand led to the appearance of a C4 component and a significant increase of the C3 component at the expense of a decrease in both the C1 and C2 components (Young et al. 1982). In another study, XPS data of wood and cellulose surfaces treated with aqueous chromium trioxide showed that at least 75% of Cr(VI) was reduced to Cr(III), and the C1s spectra showed that the surface concentration of hydroxyls decreased while the hydrocarbon component increased. It was suggested that this change occurred through oxidation of primary alcohols in the cellulose to acids, followed by decarboxylation, and also that there were possible cellulose-chromium interactions in addition to previously proposed chromium-lignin interactions (Williams and Feist 1984).

Changes in the surface chemical composition of solid residues of quaking aspen (*Populus tremuloides*) wood extracted with supercritical fluid methanol were monitored by XPS, and it was shown that the C1s peak provided information that allowed for the rapid measurement of the proportion of carbon in polyaromatics. The components of the O1s peak were also tentatively assigned to oxygen in the major wood components, and to minor extractives, recondensed material and strongly adsorbed water (Ahmed et al. 1987, 1988). In another XPS study of weathered and UV-irradiated wood surfaces, the observed increase in the O/C ratio was interpreted as an indication of a surface rich in cellulose and poor in lignin (Hon 1984, Hon and Feist 1986).

The surface composition of grafted wood fibers has also been characterized by XPS. By grafting poly(methyl methacrylate) onto wood fibers this study demonstrated the possibility of tailoring the chemical surface composition of the wood fiber for specific end uses in thermoplastic composites (Kamdem et al. 1991).

The XPS technique has also been applied to the study of the surface composition of wood pulp prepared by the steam explosion pulp (SEP), conventional chemimechanical pulp (CMP), and conventional chemithermomechanical pulp (CTMP) processes (Hua et al. 1991, 1993a,b). Based on the theoretical O/C ratios and C1 contents of the main components of the wood fibers (i.e., carbohydrates, lignin, and extractives), a ternary diagram was constructed to illustrate the relative amounts of the three components on the surface. A tentative assignment of two components of the oxygen 1s peak, O1 and O2 was also made. The O1 component, which has a lower binding energy, was assigned to oxygen in lignin, and O2 was assigned to the oxygen in the carbohydrates. The investigators suggested that the percentage of the O1 peak area could be viewed as a measure of the lignin on the fiber surface.

8.3.3 MASS SPECTROSCOPY

Surface mass spectroscopy techniques consist in the measurement of the masses of secondary ions that are ejected, in a process known as sputtering, from the surface of a specimen when a primary beam of energetic particles bombards it. The secondary ions can provide unique information about the chemistry of the surface from which they originated. Thus surface mass spectroscopy and surface electron spectroscopy complement each other. Indeed, it is common practice to use electron spectroscopy in combination with surface mass spectroscopy to characterize the surface chemistry of an unknown specimen (Perry and Somorjai 1994).

There are three main methods of surface mass spectroscopy that are commonly used; secondary ion mass spectrometry (SIMS); laser ionization mass spectrometry (LIMA); and sputtered neutral mass spectrometry (SNMS). However, SIMS is by far the most commonly used of the surface mass spectroscopic techniques. While SIMS always revolves around the sputtering process, different modes of operation of SIMS exist. The three main modes of operation are static, dynamic, and scanning (or imaging) SIMS (Johnson and Hibbert 1992).

Static SIMS (SSIMS) operates under gentle bombardment conditions, and provides information about the chemistry of the upper surface of a specimen, while causing negligible surface damage. Dynamic SIMS (DSIMS) operates under relatively harsher bombardment conditions, and provides information about the chemistry of a surface from a few nanometers to several hundred microns in depth. Scanning or imaging SIMS, using highly focused ion beams, provides detailed chemical images with spatial resolution approaching that associated with SEM (less than 100 nm).

In SSIMS, the surface of a sample maintained in a high vacuum (10^{-6}–10^{-11} torr) is bombarded with ions (Ar^+, Xe^+, Cs^+, or Ga^+) at well-defined energies in the range of 1–4 KeV. Secondary ions are sputtered from the surface, and are detected in a mass spectrometer. To keep depletion of the surface at a minimum, the primary current is held at approximately 1 nA/cm^2. Under these conditions, only a very small fraction, approximately 1%, of the uppermost monolayer is consumed. The relatively low secondary ion flux is typically detected and analyzed in a time-of-flight mass spectrometer (TOFMS). The TOFMS has high sensitivity; extended mass range and high mass resolution, and is applicable to virtually all fields of science and technology where solid surfaces and their behavior are important (Watts 1992, Winograd 1993, Benninghoven et al. 1993, Comyn 1993, Li and Gardella 1994).

The development of SIMS and its application in the study of paper surfaces is the subject of an excellent review by Detter-Hoskin and Busch (1995). Although SSIMS spectra are complicated and sometimes difficult to interpret, the SSIMS technique has been successfully applied in the study of the surface composition and topochemistry of paper surfaces. Brinen compared the information content of XPS and SSIMS for characterizing sized-paper surfaces. He observed that while XPS readily detected the presence of sizing agents, it provided little structural information. Time-of-flight secondary ion mass spectroscopy (ToF–SSIMS) on the other hand, could be used to identify the chemical structures. Brinen also showed that the spatial distribution of the sizing agents on the paper surfaces could be obtained using ToF–SSIMS (Brinen 1993). In a combination of XPS, ToF–SIMS, and paper chromatography, Brinen et al. studied the cause of sizing difficulties in sulfite paper. Using SIMS imaging, they observed concentrated islands of pitch, which were implicated in the desizing effect (Brinen and Kulick 1995).

In another study of paper surfaces, Pachuta and Staral demonstrated that ToF–SSIMS with its extended mass range and high sensitivity was a useful tool for the nondestructive analysis of colorants on paper. They observed that the use of SSIMS conditions produced no detectable alteration of the paper samples (Pachuta and Staral 1994).

SIMS has also been used in the imaging of fiber surfaces. Tan and Reeve used SIMS imaging to reveal the microdistribution of organochlorine on fully bleached pulp fibers. They observed that pulp-bound chlorine was present over the entire surface and throughout the fiber cross-section. On the fiber outer surface, the chlorine concentration was observed to be higher in some small areas. In cross-section, chlorine was observed to be concentrated in the middle of the secondary wall. They concluded that since most of the pulp-bound organochlorine is covalently linked to large carbohydrate molecules held deep within the fiber wall, it was not likely to diffuse out of the fiber even after a long time (Tan and Reeve 1992).

Wang and Peltonen 2011 used ToF–SIMS to compare the surface chemistries of pine sapwood and heat-treated spruce. They found that less oleic, arachidic acid/tripalmitin and triolein and more stearic, behenic acid and sitosterols/sterylester were present on the surface of heat-treated spruce surface compared with pine sapwood. The ToF–SIMS spectra and images were obtained with a PHI TRIFT II instrument (Physical Electronics, USA) using a 15 kV pulsed liquid-metal ion source

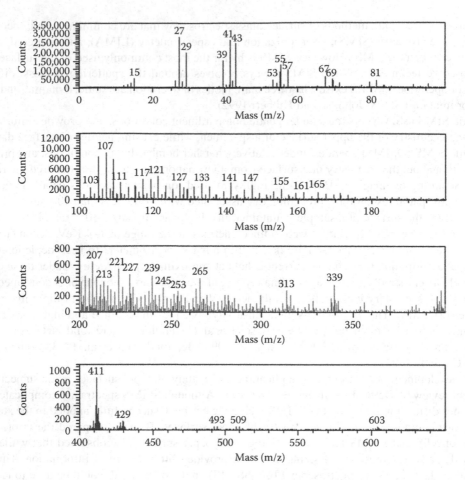

FIGURE 8.17 ToF–SIMS positive ion spectra (m/z = 0–650 amu) for pine sapwood.

(LMIS) enriched in [69]Ga[+] ions. The images were obtained using a primary beam of [69]Ga[+] LMIS with kinetic energy of 25 keV and operated in the microprobe mode. The raster size was 200 μm × 200 μm. Images were acquired for 10 min. The charge compensation was performed using an electron flood gun pulsed out-of-phase with the ion gun. The positive ToF–SIMS spectra of pine sapwood and heat-treated spruce (Thermo-S) are shown in Figures 8.17 and 8.18. All the spectral peaks used in the analysis are summarized in Table 8.3. The spectra are shown for mass to charge ratios (m/z) from 0 to 650. For wood samples, mostly fragments of hydrocarbons can be seen in the mass range below 100 m/z (CH_2, CH_3, C_2H_5, and so on). Peaks representing different functionalities in the wood components (e.g., HO-, H_3CO-, and -COO-) can also be seen in this region. However, the compounds or fragments mentioned above are very common in many organic materials, so this region is not discussed in this text. The region from 100–200 m/z was found to be the most interesting spectral range for detecting the molecular fragments of lignin and carbohydrates (Kangas 2007). In the spectra of both pine sapwood and heat-treated spruce (Figure 8.17), the dominant peaks of spruce lignin can be seen at 107 and 121 m/z. These peaks represent the *p*-hydroxyphenyl units in lignin. However, the intensities of peaks representing guaiacyl at 151, 167 m/z and syringyl units at 167, 181 m/z are relatively weak. The peaks at 115 and 133 m/z originated from xylan, while the peaks at 127 and 145 m/z were interpreted to originate from cellulose. The spectral region of 100–200 m/z for pine sapwood is similar to that of heat-treated spruce. Peaks characteristic of wood extractives are usually seen in the mass to charge ratio from 230 m/z to 650 m/z. In the pine

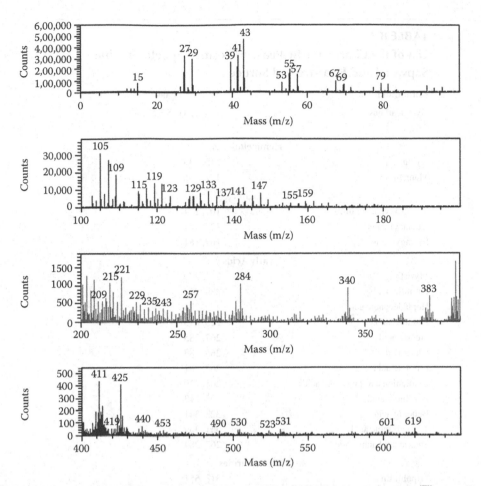

FIGURE 8.18 ToF–SIMS positive ion spectra (m/z = 0–650 amu) for heat-treated spruce (Thermo-S).

sapwood positive spectra, the presence of peaks at around 600 m/z and at 339 m/z indicates that the sample contained triglycerides. In addition, the peaks at 239 m/z originated from palmitic acid, the peak at 253 m/z was assigned to heptadecanoic acid and the peak at 265 m/z was identified as oleic acid. The peak at 313 m/z originated from arachidic acid/tripalmitin. The peaks at 411, 429 m/z originating from steryl ester were also observed in pine sapwood positive spectra. However, with respect to the extractive region for heat-treated spruce, it was seen that the most dominant peaks were those of sterol and steroid (383, 397, 411, 425 m/z) and probably stearic (284 m/z) and behenic acid anhydride (340 m/z), in addition to palmitic acid (239, 257 m/z). In order to obtain semiquantitative information about the surface compounds, the peaks of interest shown in Table 8.3 were integrated and normalized to the total intensity of the spectrum. As shown in Figure 8.19, it is interesting to note that, in comparison with pine sapwood, less oleic, arachidic acid/tripalmitin, and triolein but more stearic acid and sitosterols/sterylester were found on heat-treated spruce surface. This was probably due to some de-esterification of fats and waxes and the migration of extractives such as resins with subsequent degradation during heat treatment of spruce (Sundqvist 2004, Viitaneimi et al. 2001).

SIMS spectral libraries are still at the early stages of development, and as more spectra of biological materials, including lignocellulosic materials are added to the libraries, SIMS will become a very productive technique for characterizing the surface chemistry of wood and wood composites.

TABLE 8.3

List of the Characteristic Peaks in Positive Spectra of Pine Sapwood and Heat-Treated Spruce

Compound	Peaks
Hydrocarbons	15, 27, 41, 55, 57, 69
Cellulose	127
Hemicelluloses	
Xylan	115, 133
Mannan	127, 145
Lignin	
p-hydroxyphenyl units	107, 121
Guaiacyl units	137, 151
Syringyl units	167, 181
Fatty Acids	
Myristic	211, 229
Palmitic acid	239, 257
Heptadecanoic acid	253, 271
Pentadecanoic	225, 243
Stearic acid	267, 285
Oleic acid	265, 283
Linoleic acid	263, 281
Linolenic acid (pinolenic acid)	261, 279
Arachidic acid	295, 313
Behenic acid	323, 341
Tetracosanoic acid	351, 369
Stearic acid anhydride	267, 284
Triglycerides	
Tripalmitin	313, 551
Triolein	339, 603
Tristearin	341, 607
Triglycerides of other fatty acids	327, 335, 337, 575, 595, 599, 600, 601, 602
Resin Acids	
Dehydroabietic acid	299, 399, 301, 302, 303
Abietic acid	299, 300, 301, 302, 303
Steroids	
Ultrasitosterol	383, 397, 414, 429
Steryl esters	383, 397, 411, 412, 425/429
Sterols	
Sitosterol	397, 414, 415
Sitostanol	398, 416
Oxo-sitosterol	411, 429

Source: Adapted from Kangas H. 2007. Surface chemical and morphological properties of mechanical pulps, fibers and fines, November, KCL Communications 13; Fardim P. et al. 2005. *EngIneering Aspects* 255:91–103; Tokareva E. 2011. *Doctoral thesis – "Spatial Distribution of Components in Wood by ToF-SIMS"*. Åbo Akademi University, Turku.

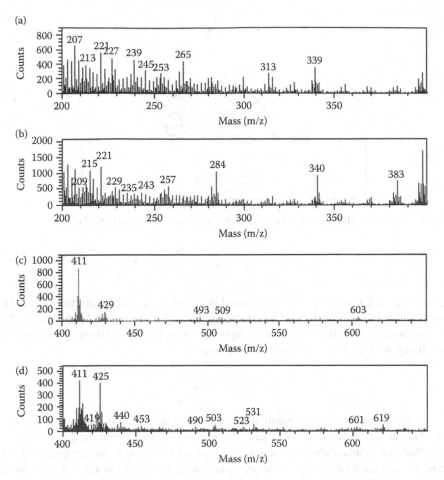

FIGURE 8.19 ToF–SIMS positive ion spectra (m/z = 200–650 amu) for pine sapwood (a, c) heat-treated spruce (b, d).

8.4 THERMODYNAMIC METHODS FOR CHARACTERIZING SURFACE PROPERTIES

Over the years, two methods for determining the surface thermodynamic properties (surface energy and acid-base characteristics) of wood have been developed. The first method, which is known as contact angle analysis (CAA), is based on wetting the wood surface with a liquid. The second technique, which is known as inverse gas chromatography (IGC), is based on the adsorption of organic vapors on the wood surface. Another major difference between the two methods is that solid samples for IGC analysis should be in a finely divided form, while samples for CAA can be in any form, including small coupons or wafers.

8.4.1 CONTACT ANGLE ANALYSIS (CAA)

There are two approaches to CAA: static and dynamic contact angle measurements. In static CAA the liquid–solid angle of a sessile drop of liquid on the surface of a wood specimen is observed and measured by means of a contact angle meter. The contact angle, θ of a drop of liquid on the surface of a specimen is shown schematically in Figure 8.20. The cosine of the contact angle (cos θ) is related to the surface energy of the specimen by the Young equation (Nguyen and Johns 1978).

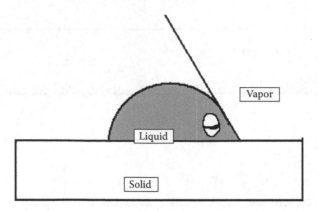

FIGURE 8.20 Schematic diagram of the contact angle, θ at the solid–liquid interface.

$$\gamma_{LV} \cos\theta = \gamma_{SV} - \gamma_{SL} = \gamma_S - \gamma_{SL} - \pi_e \tag{8.2}$$

where γ_S is the surface energy of a solid measured in vacuum, γ_{SL} and γ_{SV} are the surface free ener-gies on the interface between solid and liquid and of the solid and saturated vapor, respectively; γ_{LV} is the surface tension of the liquid against its vapor, and π_e is the equilibrium spreading pressure of the adsorbed vapor of the liquid on solid and is defined as:

$$\pi_e = \gamma_S - \gamma_{SV} \tag{8.3}$$

There are only two quantities in Equation 8.2 that can be determined experimentally, namely γ_{LV} and $\cos\theta$. The surface tension of the liquid, γ_{LV} and $\cos\theta$ can be measured by several methods (Adamson 1967), but contact angle θ can only be obtained when the surface tension of the liquid is higher than the surface-free energy of the solid (assuming γ_{SL} and π_e in Equation 8.2 are approxi-mately equal to zero). In that case, the surface energy of the solid can be estimated from Equation 8.4:

$$\gamma_S \cong \gamma_{LV} \cos\theta \tag{8.4}$$

Zisman (Adamson 1967) introduced the well-known empirical approach to estimate the surface-free energy of a solid by plotting the cosine of contact angle θ versus the surface tension of a series of liquids of known surface tension. The point at which the resulting straight line plot intercepts the horizontal line, $\cos\theta = 1$ (zero contact angle) is called the critical surface tension, γ_C. Zisman's plot is generally described by Equation 8.5.

$$\cos\theta = 1 + b(\gamma_C - \gamma_{LV}) \tag{8.5}$$

where b is the slope of the line. Zisman's critical surface tension provides one of the most convenient means of expressing the surface energy of a solid.

In dynamic CAA, the contact angle is measured by means of a dynamic contact angle analyzer (DCA), which consists of a precision balance for measuring the wetting force exerted on a specimen as it is dipped and withdrawn from a liquid of known surface tension, γ_{LV}. As shown in the schematic diagram in Figure 8.21, the weight of the solid specimen is recorded by means of a precision balance as it is immersed and withdrawn from the liquid. Immersion and withdrawal of the specimen is accomplished by raising or lowering the liquid in the cup placed on motorized elevator. The immersion graph, which is also known as a hysteresis loop is captured by a data system that is interfaced to the

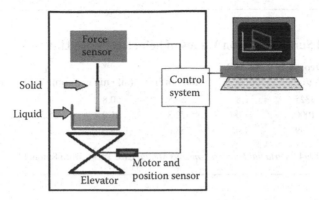

FIGURE 8.21 Schematic diagram of a dynamic contact angle analyzer.

control system of the dynamic contact angle analyzer. The immersion graph is essentially a plot of the measured wetting force versus the immersion depth of the specimen (Figure 8.22).

The wetting force is related to the contact angle by the following equation (De Meijer et al. 2000):

$$F = \gamma_L P \cos\theta - \rho_L g A d \qquad (8.6)$$

where F is the measured force, N; γ_L the surface tension of the probe liquid, mJ/m^2; P the perimeter of the specimen, m; θ the contact angle (deg) between the liquid and wood specimen; ρ_L the density of the liquid, kg/m^3; g the gravitational constant, m^2/s; A the cross-sectional area of the wood specimen, m^2; and d the depth of immersion, m.

The surface energy components are calculated from the advancing contact angles of diiodomethane, formamide, and water according to Equations 8.2 through 8.5, which are based on Equation 8.3, developed by Fowkes, van Oss, and Good et al. (Wålinder 2002, Van Oss 1994, Good 1992).

The Lifschitz–van der Waals component γ_s^{LW} is obtained from Equation 8.7 for diiodomethane

$$\gamma_s^{LW} = 0.25\gamma_L^{LW}(1+\cos\theta)^2 \qquad (8.7)$$

where γ_L^{LW} is the surface tension of diiodomethane from Table 8.4.

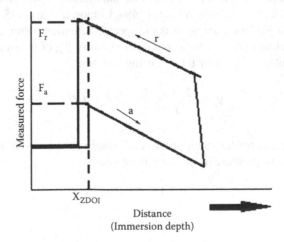

FIGURE 8.22 Schematic diagram of the immersion graph as the specimen is immersed (a) and withdrawn (r) from the liquid. The vertical broken line coincides with the point when the specimen first comes into contact with the liquid at the zero depth of immersion, $x = X_{ZDOI}$.

TABLE 8.4
Physical Data and Surface Tension Values of Selected Probe Liquids

Liquid	Density (kg/m³)	Viscosity (mPa · S)	γ_L (mJ · m^{-2})	γ_L^{LW} (mJ · m^{-2})	γ_L^+ (mJ · m^{-2})	γ_L^- (mJ · m^{-2})	γ_L^{AB} (mJ · m^{-2})
Diiodomethane	3325	2.8	50.8	50.8	0	0	0
Water	1000	1.00	72.8	21.8	25.5	25.5	51.0
Formamide	799	1.02	58.0	39.0	2.28	39.6	19.0

Source: Van Oss, C.J. 1994. *Interfacial Forces in Aqueous Media.* CRC Press, Boca Raton, FL.

Substituting the known value of γ_s^{LW} in Equation 8.8, the acid, γ_s^+, and base, γ_s^- components of the surface energy can be calculated from the advancing contact angles of water and formamide.

$$0.5\gamma_L(1+\cos\theta) = (\gamma_s^{LW}\gamma_L^{LW})^{1/2} + (\gamma_s^-\gamma_L^+)^{1/2} + (\gamma_s^+\gamma_L^-)^{1/2}. \tag{8.8}$$

The total acid-base component can be calculated from Equation 8.9.

$$\gamma_s^{AB} = 2(\gamma_s^+)^{1/2}(\gamma_s^-)^{1/2}. \tag{8.9}$$

The total surface energy of the wood can be calculated from Equation 8.10.

$$\gamma_s = \gamma_s^{LW} + \gamma_s^{AB}. \tag{8.10}$$

8.4.2 INVERSE GAS CHROMATOGRAPHY (IGC)

IGC involves packing the material to be studied, typically in a fibrous or particle form, into the column of a gas chromatograph. The retention behavior of a known vapor probe, which is injected into the carrier gas that passes through the packed column, allows the solid-vapor adsorption characteristics, and thus the surface thermodynamic properties, of the material to be quantified (Schultz and Lavielle 1989, Fowkes 1990, Kamdem and Reidl 1991, Lavielle and Schultz 1991, Felix and Gatenholm 1993, Kamdem et al. 1993, Farfard et al. 1994, Williams 1994, Chtourou et al. 1995, Belgacem 2000).

The key parameter in IGC measurements is the specific retention volume, V_g°, or the volume of carrier gas required to elute a probe from a column containing 1 g of the sample of material. V_g° is related to the experimental variables by the following equation:

$$V_g^\circ = \frac{273.15}{T_c} \cdot \frac{(t_r - t_m)}{W} \cdot F.J.C \tag{8.11}$$

where C is the correction factor for vapor pressure of water in the soap bubble flowmeter, and J is the correction factor for the pressure drop across the column.

$$C = 1 - (P_w/P_0) \tag{8.12}$$

and

$$J = 1.5\left[\frac{(P_i/P_0)^2 - 1}{(P_i/P_0)^3 - 1}\right] \tag{8.13}$$

T_c is the column temperature in K; W is the amount of material packed into the column in g; t_r is the retention time of the probe in min; t_m is the retention time of a reference probe such as methane or argon in min; F is the carrier gas flow rate in mL/min; P_w is the vapor pressure of water at the column temperature in mm Hg; P_0 is the carrier gas pressure at the column outlet (atmospheric pressure) in mm Hg; P_i is the carrier gas pressure at the column inlet in mm Hg.

8.4.3 Total Surface Energy

The total surface energy may be regarded as the sum of the contribution of two components, a nonpolar and a polar component. The nonpolar component is associated with London dispersive forces of interaction, and the polar component is associated with acid–base interactions.

8.4.3.1 Dispersive Component of the Total Surface Energy

The interaction of neutral probes such as saturated n-alkanes with the surface of the sample of the material is predominated by London dispersive forces of interaction. Under conditions of infinite dilution of the injected probe vapors, it has been shown that the dispersive component, γ_s^D of the total surface energy of the material is related to V_g° by the following equation:

$$\mathrm{RT}\ln V_g^\circ = 2N(\gamma_s^D)^{1/2}\,a(\gamma_L^D)^{1/2} + C^t \tag{8.14}$$

where N is Avogadro's number, a is the surface area of the probe molecule, γ_L^D is the dispersive component of the surface energy of the probe, C^t is a constant, which depends on the reference states.

A plot of $\mathrm{RT}\ln V_g^\circ$ versus $2N\,a(\gamma_L^D)^{1/2}$ should give a straight line with slope, $(\gamma_s^D)^{1/2}$.

8.4.3.2 Acid–Base Component of the Total Surface Energy

The interaction of polar probes with the surface of the material involves both dispersive and acid–base interactions. The free energy of desorption, ΔG°_{AB}, corresponding to specific acid–base interactions may be related to V_g° by the following equation:

$$\mathrm{RT}\ln(V_g^\circ / V_g^{\circ*}) = \Delta G^\circ_{AB} \tag{8.15}$$

where V_g° is the specific retention volume of the polar probe, $V_g^{\circ*}$ is the specific retention volume of a reference neutral n-alkane.

This equation suggests that values of $\mathrm{RT}\ln V_g^\circ$ plotted against $2N\,a(\gamma_L^d)^{1/2}$ for polar probes should fall above the straight line obtained by plotting $\mathrm{RT}\ln V_g^{\circ*}$ versus $2N\,a(\gamma_L^d)^{1/2}$ for the reference neutral n-alkane probes. The difference of ordinates between the point corresponding to the specific polar probe and the reference line gives the value of the free energy of desorption corresponding to the specific acid–base.

The free energy of desorption corresponding to the specific acid–base interactions may be related to the enthalpy of desorption, ΔH°_{AB} by the following equation:

$$\Delta G^\circ_{AB} = \Delta H^\circ_{AB} - T\Delta S^\circ_{AB} \tag{8.16}$$

where, T is the temperature in K, and ΔS°_{AB} is the entropy of desorption corresponding to the specific acid–base interactions.

A plot of $\Delta G^\circ_{AB}/T$ versus $1/T$ should yield a straight line with slope, ΔH°_{AB}.

It has been shown that the enthalpy of desorption may be used to obtain the acidic and basic constants of a substrate. The acidic and basic constants, K_A and K_B, may be regarded as the acceptor and donor numbers, respectively, and are analogous to the acceptor number, AN*, and donor number, DN of any compound as defined by the Gutmann theory of acids and bases. The enthalpy of desorption corresponding to the specific acid-base interactions is related to K_A, K_B, AN*, and DN by the following expression:

$$\Delta H^{\circ}_{AB} = K_A DN + K_B AN^*. \qquad (8.17)$$

A plot of $\Delta H^{\circ}_{AB}/AN^*$ versus DN/AN^* should yield a straight line with slope, K_A and intercept, K_B. Values of DN and AN* for various solvents are available in the literature.

IGC has been used to characterize the surfaces of cellulose fibers, birch wood meal, polyethylene and wood pulp fibers, synthetic polymers, and lignocellulosic fibers grafted with poly(methylmethacrylate). The values of the dispersive component of the surface energy of some lignocellulosic materials are summarized in Tables 8.5 and 8.6 summarizes values of acid–base properties, expressed in terms of the ratio, K_A/K_B. For acidic surfaces the value of this ratio is generally greater than 1.0.

TABLE 8.5
Values of the Dispersive Component of the Surface Energy of Lignocellulosics Obtained under Infinite Dilution Conditions

Material	Treatment	γ^D_s at (T°C), mJ/m²	Reference
Cellulose	Untreated	48.5 (25°C)	Belgacem (2000)
Wood birch	None	43.8 (50°C)	Kamdem et al. (1993)
Bleached softwood kraft pulp	Untreated	37.9 (40°C)	Belgacem (2000)
Lignin	Kraft	46.6 (50°C)	Belgacem (2000)
Wood eastern white pine	Unextracted	37 (40°C)	Tshabalala (1997)
Kenaf powder	Unextracted	40 (40°C)	Tshabalala (1997)
Wood eastern white pine	Extracted with toluene/ ethanol (2:1 v/v)	49 (40°C)	Tshabalala (1997)
Kenaf powder	Extracted with toluene/ ethanol (2:1 v/v)	42 (40°C)	Tshabalala (1997)

TABLE 8.6
Acid-Base Properties of Lignocellulosic Surfaces Obtained under Infinite Dilution Conditions

Material	Treatment	K_A/K_B	Reference
Cellulose powder, 20 µm particle size	None	24	Tshabalala (1997)
Wood birch	None	1.5	Kamdem et al. (1993)
Bleached softwood kraft pulp	Untreated	2.2–5.7	Belgacem (2000)
Wood eastern white pine	Extracted with toluene/ ethanol (2:1 v/v)	1.4	Tshabalala (1997)
Kenaf powder	Unextracted	0	Tshabalala (1997)
Kenaf powder	Extracted with toluene/ ethanol (2:1 v/v)	1.6	Tshabalala (1997)

8.5   CONCLUSIONS AND OUTLOOK

Characterization of surface properties of wood is a very complex and difficult undertaking. No single technique is adequate to completely characterize the surface chemistry of wood and related lignocellulosic materials. Rather, a combination of spectrometric techniques used together with microscopic and thermodynamic techniques can provide good insight into the chemical composition of lignocellulosic surfaces, the surface distribution, and topography of acid-base sites, and the effect of chemical composition on the reactivity and mechanical properties of lignocellulosic surfaces. Hence, as new surface sensitive spectroscopic techniques are applied to the study of lignocellulosic surfaces, it is likely that better data will be developed that should greatly enhance our understanding of the surface chemistry of wood, and its effect on surface properties of wood, including surface mechanical properties.

REFERENCES

Adamson, A. W. 1967. *Physical Chemistry of Surfaces*. 2nded. Interscience, New York.

Adusumalli, R.-B., Mook, W., Passas, R., Schwaller, P., and Michler, J. 2010a. Nanoindentation of single pulp fibre cell walls. *Journal of Materials Science* 45(10):2558–2563.

Adusumalli, R.-B., Raghavan, R., Ghisleni, R., Zimmermann, T., and Michler, J. 2010b. Deformation and failure mechanism of secondary cell wall in Spruce late wood. *Applied Physics A: Materials Science & Processing* 100(2):447–452.

Adusumalli, R.-B., Raghavan, R., Schwaller, P., Zimmermann, T., and Michler, J. 2010c. *In situ* SEM microindentation of single wood pulp fibres in transverse direction. *Journal of Electron Microscopy* 59(5):345–349.

Agarwal, U.P. 1999. An overview of raman spectroscopy as applied to lignocellulosic materials. In: *Advances in Lignocellulosic Characterization*. Dimitris S. Argyropoulos (ed.), TAPPI Press, Atlanta, Georgia.

Agarwal, U.P. and Ralph, S.A. 1997. FT-Raman spectroscopy of wood: Identifying contributions of lignin and carbohydrate polymers in the spectrum of black spruce (*Picea mariana*). *Applied Spectroscopy* 51(11):1648–1655.

Ahmed, A., Adnot, A., and Kaliaguine, S. 1987. ESCA study of solid residues of supercritical extraction of *Populus tremuloides* in methanol. *Journal of Applied Polymer Science* 34:359–375.

Ahmed, A., Adnot, A., and Kaliaguin, S. 1988. ESCA analysis of partially converted lignocellulosic materials. *Journal of Applied Polymer Science* 35:1909–1919.

Asif, S.A.S., Wahl, K.J., Colton, R.J., and Warren, O.L. 2001. Quantitative imaging of nanoscale mechanical properties using hybrid nanoindentation and force modulation *Journal of Applied Physics* 90(11):5838–5838.

Béland, M-C. and Mangin, P.J. 1995. Three-dimensional evaluation of paper surfaces using confocal micros copy. In: Surface Analysis of Paper. Terrance E. Conners and Sujit Banarjee (eds.), CRC Press, Inc. Boca Raton, Florida.

Belgacem, M.N. 2000. Characterisation of polysaccharides, lignin and other woody components by inverse gas chromatography. *Cellulose Chemistry and Technology* 34:357–383.

Benninghoven, A., Hagenhoff, B., and Niehuis, E. 1993. Surface MS: probing real-world samples. *Analytical Chemistry* 65(14):630A–640A.

Boyde, A. 1994. Bibliography on Confocal microscopy and its applications. *Scanning* 16:33–56.

Brandt, B., Zollfrank, C., Franke, O., Fromm, J., Göken, M., and Durst, K. 2010. Micromechanics and ultrastructure of pyrolysed softwood cell walls. *Acta Biomaterialia* 6(11):4345–4351.

Brinen. J.S. 1993. The observation and distribution of organic additives on paper surfaces using surface spectroscopic techniques. *Nordic Pulp and Paper Research Journal* 8:123–129.

Brinen, J.S. and Kulick, R.J. 1995. SIMS imaging of paper surfaces. Part 4. The detection of desizing agents on hard-to-size paper surfaces. *International Journal of Mass Spectrometry Ion Processes* 143:177–190.

Cheng, Y.-T. and Cheng, C.-M. 2004. Scaling, dimensional analysis, and indentation measurements. *Materials Science & Engineering R: Reports* 44(Copyright 2005, IEE):91–149.

Chtourou, H.B., Reidel, B., and Kokta, B.V. 1995. Surface characterization of modified polyethylene pulp and wood pulp fibers using XPS and inverse gas chromatography. *Journal of Adhesion Science and Technology* 9(5):551–574.

Comyn, J. 1993. Surface analysis and adhesive bonding. *Analytical Proceedings* 30:27–28.

Conley, R.T. 1996. *Infrared Spectroscopy*. Allyn and Bacon, Boston.

Danilatos, G.D. 1988. Foundations of environmental scanning electron microscopy. *AdvAnces in Electronics and Electron Physics* 71:109–250.

Danilatos, G.D. 1993. Introduction to the ESEM Instrument. *Microscopy Research and Technique*, 25:354–361.

De Hosson, J., Soer, W., Minor, A., Shan, Z., Stach, E., Syed Asif, S., and Warren, O. 2006. *In situ* TEM nanoindentation and dislocation-grain boundary interactions: a tribute to David Brandon. *Journal of Materials Science* 41(23):7704–7719.

De Meijer, M., Haemers, S., Cobben, W., and Militz, H. 2000. Surface energy determinations of wood: Comparison of methods and wood species. *Langmuir* 16:9352–9359.

Deshmukh, S.C. and Aydil, E.S. 1996. Investigation of low temperature SiO_2 plasma enhanced chemical vapor deposition. *Journal of Vacuum Science and Technology* B14:738.

Detter-Hoskin, L.D. and Busch, K.L. 1995. SIMS: Secondary ion mass spectrometry. In: *Surface Analysis of Paper*. Conners, T.E. and Banerjee, S.S. (eds), CRC Press, Inc. Boca Raton, Florida.

Donald, A.M. 2003. The use of environmental scanning electron microscopy for imaging wet and insulating materials. *Nature Materials* 2(8):511–516.

Donaldson, L. 2003. *Private Communication*. Cell Wall Biotechnology Center, Forest Research, New Zealand.

Dorris, G.M. and Gray, D.G. 1978a. The surface analysis of paper and wood fibres by ESCA I. *Cellulose Chemistry and Technology* 12:9–23.

Dorris, G.M. and Gray, D.G. 1978b. The surface analysis of paper and wood fibres by ESCA II. *Cellulose Chemistry and Technology* 12:721–734.

Farfard, M., El-Kindi, M., Schreiber, H.P., G. Dipaola-Baranyi, G., and Hor, A.M. 1994. Estimating surface energy variations of solids by inverse gas chromatography. *Journal of Adhesion Science and Technology* 8(12):1383–1394.

Fardim P., Gustafsson J., von Schoultz S., Peltonen J., and Holmbom B. 2005. Extractives on fiber surfaces investigated by XPS, ToF-SIMS and AFM, Colloids and Surfaces A: Physicochem. *EngIneering Aspects* 255:91–103.

Felix, J.M. and Gatenholm, P. 1993. Characterization of cellulose fibres using inverse gas chromatography. *Nordic Pulp and Paper Research Journal* 1:200–203.

Fischer-Cripps, A.C. 2004. *Nanoindentation*. Springer, New York.

Follrich, J., Stöckel, F., and Konnerth, J. 2010. Macro- and micromechanical characterization of wood-adhesive bonds exposed to alternating climate conditions. *Holzforschung* 64(6):705–711.

Fowkes, F.M. 1990. Quantitative characterization of the acid-base properties of solvents, polymers and inorganic surfaces. *Journal of Adhesion Science and Technology* 4(8):669–691.

Fracassi, F., d'Agostino, R., and Favia, P. 1992. Plasma—enhanced chemical vapor-deposition of organosilica thin-films from tetraethoxysilane-oxygen feeds. *Journal of the Electrochemical Society* 139(9):2936–2944.

Frisbie, C.D., Rozsnyai, L.F., Noy, A., Wrighton, M.S., and Lieber, C.M. 1994. Functional group imaging by chemical force microscopy. *Science* 265:2071–2074.

Frommer, J. 1992. Scanning tunneling microscopy and atomic force microscopy in organic chemistry. *Angewandte Chemie International Edition in English* 31:1298–1328.

Fujisawa, N. and Swain, M.V. 2008. Nanoindentation-derived elastic modulus of an amorphous polymer and its sensitivity to load-hold period and unloading strain rate. *Journal of Materials Research* 23(3):637–641.

Gacitua, W., Bahr, D., and Wolcott, M. 2010. Damage of the cell wall during extrusion and injection molding of wood plastic composites. *Composites Part A: Applied Science and Manufacturing* 41(10):1454–1460.

Gershon, A.L., Bruck, H.A., Xu, S., Sutton, M.A., and Tiwari, V. 2010. Multiscale mechanical and structural characterizations of Palmetto wood for bio-inspired hierarchically structured polymer composites. *Materials Science and Engineering: C* 30(2):235–244.

Gindl, W., Gupta, H.S., and Grunwald, C. 2002. Lignification of spruce tracheid secondary cell walls related to longitudinal hardness and modulus of elasticity using nano-indentation. *Canadian Journal of Botany* 80(10):1029–1033.

Gindl, W., Gupta, H.S., Schoberl, T., Lichtenegger, H.C., and Fratzl, P. 2004a. Mechanical properties of spruce wood cell walls by nanoindentation. *Applied Physics A (Materials Science Processing)* A79(8):2069–2073.

Gindl, W. and Schoberl, T. 2004. The significance of the elastic modulus of wood cell walls obtained from nanoindentation measurements. *Composites Part A: Applied Science and Manufacturing* 35(11):1345–1349.

Gindl, W., Schoberl, T., and Jeronimidis, G. 2004b. Erratum: The interphase in phenol-formaldehyde (PF) and polymeric methylene di-phenyl-di-isocyanate (pMDI) glue lines in wood (International Journal of Adhesion and Adhesives (2004) 24 (279–286)). *International Journal of Adhesion and Adhesives* 24(6):535–535.

Gindl, W., Schoberl, T., and Jeronimidis, G. 2004c. The interphase in phenol-formaldehyde and polymeric methylene di-phenyl-di-isocyanate glue lines in wood. *International Journal of Adhesion and Adhesives* 24(4):279–286.

Good, R.J. 1992. Contact angle, wetting, and adhesion: A critical review. *J. Adhesion Sci. Technol.* 6(12): 1269–1302.

Gray, D.G. 1978. The surface analysis of paper and wood fibres by ESCA III. *Cellulose Chemistry and Technology* 12:735–743.

Hanley, S.J. and Gray, D.G. 1994. Atomic force microscope images of black spruce wood sections and pulp fines. *Holzforschung* 48(1):29–34.

Hanley, S.J. and Gray, D.G. 1995. Atomic force microscopy. In *Surface Analysis of Paper*. Conner, T.E. and Banerjee, S.S. (eds). CRC Press, Inc. Boca Raton, Fl.

Hansma, P.K., Ellings, V.B., Marti,O., and Bracker, C.E. 1998. Scanning tunneling microscopy and atomic force microscopy: *Application and Technology. Science* 242:209–216.

Herbert, E.G., Oliver, W.C., Lumsdaine, A., and Pharr, G.M. 2009. Measuring the constitutive behavior of viscoelastic solids in the time and frequency domain using flat punch nanoindentation. *Journal of Materials Research* 24(Copyright 2009, The Institution of Engineering and Technology):626–637.

Hoh, J.H. and Engel, A. 1993. Friction effects on force measurements with an atomic force microscope. *Langmuir* 9:3310–3312.

Hon, D.N.-S. 1984. ESCA study of oxidized wood surfaces. *Journal of Applied Polymer Science* 29:2777–2784.

Hon, D.N.-S and Feist, W.C. 1986. Weathering characteristics of hardwood surfaces. *Wood Science and Technology* 20:169–183.

Hua, X., Ben,Y., Kokta,B.V., and Kalinguin, S. 1991. Application of ESCA in wood and pulping chemistry. *China Pulp and Paper* 10:52–57.

Hua, X., Kaliaguine, S., Kokta, B.V., and Adnot, A. 1993a. Surface analysis of explosion pulps by ESAC. Part 1: Carbon (1s) spectra and oxygen-to-carbon ratios. *Wood Science and Technology* 27:449–459.

Hua, X., Kaliaguine, S., Kokta, B.V., and Adnot, A. 1993b. Surface analysis of explosion pulps by ESAC. Part 2: Oxygen (1s) and sulfur (2p) spectras. *Wood Science and Technology* 28:1–8.

Jäger, A., Bader, T., Hofstetter, K., and Eberhardsteiner, J. 2011. The relation between indentation modulus, microfibril angle, and elastic properties of wood cell walls. *Composites Part A: Applied Science and Manufacturing* 42(6):677–685.

Jukes, J.E., Frihart, C.R., Beecher, J.F., Moon, R.J., Resto, P.J., Melgarejo, Z.H., Surez, O.M., Baumgart, H., Elmustafa, A.A., and Stone, D.S. 2009a. Nanoindentation near the edge. *Journal of Materials Research* 24(3):1016–1031.

Jakes, J.E., Frihart, C.R., Beecher, J.F., Moon, R.J., and Stone, D.S. 2008a. Experimental method to account for structural compliance in nanoindentation measurements. *Journal of Materials Research* 23(4):1113–1127.

Jakes, J.E., Frihart, C.R., and Stone, D.S., 2008b. Creep properties of micron-size domains in ethylene glycol modified wood across 4 1/2 decades in strain rate. In: Katti, K., Hellmich, C., Wegst, U.G.K., Narayan, R. (eds.). *Materials Research Society. Mater. Res. Soc. Symp. Proc.*, Boston, MA, USA, p. 1132Z0721.

Jakes, J.E., Lakes, R., and Stone, D.S. 2011a. Broadband Nanoindentation I: Viscoelasticity of glassy polymers. *Journal of Materials Research* 27(2):463–474.

Jakes, J.E., Lakes, R., and Stone, D.S. 2011b. Broadband nanoindentation II: Viscoplasticity of glassy polymers *Journal of Materials Research* 27(2):475–484.

Jakes, J.E. and Stone, D.S. 2011. The edge effect in nanoindentation. *Philosophical Magazine* 91(7):1387–1399.

Jakes, J.E., Yelle, D.J., Beecher, J.F., Frihart, C.R., and Stone, D.S., 2009b. Characterizing pMDI reactions with wood cell walls: 2. Nanoindentation. In: Frihart, C.R., Hunt, C.G., Moon, R.J. (eds.). *International Conference on Wood Adhesives*. Forest Products Society, Harveys Resort & Casino, Lake Tahoe, Nevada, USA (*submitted*), pp. 366–373.

Johnson, K.L. 1985. *Contact Mechanics*. Cambridge University Press, Cambridge, UK.

Johnson, D. and Hibbert, S. 1992. Applications of secondary ion mass spectrometry (SIMS) for the analysis of electronic materials. *Semiconductor Science and Technology* 7:A180–A184.

Kamdem, D.P., Bose, S.K., and Luner, P. 1993. Inverse gas chromatography of birch wood meal. *Langmuir* 9:3039–3044.

Kamdem, D.P. and Reidl, B. 1991. IGC characterization of PMMA grafted onto CTMP fiber. *Journal of Wood Chemistry and Technology* 11(1):57–91.

Kamdem, D.P., Reidl, B., Adnot, A., and Kaliaguine, S. 1991. ESCA spectroscopy of poly(methyl methacrylate) grafted onto wood fibers. *Journal of Applied Polymer Science* 43:1901–1912.

Kangas H. 2007. Surface chemical and morphological properties of mechanical pulps, fibers and fines, November, KCL Communications 13.

Konnerth, J., Buksnowitz, C., Gindl, W., Hofstetter, K., and Jager, A., 2010a. Full Set of Elastic Constants of Spruce Wood Cell Walls Determined by Nanoindentation. International Convention of Wood Science and Technology and United Nations Economic Commission for Europe–Timber Committee, Geneva, Switzerland.

Konnerth, J., Eiser, M., Jäger, A., Bader, T.K., Hofstetter, K., Follrich, J., Ters, T., Hansmann, C., and Wimmer, R. 2010b. Macro- and micro-mechanical properties of red oak wood (Quercus rubra L.) treated with hemicellulases. *Holzforschung* 64(4):447–453.

Konnerth, J. and Gindl, W. 2006. Mechanical characterisation of wood-adhesive interphase cell walls by nanoindentation. *Holzforschung* 60(4):429–433.

Konnerth, J. and Gindl, W. 2008. Observation of the influence of temperature on the mechanical properties of wood adhesives by nanoindentation. *Holzforschung* 62(6):714–717.

Konnerth, J., Harper, D., Lee, S.-H., Rials, T.G., and Gindl, W. 2008. Adhesive penetration of wood cell walls investigated by scanning thermal microscopy (SThM). *Holzforschung* 62(1):91–98.

Konnerth, J., Jager, A., Eberhardsteiner, J., Muller, U., and Gindl, W. 2006. Elastic properties of adhesive polymers. II. Polymer films and bond lines by means of nanoindentation. *Journal of Applied Polymer Science* 102:1234–1239.

Konnerth, J., Valla, A., and Gindl, W. 2007. Nanoindentation mapping of a wood-adhesive bond. *Applied Physics A: Materials Science and Processing* 88(2):371–375.

Lahiji, R.R., Xu, X., Reifenberger, R., Raman, A., Rudie, A., and Moon, R.J. 2010. Atomic force microscopy characterization of cellulose nanocrystals. *Langmuir* 26(6):4480–4488.

Lavielle, L. and Schultz, J. 1991. Surface properties of carbon fibers determined by inverse gas chromatography. *Langmuir* 7(5):978–981.

Lee, B.H., Kim, H., Lee, J.J., and Park, M.J. 2003. Effects of acid rain on coatings for exterior wooden panels. *Journal of Industrial and Engineering Chemistry*, 9(5):500–507.

Lehringer, C., Koch, G., Adusumalli, R.-B., Mook, W.M., Richter, K., and Militz, H. 2011. Effect of physisporinus vitreus on wood properties of Norway spruce. Part 1: Aspects of delignification and surface hardness. *Holzforschung*, 65(5):711–719.

Leica TCS Confocal Systems User Manual, 1999. Version 1.0, January, Leica Microsystems Heidelberg GmbH. Edited and written by EDV-Service Dr. Kehrel, Heidelberg, Germany.

Li, J-X. and Gardella, J.A. Jr. 1994. Quantitative static secondary ion mass spectrometry of pH effects on octadecylamine monolayer Langmuir-Blodgett films. *Analytical Chemistry* 66(7):1032–1037.

Libermann, M.A. and Lichtenberg, A.J. 1994. *Principles of Plasma Discharges and Materials Processing*. Wiley, New York.

Lichtman, J.W. 1994. Confocal microscopy. *Scientific American* 271:40–45.

Loubet, J.L., Oliver, W.C., and Lucas, B.N. 2000. Measurement of the loss tangent of low-density polyethylene with a nanoindentation technique. *Journal of Materials Research* 15(5):1195–1198.

Louder, D.R. and Parkinson, B.A. 1995. An update on scanning force microscopies. *Analytical Chemistry* 67(9):297A–303A.

Loxton, C., Thumm, A., Grisby, W.J., Adams, T.A., and Ede, R.M. 2003. Resin distribution in medium density fiberboard. Quantification of UF resin distribution on blowline- and dry-blended MDF fiber and panels. *Wood and Fiber Science* 35(3):370–380.

McGhie, A.J., Tang, S.L., and Li, S.F.Y. 1995. Expanding the uses of AFM. *Chemtech* 25(7):20–26.

Meredith, P., Donald, A.M., and Thiel, B. 1996. Electron-gas interactions in the environmental SEM's gaseous detector. *Scanning* 18:467–473.

Meyer, E. 1992. Atomic force microscopy. *Progress in Surface Science* 41:3–49.

Minor, A.M., Morris, J.W., and Stach, E.A. 2001. Quantitative *in situ* nanoindentation in an electron microscope. *Applied Physics Letters* 79(11):1625–1627.

Mirabella, F.M. (ed.) 1998. *Modern Techniques in Applied Molecular Spectroscopy*. John Wiley Publications, New York.

Mjörberg, P.J. 1981. Chemical surface analysis of wood fibres by means of ESCA. *Cellulose Chemistry and Technology* 15:481–486.

Nguyen, T. and Johns, W.E. 1978. Polar and dispersion force contributions to the total surface free energy of wood. *Wood Science and Technology* 12:63–74.

Oliver, W.C. and Pharr, G.M. 1992. Improved technique for determining hardness and elastic modulus using load and displacement sensing indentation experiments. *Journal of Materials Research* 7(6):1564–1580.

Oliver, W.C. and Pharr, G.M. 2004. Measurement of hardness and elastic modulus by instrumented indentation: Advances in understanding and refinements to methodology. *Journal of Materials Research* 19(1):3–20.

Oyen, M.L. 2011. Mechanics of indentation. In: Oyen, M.L. (ed.), *Handbook of Nanoindentation with Biological Applications*. Pan Stanford, Singapore, pp. 123–152.

Pachuta, S.J. and Staral, J.S. 1994. Nondestructive analysis of colorants on paper by time-of-flight secondary ion mass spectrometry. *Analytical Chemistry* 66(2):276–284.

Pandey, K.K. and Theagarajan, K.S. 1997. Analysis of wood surfaces and ground wood by diffuse reflectance (DRIFT) and photoacoustic (PAS) Fourier transform spectroscopic techniques. *Holz als Roh- und Werkstoff* 55:383–390.

Pawley, J.B. 1990. *The Handbook of Biological Confocal Microscopy*. Plenum, New York.

Perry, S.S. and Somorjai, G.A. 1994. Characterization of organic surfaces. *Analytical Chemistry* 66(7):403A–415A.

Puthoff, J.B., Jakes, J.E., Cao, H., and Stone, D.S. 2009. Investigation of thermally activated deformation in amorphous PMMA and Zr-Cu-Al bulk metallic glasses with broadband nanoindentation creep. *Journal of Materials Research* 24(3):1279–1290.

Rynders. R.M., Hegedus, C.R., and Gilicinski, A.G. 1995. Characterization of particle coalescence in waterborne coatings using atomic force microscopy. *Journal of Coatings Technology* 667(845):59–69.

Sahli, S., Segui, Y., Ramdani, R., and Takkouk, Z. 1994. RF plasma deposition from hexamethyldisiloxane oxygen mixtures. *Thin Solid Films* 250(1–2):206–212.

Schaefer, D.M., Carpenter, M., Gady, B., Reifenberger, R., Demejo, L.P., and Rimai, D.S. 1995. Surface roughness and its influence on particle adhesion using atomic force techniques. *Journal of Adhesion Science and Technology* 9(8):1049–1062.

Selamoglu, N., Mucha, J.A., Ibbotson, D.E., and Flamm, D.L. 1989. Silicon oxide deposition from tetraethoxysilane in radio frequency: Downstream reactor: Mechanism and step coverage. *Journal of Vacuum Science and Technology* B7:1345.

Schultz, J. and Lavielle, L. 1989. Interfacial properties of carbon fiber-epoxy matrix composites. In: *Inverse Gas Chromatography, ACS Symposium Series*. Lloyd, D.R., Ward, T.C., and Schreiber, H.P. (eds). American Chemical Society, Washington, D.C. 391:168–184.

Stanzl-Tschegg, S., Beikircher, W., and Loidl, D. 2009. Comparison of mechanical properties of thermally modified wood at growth ring and cell wall level by means of instrumented indentation tests. *Holzforschung* 63(4):443–448.

Stöckel, F., Konnerth, J., Kantner, W., Moser, J., and Gindl, W. 2010a. Mechanical characterisation of adhesives in particle boards by means of nanoindentation. *European Journal of Wood and Wood Products* 68(4):421–426.

Stöckel, F., Konnerth, J., Kantner, W., Moser, J., and Gindl, W. 2010b. Tensile shear strength of UF- and MUF-bonded veneer related to data of adhesives and cell walls measured by nanoindentation. *Holzforschung* 64(3):337–342.

Stone, D.S., Yoder, K.B., and Sproul, W.D. 1991. Hardness and elastic modulus of TiN based on continuous indentation technique and new correlation. *Journal of Vacuum Science & Technology A (Vacuum, Surfaces, and Films)* 9(4):2543–2547.

Stuart, B.H. 2004. *Infrared Spectroscopy—Fundamentals and Applications*. John Wiley Publications, New York.

Sundqvist B., Doctoral thesis 2004. *Colour Changes and Acid Formation in Wood During Heating*. Luleå University of Technology, Sweden.

Tan, Z. and Reeve, D.W. 1992. Spatial distribution of organochlorine in fully bleached kraft pulp fibres. *Nordic Pulp and Paper Research Journal* 6:30–36.

Tang, B. and Ngan, A.H.W. 2003. Accurate measurement of tip-sample contact size during nanoindentation of viscoelastic materials. *Journal of Materials Research* 18(Copyright 2003, IEE):1141–1148.

Tokareva E. 2011. *Doctoral thesis – "Spatial Distribution of Components in Wood by ToF-SIMS"*. Åbo Akademi University, Turku.

Tshabalala, M.A. 1997. Determination of acid-base characteristics of lignocellulosic surfaces by inverse gas chromatography. *Journal of Applied Polymer Science* 65:1013–1020.

Tshabalala, M.A., Kingshott, P., VanLandingham, M.R., and Plackett, D. 2003. Surface chemistry and moisture sorption properties of wood coated with multifunctional alkoxysilanes by sol-gel process. *Journal of Applied Polymer Science* 88(12):2828–2841.

Tze, W.T.Y., Wang, S., Rials, T.G., Pharr, G.M., and Kelley, S.S. 2007. Nanoindentation of wood cell walls: Continuous stiffness and hardness measurements. *Composites Part A: Applied Science and Manufacturing* 38(3):945–953

VanLandingham, M.R., Chang, N.K., Drzal, P.L., White, C.C., and Chang, S.H. 2005. Viscoelastic character-ization of polymers using instrumented indentation. I. Quasi-static testing. *Journal of Polymer Science, Part B (Polymer Physics)* 43(Copyright 2006, IEE):1794–1811.

Van Oss, C.J. 1994. *Interfacial Forces in Aqueous Media.* CRC Press, Boca Raton, FL.

Viitaneimi P., Jämsä S., Ek P., and Viitanen H. 2001. Method for increasing the resistance of cellulosic products against mould and decay, EP695408B1.

Vlassak, J.J., Ciavarella, M., Barber, J.R., and Wang, X. 2003. The indentation modulus of elastically anisotro-pic materials for indenters of arbitrary shape. *Journal of the Mechanics and Physics of Solids* 51(9):1701–1721.

Wålinder. M.E.P. 2002 Study of lewis acid-base properties of wood by contact angle analysis. *Holzforschung,* 56:363–371.

Watts, J.F. 1990. *An Introduction to Surface Analysis by Electron Spectroscopy.* Oxford University Press, New York.

Watts, J.F. 1992. Investigation of adhesion phenomena using surface analytical techniques. *Analytical Proceedings* 29:396–398.

Williams, D. 1994. Inverse gas chromatography. In: *Characterization of Composite Materials.* H. Ishida (ed.), Butterworth-Henemann, Boston, MA.

Williams, R.S. and Feist, W.C. 1984. Application of ESCA to evaluate wood and cellulose surfaces modified by aqueous chromium trioxide treatment. *Colloids and Surfaces* 9:253–271.

Wimmer, R. and Lucas, B.N. 1997. Comparing mechanical properties of secondary wall and cell corner middle lamella in spruce wood. *Iawa Journal* 18(1):77–88.

Wimmer, R., Lucas, B.N., Tsui, T.Y., and Oliver, W.C. 1997. Longitudinal hardness and Young's modulus of spruce tracheid secondary walls using nanoindentation technique. *Wood Science and Technology* 31(2):131.

Winograd, N. 1993. Ion beams and laser positionization for molecule-specific imaging. *Analytical Chemistry* 65(14):622A–629A.

Wang, S. and Peltonen, J. 2011. Surface characterization of wood samples by ToF-SIMS. *Private Communication.*

Wu, Y., Wang, S., Zhou, D., Xing, C., and Zhang, Y. 2009. Use of nanoindentation and silviscan to determine the mechanical properties of 10 hardwood species. *Wood and Fiber Science* 41(1):64–73.

Xing, C., Wang, S., and Pharr, G. 2009. Nanoindentation of juvenile and mature loblolly pine wood fibers as affected by thermomechanical refining pressure. *Wood Science and Technology* 43(7):615–625.

Young, R.A., Rammon, R.M., Kelley, S.S., and Gillespie, R.H. 1982. Bond formation by wood surface reac-tions. Part I: Surface analysis by ESCA. *Wood Science* 14:100–119.

Yu, Y., Fei, B., Wang, H., and Tian, G. 2010. Longitudinal mechanical properties of cell wall of Masson pine (Pinus massoniana Lamb) as related to moisture content: A nanoindentation study. *Holzforschung* 65(1):121–126.

Zickler, G.A., Schoberl, T., and Paris, O. 2006. Mechanical properties of pyrolysed wood: A nanoindentation study. *Philosophical Magazine* 86(10):1373–1386.

Part III

Wood Composites

9 Wood Adhesion and Adhesives

Charles R. Frihart

CONTENTS

9.1 GENERAL

The recorded history of bonding wood dates back at least 3000 years to the Egyptians (Skeist and Miron 1990, River 1994a), and adhesive bonding goes back to early mankind (Keimel 2003). Although wood and paper bonding are the largest applications for adhesives, some of the fundamental aspects leading to good bonds are not fully understood. Better understanding of these critical aspects of wood adhesion should lead to improved wood products. The chemistry of adhesives has been covered in detail (Pizzi 2003a–f); however, the fundamentals of adhesive mechanical performance are not well understood. This chapter is aimed at more in-depth coverage of those items that are not covered elsewhere. It will touch briefly on topics covered by other writers and the reader should examine the recommended books and articles for more details. Many of the books on adhesives and adhesion are long and complicated, but a brief but thorough book exists (Pocius 2002). Adhesives are designed for specific applications, leading to thousands of products (Rice 1990). Petrie has broken adhesives into 20 groups of synthetic structural, 11 groups of elastomeric, 12 groups of thermoplastic, and 6 groups of natural adhesives (Petrie 2000). Brief has summarized the vast number of markets for adhesives (Brief 1990).

Understanding how an adhesive works is difficult since adhesive performance is not a single science, but the combination of many sciences. Adhesive strength is defined mechanically as the force necessary to pull apart two substrates that are bonded together. Mechanical strength is dependent upon primary and secondary chemical bonds and interlocking of the polymer chains in the adhesive, wood and adhesive-wood interphase. Thus, both chemical and mechanical aspects of bond strength, and the interrelation of the two factors are all important. Because adhesive strength is a measurement of failure, the process determines where the localized stress exceeds the bond strength under specific test conditions. One concept is to illustrate the bonded assembly as being a series of links representing different domains with the failure occurring in the weakest link (Marra 1980). However, the bondline is actually more a continuum than discrete links. The localized stress is usually very different from applied stress due to stress distribution and concentration (Dillard 2002). It is generally preferred that the adhesive bond be stronger than the substrate so that the failure mechanism is one of substrate fracture.

There are generally three steps in the process of adhesive bonding. The first is usually the preparation of the surface to provide the best interaction of the adhesive with the substrate. Even though a separate treatment step may not be used in some cases, the knowledge of material science (surface chemistry and morphology) is important for understanding this interaction. Preparation of the surface can involve either mechanical or chemical treatment or a combination of the two. In some cases, the adhesive is modified to deal with problems in wetting of the surface or contamination on the surface. Surface analysis techniques are often more difficult on wood than other materials due to the complex chemistry and morphology of the wood.

The second step is that the adhesive needs to form a molecular-level contact with the substrate surface; thus, it should be a liquid so that it can develop a close contact with the substrates. This process involves both the sciences of rheology and surface energies. Rheology is the science of the deformation and flow of matter. Surface energies are influenced by both the polar and nonpolar components of both the adhesive and the substrate. Improving the compatibility by altering one or both components can lead to stronger and more durable bonds.

The third step is the setting, which involves the solidification and/or curing of the adhesive. Most adhesives change physical state in the bonding process, with the main exception being pressure sensitive adhesives that are used on tapes and labels. The solidification process depends on the type of adhesive. For hot melt adhesives, the process involves the cooling of the molten adhesive to form a solid, whether this is an organic polymer as in some craft glues, or an inorganic material as in the case of solder. Other types of adhesives have polymers dissolved in a liquid, which may be water (e.g., white glues) or an organic (e.g., rubber cement). The loss of the solvent converts these liquids to solids. The third type of adhesive is made up of small molecules that polymerize to form the adhesive, for example, super glues or two-part epoxies. Most wood adhesives involve both the polymerization and solvent loss methods. Understanding the conversion of small molecules into large molecules requires knowledge of organic chemistry and polymer science.

Once the bond is prepared, the critical test is the strength of the bonded assembly under forces existing during the lifetime of the assembly. This involves both externally applied forces and internal forces from shrinkage during the curing of the adhesive and differential expansion/contraction of the adhesive and substrate during environmental changes. Understanding the performance of a bonded assembly requires knowledge of both chemistry and mechanics. Often the strength of a bonded assembly is discussed in terms of adhesion. Adhesion is the strength of the molecular layer of adhesive that is in contact with the surface layer of the substrate, such as wood. The internal and applied energies may be dissipated at other places in the bonded assemblies than the layer of molecular contact between the adhesive and the substrate. However, failure at the interface between the two is usually considered unacceptable. Understanding the forces and their distribution on a bond requires knowledge of mechanics.

An appreciation of rheology, material science, organic chemistry, polymer science, and mechanics leads to better understanding of the factors controlling the performance of the bonded assemblies; see Table 9.1. Given the complexity of wood as a substrate, it is hard to sort out why

TABLE 9.1
Wood Bonding Variables

Resin	Wood	Process	Service
Type	Species	Adhesive amount	Strength
Viscosity	Density	Adhesive distribution	Shear modulus
Molecular weight distribution	Moisture content	Relative humidity	Swell—shrink resistance
Mole ratio of reactants	Plane of cut: radial, tangential, transverse, mix	Temperature	Creep
Cure rate	Heartwood vs. sapwood	Open assembly time	Percentage of wood failure
Total solids	Juvenile vs. mature wood	Closed assembly time	Failure type
Catalyst	Earlywood vs. latewood	Pressure	Dry vs. wet
Mixing	Reaction wood	Adhesive penetration	Modulus of elasticity
Tack	Grain angle	Gas-through	Temperature
Filler	Porosity	Press time	Hydrolysis resistance
Solvent system	Surface roughness	Pretreatments	Heat resistance
Age	Drying damage	Posttreatments	Biological resistance: fungi, bacteria, insects, marine organisms
pH	Machining damage	Adherend temperature	Finishing
Buffering	Dirt, contaminants		Ultraviolet resistance
	Extractives		
	pH		
	Buffering capacity		
	Chemical surface		

some wood adhesives work better than other wood adhesives, especially during the more severe durability tests. In general, wood is easy to bond compared to most substrates, but it can be harder to make a truly durable wood bond. A main trend in the wood industry is increased bonding of wood products as a result of the fewer old growth trees and more engineered wood products.

9.2 WOOD ADHESIVE USES

Because adhesives are used in many different applications for bonding wood, a wide variety of types are used (Frihart and Hunt 2010). Given the focus of this book on composites, the emphasis will be more on adhesives used in composite manufacturing than on those used in product assembly. Factors that influence the selection of the adhesive include cost, compatibility with the assembly process, strength of bonded assembly, and durability.

The largest wood market is the manufacturing of panel products, including plywood, oriented strandboard (OSB), fiberboard, and particleboard. Except for plywood, the adhesive in these applications bonds small pieces of wood together to form a wood-adhesive matrix. The strength of the product depends on efficient distribution of applied forces between the adhesive and wood phases. The composites (strandboard, fiberboard, and particleboard) have adhesive applied to the wood (strands, fibers, or particles); then they are formed into mats and pressed under heat into the final product. This type of process requires an adhesive that does not react immediately at room temperature (pre-mature cure), but is heat activated during the pressing operation. Given the weight adhesive (2–8%) compared to the product weight and relatively low cost of wood, adhesive cost is an issue. In addition, since the wood surfaces are brought close together, gap filling is not an important issue, but over penetration can be. On the other hand, for plywood, the surfaces are not uniformly brought in such close contact, requiring the adhesive to remain more above the surface. Light-colored adhesives are important for some applications, but many of these products have their surfaces covered by other materials. Most of the adhesives used in wood bonding have formaldehyde as a co-monomer, generating concern about formaldehyde emissions. Dunky and Pizzi have discussed many of the commercial issues relating to the use of adhesives in manufacture and the use of wood composites (Dunky and Pizzi 2002). Recently, formaldehyde emissions have become an important issue (Frihart 2011, Williams 2010).

For laminating lumber and bonding finger joints, the adhesive can either be heat or room-temperature cured. The cost of the adhesive has become more critical as the thickness of the wood decreases from glulam to laminated veneer lumber and parallel strand lumber (Stark et al. 2010). Color is sometimes an issue, but moisture and creep resistances are more important because these products are usually used in structural applications.

Adhesives used in construction and furniture assembly usually have long set times and are room-temperature cured. Furniture adhesives are light-colored, low-viscosity, and generally do not need high moisture resistance. On the other hand, construction adhesives generally have a high viscosity and need some flexibility, but color and moisture resistance are less important issues.

The movement away from solid wood for construction to engineered wood products has increased the consumption of adhesives. A wooden I-joist can have up to five different adhesives in its construction; see Figure 9.1. The wood laminates that form the top and bottom flanges may be finger joined with a melamine–formaldehyde (MF) adhesive and glued together with a phenol–resorcinol–formaldehyde adhesive. The OSB that forms the middle part (web) is often produced using both phenol–formaldehyde (PF) and polymeric diphenylmethane diisocyanate adhesives to bond the strands. The I-joist is produced by attachment of the web to the flange and bonding of the web sections together using emulsion–polymer isocyanate. Each of these adhesives has different chemistries, are bonded under different conditions of time, temperature, and pressure to a variety of wood surfaces, and are subjected to different forces during use. Thus, it is not surprising that a simple model for satisfactory wood adhesion has been difficult to derive.

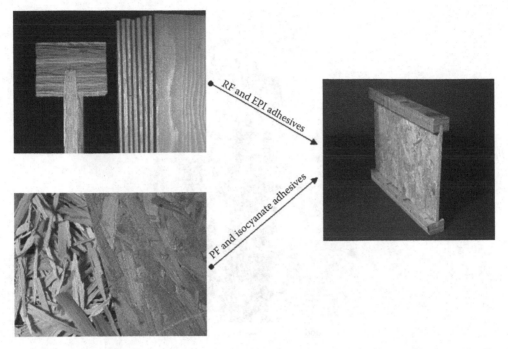

FIGURE 9.1 The importance of adhesives is illustrated by the need for different adhesives to make the flange by the bonding of laminate pieces and the oriented strandboard from the flakes and the final I-joist by attachment of the strandboard to the flange.

9.3 TERMINOLOGY

Confusion can be caused if there is no clear understanding of the terminology; this chapter generally follows that given in the ASTM Standard D 907-11 (ASTM International 2011a). *Adhesive joint failure* is "*n*—the locus of fracture occurring in an adhesively-bonded joint resulting in a loss of load-carrying capability" and is divided into interphase, cohesive, or substrate failures. *Cohesive failure* is within the bulk of the adhesive, while *substrate failure* is within the substrate or adherend (wood). The least clear failure zone is that occurring within the *interphase,* which is "a region of finite dimension extending from a point in the adherend where the local properties (chemical, physical, mechanical, and morphological) begin to change from the bulk properties of the adherend to a point in the adhesive where the local properties equal the bulk properties of the adhesive." Figure 9.2 shows the various regions of a bonded assembly. The bulk properties are the properties of one phase unaltered by the other phase.

The *assembly time* is "the time interval between applying adhesive on the substrate and the application of pressure, or heat, or both, to the assembly." This time can be "closed" with substrates brought into contact or "open" with the adhesive exposed to the air; these times are important for penetration of the adhesive and evaporation of solvent. *Set* is "to convert an adhesive into a fixed or hardened state by chemical or physical action, such as condensation, polymerization, oxidation, vulcanization, gelation, hydration, or evaporation of volatile constituents." *Cure* is "to develop the strength properties of an adhesive by chemical reaction." Note that cure is only one way for the adhesive setting step. However, because cure is a function of how it is measured, there is no universal value for an adhesive. Separating partial cure from total cure is important because they usually have very different properties, and in most bonded products, total cure is not usually obtained after long after the product is assembled while partial cure can allow the product to be handled for the next stage of manufacturing. *Tack* is "the property of an adhesive that enables it to form a bond of measurable strength immediately after the adhesive and adherend are brought into contact under low pressure." Tack is important for holding composites together during layup and pre-pressing.

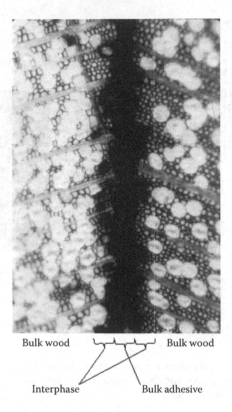

Bulk wood   Bulk wood

Interphase   Bulk adhesive

FIGURE 9.2   A transverse scanning electron microscope image of a resorcinol bond of yellow-poplar, showing the zones of bulk wood, interphase region, and bulk adhesive.

A *structural adhesive* is "a bonding agent used for transferring required loads between adherends exposed to service environments typical for the structure involved" (ASTM International 2011a). For wood products, structural implies that failure can cause serious damage to the structure, and even loss of life (Frihart and Hunt 2010), while semi-structural adhesives need to carry the structural load, but failure is not as disastrous, and nonstructural adhesives typically support merely the weight of the bonded product.

Other terms are used in different ways that can also cause confusion. The term *adhesive* can refer to either the adhesive as applied or the cured product. On the other hand, a *resin* is often used to refer to the uncured adhesive, although the ASTM defines a resin as "solid, semisolid, or liquid, usually organic material that has an indefinite molecular mass and, when solid, usually has a softening or melting range and exhibits a tendency to flow when subjected to stress" (ASTM International 2011a). Thus, a cross-linked adhesive is not a resin, but the adhesive in the uncross-linked state may be. Glue was "originally, a hard gelatin obtained from hides, tendons, cartilage, bones, etc. of animals," but is now generally synonymous with the term adhesive.

9.4   APPLICATION OF THE ADHESIVE

9.4.1   ADHESIVE APPLICATION TO WOOD

The first step in bond formation involves spreading the adhesive over the wood surface. The physical application of the adhesive can involve any one of a number of methods, including using spray, roller coating, doctor blade, curtain coater, and bead application technologies. After the adhesive application, a combination of some open and closed assembly times is used depending on the specific bonding process. Both give the adhesive time to penetrate into the wood prior to bond formation, but the open assembly time will cause loss of solvent or water from the formulation. Long open times can cause the adhesive to dry out on the surface causing poor bonding because flow is needed for bonding to the substrate. In the bonding process, pressure is used to bring the surfaces closer together. In some cases, heat and moisture are used during the bonding process, both of which will make the adhesive more fluid and the wood more deformable (Kretschmann 2010).

For any type of bond to form, molecular-level contact is required. Thus, the adhesive has to flow over the bulk surface into the voids due to surface roughness that exist for almost all surfaces. Many factors control the wetting of the surface, including the relative surface energies of the adhesive and the substrate, viscosity of the adhesive, temperature of bonding, pressure on the bondline, and so on. Wood is a more complex bonding surface than what is encountered in most adhesive applications. Wood is very anisotropic because the cells are greatly elongated in the longitudinal direction, and the growth out from the center of the tree makes the radial properties different from the tangential properties. Wood is further complicated by differences between heartwood and sapwood, and between earlywood and latewood. Adding in tension wood, compression wood, and slope of grain increases the complexity of the wood adhesive interaction. The manner in which the surface is prepared also influences the wetting process. These factors are discussed in later sections of this chapter and in the literature (River et al. 1991), but for now we will assume that the adhesive is formulated and applied in such a manner that it properly wets the surface.

9.4.2 Theories of Adhesion

Adhesion refers to the interaction of the interface between adhesive and adherend. It must not be confused with bond strength. Certainly if there is little interaction of the adhesive with the adherend, these materials will detach when force is applied. However, bond strength is more complicated because factors such as stress concentration, energy dissipation, and weakness in surface layers often play a more important role than adhesion. Consequentially, the aspects of adhesion are a dominating factor in the bond formation process, but may not be the weak link in the bond breaking process.

It is important to realize that, although some theories of adhesion emphasize mechanical aspects and others put more emphasis on chemical aspects, chemical structure and interactions determine the mechanical properties and the mechanical properties determine the force that is concentrated on individual chemical bonds. Thus, the chemical and mechanical aspects are linked and cannot be treated as completely distinct entities. In addition, some of the theories emphasize macroscopic effects while others are on the molecular level. The discussion of adhesion theories here is brief because they are well covered in the literature (Schultz and Nardin 2003, Pocius 2002), and in reality, most strong bonds are usually due to a combination of the concepts listed in each theory.

In a mechanical interlock, the adhesive provides strength through reaching into the pores of the substrate (Packham 2003). An example of mechanical interlock is Velcro; the intertwining of the hooked spurs into the open fabric holds the pieces together. This type of attachment provides great resistance to the pieces sliding past one another, although the resistance to peel forces is only marginal. In its truest sense, a mechanical interlock does not involve the chemical interaction of the adhesive and the substrate. However in reality, there are friction forces preventing detachment, indicating interaction of the surfaces. For adhesives to form interlocks, the adhesive has to wet the substrate well enough so that there are some chemical as well as mechanical forces in debonding. For a mechanical interlock to work, the tentacles of adhesive must be strong enough to be load bearing. The size of the mechanical interlock is not defined, although the ability to penetrate pores becomes more difficult and the strength becomes less when the pores are narrower. It should be noted that generally mechanical interlocks provide more resistance to shear forces than to normal forces. Also, many substrates do not have enough roughness to provide sufficient addition to bond strength from the mechanical interlock. Roughing of the substrate surface by abrasion, such as grit blasting or abrasion, normally overcomes this limitation.

If the concept of tentacles of adhesive penetrating into the substrate is transferred from the macro scale to the molecular level, the concept is referred to as the diffusion theory (Wool 2002). If there are also tentacles of substrate penetrating into the adhesive, the concept can be referred to as inter-diffusion. This involves the intertwining of substrate and adhesive chains. The interface is strong since the forces are distributed over this intertwined polymer network (Berg 2002). However, the concept can also work if only the adhesive forms tentacles into the substrate. For this to occur there has to be good compatibility of the adhesive and substrate. This degree of compatibility is not that common for most polymers. When it does occur a strong network is formed from a combination of chemical and mechanical forces.

The other theories are mainly dependent upon chemical interactions rather than truly mechanical aspects. Thus, they take place at the molecular level, and require an intimate contact of the adhesive with the substrate. These chemical interactions will be discussed in order of increasing strength of the interaction (Kinloch 1987). The strengths of various types of bonds are given in Table 9.2, along with examples of some of the bond types in Figure 9.3. It is important to remember that the strength of interaction is for just a single interaction. To make a strong bond these interactions need to be large in number and evenly distributed across the interface.

The weakest interaction is the London dispersion force (Wu 1982a). This force is the dispersive force that exists between any set of molecules and compounds when they are close to each other. The dispersion force is the main means of association of nonpolar molecules, such as polyethylene (Figure 9.3). Although this force is weak, where the adhesive and the adherend are in molecular contact, the force exists between all the atoms and can result in appreciable total strength. The ability of the gecko to walk on walls and ceilings has been attributed to this force (Autumn et al. 2002).

The other types of forces are generally related to polar groups (Pocius 2002). The weakest are the dipole–dipole interactions. For polar bonds, there is a separation of charge between the atoms; this process creates a natural, permanent dipole. Two dipoles can interact if positive and negative ends of the dipole match up with the opposite ends of another dipole. The strength of this interaction depends

TABLE 9.2
Table of Bond Strengths from Literature Bond Types and Typical Bond Energies

Type	Bond Energy (kJ mol⁻¹)
Primary bonds	
Ionic	600–1100
Covalent	60–700
Metallic, coordination	110–350
Donor-acceptor bonds	
Brønsted acid–base interactions (i.e., up to a primary ionic bond)	Up to 1000
Lewis acid–base interactions	Up to 80
Secondary bonds	
Hydrogen bonds (excluding fluorines)	1–25
Van der Waals bonds	
Permanent dipole–dipole interactions	4–20
Dipole-induced dipole interactions	Less than 2
Dispersion (London) forces	0.08–40

Source: Adapted from Fowkes, F.M. 1983. *Physicochemical Aspects of Polymer Surfaces.* Vol. 2, 583–603. Plenum Press, New York; Good, R. J. 1967. *Treatise on Adhesion and Adhesives, Volume 1: Theory,* 9–68. New York: Marcel Dekker; Kinloch, A. J. 1987. *Adhesion and Adhesives Science and Technology.* London: Chapman & Hall; Pauling, L. 1960. *The Nature of the Chemical Bond.* Ithaca, NY: Cornell University Press.

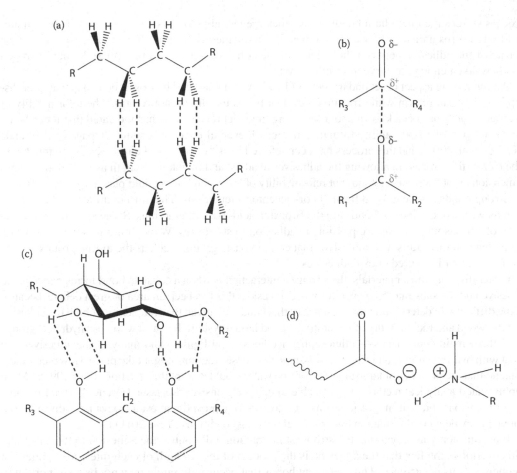

FIGURE 9.3 Examples of various types of bonds, including (a) dispersive bonds between two hydrocarbon chain, such as exist in polyethylene, (b) a dipole bond between two carbonyl group, such as in a polyester, (c) hydrogen bonds between a cellulosic segment and a PF polymer, and (d) an ionic bond between an ammonium group and a carboxylate group.

on proper alignment of the dipoles, which is not difficult for small molecules in solution, but can be very difficult between two chains because they have constrained translation and rotation (Wu 1982a). A variation of this concept is the dipole-induced dipole, but this interaction is usually weaker than the permanent dipole interaction and also suffers from the same alignment problem in polymers.

Strongest of the secondary interactions is the hydrogen bond formation. This type of bond is common with polar compounds, including nitrogen, oxygen, and sulfur groups with attached hydrogens, and carbonyl groups. This type of bond involves sharing a hydrogen atom between two polar groups, and is extremely likely with wood and wood adhesives because both have an abundance of the proper polar groups. Almost all wood components are rich in hydroxyl groups and some contain carboxylic acid and ester groups. Both of these groups form very strong internal hydrogen bonds that give wood its strength, but are also available for external hydrogen bonds. All major wood adhesives have polar groups that can form internal and external hydrogen bonds. Many bio-based adhesives depend heavily on hydrogen bonds for their adhesive and cohesive strength. Many synthetic adhesives are less dependent upon the hydrogen bond for their cohesive strength because they have internal cross-links, but most certainly form hydrogen bonds to wood. One limitation of the hydrogen bond is its ability to be disrupted in the presence of water. Water and other hydrogen bonding groups can insert themselves between the two groups that are present in the hydrogen bond. This process softens the inter-chain bonds so that they are less able to resist applied loads. Thus, a material that adsorbs and absorbs water, like wood, loses some of its strength when it is wet. The same is true of the adhesion between the wood and the adhesive—it is certainly possible that hydrogen bonds weaken enough to serve as a failure zone.

An interesting aspect of secondary bonds (dispersive, dipolar, and hydrogen bonds) is that after disruption, they can reform while fractured covalent bonds usually do not reform. The reformability of hydrogen bonds has been known about for a long time, but recent work has indicated that it can be an important part of wood's ability to maintain strength even after there is some slippage of the bonds (Keckes et al. 2003), and this process has been referred to as "velcro" mechanics (Kretschmann 2003). The role of this process in allowing the adhesives to adjust and maintain strength as the wood changes dimensions is not well understood, but reformability of hydrogen bonds could play a significant factor.

Strong bonds can be formed from donor–acceptor interactions. The most common of these interactions with wood-adhesive bonds are the Brønsted acid–base interactions. Some acid–base interactions of cations with anions are possible in adhesion to substrates. Wood contains some carboxylic acids that can form salts with adhesives that contain basic groups, such as the amine groups in MF, protein, and amine-cured epoxy adhesives.

Generally, with most materials, the strongest interaction is when a covalent bond forms between the adhesive and the substrate. However, for wood adhesion, this has been an area of great debate, because of the difficulty in determining the presence of this bond type given the complexity of both the adhesive and the wood and the difficulty of generating a good model system. Because wood has hydroxyl groups in its three main components—cellulose, hemicellulose, and lignin—and many of the adhesives can react with hydroxyl groups, it is logical to assume that these reactions might take place. However, others contend that the presence of large amounts of free water would disrupt this reaction (Pizzi 1994a). More sophisticated analytical methods were unable to definitely answer this issue (Frazier 2003). Recently, two-dimensional nuclear magnetic resonance spectroscopy proved that even isocyanate adhesives did not yield covalent bond formation under typical bonding conditions (Yelle 2011a,b).

It is commonly assumed that the strongest interaction will control the adhesion to the substrate. This overlooks the fact that the adhesion is the product of the strength of each interaction times the frequency of its occurrence. Thus, covalent bonds that occur only rarely may not be as important to bond strength as the more common hydrogen bonds or dipole–dipole interactions. Hydrogen bonds may be less significant under wet conditions than other bonds if the water disrupts these bonds. It is more important to think about forming stronger adhesion, not by a single type of bond, but by a large number of bonds of different types. Another point to consider is that the adhesive can adhere strongly to a surface and still not form a strong bond overall, due to failure within either the adhesive or the adherend interphase.

One model of adhesion that is generally not related to the bond formation step, but is observed during bond breakage, is the electrostatic model. This model assumes that adhesion is due to the adhesive or the adherend being positive while the other is the opposite charge. It is unlikely that such charges generally exist prior to bond formation, and therefore cannot aid in adhesion; however, they can occur during the debonding process.

Another model that has limited applicability to most cases of adhesion is deep inter-diffusion, which involves polymers from the adhesive and adherend mixing to form a single, commingled phase. Although it is unlikely that the wood will dissolve in the adhesive, it is quite likely that some of the adhesive molecules will be absorbed into the wood cell walls. This one-way diffusion can form one of several types of structures that more strongly lock the adhesive into the wood. This is one type of penetration, and it will be covered in Section 9.4.8. In many cases, the strength of this penetration could be as strong as covalent linkages.

Most of these adhesion models play not only a role in bond formation, but also aid the bonded assembly in resisting the debonding forces. The important part to remember is that, depending on the origin of the forces, the stresses can be either concentrated at the interface or dispersed throughout the bonded assembly. If the forces are dispersed, then the force felt at the interface may be quite small.

It is often asked which model of adhesion is correct. This question assumes that there is only a single factor dominating the interaction of the adhesive and the substrate. In reality, there is often a combination of factors that play a role to some degree. The general rule is that the more of each mode of adhesion existing at the interface, the greater the bond durability.

9.4.3 WOOD ADHESION

The comprehension of wood adhesive bonds requires both an understanding of the uniqueness of the wood structure for bond formation and an understanding of the modes of stress concentration and dissipation during environmental changes. Because adhesive strength is a mechanical property, the polymer properties of the adhesive, wood, and wood-adhesive interphase regions, are covered in the following sections. Macroscopic generalizations are difficult in the sense that wood is a nonhomogenous substrate. The adhesive needs to interact with many different types of bonding surfaces. In softwood, large longitudinal tracheids opened by vertical transwall cleavage are the main part of the surface, but parenchyma cells, various ray cells, and resin canals that are also exposed to the adhesive are also bonding surfaces with different properties. Opening the resin canals distributes resin across the surface that interferes with bonding. In hardwood, small fiber cells and large vessels form the main bonding surface, with rays and other cells also being involved. The vessel elements are split open, while the fiber cells are split in the middle lamella. Although generally for bonding studies the sapwood is used, in actual products there can be considerable amount of heartwood, which is harder to bond. Adding to the complexity, the wood can have juvenile, tension and compression wood. Adhesion studies use samples that are mainly tangential with a small slope of grain, only tiny knots, and no splits, but in commercial wood these factors are less controlled.

Most observations of adhesive interaction with wood are concentrated on scales of millimeter or larger (Marra 1992). However, the wood–adhesive interaction needs to be evaluated in three spatial scales (millimeter, micrometer, and nanometer) (Frazier 2002, Frihart 2003, 2006). The millimeter or larger involves observations by eye or light microscopy. The use of scanning electron microscopy allows observations on the micrometer or cellular level and smaller. On the other hand, the size of the cellulose fibrils, hemicellulose domains, and lignin regions are on the nanometer scale. The nanometer level is also the spatial scale in which the adhesive molecules need to interact with the wood for a bond to form. Tools, such as atomic force microscopy, developed for making observations on the nanoscale can be difficult to use with wood because its surface is rough on the micrometer scale.

To understand the adhesive interaction with the wood, we need to consider in more detail the aspects of surface preparation, types of wood surfaces, and spatial scales of wood surfaces. This provides the appropriate background for discussing the adhesive bonding as the steps of wetting the surface and solidification of the adhesive. The wood–adhesive interaction is important for the ultimate strength and durability of the bonded assembly.

9.4.4 WOOD SURFACE PREPARATION

On the larger scale, wood is a porous, cellular, anisotropic substrate. It is porous in that water and low molecular weight compounds will be rapidly absorbed and move through the wood. The elongated cells of varying size and shape with the differences between the radial and tangential directions lead to the wood being very anisotropic. A simple model cannot be developed because of the large differences between wood species in the chemistry and morphology of the wood surfaces. The cell types and sizes are dramatically different between hardwood and softwood. The individual species in each of these classes vary considerably in their ability for liquids to penetrate, the amount of extractives, as well as the distribution of the various cell types. Even within a species, there is the problem of earlywood versus latewood, sapwood versus heartwood, and juvenile, compression, and tension wood having distorted cell structures. The earlywood cells with the thinner walls should be easier to bond because of a more accessible lumen. The sapwood of a species is generally considered easier to bond than the heartwood due to changes in the extractives. The juvenile, compression, and tension wood all have distorted cell structures that should weaken the wood adhesive interphase region. To simplify the discussion, the emphasis will be placed on the wood that meets the selection criteria for standard testing.

The surface preparation has been shown to have a large effect on the quality of a wood surface (River et al. 1991). One concern is a weak boundary layer, which is a layer between the bulk materials and the true adhesive–adherend interface that is often the weak link and fails cohesively within that layer (Bikerman 1968, Wu 1982b). A classic example of a weak boundary layer was the difficulty in bonding to aluminum, due to a weak aluminum oxide surface layer, until the FPL etch was developed (Pocius 2002). Stehr and Johansson have broken down the weak boundaries of wood into those that are chemically weak and those that are mechanically weak (Stehr and Johansson 2000). The distinction is that the chemically weak layer involves extractives coming to the surface, while the mechanically weak layer involves a crushed or fractured cell layer. The role of extractives has been widely considered to be a major factor in poor adhesive strength. Certainly, low-polarity, small molecules coming to the surface can hurt the wetting process. However, it is not clear that they are normally a cause of poor bond strength. Chemically weak boundary layers are certainly an issue in oily woods, such as teak, where wiping the surface with solvent to remove the oils will solve most bonding problems. The issue of extractives should not be confused with the more general phenomena of over-dried wood. The latter case also involves chemical alteration of the wood by excessive heating that leads to poor wetting and weaker bonds (Christiansen 1990, 1991, 1994).

Wetting is an important issue, especially since most wood adhesives are water-borne. Water has such a high surface energy that wetting of many surfaces is difficult. Although surfactants can lower surface energy, they are often avoided since they can create a chemically weak boundary layer. The monomers and oligomers in the adhesive can lower surface energies, as can added low molecular weight alcohols. Wetting should be less of an issue with adhesives that have organic solvents or are 100% solids.

Mechanically weak boundary layers can be an issue with wood (River 1994a). The general problem is the crushing or excessively fracturing the wood cells during the surface preparation. Wood cells, especially earlywood, are weak in the radial and tangential direction. Crushed cells are easy to visualize by looking at cross sections microscopically. If the adhesive does not penetrate through the layer of crushed cells, then this layer will generally be the source of fracture under test or use conditions. The best method for preparing a wood surface for bonding is to use sharp planar blades. Unsharpened blades can crush cells and cause a very irregular surface (River and Miniutti 1975). The difference in penetration of an adhesive on well- and poorly planed wood surfaces is shown in Figure 9.4. Abrasively planed surfaces and saw-cut surfaces also suffer from crushed and fractured surface cells. Hand sanding is generally acceptable because it causes less damage to the cells. For laminates, ASTM prescribes that the wood surfaces be planed with sharp blades and then be bonded within 24 h to provide the most bondable surface (ASTM International 2011b), although it is generally recommended to bond surfaces immediately after preparation to provide the most durable bond.

9.4.5 WOOD BONDING SURFACE

The wood-bonding surface varies considerably both chemically and morphologically depending on how the surface is prepared and what type of wood is being used. The morphology is better characterized than the surface chemistry, and will be discussed first. Except for fiber bonding, the desire is to have sufficient open cells on the surface so that the adhesive can flow into the lumen of the cells to provide more area for mechanical interlock. The accessibility of open cells is dependent upon the tree species, types of cells, and method of preparation. When the cell wall is thin in comparison to the diameter of the cell, then there will be more longitudinal transwall fracture.

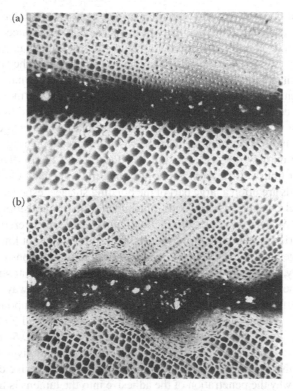

FIGURE 9.4 Bondlines show good adhesive penetration for (a) a sound wood surface, but not for (b) a crushed and matted wood surface.

Hardwood vessels and earlywood cells have thin walls that are easily split to open the lumens to the adhesive for good penetration. On the other hand, hardwood fiber cells and latewood cells have thick walls that are not easy to fracture, so cleavage often occurs more in the middle lamella providing less area for mechanical interlock (River et al. 1991). The open ends of any cells and cracks in the cell walls allow the adhesive to penetrate into the lumens. The differences in the surfaces can be large, by comparing the scanning electron microscopy pictures for southern yellow pine and hard maple (Figure 9.5), with pine having more open cells, while many of the maple's cells are closed.

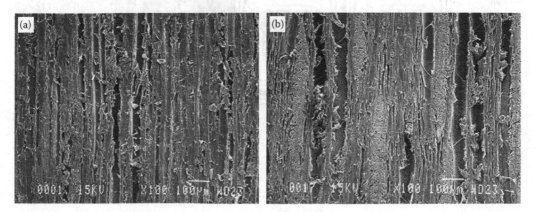

FIGURE 9.5 Scanning electron microscopy pictures of transverse sections of (a) southern yellow pine and (b) hard maple.

   The chemical composition of the wood-bonding surface is less well understood because the surface is very hard to characterize. The roughness of the surface, the presence of many different surfaces (lumen walls, middle lamella, and fractured cell walls), and the changes of the surfaces with time, heat, and moisture add to the difficulty. The main components of the wood are the cellulose, hemicellulose and lignin fractions. The interactions of PF and urea–formaldehyde (UF) polymers with cellulose have been modeled (Pizzi 1994b). Although cellulose is the main component of wood, it may not be the main component on the surface. Prior work has indicated that hemicellulose is the main site of interaction with water for hydrogen bonding because of its greater accessibility (River et al. 1991, Salehuddin 1970). The preparation of the wood surface by planing can create many types of surfaces, depending on how the cells fracture, as illustrated in Figure 9.6. If the cell walls are cleaved in longitudinal transwall fashion as desired, then the lumen should be the main bonding surface. The lumen walls are often a large part of the bonding surface, especially for earlywood cells of softwoods, and vessel elements in hardwoods. The lumen walls' compositions can vary from being highly cellulosic, if the S_3 layer is exposed, to highly lignin if they are covered by a warty layer. The middle lamella is also rich in lignin. However, for the most part we do not know when the walls are fractured if the cleavage plane runs through any of the three main fractions or between the lignin–hemicellulose boundary, which may be the weakest link in the wood cellular structure. Complicating this consideration of the bonding surface is that the typical mechanical ways of preparing binding surfaces cause a lot of fragmentation and smearing of the cell wall components. Only by careful microtome sectioning can the clean splitting of the cell walls be observed. Other methods give surfaces that are a lot less intact (Wellons 1983). As can be seen by Figures 9.5 and 9.7, there is a lot of debris on the surface even with sharp planer blades. Hardwood tends to give even more smearing of the surface. Thus, the theory of many open lumens into which the adhesive can flow is not always correct, which may be why the penetration of the adhesive into the lumens is not always that fast.

9.4.6 SPATIAL SCALES OF WOOD FOR ADHESIVE INTERACTION

Wood bonds need to be considered on three different spatial scales: millimeter and larger, micrometer, and nanometer (Frazier 2002, Frihart 2004). The millimeter and larger scale is normally used for evaluating the bonding and debonding processes of wood. The micrometer scale relates to the cellular dimensions. The nanometer and smaller scale correlates to the sizes of the cellulose, hemicellulose,

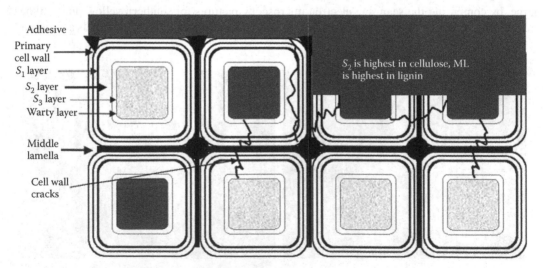

FIGURE 9.6 Illustration of a transverse section of wood showing fracture points of the wood cellular structure and surfaces available with which adhesives can interact, assuming clean fractures are occurring.

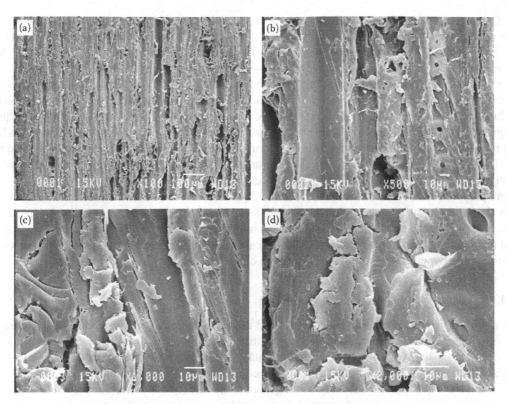

FIGURE 9.7 Scanning electron microscopy of yellow poplar surfaces at four levels of magnification, increasing from (a) to (d) as illustrated by the length bar in each image showing the extensive fracturing of the surface and generation of weakly bonded fragments even with sharp planar blades.

and lignin domains and the molecular interactions of the adhesive with the wood. Each domain size requires different observation methods and has different implications on bonding and debonding processes.

The millimeter and larger scale is the normal method for dealing with both the bonding and the debonding processes. Usually the naked eye or feel by hand touch is used to judge the smoothness of the surface for bonding. On this scale, measurement of the spread by the adhesive across the surface is typically done by contact angles. Normal examination of the adhesive bond failure is generally limited to this scale. This information is valuable for understanding bond formation and failure aspects as the first stage in evaluation of adhesive performance. However, it is important to move on to the smaller spatial scales to gain a fuller understanding of wood bonding.

The micrometer scale involves the adhesive interaction with the lumens and cell walls. While the earliest theory on the strength of wood adhesive bonds involves mechanical interlock (McBain and Hopkins 1925), others proposed that there were specific interactions of adhesives of the wood surface (Browne and Brouse 1929). Flow into the lumen of cells is still considered important as judged by many microscopic studies on penetration (Johnson and Kamke 1992). However, there has not been enough consideration of what happens to the adhesive–wood interphase as the cells and the adhesive undergo differential expansion caused by changes in moisture and temperature. Because different adhesives can interact with the cells in diverse ways, these aspects are covered in more detail in the individual bonding and debonding sections (9.4.8 and 9.6.3). The tools for looking at this level of interaction are more complicated because it is at the high end of light microscopy magnification, but it is certainly in the range of scanning electron and transmission electron microscopy (SEM and TEM, respectively).

The nanometer and smaller scale is important because it is the size of the basic domains of wood and of the adhesive–wood interactions (Fengel and Wegener 1984). The size of the cellulose fibrils, the hemicellulose portions, and the lignin networks are in the tens of nanometers. For there to be adhesion, the adhesive needs to interact with the wood at the molecular level; independent of whatever mechanism is involved. The idea of wood adhesion being more than a mechanical interlock was proposed in the 1920s with the concept of specific adhesion as being critical (Browne and Brouse 1929). The problem with understanding this specific adhesion is our lack of understanding of the composition of the wood surface. Although cellulose is the main component of wood, it may not be the main component on the wood surface. If bonding to lumen walls is important, then adhesion to lignin is important since the warty layer present in many species is high in lignin content (Tsoumis 1991). Cleavage in the middle lamella, as may occur with latewood cells, fiber cells in hardwood or fibers prepared for fiberboard, leads to a surface high in lignin content. Until we can better define how the adhesive has to interact with the wood to form durable bonds, this area is still quite speculative. Although instrumental methods, such as atomic force microscopy, surface force microscopy, and nanoindentation can look at surfaces at this scale, they work best when the surface morphology changes only by nanometers while the roughness of the wood surface varies by micrometers.

9.4.7 Wetting and Penetration in General

For a bond to form the adhesive needs to wet and flow over a surface, and in some cases penetrate into the substrate. It is important to understand that the terms mean different things even though they sound familiar. Wetting is the ability of an adhesive drop to form a low contact angle with the surface upon contact. In contrast, flow involves the adhesive spreading over that surface in a reasonable time. Flow is important because covering more of the surface allows for a stronger bond. Thus, a very viscous adhesive may wet a surface, but it might not flow to cover the surface in a reasonable time frame. Penetration is the ability of the adhesive to move into the voids on the substrate surface or into the substrate itself. The difference between flow, penetration, and transfer are illustrated in Figure 9.8.

First, we will consider the aspects of wetting, flow, and penetration that are common to most substrates. In the next section, we will discuss how these need to be modified for wood bonding. For a strong bond to form, the adhesive must intimately encounter most of the substrate surface (Berg 2002). With many plastics having low surface energies, this is a significant problem since the adhesive can find it difficult to wet the substrate. An extreme example is the bonding of Teflon, which has

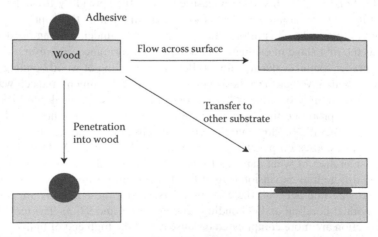

FIGURE 9.8 Adhesive wetting of wood surfaces, showing the difference between flow, penetration, and transfer.

a very low surface energy so that very few adhesives will wet it. In fact, an adhesive applied to the surface forms a bead rather than wets the surface. For bonding to many polyethylene and polypropylene materials, wetting by an adhesive is also a significant problem because of their low surface energies. Thus, a great deal of the literature places emphasis on the measurement of contact angles to determine the wetting of the surface. The contact angle is the angle at the edge of a droplet and the plane of that surface upon which it is placed. Therefore, a material with a high contact angle has poor surface wetting ability. The addition of surfactants or less polar solvents reduces the adhesive's surface energy as indicated by a decreased contact angle. With many plastics, surface treatments such as oxidation by flame or corona discharge are used to increase the polarity and surface energy of the plastic surface to improve its bondability. It is important to remember that most contact angle measurements are equilibrium values, and may not reflect the dynamics of the bonding process well. Another very important property that is closely associated with wetting is flow over the surface. Flow is dependent upon not only the contact angle, but also the viscosity of the adhesive. With a lower viscosity, the adhesive flows better and wets more of the surface.

While flow is movement across the surface, penetration is the movement into the substrate. Adhesives will not penetrate into the bulk of many substrates like metals and many plastics, but penetration is important in the sense of movement of the adhesive into the microcrevices on the surface (Berg 2002). Most surfaces have some degree of roughness, which an adhesive must penetrate. Like flow, penetration is dependent upon surface energies and adhesive viscosity, but it also depends on the size of the capillary or void that it is penetrating. For a strong bond, the adhesive must penetrate into all microscale roughness. A typical problem is a displacement of air, water, or oil on the surface. As discussed in the next section, penetration has a very different meaning for wood, due to its structure.

9.4.8 WETTING, FLOW, AND PENETRATION OF WOOD

Wood bonding faces many of the same issues as discussed in the previous section on general aspects of wetting, flow, and penetration, but there are many characteristics that are unusual about wood that require additional consideration. Wood has a relatively polar surface that allows the general use of water-borne adhesives, although some woods are harder to wet. Examples are some tropical woods that have a very oily nature, such as teak, and wood that has been treated with creosote. Wetting of the surface can be improved by removal of the oily components through solvent wiping, mechanical, or oxidation techniques. In Figure 9.9, the effect of sanding on improving the wetting of yellow birch veneer is illustrated. It has been shown that oxidation of wood surfaces by corona treatment can improve wetting and adhesion for some woods (Sakata et al. 1993). A lot of work has

FIGURE 9.9 Water droplets on a yellow birch veneer show the improved wetting by removal of surface contaminants. The photograph was taken 30 s after placing three droplets on the surface. The left drop was on an untreated surface, the middle was renewed by two passes of 320 grit sandpaper, and the right drop was renewed with four passes with the sandpaper.

been done on examining the wetting of wood; however, it is not clear what this data means. For example, studies to relate extractives with bonding have not found good correlations (Nussbaum 2001). This may be caused by wetting experiments usually being done with water at room temperature; while most adhesive bonding is done using aqueous solutions of organics with higher temperatures and pressure, all of which improve wetting of surfaces.

Understanding flow over the surface is complicated by the fact that the surface has very macroscopic roughness, and penetration is taking place at the same time. As mentioned in the previous section, penetration generally involves wetting of the micro-roughness. On the other hand, wood's cellular nature allows significant penetration of the adhesive into the substrate. A main complication is that different species of woods have different cellular structures, and therefore, adhesives will penetrate them to different degrees. This leads to problems in trying to achieve uniform penetration when bonding different species of wood, as occurs in OSB production. For a more porous wood, an adhesive can over-penetrate into the wood and not be on the surface for bonding, while the same adhesive on a less porous wood sits on the surface and may not give significant bonding. Thus, adhesives are formulated for different applications given the type of wood, the type of application, and the application conditions. An adhesive that is sprayed onto OSB tends to be much lower in viscosity for better spraying than one that is formulated for spreading on plywood that needs to sit more on the surface. Aspects of formulating adhesives are covered in later sections.

In most bonding applications, adhesive penetration into the adherend does not occur to any great degree, but penetration is very important for wood bonding. The need for the proper degree of penetration influences both the formulation of the adhesive and the bonding conditions. The proper balance is necessary in that poor bonds will result from either under- or over-penetration. In under-penetration, the adhesive is not able to move into the wood enough to give a strong wood–adhesive interaction. In contrast, with over-penetration so much of the adhesive moves into the wood that insufficient adhesive remains in the bondline to bridge between the wood surfaces, resulting in a starved joint. To solve these problems, the viscosity and composition of the adhesive can be adjusted, as well as the temperature and time for the open and closed assemblies. Some species are known to be more porous compared to other species, leading to complications when bonding mixed species. This is a significant issue for composites that usually use a wide mixture of species and a frequently changing mixture. Using mixed species certainly could lead to both over- and under-penetration and to potentially reduced bond strength. Although it is generally known that proper penetration is important to strong bonds, it is not clear whether penetration into the lumens or the cell walls is more critical.

The penetration of adhesives into wood is most often examined at the cellular level. Some lumens have openings on the surface as a result of slope of grain so that the adhesive can flow into the lumen; this is more likely with larger diameter cells in softwood. In hardwoods, most of the filling of lumens is of the larger vessels rather than the smaller fiber cells. Factors that influence the filling of the lumens can be classified into those that are

- Wood related, such as diameter of the lumen and exposure on the wood surface
- Adhesive related, such as its viscosity and surface energy
- Process related, such as assembly time, temperature, pressure, moisture level

It is normally assumed that the filling of lumens contributes to bond strength. Resin penetration into lumens has been extensively investigated in the wood bonding literature because it is easy to determine by visible light, fluorescence, and scanning electron microscopy. The problem is that these data have not been related to bond strength or level of bond failure. An example, where a filled ray cell contributed to adhesion of a coating after environmental exposure, has been shown by light microscopy (Dawson et al. 2003).

In addition to filling the lumens, an important part of wood adhesion, especially for durable bonds, might be infiltration of adhesive components into cell walls (Nearn 1965, Gindl et al. 2002).

A significant amount of lower molecular weight compounds can go into cell walls due to their ability to swell. These compounds include water, cosolvents, adhesive monomers and oligomers, but not higher molecular weight polymers. Polyethylene glycol molecules of up to 3000 g/mol were shown to penetrate into the transient capillaries or micropores in cell walls (Tarkow et al. 1965). In addition to hydrodynamic volume of an adhesive, its compatibility with the wood structure controls this infiltration. Generally, solubility parameters are widely used to determine the compatibility of adhesives and coatings to interact with surfaces (Barton 1991). Limited studies have been done trying to relate the solubility parameters of the components of wood to its ultrastructure (Hansen and Björkman 1998, Horvath 2006), which would then relate to the components' interaction with adhesives.

The observation of adhesive components in cell walls has been shown by a variety of methods. The migration of PF resins into cell walls has been shown using fluorescence microscopy (Saiki 1984), audioradiography (Smith 1971), transmission electron microscopy (Nearn 1965), scanning electron microscopy with x-ray dispersive emissions (Smith and Cote 1971), dynamic mechanical analysis (Laborie et al. 2006), and antishrink efficiency (Stamm and Seborg 1936). For polymeric diphenylmethane diisocyanate, pMDI, the presence of adhesives in cell walls has been shown by x-ray micrography and nuclear magnetic resonance spectroscopy (Marcinko et al. 1998, Marcinko et al. 2001). These and other techniques such as UV microscopy (Gindl et al. 2002) and nano-indentation (Gindl and Gupta 2002) have been used to show the presence of UF and MF (Bolton et al. 1985, 1988), while fluorescence spectroscopy has been used to show epoxy resins in the wall layers (Furuno and Goto 1975, Furuno and Saiki 1988). Because both chemical and mechanical data show the presence of adhesives in cell lumens and cell walls, it is likely that the wood portion of the interphase has very different properties than the bulk wood.

Although it has been shown that adhesive components can infiltrate into cell walls, only in one case has it been claimed to improve bond strength (Nearn 1965). Several models can be proposed as to how these adhesive components may influence bond strength. The simplest is that the oligomers and monomers are simply soluble in the cell walls, but do not react, being too diluted by the cell wall components (Laborie et al. 2006). In this case, they would maintain the cell walls in the expanded state due to steric constraint (bulking effect); thus, the process would reduce the stresses due to less dimensional change. A second model is that the adhesives react with cell wall components and possibly cross-link some of the components, thereby increasing the strength properties of the surface wood cells, as shown in Figure 9.10. A third model is that the adhesives polymerize to form molecular

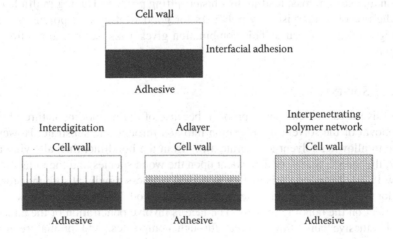

FIGURE 9.10 Modes of adhesive interaction within wood cell walls are depicted for true interfacial adhesion with no cell wall penetration, interdigitation of fingers of adhesive penetrating the microchannels, adlayer of cross-linking in the surface cell wall and interpenetrating polymer.

scale fingers of the adhesive in the wall, providing a nanoscale mechanical interlock. The fourth is that they form an interpenetrating polymer network within the wood, providing improved wall strength (Frazier 2002). All of these models have the adhesive reducing the dimensional changes of the surface cells, and therefore reducing the stress gradient between the adhesive and the wood, thereby improving the bond strength, but the degree of improvement should be different.

Knowing that adhesive components do migrate into the cell wall, the next questions is: Are they associated with any specific cell layer or the middle lamella, and are they more in the cellulose, hemicellulose or lignin domains? One study indicates that the isocyanates seem to be more concentrated in the lignin domains (Marcinko et al. 2001). Peeling experiments have shown that an epoxy adhesive gave failure in the S_3 layer while a PF adhesive resulted in failure deeper in the S_2 layer (Saiki 1984).

9.5   SETTING OF ADHESIVE

Once an adhesive is applied to wood; the adhesive needs to set for forming an assembly with high strength. Set is "to convert an adhesive into a fixed or hardened state by chemical or physical action, such as condensation, polymerization, oxidation, vulcanization, gelation, hydration, or evaporation of volatile solvent." Although the ASTM terminology uses solvent to refer to organic solvents, this chapter uses it in the more general sense of both water and organics because wood adhesives are usually water-borne. Water-borne adhesives often contain some organic solvent to help in the wetting of wood surfaces. For some of the polymeric adhesives, including polyvinyl acetate, casein, blood glue, and so on, the loss of solvent sets the adhesive. For many others, including the formaldehyde-cured adhesives, the set involves both the loss of water and polymerization to form the bond. For polymeric diphenylmethane diisocyanate, the set is by polymerization. For hot melt adhesives, cooling to solidify the polymer is sufficient. In wood bonding, all of these mechanisms are applicable, dependent upon the adhesive system that is being used.

The original wood adhesives were either hot-melt or water-borne natural polymers (Keimel 2003). These had several limitations in relation to speed of set, formation of a strong interphase region, and environmental resistance. All of the biomass-based adhesives had poor exterior resistance. The use of composites and laminated wood products has greatly expanded with the development of synthetic adhesives with good moisture resistance. Instead of being mainly polymers with limited and reversible cross-links, these adhesives have strong covalent cross-links to provide environmental resistance. In addition, these synthetic adhesives generally cure by both polymerization and solvent loss, leading to a faster setting process. Having multiple modes of set allows both the use of lower viscosity polymers for good wetting and polymers with a higher molecular weight for a faster cure. This combination gives a fast set rate that allows for higher production speeds.

9.5.1   Loss of Solvents

For many adhesive uses, solvents are a problem because of the nonporous nature of the substrate preventing removal of the solvent by migration into and through the substrate. However, wood is quite effective in allowing solvent to migrate away from the bondline, thus allowing adhesives to set. Of course, this property is very dependent upon the wood species and the moisture level of the wood (Tarkow 1979). It is not surprising that wet wood will less rapidly absorb moisture, thus making it harder for water-borne adhesives to move into the wood. The dynamics of water movement have a large effect on the bonding process. The factors involve penetration of the adhesive into the wood, rate of adhesive cure, flow of heat through composites, and premature drying of the adhesive.

Penetration of the adhesive into the wood is an important part of the bonding process. Green wood is difficult to bond with most adhesives because there is little volume into which the adhesive

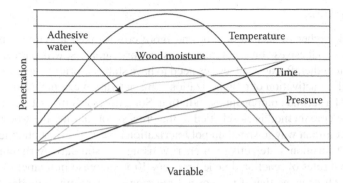

FIGURE 9.11   General effects of conditions on adhesive penetration. The temperature makes the adhesive more fluid until too much causes polymerization. At low wood moisture the water is drawn from the adhesive, while at high wood moisture the water retards the penetration. As the water content of the adhesive increases, the viscosity of the adhesive is lower and penetration increases. Both an increase in bond pressure and a longer time promote adhesive penetration.

can penetrate. (See Figure 9.11 for the generalized effect of bonding parameters on penetration.) At the other extreme, overly dry wood can also be difficult for the adhesive to penetrate because the wood surface is more hydrophobic and therefore harder to wet (Christiansen 1994). Thus, wood with a 4–10% moisture range is typically good for optimum penetration and set rates. The desire is to have the bonded product be near the normal in-use moisture condition to reduce dimensional changes and internal stress (Marra 1992). Although most of the studies on uptake of small molecules into wood have naturally concentrated on water, other solvents are also readily absorbed/adsorbed by wood.

For many of adhesives, cure rate is dependent upon the moisture content. Most bonding processes require the wood to be within a set range of moisture content to get an acceptable set rate. For the adhesive to set, the solvent needs to flow away from the adhesive into the adjoining and further removed cell walls. The sorption of the water into the nearby cell walls allows the formation of the solid, cured adhesive. Many setting reactions involve condensations that give off water; higher moisture levels can retard the reactions as expected by normal chemical equilibrium theory and from limited collisions due to dilution. The amount of water present also alters the mobility of polymer chains during the curing process, which can change the product distribution for the adhesive polymers. On the other hand, many isocyanates depend on a small amount of water to start the curing process. Thus, the isocyanates are most tolerant of higher moisture content of the wood.

A very important issue in the rate of setting is the heat flow through composites or laminates to the bond surface, especially since wood is a good insulator. In composites, water boiling in the wood near the composite surface or added steam helps transfer heat to the core of the composite. Use of core resins that cure at lower temperatures than face resins is important for fast production cycles; fast curing can be accomplished by using higher molecular weight oligomers or adding catalysts to the core resins. Controlling heat transfer and moisture levels is important for fast, reproducible composite production. Isocyanates are less sensitive to higher moisture levels in the core. The ability of resorcinol–formaldehyde (RF) and phenol–resorcinol–formaldehyde to cure rapidly at room temperature favors them over PF resins despite their higher cost where it is difficult to heat the bondline, such as laminated beams. Another way to accelerate cure is to use radiation methods, such as radio frequency curing. Polyurethanes use moisture to cure rather than heat.

With some adhesives, premature drying can be a problem if the open time is too long. This involves too much loss of solvent so that the adhesive does not flow to wet the other surface. Proper control of moisture level and penetration are accomplished by the length of open- and closed-assembly times, as well as adhesive composition.

9.5.2 POLYMERIZATION

For a strong bond, higher molecular weight and more cross-linked polymers are needed (Wool 2005). In most cases, adhesives consist of monomers and/or oligomers, which are a small number of monomers linked together. Because adhesives need to have stability prior to application, there needs to be some method for activation of polymerization. This activation can include heat, change in pH, catalyst, addition of a second component, or radiation. Sometimes a combination of methods is used for faster cure. The cure method is closely tied to the process for making the wood product.

Heat is a very common way to speed up polymerization reactions. Most chemical processes are controlled by the transition state activation energy, using the standard Arrhenius equation. The typical factor is that rates of reaction double for every 10°C increase in temperature, but this does not always apply. This means that if the normal reaction temperature is moderate, there will be appreciable reaction at room temperature and a limited storage life of the adhesive for a single component system. Since wood is a good insulator, uniform heating of the adhesive continues to be a problem for many composites and laminates. Incomplete heating gives poor bond strength as a result of incomplete formation of the adhesive polymer. To overcome this problem adhesive producers try to have the adhesive formulation in as advanced stage of polymerization as is possible while still having good flow and penetration into the wood. Having a more advanced resin means that fewer reactions need to take place to obtain the strength properties needed from the adhesive. This balance between the advancement of the resin for fast curing while still having good bonding properties has been optimized by intense study of reaction mechanisms over the years and allows for higher production rates. On the other hand, the understanding of heat and moisture levels within the composites is still being studied to allow further improvement in production rates (Winandy and Kamke 2004).

Many of the adhesive polymerization rates are sensitive to pH. This is especially true of the formaldehyde polymers, but the effect varies with the individual type of co-reactant and the different steps in the reaction. For UF resins, the initial addition step of formaldehyde to urea is base catalyzed, while the polymerization of hydroxymethylated urea is acid catalyzed. Thus, UF resins are kept at a more neutral pH for storage stability, but then accelerated by lowering the pH during the bonding process. For PF resins, there is a different pH effect with condensation reactions being faster at high pHs and very low pHs. One issue of concern is how much the pH and neutralization capacity of wood alters the adhesives' polymerization rates near the interface and within the wood. This is complicated by the fact that different woods have different pHs and buffering abilities (Marra 1992).

Another aspect that alters the polymerization rate is the addition of catalysts and accelerators. A true catalyst is one that is not consumed in the process, while an accelerator can be consumed via reaction. A number of accelerators are incorrectly termed catalysts. As mentioned in the previous paragraph, changes in pH can catalyze polymerizations. In some cases, the pH is not changed directly, but compounds are added that can generate acids, such as the ammonium chloride or ammonium sulfate accelerators for UF resins that decompose upon heat to yield hydrochloric acid or sulfuric acid, respectively (Pizzi 2003e). Certain metal ions are known to be catalysts for PF resins. Ortho esters are often described as catalysts for PF resins, but in actuality are consumed in the process, making them accelerators (Conner et al. 2002). A number of compounds have been found to speed up PF curing (Pizzi 1994c). In some cases, co-reactants, such as formaldehyde, have been referred to as accelerators, but in their general use, they serve as hardeners because they become part of the polymer.

Many adhesives are two-part products. Because the components are not mixed together until shortly before the bonding process, each component alone has a good storage life. However, the addition of a second component allows the polymerization to begin. Because the adhesive is applied at ambient temperatures and most of the polymerizations need higher temperatures, setting is slow until the composite or laminate reaches the heated press. Rapid ambient polymerizations are not desirable because they limit the adhesive's ability to wet and penetrate the wood, and to transfer when the wood surfaces are brought into contact. One area of concern is the uniformity of mixing of two components. Off-ratio mixtures do not form as strong a bond as those at optimum ratio because of the poor stoichiometry. The better the compatibility and more equal the viscosity of the two components, the better the uniformity of the product upon mixing. Most application equipment is designed to give good mixing, but this may not be as true in laboratory testing or during upsets in plant operations. A special type of two-component application is where one component is applied to one surface and the other component to the other surface and has been called a honeymoon adhesive (Kreibich et al. 1998). The two surfaces need to be brought into the proper contact to allow mixing and the two components need to have good mutual solubility for this system to work well.

A common type of cure is those that use water as a reactant in the polymerization process; this type of cure is used for curing of most one-component polyurethanes, isocyanate, and silicones adhesives (Frazier 2003, Lay and Cranley 2003, Parbhoo et al. 2002). The chemistry for the polyurethanes and isocyanates is discussed in a later section. Because these adhesives use water for curing, then water exposure needs to be prevented prior to application. In general, wood contains sufficient moisture to cause curing. In addition, because these reactions use water for curing rather than give it off as in condensation polymerizations, the adhesives are much more tolerant on bonding wood with higher moisture content than are most other adhesives.

Another method of activation of an adhesive is the use of some type of radiation. The use of ultraviolet light and electron beam radiation are common for the curing of coatings, but trying to get light into a wood adhesive bond is more difficult. However, other types of radiation can penetrate wood, including microwaves and radio frequencies, which activate curing by causing heat generation in the bondline to initiate thermal polymerization (de Fleuriot 2004).

9.5.3 SOLIDIFICATION BY COOLING

Although hot melts are a small part of the wood adhesive market, understanding the interaction of molten polymers with wood to form a strong durable interface is important for the wood–plastic composite field. Many wood adhesives used by the early civilizations were hot melts (Keimel 2003). Some hot melt adhesives have been used for bonding plastics to wood and are used in some wood assembly markets, such as cabinet construction, edge banding, window manufacturing, and mobile home construction. Because hot-melt adhesives and plastics used for composites are polymeric, they have a limited ability to flow. Heating the polymers above their softening point will allow them to flow. The lower the molecular weight of the polymer and the higher the temperature, the better the flow. However, both of these aspects can reduce the final strength and lengthen the set time. The formulation of the polymer backbone and additives can have a great effect on the set time. In fact, formulation is often used to control the set time so that the adhesive does not solidify before the two components are in place or take so long that extended clamping times are needed. Unlike other adhesives, high viscosities of hot melts limit their ability to penetrate into the wood lumens and flow across the wood surfaces. As the adhesive cools, its viscosity raises rapidly to further limit the wetting. Although the wetting of the wood is limited, there has still been reported flow into lumens (Smith. 2002). Understanding the wood-molten polymer interaction is very critical for making improved wood–plastic composites (Clemons et al. 2012).

Some of the newer hot-melt adhesives are reactive types that allow for better wetting by the adhesive and greater cured strength. Normally, hot melts need to be of high molecular weight for strength, but if the adhesive cures after application, then the initial strength is not such a critical issue. The curing also makes the adhesive a thermoset to eliminate remelting of the adhesive or flow (creep) with time. Some of these products are isocyanates so that they cure by reacting with moisture that is readily available in the wood (Paul 2002). Thus, the combination of modes of set provides benefits that are not available over adhesives with a single mode of setting.

9.6 PERFORMANCE OF BONDED PRODUCTS

Because an adhesive is used to hold two adherends together under normal use conditions, it is important to comprehend the properties of an adhesive that allow it to perform this function. The definition of an adhesive is mechanical in nature, making it important to understand the internal and external forces on the bondline and the distribution of those forces across the bonded assembly. Mechanical properties are dependent upon the chemical structure; thus, knowing the structure of the adhesive and interphases helps to understand the adhesive's performance. Bonded assemblies are usually weaker in tension perpendicular to the bondline than in shear or compression because it is easier to pull the chains apart. To understand the performance of bonded products, the structures of the wood adhesive polymers and the mechanical properties of polymers need to be appreciated. Greater strength in the bulk of the adhesive does not necessarily result in more strongly bonded assemblies because the weakest portion may still be in the interphase regions. Another factor is the need to know the forces that the bondline must withstand under normal use conditions. The effect of external forces on the bondline can be analyzed through a variety of standard tests; however, the internal forces are not as clearly determined. There are commonly accepted durability tests, but the forces that are exerted on the bondline during these tests are not well understood. The relationship of mechanical properties that are usually observed on the millimeter scale to the chemical structure that is formed under the nanoscale has to be examined.

9.6.1 BEHAVIOR UNDER FORCE

The evaluation of the integrity of a bonded object rests upon understanding the viscoelastic dissipation of energy for each of the components (bulk adherend, bulk adhesive, and adhesive–adherend interphase). A basic test is a stress–strain curve, which shows the response of a material to an applied force, usually in tension. Although the behavior of material can be measured in tension, compression, or shear, tension is usually measured because it is the most likely mode of failure.

Stress–strain data are presented for a variety of material types in Figure 9.12. A very stiff material, such as a nonductile metal or glass, does not elongate (% strain) much before the material breaks; thus, the applied force accumulates as stress until it exceeds the strength of the material, as indicated by curve A. The stiffness or modulus of A is defined as the stress divided by strain at low percent strain usually over the linear region. Plastics are represented by curve B or C in that at some point the elastic limit (when deformation is no longer reversible) is exceeded at the yield point. The

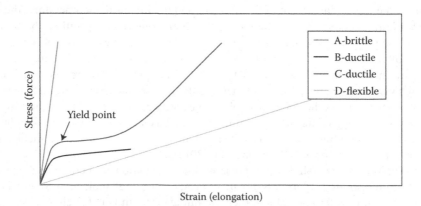

FIGURE 9.12 General stress-strain data for polymers. The rigid polymers resist applied force and build the stress showing a high modulus (stress/strain) until the material breaks. A ductile material will resist initially, but then start to flow at the yield point, with higher molecular polymers showing strain-induced crystallization. The flexible polymer will offer little resistance to the applied force, giving a high elongation.

modulus of B is the linear portion prior to the yield point. The applied force is elastically stored in the plastic prior to the yield point, but stretches inelastically after the yield point. For a lower molecular weight plastic B, at some point on this plateau the applied force exceeds what that plastic can take and the sample breaks. However, a higher molecular weight plastic C will often have a strain-induced crystallization that causes the curve to bend upward again. The last example D represents a rubber that does not store much energy as stress, but the force causes the material to elongate. The modulus in this case is much lower and hard to measure since the initial linear section is short. In addition to the stress, strain, and modulus obtained from these tensile tests, another important piece of information is the area under the curve, which is related to toughness of the material.

For wood-bonding applications, a polymer of type D is not acceptable since there is not enough rigidity in the adhesive. However, type D is excellent for caulking and sealant applications since these materials need to be flexible given the expansion and contraction of buildings. Curves B and C have large areas under the stress–strain curve giving these materials good toughness, especially for impact resistance. Curve C represents plastic used in wood–plastic composites. Some wood adhesives represented by curve B are the poly(vinyl acetate) resins, emulsion polymerized isocyanates, polyurethanes, contact cement, and hot-melt adhesives.

Curve A represents structural adhesives that have low creep, the lack of flow under force. This nonflow characteristic under normal conditions means that bonded products will retain their shape. Most wood adhesives fall into this class, including the widely used UF, PF, RF, and combinations such as melamine–urea–formaldehyde and phenol–resorcinol formaldehyde. The epoxy and fully cured polymeric diphenylmethane diisocyanate adhesives also are members of this class.

The data in these graphs represent the materials at a specific temperature. As the temperature of a material increases it softens so that a class A polymer becomes like B. The transition of going from a glassy (hard and brittle) material to a more pliable one involves going through the glass transition temperature, T_g. However, there are limits on softening for curable adhesives because they can continue to cure and become more rigid at elevated temperatures and can begin to degrade at some point, thus changing their physical properties.

Knowing the chemical structure of the adhesives allows the prediction of the general mechancial properties, but does not allow the calculation of the specific shape of the stress–strain curve. The curve D polymers are generally linear or branched organics that have low crystallinity. They also include a major nonorganic adhesive and sealant type, the silicone adhesives that are actually poly(dimethylsiloxanes) and their derivatives and copolymers. These materials will creep, that is, flow under an applied force, unless they are cross-linked. The cross-links prevent the polymer chains from continuing to flow past one another. As the number of cross-links increases, the material becomes stiffer, usually resulting in a reduction in the ultimate elongation.

For noncross-linked polymers, the properties are dependent not only upon the chemical structure, but also upon the conditions to which the material has been exposed. As would be expected, the lower the rotational energy around the bond in the backbone, the more flexible and impact resistant the product is. Thus, Si—O–Si bonds provide the most flexibility and are curve D, with C–O–C next, and then C–C–C bonds being the least flexible. Replacement of a linear structure with a cyclic group increases the stiffness of the backbone, and having an aromatic ring provides even higher stiffness. Interchain interactions, such as hydrogen or ionic bonds between chains and the formation of crystalline regions to act as reversible cross-links, usually greatly alter the properties. These interactions reduce chain mobility, and thus increase the stiffness and glass transition temperature (T_g) of the polymer. As the stiffness of the backbone increases and the number of cross-links increase, the shape of the curve goes from D to B and eventually A. However, these interactions will be weakened by heat or water exposure, reducing the strength of the polymer. Additionally, the history of the polymer affects its properties. Plastics (curves B and C) generally have a fair degree of crystallinity; this association of the molecules causes a reduction in the mobility of the polymer chains compared to more amorphous polymers. The quantity and structure of the crystalline regions depend very much on how the material solidifies. Fast cooling creates fewer and smaller crystals, resulting in a less stiff product than does slow cooling (annealing). At the interfaces, the type of adjoining surfaces influences the crystallization of the polymer.

The chemical structure and amount of cross-linking play a major role in making an A-type polymer. The backbones usually contain aromatic groups, sometimes cyclic groups, and generally few aliphatic groups, and the polymers tend to be highly cross-linked. Because many wood products are used for structural applications, it is necessary that under applied load most will not exhibit any significant elongation; thus, a high modulus is required. Unfortunately, the same factors that lead to a high modulus generally lead to brittleness in the polymer.

Cross-linking of polymer chains is required to convert a thermoplastic resin to a thermoset resin. The tying of the chains together eliminates the plastic flow of the polymers, which is necessary to eliminate creep over time. Natural rubber was known about for a long time but had little commercial utility because it softened under heat. After much research, vulcanization processes were developed which allowed rubber to retain its deformability, but eliminated the flow. As would be expected at low cross-linking levels, rubber has large segmental mobility, resulting in a very flexible product. As the cross-linking and molecular weight increases, the segments have less mobility, making the product more rigid. Unfortunately at high cross-linking levels, not only does the product become more rigid, it also becomes more brittle.

Figure 9.13 shows some idealized stress–strain curves that demonstrate the effect of increasing polymerization and cross-linking on the properties of different adhesives, and the effect of conditions on the adhesive. For thermoplastics, increasing the molecular weight mainly increases the elongation at break. This means as the adhesive cures, it is able to withstand greater force. The conversion from a thermoplastic to a thermoset will increase the stiffness at some expense of ductility. For both thermoplastics and thermosets, an increase in temperature or moisture will soften the material. However, for thermoplastics this leads to much lower strength, while for thermoset the effect upon failure properties is much less. One consequence of this softening in composite production,

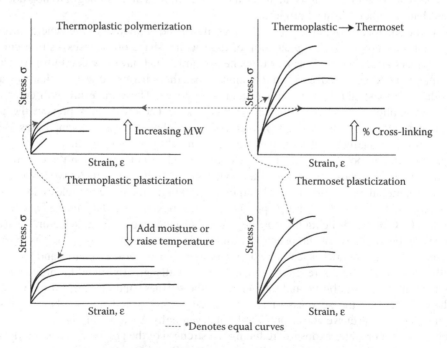

FIGURE 9.13 Effect of polymer changes on physical properties. For a thermoplastic, increasing the molecular weight leads to increases in both the stiffness and the ductility, while the thermoset loses ductility as it becomes stiffer with higher cross-linking. When the polymers are plasticized or the temperature is raised, both the thermoplastic and thermoset lose stiffness.

both the heat and moisture factors are working against the adhesive as it is trying to hold the material together to resist either a blowout (void in panels caused by steam bubbles) or excessive spring-back (tendency of compressed or bent materials to return to their original state).

Some classes of adhesives are more amenable to changing their properties by altering their formulations than are others. Polyurethanes and polyamide adhesives can go from very flexible to quite rigid depending on the formulation. PF and polymeric methanediphenyl diisocyanate adhesives do not have similar formulation flexibility. For some resins, incorporating flexible segments, which are softer than the main backbone and improve the impact resistance and reduce the brittleness of the polymer, can improve the polymer's properties.

However, the adhesive formulator does have a number of tools for varying the stress–strain behavior of these products. It should be noted that many of these additives are added for other purposes, such as lower cost, reduction of over-penetration, increase of resin tack, and improvement of wet out, but our concern here is how they affect the stress–strain behavior. The additives are divided into the classes of fillers, extenders, plasticizers, and tackifiers, see Section 9.7.13.

Fillers are common additives because they lower the cost, and thus are used at as high a level as possible to make the adhesive more economical. Fillers increase the stiffness of the adhesive, but usually also reduce its elongation and increase its viscosity. At low levels extenders have a small impact on an adhesive's properties, but at high levels they cause decreased elongation and higher viscosity. On the other hand, plasticizers soften an adhesive, resulting in a decreased modulus and T_g, and an increased elongation. For most wood adhesives the desire is to have a rigid bond; thus, plasticizers are not generally used. Tackifiers are often confused with plasticizers, but provide very different responses in raising the glass-transition temperature while decreasing the modulus. Increase in tack is often desirable with wood adhesives.

9.6.2 Effect of Variables on the Stress–Strain Behavior of Bonded Assemblies

The discussion, so far, has been on the stress–strain behavior of adhesives under one condition and in tension. It is important to understand what happens to the strength properties under other conditions. For wood adhesives, the two most important variation in conditions are changes in temperature and moisture. Additionally, it is important to consider more than just the cohesive strength of the bulk adhesive and bulk wood. Although the properties of the bonded assembly are a continuum, Marra's weakest link concept is useful in understanding failure (Marra 1980, 1992). Thus, it is important to understand the properties of the interphase, as well as the bulk adhesive and the wood. Applied forces are not going to be result in a uniform force throughout the bonded assembly for several reasons (Dillard 2002).

The differences in mechanical properties of the wood, adhesive, and interphase regions imply that stress concentrations are likely to occur in the zone of greatest change, that is, the interphase zone. Additionally, the interphase has the greatest internal stress caused by volume reduction in the adhesive upon setting. With environmental exposure, the interphase has to accommodate the large dimensional changes between the wood and the adhesive. If the applied stresses can be dispersed over the entire volume of the material, then localized stresses are reduced and higher total bond strengths obtained. The ability of the applied forces to be dissipated in certain domains without catastrophic failure can lead to higher bond strengths (the shock absorber approach). On the other hand, high internal stresses can add to the applied force and cause unexpected failure. The stresses can be concentrated such as at a flaw causing early failure (Liechti 2002).

For a bonded assembly, the overall properties are hard to predict because less is known about the properties of the interphase regions compared to the bulk properties of adhesives and adherends. The bulk mechanical properties of many wood species have been well studied (Kretschmann 2010). The bulk properties of many adhesives have also been investigated, but many of the wood adhesives form brittle, inhomogenous films that do not yield good mechanical property measurements. However, the interphase properties change from those of the bulk adhesive to those of the bulk wood. This change of properties can be gradual or sharp, and it is expected that a more gradual change should be better, as the stress concentration would be smaller. Large internal forces can be generated when the adhesive and the adherend have different responses to environmental changes, such as moisture and heat. The difference in expansion coefficient between metals and adhesives has been well studied as a cause of adhesive failure due to high internal stress. A major issue with wood is the difference in expansion coefficients with moisture changes between adhesives and wood, mainly in the radial and tangential directions. How these expansion differences are handled in the bonded assembly may be very important to its durability. The internal forces can be as significant as the applied forces for bond strength.

The strength properties of most polymers are sensitive to temperature changes (see Figure 9.13). The increased vibration and therefore mobility of polymers at higher temperatures cause the polymer to be less resistant to applied forces. However, the effect can be greatly influenced by the structure of the polymer. Thermoplastic polymers soften at the glass transition temperature (T_g) and eventually flow at the melt transition temperature (T_m). Polymers with more cyclic and aromatic character have a lower T_g. Crystalline segments will limit the effect of temperature until the T_m of the crystallites is reached. The addition of cross-links, even noncovalent cross-links, such as hydrogen bonds, can improve the resistance to softening at elevated temperatures. Covalent cross-links that exist in many wood adhesives give improved resistance to temperature changes in the bulk adhesive. However, there can be significant differences in the thermal expansion coefficients of the wood and the adhesive causing interphase stresses (Pizzo et al. 2002).

An even greater issue is the effect of moisture changes on bonded assemblies, especially in the interphase region. Some adhesives, like poly(vinyl acetate), lose much of their strength at high moisture levels, as a result of polymer plasticization. UF adhesives are known to depolymerize under high moisture environments, as shown by increased release of formaldehyde (Dunky 2003). On the other hand, wood adhesives, like PF and RF, do not change drastically in their adhesion to wood at higher moisture levels. Wood is known to weaken at higher moisture levels, and to change dimensionally in the radial and tangential directions. When an adhesive does not change dimensionally as the wood swells and shrinks, then stress concentration will occur in the interphase region.

The setting process can generate additional internal forces due to shrinkage of the adhesive. The loss of solvent/water and the polymerization process reduce the volume of the adhesive, while the surface area of the wood stays constant or even increases due to the absorption of water from the adhesive. This difference can cause significant forces that may exceed the strength of the adhesive. Weakness in the bulk of the adhesive UF has been shown to cause cracks in the adhesive (River et al. 1994c); adding flexible groups to an UF formulation reduces this deficiency (Ebewele et al. 1991), especially if those groups are of low to medium molecular weight (Ebewele et al. 1993). In other cases, the forces alone are not sufficient to cause fracture, but may be sufficient to cause fracture when combined with small applied external loads or swelling of the wood as a result of a higher combination of internal and external forces.

9.6.3 BOND STRENGTH

Adhesives are used to hold two materials together; thus, the viscoelastic dissipation of internal and external forces is the most important aspect of adhesive performance. The forces that a bond assembly has to withstand depend very much on the type of product and the use of that product. The effects of internal forces are often not considered, but such forces can be very high in wood. The most rigorous test for laminated wood is the ASTM D 2559 cyclic delamination test (ASTM International 2011b). Many adhesives that have strong wood bonds under dry conditions show significant delamination and do not pass this test; however, the extent of the delamination can be reduced by using a hydroxymethylated resorcinol primer (Vick et al. 1998) (Figure 9.14). An interesting aspect of D 2559 is that no external force is applied; swelling and shrinking forces alone

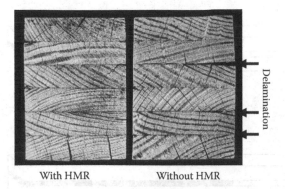

With HMR   Without HMR

FIGURE 9.14 ASTM D 2559 causes bond failure, as shown by delamination from the shrinking and swelling of the wood. The test is severe enough to cause cracking in the wood, but an acceptable adhesive gives minimal bondline failure. The same adhesive was used in both specimens, but the wood on the left that was first primed with hydroxymethylated resorcinol (HMR) resisted the delamination much better than the untreated wood on the right (Okkonen and Vick 1998) (see Section 9.7.3.2).

cause the bond failures. This test involves cycles of vacuum water soaks, followed by oven drying, with a water boil in the second cycle. The fact that dimensional changes, along with some warping of wood, are sufficient to cause substantial bondline failure shows the power of these internal forces. The problem with internal forces is that they are very hard to quantify. However, a test like the D 2559 may exaggerate these forces since the rapid drying provides sufficient force to cause extensive fracture of the wood, while under normal use conditions the moisture change in the wood is more gradual, allowing stress relaxation of the wood.

The forces on bondlines are divided into three modes: I, II, and III (Figure 9.15) (Liechti 2002). The normal force of mode I is perpendicular to the bond and is the direction in which adhesives are the weakest. On the other hand, the shearing force of mode II is the direction in which the adhesive is the strongest. The torsional forces of mode III are an intermediate test of adhesive strength. All three types of forces are common in bonded wood products. Mode I forces exist in strandboard as it resists springback from its compressed state, internal force of swelling under higher moisture, or applied force in the internal bond test. The mode II force is common in laminated veneer lumber under normal external loading or during swelling under high moisture conditions. Mode III forces exist in plywood as a result of the cross-ply construction.

The performance tests are generally covered by the ASTM and other standards (Frihart and Hunt 2010, River et al. 1991). Normally, the tests tend to be hard to pass to allow safety factors in construction. The general rule with most wood products is to have as much good bonding surface and to have as much of the force in the shear mode as possible. Knowledge of wood bond strength has generally been gained using laminated wood and plywood specimens. Distributing the adhesives as droplets on irregular surfaces of strands or fibers has been more difficult to understand the bonding for strandboard and fiberboard. The issue involves relating the data obtained for laminates and plywood that are normally tested in shear to the internal bond test data for particleboard, strandboard, and fiberboard, that involves mode I forces. Summaries of much of the work on performance testing have already been published (River et al. 1991, River 1994a).

If the bonded assembly is considered as a series of links in a chain (Marra 1992), the chain will hold unless the force exceeds the strength of one of the links. Thus, the process for making improved adhesives involves understanding what the weak link is, and why it failed. The internal forces influence the strength of the links and thus, they can vary with conditions. For example, if an adhesive, such as uncross-linked poly(vinyl acetate), softens with heat or increased moisture content, then it is likely to become the weak link under hot or wet conditions. Many adhesives give strong bonds to wood under dry conditions so that the wood is the weak link. However, under wet conditions the

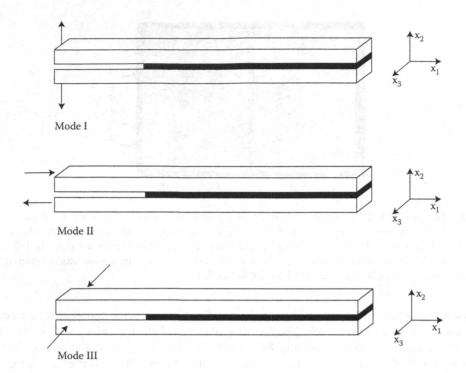

FIGURE 9.15   The force on the bondlines is often a combination of the three modes of force. Mode I is a tensile force in the normal mode and is usually the one in which the adhesive is the weakest. Mode II is the common shear force and is usually the mode in which the adhesive is the strongest. Mode III is the less common torsional force.

weak link may be in the interphase because of a greater strength loss in this link than in the bulk wood or adhesive. Epoxies exhibit a high percent wood failure when dry, but low wood failure when wet (Frihart and Hunt 2010); thus the conditions cause the weak link to change. Failure analysis has indicated that the weak link for epoxies is the interphase region (Frihart 2003), leading to the need to strengthen the epoxy or reduce the stress concentration in the interphase. Using the chain analogy can also aid in understanding why adhesives do not bond as well to dense wood species. If the strength of the bulk adhesive and the adhesive–wood interphase links are enough to hold 2000 psi and the wood strength is only 1000 psi, then the wood breaks first. If the wood strength is 3000 psi, then the fracture is in the adhesive not in the wood even though the adhesive strength has not changes. This is not to imply that more dense woods may not be harder to bond in some cases, but the data needs to be evaluated by considering the strength of the wood relative to that of the adhesive. An important aspect of this discussion is that the tests of the bonded assemblies are measuring the strength of the bondline and not the adhesive.

To make an improved adhesives and bonded assemblies, it is important to understand where failure occurs. Failures within the bulk wood and bulk adhesive are generally easy to see using the naked eye or microscopy. Failure analysis in the interphase is more complicated, especially for wood. In Figure 9.16, the different types of interfacial failure are illustrated. Understanding failure mechanism is important because it leads to better routes for improving the adhesive to solve the problem. One study showed that PF gave fracture in the S_2 layer while an epoxy gave failure in the S_3 under peel, suggesting that the PF gave deeper penetration of the cell walls (Saiki 1984).

At this time, there is insufficient knowledge to predict how well a new adhesive will hold wood pieces together without testing the bonding with the same type of wood and a similar bonding process that will be used commercially and then carrying out the performance tests. The current

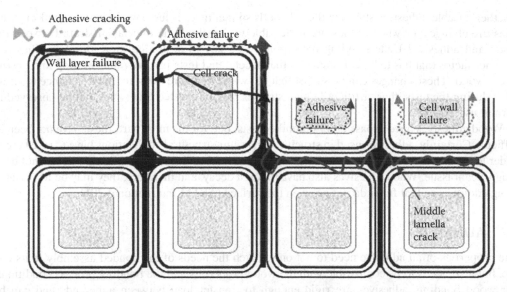

FIGURE 9.16 Failure in the interphase region of wood bonds is complicated. Besides the true interfacial failure that leads to adhesive on one surface and wood on the other, there are a number of other failure zones. The adhesive near the wood may not cure as well, leading to failure in the adhesive near the surface. The adhesive may bond strongly to the wood, but the wood itself may split between layers or within a cell wall layer.

limitations involve in understanding what is necessary about the adhesive–wood interactions to give strong durable bonds. This has been hard to examine because of the complex chemistry and morphology of wood. However, improved analysis will help to shed light on this issue.

9.6.4 DURABILITY TESTING

ASTM defines durability "as related to adhesive joints, the endurance of joint strength relative to the required service conditions" (ASTM International 2011a). Because wood products are used for many years, accelerated tests are used to estimate the long-term performance. A few studies have used field-testing to understand the performance of adhesives under some environmental conditions (River 1994b, Okkonen and River 1996). In addition, for many adhesives, there is in-use experience over many years. Several accelerated tests have been developed that give similar results on durability with the same type of adhesives (River et al. 1991). The key factor that has often been overlooked is that the failure mode must be the same for both long-term use and accelerated test results; thus, it is of paramount importance to extensively validate the accelerated aging tests. If the failure modes are different between normal use failure and an accelerated aging test, then the accelerated test is not likely to always be a reliable predictor of long-term performance.

The most common problem with wood durability is the adhesive's inability to withstand the swelling and shrinking of wood with moisture changes. Most wood products are subjected to temperature and humidity changes, but those in uncontrolled environments are subjected to greater changes. The swelling of wood can subject the bond to mode I, II, or III types of forces depending on the joint design. Swelling has normally been considered on the basis of the macroscopic changes; however, it should be considered also on the basis of the cellular (micrometer) scale. The available data indicate that the swelling of cells usually involves thickening of the cell walls outwards rather than shrinking of the lumen diameter (Skaar 1984). Thus, large forces are exerted on the adhesive at the cell wall edges (Frihart et al. 2004). One study indicates that a phenol–resorcinol–formaldehyde adhesive yields more under wet conditions, but the changes were not as large as the dimensional changes of wood during the wetting process (Muszynski et al. 2002). A key question is whether durable adhesives stabilize the cell walls so that there is less swelling and shrinking with moisture changes, or whether they are better able to distribute the interfacial strain between the wood and adhesive? Understanding these points is a key to designing more durable adhesives. Another factor that has to be considered is that accelerated tests involve rapid wetting and drying of the wood. These changes can be so fast that the wood structure does not have a chance to stress relax during the tests; thus, artificially high stresses may be created that would not be observed in normal use.

Wood adhesives have to pass other durability tests, but, for the most part, these have not been as difficult. Certainly, adhesives used in structural and semi-structural applications have to resist creep under load. Given the rigid nature of the polymer backbone and the cross-linking, this has not been a significant issue. Wood adhesives also have to resist decay, and therefore, they may be formulated using additives so that fungi do not grow on the surface (ASTM International 2011c).

9.7   ADHESIVES

The properties of an adhesive need to not only match the needs of the bonded assembly in its end use, but also needs to be compatible with the wood properties and the bonding process conditions. For wood bonding, adhesives are rigid enough to transfer load between adherends and can be divided into structural, semi-structural, and nonstructural types—see Section 9.3 (River et al. 1991). Although rigidity is often good, adhesives can be too rigid that leads to too much brittleness for some applications.

Wood adhesives can be grouped not only by their structural, semi-structural, and nonstructural use, but also by their permanence and durability. Permanent is more stable than wood under irreversible environmental conditions, while nonpermanent is less stable than wood under irreversible environmental conditions (River et al. 1991). Durable is stronger, more rigid than wood, and more stable under reversible environmental effects, while nondurable is weaker, less rigid than wood, and less stable under reversible environmental effects.

Adhesives need to be compatible with the bonding conditions used commercially. For example, heat-cured adhesives are compatible with the manufacture of panel products for the following reasons:

- They cure slowly at room temperature, allowing time for the wood components to be coated with the adhesive and brought together for assembly.
- The heat and moisture let the wood soften, allowing the adjoining wood surfaces to be brought into close contact.
- Upon heating, the adhesive cures quickly, reducing springback when the pressure is released.

However, a room temperature cure is better for thick laminates because heating the deep layers is more difficult. For manufacturing bonded products, low-cost and rapid setting of the adhesive are important factors, but for construction adhesives, a longer set time, room temperature curing, and easy dispensing from cartridges are important. In many nonwood applications, water-borne adhesives are not used because of poor surface wetting and the inability of the water to move away from the bondline. Neither of these issues is as critical for wood adhesives. However, the penetration of adhesives into wood without over-penetration is important for wood bonding, but not a factor in the bonding of most other materials.

To understand the application, setting, and performance of adhesives, some general polymer chemistry and polymer properties are covered below. The specific adhesive discussions refer back to this general discussion. The properties of polymers are controlled by the structure of the backbone and the number of cross-links, if any. In a few cases, such as polyurethanes, domain separation is also an important factor.

9.7.1 Polymer Formation

Knowledge about the structure of polymers leads to a better understanding of their properties; the properties of polymers are important both in the bonding process and in the ultimate end use performance of the bonded material. Aspects of polymers that need to be considered include use, class, type, and size.

Different applications require materials of different mechanical properties, with these being greatly influenced by the chemical structure of the polymer. It must be remembered that as discussed in Section 9.6.1, the properties of polymers are greatly influenced by the conditions under which they are measured. For example, most adhesives will tend to soften and therefore are less able to carry a load as the temperature increases. When many adhesives absorb small molecules, including water or other solvents, they will soften and, in some cases, will develop cracks that will expand and ultimately cause failure. In addition, the properties of many polymers change as they age. If a polymer is susceptible to oxidation, over time this can either make the material stiffer or depolymerize the adhesive, making it weaker. Chemicals, such as ozone, acids, and bases can also alter the performance of many adhesives.

Polymer classes are determined by how the polymer is constructed. Some polymers are homopolymers, such as polyethylene used in wood-plastic composites. This means that the polymer (AAA ...) is made up of the same individual monomer units (A). Another common type are those polymers made up of two or more components, such as A and B. One way of putting the components together is a random process where two or more monomer units form the copolymer (AAABAABBAB ...), but there is no specific order to the adjacency of the components. An example of this class is the styrene–butadiene rubber that is used in many sealants and mastics. Another way of putting the components together is an alternating copolymer (ABABABAB). Two components can also be combined by making block co-polymers where there are long stretches of monomer A that are then attached to sections of monomer B. Often the A and B components are not compatible when polymerized, so materials tend to separate into individual domains, with examples being polyurethanes and styrenated block copolymers. While the random and alternating copolymers exhibit the average properties of the homopolymers, the block copolymers often exhibit properties not obtainable with either of the homopolymers. A fourth way of reacting two monomers is a grafting process, in which monomer B is attached along the sides of a polymer A backbone. An example is the reaction of grafting of acrylate polymers onto a polyolefin backbone.

Polymer types can be used to group adhesives with different topology independent of their class. For example, the same polymer type can be either a homopolymer or copolymer. One type is a linear polymer where all the monomer units link with one another like a string of beads. Polyethylene and polypropylene are for the most part linear polymers. A second type has branches of the linear chain; the properties of the polymer change dramatically as the type and degree of branching changes. In going from the linear high-density polyethylene to the slightly branched low-density polyethylene and onto the much more branched very low-density polyethylene, there are changes in melting point, flexibility, and strength. Another type of polymer backbone involves whether the structures are linear aliphatics, such as the case with polyethylene, or whether they are cyclic structures, such as cyclohexane or aromatic rings. The cyclical nature of the monomers makes the polymers much stiffer because they have less ability to rotate around the backbone bonds. Aromatic rings make the adhesives even more rigid due to less rotation in the backbone. Many wood adhesives tend to be made from aromatic compounds, including phenol, resorcinol, and melamine, to produce much more rigid polymers with high glass transition temperatures.

Another type of morphology involves whether the polymer chains are cross-linked (thermoset) or not cross-linked (thermoplastic). Some wood adhesives are thermoplastic, including uncrosslinked poly(vinyl acetate) and hot melts. The problem with thermoplastics is that at elevated temperatures or moisture levels, they will flow, leading to creep (flow under load over time) problems. For structural and semi-structural applications creep is very undesirable. Thus the great majority of wood adhesives are thermoset. The term thermoset is used to indicate cross-linked polymers even though the setting process may not be caused by heat. Hot press adhesives are certainly thermoset because they need heat activation to develop the cross-link. On the other hand, moisture-cured adhesives, such as some polyurethanes and silicones, are cross-linked not by the heat process but by the presence of moisture, but are also considered thermosets.

Another factor in the properties of polymers is their size or molecular weight. This is an area that illustrates the two competing natures that an adhesive needs to exhibit. For bond formation, the adhesive needs to flow and penetrate into lumens well and sometimes cell walls, favoring low molecular weights. However, once the bond is formed, it is desirable that the product has great resistance to flow, which favors higher molecular weight. A higher molecular weight adhesive will tend to set faster because fewer reactions are needed to form the cured product. On the other hand, the higher molecular weight polymer can lead to solubility and stability problems for the uncured adhesive. Thus, in designing polymers to be used as adhesives, a balance is needed between low molecular weight for a good wetting of the wood and higher molecular weight for more rapid set and to resist flow once the bond is formed.

Aside from these obvious differences in formulations, changes in the curing conditions can have an effect on the properties of the resin. It is well known for epoxies that additional heating causes additional cross-linking reactions. An epoxy cured at room temperature becomes a rigid gel so that the remaining unreacted groups are not physically able to find each other. As the epoxy is heated, the mobility of the polymers increases, allowing additional groups to come into physical contact to add more cross-links in the matrix, making the product more rigid and usually more brittle. This effect has also been observed with phenolic resins, in that cure times influenced both the degree of cure and the mechanical properties (Wolfrum and Ehrenstein 1999). This is important in considering the production of composites. For particleboard, strandboard, and fiberboard, the adhesive near the surface is at a higher temperature for longer times and at a lower moisture content compared to the adhesive toward the center of the board. The gradient in the heat and moisture causes less polymerization and cross-linking to occur in the center of the composite. The primary curing problem can be reduced by using a faster reacting resin or a higher molecular weight resin in the core than in the face. However, the gradient in the reaction rates can influence the properties of the board and makes studies on the curing process exceptionally difficult.

9.7.2 Self-Adhesion

Under certain conditions wood can self-adhere, but generally adhesives are needed to give sufficient product strength. The forces working against good self-adhesion are the roughness of the surface and the lack of mobility of the wood components that inhibit wetting and interdiffusion. For good adhesion, the two surfaces have to be brought into contact at the molecular level. Obviously, this is difficult with the high surface roughness of a cellular material like wood. Contact becomes more likely if the surface cells are pressed together under high pressure and if the wood is more compliant, such as when one goes from wood laminates to chips to particles and finally to fibers. One product made with little or no added adhesive is high-density fiberboard. The adhesion of the hardboard is dependent upon hydrogen bonding and auto-cross-linking (Back 1987). Of the main wood components, the greatest likelihood for self-adhesion is with lignin and hemicellulose components. Both lignin and hemicellulose soften under high moisture and temperature conditions. Hemicellulose more readily forms hydrogen bonds to bond the adjoining fibers, while lignin more readily forms chemical bonds. The process works adequately for hardboard, but other wood products are not bonded under sufficient heat and pressure to obtain high intersurface bonding.

Another process of self-adhesion is wood welding. Vibrational welding was first demonstrated to cause bond formation (Gfeller et al. 2003). This process uses the heat and cellular distortion generated by friction to bond the wood together. The products show good adhesion under dry conditions, but so far have not been able to provide good wet strength. An even more interesting case is rotational welding. This involves driving a wood dowel into a hole in another piece of wood. With the proper conditions, a strong bond can be formed (Segovia and Pizzi 2009).

Chemical modification of wood surfaces has been shown to give improved bond strengths. A base activation of wood was found to give significant improvement in the dry strength of wood bonds, but not the wet strength (Young et al. 1985). Iron salts with hydrogen peroxides will give more durable bonds with wood particles than with unactivated wood (Stofko 1974, Westermark and Karlsson 2003), and surface activation with peracetic acid has also been used in making particle-board (Johns and Nguyen 1977).

The use of enzyme modification of wood has been shown to increase the strength of bonded wood (Felby et al. 2002, Widsten et al. 2003). Although most studies have been at the laboratory stage, at least one investigated has been done at the pilot plant stage (Kharazipour et al. 1997).

9.7.3 FORMALDEHYDE ADHESIVES

The most common wood adhesives are based on reactions of formaldehyde with phenol, resorcinol, urea, melamine, or mixtures thereof. The reactions can sometimes involve three steps of reaction with a nucleophilic center of the co-monomer with formaldehyde to form a hydroxymethyl deriva-tive, then condensation of two of these hydroxymethyl groups to form a bismethylene ether group with loss of a water molecule or the hydroxymethyl derivative can be directly attacked by a co-monomer nucleophile to form the methylene-bridged product. The methylene bridge predominates and is preferred due its greater stability. The specific chemistry is very pH sensitive. The discussion of the chemical reactions in this section is quite general and does not involve the details because these have been well covered in other books (Pizzi and Mittal 2003).

The rates of the individual reactions depend very much on the co-monomer nucleophile that is copolymerized with the electrophilic formaldehyde. All of these reactions are very pH dependent (see Figure 9.17), but the effect of pH varies depending on the co-monomer. For example, under acidic conditions, formaldehyde addition to phenol is a slower step than the condensation step to form the methylene bridged product, while the relative rates of these two reactions are reversed under basic conditions. Thus, control of the pH is very important in controlling the polymerization reactions; thus, the pH and buffering capacity of the wood may alter the curing in the interphase region. In addition to the pH, these reactions are also controlled by adjusting the temperature and adding catalysts or retarders.

The formaldehyde adhesives are usually water-borne resins so that the curing process is not only polymerization, but also the loss of the water solvent. Because the polymerization process generates water, too much water remaining in the bondline retards the reaction. On the other hand, too little water prior to polymerization not only influences wetting but also can reduce the mobility of the resins and limit collisions needed for polymerization, in addition to limiting heat transfer. Control of both the open and closed assembly times are important for controlling both the penetration and water content of the bondline.

Most wood bonding applications need an adhesive that does not creep over time, leading to the use of cross-linked or thermoset adhesives. High glass transition temperature polymers could also exhibit low creep, but they have been too expensive and hard to use for wood bonding. The formal-dehyde copolymers produce thermoset polymers by cross-linking in the later stages of curing. These reactions occur by formaldehyde bridging the reactive sites on different chains. The co-monomers used with the formaldehyde all have three or more reactive sites, leading to plentiful opportunities to cross-link. Having many available reactive sites is important due to the limited mobility of the polymer backbones, which allows close proximity between only a few locations. It is highly unlikely that every site that is converted to a hydroxymethyl group can find another group in close proximity with which to react. Longer cure times at higher temperatures will tend to push the product to a higher degree of cure. Thus, the ultimate performance of the adhesives is going to depend on the processing conditions.

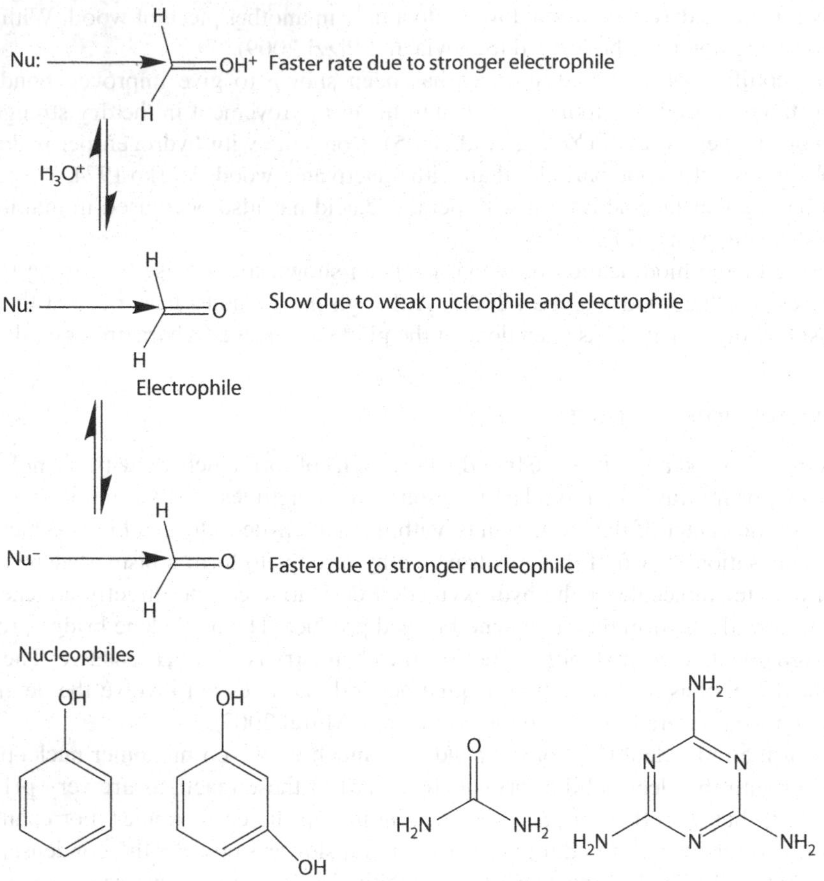

FIGURE 9.17   Reaction of formaldehyde with phenol, resorcinol, urea, and melamine. All of these compounds will copolymerize with formaldehyde, generally in an alternating fashion. The first step is the reaction of a nucleophile with an electrophilic formaldehyde that can be promoted under acidic or basic conditions.

Generally, the molar ratio of formaldehyde needs to be greater than that of the co-monomer to accommodate the need for extra formaldehyde to cross-link the chains, to compensate for formation of bismethylene ethers, and to allow for unpolymerized hydroxymethyl groups. Extra formaldehyde was therefore used to produce fast-setting adhesives with a high degree of curing. However, this caused the problem of significant formaldehyde emissions from the bonded products, mainly those made using urea as the co-monomer. The formulations needed to be adjusted to reduce the formaldehyde levels, but still give good final cures and fast set rates. This has been accomplished through a good understanding of the adhesive chemistry, but then there has been some sacrifice in operability of the bonding process and performance of the bonded assembly.

Formaldehyde copolymer adhesives are used for the production of most laminates, finger joints, and composite products, although the isocyanates are taking over some of the market share as the result of a lower sensitivity to wood moisture content and process temperatures. These formaldehyde-containing adhesives provide good wood adhesion and rigid bonds that do not creep because the formaldehyde not only forms the polymeric chain, but also provides the cross-linking group. However, the properties vary depending on the co-monomer used with the formaldehyde. UF adhesives are the least expensive of all wood adhesives, but they have poor durability under wet conditions. PF adhesives offer a good balance of cost and water resistance. Higher cost melamine-containing adhesives are used because they also provide good water resistance, and are light in color compared to the phenol resins. RF resins are useful because they cure at room temperature, but are expensive. The co-monomer or combination of co-monomer used with formaldehyde is selected depending on the costs, production conditions, and expected performance of the product.

9.7.3.1 PF Adhesives

PF polymers are the oldest class of synthetic polymers, having been developed at the beginning of the twentieth century (Detlefsen 2002). These resins are widely used in both laminations and composites because of their outstanding durability, which derives from their good adhesion to wood, the high strength of the polymer, and the excellent stability of the adhesive. In most durability testing, PF adhesives exhibit high wood failure and resist delamination. There are a vast number of possible formulations, and selection of the wrong one can lead to poor bond strength. Among the factors that can lead to poor adhesion are incomplete polymerization due to too little time at temperature; a resin with too high a molecular weight, leading to poor wetting and penetration; not enough assembly time to allow good wood penetration; or too much assembly time or pressure leading to over penetration of the adhesive and a starved bondline. In general, PF adhesives can meet the bonding needs for most wood applications if cost and heat curing times are not an issue.

For all these adhesives, phenol is reacted with formaldehyde or a formaldehyde precursor under the proper conditions to produce an oligomer that can undergo further polymerization during the setting process. There are two basic types of oligomers, novolaks that have a formaldehyde/phenol (F/P) ratio of less than 1 and are generally made under acidic conditions, and resole resins made under basic conditions with F/P ratios of greater than 1. Although at first glance the acid and base processes may seem to be similar, the chemical reactions and the polymer structures are quite different. For most wood adhesive applications, the resole resins are used because they provide a soluble adhesive that has good wood wetting properties and the cure is delayed until activated by heat allowing product assembly time.

The formaldehyde addition reaction depends on an electron-donating hydroxyl group for activation of the aromatic ring, specifically at the positions *ortho* and *para* to the hydroxyl group; these positions are nucleophilic enough to attack the electrophilic formaldehyde. Although all three sites are activated, the reaction conditions control which sites are more reactive toward the initial and subsequent modifications. The availability of three positions for reaction leads to the ability to form a polymer chain that can be cross-linked to provide good strength and durability. The chemistry described here is general because more details have been published elsewhere (Detlefsen 2002, Pizzi 2003a, Robins 1986).

Novolak resins are made using acidic conditions with typical formaldehyde to phenol ratios of 0.5–0.8 at a pH of 1–4 (Detlefsen 2002). The chemistry involves, first, the addition of the acid-activated formaldehyde to the phenol via a nucleophilic attack by the activated *ortho* or *para* positions of the phenol. This molecule can then lose a water molecule under acidic conditions due to stabilization with the phenol group. The methylene group is then reactive with another phenol group to form the methylene-bridged dimer. Continuation of this process leads to a low molecular-weight linear novolac oligomer. Under acid conditions, the linking step is faster than the addition step, which leads to polymers if the formaldehyde content is not limited. Commercial products are normally oligomers that are converted to polymers by adding more paraformaldehyde, which is usually called the hardener, just prior to application. Novolak oligomers are generally not used for wood bonding due to their low water solubility and very low pH.

On the other hand, resole resins are generally made using alkali hydroxides with a formaldehyde to phenol ratio of 1.0–3.0 at a pH of 7–13 (Detlefsen 2002, Pizzi 2003a). The chemistry involves the reaction of the base-activated phenol attacking formaldehyde, as shown in Figure 9.18. In contrast to the reaction under acidic conditions, the addition of formaldehyde to phenol under basic conditions is the rapid step, while the conversion of the hydroxymethyl derivatives to oligomers is the slow step. Thus, higher formaldehyde levels can be used without forming the final polymer until sufficient heating is applied. Some of the hydroxymethylphenols may dimerize to form a bismethylene

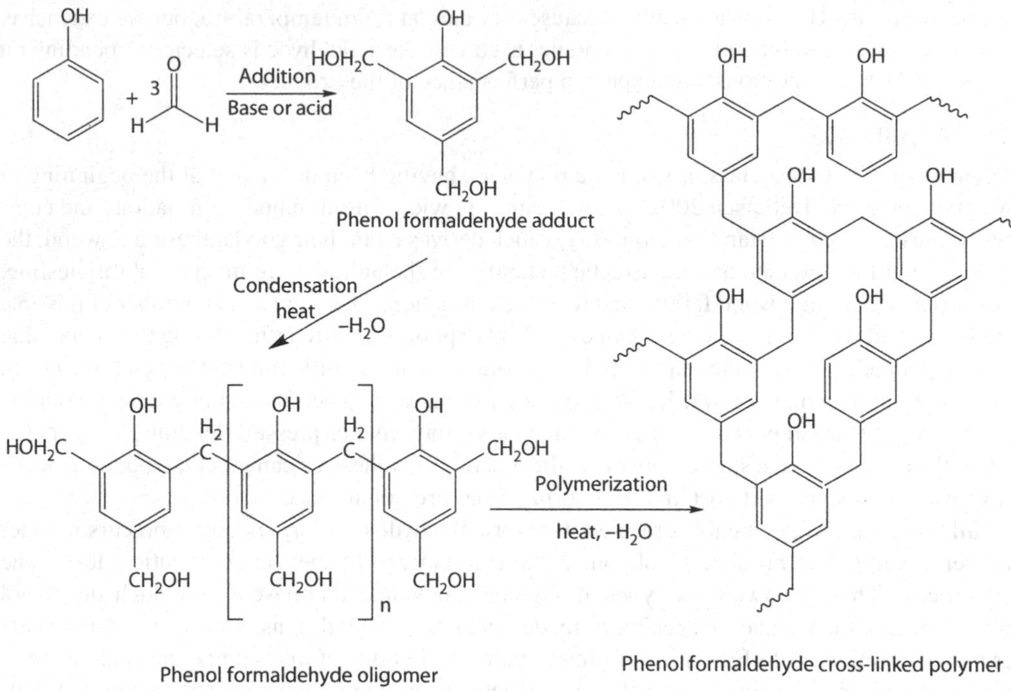

FIGURE 9.18 PF chemistry involves first formation of the hydroxymethyl group, followed by partial polymerization to the oligomer that makes up the adhesive. After applying adhesive to the substrate the polymerization is completed to form a cross-linked polymer network.

ether bridge and are then always converted to the methylene-bridged species. This process is used to generate oligomers with sufficient reactive groups to cure under the proper heating conditions without additional formaldehyde. The molecules in Figure 9.18 show the fully functionalized species, but the molar ratio of formaldehyde to phenol is usually less than 3, leading to enough groups to form the polymer backbone and some cross-linking in the cured product. Drawings often depict only one position of reaction, but it should be remembered that all the *ortho* and *para* positions are reactive, with position selectivity due to the reaction conditions. After being applied to the wood, these resins are then converted to the final adhesive by using heat and water removal conditions. The structure in Figure 9.18 shows the limited mobility of the polymer chain due to only methylene bridges between the aromatic rings and the cross-linking process. There are some hydroxylmethyl groups that cannot find a reactive site.

The PF adhesives could serve in almost all wood bonding applications, as long as the adhesive in the assembly can be heated; however, in many cases, high environmental resistance is not needed so a lower cost and more readily cured UF adhesive is used. Like most adhesives, the commercial products contain more than just the resin. The most common additive is urea to provide improved flow properties, to scavenge free formaldehyde, and to reduce cost. It is generally assumed that most of the urea does not become part of the polymer backbone due to its low polymerizability under basic conditions. For plywood, fillers and extenders are added to provide holdout on the surface and control rheology, including tack, for the specific application method.

9.7.3.2 Resorcinol and Phenol–RF Adhesives

RF resins have the advantage over PF resins of being curable at room temperature due the resorcinol to being 10 faster in reaction than phenol. Resorcinol is 1,3-dihydroxybenzene, and is very reactive because of the combined effect of the two hydroxyl groups on the aromatic ring in activating the *ortho* and *para* positions toward reaction with formaldehyde for the addition reaction, and with hydroxymethylresorcinol in the condensation step (Pizzi 2003b). Because phenol and resorcinol have three reactive sites, they are able to cross-link to form a thermosetting adhesive. The chemistry of modification and polymerization is illustrated in Figure 9.19. The resorcinol copolymerizes well with formaldehyde at room temperature. Thus, it is important to have a formaldehyde–resorcinol ratio low enough to make a noncross-linked novolac polymer, but it also requires the addition of a formaldehyde hardener just prior to applying the adhesive to wood for completing the cure.

Like the PF resins, these adhesives form very durable bonds. They are resistant to both bond failure and to degradation. The main drawback to resorcinol adhesives has been the cost of the resorcinol. To lower the cost, but to maintain the room temperature curing properties, phenol–resorcinol–formaldehyde (PRF) adhesives were developed. PRF adhesives are widely used in wood lamination and finger jointing. PRFs are covered in this section because they behave more like RFs than PFs in their cure.

Three different PRF polymers can be prepared, but all depend on the ability of the resorcinol to react at room temperature.

- A PF resole is reacted with resorcinol at the hydroxymethyl sites to form a resorcinol-terminated adhesive that is then mixed with a formaldehyde hardener just prior to bonding (most common).
- A PF resole is mixed with a resorcinol–formaldehyde hardener just prior to bonding.
- A PF resole is reacted with resorcinol at the hydroxymethyl sites to form a resorcinol-terminated adhesive that is mixed with a PF resole just prior to bonding.

The three methods give different polymer structures, and each has its own advantages and disadvantages depending on the specific application. The PRFs generally have a lengthy assembly time because of the room temperature cure. If the cure were rapid at room temperature, then there would not be enough time to mix the components, spread them on the wood, and press the wood pieces together prior to adhesive curing. The slow cure results in a longer clamping time before the adhesive has sufficient strength to allow handling of the wood pieces. Thus, a room temperature cure is desirable, to avoid heating large laminated pieces, but suffers from the long clamping times.

An interesting use of a RF resin is for making a low solids primer, called hydroxymethylated resorcinol (HMR). This primer has been found to be very useful in improving the delamination resistance of PRF adhesive to CCA treated wood (Vick 1995), epoxy bonds to Douglas-fir (Vick et al. 1998), polyurethane and epoxy to yellow birch and Douglas-fir (Vick and Okkonen 1998, Vick 1997), yellow cedar with PRF adhesive (Okkonen and Vick 1998), and epoxy to Sitka spruce (Vick et al. 1996). The original primer had to be manufactured shortly before use and had a short use time, but an improved process has solved these issues (Christiansen and Okkonen 2003).

Another type of PRF is the honeymoon adhesive, developed for finger jointing and laminating; this process circumvents the long clamping times associated with room temperature cures. In this application, the adhesive is placed on one wood surface and the activator or copolymer material is placed on the other, with the mating of these two pieces leading to the faster cures (Pizzi 2003b). One system for fast curing used an amine cure-promoter on one wood piece and a formaldehyde-based adhesive on the other, and showed that this produced rapid curing with good bonds even to green wood (Parker et al. 1997). The use of hydrolyzed soybean flour and a PRF adhesive as the two components has been shown to produce very good finger joints even with green wood (Kreibich et al. 1998).

9.7.3.3 UF and Mixed Urea Formaldehyde Adhesives

UF adhesives have several strong positive aspects: very low cost, nonflammable, very rapid cure rate, and a light color. On the negative side, the bonds are not water resistant and formaldehyde

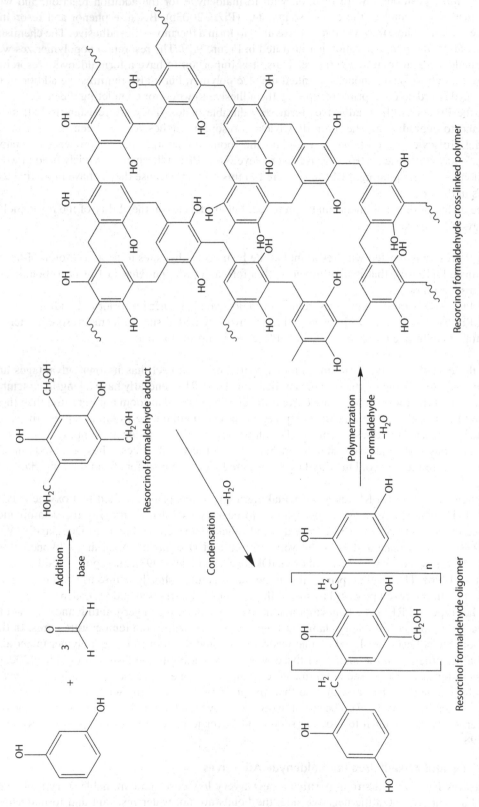

FIGURE 9.19 RF chemistry is similar to the PF in Figure 9.18, but the reaction rates are fast enough that heat does not need to be applied.

continues to evolve from the adhesive. UF adhesives are the largest class of amino resins, and are the predominate adhesives for fiberboard, particleboard some interior plywood.

The chemistry of the UF adhesives involves several steps, with the first being the addition of the formaldehyde to the urea under neutral or basic conditions (Pizzi 2003e, Updegaff 1990). Although there are only two nitrogen atoms on which the formaldehyde can add, the literature shows that the N,N,N´-tris(hydroxymethyl)urea, along with the bis- and mono-hydroxymethyl ureas are the primary products. These hydroxymethyl compounds then react under slightly acidic conditions and heat to generate oligomers, in which the urea molecules are linked by bismethylene ether or methylene bridges, see Figure 9.20. After reaching the desired molecular weight for the specific application, the polymerization is slowed by raising the pH and cooling. An additional charge of urea is added to reduce formaldehyde emissions from the resin. The UF resins are mixed with a latent acid catalyst that produces an acid catalyst during the heat cure. Latent catalysts can be salts, such as

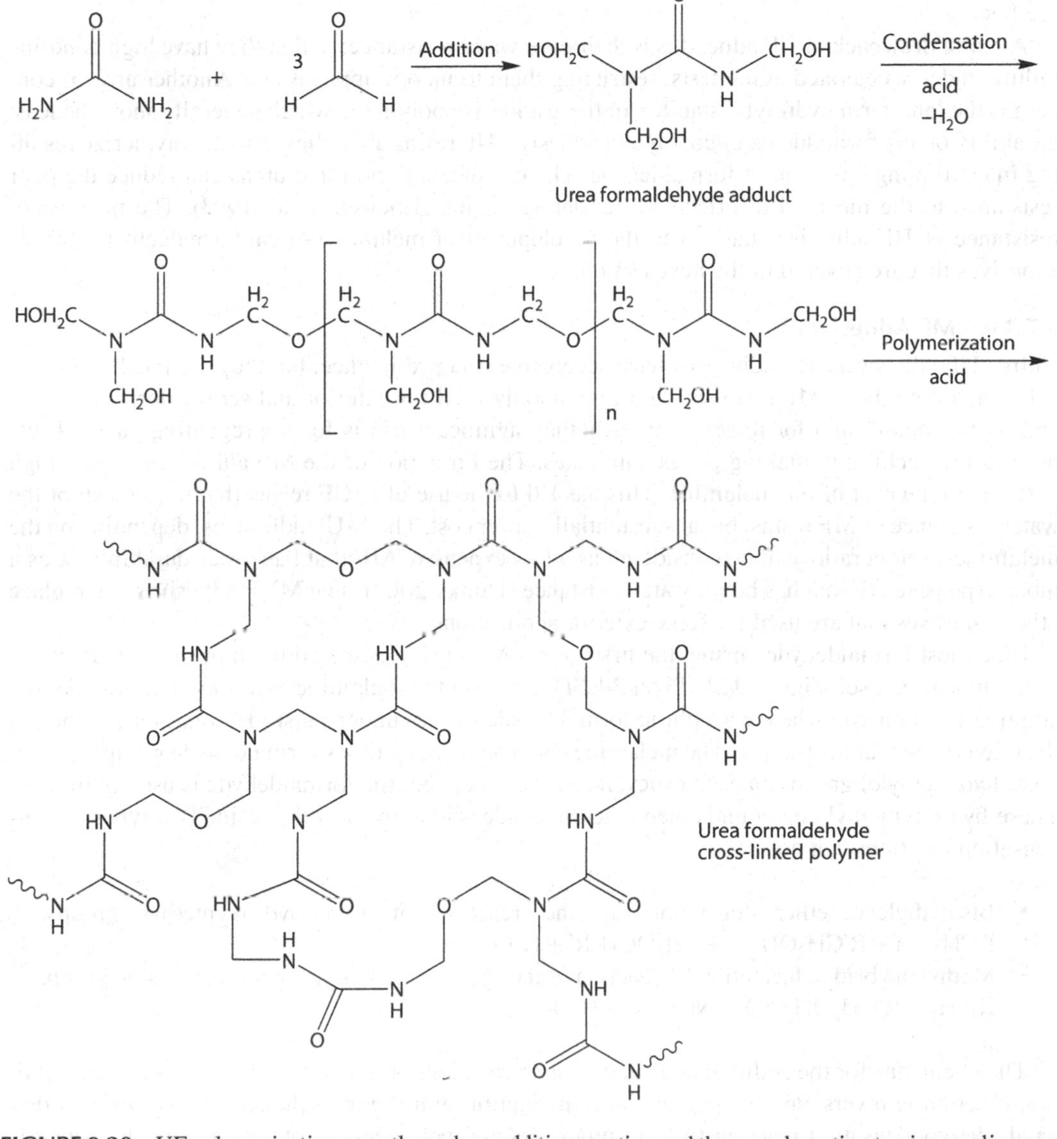

FIGURE 9.20 UF polymerization goes through an addition reaction and then condensation to give an oligomer that is applied to the wood. After application, the polymerization is completed to give a cross-linked network.

ammonium sulfate or chloride, which generate ammonia and sulfuric or hydrochloric acid, respectively. These acids and heat cause the UF to cure rapidly, giving the UF adhesive its desirable rapid setting properties. The rapid strength development leads to shorter press times than with other adhesives. The chemistry and formulation are much more complicated than there is space here to describe and understanding the chemistry has led to efficient products that are used commercially (Pizzi 2003e, Updegaff 1990).

Concern about formaldehyde emissions during production and indoor applications has led to lower formaldehyde/urea ratios, addition of melamine to the formulations, and use of scavengers in current products. However, this has not come about without some sacrifice in ultimate strength and robustness of commercial production, but was needed to meet current environmental standards. The specific UF formulation and bonding conditions are adjusted to meet acceptable formaldehyde emissions for the end product. The classes of products are more rigidly defined in Europe and the United States (Dunky 2003). The formaldehyde emissions are high initially, and decrease with time, but do not go to zero even over a long time. For further discussions and current status, see Section 9.8.

A major drawback of UF adhesives is their poor water resistance; in that they have high bondline failure under accelerated aging tests, restricting them to indoor applications. Another area of concern is the long-term hydrolytic stability of these adhesive polymers, which generally show the least durability of any formaldehyde–copolymer adhesive. UF resins are believed to depolymerize resulting in continuing emission of formaldehyde. The use of some modified ureas can reduce the poor resistance to the mechanical effects of accelerated aging (Ebewele et al. 1993). The poor water resistance of UF adhesives has led to the development of melamine–urea–formaldehyde (MUF) adhesives that are covered in the next section.

9.7.3.4 MF Adhesives

Unlike UF adhesives, MF adhesives have acceptable water resistance, but they are much lighter in color than the others. MF resins are most commonly used for exterior and semi-exterior plywood and particleboard, and for finger joints. Another significant use is for impregnating paper sheets used as the backing in making plastic laminates. The limitation of the MF adhesives is their high cost due to the cost of the melamine. This has led to the use of MUF resins that have much of the water resistance of MF resins, but at substantially lower cost. The MUF adhesives, depending on the melamine-to-urea ratio, can be considered as a less expensive MF that has lower durability or as a more expensive UF that has better water resistance (Dunky 2003). The MUF adhesives can replace other adhesives that are used for some exterior applications.

Like most formaldehyde curing, the first step in MF curing is the addition of the formaldehyde to the melamine, see Figure 9.21 (Pizzi 2003f). Because the melamine is a good nucleophile, the addition reaction with the electrophilic formaldehyde occurs under most pH conditions, although the rate is slower at neutral pH. The melamine can react with up to six formaldehyde groups to form up to two methylol groups on each exocyclic amine group, but the formaldehyde is usually limited. These hydroxymethyl compounds then react by condensation to form the resin. Two types of condensation reactions can occur:

- Bismethylene ether formation by the reaction of two hydroxymethyl groups, $RCH_2OH + R'CH_2OH \Rightarrow RCH_2OCH_2R' + H_2O$
- Methylene bridge formation by reaction of the hydroxymethyl group with an amine group, $RNH_2 + R'CH_2OH \Rightarrow RHNCH_2R' + H_2O$

The chemistry for the addition and condensation reactions is illustrated in Figure 9.21. The addition reaction is reversible, though generally the equilibrium is far to the right side. On the other hand, the condensation reaction to form oligomers and polymers is not very reversible, which is important for the water resistance of the product and makes it different from UF. It is evident from

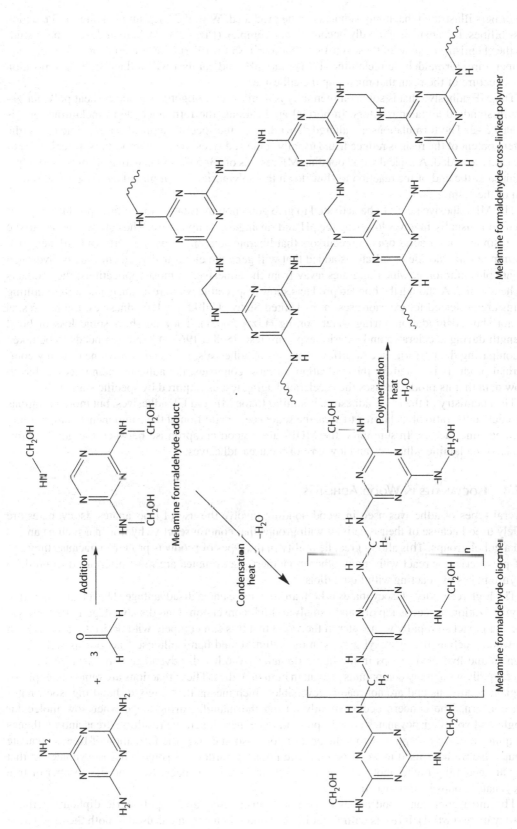

FIGURE 9.21 MF chemistry goes through similar steps as the UF in Figure 9.20.

the dimers illustrated that many isomers can be produced. With the large number sites and reaction possibilities, the chemistry rapidly becomes very complex (Pizzi 2003f). Sato and Naito have studied the chemistry of some of these reactions (Sato and Naitio 1973). The reaction conditions of time, temperature, formaldehyde/melamine (F/M) ratio, pH, and catalyst will influence the composition and structure of the resin that makes up the adhesive.

The MF adhesive that is sold commercially is a mixture of oligomers made by heat polymerization in standard agitated reactors. In normal applications, the formaldehyde to melamine ratio is about 1.5–2. The formulations are altered depending on the specific application. In some cases, the water content of the resin is reduced, and usually some additives (rheology modifiers, fillers, extenders, etc.) are added. A typical wood bonding MF resin is of 53–55% solids with a pH of 9.9–10.3 (pH is raised at the end of the reaction to slow down the polymerization of manufacturing for stabilization of the resin.

The MF adhesive needs to be activated to give good polymerization to the final product. Similar to UF, this usually involves lowering the pH and raising the temperature. The catalysts added to the MF resin are either acids or acid precursors that liberate acid upon heating. Often a hardener, such as ammonium chloride or sulfate, is added that will generate either hydrogen chloride or hydrogen sulfate plus ammonia, which migrates away from the adhesive. In most applications, the products are heat-cured. Although the bonded products show respectable water resistance, phenol-containing resins are preferred for exterior uses in the United States. Unlike the UF adhesives, the MF resins do not show degradation during water boiling (Pizzi 2003f). They do show some loss of bond strength during accelerated and exterior exposure tests (Selbo 1965). Care often needs to be taken in comparing the performance of different classes of adhesives, for usually only one of many commercial products is tested for the evaluation. In most countries, the adhesive manufacturer has to show data that its product passes the accelerated aging test as required by specific standards.

The chemistry of the MUF adhesives is similar to the MF and UF adhesives, but more variations exist due to the ratio of melamine to urea, the sequence for addition of the components, temperature, pH, and time factors. In summary, the MUFs are a good compromise between the good performance of melamine adhesives and low cost of the urea adhesives.

9.7.4 ISOCYANATES IN WOOD ADHESIVES

Several types of adhesives used in wood bonding involve the use of isocyanates. Isocyanates are widely used because of their reactivity with groups that contain reactive hydrogens, such as amine and alcohol groups. This allows great flexibility in the types of products produced because they can self-polymerize or react with many other monomers. Isocyanates are most often used to produce polyurethanes by reacting with liquid diols.

The high reactivity of isocyanates is both an advantage and a disadvantage. The advantage is that polymerization proceeds rapidly and usually to high conversion. One disadvantage is that isocyanates can react so rapidly with water in the wood that this can compete with desired reactions with the wood, such as the hydroxyl groups in the cellulose and hemicellulose fractions as well as the phenols and hydroxyl groups in the lignin domains. Another disadvantage is the isocyanates can react rapidly with many compounds present in human bodies. These reactions are rapid under physiological conditions and are not readily reversible which means that safety of handling isocyanates is a concern. The concern occurs mainly during the manufacturing stage when low molecular weight and volatile isocyanates are still present; once these react, the resulting ureas and urethanes are quite safe. An exception is that the heat of combustion causes the formation of free isocyanate groups. Isocyanates used in wood bonding are not as hazardous as some other isocyanates in that they are generally higher molecular weight so their volatility is decreased and the number of free isocyanate groups is diminished.

The most common wood adhesive is a self-curing isocyanate, polymeric diphenylmethane diisocyanate (pMDI). It reacts with water in the wood for curing and is used in both the production of composites and in the gluing of laminated wood products. The PMDI is used mainly for production of the core of oriented strandboard and is the only known adhesive that works well for the bonding of strawboard. pMDI can be used in the face of OSB, but mold release is needed to reduce sticking to the platens. Although the per pound cost of the pMDI is higher, it is taking market share away from the PF adhesives due to its rapid cure and ability to work at lower application rates.

Another common wood isocyanate adhesive is the emulsion-polymer isocyanate (EPI or API), which is a two-component adhesive (Grøstad and Pedersen 2010). These adhesives are used in bonding of oriented strandboard in engineered wood products. Mixing of the diisocyanate with a diol starts the curing process to form primarily a linear polymer with, usually, a moderate degree of cross-linking to provide more flexible products.

Another class of isocyanate adhesives is the polyurethanes, which are being used in more specialty wood-bonding applications (Dunky and Pizzi 2002, Vick and Okkonen 1998). Their advantage is wide formulation ability, given the great variety of raw materials that can be used. Polyurethanes have shown good potential for bonding green wood (Lange et al. 2001), and are more widely used in Europe.

The most common reactive adhesives contain isocyanate groups, attached to the polymer backbone (Frisch 2002, Lay and Cranley 2003). Hot-melt adhesives are very desirable in product assembly because they develop their bond strength as the molten polymer cools and transforms from the melt to a solid. Unfortunately, the bond generally has poor resistance to heat, long-term stresses, and in some cases, moisture. Moisture-cured hot-melt isocyanates behave like typical hot-melts with good initial strength, but also cross-link to yield a thermoset that can resist the effect of heat, long-term loads, and moisture (Paul 2002). These are being widely used in product assembly areas, many of which involve wood.

9.7.4.1 Polymeric Diphenylmethane Diisocyanate

Isocyanate adhesives have shown increasing use at the expense of other adhesives due to their high reactivity and efficiency in bonding. Polymeric diphenylmethane diisocyanates (pMDI) are commonly used in wood bonding and are a mixture of the monomeric diphenylmethane diisocyanate and methylene-bridged polyaromatic polyisocyanates, illustrated in Figure 9.22 (Frazier 2003). The higher cost of the adhesive is offset by its fast reaction rate, its efficiency of use, and its ability to adhere to difficult-to-bond surfaces. The pMDI forms a homopolymer, but needs water for the activation, which is not a problem with wood, but may be for bonding to other substrates. The chemistry involves several steps:

- The isocyanate first reacts with water to form a carbamic acid: $R-NCO + H_2O = R-NHCOOH$.
- The unstable carbamic acid gives off carbon dioxide to form an amine: $R-NHCOOH \Rightarrow R-NH_2 + CO_2$.
- The amine then reacts with another isocyanate group to form a urea: $R-NH_2 + OCN-R \Rightarrow R-NHCONH-R$.
- Some of the urea molecules react further with isocyanate to form a biuret: $R-NHCONH-R + R-NHCON(CON-R)-R$.

As shown by these reactions, once the isocyanate reacts with the water the rest of the process proceeds rapidly as long as there is enough isocyanate for reaction in comparison to groups with other reactive hydrogens, see Figure 9.23. Sufficient water is generally not a problem with wood given its high water content, but some other substrates need to be wetted for proper bonding. However, high water levels could potentially inhibit polymer formation by producing too many amine groups, but this has not been found to be the case in wood bonding. The carbon dioxide off gas can be a problem since it creates voids in the adhesive that can reduce the strength. Generally,

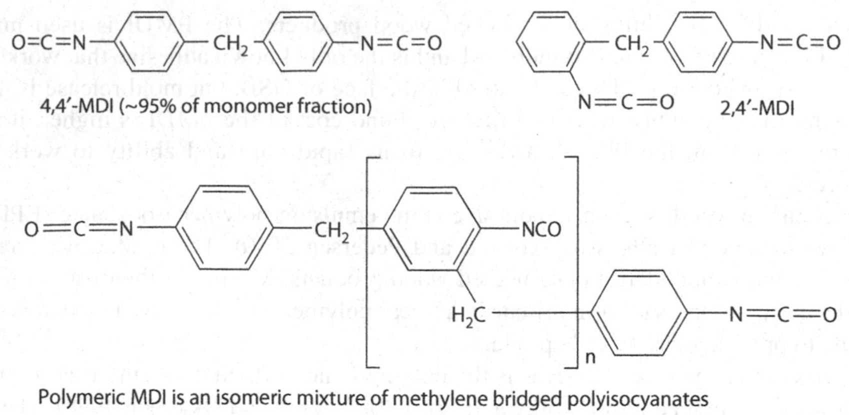

FIGURE 9.22   The polymeric diphenylmethane diisocyanates is a mixture of the monomeric and polyfunctional isocyanates.

these reactions are not reversible under normal conditions, leading to good bond integrity of the isocyanate-bonded wood.

The interaction of pMDI with the composite surface is quite different from other wood adhesives, such as PF. pMDI's low polarity and low viscosity compared to other wood adhesives leads to very rapid penetration into the wood (Frazier 2003). Normally this might lead to a starved bondline and poor strength, but this does not happen with the pMDI-bonded wood. It may be that the strength derives from the strong bridge created at the point where the wood is brought into close contact with the isocyanate adhesive. Some consider the isocyanate to be the most likely of all wood adhesives to form covalent bonds to wood due to the ease of isocyanate reacting with the hydroxyl groups of the wood to form urethane bonds (Frazier 2003). On the other hand, others believe that the fast reaction of the isocyanate with water and the large number of water groups present, especially on the

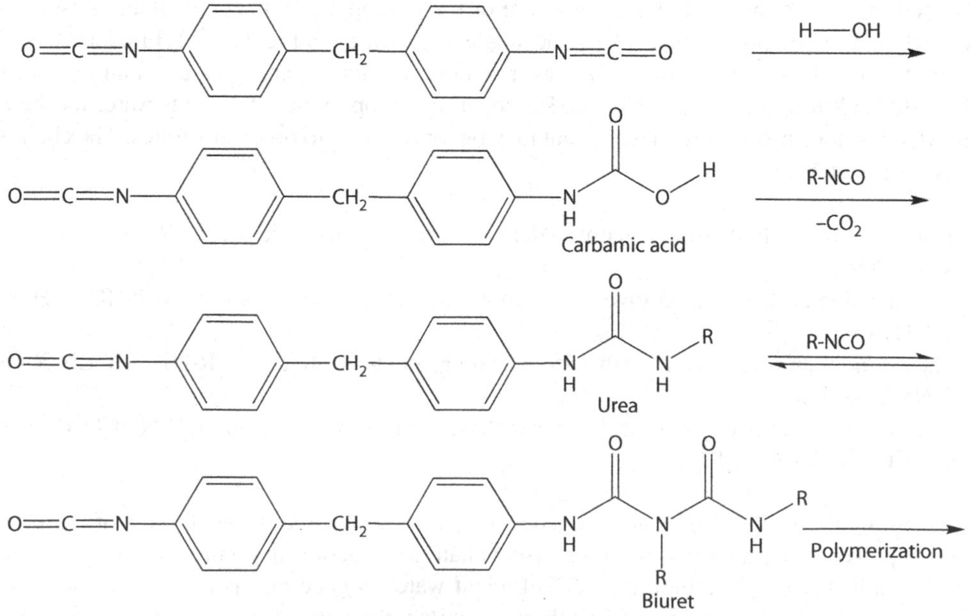

FIGURE 9.23   The isocyanate needs water to start the polymerization process. This reaction ends up forming carbon dioxide that can cause bubbles in the adhesive, but once the amine forms, self-polymerization takes place rapidly.

wood surface makes the urethane formation unlikely (Pizzi 1994a, Frazier 2003). Yelle and coworkers have proven definitely that there is no detectable reaction between the PMDI and wood polymers (Yelle et al. 2011a,b).

The unique properties of pMDI adhesives give it advantages in several markets. Its rapid polymerization and ability to form bonds in the presence of high water levels has led to its use as a core resin for OSB (Dunky and Pizzi 2002). The higher water content and lower temperatures of the OSB core section can make sufficient cure of a PF resin in the core more difficult, leading to increased use of pMDI. This ability to form bonds with high-moisture-content wood has also led to pMDI being used in bonding green or wet lumber. Low polarity allows pMDI to find cracks in the waxy coating of straw, leading to its use in strawboard, for which PF resins are unsatisfactory.

There are several disadvantages to pMDI besides cost. Unlike many wood adhesives that are poor bonders to substrates other than wood, pMDI bonds very well to other materials, including metal caul plates or press platens. Thus, this adhesive is less likely to readily displace PF from the face layer of OSB. Isocyanate's hazards have limited the use of pMDI due to the extra cost of maintaining safe operations in the plants. However, the safety issues can be addressed and no hazard exists in the bonded product due to the reaction of the isocyanate groups. PMDI is promoted as a formaldehyde-free adhesive.

9.7.4.2 Emulsion Polymer Isocyanates

Emulsion polymer isocyanates are generally two-part adhesives that are mixed prior to use and have been used for panel bonding, bonding of plastics to wood surfaces, and for bonding OSB web into the flange to make I-joists (Grøstad and Pedersen 2010). The components are a water-emulsifiable isocyanate and an emulsion latex containing polyhydroxyl functionalized molecules. The emulsion allows higher molecular weight polymers to be used while keeping a low solution viscosity for ease of application. Because the isocyanate readily disperses when mixed with the latex, it comes into contact with the hydroxyl groups as the water moves into the wood. Like all two-component systems, adequate mixing is important. As the adhesive cures, polyurethane groups are formed between by reaction of the isocyanate and hydroxyl groups. The hydroxyl functionalized pre-polymer can be varied in both its backbone structure and the number of hydroxyl groups to control the cross-linking. The variations in the latex portion allow products to be made with a wide range of stress–strain behaviors for different applications.

These adhesives can form fairly durable bonds depending on the formulation; some are known to give good water resistance. The ability to bond plastics and other non-wood substrates is an advantage of these resins over many other wood adhesives. The higher cost and the need to mix the two components prior to use are disadvantages.

9.7.4.3 Polyurethane Adhesives

Polyurethanes are widely used in coatings and adhesives, but less common in wood bonding. Polyurethanes can be either one- or two-component systems, with the selection depending on the specific application. To obtain good wetting, the components need to be low molecular weight or a solvent needs to be added to reduce the viscosity for good wetting. Low molecular weight of the isocyanate components is not desirable because it leads to excessive volatility and health problems. The one-component system is an isocyanate-functionalized polymer that has remaining isocyanate groups. These groups will react with moisture causing the generation of amines that react with other isocyanate groups to form the backbone and cross-linking connections. The two-component adhesive has an isocyanate portion and an isocyanate-reactive portion. Good mixing of these two components just prior to bonding is critical.

The market for these products has been somewhat limited in structural markets because of their marginal levels of wood failure. They are widely used in many other bonding markets due to their good strength, flexibility, impact resistance, and ability to bond many substrates. The acceptance of these products has been greater in Europe than in the United States.

9.7.5 Epoxy Adhesives

Epoxy adhesives and coatings are widely used because of their good environmental resistance and the ability to bond to a wide variety of surfaces, including wood, metals, plastics, ceramics, and concrete. They are less commonly used in wood bonding because they cost more than most wood adhesives, and in some cases, their durability is limited. On the other hand, they are structural adhesives that cure at ambient temperatures, have good gap filling ability, and do bond to many other surfaces, while most wood adhesives require heat cure, are not gap filling and do not bond well to other substrates. Thus, epoxies continue to be examined for their use in bonding wood to other materials and for in-place repair of damaged wood structural members. Besides cost, a main limitation of epoxies is their lack of acceptance for applications that require durable bonds (American Institute of Timber Construction 1990).

Although there are some cases of self-polymerization under the influence of acid or tertiary amine catalysts, most epoxies have an alternating ABABAB backbone of epoxy and hardener that is highly cross-linked, usually using a multifunctional hardener. The standard terminology is for the epoxy to be called the resin and the other component that polymerizes and cross-links the epoxy to be called the hardener. The formulation is expressed as parts per hundred resin (phr) with the weight of the epoxy as 100 and the rest of the components given relative to the epoxy weight. The hardener is anything that will react with the epoxy groups, including amines, thiols, hydroxides, and acid groups, but amines are the most common hardeners.

The most common epoxy resin is the diglycidyl ether of a bisphenol A (DGEBA), although other multifunctional epoxies can be used. The DGEBA synthesis begins by the condensation of phenol with acetone to give the bisphenol A (bisA) as illustrated in Figure 9.24. This then reacts with epichlorohydrin under basic conditions to yield the DGEBA molecule and sodium chloride. The removal of the salt is especially important in electronic applications to minimize metal corrosion by the chloride. The DGEBA epoxies vary in molecular weight due to oligomerization through the epoxy group. Another important class of the epoxies is made from the Novalak resins via the condensation of phenol with formaldehyde, as discussed in Section 9.7.3.1. Bis-F resins have also been used to impart some flexibility; these are similar to the bis-A resins except that formaldehyde is used in place of acetone for the condensation. Even more flexible resins can be made using epoxidized fatty oils and other nonaromatic epoxides. Brominated epoxies are often used for fire resistance.

The hardeners or curatives have an even wider variety of chemical structures than do the epoxies. The hardeners have an active hydrogen attached to a nucleophile and essentially add across the epoxy group. The process involves the nucleophile attacking the terminal carbon of the epoxy as illustrated in Figure 9.25, with the hydrogen then migrating to the hydroxyl anion. For less nucleophilic groups, use of a tertiary amine that interacts with the oxygen atom in the epoxy makes the epoxy ring easier to open. This continues until all the active hydrogens are reacted with epoxide or the epoxide is used up. Thus, for amine with two reactive hydrogens on a nitrogen, two epoxy groups can react, but the second addition is much slower. For formulating the ratio of hardener, the equivalent weight of the hardener is calculated by dividing its molecular weight by the number of active hydrogens and compared to a similar equivalent weight for the epoxies.

The amine hardeners are the most common for room-temperature curable epoxies; they can be divided into three different classes. The main class comprises polyamines, such as those made from the reaction of ethylene oxide and ammonia, to give products such as diethylene triamine, triethylene tetraamine, and tetraethylene pentaamine. These low molecular amines are hazardous due to their corrosivity and their ability to chelate metals. Their reasonable cost and high reactivity make them the most common curatives and are used in wood bonding and repair. Other amines used to make cured products more flexible are the amine-capped polypropylene oxide polymers and branched six carbon diamines. The other amine-containing curatives contain fatty acids that make the epoxies somewhat more flexible and hydrophobic. Polyethylene polyamines can be reacted with

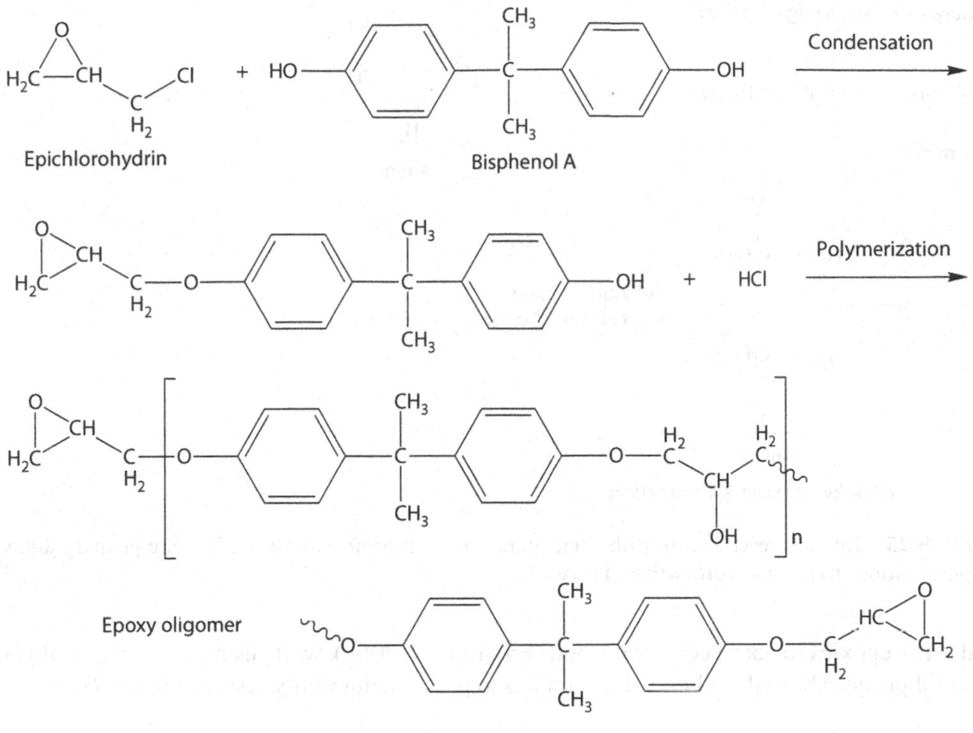

FIGURE 9.24 The epoxies are made by reaction of epichlorohydrin with phenols to produce glycidyl ethers. The bis-A epoxy is much more common than the novolak epoxies, but the latter can provide better heat resistance.

fatty acids to make amidoamines. Polyethylene polyamines can also be reacted with dimer acid, made via the dimerization of unsaturated fatty acids, producing polyamide curing agents. While the standard epoxies in home stores involve amine hardeners, those that exhibit a 5-min cure time use mercaptans hardeners. Epoxies can also be cured using anhydride hardeners or tertiary amine catalysts, but the high cure temperatures limit their use in wood products.

Although epoxies give strong and durable bonds to many substrates, they do not result in highly durable bonds to wood. There is some disagreement on the durability of epoxy bonds under wet conditions, but most standards limit epoxies for load bearing applications (American Institute of Timber Construction 1990). Considerable work has been done on supporting the use of epoxies for restoration work, but examination under severe testing shows that commercial epoxies do not pass the test requirements (Pizzo et al. 2003). Examination of the failure indicates that most of the failure is in the epoxy interphase region (Frihart 2003). Higher wood failure under wet conditions has been

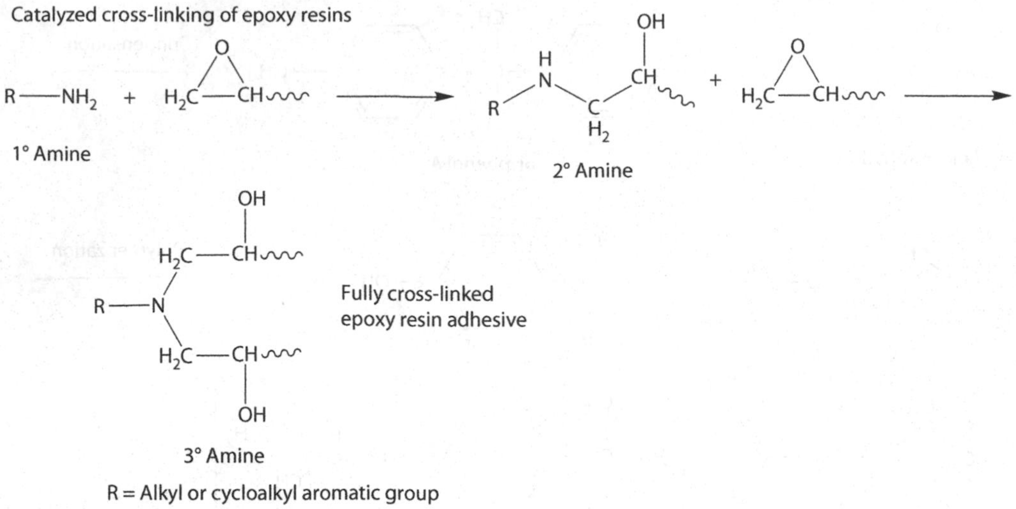

Catalyzed cross-linking of epoxy resins

1° Amine

2° Amine

Fully cross-linked
epoxy resin adhesive

3° Amine

R = Alkyl or cycloalkyl aromatic group

FIGURE 9.25 Epoxies react readily with nucleophiles, such as amines. Most amines are primary allowing multiple additions to provide a cross-linked network.

found using epoxies to bond acetylated wood (Frihart et al. 2004), while using a hydroxymethylated resorcinol primer allowed the bonded assemblies to pass the durability tests (Vick 1997).

9.7.6 VINYL ACETATE DISPERSION ADHESIVES

The water-borne adhesives poly(vinyl acetate), PVAc, and poly(ethylene-vinyl acetate), EVAc, find wide utility in the bonding of wood and paper products into finished goods. Common white glue (PVAc) has long been used in wood bonding including furniture construction. These adhesives easily set at room temperature, are cost effective and are easy to use. These waterborne adhesives set by the water being absorbed into the wood or paper product (and eventually released into the atmosphere), leading to wide use in manufacturing and construction operations involving wood. Because these products are not cured, they will lose much of their strength at high moisture levels.

The processes for making PVAc and EVAc dispersions are similar in many respects (Geddes 2003, Jaffe et al. 1990). The monomers (vinyl acetate and ethylene) are dispersed in water containing poly(vinyl alcohol), making an emulsion; the monomers in the droplets of the emulsion are polymerized to form a dispersion of an organic polymer in water. Making a stable product requires having an emulsion with small droplet sizes. Addition of the monomers is controlled to prevent overheating caused by the exothermic polymerization. The important part is to prevent the phases from separating to form the necessary fine dispersion. If there is too much surfactant, the product can have poor adhesion due to a weak boundary layer. After application, the water migrates away and the beads of adhesive coalesce to form a film, but the coalescence needs to take place on the wood surface. The polarity of the adhesive can be reduced by incorporation of ethylene in the polymerization to produce ethylene-vinyl acetate copolymers for bonding to less polar surfaces.

PVAc is a linear polymer with an aliphatic backbone; thus, it is very flexible adhesives as opposed to the rigid nature of formaldehyde copolymers normally used as wood adhesives. These PVAc adhesives being water-borne generally exhibit good flow into the exposed cell lumens, but, given their high molecular weight, they most likely do not penetrate cell walls. PVAc, with its high content of acetate groups and flexible backbone, can form many hydrogen bonds with the various fractions of the wood for good interfacial adhesion. These adhesives maintain much of their bond strength as the wood expands and contracts due to dissipation of the energy into flexing of the polymer backbone. With this dissipation of the stress into the polymer chains, there is limited stress concentration

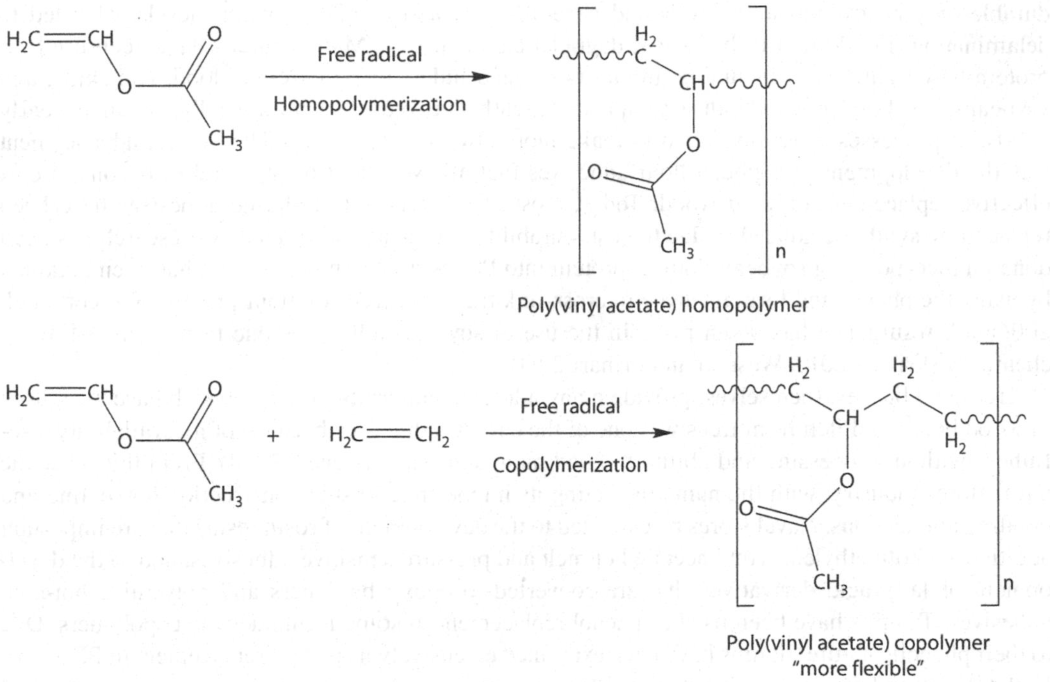

FIGURE 9.26 Polyvinyl acetate is made by the self-polymerization of vinyl acetate usually under free radical conditions. The chains can be altered by adding ethylene to form a copolymer.

at the interface. However, PVA adhesives do not work well at high moisture levels due to loss of strength or high constant stress levels because of a lack of creep resistance. A solution to these problems is to convert the thermoplastic PVA into a thermoset. This is accomplished by cross-linking the linear thermoplastic, using covalent bond formation, such as reaction with glyoxal, formaldehyde resins, or isocyanates, or using ionic bond formation, such as reaction with organic titanates, chromium nitrates, aluminum chloride, or aluminum nitrates. Other ways for making cross-linkable PVAc are to make a copolymer adding *N*-methylolacrylamide or to add phenolic resins. Cross-linked PVAc (PVAx) has improved resistance at high moisture levels, higher temperatures, and under load but is not as convenient because the cross-linker needs to be added just prior to application. However, these adhesives can be used in other applications, such as windows and door construction, for which regular PVA cannot be used.

Poly(vinyl acetate) is converted to poly(vinyl alcohol) (PVOH) by hydrolysis (Jaffe and Rosenblum 1990). PVOH products vary in degree of hydrolysis and molecular weight depending on the specific application, the main uses being textile and paper sizing, adhesives, and emulsion polymerization of vinyl acetate. Given the water solubility, these adhesives need to be either cross-linked or gelled to give some permanence under moist conditions. This conversion is done by cross-linking using a covalent bond formation, such as reactions with glyoxal, UF, and MF or using ionic bonds formations with metal salts, such as cupric ammonium complexes, organic titanates, and dichromates. The gelation is usually accomplished using boric acid or borax.

9.7.7   BIO-BASED ADHESIVES

The most common bio-based adhesives for wood bonding are protein based. In contrast to the other wood adhesives, protein glues have been used for thousands of years. Many of the early civilizations learned how to make adhesives from plants and animals (River et al. 1991, Keimel 2003). Although the original bonded wood products were made using natural protein adhesives, these bonds were durable only at low moisture levels and generally softened at high moisture levels. This led to delamination, in addition to biological degradation when wet. Many sources have been used for protein-based adhesives, including animal bones and hides, milk (casein), blood, fish skins, and soybeans. The bonded wood industry expanded greatly due to the use of these adhesives in the early 1900s, as processes were developed to make more effective adhesives. The biggest advancement was the development of soybean flour adhesives that allowed interior plywood to become a cost effective replacement of solid wood. Today, most of the original bio-based adhesives have been replaced by synthetic adhesives due to cost, durability, and availability factors. Research has been done on incorporating soybean flour or protein into PF resins, but more success has been obtained by using the phenol and formaldehyde to cross-link the denatured soy flour protein (Wescott et al. 2006a). A resurgence has taken place in the use of soybean adhesives due to new cross-linking chemistry (Frihart 2010, Wescott and Frihart 2011).

Trees and bushes, themselves, provide many adhesive materials, some of which have been used in wood bonding. Pitch from trees was one of the earliest adhesives because of its availability, usefulness without processing and ability to bond many materials (Regert 2004). From this grew the naval stores industry, with the name indicating its importance to ship construction for sealing and bonding applications. Naval stores research led to the development of rosin resins that are important additives in both ethylene-vinyl acetate hot-melt and pressure sensitive adhesives, and to the development of fatty acid derivatives that are converted to epoxy hardeners and polyamide hot-melt adhesives. Tannins have been used as phenol replacement in some formaldehyde copolymers. Due to their phenolic nature, lignins have been examined extensively as phenol replacement in PF resins. Both tannins and lignin adhesives tend to have good moisture resistance and are not readily attacked by microorganisms.

Carbohydrates have been used as adhesives, but have not found much utility in wood bonding (Conner and Baumann 2003). Starch is widely used in paper bonding, especially in the construction of corrugated board used in many packaging applications, but generally lacks the strength and water resistance needed for use in wood bonding. The cellulosic adhesives not only lose strength under wet conditions, but also support the growth of microorganisms.

9.7.7.1 Protein Glues

Once the dominant wood bonding adhesive, proteins had mainly disappeared from the market but are now seeing a resurgence (Frihart 2010, Wescott and Frihart 2011). Like most biomass materials, proteins are not uniform in composition as the source varies; thus, the processes for using these proteins and the properties of the adhesives vary as the protein source changes. To make the most effect adhesive, the native protein structure should be denatured to expose more polar groups for solubilization and bonding. The primary structure involves a polyamidoamine backbone made from the condensation of amino acids, while the secondary and tertiary structures are based on intra-chain and interchain interactions, respectively, which involve hydrogen bonds, disulfide linkages, or salt bonds. The main denaturation involves breaking the hydrogen bonds, while breaking other secondary and tertiary bonds depends on the denaturation conditions. Once the protein has been denatured, then it has the ability to come in intimate contact, and to form hydrogen bonds with the wood surface. The setting step involves reformation of the hydrogen bonds between the protein chains to establish bond strength. For prior protein adhesives, the main method of denaturation for adhesive applications was hot aqueous conditions (Lambuth 2003). The aqueous process is often done under caustic conditions and may also involve adding other chemicals to either stabilize the denatured glue or add strength to the final bonds.

Of the protein-based adhesives, soybean flour continues to be used in the largest volume; the flour is ground from the residue after the soybeans had the traditionally more valuable oil removed by extraction (Sun 2005). Traditionally, the flour was dispersed in aqueous caustic to obtain low solids dispersions that had to be used within eight hours before the adhesive starts to degrade (Lambuth 2003). These soybean protein adhesives allowed the development of the interior plywood industry in the early 1900s. The adhesives were improved to give better water resistance (Lambuth 2003), but never achieved sufficient moisture resistance to make exterior grade plywood. PF resins were slow to displace soybean adhesives due to cost and marginal performance. The need for more durable plywood adhesives during World War II led to improved and lower cost PF resins and the ultimate demise of the soybean adhesives. The upsurge in soybean use during the 1950s shows the potential for soybean adhesives on a cost basis if the water resistance, short storage stability, and inconsistency of properties can be overcome.

With the rising cost of petroleum based adhesives, soy flour based adhesives have been of interest as a partial replacement of phenol in PF adhesives. The highest replacement has been about 50% when used the resin in the face of oriented strandboard (Wescott et al. 2006a). These alkaline resins behave very similarly to the standard PF resin. However, the soy flour can provide an unusual type of phenol resin; the alkaline soy-PF can be acidified to make a stable dispersion (Wescott et al. Frihart 2006b). This dispersion is light colored and does not give caustic burns.

Another approach to more durable soybean flour adhesives is the use a poly(amidoamine)-epi-chlorohydrin (PAE) resin as a curing agent (Li et al. 2004) These protein adhesives do not need the highly alkaline conditions previously used with soybean adhesives and have good stability (Allen et al. 2010). These adhesives with no added formaldehyde provide products with very low formal-dehyde emissions not only under standard test conditions (Birkeland et al. 2010) but also at elevated temperature and humidity (Frihart et al. 2010). These adhesives have replaced UF in interior ply-wood, engineered wood flooring, and particleboard (Allen et al. 2010) to meet the new formalde-hyde emission standards discussed in Section 9.8.

None of the other protein sources are available with sufficiently low cost, large supply, and con-sistent composition as soybean flour, but they still have advantages because of their special proper-ties. Blood protein from beef and hogs has the best water resistance of any of the commercial protein adhesives but has great inconsistency (Lambuth 2003). To retard spoilage, the blood is spray dried. It has been mixed with PF adhesives for plywood bonding. Animal bone and hide glues are used in fine furniture manufacturing because they provide flexible bonds for good durability with indoor humidity changes (Pearson 2003). They have many other uses but are being replaced by synthetics, such as ethylene vinyl acetate polymers, which have lower cost and greater ability to be formulated for specific applications. Casein, like many of the protein adhesives, provides good fire resistance and is therefore used in fire doors. Each of these adhesives has its own process for dena-turation and use (Lambuth 2003).

9.7.7.2 Tannin Adhesives

Tannins are polyhydroxypolyphenolics that occur in many plant species, but only a few species have a high enough concentration to make it worthwhile to isolate them. The commercial supplies of tan-nins are limited to a few countries. Tannins are used because they are more reactive than phenol, but they are also more expensive than phenol. Extraction of the plant material and subsequent puri-fication of the isolates, followed by spray drying, yield powdered tannins (Pizzi 2003c). The purified isolates behave in many ways like a natural form of resorcinol, with their high reactivity and water-resistant bonds when polymerized with formaldehyde. Although tannin's reaction rate with formal-dehyde is quite similar to that of resorcinol, the polymer structure is quite different. Instead of multiple additions of formaldehyde to a single aromatic ring, formaldehyde adds mainly as single additions and some double additions to the connected rings of the resorcinol, pyrogallol, phloroglu-cinol and catacheol structures in tannins. Thus, the final polymer structure is very different and will have different properties than the resorcinol product due to its lower cross-link density, despite the similarity in the chemical reactions.

Three limitations of tannins compared to synthetic adhesives are their high viscosity, limited availability, and inconsistent source and therefore reactivity. Their polycyclic structure that leads to fast cure speed also makes solutions of tannins high in viscosity; using more dilute solutions to reduce viscosity leads to additional steam in the hot pressing of the composite. Tannins exist in high enough concentrations to be commercially viable in a few species, but are not available in the large quantities to compete with synthetic adhesives. Like many natural products, the composition of tannins varies depending on growing conditions; thus, making consistently performing adhesives difficult.

Tannins have been used as adhesives in South Africa, Australia, Zimbabwe, Chile, Argentina, Brazil, and New Zealand (Dunky and Pizzi 2002). The use is in composite (particleboard and medium density fiberboard) production, laminate and finger joint bonding, and for damp-resistant corrugated cardboard (Pizzi 2003c). The tannin market will always be of limited volume due to supply limitations.

9.7.7.3 Lignin Adhesives

Although lignins are aromatic compounds, they are very different from tannins—they are available in large quantities at lower cost, but they are much slower in their reaction with formaldehyde. The supply of lignin is large, being the by-product of pulping processes for papermaking; they constitute 24–33% of the woody substance in softwoods and 16–24% in hardwoods. Native lignin is a complex polymer, and this polymeric structure needs to be partially degraded to allow them to be separated from the cellulosics. For adhesive purposes, these degraded lignins need to be further polymerized to obtain useful adhesive properties. Despite being almost completely aromatic, lignins have only a few phenolic rings and no polyhydroxy phenyl rings, leading to low reactivity with formaldehyde.

The low value of lignins has led to much research in finding ways to convert the lignin into useful thermoset adhesives. Lignin from the predominant Kraft pulping process does not lead to a useful product because of the cost of separating the lignin from the pulping chemicals and the inconsistency of the lignin product. However, lignosulfonates contained in the spent sulfate liquor (SSL) from sulfite pulping of wood have been found to be a more useful feed for the production of reactive lignins (Pizzi 2003d). Because of lignin's low reactivity with formaldehyde, other curing mechanisms have been investigated, including thermal cure with acids and oxidative coupling using hydrogen peroxide and catalysts. Three methods of using SSL as the main adhesive with particleboard are to use long press times with a postheating step, to heat with sulfuric acid during bonding, and to heat with hydrogen peroxide (Pizzi 2003d). SSL has also been used as PF and UF extenders. The poor reactivity of lignin can be altered by pre-methylation with formaldehyde, and this pre-methylated lignin has been used with PF resins in plywood bonding.

9.7.8 MISCELLANEOUS COMPOSITE ADHESIVES

Understanding adhesion to wood is as important for composites where wood is the minor component as it is for composites where wood is the main component. Three product areas for wood as a minor component are wood–fiber cement board, wood–plastic composites, and wood filler for plastics. In all three cases, the nonwood component is the main phase holding the material together, but the better the adhesion of the main phase to the wood fiber, the stronger and more durable the product.

Wood-fiber-reinforced cement board competes with traditional cement board that uses other reinforcing materials such as fiberglass cloths. The reinforcement serves to reduce the fracture of these preformed panels. Making improved products involves knowing the interaction of wood with the inorganic cement. Plant fiber reinforcement is still being studied, but the market is dominated by fiberglass reinforcement.

Developing good interaction of wood with low polarity plastics is of growing importance. PPE and PP being the most significant for wood–plastic composites with a main market being a wood replacement for exterior decking. The limited interaction between wood and PE or PP is not surprising, given the large difference in polarity and the difficulty in obtaining good molecular contact between a solid and high molecular weight polymer. The most common method of addressing the polarity difference and the rheological issue is to use medium molecular weight, maleic anhydride-modified polyethylene or polypropylene that can serve as a coupling agent (Clemons et al. 2012). These agents have the polar maleic anhydride that can either react with hydroxyl groups to form esters or react with water to form organic acid groups that will form polar bonds with the hydroxyl groups in wood; thus, making the plastic more compatible with the wood. Better interaction between the hydrocarbon polymer network and the wood fiber will lead to stronger and more durable products. These products are mainly aimed at replacement of wood in decking (Morton et al. 2003), while retaining the appearance of wood.

Other areas for understanding wood–plastic interactions involve plastics filled with wood for applications such as automotive (Suddell and Evans 2003). These products are made to look like normal plastics, but wood filler is used as a partial replacement of the inorganic filler to reduce the weight of the product. The main polymer network is selected from a wide variety of polymers and more of the main fibers are agricultural (nonwood) than is the case in the previous paragraph. However, the fiber–polymer interactions are still very important, and worthy of further investigation. Poor interaction between the fiber and polymer network can cause early failure due to stress concentration. Although the plastic slows the migration of water to the fiber, under wet conditions, the fiber will eventually become saturated with water and begin to swell, putting additional stress on the interface.

9.7.9 CONSTRUCTION ADHESIVES

Construction adhesives are used for attachment of floor and wall coverings, and in assembly of buildings (Miller 1990). Most building construction still uses nails or screws for attachment of wood pieces to each other. However, the use of an adhesive can give extra rigidity to the structure if the panel products are also bonded to the frame. Because the nail or screw holds the wood together, the adhesive does not need to set rapidly. Construction adhesives are normally made to be flexible to provide lateral "give" as the various house components expand and contract with changes in moisture and temperature (Blomquist and Vick 1977). A typical application uses an adhesive that is noncuring, high in molecular weight and with a small amount of solvent to provide some flow. The adhesive is applied at room temperature from a gun to one surface as a bead. Then, the nailing or screwing provides the force necessary for transfer, spreading, and penetration of the adhesive to both surfaces. Because the surfaces are not uniformly brought into close contact, the adhesive has to have gap filling capabilities. Most standard wood adhesives are not able to be gap-filling due to void formation as the water escapes or gas bubbles form during the setting process (River et al. 1991). However, Vick made a gap-filling phenol–resorcinol resin (Vick 1973), but it would not have the flexibility needed for a construction adhesive.

Construction adhesives are usually elastomers, which provide the deformability needed for short-range movement to prevent fracture of the bondline as the wood expands and contracts. However, the adhesive is high in molecular weight to prevent long-range movement that would lead to separation of the bondline. These adhesives provide good strength for many years, but it is unlikely that many will last the lifetime of the building because most elastomers will react with oxygen and ozone, leading to embrittlement and fracture over such a long time.

9.7.10 HOT MELTS

Hot-melt adhesives are used mainly in specialty wood applications. The main applications in wood bonding are related to furniture and cabinetry assembly, although they have also been used in window construction and edge banding of decorative laminates due to their ability to form bonds quickly. Rapid bond formation is valuable for manufacturing operations because minimal clamping time is needed for assembly (Dunky 2003). Hot melts generally set by cooling that turns the molten polymer into a solid, although some hot melts can acquire additional strength by cross-linking.

Hot-melt adhesives are fully formed polymers that are molten for application, but they have such high viscosities that their ability to wet wood surfaces is limited. Upon cooling they recover their strength as the molten polymer solidifies.

The hot-melt version of ethylene vinyl acetate (EVA) coploymer sets by cooling to room temperature and is used more with paper products and nonwovens (Paul 2002) than with wood products, although there is some use in wood furniture assembly. These EVAs are made by gas phase polymerization to yield nonsolvated polymers that have a range of properties. Variation in the ethylene to vinyl acetate ratio and molecular weight of the final polymer creates the various properties needed for individual applications. These products are formulated with tackifiers and waxes (Eastman and Fullhart 1990). Although EVAs are relatively inexpensive, they often have problems with creep as the temperature increases because they contain large amounts of lower molecular weight compounds (tackifiers and waxes) in the adhesive formulations and there is limited attraction between the polymer chains.

Polyamide hot-melt adhesives are also used in wood bonding because of their stronger interactions between chains, leading to better creep resistance. These polyamides are made by the reaction of various diamines with "dimer acid," a diacid that is made from the coupling of unsaturated fatty acids at their olefinic sites (Rossitto 1990). These polyamides offer good creep and heat resistance for a thermoplastic polymer due to the strong hydrogen bonds between the chains. These interchain hydrogen bonds resist flow until enough heat is applied to break these bonds, rapidly turning the solid into a fluid. After application to the substrate, cooling then converts the melt into a strong solid, with good adhesive strength. These properties have made the "dimer acid" polyamides useful for edge banding of laminates, cabinet construction, and window assembly. The higher cost of these adhesives limits their use to high-value products that need more durable bonds.

The moisture-cured isocyanates that were discussed in Section 9.7.4 and polyesters are other hot melts that are also used in wood products. The polyesters are made by reacting aromatic diacids with aliphatic diols, where the aromatic rings provide rigidity to the polymers (Rossitto 1990).

9.7.11 PRESSURE SENSITIVE ADHESIVES

Pressure sensitive adhesives (PSAs) have been a high growth area not only for tapes and labels, but also for application of decorative laminates. PSAs are different from other adhesives in that there is no setting step in their end use. PSAs readily deform to match the topography of the surface to which they are being bonded. Because PSAs are high molecular weight polymers, and in some cases cross-linked polymers, they have limited ability to flow, though their low modulus allows enough deformation to wet the surface. Although these adhesives may not have high interfacial adhesion, most of the applied force is not concentrated at the interface, because the force is mainly expended in deformation of the elastomeric adhesive (Rohn 1999). Because rheological properties are time and temperature dependent, the development of PSAs has been strongly dependent upon dynamic mechanical analysis (DMA) measurements. DMA provides useful information about a formulation's effect on the glass transition temperature and modulus (inverse of compliance) with the small dimensional changes that occur during bonding (Satas 1999a). However, the debonding process occurs over large dimensional changes and is more dependent upon the stress–strain properties of the adhesive. PSAs offer a wide range of properties from easily removed tape or Post-It™ notes to high peel and shear strength tapes, by alterations of both the bonding ability and the energy dissipation ability in debonding.

Given that bonding involves deformation of the adhesive to conform to the substrate surface, PSAs give satisfactory bonds to most surfaces because almost all surfaces are rough on the submicrometer scale. Elastomeric polymers provide the strength for the PSA, but the formulations usually contain low molecular weight materials that are used to tackify and plasticize the polymer. Many types of homopolymers and copolymers (random and blocked) are used in PSAs; Satas' book is an excellent reference source for PSAs (Satas 1999b).

Pressure-sensitive adhesives are often used for bonding plastics (usually having information or decorations printed on them) to wood products for informational or decorative purposes. Applications for this technology range from indoor office furniture to outdoor signs.

9.7.12   OTHER ADHESIVES

Contact adhesives have been used for bonding of plastic laminates to wood. A contact adhesive is a preformed polymer dissolved in a solvent that is applied to both surfaces that are brought into contact after most of the solvent has evaporated. Thus, a countertop is produced by first coating both the particleboard base and the plastic laminate with a contact adhesive, usually neoprene dissolved in a solvent or emulsified in water; then, after the volatiles evaporate, the coated surfaces are pressed together. It is interesting to note that plastic laminates are primarily paper that has been impregnated with resin and then surface coated. Contact adhesives are mainly used in bonding plastic laminates to particleboard for countertops and furniture.

Polymerizable acrylic and acrylate adhesives are not used often for wood because of their high cost. The most common products of this type are structural acrylic (Righettini 2002) and cyanoacrylate instant adhesives (Klemarczyk 2002) that can bond to wood, but these generally require smooth surfaces. These adhesives are more often used in electronics assembly with radiation (light) curing rather than in wood bonding. They do provide rapid cure rates and high strength bonds. Light curing of adhesives does not work with an opaque substrate like wood, but the acrylates can be used for a tough finish over the paper decorative layer on paneling.

Film adhesives involve either partially cured adhesives or adhesives applied onto a carrier such as a fiberglass mat or tissue paper. They are used where applying a liquid adhesive may be difficult, such as in bonding of very thin wood veneers.

9.7.13   FORMULATION OF ADHESIVES

Adhesives are composed of several different components in addition to the base polymer. Although the other components are added for a specific purpose, they often will alter several properties of the adhesive, as applied or after setting.

Base is the polymer, either synthetic, biobased, or a combination, that provides the adhesive the strength to hold the two substrates together. This is the material from which the adhesive usually takes its name, such as PF, epoxy or casein. The base material provides the "backbone" of the adhesive, controlling its application, setting, and curing.

Solvents are liquids often used to dissolve or disperse the base material and additives in order to provide a liquid system for application to the adherends, but are removed from the adhesive in the setting step. The most common solvent for wood adhesives is water. Water is not used in many other adhesive applications due to poor wetting, low volatility, and corrosion of surfaces, but for wood it is an ideal and low-cost solvent. In some cases, the base material of the adhesive is a liquid itself and can be applied in this form without the need for solvents; for example, epoxies or pMDI. These are often referred to as "100-percent solid" adhesives. Such systems shrink less on hardening, thus reducing internal stresses in the film.

Diluents or *thinners* are liquids added to reduce the viscosity of the adhesive systems, and make them suitable for spraying or other special methods of application. However, unlike solvents, they have low volatile. A reactive diluent not only reduces the viscosity of the adhesive for application purposes, but also becomes part of the final polymeric chain.

Catalysts or *accelerators* are chemicals added in small amounts to increase the rate of chemical reaction in the curing or hardening process. True catalysts are not consumed in the reaction, while accelerators may be consumed in the reaction. An example of a catalyst is the acid catalyst generated from ammonium salts for curing UF resins (Pizzi 2003e), while an example of an accelerator is an *ortho* ester used to speed up the cure of PF resins (Conner et al. 2002).

Curing agents or *hardeners* are chemicals that actually undergo chemical reaction in stoichiometric proportions with the base resin and are combined in the final cured polymer structure. A good example is the amine component that reacts with an epoxy resin to form the final adhesive.

Fillers are solids that are added primarily to lower the cost and to give body to liquid adhesives, reducing undesired flow or over-penetration into wood. Fillers usually increase the rigidity of the cured adhesive. They may also modify the thermal expansion coefficient of the film to more nearly approximate that of the adjacent adherends, thus reducing thermal stresses in the joint formed during the cooling, following heat-curing conditions or when thermally cycled in service. Two examples of such fillers include walnut shell flour, incorporated in urea or phenolic adhesives to improve spreading or reduce penetration into open wood pores, and china clay that is sometimes added to epoxy resin systems primarily for thickening or to modify thermal expansion coefficients.

Extenders primary purpose is reducing the adhesive costs while also improving some adhesive properties. At times they can also alter other properties, such as increasing the tack of the adhesive. A good example of an extender is wheat gluten added to UF resins in making hardwood plywood for interior applications.

Stabilizers or *preservatives* are chemicals added to an adhesive to protect one or more of the components and/or the final adhesive against some type of deterioration. Preservatives are usually used for preventing biological deterioration, while a stabilizer can protect against either biological or chemical degradation. Prevention of biological deterioration can involve the use of fungicides or biocides, while chemical degradation prevention may involve the use of antioxidants or antiozonates. In some cases stabilizers are used to avoid the premature curing of an adhesive.

Fortifiers are generally other base materials added to modify or improve the durability of the adhesive system under some specific type of service. A good example is the addition of more durable melamine resins to UF resins in wood bonding to provide greater resistance to deterioration under hot and moist conditions (Dunky and Pizzi 2002).

Carriers are sometimes used to produce film-type adhesives. The carrier is usually a very thin, rather porous fabric or paper on which the liquid adhesive is applied and then dried. Examples include the use of thin tissue paper as a carrier for phenolic film adhesives in making thin hardwood plywood, where spreading the liquid adhesive on a conventional roller spreader might tear or break the thin veneers.

Adhesive formulating is an important skill, often requiring a mixture of empirical and scientific knowledge. Because there is no universal adhesive, systems must be formulated for the specific applications, for example, for a given type of joint or even for a given type of commercial bonding operation. While billions of pounds of phenolic adhesives are used each year in wood bonding, the actual adhesive formulation used in one plant may be quite different from that used in another. Additionally the adhesive formulation used within the same plant may vary with the season due to changes in temperature and humidity. The moisture content or surface roughness of the veneers or the time sequence between one operation and the next influences the actual types and proportions of additives, solvents, and resins used to make a cost effective adhesive.

9.8 ENVIRONMENTAL ASPECTS

Although wood is a natural material, some bonded wood products have caused environmental concerns. There are a number of problem areas, but the foremost area of concern has been formaldehyde emissions from the bonded products, mainly using UF resins. Formaldehyde can react with biological systems in reactions similar to those that are used for curing of adhesives. The problem can arise from both unreacted and generated formaldehyde. Unreacted formaldehyde is also a problem during the manufacturing operation and in freshly produced composites, but has been handled through formulation and engineering solutions. Formaldehyde emissions from composites decrease with time after production (Birkeland et al. 2010). The rate is high initially, but slowly decreases due to diffusion limitations. On the other hand, formaldehyde can be generated by the decomposition of some formaldehyde containing adhesives, in particular the UF adhesives (Myers 1984a,b). These adhesive bonds are more prone to hydrolysis, generating free formaldehyde. The biggest concern is with particleboard, due to the large volume of indoor usage and the high level of adhesive in the product. The formulations of formaldehyde adhesives have been altered over the years to reduce the amount of formaldehyde used and formaldehyde scavengers have been used. The reduction in formaldehyde altered the curing rate and the strength of the product; thus, the process required much research. Many of the formaldehyde concerns were addressed through adhesive reformulation (Dunky and Pizzi 2002, Dunky 2003) with the science of formaldehyde in wood products has been extensively reviewed (Marutzky 1989). The formaldehyde issue continues to be an issue as the acceptable emission levels decrease. The level in the United States was lowered by the California Air Resources Board (Williams 2010), which were used for the United States federal law.

The main concern, emissions, has focused on formaldehyde, but this is not the only compound emitted by bonded wood products. Other volatile compounds in the adhesive formulation have also been detected. In addition, a number of other volatiles are present in wood and additional ones can be generated by the heat and moisture in the production of the composite (Wang et al. 2003). Careful analysis has revealed the presence of formaldehyde, other aldehydes, methanol, and pinenes, many of which come from the wood itself rather than from the adhesive (Baumann et al. 2000).

During the use of the adhesives, volatiles from the monomers that are used to produce the polymers generate additional health concerns. Thus, free formaldehyde, phenol, methanediphenyl diisocyanate, polyethylene polyamines, and so on, are all of concern depending on the type of adhesive used. Heating certainly increases the problem because it raises the vapor pressure of these reactive chemicals. In addition, many hot pressing methods cause other chemicals to be entrained in the steam from the presses (Wang et al. 2003).

9.9 SUMMARY

Although wood bonding is one of the oldest adhesive applications, it is less understood than the most bonding applications. Many modes are possible for both bond formation and failure. Wood structure has so many variables in the different species, cell structures within a tree, and complex morphology at all spatial scales that it is hard to model the process. Despite these problems, many adhesives have been developed that are stronger and some are even more durable than the wood itself. In addition, many functional adhesives have been developed that allow a wide variety of woods and wood pieces to be glued together in a useful and cost-effective manner.

The area that is best understood is the chemistry of the adhesives, even though there are aspects, such as the effects of the composite processing dynamics that need to be more thoroughly researched. The development of the physical properties during the setting process and the interaction of the adhesive with the wood need to be better understood to allow for a more cost effective development of new adhesives.

REFERENCES

Allen, A. J., B. K. Spraul, and J. M. Wescott. 2010. Improved CARB II-compliant soy adhesives for laminates. In: *Wood Adhesives 2009*, eds. C. R. Frihart, C. G. Hunt, and R. J. Moon, 176–184. Madison, WI: Forest Products Society.

American Institute of Timber Construction. 1990. Use of epoxies in repair of structural glued laminated timber. AITC Technical Note 14. Englewood, CO AITC.

ASTM International. 2011a. D 907-11. *Standard Terminology of Adhesives*. West Conshohocken, PA: ASTM International, Vol. 15.06.

ASTM International. 2011b. D 2559-10a. Standard Specification for Adhesives for Structural Laminated Wood Products for Use Under Exterior Exposure Conditions. West Conshohocken, PA: ASTM International, Vol. 15.06.

ASTM International. 2011c. D 4300-01. *Standard Test Methods for Ability of Adhesive Films to Support or Resist the Growth of Fungi.* West Conshohocken, PA: ASTM International, Vol. 15.06.

Autumn, K., M. Sitti, Y. A. Liang, A. Peattie, W. Hansen, S. Sponberg, T. Kenny, R. Fearing, J. Israelachvili, and R.J. Full. 2002. Evidence for van der Waals adhesion in gecko setae. *Proc. Natl. Acad. Sci.* 99:12252–12256.

Back, E. L. 1987. The bonding mechanism in hardboard manufacture. *Holzforschung* 41:247–258.

Barton A. F. M. 1991. *CRC Handbook of Solubility Parameters and Other Cohesion Parameters* (2nd ed.). Boca Raton: CRC Press.

Baumann, M. G. D., L. F. Lorenz, S. A. Batterman, and G-Z. Zhang. 2000. Aldehyde emissions from particleboard and medium density fiberboard products. *Forest Products J.* 50(9):75–82.

Berg, J. C. 2002. Semi-empirical strategies for predicting adhesion. In: *Adhesive Science and Engineering—2: Surfaces, Chemistry and Applications*, eds. M. Chaudhury and A. V. Pocius, 1–73. Amsterdam: Elsevier.

Bikerman, J. J. 1968. *The Science of Adhesive Joints* (2nd ed.). New York: Academic Press.

Birkeland, M. J., L. Lorenz, J. M. Wescott, and C. R. Frihart. 2010. Significance of native formaldehyde in particleboards generated during wood composite panel production. *Holzforschung* 64, 429–433.

Blomquist, R. F. and C. B. Vick. 1977. Adhesives for building construction. In: *Handbook of Adhesives*, ed. I. Skeist (2nd ed.), Chapter 49. New York: Van Nostrand Reinhold.

Bolton, A.J., J. M. Dinwoodie, and P. M. Beele. 1985. The microdistribution of UF resins in particleboard. In: *Proceedings of IUFRO Symposium, Forest Products Research International: Achievements and the Future.* Vol. 6. 17.12-1–17.12-19. South Africa: Pretoria.

Bolton, A.J., J. M. Dinwoodie, and D. A. Davies. 1988. The validity of the use of SEM/EDAX as a tool for the detection of the UF resin penetration into wood cell walls in particleboard. *Wood Sci. Technol.* 22:345–356.

Brief, A. 1990. The role of adhesives in the economy. In: *Handbook of Adhesives*, ed. I. Skeist (3rd ed.), 641–663. New York: Van Nostrand Reinhold.

Browne, F. L. and D. Brouse. 1929. Nature of adhesion between glue and wood. *Ind. Chem. Eng.* 21(1):80–84.

Christiansen, A. W. 1990. How overdrying wood reduces its bonding to phenol–formaldehyde adhesives: A critical review of the literature, Part I: Physical responses. *Wood and Fiber Sci.* 22(4):441–459.

Christiansen, A. W. 1991. How overdrying wood reduces its bonding to phenol-formaldehyde adhesives: A critical review of the literature, Part II: Chemical reactions. *Wood and Fiber Sci.* 23(1):69–84.

Christiansen, A. W. 1994. Effect of overdrying of yellow-poplar veneer on physical properties and bonding. *Holz als Roh und Werkstoff* 52:139–149.

Christiansen, A. W. and E. A. Okkonen. 2003. Improvements to hydroxymethylated resorcinol coupling agent for durable wood bonding. *For. Prod. J.* 53(4):81–84.

Clemons, C., R. M. Rowell, and D. Plackett, 2012. *Wood-Thermoplastic Composites* (2nd ed.). Chapter 13. Boca Raton: Taylor & Francis.

Conner, A. H. and M. G. D. Baumann. 2003. Carbohydrate polymers as adhesives. In: *Handbook of Adhesive Technology*, eds. A. Pizzi and K. L. Mittal (2nd ed.), 495–510. New York: Marcel Dekker.

Conner, A. H., L. F. Lorenz, and K. C. Hirth 2002. Accelerated cure of phenol-formaldehyde resins: Studies with model compounds. *J. Appl. Polym. Sci.* 86:3256–3263.

Dawson, B., S. Gallager, and A. Singh. 2003. *Microscopic View of Wood and Coating Interaction, and Coating Performance on Wood.* Forest Research Bulletin 228. Rotorua, NZ: Forest Research.

De Fleuriot, L. 2004. Radio frequency glue bonding. http://www/highfrequency.co.nz/webfiles/Radio_Frequency_Glue_Bonding.pdf. (accessed August 20, 2012).

Detlefsen, W. D. 2002. Phenolic resins: Some chemistry, technology and history. In: *Adhesive Science and Engineering—2: Surfaces, Chemistry and Applications*, eds. M. Chadhury and A.V. Pocius, 869–945. Amsterdam: Elsevier.

Dillard, D. A. 2002. Fundamentals of stress transfer in bonded systems. In: *Adhesive Science and Engineering—1: The Mechanics of Adhesion*, eds. D. A. Dillard and A. V. Pocius, 1–44. Amsterdam: Elsevier.

Dunky, M. 2003. Adhesives in the wood industry. In: *Handbook of Adhesive Technology*, eds. A. Pizzi and K. L. Mittal (2nd ed.), 887–956. New York: Marcel Dekker.

Dunky, M. and A. Pizzi. 2002. Wood adhesives. In: *Adhesive Science and Engineering–2: Surfaces, Chemistry and Applications.* eds. M. Chadhury and A. V. Pocius, 1039–1103. Amsterdam: Elsevier.

Eastman, E. F. and L. Fullhart, Jr. 1990. Polyolefin and ethylene copolymer-based adhesives. In: *Handbook of Adhesives*, ed. I. Skeist (3rd ed.), Chapter 23. New York: Van Nostrand Reinhold.

Ebewele, R. O., B. H. River, and G. E. Myers. 1993. Polyamine-modified urea-formaldehyde-bonded wood joints, III: Fracture toughness and cyclic stress and hydrolysis resistance. *J. Appl. Polymer Sci.* 49:229–245.

Ebewele, R. O., B. H. River, G. E. Myers, and J. A. Koutsky, 1991. Polyamine-modified urea-formaldehyde resins, II: Resistance to stress induced by moisture cycling of solid wood joints and particleboard. *J. Appl. Polymer Sci.* 43:1483–1490.

Felby, C., J. Hassingboe, and M. Lund. 2002. Pilot-scale production of fiberboards made by laccase oxidized wood fibers: Board properties and evidence for cross-linking of lignin. *Enzyme Microb. Technol.* 31:736–741.

Fengel, D. and G. Wegener. 1984. *Wood: Chemistry, Ultrastructure, Reactions.* Berlin: Walter de Gruyter.

Fowkes, F. M. 1983. Acid–base interactions in polymer adhesion. In: *Physicochemical Aspects of Polymer Surfaces.* ed. K. L. Mittal, Vol. 2, 583–603. New York: Plenum Press.

Frazier, C. E. 2002. The interphase in bio-based composites: What is it, what should it be? In: *Proc. 6th Pacific Rim Bio-Based Composites Symposium & Workshop on the Chemical Modification of Cellulosics.* 1:206–212. Corvallis: Oregon State University.

Frazier, C. E. 2003. Isocyanate wood binders. In: *Handbook of Adhesive Technology,* eds. A. Pizzi, and K. L. Mittal (2nd ed.), 681–694. New York: Marcel Dekker.

Frihart, C. R. 2003. Durable wood bonding with epoxy adhesives. In: *Proceedings, 26th Annual Meeting, Adhesion Society, Inc.* 476–478. Myrtle Beach, SC, Feb. 23–26.

Frihart, C. R. 2004. Adhesive interactions with wood. In: *Fundamentals of Composite Processing: Proceedings of a Workshop,* eds. J. E. Winandy and F. A. Kamke. General Technical Report FPL-GTR-149, 29–38. Madison, WI: U.S. Department of Agriculture, Forest Service, Forest Products Laboratory.

Frihart, C. R. 2006. Wood structure and adhesive bond strength. In: *Characterization of the Cellulosic Cell Wall,* eds. D. D. Stokke and L. H. Groom, 241–253. Oxford: Blackwell Publishing.

Frihart, C. R. 2010. Soy protein adhesives. In: *McGraw Hill Yearbook of Science and Technology 2010.* 354–356. New York: McGraw Hill.

Frihart, C. R. 2011. Wood adhesives vital for producing most wood products. *Forest Prod. J.,* 61(1):5–11.

Frihart, C. R., R. Brandon, and R. E. Ibach. 2004. Selectivity of bonding for modified wood. In: *Proceedings, 27th Annual Meeting of The Adhesion Society, Inc.* pp. 329–33, Wilmington, NC, February 15–18.

Frihart, C. R. and C. G. Hunt. 2010. Adhesives with wood materials: bond formation and performance. In: *Wood Handbook: Wood as an Engineering Material,* Chapter 10. Madison, WI: U.S. Department of Agriculture, Forest Service, Forest Products Laboratory.

Frihart, C. R., J. M. Wescott, M. J. Birkeland and K. M. Gonner 2010. Formaldehyde emissions from ULEF- and NAF-bonded commercial hardwood plywood as influenced by temperature and relative humidity. In: *Proceedings of the International Convention of Society of Wood Science and Technology and United Nations Economic Commission for Europe – Timber Committee.* WS-20, Madison, WI: Society of Wood Science Technology.

Frisch, Jr., K.C. 2002. Chemistry and technology of polyurethane adhesives. In: *Adhesive Science and Engineering—2: Surfaces, Chemistry and Applications,* eds. M. Chadhury, and A.V. Pocius, 759–812. Amsterdam: Elsevier.

Furuno, T. and T. Goto. 1975. Structure of the interface between wood and synthetic polymer, VII: Fluorescence microscopic observation of glue line of wood glued with epoxy resin adhesive. *Mokuzai Gakkaishi* 21(5):289–296.

Furuno, T. and H. Saiki. 1988. Comparative observations with fluorescence and scanning microscopy of cell walls adhering to the glue on fractured surfaces of wood-glue joints. *Mokuzai Gakkaishi* 34(5):409–416.

Geddes, K. 2003. Polyvinyl and ethylene-vinyl acetates. In: *Handbook of Adhesive Technology,* eds. A. Pizzi, and K. L. Mittal (2nd ed.), 719–729. New York: Marcel Dekker.

Gfeller, B., M. Properzi, M. Zanetti, A. Pizzi, F. Pichelin, M. Lehmann, and L. Delmotte 2003. Wood bonding by mechanically induced *in situ* welding of polymeric structural wood constituents. *J. Appl. Polymer Sci.* 92(1):243–251.

Gindl, W., E. Dessipri, and R. Wimmer. 2002. Using UV-microscopy to study the diffusion of melamine urea–formaldehyde resin in cell walls of spruce wood. *Holzforschung* 56:103–107.

Gindl, W. and H. S. Gupta. 2002. Cell wall hardness and Young's modulus of melamine-modified spruce wood by nano-indentation. *Composites: Part A* 33:1141–1145.

Good, R. J. 1967. Intermolecular and interatomic forces. In: *Treatise on Adhesion and Adhesives, Volume 1: Theory,* ed. R. L. Patrick, 9–68. New York: Marcel Dekker.

Grøstad, K. and A. Pedersen. 2010. Emulsion polymer isocyanates as wood adhesive: a review. *J. Adhes. Sci. Tech.* 24:1357–1381.

Hansen, C. M. and A. Björkman 1998. The ultrastructure of wood from a solubility parameter point of view. *Holzforschung* 52(4):335–344.

Horvath, A. L. 2006. Solubility of structurally complicated materials: I. wood. *J. Phys. Chem. Ref. Data* 35(1):77–92.

Jaffe, H. L. and F. M. Rosenblum 1990. Poly(vinyl alcohol) for adhesives. In: *Handbook of Adhesives*, ed. I. Skeist (3rd ed.), Chapter 22. New York: Van Nostrand Reinhold.

Jaffe, H. L., F. M. Rosenblum, and W. Daniels. 1990. Poly(vinyl acetate) emulsions for adhesives. In: *Handbook of Adhesives*, ed. I. Skeist (3rd ed.), 381–400. New York: Van Nostrand Reinhold.

Johns, W. E. and T. Nguyen. 1977. Peroxyacetic acid bonding of wood. *Forest Products J.* 27(1):17–23.

Johnson, S. E. and F. A. Kamke. 1992. Quantitative analysis of gross adhesive penetration in wood using fluorescence microscopy. *J. Adhesion* 40:47–61.

Keckes, J., I. Burgert, K. Frühmann, M. Müller, K. Kölln, M. Hamilton, M. Burghammer, S. V. Roth, S. Stanzl-Tschegg, and P. Fratzl. 2003. Cell-wall recovery after irreversible deformation of wood. *Nat. Mater.* 2:810–814.

Keimel, F. A. 2003. Historical development of adhesives and adhesive bonding. In: *Handbook of Adhesive Technology*, eds. A. Pizzi, and K. L. Mittal (2nd ed.), 1–12. New York: Marcel Dekker.

Kharazipour, A., A. Hüttermann, and H. D. Lüdemann 1997. Enzymatic activiation of wood fibres as a means for the production of wood composites. *J. Adhesion Sci. Technol.* 11(3):419–427.

Kinloch, A. J. 1987. *Adhesion and Adhesives Science and Technology*. London: Chapman & Hall.

Klemarczyk, P. 2002. Cyanoacrylic instant adhesives. In: *Adhesive Science and Engineering—2: Surfaces, Chemistry and Applications*, eds. M. Chadhury, and A.V. Pocius, 847–867. Amsterdam: Elsevier.

Kreibich, R. E., P. J. Steynberg, and R. W. Hemingway. 1998. End jointing green lumber with SoyBond. In: *Wood Residues into Revenue: Residual Wood Conference Proceedings*. November 4–5, 1997, 28–36. Richmond, British Columbia: MCTI Communications Inc.

Kretschmann, D. 2003. Velcro mechanics in wood. *Nat. Mater.* 2:775–776.

Kretschmann, D. E. 2010. Mechanical properties of wood. In: *Wood Handbook: Wood as an Engineering Material*, Chapter 5. Madison, WI: U.S. Department of Agriculture Forest Service, Forest Products Laboratory.

Laborie M. P. G., L. Salmén, and C. E. Frazier. 2006. A morphological study of the wood/phenol-formaldehyde adhesive interphase. *J. Adhesion Sci. Technol.* 20(8):729–741.

Lambuth, A. L. 2003. Protein adhesives for wood. In: *Handbook of Adhesive Technology*, eds. A. Pizzi, and K. L. Mittal (2nd ed.), 457–477. New York: Marcel Dekker.

Lange, D. A., J. T. Fields, and S. A. Stirn. 2001. Finger joint application potentials for one-part polyurethanes. In: *Wood Adhesives 2000*, pp. 81–90. Madison, WI: Forest Products Society.

Lay, D. G. and P. Cranley. 2003. Polyurethane adhesives. In: *Handbook of Adhesive Technology*, eds. A. Pizzi, and K. L. Mittal (2nd ed.) 695–718. New York: Marcel Dekker.

Li, K., S. Peshkova, and X. Gen. 2004. *J. Am. Oil Chemists Soc.* 81:487.

Liechti, K. M. 2002. Fracture mechanics and singularities in bonded systems. In: *Adhesive Science and Engineering—1: The Mechanics of Adhesion*, eds. D. A. Dillard and A. V. Pocius, 45–75. Amsterdam: Elsevier.

Marcinko, J. J., S. Devathala, P. L. Rinaldi, and S. Bao. 1998. Investigating the molecular and bulk dynamics of pMDI/wood and UF/wood composites. *Forest Products J.* 48(6):81–84.

Marcinko, J., C. Phanopoulos, and P. Teachey. 2001. Why does chewing gum stick to hair and what does this have to do with lignocellulosic structural composite adhesion. In: *Wood Adhesives 2000*, 2000, pp. 111–121. Madison, WI: Forest Products Society.

Marra, A. A. 1980. Applications in wood bonding. In: *Adhesive Bonding of Wood and Other Structural Materials*, eds. R. F. Blomquist, A. W. Christiansen, R. H. Gillespie, and G. E. Myers. University Park, PA: Pennsylvania State University. Chapter 9.

Marra, A. A. 1992. *Technology of Wood Bonding: Principles in Practice*. New York: Van Nostrand Reinhold.

Marutzky, R. 1989. Release of formaldehyde by wood products. In: *Wood Adhesives Chemistry and Technology*. Vol. 2. ed. A. Pizzi, Chapter 10. New York: Marcel Dekker.

McBain, J. W. and D. G. Hopkins. 1925. On adhesives and adhesive action. *J. Phys. Chem.* 29:188–204.

Miller, R. S. 1990. Adhesives for building construction. In: *Handbook of Adhesives*, ed. I. Skeist (3rd ed.), Chapter 41. New York: Van Nostrand Reinhold.

Morton, J., J. Quarmley, and L. Rossi. 2003. Current and emerging applications for natural and woodfiberplastic composites. *Seventh International Conference on Woodfiber-Plastic Composites*, 3–6. Madison, WI: Forest Products Society.

Myers, G. E. 1984a. How mole ratio of UF resin affects formaldehyde emission and other properties: A literature critique. *Forest Prod. J.* 34(5):35–41.

Myers, G. E. 1984b. Effect of ventilation rate and board loading on formaldehyde concentration: A critical review of the literature. *Forest Prod. J.* 34(10):59–68.

Muszynski, L., F. Wang, and S. M. Shaler. 2002. Short-term creep tests on phenol-resorcinol-formaldehyde (PRF) resin undergoing moisture content changes. *Wood Fiber Sci.* 34(4):612–624.

Nearn, W. T. 1965. Wood-adhesive interface relations. *Off. Dig., Fed. Soc. Paint Technol.* 37(June): 720–733.

Nussbaum, R. 2001. Surface interactions of wood with adhesives and coatings. Doctoral thesis, KTH—Royal Institute of Technology, Stockholm, Sweden.

Okkonen, E. A. and B. H. River. 1996. Outdoor aging of wood-based panels and correlation with laboratory aging: Part 2. *Forest Products J.* 46(3):68–74.

Okkonen, E. A. and C. B. Vick. 1998. Bondability of salvaged yellow-cedar with phenol-resorcinol adhesive and hydroxymethylated resorcinol coupling agent. *Forest Prod. J.* 48(11/12):81–85.

Packham, D. E. 2003. The mechanical theory of adhesion. In: *Handbook of Adhesive Technology*, eds. A. Pizzi and K. L. Mittal (2nd ed.), 69–93. New York: Marcel Dekker.

Parbhoo, B., L.-A. O'Hare, and S. R. Leadley. 2002. Fundamental aspects of adhesion technology in silicones. In: *Adhesive Science and Engineering—2: Surfaces, Chemistry and Applications*, eds. M. Chadhury and A.V. Pocius, 677–709. Amsterdam: Elsevier.

Parker, J. R., J. B. M. Taylor, D. V. Plackett, and T. D. Lomax. 1997. Method of joining wood, U.S. Patent 5,674,338.

Paul, C. W. 2002. Hot melt adhesives. In: *Adhesive Science and Engineering – 2: Surfaces, Chemistry and Applications*, eds. M. Chadhury, and A.V. Pocius, 712–757. Amsterdam: Elsevier.

Pauling, L. 1960. *The Nature of the Chemical Bond*. Ithaca, NY: Cornell University Press.

Pearson, C. L. 2003. Animal glues and adhesives. In: *Handbook of Adhesive Technology*, eds. A. Pizzi, and K. L. Mittal (2nd ed.), 479–494. New York: Marcel Dekker.

Petrie, E. M. 2000. *Handbook of Adhesives and Sealants*, Chapter 8. New York: McGraw-Hill.

Pizzi, A. 1994a. *Advanced Wood Adhesives Technology*, 275–276. New York: Marcel Dekker.

Pizzi, A. 1994b. *Advanced Wood Adhesives Technology*, 31–39 and 92–98. New York: Marcel Dekker.

Pizzi, A. 1994c. *Advanced Wood Adhesives Technology*, 119–126. New York: Marcel Dekker.

Pizzi, A. 2003a. Phenolic resin adhesives. In: *Handbook of Adhesive Technology*, eds. A. Pizzi, and K. L. Mittal (2nd ed.), 541–571. New York: Marcel Dekker.

Pizzi, A. 2003b. Resorcinol adhesives. In: *Handbook of Adhesive Technology*, eds. A. Pizzi, and K. L. Mittal (2nd ed.), 599–613. New York: Marcel Dekker.

Pizzi, A. 2003c. Natural phenolic adhesives I: Tannin, In: *Handbook of Adhesive Technology*, eds. A. Pizzi, and K. L. Mittal (2nd ed.), 573–587. New York: Marcel Dekker.

Pizzi, A. 2003d. Natural phenolic adhesives II: Lignin. In: *Handbook of Adhesive Technology*, eds. A. Pizzi, and K. L. Mittal (2nd ed.), 589–598. New York: Marcel Dekker.

Pizzi, A. 2003e. Urea-formaldehyde adhesives. In: *Handbook of Adhesive Technology*, eds. A. Pizzi, and K. L. Mittal (2nd ed.), 635–652. New York: Marcel Dekker.

Pizzi, A. 2003f. Melamine-formaldehyde resins. In *Handbook of Adhesive Technology*, eds. A. Pizzi, and K. L. Mittal (2nd ed.), 653–680. New York: Marcel Dekker.

Pizzo, B., P. Lavisci, C. Misani, and P. Triboulot 2003. The compatibility of structural adhesives with wood. *Hoz als Roh- und Werkstoff* 61:288–290.

Pizzo, B., G. Rizzo, P. Lavisci, B. Megna, and S.Berti, 2002. Comparison of thermal expansion of wood and epoxy adhesives. *Hoz als Roh- und Werkstoff* 60:285–290.

Pocius, A. V. 2002. *Adhesion and Adhesives Technology: An Introduction* (2nd ed.). Munich: Hanser.

Regert, M. 2004. Investigating the history of prehistoric glues by gas chromatography-mass spectroscopy. *J. Sep. Sci.* 27:244–254.

Rice, J. T. 1990. Adhesive selection and screening testing. In: *Handbook of Adhesives*, ed. I. Skeist (3rd ed.), Chapter 5. New York: Van Nostrand Reinhold.

Righettini, R. F. 2002. Structural acrylics. In: *Adhesive Science and Engineering—2: Surfaces, Chemistry and Applications*, eds. M. Chadhury, and A.V. Pocius, 823–845. Amsterdam: Elsevier.

River, B. H. 1994a. Fracture of adhesively-bonded wood joints. *Handbook of Adhesive Technology*, eds. A. Pizzi, and K. L. Mittal (2nd ed.), 325–350. New York: Marcel Dekker.

River, B. H. 1994b. Outdoor aging of wood-based panels and correlation with laboratory aging. *Forest Products J.* 44(11/12):55–65.

River, B. H., R. O., Ebewele, and G. E. Myers. 1994c. Failure mechanisms in wood joints bonded with urea-formaldehyde adhesives. *Holz als Roh- und Werkstoff* 52:179–184.

River, B. H. and V. P. Miniutti. 1975. Surface damage before gluing—weak joints. *Wood Wood Prod.* 80(7):35–36.

River, B. H., C. B. Vick, and R. H. Gillespie. 1991. Wood as an adherend. In: *Treatise on Adhesion and Adhesives*, ed. J. D. Minford, Vol. 7. 1–230. New York: Marcel Dekker.

Robins, J. 1986. Phenolic resins. In: *Structural Adhesives, Chemistry and Technology*, ed. S. R. Hartshorn, 69–112. New York: Plenum Press.

Rohn, C. L. 1999. Rheology of pressure sensitive adhesives. In: *Handbook of Pressure Sensitive Adhesive Technology*, ed. D. Satas (3rd ed.), Chapter 9. Warwick, RI: Satas & Associates.

Rossitto, C. 1990. Polyester and polyamide high performance hot melt adhesives. In: *Handbook of Adhesives*, ed. I. Skeist (3rd ed.), 478–498. New York: Van Nostrand Reinhold.

Saiki, H. 1984. The effect of the penetration of adhesives into cell walls on the failure of wood bonding. *Mokuzai Gakkaishi* 30(1):88–92.

Sakata, I., M. Morita, N. Tsurata, and K. Morita. 1993. Activation of wood surface by corona treatment to improve adhesive bonding. *J. Appl. Polymer Sci.* 49:1251–1258.

Salehuddin, A. 1970. A unifying physico-chemical theory for cellulose and wood and its application in gluing. PhD thesis, North Carolina State University at Raleigh.

Satas, D. 1999a. Dynamic mechanical analysis and adhesive performance. In: *Handbook of Pressure Sensitive Adhesive Technology*, ed. D. Satas (3rd ed.), Chapter 10. Warwick, RI: Satas & Associates.

Satas, D. 1999b. *Handbook of Pressure Sensitive Adhesive Technology* (3rd ed.). Warwick, RI: Satas & Associates.

Sato, K. and T. Naitio. 1973. Studies on melamine resin 7. Kinetics of acid catalyzed condensation of dimethylolmelamine and trimethylolmelamine. *Polymer. J.* 5(2):144–157.

Schultz, J. and M. Nardin. 2003. Theories and mechanisms of adhesion. In: *Handbook of Adhesive Technology*, eds. A. Pizzi, and K. L. Mittal (2nd ed.), 53–67. New York: Marcel Dekker.

Selbo, M. L. 1965. Performance of melamine resin adhesives in various exposures. *Forest Products J.* 15(12):475–483.

Segovia, C. and A. Pizzi 2009. Performance of Dowel-welded wood furniture linear joints *J. Adhes. Sci. Tech.* 23:1293–1301.

Skaar, C. 1984. Wood-water relationships. In: *The Chemistry of Solid Wood*, ed. R. Rowell, Chapter 3. Washington, DC: American Chemical Society.

Skeist, I. and J. Miron. 1990. Introduction to adhesives. In: *Handbook of Adhesives*, ed. I. Skeist (3rd ed.), Chapter 1. New York: Van Nostrand Reinhold.

Smith, L. A. 1971. Resin penetration of wood cell walls—Implications for adhesion of polymers to wood. PhD thesis, Syracuse University, Syracuse, NY.

Smith, L. A. and W. A. Cote. 1971. Studies on the penetration of phenol-formaldehyde resin into wood cell walls with SEM and energy-dispersive x-ray analyzer. *Wood Fiber J.* 56–57.

Smith, M. J., H. Dai, and K. Ramani. 2002. Wood-thermoplastic adhesive interface—Method of characterization and results. *Int. J. Adhesion Adhesives* 22:197–204.

Stamm, A. J. and R. M. Seborg. 1936. Minimizing wood shrinkage and swelling. Treating with synthetic resin-forming materials. *Ind. Eng. Chem.* 28(10):1164–1169.

Stark, N. M., Z. Cai, and C. Carll. 2010. Wood-based composite materials panel products, glued-laminated timber, structural composite lumber, and wood–nonwood composite materials. In: *Wood Handbook: Wood as an Engineering Material*, Chapter 11. Madison, WI: U.S. Department of Agriculture, Forest Service, Forest Products Laboratory.

Stehr, M. and I. Johansson. 2000. Weak boundary layers on wood surfaces. *J. Adhesion Sci. Technol.* 14:1211–1224.

Stofko, J. 1974. The autoadhesion of wood. PhD thesis, University of California, Berkeley.

Suddell, B. C. and W. J. Evans. 2003. The increasing use and application of natural fiber composite materials within the automotive industry. *Seventh International Conference on Woodfiber-Plastic Composites*, pp. 7–14. Madison, WI: Forest Products Society.

Sun, X. S. 2005. Isolation and processing of plant materials. In: *Bio-Based Polymers and Composites*, eds. R. P. Wool and X. S. Sun, 33–55. Burlington, MA: Elsevier-Academic Press.

Tarkow, H. 1979. Wood and moisture. In: *Wood: Its Structure and Properties*, ed., F. F. Wangaard, 155–185. University Park, PA: Pennsylvania State University.

Tarkow, H., W. C. Feist, and C. F. Southerland, 1965. Interpenetration of wood and polymeric materials, II: Penetration versus molecular size. *Forest Products J.* 16(10):61–65.

Tsoumis, G. 1991. *Science and Technology of Wood: Structure, Properties and Utilization*. New York: Van Nostrand Reinhold.

Updegaff, I. V. 1990. Amino resin adhesives. In: *Handbook of Adhesives*, ed. I. Skeist (3rd ed.), Chapter 18. New York: Van Nostrand Reinhold.

Vick, C. B. 1973. Gap-filling phenol-resorcinol resin adhesives for construction. *Forest Products J.* 23(11):33–41.

Vick, C. B. 1995. Coupling agent improves durability of PRF bonds to CCA-treated southern pine. *Forest Products J.* 45(3):78–84.

Vick, C. B. 1997. More durable epoxy bonds to wood with hydroxymethylated resorcinol coupling agent. *Adhesives Age* 40(8):24–29.

Vick, C. B., A. W. Christiansen, and E. A. Okkonen, 1998. Reactivity of hydroxymethylated resorcinol coupling agent as it affects durability of epoxy bonds to Douglas-fir. *Wood Fiber Sci.* 30(3):312–322.

Vick, C. B. and E. A. Okkonen. 1998. Strength and durability of one-part polyurethane adhesive bonds to wood. *Forest Products J.* 48(11/12):71–76.

Vick, C. B., K. H. Richter, and B. H. River. 1996. Hydroxymethylated resorcinol coupling agent and method for bonding wood. U.S. Patent 5,543,487.

Wang, W., D. J. Gardner, and M. G. D. Baumann. 2003. Factors affecting volatile organic compound emissions during hot-pressing of southern pine particleboard. *Forest Products J.* 53(3):65–72.

Winandy, J. E. and F. A. Kamke. 2004. *Fundamentals of Composite Processing: Proceedings of a Workshop*, General Technical Report FPL-GTR-149. Madison, WI: U.S. Department of Agriculture, Forest Service, Forest Products Laboratory.

Wellons, J. D. 1983. The adherends and their preparation for bonding. In: *Adhesive Bonding of Wood and Other Structural Materials*, eds. R. F. Blomquist, A. W. Christiansen, R. H.Gillespie, and G. E. Myers, Chapter 3. University Park, PA: Pennsylvania State University.

Westermark, U. and O. Karlsson. 2003. Auto-adhesive bonding by oxidative treatment of wood. In: *Proceedings, 12th Intl. Symp. Wood and Pulping Chem.* Vol. 1. 365–368. Madison, WI, June 9–12.

Wescott, J. M., C. R. Frihart, and A. E. Traska. 2006a. High-soy-containing water-durable adhesives. *J. Adhes. Sci. Technol.* 20(8):859–873.

Wescott, J. M., C. R. Frihart, A. E. Traska, and L. Lorenz. 2006b. in: *Wood Adhesives 2005*, 263–269. Madison, WI: Forest Products Society.

Wescott, J. and C. Frihart. 2011. Sticking power from soya beans. *Chemistry & Industry*. February 7, 21–23.

Widsten, P., J. E. Laine, S. Tuominen, and P. Qvintus-Leino. 2003. Effect of high defibration temperature on the properties of medium-density fiberboard (MDF) made from laccase-treated hardwood fibers. *J. Adhesion Sci. Technol.* 17(11):67–78.

Williams, J. R. 2010. The CARB rule: Driving technology to improve public health. *In: Proceedings of the International Conference on Wood Adhesives 2009*, eds. C. R. Frihart, C. G. Hunt, and R. J. Moon, pp. 12– 16. Madison, WI: Forest Products Society.

Wolfrum, J. and G. W. Ehrenstein. 1999. Interdependence between the curing, structure, and the mechanical properties of phenolic resins. *J. Appl. Polymer Sci.* 74:3173–3185.

Wool, R. P. 2002. Diffusion and autoadhesion. In: *Adhesive Science and Engineering—2: Surfaces, Chemistry and Applications*, eds. M. Chadhury and A. V. Pocius, 351–401. Amsterdam: Elsevier.

Wool, R. P. 2005. Fundamentals of fracture in bio-based materials. In: *Bio-Based Polymers and Composites*, eds. R. P. Wool and X. S. Sun, 149–201. Burlington, MA: Elsevier-Academic Press.

Wu, S. 1982a. *Polymer Interface and Adhesion.* 29–65. New York: Marcel Dekker.

Wu, S. 1982b. *Polymer Interface and Adhesion.* 449–463. New York: Marcel Dekker

Yelle, D. J., J. Ralph, and C. R. Frihart 2011a. Delineating pMDI model reactions with loblolly pine via solution-state NMR spectroscopy. Part 1. Catalyzed reactionswith wood models and wood polymers. *Holzforschung* 65:131–143.

Yelle, D. J., J. Ralph, and C. R. Frihart. 2011b. Delineating pMDI model reactions with loblolly pine via solution-state NMR spectroscopy. Part 2. Non-catalyzed reactions with the wood cell wall. *Holzforschung* 65:144–154.

Young, R. A., M. Fujita, and B. H. River. 1985. New approaches to wood bonding: A base-activated lignin system. *Wood Sci. Technol.* 19:363–381.

10 Wood Composites

Mark A. Irle, Marius Catalin Barbu, Roman Reh,
Lars Bergland, and Roger M. Rowell

CONTENTS

10.1 INTRODUCTION

The principal aim of this chapter is to provide the reader with a brief overview of the manufacturing technologies used to make wood-based panel (WBP) products. Some excellent reviews on WBP manufacture exist, for example, Maloney (1993), Moslemi (1974), Schniewind et al. (1989), Walker (1993, 2006), and Youngquist (1999). WBP manufacturing technology moves fast in response to ever-changing markets and regulations and so new texts are very necessary. Much of the text of this particular chapter was initially published as a chapter for a state-of-the-art report prepared by a European network of researchers called COST Action E49: Processes and Performance of Wood-Based Panels, see Irle and Barbu (2010).

Another good source of information on current manufacturing practices is the trade magazine *Wood-Based Panels International*. Practically every issue of this journal includes at least one report on a particular factory. The reports describe the products made and the manufacturing line in detail. Therefore, the motivated reader can obtain a great deal of detailed information on current manufacturing practices by looking back over the issues published in recent years.

WBP products are made with fibers, particles or veneers (see Figure 10.1). Each of these three raw materials is discussed in the following chapters on particleboard, fiberboard, and plywood manufacture.

10.2 WHY MAKE WOOD-BASED PANELS?

It is often quoted that the manufacture of WBP products has been brought about by the ever-increasing cost of logs and lumber, which in turn, has caused the managers of the world's forest resource to investigate ways and means of using trees more efficiently. This is certainly true, as many wood composites can utilize low-grade logs such as thinning, bowed, and twisted logs. They can also use wood by-products and recycled materials. All sawmills produce large quantities of residues in the form of chips, sawdust, and slabs. Even the most efficient sawmiller is unlikely to convert more than 65% of each log in his yard. These residues can be used to manufacture some of the many kinds of particleboards and fiberboards that are made today.

Even disregarding the economic advantages, panel products would still be manufactured because of man's general desire for better building materials. As a building material, wood has not only a great number of advantages but also some disadvantages, the main one being its variability. Wood is a very variable material both between and within species, and not just in appearance but, more importantly, in density, strength, and durability. Tables 10.1 and 10.2 show that wood composites can be manufactured to have much more uniform properties. Although the strength properties of wood composites are generally lower than solid timber they are more consistent. This means that they can support loads with smaller safety margins, which in effect reduces the apparent difference in strength between solid wood and composites.

Bacteria, fungi, and insects readily decay wood, especially when it is wet. Some panel products are better in this respect, particularly in the case of insect attack. Cement-bonded composites have been found to be extremely resistant to degradation by fungi and even termites.

Other benefits of wood composites come from the fact that their properties can be engineered. Lumber is limited to a large extent by size, width in particular. It is difficult to obtain wood wider than 225 mm and thicker than 100 mm. The dimensions of typical panels vary from market to market but are usually 2–2.5 m long and 1–1.5 m wide, but it is possible to buy panels in much larger sizes if necessary. The majority of houses built today have particleboard floors because wooden floors are more expensive to buy and lay. The comparatively large size of a tongue and grooved particleboard floor panel (2440 × 660 mm in the UK) enables a floor to be laid down far faster and produces a less "creaky" and flatter result than a traditional timber floor. Wood composites can be made to have special properties such as low thermal conductivity, fire resistance, better bio-resistance, or have their surfaces improved for decorative purposes.

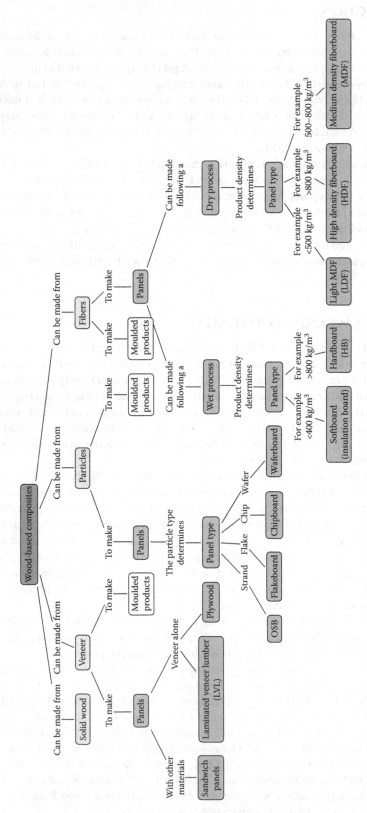

FIGURE 10.1 A map summarizing the wide range of wood composites that can be made. This chapter is concerned with the panel products named in the last row of each branch of panels.

TABLE 10.1

Dimensional Stability of Timber and Boards. Change in Dimensions from 30% to 90% Relative Humidity

	Direction to Grain or Board Length		
	Parallel (%)	Perpendicular (%)	Thickness (%)
Solid Timber			
Douglas fir	Negligible	2.0–2.4	2.0–2.4
Beech	Negligible	2.6–5.2	2.6–5.2
Plywood			
Douglas fir	0.24	0.24	2.0
Particleboard			
UF bonded	0.33	0.33	4.7
PF bonded	0.25	0.25	3.9
MF/UF bonded	0.21	0.21	3.3
Fiber-Building Board			
Tempered	0.21	0.27	7–11
Standard	0.28	0.31	4–9
MDF	0.24	0.25	4–8

Source: Adapted from Dinwoodie, J.M. 1981. *Timber Its Nature and Behaviour.* Van Nostrand Reinhold Company, London.

TABLE 10.2

Strength Properties of Timber and Boards

	Thickness (mm)	Density (kg/m³)	Bending Strength (MPa)		Bending Stiffness (MPa)	
			Parallel	Perpendicular	Parallel	Perpendicular
Solid Timber						
Douglas fir	20	500	80	2.2	12700	800
Plywood						
Douglas fir	4.8	520	73	16	12090	890
Douglas fir	19	600	60	33	10750	3310
Chipboard						
UF bonded	18.6	720	11.5	11.5	1930	1930
PF bonded	19.2	680	18.0	18.0	2830	2830
MF/UF bonded	18.1	660	27.1	27.1	3460	3460
Fiber-Building Board						
Tempered	3.2	1030	69	65	4600	4600
Standard	3.2	1000	54	52	—	—
MDF	9–10	680	18.7	19.2	—	—

Source: Adapted from Dinwoodie, J.M. 1981. *Timber Its Nature and Behaviour.* Van Nostrand Reinhold Company, London.

10.3 MANUFACTURE OF PARTICLEBOARDS: A SHORT OVERVIEW

10.3.1 DEFINING PARTICLEBOARD

Particleboard is often used as a generic term for any panel product that is made with wood particles. Of course, there is a great range of particle shapes and sizes used to make particleboards. The type of particle is therefore used to define the type of particleboard product. For example, following English terminology, chipboard* is made with chips, a flakeboard with flakes, oriented strand board (OSB) with strands and so on (see Figure 10.1 for examples).

Another aspect of particleboards is that the wood particles are bonded together by adding a synthetic adhesive and then pressing them at high pressures and temperatures. This is important as the manufacturing method of these panel products has a marked influence on their subsequent properties.

10.4 WOOD AS A RAW MATERIAL

Approximately 95% of the ligno-cellulosic material used for particleboard production is wood. The rest consists mainly of seasonal crops such as flax, bagasse, and cereal straw. A discussion on these various seasonal crops is beyond the scope of this chapter.

Some wood species are more suitable for particleboard production than others.

10.4.1 WORKABILITY

A material that is difficult and costly to break into particles is not suitable. Once produced, the chips should have smooth surfaces and a minimum of end grain, otherwise the particle will absorb too much adhesive to be cost effective. The characteristics of a chip are, to a certain extent, dependent on the anatomy of the wood. Softwoods are preferred to hardwoods because they tend to be easier to cut and the vessels present in hardwoods cause the chip to have a rough surface.

10.4.2 DENSITY

In order to manufacture a board of adequate strength the particles must be compressed to at least 5% above their natural density. In practice, the raw material is usually compressed to nearer 50% of its natural density; so if a raw material of about 400 kg/m³ is used then the finished board will have a density of approximately 600 kg/m³. This degree of compression is needed to achieve sufficient chip-to-chip contact to enable the adhesive to bond the particles together. The need for compression is shown in Figure 10.2 where the very outer, rough edge of the panel consists of loose particles that are easily brushed off of the edge, even though these particles have the same amount of glue as particles in the middle and have been hot-pressed for the same amount of time.

Figure 10.3 shows the relationship between raw material density and particleboard physical properties. From this graph it is clear that for a single species bending strength increases with compaction ratio. This is as expected because the more the particles are squashed the greater the contact between them. The graph also shows that for a specific bending strength the required particleboard density increases with the density of the raw material. Most regulations specify minimum strengths, consequently a manufacturer using say Birch (*Betula* spp.) will have to produce boards of much higher density than another using Spruce (*Picea* spp.) to attain these minima. The heavy boards will require a stronger press for manufacture, incur greater transport costs, and be more difficult to cut and handle. Therefore, it is not surprising that low-density woods are preferred.

---

* Most European countries use the term particle rather than chip and therefore particleboard as a term for chipboard. To avoid confusion over whether the text is referring to particleboard in the generic sense or particleboard in the specific product sense, the name chipboard is being used in this text for the specific product.

FIGURE 10.2 A laboratory made particleboard approximately 600 by 600 mm. The impression of a frame on the surface is clearly visible. The frame densifies the edges and helps to raise the steam pressure inside the mattress during hot-pressing.

Practically all the physical and mechanical properties of particleboards are related to density. A manufacturer who attempts to use a number of different density species over a production period will produce boards with inconsistent properties unless he mixes the chips from the different species to form a uniform furnish for the press.

10.4.3 pH of Wood

The curing rates of formaldehyde-based resins, for example, urea formaldehyde (UF), which is used to make most of the world's particleboard production, are very dependent on the pH of the environment in which they cure. All wood species have a pH, if a near neutral species is used then a resin may not cure sufficiently, or if an acidic species is used then precure may result. When adhesive precure occurs the board's surface layer is weak and flaky. This is because the adhesive cures before

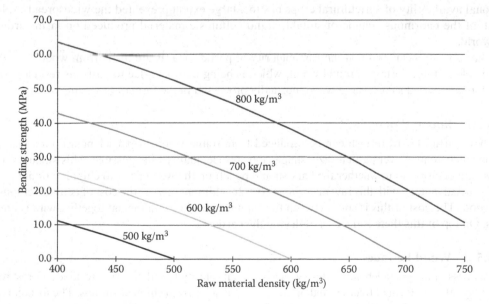

FIGURE 10.3 Relationship between raw material density and particleboard bending strength, as predicted by a computer model.

the particles have been fully compressed and so when the press closes further the precured resin bonds are broken. The pH of the raw material is not usually a problem if it remains fairly constant, but if it fluctuates then the quantities of hardeners and buffers added to the adhesives must be continually altered to suit the wood in use. Differences occur between species, within species depending on where the tree grew, and within the tree (principally a difference between heart and sapwood). In addition, the pH and buffer capacity of wood can change with storage time and conditions (Elias and Irle 1996).

10.4.4   PERMEABILITY

It is postulated that highly permeable species of wood will produce poor-quality particleboard because chips from such wood will absorb the adhesive applied to it thus creating starved joints. The way a wood breaks during chipping is probably more important than its permeability as this will determine the amount of exposed end grain. The amount and type of extractives present might be even more important still as these will influence the contact angle that the adhesive makes with the wood after blending. It is practically impossible to show the effect of wood permeability on particleboard properties, because when two boards made with different species are compared many more factors other than just wood permeability will be different. For example, anatomy will affect surface roughness and possibly the geometry of the particles and these two factors could be more important than particle permeability.

10.4.5   WOOD SOURCES

In addition to having the physical characteristics described above the raw material should also be inexpensive and available in sufficient quantities to support sustained production over many years. Conventional particleboards generally compete with each other on price, because there are many manufacturers producing very similar products. Therefore, to be competitive, cheap raw materials must be used. Other forms of particleboard such as OSB require much higher-quality raw materials because they need engineered particles (strands). The extra cost involved in manufacturing these chips is offset by the greater selling price of the finished product. The seasonal availability of agricultural crops has to a large extent prevented the widespread exploitation of the enormous volume of suitable ligno-cellulosic material produced on farms around the world.

It has already been said that the vast majority of particleboards are made from wood. Wood is obtainable in three forms: as round wood, which is being used less due to costs, as residues from other processes and as recovered wood (normally processed in the form of large chips).

10.4.5.1   Round Wood

The best particleboard furnish can be produced from round wood. A particleboard manufacturer has as much control over particle size, shape, and surface quality as is possible. This situation also allows the decision as to whether the bark should be left on the wood prior to chipping or not.

The disadvantage with this raw material is cost. In addition, there is the cost of actually chipping the wood. The cost of this is split between the capital cost of the equipment together with running costs. On top of this there is the cost of drying the particles.

10.4.5.2   Wood Residues

The success of the particleboard industry stems from its ability to utilize wood residues. Forest residues may take the form of treetops and branches, or particles from chipped stumps. The former have not proved popular because they contain a high quantity of bark a needles.

Sawmill residues are preferred since the slabs, edge trimmings, or chips, if a chipping head rig is used, are usually debarked. The larger residues have to be chipped, so they have some of the advantages and disadvantages of the round wood described above. However, sawmill residues are likely to have a lower moisture content than round wood, so the chips produced should require less drying. Further down stream there are the joinery manufacturers who produce vast amounts of shavings and sawdust. Of the two, shavings are preferred; however, since they are used for pellet manufacture and animal bedding the demand for them is high, so the cost of shavings is usually prohibitive. The prime advantage of sawdust is that it is cheap. Joinery mill residues are usually dry and in a particulate form so only secondary breakdown is needed to produce particleboard furnish. However, much less control over particle geometry is possible and the residue is likely to contain a range of different wood species.

Sawdust, which many particleboard manufacturers buy and use, is increasing in cost rapidly as it is being used for pellet manufacture. When used as a surface layer furnish it helps to produce a hard smooth dense surface which many furniture manufacturers prefer. Tensile strength perpendicular to plane of board, often called internal bond (IB) strength, is improved by the addition of sawdust. This is probably due to increased inter-particle contact. Other physical properties are lowered by sawdust addition.

Plywood mill veneer cores are an excellent raw material for particleboard manufacture. They are all the same size, inexpensive, and of course bark free. Of course, the volumes available are limited.

10.4.5.3 Recovered Wood

Competition is very fierce in most sectors of the panels industry and so many manufacturers concentrate on reducing manufacturing costs as much as possible. Most manufacturers of particleboard use recovered wood to reduce their manufacturing costs because it is often a cheap alternative to other sources of wood and it is generally drier than other sources and so there is a considerable saving in energy during the drying stage of panel production.

Using recovered wood seems to be an environmentally friendly thing to do and makes economic sense, but it does not come without its own problems. The particleboard industry has always used a lot of "wastes" as raw materials for its products. These have included using trimmings, sanding dust and reject boards within the production line, sawmill wastes, secondary processing residues, for example, off-cuts, shavings and sawdust, and forest residues. Many of these are classified as being pre-consumer sources of recovered wood. In other words, they are residues generated as a result of making a product, for example, wooden furniture, and have not been used for a particular purpose prior to being used as a raw material for another product, in this case, particleboard.

The trend of greater use of recovered wood has come from a greater use of postconsumer sources of recovered wood such as demolition timbers, old furniture, and pallets and packaging.

Many sources of recovered wood, but especially post-consumer sources, contain contaminants that must be removed. These include: minerals, for example, stones, concrete, soil, and so on; ferrous metals, for example, iron and steel; non-ferrous metals, for example, aluminium, lead, brass, and so on, and organic materials such as plastics, paints, rubber, and fabrics. Sophisticated cleaning systems are available but these require significant capital expenditure. Generally, the economic benefits of using recovered wood justify the investment required and so a considerable increase in the use of recovered wood is anticipated for particleboard manufacture over the next couple of years.

The limitation in the use of recovered wood in many countries is the lack of infrastructure to collect, process, and deliver it. Another potential limitation is competition for this resource from the new bioenergy generation plants, which are often established with the help of state grants and subsequently supported through the receipt of higher than market prices for each unit of energy produced.

The use of recovered wood in particleboard provides manufacturers with an opportunity to market their products as being environmentally friendly, in much the same way as the paper industry has successfully promoted recycled paper products.

10.4.5.4 Bark

Bark is chemically, physically, and mechanically different to wood fiber plus it tends to have a high mineral contamination. So ideally bark should not be included in a particleboard furnish as it reduces board strength properties and increases resin demand. The removal of bark is even more important for other types of WBP and especially plywood and so this is discussed more fully in the plywood section.

Removal of bark helps to extend the lifetime of equipment in subsequent processing steps, for example, chipper knives, plug screw, refiner plates, and so on, and improves the mechanical properties as well as visual appearance of the board. During the manufacture of densified WBP bark is most often removed from small-diameter logs with either a drum or a roller debarker.

The operating principle of a drum debarker (see Figure 10.4), is that the logs rub and knock against each other during the tumbling action of the rotating drum causing the bark to be sheared off. The drum debarker is the most common used debarker in the paper and WBP industry, due to the excellent debarking efficiency (up to 99%) and high achievable capacities (up to 350 m³/h). The drum has a pre-set inclination which causes the logs to travel through by gravity and it also determines the log retention time. The drum is normally supported by rubber tyres mounted on truck axes that also drive unit. There are bars fixed to the inner surfaces of the drum shell that lift the logs as the drum rotates. Eventually, the logs fall down and move forward because of the inclination.

Logs that are shorter than the drum diameter will tend to tumble in the drum whereas logs that are longer than the drum diameter will tend to lose their bark by friction and impacts from other logs. At the discharge of the drum a hydraulically operated, adjustable gate provides some control over the log retention time. Drum diameters range from 4–6 m and their lengths from 15–30 m. Bark as well as stones and sand can be discharged from the drum by slots in the drum shell. After the drum a set of rollers allows loose bark to be separated completely from the logs. Special roller conveyors exist for the removal of stringy bark, for example, eucalyptus, which can roll up to form huge bundles.

A rotor debarker has a 4–5 of parallel, toothed rotors that transport the logs and remove their bark (see Figure 10.5). Retention time is controlled using a frequency converter to control rotor rotation speed. The debarked logs are conveyed forward to the discharge-end of the rotor debarker,

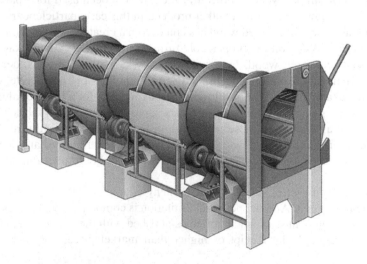

FIGURE 10.4 Rubber Tyre supported debarking drum. (Courtesy of Andritz.)

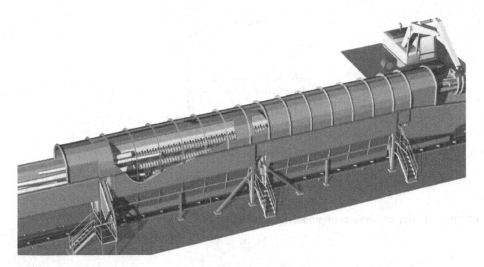

FIGURE 10.5 Rotor debarker. (Courtesy of Andritz.)

while loose bark falls through the openings between the rotors. Rotor debarker lines are designed for capacities from 30 up to 300 m³/h.

The advantages of the rotor debarker include: its ability to debark even frozen logs (due to the tools mounted on the rotors), low wood losses and low energy consumption compared to a drum debarker of the same capacity. When using a drum debarker in cold areas where the logs are frozen a de-icing unit (using hot water or steam) in front of the debarker is necessary.

Bark is collected by chutes underneath the debarkers which then falls onto a debris conveyor, this line is normally equipped with a disc screen and over size shredder before feeding to a boiler. The bark is usually burnt in the factory's boilers as it has a high calorific value; often providing up to 30% of a factory's thermal energy need.

During wood chipping, and subsequent drying and sorting operations, some of the bark that remains is reduced to a fine dust. It is thought that the adhesive is absorbed by the dust, because of its high surface area-to-volume ratio, thus lowering the amount of adhesive available for inter-particle bonding. Including bark also tends to increase thickness swelling, linear expansion and decrease IB strength.

Some people regard the darker board color which results from including bark depreciates its marketability. Despite these disadvantages many manufacturers do not debark their logs before further processing for cost reasons. In such cases, about half of the bark is removed from the wood furnish, at two stages: primary chipping, where large pieces of bark fall off and are removed; after drying where bark dust makes up a high proportion of the dust collected at the air-cleaning cyclones.

10.5 PARTICLE PRODUCTION

The way particleboard furnish is produced is dependent on the raw material. Figure 10.5 illustrates some of the many routes which are possible. Details of the machines themselves are shown in Figures 10.6 through 10.11. Fischer (1972) covers the main operational aspects of many of the wood reductionizers used in modern particleboard mills. Although this reference is rather old the general principles are still relevant in today's factories.

There are many different ways to generate particleboard furnish. For example, if the raw material were round wood a manufacturer may decide to use a drum flaker only, sort the produced flakes and sending those that are too small to a boiler. Alternatively he may decide to use a hacker to initially break the wood down, then use a knife ring flaker to reduce the hacker chips to particles.

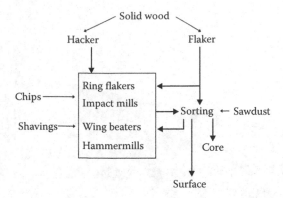

FIGURE 10.6   Different methods of producing particleboard furnish.

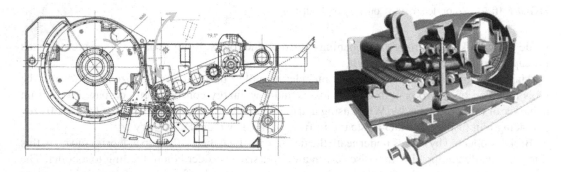

FIGURE 10.7   The inside of a large drum hacker. (Courtesy of Bruks-Klöckner and Pallmann.)

FIGURE 10.8   A drum flaker capable of cutting fixed length, random diameter logs. (Photo by Mark Irle.)

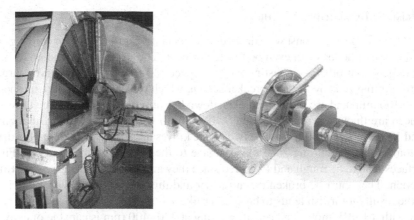

FIGURE 10.9 Vertical disc flakers that can cut random length logs. (Photo by Mark Irle and diagram on right by courtesy of Metso Panelboard.)

FIGURE 10.10 Knife ring flaker. (Courtesy of Pallmann.)

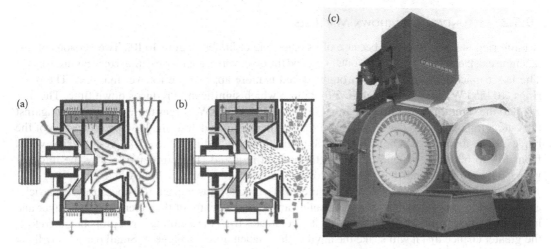

FIGURE 10.11 An impact mill double stream. (Courtesy of Pallmann and Ressel.)

10.5.1 PRIMARY BREAKDOWN MACHINES

Hackers tend to be large and robust with drum diameters up to 2.4 m. They can chip logs of random length and diameter; the logs can even be twisted or bent. The logs are fed end on to the drum so chip size is largely controlled by adjusting the feed speed (20–36 m/min). However, large particles are held in the cutting zone by a heavy breaker screen, which typically has square meshes twice the chip length (<100 mm) and are therefore broken down by subsequent passes of the knives. The particles produced are thick (<10 mm) and long (<40–80 mm) with a wide size distribution: 15% fine (<5 mm) and 10% oversize (>50 mm). The two to five knives are spring loaded into the drum so that if a large stone enters the cutting zone then damage to the knife is minimized (see Figure 10.6). Particle surfaces tend to be rough and fractured since they are produced more by a splitting than by a cutting action. They must be broken down further and this is often done with knife ring flakers, see below, the resultant furnish tends to be splinter like.

A hacker with a 1 MW motor and a shaft rotating at 300–400 rpm is capable of converting logs up to 1 m in diameter. It can produce around 150 tons wet (or 70 tons o.d.) of chips 35–50 mm long per hour. A hacker provides versatility as it can process a wide range of woods in various forms. Not all chipboard manufacturers have chosen not to install hacker/knife ring combinations because the quality of furnish produced is not as good as that possible with flakers. In addition, recent developments in flaking technology, which now allow some versions of both disc and drum flakers to utilize random length and diameter logs, have reduced the versatility gap between hackers and flakers.

Round wood can also be chipped with drum or disc flakers. Early drum flakers were limited to logs that were cut to length (dependent on length of drum, see Figure 10.6) and fairly straight. Current drum and disc flakers can use random length logs of poorer quality. In all drum flakers, regardless of the type, the knives are set at an oblique angle to the axis of the drum. This is to reduce vibration and strain applied to the drum bearing as the knife impacts on the log; it also causes a slicing action.

The development of disc flakers has followed a line similar to that of drum flakers in that early versions were limited to logs of a specific length. For example, the first flaking system installed at the OSB factory in Inverness in 1986, used horizontal disc flakers which accepted logs cut to approximately 1 m in length. Cutting the logs into short lengths reduces the problems associated with bent and twisted logs. In 1994, the horizontal disc flakers were replaced with vertical disc flakers similar to that shown in Figure 10.9. These flakers can convert logs of mixed lengths and diameters to flakes of predefined lengths and thicknesses. The installed power of 500 kW allows using a 2 m diameter disc with six knives an output of 50 tons wood/hour at a belt conveyer in feed speed of 76 m/min.

10.5.2 SECONDARY BREAKDOWN MACHINES

A knife ring flaker is so called because of its outer ring of knifes (Figure 10.10). Two versions of this machine exist: one with a stationary outer ring and the other with a counter-rotating ring (Figure 10.10b). The latter machine has a higher throughput and is more appropriate for wet material. The outer rings (10–55 kW) can be replaced in 5 to 25 min which significantly reduces down time. The diagram to the right illustrates how the inner impeller (100–630 kW) forces the wood particles against the outer ring of knives. The thickness of the resultant chip is determined by the protrusion of the blades. Ring diameter can be between 600 to 2000 mm including 28 to 92 knives and ring width from 140 to 600 mm, depending on the throughput required. The circulated air by the impeller varies between 4000 and 18,000 m^3/h allowing the processing of 2 to 30 t oven dry chips.

The machines of the type shown in Figure 10.11 are known as impact mills because the particles are reduced by them striking against the solid "anvil" in the centre of the outer ring. The larger and heavier a particle is the greater the moment it accumulates from the spinning inner propeller, therefore, the greater chance that it will strike the anvil with sufficient force to break it. Small particles will not accumulate much momentum from the spinning propeller and are much more likely to follow the air

FIGURE 10.12 A hammer mill. (Courtesy of Maier and Pallmann.)

flow (100 m/s) through the holes of the mesh and not be reduced further. These machines are efficient as in general only those particles which require reducing are broken down. Such machines are often used for the preparation of surface particles. Typically, these machines are 800 to 1800 mm in diameter, have a grinding track width of 150 to 500 mm, powered by motors of 100 to 1000 kW, creating an airflow of 6000 to 30,000 m³/h and have an output of 1 to 10 t chips o.d./h.

Wing beaters are similar to impact on mills except that the outer screen does not have a solid region. Particles that are too large to pass through the screen are ground down to the appropriate size by successive strikes from the rotating impellor. Again larger particles will accumulate more moment and strike the outer mesh ring with more force.

The hammers of a hammer mill (see Figure 10.11b) are attached to the central shaft by hinges which allow the hammers to swing back if they collide with a large particle. Large particles are therefore broken down by a series of blows. Particle size is determined by the size and shape of the holes in the screen. The robust construction of these machines enables them to break down solid wood into splinter-like particles. Hammer mills are available in a very wide range of sizes from laboratory scale to huge machines like that shown in Figure 10.12b, which are used to produce particles from recovered wood. Consequently, rotor diameter can vary from 230 to 1800 mm, rotor lengths from 250 to 2000 mm, motor size from 160 to 500 kW, producing 30–80 m³ o.d. chips/h.

10.5.3 PARTICLE GEOMETRY

Wood composite properties can be engineered, to a certain extent, by adjusting particle geometry. Unfortunately, there is a lot of contradictory evidence in this area. The clearest patterns emerge when board properties are compared to the slenderness ratio of particles. Slenderness ratio is calculated by dividing the particle's length by its thickness:

$$\text{Slenderness ratio} = \frac{\text{Length}}{\text{Thickness}}$$

For the majority of properties long thin chips seem the best. However, surface quality (smoothness and hardness) and IB strength are enhanced small particles. This is why most manufacturers attempt to classify their furnish using the fine particles for the surface layers and the larger particles for the core.

10.6   PARTICLE DRYING

Once the particles have been cut their moisture content must be reduced to between 2% and 8%, depending on the adhesive system to be used to make the WBP. Such low moisture contents are required because residual moisture is converted to steam in the hot pressing stage, if too much steam is generated then, when the press opens, the board is likely to be delaminated by the sudden release of steam pressure.

There are a number of different dryer types on the market as shown in Figure 10.13. However, chapter will just consider two of the most common, the three-pass and single-pass dryers.

Figure 10.14 illustrates the basics of a three-pass dryer. The particles are introduced into the central tube which will often be heated by a direct flame fuelled by gas, oil, or wood residue. The majority of dryers are capable of being heated by more than one heat source thus ensuring that particle drying can occur all year round. The conditions in the central tube are quite harsh; temperatures range from 250°C to 850°C (700°C is typical) and the air speeds are often as high as 8 m/s. In the second tube the air flow is reversed. A combination of water evaporation and greater tube volume causes the air temperature and speed to fall. In the final outer tube the air flow is again reversed and the air temperature will have fallen to between 60°C and 100°C.

The outer tube rotates, typically at 8 rpm, thus causing the particles to tumble which aids in their passage through the dryer. These dryers are good in that they are compact for a given evaporation rate. Typical dryers of this type maybe 30 m long, 4.5 m in diameter and have evaporation rates of 7 tons per hour, reaching a drying capacity of 25 tons furnish per hour.

The conditions inside a single-pass dryer (see Figure 10.15), tend to be moderate in comparison with the three-pass dryer. For example, the inlet temperature will usually be around 500°C and the average particle dwell time is increased to compensate for the lower temperatures. The dwell time can range from 20 min to as much as 60 min. The particles are helped through the dryer by heated paddles and the rotation of the drum itself.

An increase in demand for particleboard has caused many factories to expand which, in turn, has created the need for higher throughputs. The current trend is toward the installation of a single large dryer including a flash tube predryer (500°C) with a high evaporation rate of maybe 50 tons/h whereas previously 5–10 tons/h would have been the norm. In recent years, most new dryers that

Dryer type	Scheme	Temp. range	Drying time	Capacity
Rotary bundle dryer	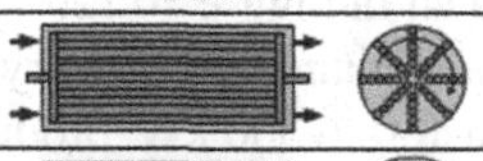	up to 200°C	≤20 min	1 ... 9 t/h
Tube bundle dryer	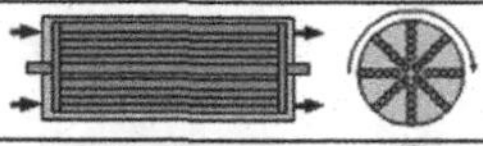	up to 160°C	n.a.	10 ... 18 t/h
Single-pass drum dryer	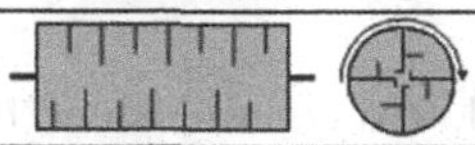	up to 450°C	20–30 min	≤40 t/h
Three-pass drum dryer	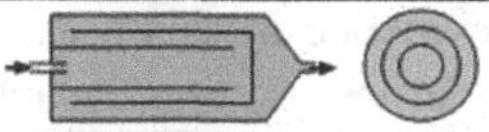	up to 400°C	5–7 min	≤25 t/h
Flash tube pre-dryer		up to 500°C	≈20 s	2 ... 14 t/h
Jet tube dryer	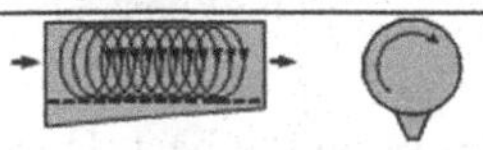	approx. 500°C	≈0.5–3 min	≤10 t/h

FIGURE 10.13   The characteristics of different chip dryer types. (From Deppe, H. and Ernst, K. 2004. Taschenbuch der Spanplattentechnik. DRW Verlag Weinbrenner. ISBN: 978-3871813498 and after Ressel, J. 2008. Presentation during the *3rd International Wood Academy*, Hamburg.)

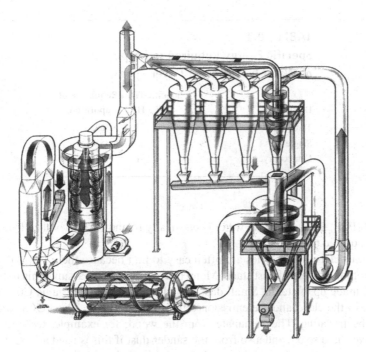

FIGURE 10.14 A three-pass dryer. (Courtesy of Metso Panelboard.)

have been installed are of the single pass type with a predrying step and all are large with drying capacity rates around the 20 tons of furnish or more per hour.

The amount of energy required to evaporate a unit of water from a wood chip is dependent on the way the chip is heated, the particle geometry and the moisture content of the wood. Heat transfer by conduction is more rapid than by convection, so those dryers that heat the particles mainly by conduction, that is, by touching the dryer sides or plates in the dryer, are likely to use less energy (see Table 10.3). Water evaporates more readily at high wood moisture contents, particularly above fiber

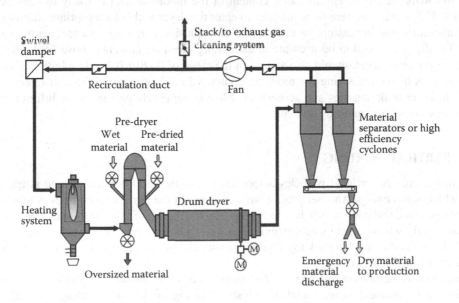

FIGURE 10.15 A single-pass dryer with dust extraction system. (Courtesy of Büttner.)

TABLE 10.3
Specific Energy Required to Evaporate 1 kg
of Water by Different Dryer Types

Dryer Type	Specific Heat Requirement (kJ/kg H_2O Evaporated)
Three pass	3350 to 3675
Single pass	3255 to 3550
Contact dryer	3150

saturation point (FSP). Below FSP the water is physically bound to the wood cell wall thus increasing the energy required to evaporate it.

Operational particleboard factories are often easy to find because there is usually a large white cloud coming from the dryer exhaust stack. Many people incorrectly assume this to be smoke but in fact it is mainly made up of steam. Small dust particles and Volatile Organic Compounds (VOCs) are also present in the cloud and the emission of these is restricted. The VOCs can have a strong smell and can be irritating. They emanate from the wood, for example, terpenes, waxes, other organic extractives, and so on, and also from the sander-dust if this is used as a fuel.

Emissions from dryers are limited by regulation. In general, manufacturers are not permitted to achieve these limits by dilution, instead they must limit the production of emissions or install a treatment process, for example, a scrubber (WESP—wet electrostatic precipitator). These restrictions may cause a move toward lower drying temperatures (inlet temperatures of <400°C) coupled with additional exhaust cleaning.

Drying small particles of wood to very low moisture contents is obviously a hazardous operation. Consequently, all modern dryers have sophisticated fire detection and extinguishing systems. Many dryers also have spark detectors and automatic-controlled sprinkler nozzles that are designed to detect a potential fire or explosion hazard before either occurs.

For efficient drying it is very important to determine the moisture content of the furnish as it enters and exits the dryer. The moisture content of the input material is likely to vary between 12–150%. If dry material were to be allowed to enter the dryer without appropriate control (lowering termperature and increasing gas circulation) an explosion may result. Consequently, moisture meters for the input need to be accurate at estimating moisture contents above FSP. The output meters, on the other hand, should be accurate in the range of 0–20 ± 0.2%. In addition, the meters should be capable of measuring the moisture content of moving material from different species, having different bulk densities and geometries. Most systems rely on the use of infrared light or microwaves to measure moisture content.

10.7   PARTICLE SORTING

Some manufacturers sort their particles before drying, so those outside the desired range are not dried, which saves energy Another goal of wet screening in the particleboard lines is to adjust the ratio between face and core particle before drying. The drying of particles process can be controlled more accurately when particle size distribution is known before. But wet particles are difficult to sort efficiently as they tend to stick together. Consequently, in most particleboard factories the particles are classified after drying.

There are two methods of sorting particles: mechanical sieves and air classifiers. There are three types of the mechanical sieves found in industry (see Figure 10.16): vibrating inclined screen, vibrating horizontal screen, and gyratory screen, which are readily recognizable from their housings.

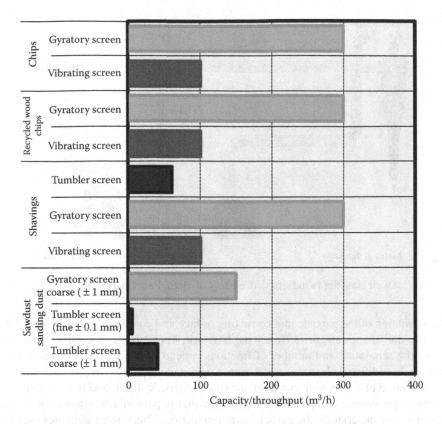

FIGURE 10.16 Performances of different screening systems. (From Ressel, J. 2008. Presentation during the *3rd International Wood Academy*, Hamburg.)

All of these work on the same principle in that the particles are fed over a series of wire meshes, the particles either fall through or are passed to a collecting bin (see Figure 10.17).

Air classifiers sort particles by air flow. Particles are introduced into a counter-flowing air stream so that the small particles are taken away by the air flow and the heavier particles fall to the bottom where they are removed by mechanical means (see Figure 10.18). A number of these may be joined together in series, each with a different air flow calculated to sift out a particular size. These are most commonly used in conjunction with a sieve system.

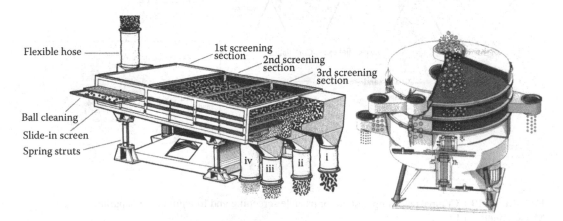

FIGURE 10.17 Gyratory rectangular and tumbling screen sieve. (Courtesy of Algeier.)

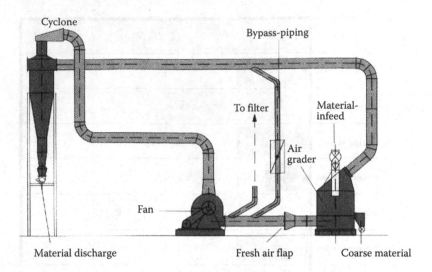

FIGURE 10.18 An air classifier (wind sifter). (Courtesy of Metro Panelboard.)

For the formation of the particle mat consisting of face and core layers, the sorting of dried particles into size and shape categories (fractions) like dust, fines, coarse, over-size is necessary. The efficiency of the separation and number of fractions depend on the number of screen decks. This equipment needs additional cleaning, maintenance, and noise reduction measures.

The combination of a screening conveyer, air sifter, magnetic drum, and heavy particle separator in one machine provides a compact cleaning system that is particularly effective when recovered wood is used as raw material for the particleboard production. This system separates oversize, heavy particles, ferrous and partially nonferrous matter and mineral impurities and dust (Figure 10.19).

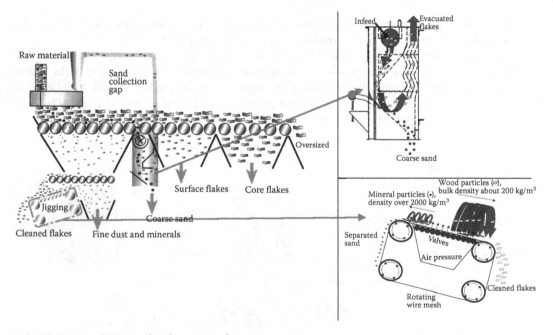

FIGURE 10.19 Compact cleaning system for particle screening and foreign matter separation. (Courtesy of Metso Panelboard.)

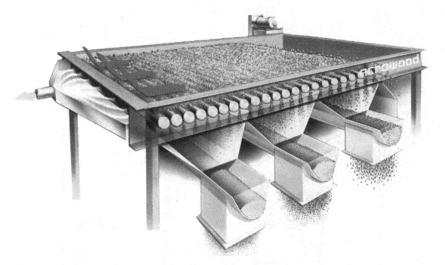

FIGURE 10.20 A roller system for sorting particles generated from recovered raw material. (Courtesy of Acrowood.)

Particle size classification can also be achieved using dynamic screens of rotating rollers as shown in Figure 10.20. The advantage of this method is that the screens are largely self-cleaning. The texture of the roller surface and the distance between the rollers permits highly accurate separation of particles into face and core fractions and also the elimination of sand, soil, and foreign particles.

10.8 RESIN METERING AND BLENDING

Liquid raw adhesives are often purchased as water-based solutions, containing approximately 50% (PF) to 65% (UF) solids. These are thermosetting adhesives in which the curing process (condensation reaction) has been interrupted before delivery of the solution, thus storage duration is limited to several weeks depending on season, transportation and storage temperature. Adhesives can be purchased in powder form for longer storage, but, this is rarely used in Europe.

Before application on the dried furnish the adhesive solution must be blended, according to proven recipes, with water and additional additives, for example, hardeners, colors, fire retardants, preservatives and so on. Different adhesive formulations may be used for the surface and core layers. The amount of adhesive mix to be added is calculated on a solid adhesive substance to oven dry wood basis.

The hardener solution, which is added to catalyze the resin curing reaction is introduced as a percentage of solid hardener substance to solid resin basis, is added as late as possible to avoid premature curing due to production line stoppages. The surface and core layers experience different curing conditions during hot pressing and so different adhesive mixes are used for surface and core layer furnishes.

Resination of the furnish is a continuous process, which requires constant and continuous mass and/or volume flow control (furnish weight, resin weight/volume) to guarantee accurate blending and uniform panel properties. The mixing of the adhesive recipe on the other hand before application to the furnish fractions is done either continuously or batch wise. The type and amount of adhesive depend on panel type (interior or exterior application), particle size, hot pressing conditions, and so on. The data in Table 10.4 are just for general information and may vary.

TABLE 10.4
Typical Resin Addition Levels for Different Panel Types

Panel	Resin	Level	Application
Particle board	UF	4-10%	→Surface layer 8-14%
			→Core layer 4-8%
	PF	6-8%	→Surface layer 8-12%
			→Core layer 6-9%
	MDI	2-6%	→Surface layer 6-8%
			→Core layer 2-4%
OSB	PF	6-8%	
	MDI	2-6%	
MDF	UF	8-14%	→In blow-line resin application
	UF	6-10%	→Resin application to dry fibres
	MUF	8-12%	→For HDF as flooring quality
	MDI	≈4-10%	
All panel types	Wax	0.3-2%	Applied as micro-crystalline wax emulsion or liquid paraffin

UF bonded particleboard for interior application

UF resin (66% solid content)	100 kg		
Hardener (15%)	10 kg	SL	5% $(NH_4)_2SO_4$
Paraffin emulsion (50%)	10 kg		10% NH_3
Water	12 kg		85% H_2O
Adhesive solution	132 kg	CL	10% $(NH_4)_2SO_4$
			5% NH_3
			85% H_2O

Hardener solution

PF bonded particleboard for exterior application

PF resin (45% solid content)	100 kg		
Hardener (50%)	5 kg	SL	None
Paraffin emulsion (50%)	15 kg	CL	50% K_2CO_3
Water	0 kg		50% H_2O
Adhesive solution	120 kg		

Hardener solution

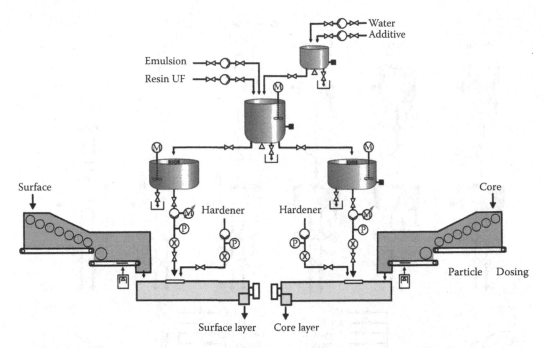

FIGURE 10.21   Batch wise adhesive blending of particle. (Courtesy of Metso Panelboard.)

Batch-wise adhesive blending is used for small and medium-capacity particleboard plants, operating with long production series without changing the adhesive formulation (see Figure 10.21). During batch-wise adhesive blending all of the adhesive formulation components, with exception of the hardener, are mixed in a tank. Hardener solution is first added to the furnish in the particle blender.

Continuous or in-line adhesive blending is used for large capacity particleboard plants (Figure 10.22). Adhesive blending and the recipes are normally controlled and managed by a computer system. Facilities that use MDI are required to take special precautions. With in-line blending all components are added simultaneously into the particle blender. Pre-conditions for reliable adhesive blending and thus uniform product quality, but also high flexibility in changing the production programme, are reliable and accurate measurement and metering. The benefits of such highly sophisticated systems are: significant raw material savings of up to 5%, optimum adjustment to any type of raw material and requirements through integrated metering on the basis of flow meters and metering scales, high flexibility due to individual simultaneous metering.

Different methods for measuring and metering of the adhesive components:

- Mass flow measurement according to the Coriolis principle, which is suitable for all media with variable density, viscosity, or conductivity. The system measures the real mass flow.
- Magneto-inductive flow measurement uses the electric conductivity of the medium as its measuring method and therefore is applicable to aqueous solutions with constant conductivity.
- Mass flow measurement with differential metering scales is a proven and highly accurate measuring and metering method. It is equally suitable for liquid and powdery adhesives and additives. It is easy to calibrate and to check as well as a uniform technology, suitable for all media and flow rates.

Blending is the mixing of dried wood particles and adhesive formulation with the aim of achieving a uniform distribution of drops of resin on each and every particle. The additives may be

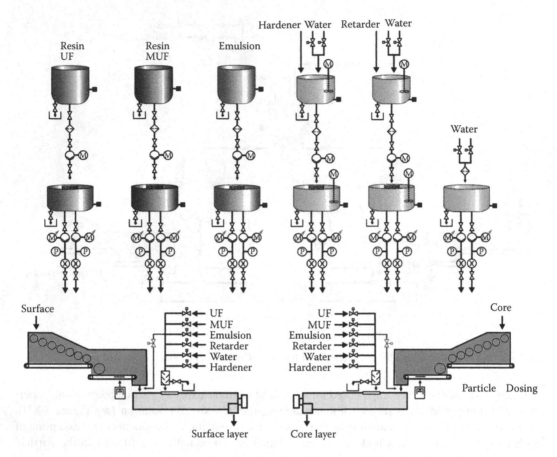

FIGURE 10.22 Continuous adhesive blending of particle. (Courtesy of Metso Panelboard.)

atomized by either hydraulic or air spraying nozzles. For the former, the droplet size is dependent on nozzle design, liquid viscosity, and liquid pressure (5–17 MPa). The disadvantages of these sprayers are the high pressures required and nozzle clogging. The droplet size for air atomizing spraying nozzles is also affected by the above factors (although the liquid pressures are very much lower). In addition, the pressure of the air impacting on the liquid jet also influences droplet size. The ideal droplet size is reported to be less than 35 μm in diameter (Moslemi 1974). In practice, the size varies in the range of 30–100 μm. The main reason for this is that formaldehyde adhesives are colloidal suspensions, and suspensions do not atomise readily.

Viscosity and ambient temperature affect resin penetration in wood. Ideally, the resin formulation should remain on the particle surface so that it is available for bonding. The smallest particles absorb five times more resin than the largest on a weight-for-weight basis. Some mills add the finest particles just at the end of the blender in order to minimize resin absorption.

Particle dosing is required to ensure a controlled furnish flow into the blender and to achieve reliable constant resination. Particle dosing bins are built with either one or two belt tables. Additionally, they are equipped with dust suction nozzles and belts with scrapers to keep the dosing bin belt clean. The weight scale controls furnish discharge volume. If two separate belt tables operate, the bin itself is separated from the metering belt to allow high bin volume and to ensure high measuring accuracy. On the first belt table, the level of the bin is measured and furnish flow is adjusted. The second belt conveys the flakes via online weight scale, controlling the furnish discharge mass (Figure 10.23).

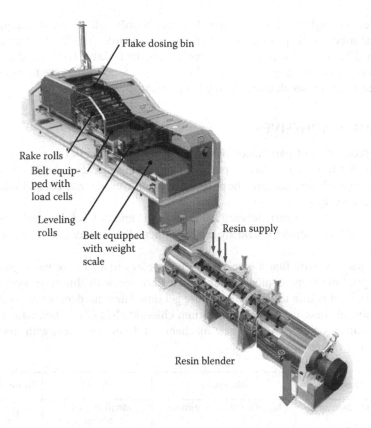

FIGURE 10.23 Furnish dosing bin and blender. (Courtesy of Metso Panelboard.)

Most modern mills use continuous blenders to add resin and other additives such as wax to the furnish. The spraying of adhesives onto dried particles is often termed as blending. Blenders are classified in two types: short and long retention time. From the 1950s onwards, the short retention time (2–3 seconds) blenders have been preferred. These are 2–3 m long and 600 mm diameter. They have a short concentrated spraying area, so some particles inevitably receive more resin than others. Resin redistribution is improved in the rest of the blender by violent tumbling (600–1000 rpm) in the bulk of the blender (Figure 10.24).

A long retention blender is a large drum ($L < 5$ m) requiring minutes for the same throughput and job as the short one. It treats the furnish gently and reaches an efficient resin distribution for inhomogeneous one.

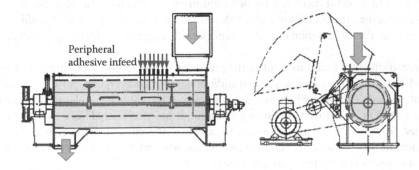

FIGURE 10.24 A short retention time blender. (Courtesy of Binos and Metso Panelboard.)

Manufacturers of single layer and graduated density boards will often use a single blender. Two blenders will be used for the production of multi-layered boards, one for the core and one for the surface furnish. This allows the possibility of optimizing the additive levels for the different layers. For example, more hardener may be added to the core layer to increase cure rate or more water for the faces to cause a steam shock effect during hot pressing.

10.9   COMMON ADHESIVES

For economic production of particleboards the adhesive must cure in the press very quickly, say within 1–5 min, but it must also have a potlife of something in excess of 20–30 min so that the adhesive does not cure before entering the press. Longer potlives are desired, and achieved by many, to allow for line stoppages.

The most commonly used particleboard adhesive is UF, and this is followed by melamine formaldehyde (MF) and then phenol formaldehyde (PF). Other adhesives have only minor importance in the global view.

The advantages of UF are that it is relatively cheap, cures to a clear or white glueline, provides good dry strength, and there is much production experience with this resin system. In order to achieve a high ratio of potlife to high temperature gel time latent hardeners are used (Meyer 1981). The most popular of these used to be ammonium chloride (NH_4Cl). When added to UF resin it produces hydrochloric acid by two different mechanisms. In one, it reacts with free formaldehyde and chain end methylol groups thus:

Urea	Melamine	Phenol	Formaldehyde
Mono methylol urea	Mono methylol melamine	Mono methylol phenol	

$$4NH_4Cl + 6CH_2O \rightleftharpoons 4HCl + (CH_2)_6N_4 + 6H_2O$$

Like similar chemical reactions the reaction rate increases with temperature. So when the ammonium chloride is first added there is a sudden drop in pH, as result of the high availability of free formaldehyde, but then the pH falls at a much slower rate until heat is applied. The additional source of acid arises from the dissociation of ammonium chloride to ammonia (NH_3) and hydrochloric acid on heating.

A fast transition from liquid to solid in the press can be obtained by adding 1 g of ammonium chloride to 100 g of resin (solids basis). Such addition levels, however, can cause precure, so NH_3 or hexamine are added. The former is better in that it reduces formaldehyde release. The use of ammonium chloride and other halogen containing compounds has declined. This is because many mills burn the dust from their sanding operations to help heat the chip dryers. Chlorides are therefore released when the dust is burnt because it contains adhesive and the chlorides attack the metal components of the dryers leading to severe corrosion.

Ammonium chloride has been largely replaced with ammonium sulphates and ammonium nitrates, both of which avoid corrosion problems in the dryer.

Using UF resins is not without its disadvantages. UF is not weather resistant, which precludes exterior uses. It also releases formaldehyde and since there are regulations which limit the maximum concentration of formaldehyde in the air, this can restrict the number of interior uses for boards bonded with this resin.

The moisture resistance of UF can be improved by fortifying it with melamine to form a melamine-urea formaldehyde resin (MUF). These adhesives are clear and strong, but they are more expensive. The price difference between a fortified and pure UF resin depends on the amount of melamine added, but a pure MF resin is about three times more expensive than pure UF. Although not quite such a problem, MF and MUF resins still emit formaldehyde.

Phenol formaldehyde resins on the other hand are much more weather resistant and do not have a formaldehyde problem. Admittedly, they are more expensive, being approximately twice the price of UF resins. Some people consider the characteristic dark-red color of cured PF resin detracts from the board's appearance, but since most boards are covered in use, for example, melamine or veneer coverings, this is probably not a major disadvantage. However, in use some additional production problems are encountered with this adhesive. Higher temperatures and longer time periods are required to cure PF. Not only does this reduce productivity but it can also lead to significant penetration of the adhesive into the wood chip; if the glue has been absorbed by the particles then it is not available to bond them together and poor board strength results.

Isocyanate-based adhesives have been used for commercial production of particleboards, MDF, and OSB. Relative to the volumes of UF adhesives used, however, the isocyanate adhesives are small. These adhesives have become known as MDI, which stands for methylene diphenyl diisocyanate. In actual fact, the isocyanates used consist of polymeric forms of MDI. Although more expensive than formaldehyde-based adhesives, MDI performs so well that a particleboard with adequate properties can be made with much less resin than is possible with formaldehyde resins. The first isocyanates caused production problems as they tended to cause the board to stick to the metal platens. This has largely been solved using release agents, see Galbraith et al. (1983). There are versions of isocyanates that can be mixed with water to form an emulsion which considerably eases the difficulty of spraying very small quantities of resin onto the furnish.

The binder cost as a proportion of total manufacturing costs is around 15–25% of total. When the amount used is considered (2–10% of wood weight), then it can be seen that a small change in use or cost can have a significant effect on profit. A great deal of work has been done on wood adhesives. Much of this information has been coherently reviewed by Pizzi (1983) and in the two state-of-the-art reports written by members of COST Action E31 (Dunky et al. 2002, Johansson et al. 2002).

10.10 MATTRESS FORMING

It is vital that this stage is carried out properly. Board density directly affects properties; consequently, the preparation of a consistent mattress is critical in producing a board with suitable properties. The mattress must be laid down uniformly across its length and width, but not necessarily through its thickness. Instead the density profile through the board's thickness must be symmetrical about the board's centre. If this is not achieved then the board will probably warp in service.

Mattress formers are either segregating or nonsegregating. They may operate as either batch or continuous formers, the latter being preferred in modern mills. In some small and old mills the mattress is laid onto caul plates, which are metal plates 4 to 6 mm in thickness and the size of the press platens. The caul plate is used to transfer the mattress to other process stages, including pressing. The high-capacity mills of today do not use caul plates because of the capacity limitations they impose, capital outlay, maintenance costs, and additional handling equipment required for example, caul separators and return lines. Instead, in these mills, the mattress is laid directly onto a conveyor belt.

Nonsegregating formers tend to be used in the production of single- or multilayered boards. There are many different designs from various manufacturers but they all aim to randomly orientate the particles so that the board strength properties are similar with and across the machine direction. The motion of the conveyor belt, or the forming machine, does, however, tend to cause some particle orientation in favor of the machine direction.

Segregating formers work by one of three principles: air jets; throwing rollers and textured separating rolls (see Figure 10.25). In the first machine, particles are dropped in front of a column of air jets. The horizontal air flow created by the air jets causes the small light particles to land on the conveyor belt away from the column; heavier particles tend to settle at the column base. The other former classifies the particles by dropping them onto a spinning ribbed roller. The chips are thrown off the roller; the heavy particles having more momentum than the small, tend to travel further. The overall effect of both machines is to produce a board with small fine particles on the surfaces (good for smoothness and hardness) and larger chips in the middle (good for bending and impact strength).

Figure 10.26 shows an example of the relatively new textured roller system that uses a bed of rollers to classify particles. The roller surfaces have a range of patterned surfaces, of which there are many designs with new patterns appearing with further development. The regular patterns have 3D shapes like diamonds/pyramids of 0.5 to 3.0 mm height which facilitate separation while minimizing further particle size reduction. The diameter of the rollers is around 80 mm and some rotate

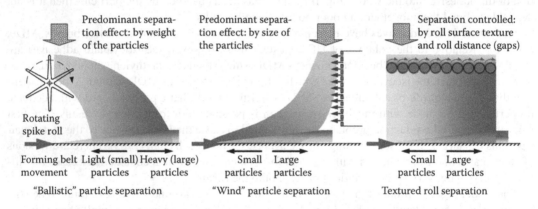

FIGURE 10.25 Principle of classifying and forming of particle mattresses. (From Ressel, J. 2008. Presentation during the *3rd International Wood Academy*, Hamburg.)

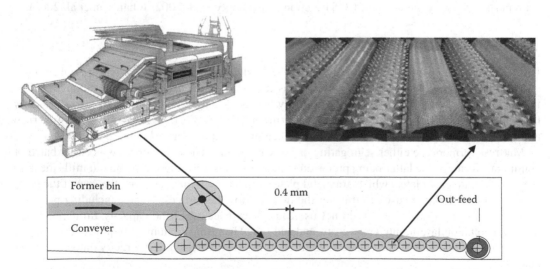

FIGURE 10.26 Roller (crown) former line for high production rate (3,100 m³/day). (Courtesy of Metso Panelboard.)

rapidly (80 rpm) and have a coarse pattern whereas others rotate more slowly (30 rpm) and have a fine pattern. This helps to uniformly distribute the furnish across the whole machine width. The gap between the rollers at one end of the bed is about 0.1 mm and this increases to 2 mm along the length of the bed. Consequently, fine particles are able to pass through at the start of the bed and the larger particles are transported down the bed where the gap increases thus allowing more particles to fall through. Therefore, one bed of rollers is able to lay one half of a panel or one layer if the panel is of a 3-layer structure. A significant advantage is that with proper adjustment of the beds there are no abrupt changes in particle size and densities between layers and this leads to better IB values.

The roller system is very compact compared to traditional formers. This is most obvious when a roller system is retro-fitted to an existing line where, because it requires less space than the previous former, there are large gaps around the new former. The roller system does not require air flow and so there is less dust emission and particle drying so trimmings from mattress can be recirculated back into the former and still have similar properties to fresh particles. Also it is more able to form low-density mattresses and fine and uniform surfaces.

A significant advantage is that, unlike sieve-based systems, the roller system is largely self-cleaning. Compared to conventional former the investment is low because the building height is lower, the platforms and supportive structure are simple, and no fans and exhaust pipe are necessary. The equipment can eliminate oversize foreign matter and particles and thus protect the steel belts of the continuous press.

10.11 MATTRESS PREPRESSING AND PREHEATING

The press is the most expensive single piece of equipment in a particleboard factory (approximately 15% of investment) so it must be operated as efficiently as possible. Many presses produce boards in batches whereas the rest of the mill, such as flakers, dryers, and formers, operate continuously. As a result, most modern mills have cold prepresses that compress the mattress to about 50–70% of its formed height (see Figure 10.27). Since a prepressed mattress is thinner the press does not have to open so much in order to accommodate it. It therefore follows that the press does not have to close so much to compress the mattress to the desired thickness, which also helps to reduce the chance of precure. A press can compress a prepressed mattress far quicker than one that has not been prepressed because the prepress squeezes out much of the air in the mattress. This air can sometimes cause particles to be blown out of the side of the mattress if it is pressed too quickly. Consequently, the overall press cycle

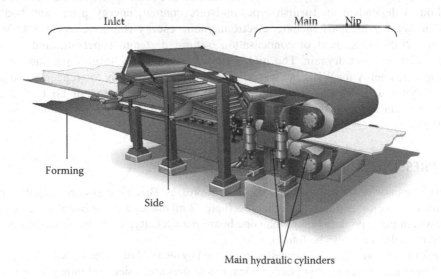

FIGURE 10.27 A prepress. (Courtesy of Metso Panelboard.)

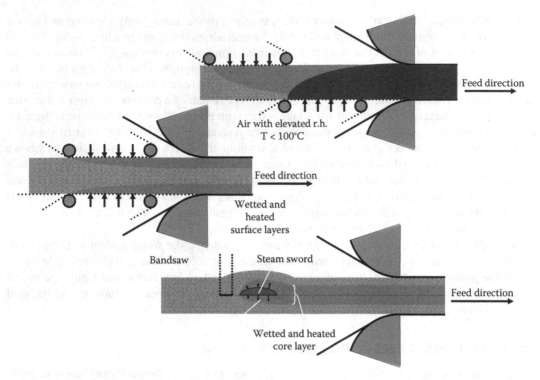

FIGURE 10.28 Preheating of mattresses using steam or hot air via surface or core. (From Thoemen, H. 2008. Pressing of panels. Presentations during the 3rd International Wood Academy, Hamburg.)

is reduced. A good example of how prepresses may increase production was shown by Kronospan who raised their production from 180,000 m³/annum to 216,000 m³, a 20% increase, just by adding prepresses to their manufacturing lines at the end of the 1970s. All modern lines include a prepress.

Pre-presses are also used on production lines that have continuous presses. Many of the benefits given above are also of relevance to continuous presses.

After cold pre-pressing, and especially for thicker mattresses, mattress preheating is used to increase mat temperature (ideally the core) sometimes. The most appropriate choice of the heating method is dependent on furnish type, moisture content, energy price, and board type. Mattress preheating methods include: electromagnetic energy (high frequency <45 MHz and microwave >2000 MHz), heat of condensation (saturated steam, supersaturated steam, high humid hot air), and hot dry air. The high costs of power today largely preclude the use of electromagnetic energy for WBP, but it is still used for the preheating for veneer-based products (moulded plywood, LVL, PSL etc.) and for heating during pressing of GLT. For particleboard and especially for dry process fiberboard, preheating typically uses steam or hot air (see Figure 10.28).

10.12   PRESSING

Particleboard can be made with batch or continuous presses. Batch presses are available in single- or multi-daylight forms. A single-daylight press (upto 52 m) makes a single board at a time, whereas a multi-daylight press produces more than one board per cycle (typically 16 for particleboard, but, much larger numbers can also be found).

The majority of commercial size presses are heated by steam, hot water, or hot oil, the latter two being preferred because consistent platen wide temperatures are easier to obtain. The use of electric heating elements is restricted to laboratory scale presses because of the high cost and large temperature variations likely in large platens. Other heating systems have been proposed (Moslemi 1974, Axer 1975).

Typical platen temperatures range from 200°C to 220°C. Such high temperatures are required to achieve rapid cure of the adhesive. The specific pressing pressure ranges from 2 to 4 MPa, principally depending on final panel density, but raw material density and panel thickness also have an impact.

10.13   BATCH PRESSES

10.13.1   SINGLE-DAYLIGHT PRESSES

In order to attain economic capacities and to maximize the advantages of single daylight, these presses tend to be designed to manufacture large boards (Figure 10.29). Most presses installed today are 25 m or longer. Egger (UK) Ltd. had until 2009 a 52 m press, which has now been replaced by a continuous press. Although long, the widths of the boards that these presses are invariably not more than 2.8 m. This is to allow steam to escape from the mattress during pressing. Similar widths are found in multi-daylight presses for the same reason.

The main advantage of single-daylight-type presses is related to the size of product produced. All densified wood composites must be trimmed after pressing, this is not just to provide a straight edge, but also to remove, or reduce the amount of, edge effects. These edge effects are caused by the inherent temperature and pressure variations experienced by particles in the outer edges of a mattress. As a result, the outer portions tend to have inferior physical properties to the bulk of the board, see Section 3.10.2. The amount of "waste" removed as a proportion of board size is much smaller for large boards than for small ones. A similar reduction in waste tends to occur when non-standard sizes are being cut from large production panels.

Better thickness tolerances and shorter pressureless waiting time have contributed to improved sanding allowances of 0.7–1.4 mm. The steam injection press is a special single-daylight press that can make very thick panels (>50 mm) or panels with a homogeneous density profile.

10.13.2   MULTI-DAYLIGHT PRESSES

Typical multi-daylight presses produce between 4 and 16 boards, 5 to 7 m long and up to 2.5 m wide. Simply multiplying the number of daylights by panel dimensions demonstrates that multi-daylight

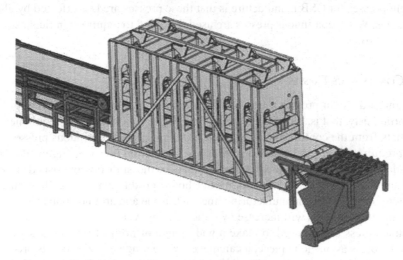

FIGURE 10.29   A single daylight (opening) press. (Courtesy of Dieffenbacher.)

FIGURE 10.30 The simultaneous opening and closing mechanism of a multi-daylight press. (Photo by Mark Irle.)

presses can have much higher capacities than single-daylight presses for a given footprint; for example, a multi-daylight press producing 4 boards 5 m long per cycle will produce the equivalent of a single-daylight press that is 20 m long.

An essential part of using a multi-daylight press efficiently is that all the boards of a press load must enter and exit the press at the same time. However, in the press they must also be subjected to the same temperature and pressure regimes. If this is not done then there will be significant variation between the boards of a single press load. Multi-daylight presses therefore have loading and unloading cages attached to them which introduce and extract the boards. Even so, the loading and unloading typically takes around 30 s and so productivity falls with panel thickness. Figure 10.30 illustrates the system of arms and levers employed to ensure the simultaneous closing and opening of all daylights. This equipment is not required for single-daylight presses.

A major disadvantage of multi-daylight pressing is the large sanding allowance of 1.0–2.5 mm depending on thickness and product type. Low thickness tolerances and softer surfaces due to some pressureless closing time are the main causes of this.

Recent OSB projects have been based on multi-daylight systems, have reached a design capacity of 900,000 m³/a with 16 daylights of 3.66 × 7.93 m. The main reason for the re-gained success of multi-daylight presses for OSB manufacture is that these presses are less affected by glue spots and mat inaccuracies. When continuous presses are used they tend to require significant steel belt repair when used for OSB.

10.13.3 CONTINUOUS PRESSES

Single- and multi-daylight presses make boards in batches and yet the rest of the production line operates continuously, that is, blending, forming, prepressing. Consequently, equipment is required to transfer mats from the continuous to the batch pressing stages. Continuous presses, on the other hand, do not require such equipment and so simplify production flow (see Figure 10.31).

The popularity of continuous presses can be seen in the statistics published in Wood Based Panels International (Wadsworth 2001) on particleboard production. These show that continuous presses account for 36% of the manufacturing lines in Europe and around about 50 % of the production volume and these figures will increase over the next few years.

A continuous press can be used to make a wide range of product thicknesses and because the panel leaves the press as one long piece, it can make a wide range of sizes too. Its production capacity is easily calculated by multiplying its width by the panel exit speed. Thin panels travel through

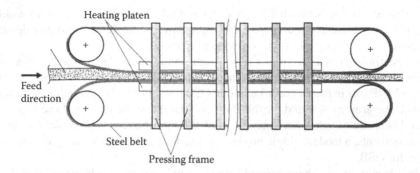

FIGURE 10.31   A diagram of the principles of a continuous press. (From Thoemen, H. and Humphrey, P.E. 1999. In *Proceedings of the Third European Panel Products Symposium*, The BioComposites Centre, University of Wales, Bangor.)

the press much faster (i.e., 80–90 m/min for 2 mm) than thick ones (i.e., 3–5 m/min for 38 mm), because there is less material to heat. In theory then, the production capacity of the press is the same regardless of panel thickness. This does not quite happen in practice. They are, however, more efficient than batch presses at making thin panels. A batch press must open, be loaded, and then close. This takes a finite amount of time. So when making thin panels, this opening and closing takes a significant proportion of the production time and overall capacity falls.

Another advantage caused by the fact that a continuous press remains "closed" the whole time, is that the panels produced have very close thickness tolerances of 0.3–1.8 mm depending on the board thickness. This in turn, reduces the amount of sanding required. Tremendous amounts of money can be saved by minimizing sanding losses.

The operating principle of a continuous press is that the mat is carried through the continuous press by an upper and a lower steel belt (see Figure 10.32). The temperature is transferred from the heating platens over rolling rods. One older system utilized an oil film instead of rolling rods which improved heat transfer but too much oil was lost per m³ of production to make it profitable. Nowadays all continuous presses use rolling rods between the heated press platens and steel belts to decrease the friction and reduce the wear of the steel belts.

Friction between the heating platens and the rolling rods can be reduced, and their subsequent wear, if additional extra hardened protection plates are placed between the heating platens and rolling rods. Another method is the direct hardening of the respective surface of each heating platen.

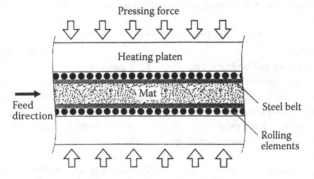

FIGURE 10.32   A diagram of the mechanism employed to minimize the friction between the stationary hot platens and the moving steel belts that carry the mat through the press. (From Thoemen, H. and Humphrey, P.E. 1999. In *Proceedings of the Third European Panel Products Symposium*, The BioComposites Centre, University of Wales, Bangor.)

Typical press widths are between 1.2 m and 3.0 m. Only a few plants exist with widths greater than 3.0 m. Pressure distribution, excessive vapor pressure, frame design, and mat de-aeration are the main reasons why wider presses are not made frequently.

There is a continuing trend for greater press lengths and speeds. The longest continuous presses for particleboard and MDF are in the 50 m range and for OSB in the 60 m range (the current record is 77 m for an MDF line in Brazil and 70 m for an OSB line in Canada). Press line speeds have been increased with the growing demand for thin MDF (<3 mm) and the ongoing lowering of minimum thickness of this product (<1.5 mm). Line speeds of 120 m/min are now possible. As a consequence of these developments, a modern single production line for particleboard can make 300 m^3/day and 2000 m^3/day for OSB.

The hydraulic systems are characterized by precise pressure control by proportional valve technology. The high line speeds demand fast control systems, which are realized by PLC, local numerical control systems, or by central microprocessor control.

For press gap control, specialized thickness metering systems are needed. Depending on the basic press design these instruments may have to withstand quite high working temperatures. The product thickness profile is fed back to automated control loops to achieve good accuracy.

A continuous presses the formed mat along its whole length under varying pressures and temperatures (Figure 10.33). The press length is subdivided in to different pressure and heating zones, each one consisting of a large number of sequential press frames. The number of pressure zones is determined by the total length of the press but every press always has at least one zone of high pressure followed by one zone of medium pressure. Every continuous press has a lengthwise and a crosswise pressure control. Some continuous presses are constructed with two big main pressure cylinders in combination with two pushback cylinders at the right and at the left side of each press frame. These cylinders push the heating platen downwards and are responsible for the precise lengthwise pressure control. Additional cylinders are seated below the lower heating platen pushing upward for an exact crosswise pressure control to obtain an excellent cross profile of the panel. Other continuous presses are supplied with a high number of pressure cylinders at the top of the upper heating platen. By pushing downward, the pressure is controlled lengthwise as well as crosswise in one vertical direction. With both solutions consistent board properties can be realized over the complete product width.

Furthermore, continuous presses have controllable heating zones down their length. This adds another variable for manufacturers to use to manipulate board properties, in particular the density profile, so that the product is more suited to a particular end-use, for example, flooring as opposed to furniture applications, all on the same press. Roll et al. (2001) presented some results from a continuous press that has a cooling zone at the end of the pressing section. It would appear that this helps reduce "blows" and, consequently, the mattress can enter the press with a higher moisture content which helps heat transfer through the mattress and speeds production.

The main advantages and disadvantages of continuous presses are summarized as:

Advantage	Disadvantage
Raw Materials: Reduced trimming and sanding losses.	High initial cost.
Production: Lower press factors compared to batch presses. Lower energy requirement per unit volume produced. Similar production capacity at different product thicknesses production capacity of batch presses is reduced when making thin boards. Thickness changes are easy to perform and quick.	*Steel belt:* Need for careful belt tracking control. Belt easily damaged by hard objects on/near mattress surface, so excellent metal detection in mattress required. Extensive wear of belt requires expensive replacement after 3–8 years of operation.
Properties: Better density profile control permitting a higher IB for a given density and reduced tool ware. Reduced precure and no asymmetry so there is scope for resin development.	Poor heat transfer from platen to mattress. Not so appropriate for pressing thick (>40 mm) boards.

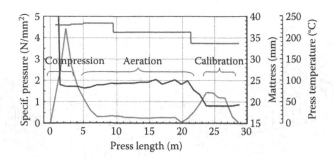

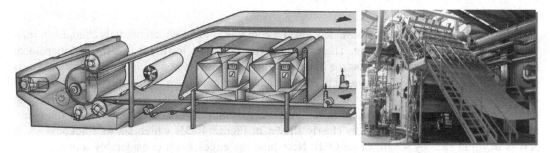

FIGURE 10.33 Change of parameters during continuous hot pressing. (After Thoemen et al. 2006. Pressing from a fundamental point of view: Actions to increase press performance. *Proceedings of 4th "Press Users Club" Seminar*, Shanghai, China, October 23–26, 2006.)

FIGURE 10.34 Calendar press with microwave preheater and simultaneous one face coating (detail with a modern calendar press). (Courtesy of Bison and Binos.)

The *calendaring presses* were the first systems able to make WBP continuously. They were introduced to the market in 1971. Due to the limited heat transfer and pressure capabilities the calendar presses are limited to thin particleboard and MDF (<4 mm). They often have microwave preheaters (Figure 10.34). Around hundred lines of this type are still operating. Limitations in capacity (300 m³/day) caused by the drum diameter (3–4 m) and length (<2.5 m) and of the density of the product limits development possibilities.

10.14 MATTRESS CONDITIONS DURING PRESSING

Prior to pressing a mattress should have the same temperature through out and have a uniform density (symmetrical for multilayer boards). The board's moisture content should also be consistent throughout unless a manufacturer is using the steam shock technique, more of which later. In addition, there will not be any vapor pressure or internal stresses present in the mattress before pressing. Once pressing starts, however, all these factors will begin to change at various rates in different parts of the mattress. Consequently, the geographical position of a wood particle will determine the conditions it is subjected to, which in turn will give rise to the board possessing inconsistent physical properties.

The minimum press time possible is determined by the time required to heat and cure the adhesive in the core layer. Heat is transferred in the mattress by radiation, conduction, and convection (water vapor movement). The latter transfer mechanism is predominant, particularly in the early stages of pressing (Bolton et al. 1989). The main reasons for this are that wood is a good insulator and vapor flow will occur readily as soon as a vapor pressure gradient is created (by surface heating). Note that the surface moisture does not have to turn to steam, which is water above its boiling point, in order to move through mattress since any vapor pressure gradient will induce some flow.

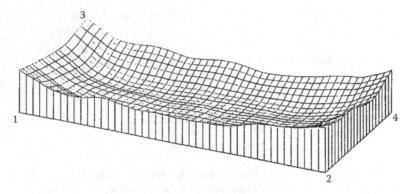

FIGURE 10.35 Observed thickness swelling across a daylight (4×2.5 m) of OSB. Note swelling at lowest point is approximately 10 and 24% at highest point.

Humphrey and Bolton (1989) have developed a computer model which predicts changes in mattress environment during pressing. Their model is unlike others in that it is a three-dimensional model; predicting conditions across the board's width and length and not just through its thickness like most models. Their model predicts that particles at the edges of a mattress will experience lower vapor pressures and temperatures during the press cycle. These two factors are likely to reduce particle stress relaxation and hence increase the internal stresses remaining in the board after pressing. The effect of this is clearly shown in Figure 10.35, which shows thickness swell across daylight (2.5 by 4.5 m) of an OSB. Note how the edges swell considerably more than the central region.

The way a board is pressed will determine its density profile, which is the change in density from one surface to the other. A board's density profile can significantly affect its strength properties and therefore its end-use.

When a board is first placed into a press its surfaces begin to warm. Since wood temperature is negatively correlated to strength then this surface heating will cause the surface layers to squash more readily than the core. As a result, even single layer commercially pressed boards will have higher-density surfaces. This is often seen as an advantage because a board with high-density surfaces will have a higher bending strength than a board which does not. In addition, the high density makes the surfaces scratch resistant and less prone to absorb paints and adhesives applied to the surface. However, particleboards are usually made to a set thickness and density, so if the surface layers are dense then it follows that the core layers must have a low density. It has already been mentioned that board physical properties are closely related to the density of the product, see Section 3.2.2. A low-density core will therefore equate to a low IB strength.

Studies have shown that this variation in density profile can be reduced by closing the press very slowly, for example, seven or more minutes (Moslemi 1974). Such long closing times inevitably lead to precure in the faces and loss of production. The alternative method would be to close the press extremely quickly, say within a few seconds, so that the surfaces do not have sufficient time to heat up and plasticize, but this would require very large and powerful hydraulic pumps. The relatively new technique of steam injection pressing (Geimer 1982) significantly reduces density variation and, at the same time, greatly reduces total press time. This is achieved by injecting steam through a perforated platen as the board is being pressed. The density variation is reduced by the fact that the board is more uniformly heated through out during compression. While the press cycle may be reduced because the core layer is heated more quickly the adhesive starts to cure sooner. Many people see this as steam injection's main advantage, but there are other benefits. Instead of producing particleboards with conventional properties in half the time, maybe this technique could be used to make much more stable boards using standard press time cycles. Much of the inherent instability of densified wood composites comes from particle densification reversal. When particleboard mattresses are pressed for long the stresses induced by compression are relaxed through wood creep; consequently, if stress relaxation is sufficient then spring back through stress reversal becomes much less important. The development of steam injection pressing will also allow the production of much thicker products, 100 mm or more. Such thicknesses cannot be made by conventional means because of the difficulty in heating the core layer.

Mattress moisture content and distribution will also affect density profile. Some manufacturers apply a fine spray of water to the mattress faces with the idea that during pressing this extra water will evaporate, migrate to the core and therefore heat the centre more quickly. This is known as the steam shock technique. There is evidence that this does indeed happen. The addition of water to the faces has a secondary effect; it plasticizes the surface layers so that surface density is increased and density profile variation is enlarged (Maloney 1993).

As a press closes stresses are induced in the wood particles. These stresses act in opposition to the closing of the press and so the level of stress is often termed as counter pressure. A simple press will operate by pumping hydraulic oil into the press rams until a certain pressure is reached. Board thickness is governed by metal stops. When such a press has "closed to stops" the stops will bear the load from the press ram. The stresses in the mattress will begin to fall off with time. Panel thickness can be controlled by using distance transducers and adjusting hydraulic pressure in the rams. Again, the hydraulic pressure will fall during the cycle. The level of stress remaining in the board once the press opens will be dependent on temperature, moisture content, particle geometry, density of raw material and final product, and total press time.

10.15 PROCESSING STEPS IMMEDIATELY AFTER PRESSING

All densified wood composites must be trimmed after pressing, this is not just to provide a straight edge, but also to remove, or reduce the amount of, edge effects (Figure 10.34). These edge effects are caused by the inherent temperature and pressure variations experienced by particles in the outer edges of a mattress. As a result, the outer portions tend to have physical properties inferior to the bulk of the board. The amount of "waste" removed as a proportion of board size is much smaller for large boards than for small ones. A similar reduction in waste tends to occur when nonstandard sizes are being cut from large production panels. This explains why, initially large single-daylight presses, and then continuous presses gained in popularity. Figure 10.36 shows the main processing steps that are found immediately after pressing.

10.15.1 BLISTER DETECTION

If the internal gas pressure in the mattress at the end of the press time exceeds the internal bond strength in the hot state (at 100°C in core), blisters (delaminations) occur. They tend to be concentrated in the middle because this is where gas pressure is highest. Blisters can be found using ultrasound transmitters and receivers distributed across the width of the production line immediately after the boards exit the press (Figure 10.37). The earlier this information is available, the easier it is to adapt production parameters. The press operator initially changes the press speed (lowering), then may try reducing the moisture content of the mattress at press entrance or improve the furnish bond by increasing adhesive content; all of which add to production costs. If no blisters are detected and the IB stays within the safety margins, then the operators usually try to gradually increase the line speed to maximize production capacity.

If the hot panels contain blister they are immediately eliminated from the flow. They may be used as packaging material or rechipped to particles with a hammermill and reused as furnish or as fuel.

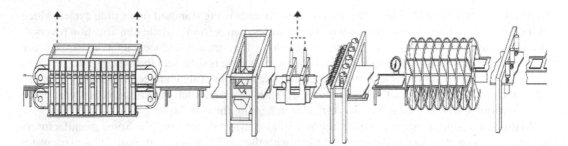

FIGURE 10.36 Primary board finishing after hot pressing: blister detection, trimming, cutting to length, thickness determination, weighing, and cooling. (Courtesy of GreCon.)

FIGURE 10.37 Blister detection in the hot panels after press. (Courtesy of GreCon.)

10.15.2 TRIMMING AND CUTTING TO LENGTH

An endless hot board exiting a continuous press is cut to length (normally no more than 6.5 m) to a so-called master panel with a diagonal or flying saw composed of two or more circular saws (Figure 10.38). The correlation between the press speed and the traversing circular saws is computer controlled. The minimum possible master panel length is dependent on the current production speed and the number of saw units available for cutting. For thin panels, which are pressed at high speeds, all available circular saws operate. Not only does the flying saw have operate quickly but it must also be precise achieving the desired panel length within ±2 mm, a squareness of ±1.5 mm/m, and a straightness ±1.5 mm/m.

FIGURE 10.38 Diagonal or flying saw for cutting master panels after continuous hot pressing. (Courtesy of Dieffenbacher.)

10.15.3 WEIGHING AND PANEL THICKNESS MEASUREMENT

Each individual trimmed panel is controlled in terms of thickness and weight. Usually, the thickness is measured continuously at three or more traces over the whole panel using roller pair gauges (Figure 10.39). Between measurements the roller pairs contact each other and calibrate themselves in order to assure a high measurement precision (0.02 mm). For high line speeds (>100 m/min) this type of calibration is not possible and a movable C-frame like that for the ultrasound equipment permits external calibration of the equipment.

Normally, the weight of each panel is measured by a scale installed under a separate belt conveyer. The weight cells need 2 s to obtain an accurate value. For new production lines with high line speeds classic scales do not have enough time to accurately weigh the panel. Also the accuracy of this weighing system is not very good for thin panels because the weight of the panel is low relative to the weight of the scale itself. Dimensional data from the saws and thickness gauges and the weight of each panel are used to calculate panel density. This can also be achieved using an X-ray system, which directly measures the density and can also detect foreign matters, furnish balls and other defects, but it is relatively expensive.

10.15.4 COOLING

Particleboards bonded with UF or MUF must be cooled after the have been taken out of the press. If the boards were stacked together immediately, the residual heat would cause thermal degradation of the glue. On exiting the press the boards are placed in a star cooler, which looks like a paddle boat wheel (Figure 10.40). The cooler is usually large enough to accommodate a sufficient number of boards so that as boards leave the cooler their surface temperature will have fallen to about 40°C. From the graph shown in Figure 10.41, it can be seen that the internal temperatures are likely to be much higher. Depending on the stack size and season, the temperature in the centre can be more than 55°C after 3 days of storage.

Phenolic bonded boards can be hot stacked. In fact, hot stacking is recommended for PF bonded boards as their properties are usually seen to improve by this process. It is said that the improvement is due to increased cure of the adhesive, which seems very plausible. However, some of the improvement may also be due to continued internal stress relaxation permitted by the high temperatures in the stack.

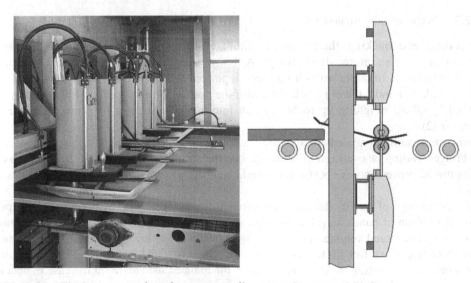

FIGURE 10.39 Thickness gauge based on contact roller pairs. (Courtesy of GreCon.)

FIGURE 10.40   A star cooler. (Photo by Mark Irle.)

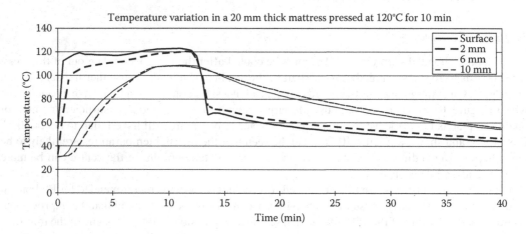

FIGURE 10.41   Temperature change in a chipboard mattress during and after pressing.

10.15.5   INTERMEDIARY STORAGE

After cooling and stacking, the panels are stabilized through additional cooling, curing, and rehumidification in an intermediary storage. Although transportation based on fork lift trucks is still common place, in larger mills it has been replaced with fully automated systems (robots), which are able to handle the big stack sizes with the master formats (press width × master board length × 1.5–4.0 m, weighing up to 50 t) to an intermediary storage of some 0.5 ha or more (Figure 10.42).

There are several automated storage systems. One type of such system in based on stack care crane bridges, rolling on overhead rails installed on the building frame (Figure 10.43). Both cross- and lengthwise movements are performed simultaneously, which permits high transfer rates of up to 3 m/s.

Some advantages of such modern and full automatic intermediary storage system are: the perfect horizontal position of panels in piles, complete protection of edges and surfaces, continuous and precise records of stored volume, minimal energy consumption, low pollution and dust generated relative to the traditional fork lift trucks, and no supervision needed by operators. The disadvantages are high investment costs, high skills required for the maintenance team, in the case of accidental stop, complete downtime for the whole line.

FIGURE 10.42 Full automatic intermediary storage system based on chariots. (Courtesy of Dieffenbacher.)

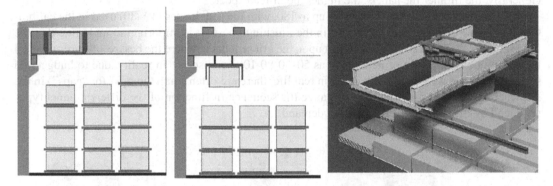

FIGURE 10.43 Full automatic intermediary crane based storage system. (Courtesy of Metso Panelboard.)

10.15.6 SANDING AND CUT-TO-SIZE

There are three reasons why the panels are sanded after hot pressing and intermediary storage:

1. To remove precured surfaces that are often weak, fibrous, and porous
2. To calibrate panel thickness because the panels spring back on release from the press and this varies from panel to panel and also within a panel
3. To improve the surface quality for subsequent value-added applications such as laminating, painting, varnishing, and so on

The sanding allowances depend mainly on panel thickness and press type: it is normally lower for thin panels and higher for old multi-daylight presses. For thin panels (<6 mm), which require a high tolerance, a 0.3–0.5 mm sanding allowance for both sides is common. The thicker the panel is, the higher the requested sanding allowance (0.8–1.2 mm). Calculated for the whole panel production (PB and MDF) approximately 3% of the line output is processed to sanding dust. Generally, this dust results from the high-densified (600–1000 kg/m³) and high-resinated (i.e., 10–12% UF resin) face layers of the boards. The sanding dust is air conveyed to a specialized bunker together with trimmings and cut to size off cuts and injected in to the upper part of the burning chamber of the power plant, often generating about one-third of the thermal energy requirement of the factory.

A modern sanding line typically consists of 8 to 10 sanding units (sometimes 12 and rarely up to 14), arranged in opposed pairs, each equipped with the same type of sanding heads and abrasive belts. Whenever possible, it is a good rule to separate the calibration phase (in which up to 75% of the over thickness is removed) from the finishing phase. It is quite common to have 4 contact rolls (or drums) for calibration and 4–6 sanding shoes (or platens) for finishing. A combined roll-platen sanding head can also be used in between to provide a certain degree of flexibility to the line, enabling it to calibrate more, or finish more, depending upon the production needs (Figure 10.44).

Sometimes different types of sanding machines are also added to the line to provide an even smoother surface, like brushing machines or high-speed cross-belt sanders (Figure 10.45). While the first ones use a couple of ScotchBrite® rolls, the latter use narrow conventional abrasive belts rotating perpendicularly to the feeding speed and thus "cross" the board.

In recent years, the use of electronic systems to control the sanding process has increased significantly. These systems combine mechanical components, sensors, and software (with graphical interface, machine status and recipes), to precisely position each working unit, thus reducing the workload of the operators and helping to improve the consistency of production quality.

Panel stacks after 2–3 days intermediary storage have a temperature of below 40°C and their surfaces are firm and the resin is completely cured. The panel conveyers and rollers feed the panels at 40 to 100 m/min depending on the sanding allowance, thickness, and sanding quality target. Generally, the thinner the panels, the higher the in-feed speed.

Abrasive belt grits can range from 40 up to 180, with the rough ones (40–80) used in the calibration phase and the finer ones (100–180) in the finishing phase. It is important to underline that the final sanding quality is strictly dependent upon the grit sequence, where the best results are achieved by a continuous sequence of belts, such as 50-60-80-100- and so on. In reality, due to budget and space limitations which normally occur in real life, there are often small "jumps" (or "gaps") in the sequence. Several trials are used to optimize the sequence in function of the different panel types produced by the plant, as well as market demand.

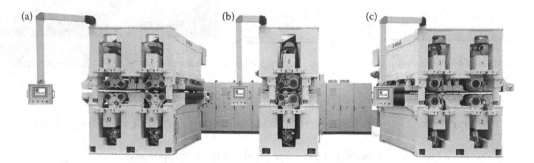

FIGURE 10.44 A 10-unit sanding line, showing four contact rolls (a) two roll-platens (b) and four platen heads (c). (Courtesy of IMEAS.)

FIGURE 10.45 A modern cross-belt sander. (Courtesy of IMEAS.)

To assure a proper processing of the whole surface, the sanding paper width slightly exceeds that of the panel, typically by 50 mm (or 2 inches), which compensates for the natural oscillation of the belt during its rotation. To optimize sanding costs, modern sanding machines can manage belts of different widths. The speed of the sanding paper is typically in the range of 25–40 m/s, higher for calibration rolls and slower for the platen heads.

The wear of the sanding paper depends on the resin type and amount, the furnish quality (especially recycled wood proportion), sanding temperature and dust extraction efficiency. The power increase of the sander motors together with surface quality changes, for example, discoloration, is the first sign of paper wear, which occurs in hours. For coating with melamine paper, a grit of 120 is sufficient, for lacquering or direct printing 150 grit or even higher is needed. The high in-feed speed makes it impossible to control by eye alone. Typically defects after sanding are dark spots (caused by glue balls and foreign bodies), light spots (caused by dust agglomeration and nonresinated furnish), longitudinal traces (caused by broken grits or agglomerations on the paper), transverse traces (caused by grit delamination from the paper or deterioration of the paper joint area). Major sanding defects are differences in thickness in the cross sections (caused by cylinder misalignment) and along the panels (caused by the vibration of cylinders due to machine construction or foundations).

The thickness of each panel should be measured online at two points or more before and after sanding in order to allow a continuous machine control and setting. Only big dark spots or major defects can be recognized by the operator. Modern online self-learning video control of both surfaces simultaneously can assure a continuous and reliable check. An important aspect for process security is the installation of spark detection as sparks may be generated during sanding by the overheating of the grit or on contact with metals. The mixture of dry fine dust and air in the exhausting pipes or air filter bags increases the explosion risks. The dust filtering from the exhaust air remains an expensive and time-intensive maintenance task.

Main producers of sanding equipment for raw panels are Steinemann, Imeas, and Binos.

Cut-to-size equipment consists of de-stacking units, tables, conveyers and aligning units, stationary and movable saws, racks for intermediary storage of small size piles, pilling and packaging units (Figure 10.46). The theoretical minimum size could be an A4 paper, but in practice, are multiples of doors, furniture size (table, walls, or fronts), and flooring elements. The height of a mini pile which can be processed in one cut does not normally exceed 20 cm and requires a pair of circular saws (under and over the table) so that relatively thin blades can be used to reduce material loss.

The amount of sawdust including the panel margins which results from cutting the master panels to size should not exceed 1–3%. To reach this target, an optimization of the master panel size based on the order of small parts is necessary before production. Small-sized parts obtain a better price, access an important market segment, and allow a flexible use of stored panels. It needs a

FIGURE 10.46 Cut to size center. (Courtesy of MDF Hallein.)

well-prepared production plan, a good organization of the storage areas, and many intermediary packet handling zones. The capacity of such a cut-to-size center could cover 30% to 50% of the production. Main producers of such systems are Anton, Giben, and Schelling.

10.16 MANUFACTURE OF ORIENTED STRAND BOARD: A SHORT OVERVIEW

10.16.1 Introduction

Oriented Strand Boards (OSB) are typically three-layered panels. The strands in the outer layers are aligned parallel to the long edge of the board and to the production line, whereas the strands in the core layer are often smaller and aligned at right angles to the strands in the face layers. The alignment in the core layer is less precise than the outer layers, that is, the orientation of strands is more random. The layering gives OSB panels anisotropic mechanical properties in that the bending strength parallel to the long edge of the board is higher than that parallel to the short edge. Some manufacturers print arrows on the faces to help ensure that the panels are installed correctly.

Placing the largest strands on the surface increases the bending properties of the panels relative to their densities. It also gives OSB a distinctive and easily recognizable "look." The large surface strands can make the panel surface rough in comparison with other panels. Although sanding is possible, only a small percentage of production is sanded as it significantly changes the appearance of the panel.

OSB was first developed in USA based on patents dating from 1935 and later wood panels based on "veneer strips crosswise oriented." The first pilot plant in USA started experimental production in 1963 and the first commercial plant in Europe started in 1978. The growth in demand for OSB is second only to MDF. In only 35 years, the market acceptance is complete and around 100 production lines with a peak in world production capacity of over 40 million m³/year in 2008. The financial crisis caused a significant reduction in the number of production plants, especially in North America. At the end of 2010, the North American production capacity was reported to be just less than 25 million m³/year and just over 5 million m³/year in the rest of the world (Anon. 2011).

About 75% of OSB is used in construction, 20% for packaging and the rest for decorative and various uses. The European standards EN 300 and EN 13986 classify OSB into four classes:

OSB/1 is for general purposes in dry conditions (service class 1)
OSB/2 is load-bearing in dry conditions (service class 1)
OSB/3 is also for load-bearing but in humid conditions (service class 2)
OSB/4 is for heavy duty construction in humid conditions (service class 2)

OSB is available in thicknesses ranging from 9 to 32 mm.

10.17 MANUFACTURE OF OSB

The strands used to make OSB are very large compared to the particles used to make chipboard. Consequently, the strands are invariably cut from fresh wood. Other forms of wood, especially recovered wood, are less suitable because they are too small to efficiently produce large strands and are too dry. Spruces and pines in Europe and aspen in North America in the form of thinning and tree tops are the most common sources of wood.

A typical production line is shown in Figure 10.47. Production starts with drum debarking followed by stranding with drum flakers (Figure 10.8) or disc flakers (Figure 10.9). The flakers will be set to cut strands at a particular thickness (0.3–0.8 mm) and length (50–150 mm). Unfortunately, strand width cannot be controlled because wood splits easily down the grain. In fact, subsequent processing steps are designed to minimize further strand splitting. Obviously, cutting long, thick strands will tend to produce more wide strands. However, the variation is enormous with some strands having widths of less than 1 mm while others could be several tens of millimeters.

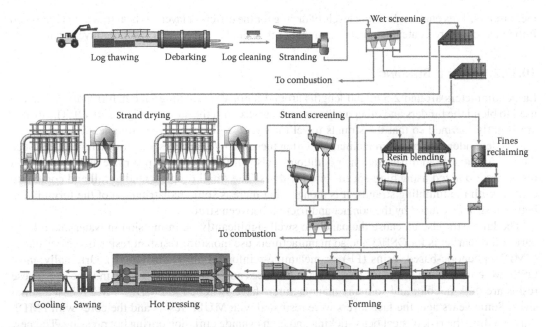

Wet screening

Log thawing   Debarking   Log cleaning   Stranding

To combustion

Strand drying   Strand screening

Fines reclaiming

Resin blending

To combustion

Cooling   Sawing   Hot pressing   Forming

FIGURE 10.47   An OSB production line. (Courtesy of Metso Panelboard.)

10.17.1   STRAND PREPARATION

Strand grading is often achieved in large rotating drums (Figure 10.48) that rotate at low speeds in order to avoid further strand breakage. Oscillating screens are also sometimes used for strand sorting. Alternatively, a disc screen based on interlocked rotating discs can separate the wet strands into different fractions (Figure 10.48). Such equipment minimizes additional strand breakage compared to sieve-type screens, does not require air cleaning, and requires lower overall energy input but it is relatively expensive. As mentioned above, strand sorting is used to ensure that the strands of the appropriate sizes are sent through the core and surface preparation systems.

The drying technologies used are the same as those described for chipboard manufacture. Examples of both 3-pass (Figure 10.14) and single-pass (Figure 10.15) dryers can be found in existing factories. Some manufacturers use two separate single-pass dryers; one for the core and the other for

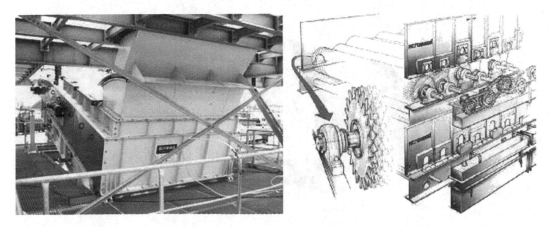

FIGURE 10.48   Disc screens for the cleaning of wet strands. (Courtesy of PAL and Acrowood.)

face strands. This enables different levels of drying for the different layers as heat transfer is improved if the face moisture content is greater than that of the core layer, but requires more investment.

10.17.2 STRAND BLENDING

Large (diameters around 2.5 m and lengths around 8 m), slow rotating (about 100 rpm) drums are used to blend the binders and other additives such as wax with the strands (Figure 10.49). The drums are slightly inclined so that the strands travel through the blender as they tumble. Their retention time is dependent on the drum inclination. Often there are two blenders, one for each layer, because the formulation of additives and their addition levels is different between surface and core. Liquid resins are atomized using high-speed spinning discs. The small droplets tend to fall and come into contact with the tumbling strands. It is thought that there is some redistribution of the formulation between strands caused by the contact and friction between strands.

The large strands can cause the panels to swell significantly on immersion in water and a large part of the market is for OSB/3 and so manufacturers use moisture-resistant resins like isocyanates (PMDI), phenolic-based resins (PF), or melamine-reinforced UF resins (MUF). Originally, most OSBs were made with PF resins (Novalacs in the faces and fast reacting resoles in the core). These resins are dark in color and so many manufacturers have switched to PMDI- or MUF-based adhesives. Some years ago, the face layers were resinated with MUPF resins and the cores with PMDI, thus avoiding the risk of steel belts sticking and high cyanide emission during hot pressing. The new trend in Europe is to glue the faces and core with 3–6% and 4–10% PMDI, respectively (depending on OSB type). The advantages are: very low formaldehyde emission, good weathering resistance, and high line speeds (4–8 s/mm). Whereas the disadvantages are: high costs, the need for permanent application of release agent spraying by nozzles or roller application on the steel belt or mat surface, and controlled exhaustion in the press and cooling star area.

10.17.3 MATTRESS FORMING

The degree of orientation and proportion of long strands in the face layers has a marked effect on the level of bending properties in the two board directions. If the surface and core layers have a high degree of orientation, then the panel properties are too anisotropic and delamination can occur between the layers. Experience has shown that the degree of orientation in the surfaces should be around 80% and much less for the core. The degree of orientation is adjusted by changing the distance between the bed of orientating discs and the mattress top; the smaller the distance the greater the degree of orientation because the strands have less chance to reorient themselves as they fall onto the mattress.

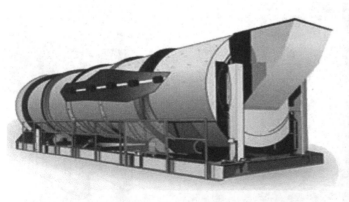

FIGURE 10.49   Drum resination of strands using atomizing discs. (Courtesy of Metso Panelboard.)

FIGURE 10.50   A bed of rotating discs for the face layers (b) and the core layer is formed by a rotating star (a). (Photos by Mark Irle.)

The distribution of strand sizes is achieved by adjusting the space between the discs. The discs are close together where the surface layer meets the core and the gap between the discs is gradually increased to the outer face (Figure 10.50). In the core the strands are laid across the machine direction, but, the orientation is less well defined and in some cases random. The strands are fed in to pockets that rotate and because the strands are long they tend to orientate themselves in the "V" of the pocket.

10.17.4   Mattress Leveling and Pressing

The large strands are quite resistant to compression and so cold prepresses are not very effective. The mattresses are more easily fed in to the hot press if they are first leveled which is achieved with a heavy steel drum (Figure 10.51).

FIGURE 10.51   An OSB leveling roller as prepress unit and trimming/metal detection in the mattress before pressing. (Courtesy of Dieffenbacher.)

Multi-daylight presses are still very common in the OSB industry although examples of both single-daylight and continuous press technology can be found. Some mills use a screen in the hot press to make nonslip surfaces on their panels.

10.18   MANUFACTURE OF MEDIUM-DENSITY FIBERBOARD: A SHORT OVERVIEW

10.18.1   INTRODUCTION

Medium-Density Fiberboard (MDF) was first developed in USA from hardboard manufacture. Hardboard is made via a wet process that is similar to paper manufacture and therefore produces a lot of polluted water that requires treatment before disposal. Semi-dry processes were developed in the 1950s and this led to a fully dry-process method. The first dry-process MDF factory was built in Deposit, USA, in 1965 and the first MDF factory in Europe is thought to be that built in the former German Democratic Republic at Ribnitz-Damgarten in 1973 (Williams 1995).

Since these small beginnings there has been exponential growth in the installed production capacity of MDF across the world (see Figure 10.52). By the early 1990s, it was being hailed as a "... one of the most exciting new products to come along over the last 50 years" (Anon 1995). There was tremendous optimism in the MDF industry in the early 1990s, which was fuelled by the rapid expansion of both the market and supply. The optimism turned out to be founded and growth continues at an exponential rate.

MDF is an engineered wood product composed of fine lignocellulosic fibers, combined with a synthetic resin and joined together under heat and pressure to form panels. The most commonly used lignocellulosic fibers are wood, but other plant fibers can be used, for example, bagasse and cereal straws.

MDF is available in a range of thicknesses from 2 to 100 mm and a very wide range of panel sizes. The density of these panels varies from about 500 to 900 kg/m^3; panel density tends to increase as panel thickness decreases. Panels have smooth, high-density faces and are pink brown to dark brown in color, unless a dye has been added during manufacture.

The European definition of MDF is provided by EN 316:2009. The main grades of MDF are: General purpose, dry (MDF), General purpose, humid (MDF.H), Load-bearing, dry (MDF.LA), Load-bearing, humid (MDF.HLS). The minimum requirements for these panels are specified in EN 622–5:2009.

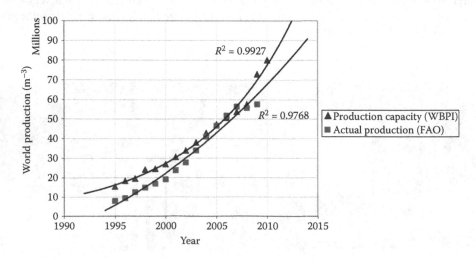

FIGURE 10.52   The world-wide installed production capacity for MDF. (Data from FAO and Wood Based Panels International.)

10.19   MANUFACTURE OF MDF

MDF can be produced from a wide range of lignocellulosic fibers including agrofibers (Suchsland and Woodson 1986) and recycled wood (Anon. 1995). An overview of a typical MDF manufacturing is shown in Figure 10.53. MDF fibers are normally made by using a thermomechanical pulping (TMP) process. This process uses the combined action of heat and mechanical energy to break the bonds between the cells that make up wood. Wood cells are joined by a region called the middle lamella, which is rich in lignin. Lignin is an amorphous polymer that can adsorb small quantities of water, and so, its softening temperature is moisture content dependent. The high temperatures (170–195°C) and humidities (60–120%) used in the TMP process, therefore, cause significant reductions in the strength of the middle lamella region and this increases the likelihood of failure occurring in the middle lamella when mechanical energy is applied during the refining process.

The main process steps for the production of MDF are described in more detail below:

10.19.1   CHIPPING

Wood is chipped in to square particles with sides of approximately 25 and 5 mm-thick (Standard: TAPPI UM 21-SCAN CM 40:88) using drum chipper (Figure 10.8), a disc chipper (Figure 10.54) or as residues from modern saw-mills using a chipper canter. Many other sizes are used, including sawdust, depending on the source of the raw material. Whatever the size, for the best digestion/cooking results the chip geometry should be homogenous.

The particles are screened to remove under (<2 mm) and oversized (>50 mm) particles.

10.19.2   CHIP STORAGE

Depending on climate region and availability of raw materials there are many ways to buffer the amount of chips from 2 weeks to 2 months of production. The easiest and cheapest option is to store the chips outdoors on platforms, but, this brings the potential of further "contamination" with soil, stones and foreign mater, freezing, or bio-degradation. Another is to use, round steel or concrete

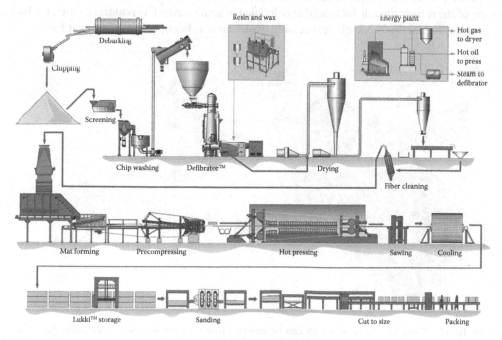

FIGURE 10.53   The main process stations in an MDF production line. (Courtesy of Metso Panelboard.)

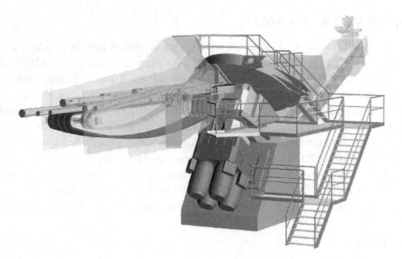

FIGURE 10.54 A horizontally fed disc chipper. (Courtesy of Andritz.)

silos with or without insulation having a volume between 5,000 to 30,000 m³ and diameters of 25 to 42 m (Figure 10.55). The advantages compared to the outdoor platforms are: reliable and uniform line feeding (based on screw reclaimer discharge unit), first in—first out discharge, lower moisture content because the stock is protected from rain and snow and wind blow from the stock is avoided.

A linear storage system can be closed or open. The closed type is for volumes up to 50,000 m³ (i.e., bark as fuel for energy generation) and open type for volumes up to 200,000 m³. The chips are reclaimed by a conical screw of 10 to 18 m. Also a boom type stacker with crane-type belt reclaimer and reversible conveyer for in and out feed typical for the paper industry could be used for high-capacity plants.

10.19.3 WASHING

Chip washing is needed to remove bark, soil, sand, and other abrasive contaminants. In this way the life time of the refiner discs is increased and doubled in some cases. Current trends are also looking into dry cleaning methods because processing waste water is becoming a larger cost factor.

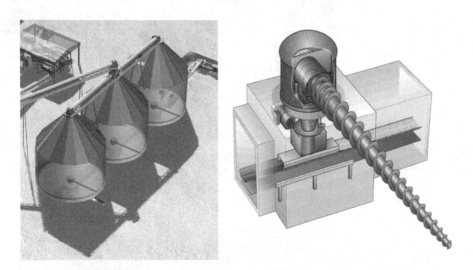

FIGURE 10.55 Chip storage silos which can be emptied using a rotating screw reclaimer (for silos up to diameter 25 m). (Courtesy of Andritz.)

FIGURE 10.56 Integrated chips washing unit (left) before preheating, steaming and refining equipment. (Courtesy of Andritz.)

The water can be heated to better clean species with high resin or gum content and to defrost chips stored outdoors in winter. Additionally, this protects conveyer, screws, and dryers and energy is saved. The water is kept in a closed loop and cleaned in the same circuit as the water emanating from the chip plug screw. The surface wetting caused by washing is thought to improve steaming and refining, but also increases the steam demand in the presteaming bin.

Whereas the chip washers are very similar to each other, the process water cleaning methods are different. The tank system (Figure 10.56) is more often used. For very sandy feed stock a hydro cyclone can be used to separate the sand from the chips before being dewatered.

10.19.4 PREHEATING

One decade ago preheating was an optional stage where the washed chips are heated to 40–60°C at atmospheric pressure in a surge bin (Figure 10.57b). Today this step is seen as being essential as it softens the particles so that a better plug is formed in the conical screw feeder (Figure 10.57c), which is a device that allows the continuous compression (2:1 rate) of chips into the main steaming tube (Figure 10.57a), which is also known as a preheater or digester, while still maintaining a high pressure in the tube.

The plug screw squeezes out some of the free water in the chips. The volume removed can be some m^3/h depending on line capacities. This water and a part of steam condensates must be treated as it has a high biochemical oxygen demand (BOD) and a pH between 5.5 and 6.5. This has driven the demand for plants that can operate at lower feed moisture contents, necessitating improved screw designs and dewatering casings.

Traditionally, waste water is precleaned and discharged into a sewage plant. The increasing costs of water, in particular waste water has resulted in different cleaning methods. Mechanical cleaning with filtering and reverse osmosis (ESMIL) or evaporation plants (Andritz, Schrader). Modern systems have the advantage that the waste water can be easily used as "make up" water (demineralized) for the boiler system with limited conditioning (pH adjustment).

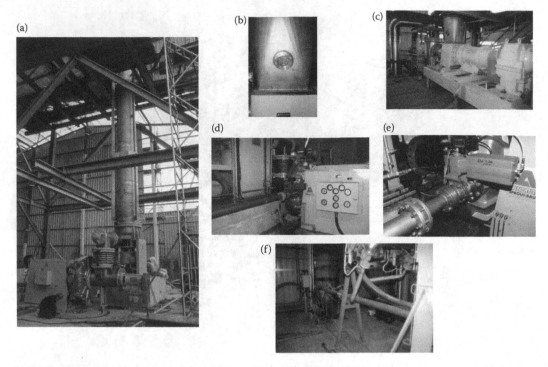

FIGURE 10.57 The main parts of a defibrator. (Photos by Mark Irle.)

10.19.5 STEAMING

The compressed and squeezed chips are heated with saturated steam at 6 to 10 bar, which creates an internal temperature of 175–195°C for 3 to 7 min depending of the digester size (5–18 m³), wood species, chip size, and required fiber quality. The retention time of the chips in the preheater influences fiber color and quality and is determined by the flow rate (screw feeder speed). The digester level is controlled by a radio-active measuring device. The discharge is accomplished by a ribbon screw unit (Figure 10.58).

The steam-heated digester is controlled by the level and the adjusted steam pressure. Different wood species need different settings to gain best results. At the moment efforts are being made to reduce the moisture content after the digester prior to the refiner as this would reduce the thermal energy demand at the dryer.

The chips are discharged from the digester via a discharge screw to a fast-rotating ribbon feeder. The ribbon feeder is an integral part of the refiner. The ribbon feeder rotates rapidly and injects the chips into the refiner; to increase the feeding efficiency the screw is designed with a double thread (Figure 10.58). The ribbon is fixed via bars to the screw shaft and so leaving an open space between ribbon and shaft to guide the surplus steam generated in the refiner back to the digester via a steam balancing pipe.

10.19.6 REFINING

The refiner converts steamed chips into fiber bundles. The refiner housing is pressurized with saturated steam (8–10 bar). Most refiners have two discs (diameter 44–72 inch), one stationary and another that rotate at about 1500 rpm by powerful motors (typically 5–12 MW depending on the diameter of the discs, see Figure 10.57). Wood chips are fed into the gap (about 1 mm) between two discs via a hole in the middle of the stationary disc.

The surfaces of the discs are equipped with exchangeable refining segments which have special grooves and dams whose pattern has a great influence on the efficiency of the refining process and

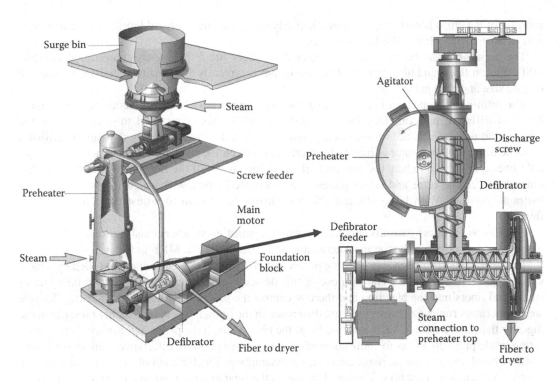

FIGURE 10.58 A typical pressurized disc refiner-based system with preheating unit and digester tube suitable for the production of thermo-mechanical pulp (TMP). (Courtesy of Metso Panelboard.)

the quality of the final fiber. At the center of the disc the breaker bar pattern is coarse and at the periphery of the disc the bars are much finer (Figure 10.59). Consequently, as the wood is driven across the radius of the disc by centrifugal forces, it is gradually broken down into its constituent fibers and fiber bundles. Pattern design depends on wood species and chip size. The refiner housing must be opened to change the refiner segments as well as servicing and maintenance reasons. The housing can be a cap design (horizontally split casing) or with a swing-door (Figure 10.58). Cap designs have the disadvantage that access to the housing is limited and additionally segment holders are required to fix the refining plates in the machine. Modern machines are designed with a hinged door allowing swift access to the machine.

The heart of the refiner is the bearing assembly consisting of the radial and axial thrust bearings. The axial thrust resulting from the pressure in the system and the steam pressure generated by the

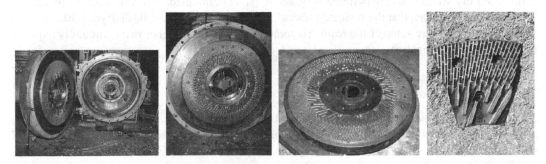

FIGURE 10.59 Photographs of the discs found inside of a typical pressurized refiner used to make MDF fiber. (Photos by Mark Irle.)

refining of the chips is best supported by a hydrodynamic (permanently oil lubricated) bearing system to assure a long life of the bearing assembly.

The adjustable gap between the two discs determines the refiner energy consumption (150–400 kWh/ton fiber) and fiber quality. The throughput is typically 15–70 t/h depending on disc and engine size and chip geometry.

The refining plate pattern is the key to a good refining result. All manufacturers carry a large range of different patterns. Despite the efforts to find a computer model to design plates, the selection at the end is based on experience and onsite testing. For extremely dry, small furnishes or some annual plants water is injected directly in between the refining plates to achieve acceptable fiber quality and reduce the carbon built up in the groves of the refining segments. The distance between the rotor and stator plates can be adjusted either with an electro-mechanic or a hydraulic system. With large plants (>40 t/h) a hydraulic system to adjust the refining gap is favored.

Double-disc refiners (two discs counter rotating against each other) exist; however, the single disc refiner has proven to be the most economic and efficient for the MDF industry.

The fibers exit via a discharge opening positioned in the refiner housing and controlled by a blow valve (Figure 10.57d and e). The purpose of this device is to continuously adjust the flow rate of steam and fibers into the blowline and thereby control the pressure in the refiner housing. To save steam operators run at a negative pressure difference in the refiner, this means higher pressure in the digester than in the refiner. Such a setting helps the fiber through the plates and is particularly interesting while producing fiber for flooring products. A positive pressure difference reduces the shives content, but does in return increase the energy consumption. The function of the blow valve is very simple: constriction of the flow. The problem lays in the wear and the tendency to stick when left for a long time in one position.

A diverter valve (Figure 10.57f) after the blow valve can divert the fiber flow to the start-up cyclone or to the blowline. The start-up cyclone guarantees a safe start-up and shut-down of the system (under steam pressure). Only if the fiber quality and resination are satisfactory and the process parameters constant (dryer), may the valve be switched to production (blowline).

10.19.7 RESINATING (BLOWLINE)

The fibers are discharged from the refiner down a pressurized blow line, a pipe of 80 to 120 mm diameter, which conveys the hot, wet fibers, and steam to the dryer. Blending of the fiber in the blowline under high pressure facilitates a uniform resin distribution due to the steam expansion, which cause high turbulent flow, rapid accelerations (that separate fibers) to high velocities of up to 340 m/s (the speed of sound!). The adhesive, usually a formaldehyde-based resin (UF), is injected at high pressure (12–14 bars) in to the blowline with other additives, for example, water, hardener, repellents, colors, and so on. The amount of adhesive added will vary between 8 and 15% (resin solids/oven dry wood basis), depending on grade of panel being made. Disadvantage of this blowline blending is the fact that the resinated fibers have to pass through the flash dryer (110...140°C) which tends to precure some of the resin. To compensate for this a higher resin amount is required (compared to dry resination).

10.19.8 DRY BLENDING

A new approach to fiber blending has two stages. In the 1st stage the resin is applied via the blowline at an addition level of only 3–5%. The wax is applied as usual via refiner discharge screw. The 2nd resination stage takes place (like 30 years ago) in a separate blender (Figure 10.60) after fiber drying. Resin is applied by air-nozzles, mounted on a ring in front of the open in-feed (negative air pressure because of exhausting funs) of the drum blender. The internal paddles generate a turbulent

FIGURE 10.60   Dry fiber blender with open in-feed and spraying nozzles ring. (Courtesy of IMAL.)

fiber flow ensuring effective blending. The blender and out-feed pipe are water cooled to avoid resin and fiber deposits. After 2nd dry fiber resination a gentle drying to the target moisture content is recommended by IMAL. It is claimed that resin savings of up to 35% compared to conventional blowline are possible.

10.19.9   DRYING

The wet resinated fibers are blown through a flash tube dryer at around 30 m/s. The dryer is often 1 to 3 m in diameter, over 100 m long. Cyclones (3–5 m diameter) at the end of the dryer separate the dried fiber from the steam. The moisture content of the fibers after drying is approximately 8–12%. The flash dryers are often heated directly with cooled gases (<200°C) from the energy plant, in order to avoid an extended precurring of the resin. They are designed as one stage dryer (longer and less temperature control) or two stages dryers, which requires two cyclones, two fans (600 and 300 kWh), more energy, but improved drying and lower precurring (Figure 10.61). The moisture content of fibers at the end of 1st stage is around 40% and the gas temperature in 2nd stage is low (<100°C). Waste air from 2nd stage can be recirculated to 1st stage and mixed with fresh hot gases due its lower humidity.

10.19.10   FORMING

The dry fiber is pneumatically conveyed to mat formers via a system of classifiers (sifter) and filters (Figure 10.62); these latter devices are designed to remove any fiber clumps. The mat formers are designed to distribute an even layer of fibers (no layering) onto a continuously moving belt. The belt speed varies with the panel thickness being made. The fiber mat forming requires a dosing bin with integrated weight scale, and a spreader system. Due to the fiber consistency, handling and spreading are different to that for particleboards. Only one forming head is used to spread the homogenous fiber mattress. The dosing and metering bin has to assure a constant fiber flow to avoid compressing at the spreader head. A back-rake conveyor built on the top of the bin keeps the fibers at a constant height when they reach the doffing rollers.

A typical mat for an 18 mm-thick MDF is 680 mm high and has a bulk density of 23 kg/m^3.

10.19.11   PREPRESSING

The mat is then passed to a continuous prepress. Here the mat is squashed to reduce its air content and to increase its density. This reduces the time required for hot pressing and avoids fiber

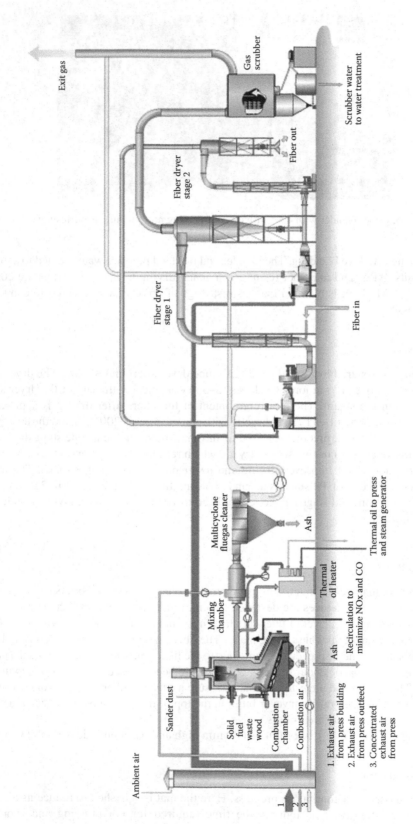

FIGURE 10.61 A two stage dryer with waste air recirculation. (Courtesy of Metso Panelboards.)

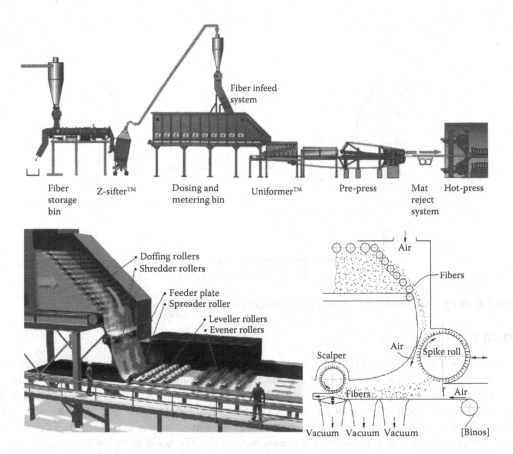

FIGURE 10.62 Fiber weighing, sifting, mat forming (mechanical) and prepressing until continuous press in-feed and air forming with spike roll. (Courtesy of Metso Panelboards and Binos.)

dislocations at the press in feed. Before prepressing a 38 m-thick MDF has a mattress height of about 1 m and after the mat height is around 350 mm. Mattresses over 200 mm height (>18 mm panels) are often preheated.

10.19.12 HOT PRESSING

The hot press applies a combination of heat (180–210°C) and pressure (0.5–5.0 MPa) to consolidate the mat and convert it to MDF. The pressing stage is not a simple one and can be used to manipulate panel properties by altering the panel's density profile, which is the variation in density through the thickness of the board (Figure 10.63).

Immediately after pressing the panels are stood on their edges to expose their faces to ambient air to facilitate rapid cooling.

Finishing entails trimming, sanding, and cut to size. Various added-value steps such as laminating, profiling, and painting may be applied.

10.20 POLLUTION CONTROL

As with all industrial processes, MDF production generates wastes in the form of solids (dust and ash), liquids (water), and gas (containing volatile organic compounds—VOC).

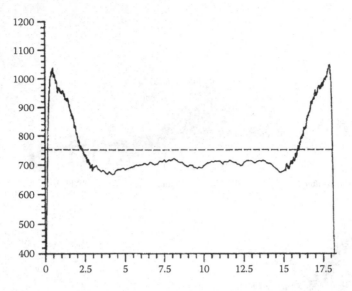

FIGURE 10.63   A typical density profile for an 18 mm-thick MDF panel.

10.20.1   WASTE WATER

The waste water of modern MDF plants is mostly generated at the plug screw when it compresses the washed wood chips. Improvements in washing and dewatering have reduced the volume of water generated per day from 500 to 350 m³/day. These waste waters have a high and variable organic loading (Table 10.5) with a typical COD (chemical oxygen demand) of 7000 mg/L. The extraction of organic acids from the (soft) wood makes the waste water acid (pH = 5.5). The nitrogen load is generally too low to support an adequate biocenosis. The solid matter of this waste water is about 3 g/L and contains mostly fine wood particles (Thompson et al. 2001).

10.20.2   WASTE AIR

The drying of steamed and resinated wood fibers, which have moisture contents in excess of 100%, results in large quantities of waste air; a typical dryer may generate 500,000 m³/h, which is some improvement on previous generations of 700,000 m³/h. As with water, the organic content varies widely and rapidly. The dust collected from unfiltered plumes contains 45% wood dust and 55% inorganic flue ash from the energy plant because of direct heating of dryer. The gaseous pollutants

TABLE 10.5
Control Parameters of the Waste Water Treatment Plant

Parameter	Waste Water	Output Biology	Output Flotation	Permeate
CCO (mg/L)	7000	3000	1000	30
HCHO (mg/L)	<1	<1	<1	<1
N-NH$_4$ (mg/L)	2	8	6	n.n.
P-PO$_4$ (mg/L)	22	16	10	n.n.
Conductivity (µs/cm)	1400	2400	2800	250
pH	5.5	6.9	4.5	4.5

Source: Adapted from Portenkirchner, K., Barbu, M., and Stassen, O. 2004. *Proceedings of the 7th European Panel Products Symposium*, Bangor, ISBN 18-422-0057-7, 201–207.

are mainly pyrolysis products, for example, formaldehyde, carbon acids, and so on, and vaporized VOC, for example, terpenes (Barbu et al. 2008).

10.20.3 TREATMENT PLANT DESCRIPTION

The key ideas of a modern waste air/water treatment plant include (Figure 10.64):

- Grouping the waste air and waste water treatment plants by combining the water cycles in an aerobic, thermophilic activated sludge process
- Closure of all water cycles by combining a biological cleaning stage with a reverse osmosis and usage of the permeat for process steam generation
- Combination of various waste air cleaning systems for the elimination of gaseous pollutants (quenche/scrubber) and aerosols (wet electrostatic precipitator) (Portenkirchner et al. 2004)

Moreover, the activated sludge suspension is also used as washing water for the bio-scrubber, where it is directly injected into the waste air flow. The washed-out gaseous pollutants are then biologically degraded in the activated sludge tank. Therefore, the temperature of the activated

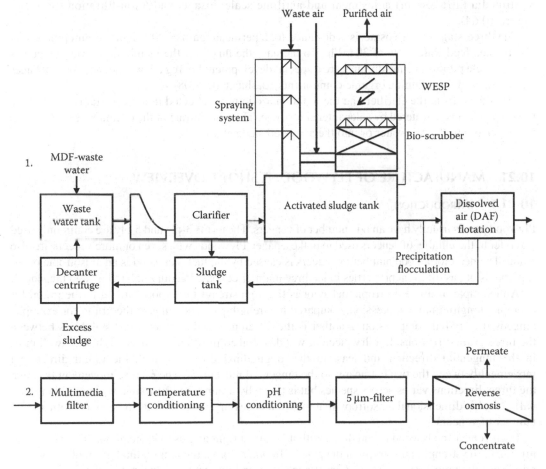

FIGURE 10.64 Combined waste water and waste air treatment plant for a MDF plant. (Adapted from Portenkirchner, K., Barbu, M., and Stassen, O. 2004. *Proceedings of 7th European Panel Products Symposium*, Bangor, ISBN 18-422-0057-7, pp. 201–207.)

sludge regulates itself approximately to the adiabatic saturation temperature of the waste air (45–60°C), which is quite high for a biological cleaning system.

All waste water flows of the plant are collected in a central tank. From there the excess waste water is transported over a bow screen and a clarifier (separation of the wood particles) into an activated sludge tank. In this aerated tank (dissolved oxygen, DO = 1–2.5 mg/L) a stable, aerobic eco-system develops. The system is particularly effective at the degradation of organic components without sludge recycling. The elimination of nitrogen or phosphorus is a bonus but was never been an objective because of the low nutrient content of the waste water.

Generally the waste air treatment plant consists of three stages:

- Spraying system (rough dust and formaldehyde removal, air cooling)
- Bio-scrubber (formaldehyde removal)
- Wet electrostatic precipitator (dust and aerosol removal) (Portenkirchner et al. 2004, Barbu et al. 2008)

The activated sludge suspension is also used for the bio-scrubber and WESP.

After the biological treatment the waste water quality still does not satisfy the requirements for the feed water of a reverse osmosis. The additional steps required to achieve a stable condition for reverse osmosis are: chemical precipitation and flocculation, dissolved air flotation, multimedia filtration, pH adjustment and antilime scale dosage, and 5 μm-filtration (part 2 of Figure 10.64).

The three-stage reverse osmosis is designed for a permeate gain of 90%, a maximum pressure of 30 bar (i.e., feed water flow of 30 m³/h). To increase the flux over the membranes, every stage has its own cycle pump. The membranes are a special development having a low tendency for blockages and optimized for retaining organic components (Barbu et al. 2008).

The sludge from the clarifier and the flotation are jointly collected in a stirred tank. From there the suspension is dewatered by a decanter centrifuge. The concentrate of the decanter centrifuge can be recycled into the raw water or into the activated sludge tank.

10.21 MANUFACTURE OF PLYWOOD: A SHORT OVERVIEW

10.21.1 INTRODUCTION

Plywood panels usually have an odd number of veneers. The terms 3-ply and 5-ply are commonly used and refer to the number of veneers used to make a panel. Plywoods with seven or more veneers are also made. The need for an odd number of veneers is caused by the fact that wood is an anisotropic material, that is, it has different properties in its three main directions, longitudinal, radial, and tangential.

An example of this anisotropic behavior is the high strength of wood parallel to the grain, for example, longitudinal compression, compared to strength perpendicular to the grain, for example, tangential or radial compression. Another is the difference in the dimensional stabilities between the three grain directions. If a dry piece of wood is soaked in water it will swell. It will swell most in the tangential direction and least in the longitudinal direction, with the radial directional movement between the two but nearer to the tangential movement. The relative movement between the three directions varies across species but is typically around 20:12:1 (tangential:radial:longitudinal). It is this dimensional anisotropy that makes it necessary for plywoods to be made with an odd number of veneers.

Each veneer in plywood is laid down with its grain at right angles to its neighbor. This is done to minimize the strength anisotropy in the panels. In other words, the longitudinal grain of one veneer reinforces the tangential grain of adjacent veneers. Bonding just two veneers together, so that their grains are at right angles to one another as in plywood, dramatically reduces strength anisotropy but such a panel would be dimensionally unstable. To visualize this, imagine the cross-section of a 2-ply

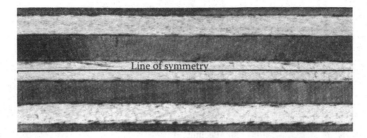

FIGURE 10.65 The cross-section of a 7-ply plywood with the line of symmetry shown.

panel, with the top veneer showing longitudinal grain and the bottom transverse grain. When placed in water, the bottom veneer, in this cross-section, will swell 20 times more than the top veneer, because of the ratio given above, and so the panel will cup.

To make a stable panel, therefore, a third veneer must be added. This creates a panel that is symmetrical through its thickness, that is, the bottom half is a mirror image of the top half. So when a 3-ply panel is placed in water, the veneers still try to swell, but this time the swelling forces are balanced and the panel remains flat. Plus, the amount of swelling is limited by the longitudinal swelling of adjacent veneers, that is, generally <1%, and so plywood panels are very stable. Additional veneers must be added in pairs so as to maintain panel symmetry.

Figure 10.65 shows a cross-section of a 7-ply plywood with the line of symmetry marked. Any dimensional movement in the top half should be counteracted by similar movement in the bottom half.

There are two ways of making thicker plywood panels. One is to use thicker veneers and the other is to use more veneers. The advantage of the latter method is that it produces more homogeneous panels. This is because the ratio of veneers in the two directions tends to one as the number of veneers increases. For example, in a 3-ply the ratio is 2:1, in a 5-ply it is 3:2, in a 7-ply 4:3, and so on. Unfortunately, increasing the number of veneers also increases manufacturing costs.

10.22 MANUFACTURING STEPS

The main manufacturing steps found in a typical European plywood factory can be seen in Figure 10.66.

10.22.1 LOG PREPARATION

The quality of plywood and certainly its ease of manufacture depend to a great extent on the quality of the logs (Mahut and Reh 2003). Some log defects of are not acceptable for plywood production, for instance, excessive conicalness and bow, rotten and defective knots, splits, borer holes, and decay, and they should be completely or partially eliminated during log sawing.

Anomalies and defects in logs can be classified into the following groups (Figure 10.67):

Shape Defects	Knots and Knobs	Other Defects	Damages by Various Organisms
Conicalness	Sound knots	Included bark, galls, blisters, burls, thorns, small knots	Insect, for example, pin holes
Bow	Rotten and defective knots	Splits, cracks, breaks, and wind shakes	Heart rot
Flattened section on a log	Sound knobs	Cup shakes	
Buttresses		Abnormal heart	
Humps		Spiral grain, entangled grain, and local deviation of the grain	

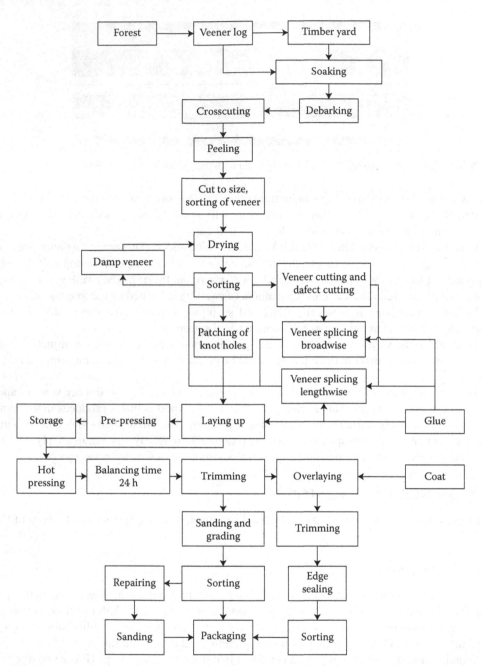

FIGURE 10.66 An example of a plywood production line flow. (Data from Kaspars Zudrags, personal communication.)

Veneer logs can be kept in good condition for some time with appropriate storage. Poor storage conditions lead to deterioration by drying and cracking at the ends. The exposed wood develops blue stain, decay, and oxidation stain, and in an advanced stage could be attacked by insects.

Log end splits can occur with susceptible species, like dense hardwoods, in just one hot dry windy day, particularly if sunlight falls directly on the log end. Blue stain and mould can occur within a week to 10 days in the sapwood in humid weather. This defect is particularly noticeable in light-colored species. Decay generally requires weeks or months to develop. Oxidation stain which

(a)

(b)

FIGURE 10.67 An example of a log shape defects. (a) Buttresses (Photo by Roman Reh.). (b) Log flat (Photo by Mark Irle.).

lowers the value of white sapwood of light species may occur through the ends of unprotected logs after several weeks. Insects may attack a log within hours after felling. To minimize insect attack, logs stored in warm weather should be used within two weeks after felling, treated with an approved chemical, or stored under water.

The sapwood of many species is subject to attack by anaerobic bacteria even though the wood is kept wet. This causes an objectionable odor, particularly from tropical hardwoods. The best way to control bacterial action is processing felled trees within one month. Spraying with chemicals may help if the bacteria have not already entered the wood.

Veneer log storage should be kept to a minimum. Log yards should be organized to ensure that the first logs put into storage are the first to be processed into veneer.

Long-term storage is possible if the log ends are properly sealed and the bark is kept intact. This ensures that the log is maintained at a high humidity, which reduces the incidence of cracks due to drying, and it reduces the chance of infection as the seal and bark act as physical barriers. Another often used practice is to submerge the logs or completely in cold water.

The quality of plywood is largely dependent on the quality of veneer. Cutting smooth-faced veneers requires well-maintained equipment and the wood to be as easy to cut as possible. The strength of wood is dependent on its moisture content and so the logs should be kept wet and certainly above their fiber saturation point. Another reason to keep the logs wet is to prevent them from cracking as a split in a log will not allow a continuous ribbon of veneer to be produced.

Before veneer can be cut the logs must be prepared so that the wood can be cut efficiently and produce a smooth, even veneer. The peeling step uses a sharp knife that is easily damaged by hard objects like stones, grit, nails, and so on. The bark often has such hard objects in it as a result of tree felling. Since the bark cannot be cut in to useful veneer it is best to remove the bark before the peeling step so as to protect the knife.

The adhesion of the bark to the wood varies between species and time since felling. Softwoods tend to be easier to debark than hardwoods, but there are many exceptions. Fresh logs also tend to be easier to debark.

A number of different systems have been used for debarking veneer logs. These include hand tools, bark saws, high-pressure water jets, flailing chains, and drum debarkers. Some mills use an old lathe to debark and round bolts. At present, however, two methods are by far the most common for debarking veneer logs: the cambio-shear or ring debarker (Figure 10.68) and rosser head debarker (see Figure 10.69).

A Rosser head debarker has two rows of wheels that support and rotate the log. The rotating speed can be regulated steplessly. The milling arm is attached to a movable carriage running along the guide placed next to the wheels and its movement is synchronized with the rotation of the log. With this type of control, the fore-movement of the milling arm is easily regulated while its automatic back stroke and repositioning has a high fixed speed. All the controls are suitably protected

FIGURE 10.68 Knife ring debarkers are often used to clean logs prior to peeling. (Photos by Mark Irle.)

against bark and possible accidental falls of the log during loading and unloading. This type of debarker can be equipped with suitable conveyors to load and unload the logs. The main advantage of this type of debarker is that one machine can cope with a very wide range of log diameters.

There are also combination machines which can use either cambio-shear or rosser-head, or both. Some factors to consider in choosing a debarker include equipment cost, maintenance cost, species to be debarked, volume of logs to be debarked, wood loss, pollution, easy operation, and maintenance.

The rosser-head debarker has a lower initial cost, lower maintenance cost, is easier to adjust, and is more adaptable for logs of a wide range of diameters. The rosser-head is generally preferred for debarking rough logs which vary widely in diameter. The cambio-shear or ring debarkers are generally preferred by plants processing logs with relatively uniform diameters and where high production and low wood loss are important.

Most plywood factories will cut their logs in to two lengths; a longer one of around 2.7 m and a shorter one of around 1.3 m. The longer logs provide veneer for the plywood faces, where the

FIGURE 10.69 Rosser head debarker. (Photo by Mark Irle.)

grain direction is normally parallel to the longest edge of the panel, and internal veneers where necessary.

Logs are cut to bolt lengths primarily with large circular saws or with chain saws. In both cases it is important that the log and saw be positioned so the cut is at a right angle to the axis of the log.

Factors to be considered in preparing bolts include: sweep in the log, end trim, presence of large defects like knots and splits, and the length of the bolts required. If possible, sweep in the bolt should be minimized as it results in excessive roundup and short grain in the veneer. Thus, even though long bolts are generally more valuable than short bolts, a log with excessive sweep would probably be more valuable if cut into two or more bolts to minimize the sweep. Logs that have been end coated or have been checked due to drying and should be end trimmed. Crosscutting with a hand-held saw can result in irregular bolt ends, which in turn can reduce the surface engaged by the lathe chucks and also cause the veneer to vary in length or require excess spurring at the lathe.

Often, but not always, bolts are heated prior to peeling. The most obvious effect of heating is that it makes it possible to cut tighter veneer, that is, with shorter, less frequent peeling checks, than that from unheated fresh logs. Tighter cutting means greater strength in tension cross to the veneer grain, so there is less splitting and checking of face veneer in handling and service. A second effect of heating is that it softens knots, thereby reducing nicks in the lathe knife. The sharper knife in turn helps to produce smoother veneer surfaces. Other possible benefits of heating include less power consumption by lathe, improvement of color by decreasing oxidation stain in the sapwood, and reduced veneer drying time.

Most disadvantages of heating can be attributed to the use of too high a heating temperature or too long a heating time. Overheating may cause excessive end splits in bolts for species like oak, fuzzy surfaces in earlywood in which fibers are pulled out instead of being cut and glossy surfaces on latewood, shelling or separation of earlywood and latewood during cutting, unwanted darkening of the veneer, increased spinout of bolts due to over soft end grain, or increased shrinkage.

Most heating vats or steam chambers are made from reinforced concrete. The vats should be designed to allow a good circulation of the heating medium. The steam pipes should be placed so that steam does not blow directly on the log ends as this overheats them and accentuates log end splits. If logs that float are to be heated in hot water, then the tank should have hold-downs that hold the logs under water during entire heating process as this will ensure that all the logs are heated evenly. The doors on steam chests or covers on water vats should be tight and preferably insulated. In many industrial operations as much heat is lost to the atmosphere as is used to heat the wood! Temperature-sensing devices should be placed in several locations in the vats or steam chambers. These in turn should automatically control the heating of the water vats or steam chambers. With an appropriate system it is possible to keep the temperature variation within 1–2°C. When using hot water vats it is best to pump the water from one vat to a second vat which has just been loaded with unheated logs. The process is repeated and the hot water is recycled from tank to tank.

The wood is not always heated before peeling, but, it is necessary for high-density species, or when cutting thick veneers, and when the logs are frozen. Warming a log lowers its strength and reduces wear on knife and especially knots which can cause nicks in knife. As will be shown later, warming the wood also increases its flexibility enabling it to survive some of the strains that are applied to it during peeling.

Logs are heated with steam, hot water, or a combination of the two. Steaming is done by heating in pits or vats either directly or indirectly. Direct steaming carries the risk that the log will begin to dry and split, whereas the use of hot water avoids this. The temperature that a log is heated to depends on the species. Generally, the higher a log's specific gravity the higher the temperature. Certainly, low-density species, for example, poplar, do need to be heated, whereas Okumé, Beech, and Oak are heated to 65, 80, and 85°C, respectively.

A side effect of heating is color change. Perhaps that most well known is the pink color that is caused when beech is steamed whereas unsteamed beech has yellow/straw color. The color change

FIGURE 10.70 A photograph showing log ends in which chuck spin-out has occurred. (Photo by Mark Irle.)

is associated with chemical changes in the extractives present. Nearly all sapwood of species darken on heating as do many heartwoods and sometimes the change may not be desirable, especially if it is nonuniform and the veneer is wanted for face veneers.

The heat should be applied carefully. High water temperatures combined with short heating periods cause temperature differentials where the outside and ends of the log are softened but inside is still cold and hard. This situation will cause the veneer to break when the knife hits the cold zone of the log. In extreme cases it can cause chuck spin-out, where the chucks which spin the log at its ends continue to turn when the log stops on the knife edge (see Figure 10.70).

The heating time should be long enough to give an even temperature throughout log, that is, the temperature differential between the outside and the innermost part of the log should not exceed 6°C. Clearly the heating time is proportional to the diameter squared of the log, so if the diameter is doubled then heating time is quadrupled.

10.22.2 PEELING

The principal aim of veneering is to form continuous sheets. Veneering knives tend to have high rake angles (typically around 70°) and sharpness angles of around 20° (Figure 10.71). It goes without saying that the knife should be sharp to ensure that the knife geometry is maintained all the way to the cutting tip. Knife sharpening represents the largest maintenance cost in veneer peeling and consequently it is worthwhile to use good purchasing specifications and to take care in grinding and setting the knives in the lathe. An ideal knife should have a maximum stiffness, toughness, corrosion, and wear resistance.

The most common knife thickness for lathes is 16 mm. Thinner knives such as 13 mm are sometimes used because they are cheaper but they are also less stiff and so veneer thickness variation can be greater. The veneer knife should be thicker when cutting thick veneer or peeling hardwood species.

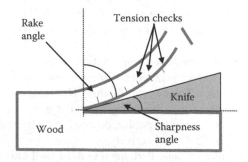

FIGURE 10.71 A diagram showing the main aspects of veneering.

When cutting thin veneer or softwood species, thinner knives can be used if there are properly supported. The hardness of the knife should be specified and can readily be tested. A soft knife can be easily honed and is tough but also wears rapidly. A hard knife is difficult to hone, is more likely to chip if it hits something hard, for example, stone, nail, and so on, but holds a sharp edge for longer.

Knife wear occurs in three ways: by impact, abrasion, and corrosion. Impact and abrasion are mechanical phenomena while corrosion has a chemical nature.

Mechanical impact is more obvious when a hard object such as a small stone makes a nick in the knife edge. Damage due to mechanical impact may also occur when the knife hits hard, unheated knots. Once damaged, all the subsequent veneer that is cut will show this small defect. Woods containing 1% or more of silica or calcium carbonate are abrasive and rapidly create a rough edge to the veneer knife. In such circumstances the use of a tough tool steel rather than brittle steel may help reduce the damage due to mechanical impact.

The third way of knife wear is corrosion. Wood is naturally acidic and the acids and polyphenols present in some wood species react with the steel knife and corrode it. This reaction makes the common blue iron stain that is objectionable on face veneer, as well as causing wear of the knife.

The purpose of grinding veneer knives is to restore a straight, sharp, tough edge. In order to grind a straight edge, it is necessary to start with a rigid level grinder. The most satisfactory veneer knife grinders have a fixed bed for mounting the knife and a travelling grinding wheel. The abrasive may be a solid cup wheel or may be segmented. The surface of the knife that goes against the grinder bed must be checked for bumps or other rough spots that will prevent the knife from lying perfectly flat. If necessary, the back of the knife should also be ground to restore a flat surface.

Information on setting the knife and bar in a lathe assumes that the knife and bar frames (holders) of the machine are in proper alignment with the center of rotation of the spindles. It is further assumed that there is a minimum of play in the moving parts of the lathe, and that the machine parts are at the same temperature will be attained in use. If these conditions are not met, the careful setting of the knife and bar on the static machine may be changed so much in the dynamic cutting condition that poor-quality veneer will be produced.

A correctly ground flat knife with a straight cutting edge is the first requirement. If a knife holder is used, it must also be clean and flat. A clean, flat bed on the lathe is the second requirement. If these conditions are not met, it is difficult or impossible to correctly set the knife. The knife or knife holder is then set with two end adjusting screws. The clamping screws are tightened by hand and so the knife is flat against the bed but free to move.

Positioning the veneer knife correctly plays an important role in producing high-quality veneer. It also determines uniformity of veneer thickness. For best results the lathe knife tip will set at or very near the horizontal centre line of the lathe spindles. Setting the knife above the spindle center line decreases the knife cutting angle in relation to the same knife position at the center line. Setting the knife tip below the spindle center line increases the knife angle in relation to the peeling block.

FIGURE 10.72 A magnified view of edge of a plywood showing tension checks (small splits).

The veneer is subject to tremendous stresses as it is cut away from the bolt. These stresses cause the veneer to split at regular intervals on the knife side of the veneer. These splits are called tension or lathe checks. Their presence changes the properties of the surface, particularly in terms of permeability. Consequently, the face without the tension checks is called the "tight side" and the face with the checks is the "loose side."

In general, the thicker the veneer, the greater the chance of tension checks being formed. This is because the minimum radius of curvature for a thick veneer is larger than that for a thin veneer or in other words, thin veneer is more pliable than thick veneer. If the checking is very bad, then the veneer can break.

To help minimize tension checks, the wood is kept saturated and often heated to ensure that it is as pliable as possible. A compression force applied just ahead of the knife tip with a nose bar can also reduce tension checking. The pressure applied by the bar and its distance from the knife tip are on the species of wood being cut.

The nosebar is important for controlling thickness, smoothness, and depth of tension checks in the veneer. It compresses the wood just ahead of the knife and so allows the knife to cut rather than split the veneer from the bolt. By keeping force between the knife carriage and the bolt, the nosebar always takes up slack in the machinery in the same direction, and so aids control of the veneer thickness. The most common type of nosebar is a fixed nosebar. Other examples include: a roller nosebar, which reduces the friction between the bolt and nosebar and a double-surfaced nosebar.

The two factors to consider when selecting a fixed nosebar are its stability and wear resistance. The most common metals are tool steel, stellite, and stainless steel. A tool steel bar is relatively stable, machines easily, and is relatively inexpensive. A stellite bar is more expensive, harder to grind, and less stable. However, the stellite bar will wear many times longer than tool steel. Stainless steel is easier to grind than stellite and like stellite, does not stain the veneer.

The second major type of nosebar, is the roller bar for peeling of thick softwood veneer (>2.5 mm) for engineered products, that is, LVL. It is commonly of bronze, generally 16 mm in diameter if it is a single bar and 12.7 mm in diameter if it is a double roller bar. The single roller bar is driven directly while the double roller bar is driven with a backup roll. With the double roller bar, the drive roller can be larger so that there is less breakage of the rollers and the knife and nosebar can advance very close to the chucks, permitting peeling to smaller diameter cores (50 mm) than with a single roller bar. The drive chain for a single roller bar may protrude up to 25 mm beyond the surface of the roller bar. Roller bars are generally lubricated with 1% vegetable oil mixed in water and introduced through holes in the cap that holds the bar.

The setting of the nosebar as well the setting of the knife assumes the lathe is in good mechanical condition with a minimum of looseness in moving parts. The knife, nosebar, and surrounding metal parts on the lathe should be at the approximate temperature they will attain during cutting.

The position chosen for the knife and nosebar has been studied thoroughly over many years. Initially peelers had a fixed vertical knife with the flat side oriented to the log; both tools (knife and nosebar), were set to the horizontal center line of the log. This construction was unsatisfactory since the flat side of the knife was exposed to wear. In this case regrinding became highly uneconomical. Today, the knife is inclined to the appropriate rake angle and is supported on the back almost to the tip. Thus, sharpening is minimized and the knife is securely supported making it more stable during cutting.

The peeling of veneer is achieved by rotating a log, or more correctly a bolt, against a sharp knife. The bolt is rotated by "chucks" inserted in its ends (see Figure 10.73). The chuck teeth, of which there are many designs, provide the torsion resistance needed to rotate the bolt with enough force to cut the veneer from it.

Profitable plywood manufacture is dependent on maximizing the yield of veneer, and preferably in full-size sheet form, from the bolt. Since logs are not perfectly round and often contain defects, simply inserting the chucks that spin the bolt in to the centre might not be the most efficient solution.

(a)   (b)

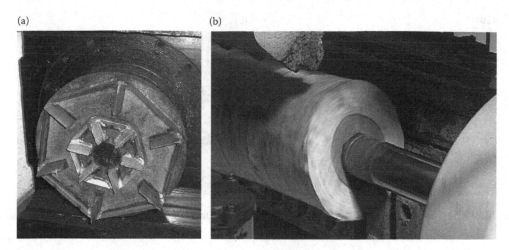

FIGURE 10.73   (a) One of many chuck designs and (b) the chucks inserted in the log ends.

Centering the log in the lathe can be achieved mechanically, optically or by using laser scanning combined with using X-Y block chargers.

Mechanical centering devices are usually combined with a charging device. These are made from robust structural steel, comprising two gripping elements, each with three or four jaws which automatically lock and center the log. The centering is determined by the geometrical configuration of the jaws. The centering and charging device receives the logs from a chain conveyor. The log to be inserted in the peeling lathe is picked up from a storage platform by means of two arms and placed in the gripping elements, where it is locked by the jaws. During this phase the log is lined up to the spindles of the peeling lathe. As soon as the preceding log has been peeled and the core removed, the charging device loads another one to the peeling lathe.

The block is measured only at 6 or 8 points. This way of mechanical centering does not consider the block shape in detail with its entire defects, like crooks, irregularities, excessive growth around knots, conicalness, splits, and so on. Because of all round-ups of such an inaccurately centered block, a substantial part of the useful and valuable raw material ends up in waste and a considerable amount of defective parts must be cut out from the peeled veneer after.

The work cycle of the centering and of the charging device is usually automatic and takes place while the peeling lathe is in operation.

Optical log centering and charging devices are usually an indispensable complement to large or medium diameter peeling lathes. They have a robust construction composed of two V-shaped lifting forks which receive the logs from a transversal chain conveyor. The forks can be moved either vertically or horizontally and simultaneously or independently of each other. The log is positioned accurately by concentric rings of light, which are projected by two powerful light sources at either ends (see Figure 10.74). A convex mirror enables the operator to centre the side opposite the log section. The log once centered with respect to the axis of the peeling lathe spindles, is gripped by an overhead carriage driven by hydraulic cylinders and controlled by electro-servos. The gripped log remains stationary in a waiting position until the previous log has been peeled, or it can be placed directly onto the peeling lathe. Once the log has been clamped between the spindles of the peeler, the carriage moves back to the initial position to grip another log which has already been positioned and centered by the lifting forks. Optical centering and charging devices work to a maximum diameter of 2 m and maximum length of 3.6 m, and are ideal for tropical species.

Modern mills tend to use laser scanning technology which produces a 3D image of the log that a software optimization program uses to decide where best to place the chucks. This method tends to give the best yield of veneer, that is, more full sheets, fewer strips and fishtails. As in mechanical

(a)   (b)

FIGURE 10.74   Two methods of aligning logs: (a) a shadow mask and (b) a laser scan of log as it is slowly rotated.

centering, the block is first precentered at 3 points at both ends. Once the geometry precentering operation has taken place, an electronic centering group optimizes centering. This group is composed of a pair of spindles which are positioned by motor control according to the log length. The spindles clamp onto the log and rotate it. While the block is rotated, its geometry is measured simultaneously at each 0.5 m step by laser scanners or CCD cameras. The scanning system transmits the data to the optimizing computer for processing to determine the axis of the optimal cylinder within the block which would give the maximum veneer yield. In addition, based on these data the computer calculates the best possible moment for the lathe for changing over from round-up to peeling, and in this way enables the peeling line to be run under full automatic control. Computer-controlled hydraulic servo-cylinders will position both ends of the block in the X and Y directions. The block is taken in its optimum position by the transfer arms to the lathe. When peeling is over, the lathe spindles will open and the centered block will be transferred to the lathe at high accuracy without major stops in the process.

The size of the chuck also has an effect on yield. On the one hand, a large chuck is needed to avoid chuck spinout and at the start when a log is heavy. On the other, a small chuck diameter allows the log to be peeled to a small spindle. Most modern mills achieve both by using retractable chucks which consist of a central chuck surrounded by an outer ring. Initially, both the ring and chuck are inserted thus providing a large diameter chuck. As the log diameter diminishes the outer ring is retracted thus permitting peeling to a smaller diameter.

As a bolt's diameter decreases then its resistance to the bending forces exerted by the knife and nose bar falls rapidly because bending resistance is proportional to the diameter cubed. If the log is not supported in some way then the bolt will deflect away from the knife, causing thicker veneer to be cut from the log ends than in the middle. The resultant spindle will be barrel shaped resulting in the loss of veneer.

This situation can be mitigated by supporting the back of the bolt with a backing roll. Motorized backing rolls exist that also help to rotate the log permitting even smaller chuck diameters.

Spindless peelers can peel down to 50 mm cores, which can significantly reduce waste when peeling small diameter logs. The technology is not widely used; however, it is suitable in combination with conventional lathes for processing the cores after an initial peeling step.

10.22.3   VENEER PROCESSING

The veneer ribbon is often wound on to a bobbin for processing, that is, clipping to size, dry, and so on at a later date (Table 10.6). This method is often known as "peel and reel" (see Figure 10.75). Although still very common in traditional mills around the world, modern mills tend to favor processing the veneer ribbon immediately after it leaves the lathe (Baldwin 1981).

TABLE 10.6

Three Different Approaches to Veneer Processing

Ribbon Processed Later	Ribbon Processed Immediately	Ribbon Dried Immediately
1. Peel and reel	1. Peel to ribbon	1. Peel to ribbon
2. Clip	2. Clip	2. Dry
3. Sort	3. Sort	3. Clipping
4. Dry	4. Dry	4. Veneer grading
5. Veneer grading	5. Veneer grading	5. Jointing
6. Clipping	6. Clipping	
7. Jointing	7. Jointing	

Mills that tend to peel small diameter logs, for example, birch in Scandinavia, send the veneer directly to a continuous dryer that is close coupled to the lathe. The advantage of this approach is that the veneer is subject to only one clipping and grading step, whereas, for all other methods there is a clipping and grading step before drying followed by a regarding and possible reclipping step. This approach is not used widely because it is technically difficult to dry a continuous ribbon of veneer. The main difficulty is to ensure that speed of the belt carrying the veneer through the dryer adjusts for the shrinkage that takes place. The shrinkage effectively shortens the length of the veneer and therefore the speed that it should be conveyed through the dryer.

Reeled veneer is stored in racks and unreeled just ahead of the clipper. A modern clipper has a sensing and measuring device so veneer can be clipped to normal or random widths.

Random widths may be generated when defects such as knots and splits are clipped from the veneer ribbon. An accurate sensing device coupled with the clipper soon pays for itself by greater yield of usable veneer. The green veneer is then graded/sorted by widths, defects and possibly by sapwood and heartwood before drying. Sizes of veneer sheets depend on the full size of plywood plus oversize allowances for shrinkage and further processing.

One limitation of reeled veneer is that if it is cut from hot bolts, it should be clipped before the veneer cools and sets in a curved shape. Veneer stored on trays is fed to one or more clippers. In a typical installation, with six trays from a lathe, three trays would feed to one clipper and the other three to a second clipper.

Vertical guillotine clippers are used when the feed speed of the veneer is relatively slow:

1. The most efficient clipper is operated by compressed air. The operating pressure is about 0.6 to 0.8 MPa depending upon the width and thickness of the veneer. The time for a cut is extremely short (about 1/10 to 1/20 s). The clipper may be used for wet or for dry clipping. The veneer draw-in speed is adjustable between 0 and 60 m/min^{-1}.

(a)   (b)   (c)

FIGURE 10.75 The "peel and reel" method. (a) Veneer being re-reeled on a bobbin as it exists the lathe. (b) Reels of veneers waiting to be processed. (c) Veneer being fed from a bobbin in to a clipper.

(a)   (b)

FIGURE 10.76   A rotary clipper in action. (a) Just after a cut. (b) During a cut as viewed down stream of the production line. (Photos by Fabien Clement.)

2. An electric clipper suitable for heavy cuts of veneers of poor quality which cannot be cut at high speed. Time for a cut is approximately 1/3 s.
3. Hydraulic clippers are used for lathe/veneer speeds up to 20 m/min.

For faster speeds a rotary clipper is used. The new generation of clippers is rotary clippers with sensing devices which have a clipping speed from 150 to 300 m/min. The name comes from the fact that the knife rotates in the same direction as the feed of the veneer and its rotation speed is synchronized with the feed speed (see Figure 10.76). For these clippers it is not necessary to use a bobbin or tray system after peeling as the veneer can be conveyed directly to the rotary clipper where it is clipped, and then transported to the full sheet and random sheet handling system, where it is selected, graded, and stacked.

The grading and stacking of green veneer is physically demanding because it requires full concentration, rapid reaction, and the veneer sheets are relatively heavy when handled with care. Too often employee performance does not achieve the results needed or expected. The length of the green chain, its height, and the position of the veneer chart should be considered when designing the work layout. The work area must be well lit. Most sorting chains range in length from 20 to 40 m; the length will vary with the sorting requirements, the number of pullers, and the chain speed. The height should equal the distance from the floor to the veneer cart plus the desired load height at the dryers.

Elevated walkways are frequently used to obtain tall veneer stacks while maintaining the right pulling height. Mechanical stacking systems ease the work and reduce the crew. The possible disadvantage of mechanical stacking devices is increased damage to the veneer sheets.

Automated stacking has evolved around two sheet handling concepts—mechanical and vacuum. The mechanical system is routinely used in mills having a high percentage of full sheets. Vacuum stackers are the best solution; these are usually individually designed for each customer. For the standard dimensions the actual best solution is a stacker with fixed suction gaps.

10.22.4   DRYING

The amount of drying required varies depending on the end-use of the veneer. For example, minimal drying (to about 20%) for veneer intended for moulding in to baskets and fruit containers whereas veneers that are to be glued with PF resins and then hot pressed must be dried to about 5%. Considering that the veneer has a moisture content well above FSP after peeling, this step requires a lot of energy. In between are such products as commercial hardwood veneers that are to be glued with UF resins, in which case the desired moisture content of veneers is 6–8%. In all cases, drying at minimum cost is paramount. The initial moisture content of the veneer (over FSP) after peeling depends to a great extent on species, time of harvesting, transportation conditions, storing, and bolt heating method.

Some desirable characteristics of the dried veneer are the uniform moisture content; drying without buckled or end waviness; free of splits; easy gluable surface; appropriate or minimal color change; low shrinkage, avoiding collapse and honeycomb and minimum of casehardening of veneers.

Factors that affect drying of veneer include both the wood itself and the drying conditions. An obvious factor is the thickness of the veneer. Thicker veneers dry more slowly than thin veneers. A modification of this is variation in veneer thickness from the nominal thickness. Commercial 3.2 mm veneer will often vary 0.2 mm or more in thickness. The thicker portions of the veneer take longer to dry than the thinner parts and contribute to a nonuniform final moisture content.

A second factor is the grain direction on the surface of the veneer. End grain dries several times faster than tangential (flat) grain. End-grain drying is significant at the ends of all veneer sheets, which tend to dry faster than the bulk of the sheet. It may also be a factor in curly grained or other figured veneer where at least partial end grain is exposed on the broad surface of the veneer. As these areas dry faster than surrounding straight-grain areas, they can cause stresses and buckling in the veneer sheet. The difference in drying rates between radial and tangential surfaces is small, but may show up, in that quarter-sliced veneer will take slightly longer to dry than rotary-cut veneer of the same thickness, and flat-sliced veneer may dry slower on the near-quarter edges than in the flat-grain area at the center of the sheet.

Factors in the dryer that can affect the rate of drying include temperature, air velocity across the surface of the veneer, the relative air humidity, and the related equalized moisture condition for the wood.

Various drying defects will occur in veneer unless it is dried carefully. Wood is very weak perpendicular to the grain and so veneers often split. It is surprising to learn, perhaps, that splitting is not generally caused by drying but by rough handling as they are splits initiated in green state, perhaps in the tree, or during veneering.

Veneer often has a buckled or wavy appearance. This may be caused by reaction wood or improper cutting. It is undesirable because of the difficulty of laying up a board with veneer that does not lie flat.

Even though veneer is relatively thin collapse and honeycombing defects are still possible if the veneer is dried too fast and at too high a temperature. The rippled surface can sometimes be removed by steam reconditioning. Likewise, case hardening, again caused by drying too fast which generates stresses in the veneer surface.

Drying can also alter the veneer surface so that glue does not wet it so well. For example, extractives can migrate to the surface and alter the contact angle between the glue and veneer surface. Alternatively, the thermal energy can cause cross-linking of cellulose.

10.22.4.1 Dryers

The two most common dryers are the roller dryers and belt dryers (see Figure 10.77). The latter allows veneer to be fed into the dryer with its grain angle perpendicular to the sense of feed. Consequently, they can be used to dry continuous ribbons of veneer. Veneer on reels or on trays is sometimes fed to the dryer in a continuous ribbon. As the veneer comes from the dryer, it is clipped to size. This system reportedly results in less waste and split veneer, but it is convenient only to species which do not tend to buckle.

It is much more common to dry sheets of veneers using roller dryers which are energetically efficient because the rollers are heated with steam or hot water and since these carry the veneer through the dryer the direct contact with the roller permits heating through conduction, which is more efficient than by convection. The rollers are typically 100 mm in diameter.

Very high air speeds of 600–3000 m/min are used in the dryers to minimize the thickness of the boundary layer of air surrounding wood because it is the thickness of this layer that determines drying rate. The air is blown vertically onto the veneer faces facilitating fast drying.

(a)

(b)

FIGURE 10.77   (a) Feeding a roller dryer. (b) In feed of a belt dryer.

Dryers are split into sections. The temperatures and humidities can be controlled independently and used to help ensure uniform drying. The first stage tends to have the highest temperature of around 150–200°C. This is acceptable during initial drying because of the high moisture content of the veneer. Subsequent stages are cooler to produce uniform dried veneer. As a matter of interest, the calculated equilibrium moisture content of wood in saturated steam at 105°C is about 11% whereas at 115°C it is about 5%. Drying veneer to controlled final moisture content should reduce degradation and shrinkage, and provide a superior surface for gluing.

Most of the dryers are equipped with a separate cooling section. The conveyor system in the cooling section is usually independent of that in the main dryer. High cooling capacities are computer calculated on the basis of drying temperature, veneer temperature, wood species, veneer thickness, and throughput speed. With the continuous contact pressure during cooling, even better fixing is achieved in terms of flatness and flexibility compared to the press dryer.

Modern veneer dryers have energy-efficient constructions. Good sealing as well as thick mineral insulation in the floor as well of other parts not only achieves energy savings but also stabilizes conditions in the dryer, facilitating more precise process control possibilities. The smooth inner surfaces of the dryer together with strategically placed deflectors ensure turbulent free air circulation. This reduces the demand on the blowers and facilitates greater nozzle velocity at the jet boxes for better heat transfer to the veneer. Throttle baffles ensure constant pressure and thus nozzle velocity equalization in the jet boxes, allowing uniform drying throughout the dryer decks. The nozzle is profiled for minimum air resistance and maximum air speed, but recessed to eliminate interference with the veneer.

10.22.5 Veneer Preparation

Veneer grading follows drying. In simple terms, sheets are graded as being best face quality, back face quality and interior quality, that is, veneers to be used within the panel. Actual grading rules contain many more grades and vary from species to species and type of plywood being made.

Quantitatively five factors of veneer should be checked at regular intervals: stain, uniformity of thickness, surface roughness, veneer breaks, and buckle or other distortions of the veneer.

Most veneer is readily dried satisfactorily for the intended end use, but since veneer is easy to dry, potential problems are sometimes overlooked. Some veneer drying problems are nonuniform moisture content in the veneer as it emerges from the dryer, buckle and end waviness of veneer sheets, splits and checks in the veneer, surface inactivation (making it difficult to bond), scorched veneer surfaces, veneer that shows signs of collapse, honeycomb or casehardening, excessive veneer shrinkage, and undesirable color.

The appearance of lower grade plywood can be greatly improved by removing the larger and unsound knots and filling the resultant spaces with plugs of sound veneer of similar color and texture to that of the veneer surroundings. This is achieved by stamping out the defect, for example, a

(a) (b)

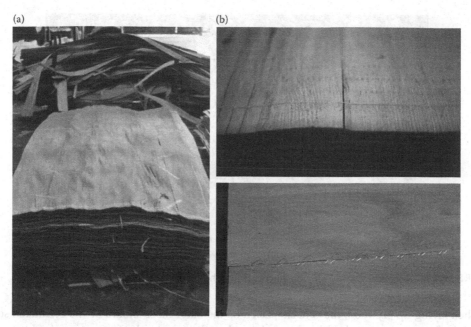

FIGURE 10.78 Examples of veneer jointing methods. (a) Adhesive paper tape. (b) String impregnated with hot-melt glue. (Photos by Mark Irle.)

knot, included bark, dark stain, and so on, with a standard-sized punch. The hole is then filled with a patch with the same shape and size that has been previous punched out of clear veneer.

Even more veneer is jointed together to form full size sheets. There are many different joint methods: adhesive tape, hot-melt impregnated string, stitching, edge gluing, scarf jointing to name a few (see Figure 10.78). Modern mills use compositors that combine veneer pieces of varying widths in to one long piece that is then cut in to full size sheets.

Many people are surprised to learn that board strength is not significantly affected by the use of veneer sheets that have been joined together. This is because the strength of wood perpendicular to the grain is low and so the presence of a jointed piece is little different to a whole piece.

10.22.6 BOARD LAYUP

Despite huge advances in computer technology and automation, plywood panels are still essentially compiled by hand. This is because of the large variation seen in the form and color of veneers; they are not flat, not particularly flexible and are often split. These factors make it difficult for an automated system to cope with veneer. Humans can recognize veneer defects and can modify the layup accordingly.

The most common method of applying adhesive is with a roller coater which applies a known amount of adhesive to both faces. Adhesive is applied to both faces of alternate veneers and not every veneer. For example, in a 3-ply panel adhesive is only applied to the middle veneer and for a 5-ply only to veneers 2 and 4. Other adhesive application systems are used including: curtain coating, spraying, and extrusion. Discussion of these is beyond the scope of this short review and readers are recommended to read Baldwin (1981).

The grammage of adhesive applied varies with grade of panel, type of adhesive, wood species and, in particular, the surface smoothness of the veneer. Less adhesive is necessary for smooth faced veneers. A typical grammage will range from 140 to 240 g/m^2 per face (or glueline). It is sometimes quoted as a double face measure (and so double the figures given in the previous sentence) because the roller coaters apply adhesive to both faces at once.

FIGURE 10.79   A typical plywood lay-up station. (Photo by Fabien Clement.)

When laying-up plywood, the veneers placed on the outer faces should be the orientated so that their tight faces are exposed. If this is not done, then there is a risk of excessive checking on the surface with time as the veneer moves with changing atmospheric conditions. Some mills have semi-automatic lay-up lines where surface veneers are handled by machine.

As an example, a station making a 5-ply panel may function as follows (Figure 10.79). Two operators will lay up the panels on the discharge side of the roller coater that will apply the adhesive. They will have within easy reach two piles of veneer: one of face veneers and another of core veneers. A third pile of core quality veneers will be placed on the in-feed side of the roller coater. One operator will place a face veneer, face down on the pile. A core-stock veneer is then passed through the roller coater and an even layer of glue applied to both sides, and then placed on top of the face veneer. The other operator will then take a dry core veneer and place it on the adhesive coated veneer. A second adhesive-coated veneer will then be added and finally, another dry face veneer. The process will then repeat until enough panels have been made to fill all the daylights of the press.

Core veneers may be in single sheets of similar size to the outer plies, or made up of several narrow pieces to make up the same area. In this case the greatest care must be taken to ensure that pieces are carefully butted together and that one veneer does not overlap another. Gaps between the veneers must also be avoided. Any glue removed from the veneer or displaced during handling must be carefully respread.

10.22.7 PRESSING

Panels are generally cold prepressed after lay-up. The aim is to flatten the panels and allow the adhesive to develop tack. Both factors make it easier to feed the panels in to the hot-press. The prepress operates cold and the panels are pressed as a bundle in a single-daylight press (see Figure 10.80).

The advantages of cold prepressing are:

- It gives more flexibility in the handling of panels between the lay-up and hot-pressing. The laid-up panels can be stored before hot-pressing from 2 to 8 h.
- Individual panels are easier to load into the hot press.
- It improves the plywood quality because the moisture content of the glued veneer is more even.
- The degree of compression of plywood during hot pressing is reduced.
- It enables the use of shorter hot-press cycles.

FIGURE 10.80   (a) a pack of plywoods being cold prepressed. (b) Prepressed plywoods being fed in to a loading cage of a multi-daylight hot press. (Photos by Fabien Clement.)

Prepressing time is usually 5–20 min depending of different factors: wood species, veneer thickness, moisture content, glue type and so on. The specific pressure applied is typically between 0.6 and 1 MPa.

Prepressing is synchronized with the hot press. Modern mills with large multi-daylight presses often use an automatic loading and unloading cages to minimize dead time between cycles and improve overall production capacity. The cages also help to ensure that all panels are subjected to the same temperature histories. The presses in such mills also tend to have simultaneous daylight closing mechanisms to ensure that each panel is pressed to the same pressure and for the same time. This results in more consistent panels.

Panels are fed into presses by hand in older mills with smaller presses and these presses tend to close by pressing one daylight against another. So for an upward stroke press, the bottom daylight will close first followed by the second and so on. On opening it is the top daylight that opens first and so it is clear that the panel in the top daylight is pressed for a shorter time than the bottom panel. Plywood press cycles are much longer than those for particle and fiberboards and so the difference in the pressing cycle of the plywood in the top daylight compared to the one in the bottom is relatively small in percentage terms, but, it is different and does not help panel consistency.

The press temperature for panels made using amide-based glues, for example, UF and MF, is between 100°C and 120°C. This is much lower than the temperatures used for particle and fiberboards using similar adhesives. The lower temperatures are associated with lower internal steam pressures and so less risk of blisters, lower energy costs but also longer press times. The panels are normally cooled briefly before being stacked ready for further processing.

Many types of plywood are made with phenolic-based adhesives, which require more energy to initiate cure. This is achieved using higher press temperatures, typically 160°C and "hot stacking" immediately after pressing so that the panels remain hot for many hours.

The length of a cycle can be calculated using a simple "rule of thumb" which is 1 min plus 30 s for each millimeter of panel thickness. For example, 10 mm thick panel would have a press cycle of around 6 min. As mentioned above, this is much longer than for particle and fiberboards, which are pressed between 4 and 10 s/mm of panel thickness. So a 10 mm-thick particleboard may be pressed every 50 s.

The specific pressure applied to the panels is around 1 MPa; a little lower for low-density plywoods, for example, poplar and higher for high-density plywoods, for example, beech. Once again, this is quite different of particle and fiberboards, which are pressed at around 3 MPa. Pressure is applied in any gluing situation to ensure adequate contact between the bonding surfaces. Since plywood is made of "flat" sheets a low pressure is sufficient to achieve the contact required; particle and fiberboards on the other hand are made of elements with varying sizes and shapes that are randomly orientated so much higher pressure is required to force the elements close enough for the adhesive to bond them together.

The press pressure peaks near the start of the cycle when the wood is cool. The pressure is reduced during the cycle to minimize induced compression and to maintain panel thickness.

10.22.8 FINISHING

Plywood panels on exit from the hot press are dimensionally unstable. The outer plies are overdried, but the core is not. This produces temperature and moisture content gradients which must be reduced or removed and is most easily achieved with storage. Problems can occur if panels are shipped out to customers too soon.

Boards are trimmed so that they have straight edges and 90° corners. Often the panels are sanded to ensure panel thickness calibration.

Repairs may be necessary to panel faces and edges. Face repair is most often needed when using a species with dark knots. These can be routered out and the hole filled with a colored hot-melt plastic. The plastic cools and solidifies very quickly and once sanded it is difficult to see the repair.

Plywood manufacturers have developed a wide range of finishes for their products in order to increase their value and to differentiate themselves from other producers. The finishes include: nonslip surfaces, phenolic impregnated paper finishes for concrete form work and various colors.

10.22.9 PLYWOOD GRADING

There are two broad classes of plywood, each with its own grading system.

One class is known as for construction and/or industrial use. Plywood in this class are used primarily for their strength and are rated by their exposure capability and the grade of veneer (1/10–1/8″ or 2.5 to 3.2 mm) used on the face and back. Exposure capability may be interior or exterior, depending on the type of glue. Veneer grades may be N, A, B, C, or D. N grade has very few surface defects, while D grade may have numerous knots and splits. For example, plywood used for subflooring in a house is rated "interior C-D." This means it has a C face with a D back, and the glue is suitable for use in protected locations. The inner plies of all construction and industrial plywood are made from grade C or D veneer, no matter what the rating.

The other class of plywood is known as hardwood and decorative. Plywoods in this class are used primarily for their appearance and are graded in descending order of resistance to moisture as technical (exterior), type I (exterior), type II (interior), and type III (interior). Their face veneers are virtually free of defects.

According to US Plywood Standard (APA PS1–95 Construction and Industrial Plywood), plywood can be classified into several grades, such as A-A, A-B, A-C, B-B, B-C, C-C, C-D. The grade refers to the grade of surface covering. The first letter refers to the grade of face veneers and the second letter refers to the grade of back veneers. The better is the grade, the nicer is the surface.

Plywood sheets range in thicknesses from 1.6 to 50 mm. The most common thicknesses are in the 6.4–19.0 mm range. The most common size for plywood sheets used in building construction is 1.2 m wide by 2.4 m long. Other common widths are 0.9 and 1.5 m. Lengths vary from 2.4 to 3.6 m in 0.3 m increments. Special applications like boat building may require larger sheets.

10.23   A HISTORY OF WOOD-BASED COMPOSITES

Veneers were initially produced manually, by sawing, and later, mechanized, by means of large diameter rotating saws and special veneer frame saws. It was only in 1818, in France, that the first rotary cutter (lathe) was patented for the production of veneers. The first slicer for decorative veneers was patented somewhat later in 1870. These two inventions have been modernized and upgraded ever since, but they are still in use today. In 1934, the waterproof synthetic resins (PF glues) were produced, which made possible the production of plywood for exterior applications.

Particleboards originated in Germany. The first mention of the manufacture of such boards dates back to 1887, when Hubbard made a so-called "artificial wood" of wood flour and albumin-based glues, consolidated under high temperature and pressure. In 1889, Kramer obtained a German patent for his method of gluing of wood shavings onto a flax fabric. Then he layered the fabrics in a similar way to plywood (alternatively cross-oriented). In 1905, Watson (USA) developed a method to produce boards of thin square particles (flakes). Today, his patent is still at the centre of flakeboard and OSB manufacturing.

Beckmann (Germany) suggested in 1918 a new technique for the production of a layered board with a core of compressed particles or wood flour and veneer faces; it was the forerunner of the products known at present as Com-Ply, a veneered particleboard or plywood with a particle core. Freudenberg (Germany) mentions in 1926 the use of planer shavings glued with resins of the epoch, to produce boards. In 1934, Nevin (USA) recommended the mixing of coarse sawdust and shavings with resins and to harden them by hot-pressing for board manufacturing. Antoni (France) obtained, in the same year, boards from a combination of wood fibers and particles glued with urea- or phenol-formaldehyde resins.

The year 1935 was a very innovative as researchers in France, Germany, Japan, and USA suggested a range of new WBP products. Samsonow (France) suggested the use of cross-alternating veneer strips for a particleboard, similar to the present products OSB. Satow (Japan) obtained an American patent for the manufacture of boards made of 75 mm long chips, randomly arranged in order to reduce board warpage. Roher (Germany) outlined the possibility of gluing wood particles onto the surfaces of a plywood core within a single pressing operation.

In 1936, Loetscher (USA) presented a patent concerning the automated manufacturing of particleboards. In 1937, Chappuis (Switzerland) described the manufacture of particleboards from dry particles, by applying powder resins (Bakelite).

Another Swiss, Phol, presented in 1936 his patent for the alternative use of long veneer strips (50–200 mm), now used for making load-bearing structures from veneer-based composites, that is, LVL.

During World War II, when the production of synthetic resins was improved, the first attempts for the industrial manufacturing of particleboards were made. Between 1938 and 1940, the German company Torfit obtained two patents for the production of particleboards; liquid resins were used for gluing particles and formed in to a particle mat which was subsequently hot-pressed.

The same company built the first plant for the industrial manufacturing of particleboards in 1941 in Bremen (Germany). Fahrni obtained in 1943 a French patent for the production of particleboards. He later developed equipment (Novopan) for particleboard plants. Kreibaum (Germany) produced between 1947 and 1949 the first extruded, as opposed to flat-pressed, particleboards.

Although the idea of producing wood fibers was one-and-a-half-century old, it was only in 1844 in Germany that a practical method for producing wood fibers by mechanical processing in the so-called "mills" could be developed. In England, in 1851, a chemical system was invented for wood refining, in order to obtain fibers for paper production. In its competition with other raw materials used to produce paper, wood became the most economically profitable material, due to its low cost and availability. Fiberboard structure is closer to that of paper products than of other wood-based products.

The first insulation board plant built in 1898 in England answered the efforts to capitalize the large quantities of oversized fiber bundles removed from pulp.

In 1914, Carl Muench made a softboard from wood fibers, which had insulating properties and this explains why this type of product is often called insulation board. "Insulite" is the trade name of this fiberboard, produced in a large plant on the same location since 1916 (USA). A second plant for insulation fiberboards was built in 1931. The manufacturing process is a modified paper making method. A large tank stored the mixture of fibers and water. This mixture was then "dehydrated" on a sloping and continuously moving screen/sieve, thus forming a fiber mat. By drying the wood fiber mat in an oven, porous fiberboards with densities between 160 and 400 kg/m^3 could be made. This wet fiberboard process has been improved over the years and is an established manufacturing method for fiberboards. It has relatively low production costs because of the relatively small amounts of binders present in the product. Muench investigated alternative sources of fibers and also used agriculture by-products like corn, wheat straw, bagasse, and so on. The first plant for insulation fiberboards made with non-wood ligno-cellulose raw materials was built in 1920.

An alternative method for pulping wood using the steam explosion technique invented by William Mason was introduced in 1924. Wood chips are introduced in to a vessel called a "gun," which was then pressurized with steam to pressures as high as 8 MPa causing a temperature of around 290°C. The pressure is suddenly released causing the softened chips to pass rapidly through a grid which breaks up the chips into coarse fibers. In 1926, in USA, the Mason Fibre Company started manufacturing the first hardboard using the Masonite process.

Asplund (Sweden) proposed the thermomechanical pulping process in 1934. This technology has been subsequently widely adopted for making fibers both in the paper and in building board industries. The principal steps of this method are described in Section 5.2.

The excellent properties of the fiberboards are the result of the bonds created due to lignin reactivation under high-pressure, -temperature, and -moisture conditions. In order to improve the fiberboard properties when used in contact with water, additional water-repellent substances (paraffin/wax) or synthetic binders (phenol-formaldehyde resins <3%) can be added. This technology is widely known as the wet process and the fireboards thus manufactured are internationally known as hardboard when panel densities exceed 800 kg/m^3 or softboards if the density is less than 400 kg/m^3 (see Figure 10.1). The particularity of this process lies in that water is used to convey fibers and to form the fiber mat. Another specific aspect of the wet process is that a screen is used in the press, at the bottom of the fiber mat, so that water and steam can readily escape during hot pressing. As a result, the screen pattern is embossed on the backside of the finished hardboard. The production of fiberboards using the wet process has decreased drastically during recent decades. The main reason for this is that the process is inefficient for making thick panels (essentially they are limited to panels of 12 mm or less). In addition, the strict environmental requirements imposed on water pollution in highly industrialized countries have significantly increased manufacturing costs. This technology has patent environmental shortcomings regarding the recycling and treatment of waste waters, which contain large amounts of fibers and organic substances (cellulose, lignin, wax, etc.). Nevertheless, softboards still hold an important position in the market of insulation boards (i.e., made of polyurethane foams or of mineral-bonded fibers).

Research dating back to 1945 aimed at replacing the wet forming system with a dry forming process using an airflow. Much of the research was conducted in USA and it ultimately resulted in the development of Medium-Density Fibreboard (MDF).

The first two dry-process hardboard plants were built in 1952 in USA as an adaptation to industrial production of the semi-dry process. Dry-process hardboards have lower mechanical properties than those made using the wet process, due to the high water absorption capacity of the remaining "unbound" lignin and hemicellulose, but have the advantages of smooth faces on both sides, lower densities, thickness variability, and a high workability.

The first MDF plant was built in 1965 in Deposit, USA. The first MDF factory in Europe is thought to be that built in the former German Democratic Republic at Ribnitz-Damgarten in 1973. Other factories were built in Eastern Europe: 1975 Busovaca (today Bosnia-Herzegovina) and 1978 Illirska Bistrica (today Slovenia).

During the last decade, a distinct category of products have appeared, namely the Low-Density Fibreboards (400–600 kg/m^3).

The use of multiopening and today, continuous presses, was the next stage in developing the manufacturing technologies for particleboards, MDF and OSB. These three categories of WBP now dominate the sector in terms of production volume (world 70%, Europe 90%). Equipment production and technology improvement, as well as WBP development, has remained until the late 1990s in German hands, which has fitted out almost the entire WBP factories all over the world. The first continuous press made outside of Germany was produced in China and installed in the Czech Republic.

The field of wood-based composites has dramatic changes ever since the 1950s in terms of production capacities and technological developments. The prospect of producing particles and fibers with various dimensions and shapes, the use of new types of synthetic resins, of modern technologies and specially designed, reliable, partially or completely automated equipment have given a boost to WBP development. It was the beginning of a new age, not only in the production of WBP, but of other wood-based products as well.

10.24 CONCLUSIONS

This chapter covers only a tiny part of the wealth of literature and knowledge on WBP. For example, no mention has been made of durability, testing, or to any extent how WBP may be improved.

The main aim of this chapter is to provide enough background information so that the reader may more readily understand the other chapters in this book. It is further hoped that this chapter encourages researchers to contribute research effort to the WBP sector.

10.25 POTENTIAL FOR NEW WBP

As concluded, the current forest products industry relies heavily on particleboards, MDF, OSB, and plywood. It is helpful to consider fiberboard products in a wood composites context, since this puts a stronger emphasis on the engineering materials nature of WBP. They are often used in the building industry, where many structures are subjected to significant loads, which can be both static and dynamic. By using the term *new* in the heading of this section we underline that the scope is to consider the potential to considerably extend properties beyond those of traditional WBP. The experimental data reported in the present section are from laboratory-scale trials. The work should be considered as a source of inspiration for the development of new WBP-type composites for new applications.

A widening of the perspective for WBP may help to develop new materials for new applications. Also, a context of composite materials rather than forest products is helpful since the science and engineering of materials puts a strong focus on the microstructural organization of material components. Focus is on relationships between microstructural composition and properties. Material components such as fibers, nanofibers, and polymers are subjected to processing and combined into a material with a certain microstructure. During processing, the material may also be molded into a given geometrical shape. Common examples of shapes in the context of structural mechanics include plates, beams, and cylinders. A material of a given shape can serve simple or complex

TABLE 10.7
Examples of Wood-Based Panels of Composites Nature

Wood Composite Category	Specific Wood Composite Material	Description	Example of Applications
Laminated wood	Plywood	Veneer layers are laminated and bonded with a certain veneer orientation distribution	Building industry, furniture
Strands or particles with adhesive	Particle board	Large wood particles are coated by adhesive and hot-pressed to porous particleboard	Furniture, building industry
	Oriented strand board (OSB)	Anisotropic strands are coated by adhesive and compressed to oriented high-density boards	Competes with plywood at lower cost
Porous wood fiber networks	High density fiberboard (HDF) (850–1100 kg/m^3)	Mechanical or masonite pulps are hot-pressed and bonded by lignin or adhesive	Flooring, siding, wall panels, furniture
	Medium density fiberboard (MDF) (600–800 kg/m^3)	Mechanical pulp is combined with an adhesive and hot-pressed	Furniture, cupboards, doors flooring
	Paper (chemical pulp paper 600–800 kg/m^3)	Wood pulp is filtrated and dried into network	Printing, packaging
	Paperboard	Typical thickness >0.25 mm	Packaging
Impregnated wood fiber networks	Paper laminates	Paper is impregnated by resin and polymerized	Electric insulation boards, flooring

functions: transmitting loads, heat, ability to survive repeated folding or storing mechanical energy at minimum weight.

Common components for mechanical reinforcement include fibers, clay platelets, and inorganic particles. Air is also an important constituent, since the density of the material can be reduced. Foams and porous fiber networks can therefore be classified as composites. Table 10.7 presents some examples of current material categories. We have made the classification according to the microstructural characteristics and the scales of constituent size or constituent type. Some application examples are also presented.

The orientation distribution of the reinforcement component as well as its size is important for the mechanical properties of the composite. Larger components tend to result in materials with larger defect size, and therefore lower strength. Oriented reinforcement provides higher strength in the orientation direction, and this is one of the main advantages of composite materials. It makes it possible to tailor the anisotropy (orientation dependency) of the material properties.

The typical densities and mechanical properties of different wood composites are listed in Table 10.8. Density is important to consider in materials comparisons since mechanical properties show strong dependence on density. Since wood composites tend to be porous, the correct parameter in a micromechanics context is the relative density or volume fraction.

Comparing the different materials in Table 10.8, many materials are limited by the restricted geometric shape possible. The anisotropy and high porosity also result in locally weak regions. Laminated structures such as plywood sheets often show high strength due to the thin lamellae. In addition, the orientation distribution of the lamellae can be controlled. However, there is little freedom with respect to shape. The comparison between high-density fiberboard and a linerboard is of some interest. Kraft linerboard is a paperboard made of chemical pulp fibers and used as the surface ply in corrugated boards. It has higher strength than the fiberboard, despite lower density. Most likely, the high-density fiberboard is weaker since it has a heterogeneous structure with locally weak regions.

TABLE 10.8

Typical Properties of Wood Composites

Wood Composite	Density (kg/m³)	Elastic Modulus (GPa)	Bending Strength (MPa)	Tensile Strength (MPa)
Spruce plywood	≈550	6.9–13	21–48	6.9–13
Oriented strand board (OSB)	≈550	4.8–8.3	21–28	6.9–10.3
High-density fiberboard (HDF)	850–1100	2.8–5.5	31	15
Medium-density fiberboard (MDF)	600–800	2	20	n.a
Particleboard	550–750	2–4 (from bending)	15–25	n.a
Kraft linerboard	600–800	2.8–4.1	n.a	25 (4% fracture strain)
Nano-paper	1300	13	n.a	200–300 (10% fracture strain)

The wood fiber itself has attractive characteristics including high aspect ratio (the length-to-diameter ratio), high axial strength and elastic modulus in the fiber wall as well as favorable fiber network forming characteristics. Networks made of strong wood fibers or chemically tailored fibers, can for instance be used in new fiber architectures of designed orientation distributions, and combined with new polymer matrices, foams or other porous materials to form new types of wood composites. Interesting functions include thermal insulation and mechanical performance. Wood fiber/polymer composites could also provide new opportunities with respect to molding of intricate geometrical shapes. The *nanopaper* material in Table 10.8 represents some of the advantages that can be obtained with cellulosic *nanofibers*. Their dimensions are three orders of magnitude smaller than regular wood fibers. The modulus is 13 GPa and tensile strength exceeds 200 MPa, due to the fine structure of the material.

The wood products industry supplies many different types of wood fibers for different applications. It is interesting to consider the suitability of different fibers for new types of wood fiber biocomposites. Table 10.9 lists the most common types of wood fibers, with comments on their

TABLE 10.9

Examples of Some Common Wood Fiber Types

Wood Fiber Type	Chemical Composition	Characteristics
Saw dust, wood flour	Similar to wood	Short aspect ratio, large size since the particles often are tracheid bundles
Mechanical wood pulp	Similar to wood (high yield)	Individualized fibers 20–30 μm in width and 1mm (hardwood) to 3 mm (softwood) in initial length, mechanically cut or damaged, little fiber collapse
Chemi-thermomechanical pulp (CTMP)	Low extractives content, otherwise similar to wood (yield 90–93%)	Less mechanical damage compared with TMP
Bleached chemical sulfate pulp (kraft)	Low lignin conc., typically 15% hemicellulose, 85% cellulose	Individualized, mechanically intact, collapsed after refining
Chemical sulfite pulp	Low lignin conc., 4–15% hemicelluloses, 85–96% cellulose	Individualized, mechanically intact, collapsed after refining, lower cellulose molar mass than kraft pulp
Nano-fibrillated cellulose (NFC)	Typically 4–15% hemi-celluloses, depending on pulp source	5–15 nm in width and several micrometers in length

characteristics. Due to the low cost, saw dust and other particles from saw mills and in general machining of wood are in widespread use in particleboards and melt-processed thermoplastic wood composites. However, since the typical aspect ratio of these particles is low (≤10), the reinforcement potential of the stiff wood fiber is not utilized. In contrast, mechanical pulps such as TMP and CTMP have much higher aspect ratio and should have better reinforcement efficiency in bio-composites. The chemical composition in TMP and CTMP is fairly similar to wood. This may cause problems with odor (thermal degradation of hemicelluloses and lignin) or discoloration problems (extractives) in biocomposites that are processed at high temperature (e.g., melt-processing or compression molding). Bleached kraft pulps are more stable. They can also have high molar mass of cellulose, which gives good fiber strength. The hemicellulose content is usually fairly high, and hemicelluloses may degrade by hydrolysis in high-temperature processing. Sulfite pulps can have very low hemicellulose content, but at the same time the average molar mass of the cellulose is lower than in kraft pulps.

10.25.1 COMPRESSION MOLDING OF BINDERLESS CHEMICAL PULP

WBPs are essentially composite materials, although the components are sometimes limited to wood fiber and air. Existing WBP have limitations such as low strength, low strain-to-failure, and moisture sensitivity. In addition, odor and discoloring problems can arise due to thermal decomposition of hemicelluloses and lignin during processing. The chemical and structural heterogeneity of WBP is therefore the most important reason for the property limitations. Chemical wood pulp can actually be used in a simple compression molding procedure to form materials of much improved stiffness, strength, and toughness (Nilsson et al. 2010). The processing route is explained in Figure 10.81. Cellulosic wood pulp of high cellulose content is subjected to high pressure in the moist state. As the cellulose is dried at elevated temperature under pressure, the cellulose aggregates in the pulp fiber cell wall bond very strongly to each other. Also the fibers bond to each other since

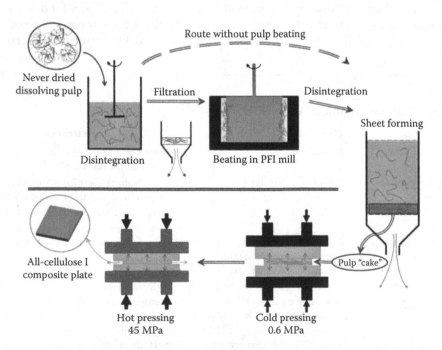

FIGURE 10.81 Schematic representation of the processing route from a wet dissolving pulp. (After Nilsson, H. et al. 2010. *Composites Science and Technology* 70: 1704–1712.)

they are brought in close contact due to the high pressure. With this method it is possible to mould components of complex geometrical shape. A fibrous material of comparably high density is formed. The modulus can be 13 GPa and the strength 76 MPa at a density of around 1300 kg/m^3. This should be compared with the typical 5.5 GPa modulus and 15 MPa tensile strength of high-density fiberboard at 1100 kg/m^3 density. Linerboard for paperboard boxes has a tensile strength of 25 MPa at lower density ($\approx$800 kg/m^3).

Moulded chemical pulp is first of all interesting due to the possibility to mold complex shapes. Secondly, the increased strength and stiffness compared with high-density fiberboard is more dramatic than expected. The increased stiffness is related to the better organization of the fibers. They are random-in-the-plane, but the extent of out-of-plane oriented fibers is probably fairly limited due to the nature of the manufacturing process. A wet "paper-making" approach in the first stage ensures a well-ordered planar arrangement. In addition, the inherent strength of chemical pulp fibers is probably higher than for the fiberboard fibers. Hot-pressing is also likely to provide strong interfiber bonding, so that fibers actually are fractured rather than debonded during mechanical loading to failure.

10.25.2   Wet-Processed High-Density Fiberboards—A Combination of Chemical Pulp and Nanofibrillated Cellulose (NFC)

The compression moulded chemical pulp (sulfate or sulfite) can obviously also be produced as a fiberboard type of material. This was done by the use of a simple wet vacuum filtration procedure, see the process demonstrated in Figure 10.82. At a density of about 850 kg/m^3, the tensile strength was almost 100 MPa, with a strain-to-failure of 2.7% and a modulus of 8 GPa. Encouraged by the insight from the molded pulp work that the fiber-fiber bonding is a critical property, this "chemical pulp fibreboard" was combined with up to 10% cellulose nanofibers (NFC) during processing (Sehaqui et al. 2011a,b). The NFC served as a small-scale cellulosic binder (see Figure 10.83). The average pore size is reduced in the fiber network and one can expect that fiber-fiber bonding is improved. These expectations were fully met. The chemical pulp with 10% NFC showed a tensile strength of 160 MPa, a strain-to-failure of 4.2% and a modulus of 10 GPa at a density of about 1000 kg/m^3. Tempered hardboard at similar density can have a tensile strength of 33 MPa. The effect of NFC on failure properties is probably quite complex. Possibly, the NFC links individual fibers up to fairly high strain, whereas regular hardboard materials show early fiber-fiber debonding.

The use of NFC nanofibers can apparently address the shortcomings of WBP. Perhaps the most interesting form of nanocellulose for large-scale industrial use is native cellulose NFC from pulp

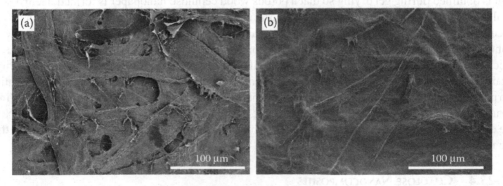

FIGURE 10.82   Surface texture of the high-density fiberboard reference based on chemical pulp (a) and the corresponding surface texture when 10% nanofibrillated cellulose (NFC) has been added (b). Densities are around 950 kg/m^3. (After Sehaqui, H. et al. 2011a. *Composites Science and Technology* 71: 382–387.)

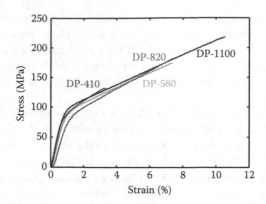

FIGURE 10.83 Stress-strain curve showing the effect of adding nanofibrillated cellulose to chemical pulp fiberboard.

fibers. Cellulose molecules are in extended chain conformation and form highly ordered fibrous crystallites. Experimental data (x-ray diffraction measurements of crystal displacement during tensile loading of fibrous plant cells) led to an estimated axial cellulose crystal modulus of 130–140 GPa. It is interesting to note that the modulus of native cellulose is similar to the modulus of aramid fibers. The advantage of cellulose, however, is in its bio-based origin and the potential to use it in nanofibrillar form. Nanostructured cellulose networks appear to offer a general reinforcement mechanism.

Wood NFC has the potential to be widely used due to the low cost associated with recently introduced economical disintegration procedures. Wood fibers cooked in a chemical solution to yield high cellulose content can be used as starting materials. As these fibers are subjected to mechanical homogenization (as used in the food industry for tomato soup and orange juice), the fiber cell wall disintegrated into cellulose nanofibers. Nakagaito and Yano (2005) used such nanofibers in composites. Enzymatic pretreatment was developed in Sweden in order to reduce energy requirements during nanofiber disintegration (Henriksson et al. 2007), although chemical pretreatments have some advantages in terms of facile disintegration (Saito et al. 2009). The lateral dimension of enzyme-treated nanofibers is around 15 nm and corresponds to the cellulose aggregate scale in the pulp fiber cell wall.

10.25.3 CELLULOSE NANOPAPER

After disintegration from wood fibers, cellulose nanofibers are available as a dilute (≈1%) water suspension. Such a water suspension can be vacuum-filtered or cast, and strong interfibril interaction is obtained during gel drying so that a porous wood cellulose nanopaper is formed, Henriksson et al. (2008). The nanopaper is a fibrous network analogous to conventional paper. The main difference is that the nanopaper is structured at the nanoscale rather than at the microscale of conventional paper. The fine structure and improved properties of the fibrous unit leads to interesting mechanical properties, including 250 MPa tensile strength and high work-to-fracture. The high strain-to-failure is most remarkable and is due to a nanofiber slippage mechanism. For comparative purposes, typical strain-to-failure data for conventional high-strength microscale paper (kraft liner board) are 3–4%, rather than the 10% achievable with nanopaper (Henriksson et al. (2008)). Note the data for the most recent form of nanopaper, where the drying method has been controlled so that the material becomes very tough (see Figure 10.84).

10.25.4 CELLULOSE NANOCOMPOSITES

As the potential of wood-based cellulose nanocomposites is discussed, moisture sensitivity is an issue. In the absence of amorphous polysaccharides also present in plant cell walls, highly

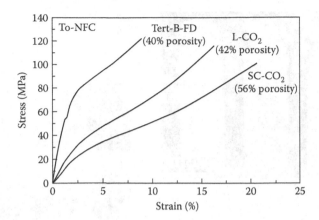

FIGURE 10.84 Stress–strain curves for cellulose nanopaper of high specific surface area. Note the toughness (large area under stress-strain curve). (After Sehaqui, H. et al. 2011b. *Biomacromolecules*, 12(10): 3638–3644.)

crystalline cellulose has very low moisture adsorption. This is supported by detailed cellulose-moisture interaction studies by molecular dynamics simulations, where it has been shown that the interior of the crystal is inaccessible to water. It is also possible to chemically modify cellulose nanofibers (i.e., by acetylation) in order to reduce hygroscopicity. Since fibril surface and disordered cellulose adsorb more moisture, those regions are most likely to be influenced by acetylation.

The tensile strength, Young's modulus, and work-of-fracture of cellulose network nanocomposites depend strongly on the cellulose content. High-volume fraction composites have been prepared by impregnation of porous nanopaper structures with monomers, followed by polymerization to form a thermoset composite. These materials show high modulus and flexural strength but are quite brittle. In brittle materials, flexural strength is controlled by defect size and depends strongly on specimen size and surface smoothness. For applications where ductility and toughness are required, more ductile polymer matrices are of interest.

Wood cellulose nanofiber networks also form if a starch/glycerol mixture containing suspended nanofibers is cast (Svagan et al. 2007). If the matrix is highly plasticized and nanofibers are well dispersed and of high content, the resulting biocomposite shows an attractive combination of strength, modulus, and work-to-fracture. The reason is high damage tolerance due to the fine nano paper structure in a highly ductile matrix. Again, nanofiber slippage is an important deformation mechanism, providing ductility.

Work by Nogi et al. (2006) demonstrates additional interesting characteristics of cellulose nanocomposite properties such as optical transparency and low thermal expansion. The reason is low in-plane axial thermal expansion of cellulose crystals combined with the random-in-the-plane network structure. It suggests that the in-plane hygroscopic expansion is also extremely low. Pure cellulose is thermally stable, with degradation typically commencing at temperatures exceeding 300°C. Recently, it was also shown that thin films of wood cellulose nanofibers can be both transparent and have low oxygen permeability, and this was utilized in PLA film coatings (Fukuzumi et al. 2009).

10.25.5 Clay Nanopaper-Based Fiberboards

WBP with fire retardant characteristics are of great industrial interest. Previous materials, such as cement-bonded WBP tend to have high density and be very brittle. One way to address the problem is to reduce the particle size, as already discussed for cellulose nanopaper. Thin silicate platelets from bentonite clay were therefore combined with NFC and processed by a wet vacuum-filtration route akin to paper-making (see Figure 10.85b). First, the clay was subjected to mechanical disintegration and the

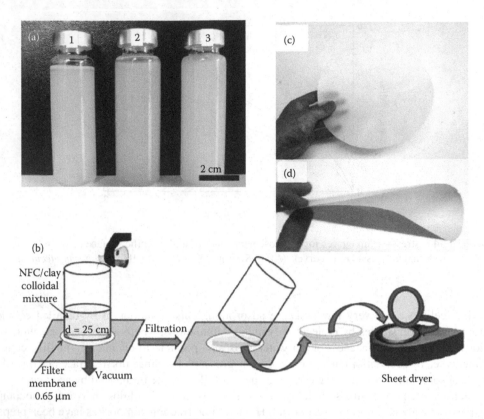

FIGURE 10.85   Hydrocolloid mixture of NFC and clay of high stability (a) processing route (b) and optical characteristics of clay nanopaper (c) and flexibility of clay nanopaper (d). (After Liu, A. et al. 2011. *Biomacromolecules*, 12: 633–641.)

clear fraction was primarily used. This fraction contains a large proportion of exfoliated silicate layers, single layers disintegrated from the clay particles. As this clay particle hydrocolloid was combined with NFC, stable dispersions were obtained (see Figure 10.85a). This was filtered through a membrane, the wet cake was dried and clay nanopaper was obtained. The paper structure showed some transparency, and the paper could be folded despite an inorganic content of 50–80 wt%. Also, the material was quite tough and strong. The fire retardance characteristics were quite interesting. Most likely, the clay platelets limited the oxygen diffusion and thus slowed down the burning process.

10.25.6   Cellulose Biofoams and Aerogels

Biofoams based only on biological polymers are interesting as replacements of petroleum-based polymer foams. Cellulose nanofibers are small enough to reinforce the cell walls in such foams. Efforts to include microcrystalline cellulose in starch foams processed by industrial methods show limited success in terms of property effects. In contrast, freeze-dried starch biofoams show the potential of cellulose nanofiber reinforcement (Svagan et al. 2008). The energy absorption, which is estimated as the area under the stress-strain curve, is more than doubled as 40wt% of cellulose nanofibers are added to the starch. This is highly relevant in the context of packaging materials. The biofoam shows as good energy absorption as expanded polystyrene (EPS) from petroleum-source, even at ambient conditions. The hygroscopic starch cell wall of the biofoam is stabilized by NFC, also at moisture contents of about 10%.

Aerogels is another class of materials where the liquid phase of a gel is removed without substantial shrinkage. One limitation of low-density aerogels is fragility and brittleness. Freeze-drying of

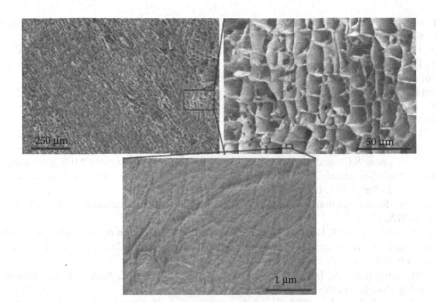

FIGURE 10.86 Foam structure of very-low-density cellulose foams at different scales. The lowest image shows the nanopaper structure of the cell wall. (After Sehaqui, H. et al. 2010. *Soft Matter* 6: 1824–1832.)

wood cellulose nanofiber gels results in low-density aerogels or foams with considerable mechanical robustness. These NFC aerogels consist of either nanofiber networks or nanopaper cells (see Figure 10.86), depending on how the drying is performed. They can also be functionalized by a secondary modification step, for instance with a conducting polymer coating (Pääkkö et al. 2008). These efforts and others such as flexible displays illustrate an interesting trend of industrial cellulose utilization of NFC expanding to high-tech applications.

Substantially improved performance of WBP requires biocomposite engineering, including fiber length and orientation control, fiber-matrix interface optimization, and reduced moisture adsorption, which can be achieved through chemical fiber modification. Chemical and microstructural heterogeneity can be reduced through the use of more highly purified cellulosic fibers. For the future, cellulose nanofibers from wood pulp fiber disintegration offer great promise as low-cost nanofiber reinforcements. This is based on a consistent raw material source with an established industrial infrastructure. The nanopaper and nanocomposites based on wood cellulose nanofiber networks show high strength, high work-of-fracture, low moisture adsorption, low thermal expansion, high thermal stability, high thermal conductivity, high optical transparency, and also exceptional barrier properties. Finally, bioinspired foams and low-density aerogels have potential in high-technology applications, although more widespread use of cellulose nanotechnology requires the development of new concepts for large-scale manufacturing.

REFERENCES

Anon. 1995. Magic number in sight for world production. *Wood Based Panels International* 17(October): 10–11.
Anon. 2011. Focus on OSB. *Wood Based Panels International* 31(2): 16–28.
Axer, J. 1975. New concepts for the production of particleboard. In *Proceedings of the 9th International Particleboard Symposium*, Pullman, WA, USA.
Baldwin, R.F. 1981. *Plywood Manufacturing Practices*. Miller Freeman Publications Inc., San Francisco.
Barbu, M.C., Stassen, O., and Frühwald, A. 2008. Impact of wood-based panels industry on the environment. *Proceeding of the SWST Annual Convention*, Concepción, ISBN 978-0-9817876-0-2, pp. 1–10.
Bolton, A.J., Humphrey, P.E., and Kavvouras, P.K. 1989. The hot pressing of dry-formed wood-based composites. Part III. Predicted Vapour Pressure and Temperature Variation with Time, Compared with Experimental Data for Laboratory Boards. *Holzforschung* 43.

Deppe, H. and Ernst, K. 2004. *Taschenbuch der Spanplattentechnik.* DRW Verlag Weinbrenner. (ISBN: 978-3871813498).

Dinwoodie, J.M. 1981. *Timber Its Nature and Behaviour.* Van Nostrand Reinhold Company, London.

Dunky, M., Piaai, A., and Van Leemput, M. (Editors) 2002. *Wood Adhesion and Glued Products. Working Group 1: Wood Adhesives.* State-of-the-art report Volume 1, COST Action E31.

Elias, R. and Irle, M.A. 1996. The acidity of stored Sitka spruce chips. *Holz als Roh-und Werkstoff* 54(1): 65–68.

Fischer, K. 1972. Modern flaking and particle reductionizing. In *Proceedings of the 6th International Particleboard Symposium,* Pullman, WA, USA.

Fukuzumi, H., Saito, T., Wata, T., Kumamoto, Y., and Isogai, A. 2009. Transparent and high gas barrier films of cellulose nanofibers prepared by TEMPO-mediated oxidation. *Biomacromolecules* 10: 162–165.

Galbraith, C.J., Cohen, S.C., and Ball, G.W. 1983. Self-releasing emulsifiable MDI isocyanate: An easy approach for all-isocyanate bonded boards. In *Proceedings of the 17th International Particleboard Symposium,* Pullman, WA, USA.

Geimer, R.L. 1982. Steam injection pressing. In *Proceedings of the 16th International Particleboard Symposium,* Pullman, WA, USA.

Henriksson, M., Henriksson, G., Berglund, L.A., and Lindström, T. 2007. An environmentally friendly method for enzyme-assisted preparation of microfibrillated cellulose (MFC) nanofibers. *European Polymer Journal* 43: 3434–3441.

Henriksson, M., Berglund, L.A., Isaksson, P., Lindstrom, T., and Nishino, T. 2008. Cellulose nanopaper structures of high toughness. *Biomacromolecules* 9, 1579.

Humphrey, P.E. and Bolton, A.J. 1989. The hot pressing of dry-formed wood-based composites. Part II. A simulation model for heat and moisture transfer, and typical results. *Holzforschung* 43(3): 199–206.

Irle, M.A. and Barbu, M.C. 2010. Wood-based panel technology. In Thoemen, H., Irle, M.A. and Sernek, M. (Editors). *Wood-Based Panels: An Introduction for Specialists.* Brunel University Press, London, England. (ISBN: 978-1-902316-82-6).

Johansson, J., Pizza, A., and Van Leemput, M. (Editors) 2002. Wood adhesion and glued products. Working group 2: Glued wood products. State-of-the-art report Vol. 2, COST Action E31 (ISBN 92-894-4892-X).

Liu, A. Walther, A. Ikkala, O. Belova, L., and Berglund, L.A. 2011. Clay nanopaper with tough cellulose nanofiber matrix for fire retardancy and gas barrier functions *Biomacromolecules,* 12: 633–641.

Mahut, J. and Reh, R. 2003. Plywood and decorative veneers. Technical University in Zvolen, 240 p. ISBN 80-228-1282-X

Maloney, T.M. 1993. *Modern Particleboard and Dry-Process Fibreboard Manufacturing.* Miller Freeman Publications, San Francisco. (TS 875.M3).

Meyer, B. 1981. *Urea Formaldehyde Resins.* Addison-Wesley Publishing Co., Inc., MA.

Moslemi, A.A. 1974. *Particleboard. Vol. 1—Materials. Particleboard. Vol. 2—Technology.* Southern Illinois University Press. (TS 875.M6).

Nakagaito, A.N. and Yano, H. 2005. Optically transparent composites reinforced with plant fiber-based nanofibers. *Appl. Phys.* A 80: 155–159.

Nogi, M. Ifuku, S. Abe, K. Handa, K., Nakagaito, A.N., and Yano, H. 2006. Fiber-content dependency of the optical transparency and thermal expansion of bacterial cellulose. *Appl. Phys. Lett.* 88: 13.

Nilsson, H., Galland, S., Larsson, P.T., Gamstedt, E.K., Nishino, T., Berglund, L.A., and Iversen, T. 2010. A non-solvent approach for high-stiffness all-cellulose biocomposites based on pure wood cellulose. *Comp Sci and Techn* 70: 1704–1712.

Pääkkö, M., Vapaavuori, J., Silvennoinen, R., Kosonen, H., Ankerfors, Lindstrom, T., Berglund, L. A., and Ikkala, O. 2008. Long and entangled native cellulose I nanofibers allow flexible aerogels and hierarchically porous templates for functionalities. *Soft Matter* 4(12): 2492–2499.

Pizzi, A. 1983. *Wood Adhesives Chemistry and Technology.* Marcel Dekker, Inc., New York.

Portenkirchner, K., Barbu, M., and Stassen, O. 2004. Combined waste air and water treatment plant for the wood panel industry. *Proceedings of 7th European Panel Products Symposium,* Bangor, ISBN 18-422-0057-7, 201–207.

Ressel, J. 2008. Presentation during the *3rd International Wood Academy,* Hamburg.

Roll, H., Barbu, M., Beck, P., Hoepner, D., Kaiser, U., and Lerach, K. 2001. Continuous hot press with cooling section for MDF. In *Proceedings of the 5th European Panel Products Symposium,* October 11–12, 2001, Llandudno, Wales, UK.

Saito, T., Hirota, M., Tamura, N., Kimura, S., Fukuzumi, H., Heux, L., and Isogai, A. Individualization of nano-sized plant cellulose fibrils by direct surface carboxylation using TEMPO catalyst under neutral conditions. 2009. *Biomacromolecules* 10(7): 1992–1996.

Schniewind, A.P., Cahn, R.W., and Bever, M.B. 1989. *Wood and Wood-Based Materials—Concise Encyclopedia.* Pergamon Press Plc, Headington Hill Hall, Oxford OX3 0BW, England.

Sehaqui, H., Salajková, M., Zhou, Q., and Berglund, L.A. 2010. Mechanical performance tailoring of tough ultra-high porosity foams prepared from cellulose I nanofiber suspensions. *Soft Matter* 6: 1824–1832.

Sehaqui, H., Allais, M., Zhou, Q., and Berglund, L.A. 2011a. Wood cellulose biocomposites with fibrous structures at micro- and nanoscale. *Composites Science and Technology* 71: 382–387.

Sehaqui, H., Zhou, Q., Ikalla, O., and Berglund, L.A. 2011b. Strong and tough cellulose nanopaper with high specific surface area and porosity, *Biomacromolecules,* 12(10): 3638–3644.

Shumate, R.D., Schenkmann, A.H., and Sloop, J.E. 1976. Experiences with air suspension classifiers in particleboard manufacture. *Proceedings of the 10th International Particleboard Symposium*, Pullman, WA, USA.

Stipek, J.W. (Translated by R.J. Vance) 1982. The present and future technology in wood particle drying. *Proceedings of the 16th International Particleboard Symposium*, Pullman, WA, USA.

Suchsland, O. and Woodson, G.E. 1986. *Fibreboard Manufacturing Practices in the United States.* U.S. Department of Agriculture, Forest Service, Agriculture Handbook No. 640.

Svagan, A.J., Samir, M.A.S.A., and Berglund, L.A. 2007. Biomimetic polysaccharide nanocomposites of high cellulose content and high toughness. *Biomacromolecules* 8: 2556–2563.

Svagan, A.J., Samir, M.A.S.A., and Berglund, L.A. 2008. Biomimetic foams of high mechanical performance based on nanostructured cell walls reinforced by native cellulose nanofibrils. *Adv. Mater., Adv. Mater.* 20: 1263–1269.

Thoemen, H. 2008. Pressing of panels. Presentations during the 3rd International Wood Academy, Hamburg.

Thoemen, H. and Humphrey, P.E. 1999. The continuous pressing process for wood-based panels: An analytical simulation model. In *Proceedings of the Third European Panel Products Symposium*, The BioComposites Centre, University of Wales, Bangor.

Thoemen, H., Meyer, N., and Barbu, M. 2006. Pressing from a fundamental point of view: Actions to increase press performance. *Proceedings of 4th "Press Users Club" Seminar*, Shanghai, China, October 23–26, 2006.

Thompson, G., Swain, J., Kay, M., and Forster, C.F. 2001. The treatment of pulp and paper mill effluent. *Bioresource Technol.* 77: 275–286.

Wadsworth, J. 2001. Focus on particleboard. Part 1: Europe and North America revealed. *Wood Based Panels International* 21(5): 9–25.

Walker, J.C.F. 1993. *Primary Wood Processing: Principles and Practice.* Chapman & Hall. London.

Walker, J.C.F. 2006. *Primary Wood Processing: Principles and Practice*, 2nd Edition. Springer, Amsterdam (ISBN 1402043929).

Williams, W. 1995. The panel pioneers: An historical perspective. *Wood Based Panels International* 17(September):6.

Youngquist, J.A. 1999. Wood-based composites and panel products. In *Wood Handbook—Wood as an Engineering Material.* Gen. Tech. Rep. FPL–GTR–113. Madison, WI: U.S. Department of Agriculture, Forest Service, Forest Products Laboratory. 463 p.

11 Chemistry of Wood Strength

Jerrold E. Winandy and Roger M. Rowell

CONTENTS

The source of strength in solid wood and engineered wood composites is the wood fiber. Wood is basically a series of tubular fibers or cells cemented together. Each fiber wall is composed of various quantities of three polymers: cellulose, hemicelluloses, and lignin. Cellulose is the strongest polymer in wood and, thus, is highly responsible for strength in the wood fiber because of its high degree of polymerization and linear orientation. The hemicelluloses act as part of a matrix for the cellulose and increase the packing density of the cell wall; hemicelluloses and lignin are also closely associated and make up the cell wall matrix in which the cellulose is imbedded. The actual role of hemicelluloses relative to the strength of virgin wood has recently been shown to be far more critical toward the overall engineering performance of wood than had previously been assumed. It is now accepted that one important role of hemicelluloses is to act as highly specific coupling agents capable of associating both with the more random areas (i.e., noncrystalline) of hydrophilic cellulose and the more amorphous hydrophobic lignin. Lignin not only holds wood fibers themselves together, but also helps bind carbohydrate molecules together within the cell wall of the wood fiber. The chemical components of wood that are responsible for mechanical properties can be viewed from three levels: macroscopic (cellular), microscopic (cell wall), and nano-molecular (polymeric) (Winandy and Rowell, 1984). Mechanical properties change with changes in the thermal, chemical, and/or biochemical environment. Changes in temperature, pressure, moisture, pH, chemical adsorption from the environment, UV radiation, fire, or biological degradation can have significant effects on the strength of wood.

Cellulose has long thought to be primarily responsible for strength in the wood fiber because of its high degree of polymerization and linear orientation. Hemicelluloses seem to act as a link between the fibrous cellulose and the amorphous lignin. Lignin, a phenolic compound, not only holds the fibers together, but also acts as a stiffening agent for the cellulose molecules within the fiber cell wall. All three cell wall components contribute in different degrees to the strength of wood. Together the tubular structure and the polymeric construction are responsible for most of the physical and chemical properties exhibited by wood.

The strength of wood can be altered by environmental agents. The changes in pH, moisture, and temperature; the influence of decay, fire, and UV radiation; and the adsorption of chemicals from the environment can have a significant effect on strength properties. Environmentally induced changes must be considered in any discussion on the strength of treated or untreated wood. This susceptibility of wood to strength loss, and the magnitude of that degrade, is directly related to the severity of its thermal/chemical/biochemical exposure.

The strength of wood can also be altered by preservative and fire-retardant compounds used to prevent environmental degradation. In some cases, the loss in mechanical properties caused by these treatments may be large enough that the treated material can no longer be considered the same as the untreated material. The treated wood may now resist environmental degradation but may be structurally different than untreated material. An integrated series of studies over the last 20 years specifically focused on this problem has now helped engineers account for these potential alterations from untreated wood in the structural design process. The approach is based on a cumulative-damage approach relating thermal- and chemical-degradation of the polymers responsible for wood strength to kinetic- or mechanical-based models (Lebow and Winandy, 1999a; Winandy and Lebow, 2001; Green et al., 2005; Green and Evans, 2008). With preservative-treated wood, a large amount of work was undertaken in the late 1980s and early 1990s to address treatment-related concerns in the structural design process. This work developed an understanding of the thermochemical issues that primarily control preservative-related strength loss (Winandy, 1996a) and was then applied by limiting treatment-processing levels allowed in standards, especially posttreatment kiln-drying temperatures (Winandy, 1996b). Similarly, with fire-retardant treated wood, an integrated series of studies define the pertinent issues related past problems with in-service thermal degradation of FRT wood exposed to elevated in-service temperatures (Winandy, 2001).

This chapter presents a theoretical model to explain the relationship between the mechanical properties and the chemical components of wood. This model is then used to describe the effects of altered composition on those mechanical properties. Many of the theories presented are only partially proven and just beginning to be understood. These theories should be considered as a starting point for dialogue between chemists and engineers that will eventually lead to a better understanding of the chemistry of wood strength.

11.1 MECHANICAL PROPERTIES

Even wood that has no discernible defects has extremely variable properties as a result of its heterogeneous composition and natural growth patterns. Wood is an anisotropic material in that the mechanical properties vary with respect to the three mutually perpendicular axes of the material (radial, tangential, and longitudinal). These natural characteristics are compounded further by the environmental influences encountered during the growth of the living tree. Yet wood is a viable construction material because workable estimates of the mechanical properties have been developed.

Mechanical properties relate a material's resistance to imposed loads (i.e., forces). Mechanical properties include: (1) measures of resistance to deformations and distortions (elastic properties), (2) measures of failure-related (strength) properties, and (3) measures of other performance-related issues. To preface any discussion concerning mechanical properties, two concepts need to be explained: stress (σ) and strain (ε).

Stress is a measure of the internal forces exerted in a material as a result of an application of an external force (i.e., load). Three types of primary stress exist: tensile stress, which pulls or elongates an object (Figure 11.1a); compressive stress, which pushes or compresses an object (Figure 11.1b); and shear stress, which causes two contiguous segments (i.e., internal planes) of a body to rotate (i.e., slide) within the object (Figure 11.1c). Bending stress (Figure 11.1d) is a combination of all three of the primary stresses and causes rotational distortion or flexure in an object.

Strain is the measure of a material's ability to deform—that is, elongate, compress or rotate—while under stress. Over the elastic range of a material, stress and strain are related to each other in a linear manner. In elastic materials, a unit of stress (σ) will cause a corresponding unit of strain (ε). This elastic theory yields one of the most critical engineering properties of a material, the elastic modulus (E). The theory is commonly known as Hooke's law (Larson and Cox, 1938):

$$E = \sigma/\varepsilon \qquad\qquad (11.1)$$

It applies to all elastic materials at points below their elastic or proportional limits.

Elastic theory relates a material's ability to be deformed by a stress to its ability to regain its original dimensions when the stress is removed. The criterion for elasticity is not the amount of deformation, but the ability of a material to completely regain its original dimensions when the stress is

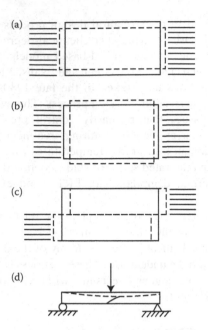

FIGURE 11.1 Examples of the three axil and one flexural types of stress. (a) Tension, (b) compression, (c) shear, and (d) bending.

removed. The opposite quality is viscosity, which can also be thought of as plasticity. A perfectly plastic body is one that makes no recovery of its original dimensions upon the removal of a stress. Wood is not ideally elastic; it will not completely recover deformation immediately on unloading, but in time, residual deformations tend to be recoverable. Wood is considered a viscoelastic material. This viscoelasticity explains the creep phenomenon in which a given load will induce an immediate deformation and if that load is allowed to remain on that piece, additional secondary deformation (i.e., creep) will continue to occur over long times. However, for simplicity's sake and because the engineering community often also assumes such, wood will be considered as an elastic material in this chapter.

The two main elastic moduli are modulus of elasticity, which describes the relationship of load (stress) to axial deformation (strain), and modulus of rigidity or shear modulus, which describes the internal distribution of shearing stress to shear strain or, more precisely, angular (i.e., rotational) displacement within a material.

Strength values are numerical estimates of the material's ultimate ability to resist applied forces. The major strength properties are ultimate or limit values for the stress–strain relationship within a material. Strength, in these terms, is the quality that determines the greatest unit stress a material can withstand without fracture or excessive distortion. In many cases the unqualified term strength is somewhat vague. It is sometimes more useful to think of specific strengths, such as compressive, tensile, shear, or ultimate bending strengths.

The American Society for Testing and Materials (ASTM) is the ISO-accredited organization in North America that standardizes testing procedures to provide reliable and universally comparable estimates of wood strength. Several ASTM Standards for wood (see ASTM 2009a,b,c,d,e) outline procedures for determining basic mechanical properties and deriving allowable design stresses. In performing a test, a load is applied to a specimen in a particular manner and the resulting deformation is monitored. The load information and the geometric configuration of the specimen allows the internal forces within the specimen (stress) to be calculated. The deformation information allows the internal distortion (strain) to be calculated when accepting specific assumptions. When stress

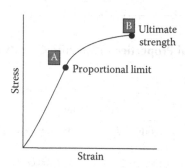

FIGURE 11.2   A typical stress–strain diagram for wood.

and strain are plotted against each other for a viscoelastic material like wood, an initially linear, then nonlinear stress–strain diagram is often developed (Figure 11.2).

The unit stress corresponding to the upper limit of the linear segment of the stress–strain diagram is known as the proportional limit (Figure 11.2A). This proportional or elastic limit measures the boundary of a material's completely recoverable strength. At stress levels below the proportional limit, a perfectly elastic material will regain its original dimensions and form. At stress levels in excess of the proportional limit, an elastic material will not regain its original shape; it will be permanently distorted.

The unit stress represented by the maximum ordinate is the ultimate (maximum) strength (Figure 11.2B). This point estimates the maximum stress at the time of failure. Many of the mechanical properties of interest to the engineer, such as maximum crushing strength or ultimate bending strength, describe this point of maximum stress.

11.2   FACTORS AFFECTING STRENGTH

11.2.1   MATERIAL FACTORS

11.2.1.1   Specific Gravity

Specific gravity is the ratio of the weight of a given volume of wood to that of an equal volume of water. As specific gravity increases, strength properties increase (USDA, 2010) because internal stresses are distributed among more molecular material. Mathematical approximations of the relationship between specific gravity and various mechanical properties are shown in Table 11.1.

11.2.1.2   Growth Characteristics

As a fibrous product from living trees, wood is subjected to many environmental influences as it is formed and during its lifetime. These environmental influences can increase the variability of the wood material and, thus, increase the variability of the mechanical properties. To reduce the effect of this inherent variability, standardized testing procedures using small, clear specimens of wood are often used. Small, clear specimens do not have knots, checks, splits, or reaction wood. However, the wood products used and of economic importance in the real world have these defects. Strength estimates derived from small clear specimens are reported because most chemical treatment data have been generated from small clear specimens. Further, comparative analyses of chemical treatment-related effects has clearly shown that clear wood material is affected more than with material containing knots and voids (Winandy, 1996a). Thus, it is now commonly assumed within the wood-engineering community that treatment effects are greater, the more defect-free the wood material and the straighter its grain.

TABLE 11.1
Functions Relating Mechanical Properties to Specific Gravity of Clear, Straight-Grained Wood

	Specific Gravity–Strength Relationship			
	Green Wood		Wood at 12% Moisture Content	
Property	Softwoods	Hardwoods	Softwoods	Hardwoods
Static bending				
MOR (KPa)	109,600 $G^{1.01}$	118,700 $G^{1.16}$	170,700 $G^{1.01}$	171,300 $G^{1.13}$
MOE (MPa)	16,100 $G^{0.76}$	13,900 $G^{0.72}$	20,500 $G^{0.84}$	16,500 $G^{0.7}$
WML (KJ/m³)	147 $G^{1.21}$	229 $G^{1.51}$	179 $G^{1.34}$	219 $G^{1.54}$
Impact bending (N)	353 $G^{1.35}$	422 $G^{1.39}$	346 $G^{1.39}$	423 $G^{1.65}$
Compression parallel (KPa)	49,700 $G^{0.94}$	49,000 $G^{1.11}$	93,700 $G^{0.97}$	76,000 $G^{0.89}$
Compression perpendicular (KPa)	8800 $G^{1.53}$	18,500 $G^{2.48}$	16,500 $G^{1.57}$	21,600 $G^{2.09}$
Shear parallel (KPa)	11,000 $G^{0.73}$	17,800 $G^{1.24}$	16,600 $G^{0.85}$	21,900 $G^{1.13}$
Tension perpendicular (KPa)	3800 $G^{0.78}$	10,500 $G^{1.37}$	6000 $G^{1.11}$	10,100 $G^{1.3}$
Side hardness (N)	6230 $G^{1.41}$	16,550 $G^{2.31}$	8590 $G^{1.49}$	15,300 $G^{2.09}$

Source: Adapted from U.S. Department of Agriculture, 2010. *Forest Service. Wood Handbook*. USDA General Technical Report GTR-190, Madison, WI, 508 p.
Note: G = specific gravity.

Because strength is affected by material factors such as specific gravity and growth characteristics, it is important to always consider property variability of clearwood. The coefficient of variation is the statistical parameter used to approximate the variability associated with each strength property. The estimated coefficient of variation of various strength properties can be found in Table 11.2. The variability of lumber and materials allowing knots, sloped grain and other naturally occurring growth characteristics is higher.

TABLE 11.2
Average Coefficients of Variation for Some Mechanical Properties of Clear Wood

Property	Coefficient of Variation (%)
Static bending	
Modulus of rupture	16
Modulus of elasticity	22
Work to maximum load	34
Impact bending	25
Compression parallel to grain	18
Compression perpendicular to grain	28
Shear parallel to grain	14
Maximum shearing strength	14
Tension parallel to grain	25
Side hardness	20
Toughness	34
Specific gravity	10

Source: Adapted from U.S. Department of Agriculture, 2010. *Forest Service. Wood Handbook*. USDA General Technical Report GTR-190, Madison, WI, 508 p.

11.2.2 ENVIRONMENTAL FACTORS

11.2.2.1 Moisture

Wood, which is a hygroscopic material, gains or loses moisture to equilibrate with its immediate environment. The equilibrium moisture content (EMC) is the steady-state level that wood achieves when subjected to a particular relative humidity and temperature. The eventual EMC of two similar specimens will differ if one approaches EMC under adsorbing conditions and the other approaches EMC under desorbing conditions. For example, if the relative vapor pressure of the environment is 0.65 (i.e., relative humidity of 65%), two similar specimens exposed under either adsorbing or desorbing conditions will equilibrate at moisture contents of approximately 11% and 13%, respectively (USDA, 2010) (Figure 11.3).

Wood strength is related to the amount of water in the wood fiber cell wall (Wilson, 1932; Gerhards, 1982; USDA, 2010). At moisture contents from oven-dry (OD) to the fiber-saturation point, water accumulates in the wood cell wall (bound water). Above the fiber-saturation point, water accumulates in the wood cell cavity (free water), and there are no tangible strength effects associated with a changing moisture content. However, at moisture contents between OD and the fiber-saturation point, water does affect strength. Increased amounts of bound water interfere with and reduce hydrogen bonding between the organic polymers of the cell wall (Rowell, 1980; USDA, 2010), which decreases the strength of wood. The approximate relationships are shown in Tables 11.3 and 11.4.

Not all mechanical properties change with moisture content. The performance of wood under dynamic loading conditions is a dual function of the strength of the material, which is decreased with increased moisture content, and the pliability of the material, which is increased with increased moisture contents. Changes in strength and pliability somewhat offset one another and, therefore, mechanical properties that deal with dynamic loading conditions are not nearly as affected by a changing moisture content as are static mechanical properties.

11.2.2.2 Temperature

Strength is related to the temperature of the working environment (Gerhards, 1982; Barrett et al., 1989; USDA, 2010). At constant moisture content, the immediate effect of temperature on strength is linear (Figure 11.4) and usually recoverable when the temperature returns to normal. In general, the immediate strength of wood is higher in cooler temperatures and lower in warmer temperatures. However, permanent (nonrecoverable) effects can occur. This relationship of permanent strength loss during extended high-temperature exposure can be dramatically influenced by higher moisture contents.

The immediate effects of increased temperature are an increase in the plasticity of the lignin and an increase in spatial size, which reduces intermolecular contact and is, thus, recoverable. Permanent

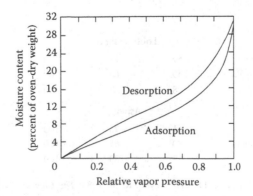

FIGURE 11.3   Adsorption–desorption isotherms for water vapor by spruce at 25°C.

TABLE 11.3
Approximate Change in Mechanical Properties of Clear Wood with Each 1% Change in Moisture Content

Property		Change per 1% Change in Moisture Content
Static bending	Fiber stress at proportional limit	5
	Modulus of rupture	4
	Modulus of elasticity	2
	Work to proportional limit	8
	Work to maximum load	0.5
Impact bending	Height of drop causing complete failure	0.5
Compression parallel to grain	Fiber stress at proportional limit	5
	Maximum crushing strength	6
Compression perpendicular to grain	Fiber stress at proportional limit	5.5
Shear parallel to grain	Maximum shearing strength	3
Hardness	End	4
	Side	2.5

Source: Adapted from Winandy, J.E. and Rowell, R. M. 1984. *The Chemistry of Solid-wood.* ACS Sym Series #208, Washington, DC, 211–255.

effects manifest themselves as an actual reduction in wood substance or weight loss via degradative mechanisms, and are thereby nonrecoverable. This permanent thermal effect on wood strength has been extensively studied (LeVan et al., 1990; Winandy, 2001; Green et al., 2003, 2005; Green and Evans, 2008) and predictive kinetic-based models have been developed (Woo, 1981; Lebow and Winandy, 1999a; Green et al., 2003). The reasons for these permanent thermal effects on strength

TABLE 11.4
Relationship between Some Mechanical Properties and Moisture Content (Property at 12% Moisture Content Set at 100%)

	Moisture Content				
Property	**Green**	**19%**	**12%**	**8%**	**Oven-dry**
		Douglas-Fir			
Modulus of rupture	62	76	100	117	161
Compression parallel to grain	52	68	100	124	192
Modulus of elasticity	80	88	100	108	125
		Loblolly Pine			
Modulus of rupture	57	72	100	121	175
Compression parallel to grain	49	66	100	127	203
Modulus of elasticity	78	87	100	109	128
		Aspen			
Modulus of rupture	61	75	100	118	165
Compression parallel to grain	50	67	100	126	199
Modulus of elasticity	73	83	100	111	137

Source: Adapted from Winandy, J.E. and Rowell, R. M. 1984. *The Chemistry of Solid-wood.* ACS Sym Series #208, Washington, DC, 211–255.

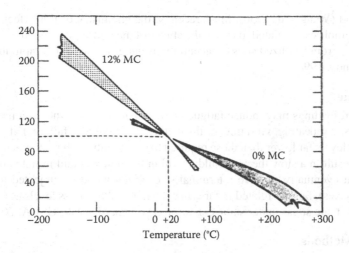

FIGURE 11.4 Immediate effect of temperature on strength properties expressed as a percent of value at 20°C. Trends illustrated are composites from several studies on three strength properties (MOR, $T_{\parallel}$, and $C_{\parallel}$).

relate to changes in the wood polymeric substance and structure and predictive models have been developed (Winandy and Lebow, 2001).

11.2.3 LOAD FACTORS

11.2.3.1 Duration of Load

The ability of wood to resist load is dependent upon the length of time the load is applied (Wood, 1951; Gerhards, 1977; USDA, 2010). The load required to cause failure over a long period of time is much less than the load required to cause failure over a very short period of time. Wood under impact loading (duration of load >1 s) can sustain nearly twice as great a load as wood subjected to long term loading (duration of load >10 years). This time-dependent relationship (Wood, 1951) can be seen graphically in Figure 11.5. Hydrolytic chemical treatments have long been known to incur

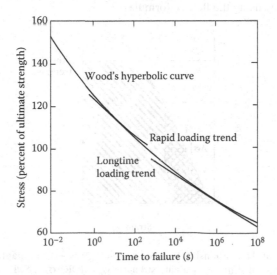

FIGURE 11.5 Hyperbolic load–duration curve with rapid loading and longtime loading trends for bending.

brashness in wood (Wangaard, 1950). More recently, the chemistry of this effect on treated wood was shown to be empirically related to loss in the ability of the hydrolyzed wood material to dissipate strain energy away from localized stress concentrations under impact-type, rapid loadings (ultimate in <1–2 s) (Winandy, 1995b).

11.2.3.2 Fatigue

Cyclic or repeated loadings may induce fatigue failures. Fatigue resistance is a measure of a material's ability to resist repeating, vibrating, or fluctuating loads without failure. Fatigue failures often result from stress levels far lower than those required to cause static failure. Repeated or fatigue-type stresses usually result in a slow thermal buildup within the material and initiate and propagate tiny micro checks that eventually grow to a terminal size. When wood is subjected to repeated stress (e.g., 5.0×107 cycles), fatigue-related failures may be induced by stress levels as low as 25–30% of the anticipated ultimate stress under static conditions (Wangaard, 1950; USDA, 2010).

11.2.3.3 Test Methods

To design with any material, mechanical property estimates need to be developed. ASTM standard test methods detail the procedures required to determine mechanical properties via stress–strain relationships (see ASTM 2009a,b,c,d,e).

11.2.4 FLEXURAL LOADING PROPERTIES

Flexural (bending) properties are important in a wood design. Many structural designs recognize either bending strength or some function of bending, such as deflection, as the limiting design criterion. Structural examples in which bending-type stresses are often the limiting consideration are bridges, floors or bookshelves. Five mechanical properties are derived from the stress–strain relationship of a standard bending test: modulus of rupture (MOR), fiber stress at proportional limit (FSPL), modulus of elasticity (MOE), work to proportional limit (WPL), and work to maximum load (WML).

11.2.4.1 Modulus of Rupture

The MOR is the ultimate bending strength of a material. Thus, MOR describes the load required to cause a wood beam to fail and can be thought of as the ultimate resistance or strength that can be expected (Figure 11.2, Point B; Figure 11.6, Point B) from a wood beam exposed to bending-type stress. MOR is derived by using the flexure formula

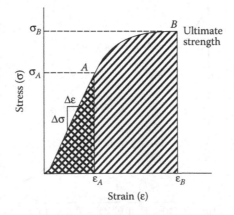

FIGURE 11.6 Examples of the relationship between a typical stress–strain diagram and some mechanical properties. Key A, proportional limit; B, ultimate strength; σ_B, MOR; σ_A, FSPL; $\Delta\sigma/\Delta\varepsilon$ (from origin to A), MOE; $\int\sigma_A\,d\varepsilon$, WPL; and $\int\sigma_B\,d\varepsilon$, WML.

$$MOR = Mc/I \tag{11.2}$$

which assumes an elastic response although that assumption is not exactly true, where M is the maximum bending moment, c is a measure of distance from the more highly stressed flanges of the beam to its neutral axis (i.e., the plane where internal stress becomes zero as those stresses change from tension on one flange to compression on the opposite), and I is the moment of inertia, which relates the bending moment to the geometric shape of the beam. Engineers often simplify these geometric factors (c/I) to a parameter known as S, the section modulus

$$S = I/c \tag{11.3}$$

For a rectangular or square beam in bending under center point loading, the flexure formula is varied to reflect loading conditions and beam geometry:

$$MOR = (1.5 * P * L)/(b * h^2) \tag{11.4}$$

where P is the ultimate load, L is the span of beam, b is the width of beam, and h is the height of beam.

11.2.4.2   Fiber Stress at Proportional Limit

The FSPL is the maximum bending stress a material can sustain under static conditions and still exhibit no permanent set or distortion. It is by definition the amount of unit stress on the y-coordinate at the proportional limit of the material (Figure 11.2, Point A; Figure 11.6, Point A). FSPL is also derived using the flexure formula where M is the bending moment at the proportional limit and S is the section modulus.

11.2.4.3   Modulus of Elasticity

The MOE quantifies a material's elastic (i.e., recoverable) resistance to deformation under load. The MOE corresponds to the slope of the linear portion of the stress–strain relationship from zero to the proportional limit (Figure 11.6). Stiffness (MOE * I) is often incorrectly thought to be synonymous with MOE. However, MOE is solely a material property and stiffness depends both on the material and the size and geometry of the beam. Large and small beams of similar material would have similar MOEs but different stiffnesses. The MOE can be calculated from the stress–strain curve as the change in stress causing a corresponding change in strain.

11.2.4.4   Work to Proportional Limit

The WPL is the measure of work performed, that is, energy used, in going from an unloaded state to the elastic or proportional limit of a material (Figure 11.6). For a beam of rectangular cross section under center point loading, WPL is calculated as the area under the stress–strain curve from zero to the proportional limit.

11.2.4.5   Work to Maximum Load

Work to maximum load (WML) is the amount of work needed to actually fracture or fail a material. It is a measure of the amount of energy required to fracture the material. Toughness or work to total load are analogous properties, but their final limit state also includes energy absorbed beyond the ultimate failure. WML is calculated as the area under the stress–strain curve from zero to the ultimate strength of the material (Figure 11.6). Because WML is a measure of work both below and beyond the proportional limit it is derived by either graphical approximations or by means of calculus.

11.2.5 Axial Loading Properties

Because of the anisotropic and heterogeneous nature of wood, there can be profound differences in the strength in various directions. Wood is stronger along the grain (parallel to the longitudinal axis of the log or longitudinal axis of the wood cell) than perpendicular to the grain (at right angles to the longitudinal axis). Axial loads describe forces that have the same line of action and are, thus, both parallel and concurrent. Because there is no eccentricity in the application of these forces, they do not induce flexure or bending moments.

Mechanical properties dealing with axial loading conditions are maximum crushing strength (compression parallel to the grain), compression perpendicular to the grain, tension parallel to the grain, and tension perpendicular to the grain.

11.2.5.1 Compression Parallel to the Grain

If wood is considered a bundle of straws bound together, then a compression parallel to the grain ($C_\parallel$) can be thought of as a force trying to compress the straws from end to end. The distance through which compressive stress is transmitted does not increase or magnify the stress, but the length over which the stresses are carried is important. If the length of the column is far greater than the width, the specimen may buckle. This stress is analogous to bending-type failure rather than axial-type failure. As long as specimen width is great enough to preclude buckling, $C_\parallel$ is solely an axial property.

Examples loading conditions inducing compression parallel to the grain are wooden columns or the top chord of a roof truss. Compression-parallel-to-the-grain strength or the maximum crushing strength is derived at the ultimate limit value of a standard stress–strain curve. The strength of wood in $C_\parallel$ is derived by

$$C_\parallel = P/A \tag{11.5}$$

where $C_\parallel$ is the stress in compression parallel to grain, P is the maximum axial compressive load, and A is the area over which load is applied.

11.2.5.2 Compression Perpendicular to the Grain

Compression perpendicular to the grain ($C_\perp$) can be thought of as stress applied perpendicular to the length of the wood cell. Therefore, in our straw example, the straws (or wood cells) are being crushed at right angles to their length. Until the cell cavities are completely collapsed, wood is not as strong perpendicular to the grain as it is parallel to the grain. However, once the wood cell cavities collapse, wood can sustain a nearly immeasurable load in $C_\perp$. Because a true ultimate stress is nearly impossible to achieve, maximum $C_\perp$ in the sense of ultimate load-carrying capacity is undefined and discussions of $C_\perp$ are usually confined to stress at some predetermined limit state such as the proportional limit or 4% deflection.

Compression-perpendicular-to-the-grain stresses are found whenever one member is supported upon another member at right angles to the grain. Examples of loading conditions inducing compression perpendicular to the grain stress are the bearing areas of a beam, truss, or joist.

The $C_\perp$ strength is derived by

$$C_\perp = P/A \tag{11.6}$$

where $C_\perp$ is the stress in compression perpendicular to the grain, P is the proportional limit load, and A is the area.

11.2.5.3 Tension Parallel to the Grain

A tension parallel to the grain ($T_\parallel$) stress is a force trying to elongate the wood cells, or straws in our straw example. Wood is extremely strong in $T_\parallel$. The distance through which tensile stress is transmitted does not increase the stress. The $T_\parallel$ is difficult to measure because of the difficulty in securely gripping the tensile specimen in the testing machine, especially with clear straight-grained wood. Often $T_\parallel$ of clear straight-grain wood is conservatively estimated by the MOR (ultimate strength in bending). This conversion is accepted because bending failure of clear wood often occurs on the lower face of a bending specimen where the lower face fibers are under tensile-type stresses. An example of a loading condition inducing tension parallel to the grain stress would be the bottom chord of a truss that is under tensile stress. The $T_\parallel$ strength of wood is derived by the formula:

$$T_\parallel = P/A \qquad (11.7)$$

where $T_\parallel$ is the stress in tension parallel to the grain, P is the maximum load, and A is the area.

11.2.5.4 Tension Perpendicular to Grain

Tension perpendicular to the grain $(T_\perp)$ is induced by a tensile force applied perpendicular to the longitudinal axis of the wood cell. In this case, the straws (or wood cells) are being pulled apart at right angles to their length. The $T_\perp$ is extremely variable and is often avoided in discussions on wood mechanics. However, $T_\perp$ stresses often cause cleavage or splitting failures along the grain, which can dramatically reduce the structural integrity of large beams. Failures from $T_\perp$ are sometimes found in large beams that dry while in service. For example, if a beam is secured by a top and a bottom bolt at one end, shrinkage may eventually cause cleavage or splitting failures between the top and bottom bolt holes. Wood can be cleaved by $T_\perp$ forces at a relatively light load. It is this weakness that is often exploited in karate and other demonstrations of human strength. The $T_\perp$ strength of wood is derived by

$$T_\perp = P/A \qquad (11.8)$$

where $T_\perp$ is the stress in tension perpendicular to the grain, P is the maximum load, and A is the area

11.2.6 OTHER MECHANICAL PROPERTIES

11.2.6.1 Shear

Shear parallel to the grain (γ) measures the ability of wood to resist the slipping or sliding of one plane past another parallel to the grain. Shear strength is derived in a manner similar to axial properties, by using the equation:

$$\gamma = P/A \qquad (11.9)$$

where γ is the shear stress parallel to grain, P is the shearing load, and A is the area of the shear plane through the material.

11.2.6.2 Hardness

Hardness is used to represent the resistance to indentation and/or marring. Hardness (ASTM, 2009a) is measured by the load required to embed a 1.128-cm steel ball one-half its diameter into the wood. While a material may be softer or harder in a common vernacular, hardness, in engineering terms, is a material property that is measured using specified methods detailing sizes, sources and test speeds. Beyond those specific test conditions, the term hardness when used in common language may have widely differing meanings to different people.

11.2.6.3 Shock Resistance

Shock resistance or energy absorption is a function of a material's ability to quickly absorb and then dissipate energy via deformation. This is an important property for baseball bats, tool handles, and other articles that are subjected to frequent shock loadings. High shock resistance on energy absorption properties requires both the ability to sustain high ultimate stress and the ability to deform greatly before failing.

Shock resistance can be measured by several methods. With wood, three of the most often used methods are work tests (to maximum load), impact bending tests and toughness tests. Both of the later two test methods yield measures of strength and pliability, mutually referred to as energy absorption. These two measures of shock resistance are similar but are not particularly relative to one another. Impact bending is tested by dropping a weight onto a beam from successively increasing heights (ASTM, 2009a). All that is recorded is the height of drop causing complete failure in a beam such that for a different sized beam or a different mass of weight the measured value would most certainly change. Toughness is the ability of a material to resist a single impact-type load from a pendulum device (ASTM, 2009a). Thus, toughness is similar to impact bending in that both are measures of energy absorption or shock resistance. Yet, critical differences exist. Toughness uses a single ultimate load and impact-type bending, whereas Impact Bending uses a series of progressively increasing, multiple loads in which the earlier load history can certainly alter the eventual result. While each test method defines a material characteristic, each measured property should only be compared within the limited definitions of that method. They should not be compared on a method-to-method basis, nor compared if tested on differing sized or conditioned materials.

11.3 COMPONENTS OF STRENGTH

11.3.1 RELATIONSHIP OF STRUCTURE TO CHEMICAL COMPOSITION

The chemical components responsible for the strength properties of wood can be theoretically viewed from three distinct levels: the macroscopic (cellular) level, the microscopic (cell wall) level, and the nano-molecular (polymeric) level.

11.3.1.1 Macroscopic Level

Wood with its inherent strength is a product of growing trees. The primary function of the woody trunk of the living tree is to provide support for the phototropic energy factory (i.e., leaves or needles) at the top and to provide a conduit for moving water and nutrients moving up to those leaves or needles. The phototropic sugars produced by the leaves or needles mostly move down the stem via the bark tissues. Woody tissues, interior to the bark, exist as concentric bands of cells oriented for specific functions (Figure 11.7). Thin-walled earlywood cells act both as conductive tissue and support; thick-walled latewood cells provide support. Each of these cells is a single fiber. Softwood fibers average about 3.5 mm in length and 0.035 mm in diameter. Hardwood fibers are generally shorter (1–1.5 mm) and smaller in diameter (0.015 mm). Alternating layers of earlywood and latewood fibers comprise a series of composite bands, bonded together by a phenolic adhesive, lignin. Each band is termed a growth ring and is anisotropic in character and is reinforced in two of the three axial directions by longitudinal parenchyma and ray parenchyma cells, respectively These parenchyma cells function as a means of either longitudinal or radial nutrient conduction and as a means of providing lateral support by increased stress distribution (Figure 11.8a,b). Because wood is a reinforced composite material, its structural performance at the cellular level has been likened to reinforced concrete (Freudenberg, 1932; Mark, 1967).

The macroscopic level of consideration takes into account fiber length and differences in cell growth, such as earlywood, latewood, reaction wood, sapwood, heartwood, mineral content, extractive

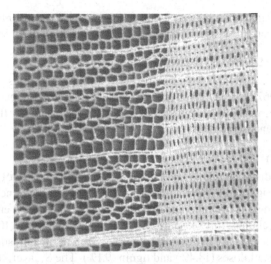

FIGURE 11.7   Scanning electron micrograph showing cellular structure with thin-walled earlywood (left) and thicker-walled latewood (right) at ×120.

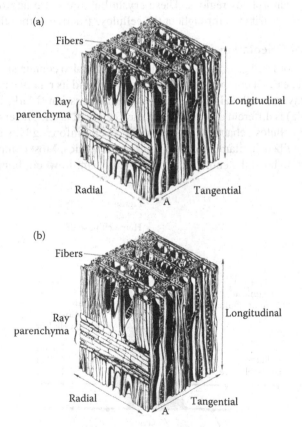

FIGURE 11.8   (a) Cube of hardwood and (b) cube of softwood.

chemicals, resin content, and so on. Differences in cellular anatomy, environmental-controlled growth patterns, and chemistry can cause significant differences in the strength of wood.

11.3.1.2 Microscopic Level

At the microscopic level, wood has been compared to multipart systems, such as filament-wound fiber products (Mark, 1967). Each component complements (i.e., reinforces) the other in such a manner that, when considering the overall range of physical performance, the components together outperform the components separately.

11.3.1.3 Composition

Within the cell wall are distinct regions (see Figure 3.12, Chapter 3) each of which has distinct composition and attributes (Figure 11.9). For a typical softwood the middle lamella and primary wall are mostly lignin (8.4% of the total weight) and hemicellulose (1.4%), with very little cellulose (0.7%). The S_1 layer consists of cellulose (6.1%), hemicellulose (3.7%), and lignin (10.5%). The S_2 layer is the thickest cell wall layer and has the highest carbohydrate content; it is mostly cellulose (32.7%) with lesser quantities of hemicelluloses (18.4%) and lignin (9.1%). The S_3 layer, the innermost layer, consists of cellulose (0.8%), hemicelluloses (5.2%), and very little lignin. One interesting caveat is that while the relative lignin ratio is low within the S_2 layer, the largest amount of lignin exists within this layer because of its large overall mass.

The large number of hydrogen bonds existing between cellulose molecules results in such strong lateral associations that certain areas of the cellulose chains are considered crystalline. More than 60% of the cellulose (Stamm, 1964) exists in this crystalline form, which is stiffer and stronger than the less crystalline or amorphous regions. These crystalline areas are approximately 60 nm long (Thomas, 1981) and are distributed throughout the cellulose fraction within the cell wall.

11.3.1.4 Microfibril Orientation

Microfibrils are highly ordered groupings of cellulose that may also contain small quantities of hemicellulose and lignin. The exact composition of the microfibril and its relative niche between the polymeric chain and the layered cell wall are subjects of great discussion (Mark, 1967). The microfibril orientation (fibril angle) is different and distinct for each cell wall layer. The entire microfibril system is a grouping of rigid cellulose chains analogous to the steel reinforcing bars in reinforced concrete or the glass or graphite fibers in filament-wound-reinforced plastics. Most composite materials use an adhesive of some type to bond the entire material into a system. In wood, lignin fulfills the function

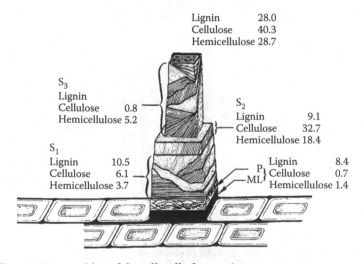

FIGURE 11.9   Chemical composition of the cell wall of scots pine.

FIGURE 11.10 Scanning electron micrograph of softwood fibers embedded in lignin (×400).

of a matrix material. Yet, it is not truly or solely an adhesive and by itself adds little to strength (Lagergren et al., 1957). Lignin is a hydrophobic phenolic material that surrounds and encrusts the carbohydrate complexes (Figures 11.10 and 11.11). It aids in holding the cell components together at the microscopic level. Lignin also seems to be responsible for part of the stiffness of wood and it most certainly is primarily responsible for the exclusion of water from the moisture-sensitive carbohydrates. Rubbery wood, a viral disease of certain varieties of apple (*Malus* spp.), is characterized by extremely flexible wood. The affected wood has been shown (Prentice, 1949) to have cells rich in cellulose but low in lignin.

11.3.1.5 Nano-Molecular Level

At the nano-molecular level the relationship of strength and chemical composition deals with the individual polymeric components that make up the cell wall. The physical and chemical properties

FIGURE 11.11 Scanning electron micrograph of delignified softwood fiber wall (×16,000).

of cellulose, hemicelluloses, and lignin play a major role in the chemistry of strength. However, our perceptions of wood polymeric properties are based on isolated polymers that have been removed from the wood system and, therefore, possibly altered. The three individual polymeric components (cellulose, hemicelluloses, and lignin) may be far more closely associated and interspersed with one another than has heretofore been believed (Attalla, R. personal communication, June 2002). Recent theories speculate that the crystalline and amorphous regions of the cellulose chains are more diffuse and less segregated between themselves than may have earlier been believed (Attalla, 2002).

Cellulose is anhydro-D-glucopyranose ring units bonded together by (β-1–4-glycosidic linkages. The greater the length of the polymeric chain, the higher the degree of polymerization, the greater the strength of the unit cell (Ifju, 1964; Mark, 1967) and, thus, the greater the strength of the wood. The cellulose chain may be 5000–10,000 units long. Cellulose is extremely resistant to tensile stress because the unit stiffness derived from the covalent bonding within the pyranose ring and mutual reinforcement shared between the individual units when grouped. Hydrogen bonds within the cellulose provide rigidity to the cellulose molecule via stress transfer and allow the molecule to absorb shock by subsequently breaking and reforming. Past theories have speculated that the cellulose is the predominate factor in wood strength. It clearly is the individually strongest component, but recent work has clearly shown that when the combined three-part system of cellulose–hemicellulose–lignin is considered, any hydrolytic or enzymatic action upon the hemicelluloses always seems to manifest themselves in the earliest levels of strength loss in woody materials (Levan et al., 1990; Lebow and Winandy, 1999a; Winandy and Lebow, 2001).

The hemicelluloses are a series of carbohydrate molecules that consist of various elementary sugar units, primarily the six-carbon sugars, D-glucose, D-galactose, and D-mannose, and the five-carbon sugars, L-arabinose and D-xylose. Hemicelluloses have linear chain backbones (primarily glucomannans and xylan chains) that are highly branched and have a lower degree of polymerization than cellulose. The sugars in the hemicellulose structure exhibit hydrogen bonding both within the hemicellulose chain as well as between other hemicellulose and amorphous cellulose regions. Most hemicelluloses are found in interspersed within or on the boundaries of the amorphous regions of the cellulose chains and in close association with the lignin. Hemicellulose may be the connecting material between cellulose and lignin. The precise role of hemicellulose as a contributor to strength has long been a subject of conjecture. Recent work on hydrolytic chemical agents and enzymatic decay has indicated that early degradation of hemicellulose(s), especially degradation of the shorter branched monomers of D-galactose and L-arabinose along the hemicellulose main chains, seem primarily responsible for the earliest portions of strength loss in wood exposed to severe thermal, chemical or biological exposures (Winandy and Morrell, 1993; Winandy and Lebow, 2001; Curling et al., 2002). Other work, has speculated that the actions of hydrolytic agents at the aforementioned sheath of hemicellulose along the boundary areas of the cellulose microfibrils (Figure 11.12) may account for the radial degradation fissures (i.e., interwall cracks of the S_2 layer often seen during chemical or biological decay (Larsen et al., 1995).

Lignin is often considered nature's adhesive. It is the least understood and most chemically complex polymer of the wood-structure triad. Its composition is based on highly organized three-dimensional phenolic polymers rather than linear or branched carbohydrate chains. Lignin is the most hydrophobic (water repelling) component of the wood cell. Its ability to act as an encrusting agent on and around the carbohydrate fraction, and thereby limit water's influence on that carbohydrate fraction, is the cornerstone of wood's ability to retain its strength and stiffness as moisture is introduced to the system. Dry delignified wood has nearly the same strength as normal dry wood, but wet delignified wood has only approximately 10% of the strength of wet normal wood (Lagergren et al., 1957). Thus, wood strength is due in part to lignin's ability to limit the access of water to the carbohydrate moiety and thereby lessen the influence of water on wood's hydrogen-bonded structure.

Another cell wall model that shows the orientation of the cell wall polymers is shown in Figure 11.13. This model clearly shows little or no association between lignin and cellulose. The hemicelluloses and lignin are closely associated with each other and bonded together in a few places.

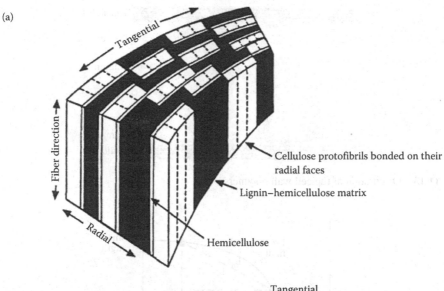

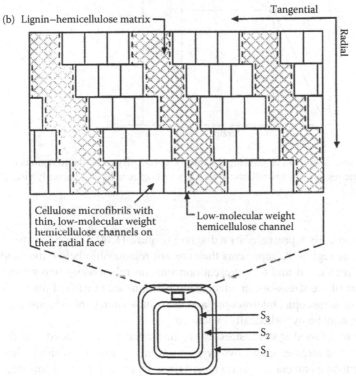

FIGURE 11.12 Representation of proposed ultrastructural models of the arrangement of lignin, cellulose, and hemicellulose in S_2 layer of the wood cell wall. (a) Three-dimensional view (Adapted from Kerr, A.J. and Goring, D.A.I. 1975. *Cellul. Chem. Technol.* 9:563–573.) and (b) (Adapted from Larsen, M.J., Winandy, J.E., and Green, F. III.1995. *Materials and Organism*, 29(3):197–210.)

11.4 RELATIONSHIP OF CHEMICAL COMPOSITION TO STRENGTH

To relate chemical composition to strength properties, to work and toughness properties, and eventually to elastic parameters, a preliminary model or hypothesis must be developed to aid in conceptualizing the relationship between strength and wood composition. The theoretical relationship of

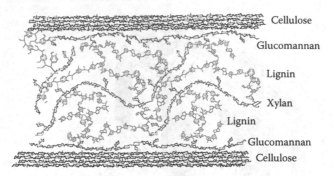

FIGURE 11.13 Orientation of the cell wall polymers.

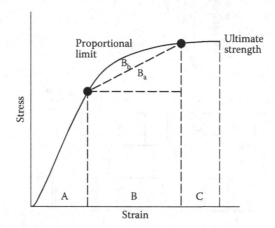

FIGURE 11.14 Typical stress–strain curve showing the three theoretically identifiable regions of mechanical behavior. Key: (A) elastic region; (B) viscoelastic—partially viscoelastic B_a and partially viscoplastic B_b; and (C) plastic region.

stress to strain can be graphically represented by a diagram (Figure 11.14). If wood is assumed to be an elastic material, a linear region, A, represents the constant relationship below the proportional limit, and the nonlinear regions, B and C, represent nonconstant relationship beyond the proportional limit. If each region of the stress–strain relationship is examined at each of the three distinct levels of wood structure (macroscopic, microscopic, and nano-molecular), the relationship between strength and composition may be hypothetically explained.

As loads are applied to a wood system, stresses are immediately introduced and distributed throughout the material. The stresses cause two types of strain or distortion: immediate, under which wood can be described as an elastic material, and time dependent. Even at low stress levels, permanent set or distortion will eventually be induced in a wood member. This phenomenon, known as creep, explains why wood is considered viscoelastic. But, for purposes of simplification, the ensuing discussion will be confined to immediate strain or distortion and will consider wood as an elastic material. Immediate strain or distortion can be conceptualized at each of the three distinct levels of wood structure.

11.4.1 BELOW PROPORTIONAL LIMIT (ELASTIC STRENGTH)

When a load is applied to a piece of wood, at the nano-molecular level, hydrogen bonds between and within individual polymer chains are breaking, sliding (uncoiling), and subsequently reforming

(Figure 11.15a–c, respectively); C–C and C–O bonds are elastically distorting within the ring structures (Figure 11.16).

At the microscopic level, hydrogen bonds between adjacent microfibrils are breaking and reforming (Figure 11.15a–c), to allow the micro-fibrils to slide by one another with only the disruption of the hydrogen bonds that are subsequently reformed. Additionally, the individual cell wall layers are distorting in relation to each other, but no permanent set or distortion is occurring between these individual cell wall layers.

At the macroscopic level, there is distortion with its inherent build-up of elastic strain energy between the individual cells, but it is not permanent because the stresses are being distributed between the individual cells such that no permanent translocation or set is introduced.

Within the limits of the elastic model, all strain or distortion resulting from the accumulation of stress in this material has been recoverable up to this point. As the proportional limit is approached, the wood material can no longer distribute the stress in a linear elastic manner.

11.4.2 Beyond Proportional Limit (Plastic Strength)

As the proportional limit is exceeded (Figure 11.14, Region B), the stress–strain relationship is no longer linear. Stresses are now great enough to induce covalent bond rupture and permanent distortion at all three structural levels.

At the nano-molecular level, the limit of reversible or recoverable hydrogen bonding has been exceeded. Covalent C–C and C–O bonds are breaking, thus reducing larger molecules to smaller ones. This reduction in degree of polymerization by covalent bond scission is nonrecoverable.

At the microscopic level, stresses develop within the microfibril region of the carbohydrates. Failure of the microfibril from stress overload causes actual covalent bond rupture and excessive microfibril disorientation. Additionally, the cell wall layers distort such that permanent micro cracks occur between the various cell wall layers. Separation of the cell wall layers is soon noticeable.

At the macroscopic level, entire fibers actually distort in relation to one another, such that recovery of original position is now impossible. The wood cells or wood fibers are actually failing either by scission of the cell, in which the cell actually fails by tearing into two parts to give a brash type of failure, or by cell-to-cell withdrawal (middle lamella failure) where the cells actually pull away from one another to give a splintering type of failure.

Permanent set is now evident at all levels of consideration, and eventual failure is imminent. In approaching the ultimate strength (Figure 11.14, Region C), nano-molecular level failures occur by C–C and C–O bond cleavage. Stress redistribution within the individual polymers is now impossible. At the microscopic level, the cell walls are distorting without additional stress. These walls are actually deforming at such an exaggerated rate that they can be thought of as being completely viscous or plastic, and they continue to deform and absorb strain energy but they can no longer handle additional stress. The cell wall is being sheared or torn apart. At the macroscopic level, failure is related to cell wall scission or cell-to-cell withdrawal.

11.5 RELATIONSHIP OF STRUCTURE TO STRENGTH

The mechanism of strength, as it relates to wood composition, has been discussed as a theoretical elastic model. To better understand this proposed model, we will look at what may be happening at each of the three structural levels.

11.5.1 Nano-Molecular Level

At the nano-molecular level, strength is elastic or recoverable because the polymeric structure can flex and, thus, absorb energy without fracturing the important covalent bonds (Figure 11.14, Region A).

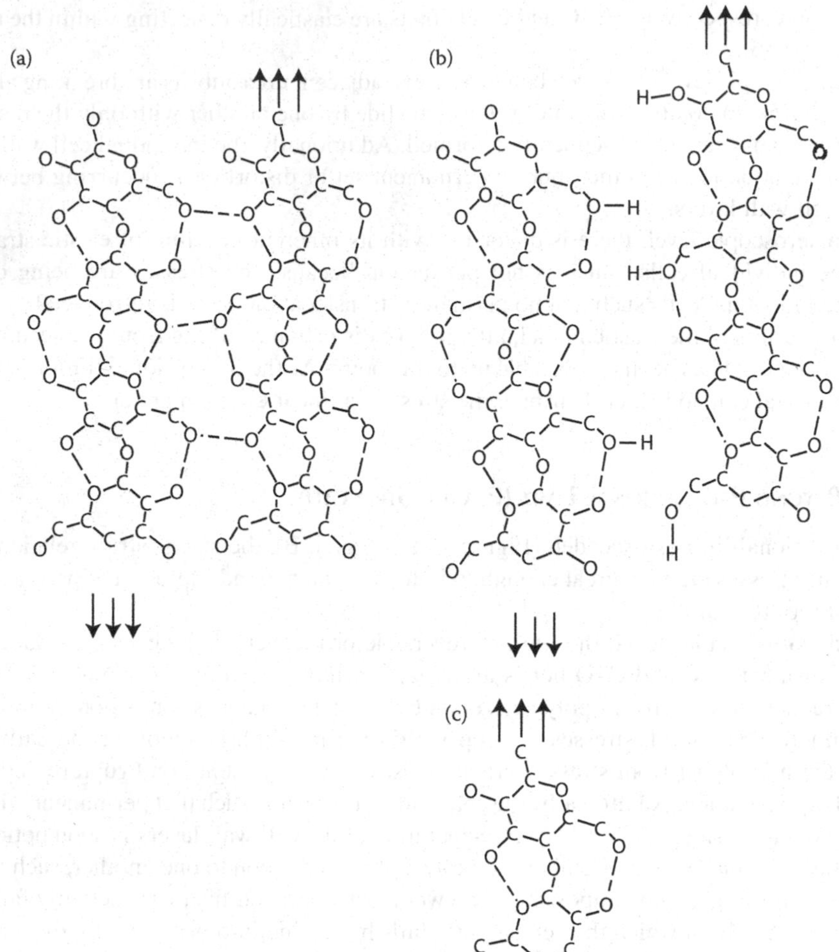

FIGURE 11.15 (a) Hydrogen bonding (bonded) between polysaccharide chains under shear forces; (b) Hydrogen bonding (sliding, unbounded) between polysaccharide chains under shear forces; and (c) Hydrogen bonding (rebounded) between polysaccharide chains under shear forces.

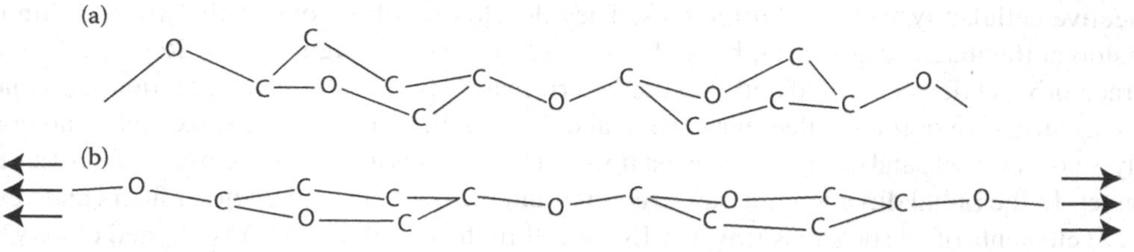

(a)

(b)

FIGURE 11.16 Flexing and elongation of polysaccharide molecules under tensile force. (a) No tensile force, no elongation; and (b) tensile force, elongation.

The second region of the stress–strain diagram (Figure 11.14, Region B) consists of Region B_a which is indicative of residual elastic strength such as represented in Region A, and Region B_b, which represents plastic strength or, more appropriately, strength associated with initial permanent set or distortion.

Section B_b is representative of C–C and C–O cleavage at the intra-polymer level which cannot be recovered. Examples of C–C bond breakage are lignin–hemicellulose copolymer separation, hemicellulose depolymerization, and amorphous cellulose depolymerization.

In Region C (Figure 11.14), elastic deformation essentially ends; there is now nearly pure plastic flow in the stress–strain relationship. Strain is continuing with little additional increase in stress and ultimate failure is imminent. This region is characterized by all the same mechanisms as in B_b, but a new and terminal intrapolymeric factor is introduced—the possibility that crystalline cellulose fails. It is more likely that the failure occurs in the noncrystalline regions of the cellulose polymer since the specific modulus of crystalline cellulose is 137 GPa. The specific modulus of whole clear wood is 12 GPa and the wood cell wall 35 GPa so the cell wall will fail long before the crystalline cellulose. However, if crystalline cellulose failure occurs, the main framework of the wood material at the nano-molecular level is disintegrating.

11.5.2 MICROSCOPIC LEVEL

At the microscopic level, the strength of the phenolic matrix is usually great enough that the cell wall stress reaches failure level in the carbohydrate framework. The S_1 layer microfibrils are oriented in both a right-hand (S helix) and a left-hand (Z helix) arrangement whereas the S_2 and S_3 have only the S-helix arrangement. The S_3 layer can be bihelical or monohelical, but, for the purpose of simplification, it has been assumed to be monohelical in this example. Because of the different linear elongation of the bihelical S_1 layer as compared to the monohelical arrangement of the S_2 and S_3 layers, the cell wall initially fails by S_1–S_2 separation (Siau, 1969). As S_1–S_2 separate, the S_2–S_3 layers assume the transferred stresses, and sustained stress increases, which will eventually cause either a brash-type failure (carbohydrate covalent bond failure) or a slow buildup to ultimate stress yielding a fibrous-type failure (phenolic covalent bond failure).

Below the proportional limit (Figure 11.14, Region A), there is elastic transfer of stresses between the S_1–S_2–S_3 cell wall layers. As Region B is entered, stress is still transferred between the S_1–S_2–S_3 cell wall layers as characterized by Section B_a. But S_1–S_2 separation is initiating, causing a sizeable transfer of stresses to the S_2–S_3 layers characterized by Section B_b. In Region C, ultimate strength is now dictated by the S_2–S_3 cell wall layer's ability to sustain additional stress until eventual failure of the substantial S_2 layer.

11.5.3 MACROSCOPIC LEVEL

At the macroscopic level, it is necessary to consider wood a viscoelastic material. As stress is applied to a wooden member, minute cracks initiate, propagate, and terminate throughout the collective cellular system in all directions. They develop in all regions of the stress–strain relationship at the macroscopic level, but only in the elastic region (Figure 11.14, Regions A and B_a) is crack propagation controlled and eventually terminated. In the tangential direction, the concentric ring structure of thin-walled early wood and thick-walled latewood in softwoods, and porous early-season vessels and dense late-season fibers in hardwoods act as the elements of elastic stress transfer. In the radial direction, the ray structures and the linear arrangement of fibers and vessels are the elements of elastic stress transfer. Every cell in the radial direction is aligned closely with the next cell because each cell in the radial direction has originated from the same cambial mother cell. Thus, the material can transfer stress elastically until an induced crack or a natural growth defect interrupts this orderly cellular arrangement. As stresses are built up within the material, cracks are initiated in the areas where elastic stress transfer is interrupted. These cracks continue to propagate until they are either terminated via dispersion of the energy away from the crack by the structural elements of stress transfer, or by eventual terminal failures as graphically characterized by Regions B_b and C (Figure 11.14).

11.6 ENVIRONMENTAL EFFECTS

When wood is exposed to environmental agents of deterioration, such as chemical treatments or elevated temperatures, each mechanical property reacts differently. Most commonly, ultimate strength properties are reduced and properties dealing with the proportional limit initially show little or no effect. However, the strain-to-failure (strain rate) is often considerably reduced, which, due to embrittlement of the fibers, is reflected as a reduction in pliability and energy-related properties such as work, toughness, and so on.

As individual wood components are altered in size, stature, or composition, the strength of the wood material is dramatically affected. Hypothetically, when ultimate stress is reduced 5% (Figure 11.17, U_1–U_2) and the proportional limit is not affected, the properties dealing with proportional limit (FSPL, MOE, WPL) reflect this in that they too are unaffected. The mechanical properties dealing with the point of ultimate stress (MOR in bending tests, $C_\parallel$ in axial-type compression tests, and $T_\parallel$ and $T_\perp$ in axial-type tensile tests) are reduced 5%. Work to maximum load can be reduced 33% because it is a function of both stress and strain. If the stress level at the proportional limit is reduced, and both the ultimate stress and strain levels are significantly reduced (Figure 11.18), larger decreases will occur in proportional limit properties (MOE, and FSPL, WPL), ultimate strength properties (MOR, $C_\parallel$, $T_\parallel$, $T_\perp$) and in work to maximum load. The examples in both Figures 11.17 and 11.18 are intended to show why energy-related properties such as that WML, WPL and related properties, such as toughness and impact bending, are usually

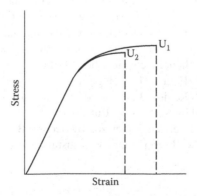

FIGURE 11.17 Hypothetical example of the effect of no change in proportional limit and a 5% reduction in ultimate bending strength on a few mechanical properties: MOE is not affected, MOR is reduced 5%, but WML is reduced by 33% because it is a dual function of stress and strain.

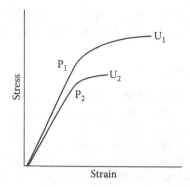

FIGURE 11.18 Hypothetical example of the effect of a 20% reduction in proportional limit and a 30% reduction in ultimate bending strength on a few mechanical properties: MOE is reduced 20%; MOR is reduced 30%, and WML is reduced 50%, because it is a dual function of stress and strain.

affected long before the other properties dealing with ultimate strength and the proportional limit are significantly affected.

What causes the phenomenon of stress and strain reduction and why is the reduction in impact and work properties so visible at small or negligible changes in elastic modulus and ultimate strengths? As discussed previously, mechanical properties deal with stress and strain relationships that are simply functions of chemical bond strength. At the nano-molecular level, strength is related to both covalent and hydrogen intrapolymer bonds. At the microscopic level, strength is related to both covalent and hydrogen interpolymer bonds and cell wall layer bonds (S_1–S_2 and S_2–S_3). At the macroscopic level, strength is related to fiber-to-fiber bonding with the middle lamella acting as the adhesive. Thus, any chemical or environmental agent that affects those bonds, also affects strength.

In considering the structural performance of the polysaccharide and phenolic polymers in wood fiber, the chemical environment of the fiber is of great importance. Chemicals can swell, hydrolyze, pyrolyze, oxidize, and, in general, depolymerize wood polymers, causing a loss in strength properties due to wood fiber network degradation. Other environmental agents such as UV light, heat, and biological organisms have a similar influence in changing strength properties.

11.6.1 ACIDS AND BASES

The average pH of wood is between 3 and 5.5 (Stamm, 1964) due to the acetyl content, the presence of acid extractives, and the adsorption of cations that comprise the ash. Even after several hundred years, this naturally mild acidic state does not induce any appreciable strength losses as long as the wood is protected from biological attack (USDA, 2010).

If the pH of the environment substantially changes or if temperatures increases to a point where the pH of the environment changes, strength properties can be reduced. These effects are further compounded by time and moisture. In general, the longer the time or the higher the temperature at which wood is exposed to an acid or a base, the greater is the degradative effect on strength (USDA, 2010; Wangaard, 1950; Stamm, 1964). Heartwood is generally more resistant to acid than is sapwood, probably because of heartwood's lower permeability and higher extractives content. Hardwoods are usually more susceptible to degradation by either acids or alkali's than are softwoods. This may be due to hardwood's lower lignin content and higher proportion of pentosan hemicelluloses. Oxidizing acids, such as HNO_3, have a greater degradative action on wood fiber than do nonoxidizing acids. Alkaline solutions are more destructive to wood fibers than are acidic solutions because wood adsorbs alkaline solutions more readily than acidic solutions. Acids with pH values above 2 and bases with pH values below 10 do not degrade the wood fiber greatly over short periods of time at low temperatures

(Kollman and Cote, 1968). Again at low temperatures mild acids, such as acetic acid, have little effect on strength, whereas strong acids, such as H_2SO_4, cause extensive strength losses (Alliott, 1926).

11.6.2 ADSORPTION OF ELEMENTS

Chemicals other than acids and bases can also be adsorbed and can cause degradation of the wood fiber. For example, fibers of southern pine exposed to the ocean air can be degraded badly (Figure 11.19). Salt crystals deposited in the void structure (Figure 11.20) can cause extensive chemical and physical damage. This chemical damage is due, in part, to the salt catalyzing hydrolysis reactions where as mechanical damage is related to the hydroscopic salts promoting greater shrinkage and swelling.

Other materials can be adsorbed from the environment if a hydrolytic solvent (e.g., water) is available. When water is available, wood will adsorb iron from oxidized metal (rust) and cause

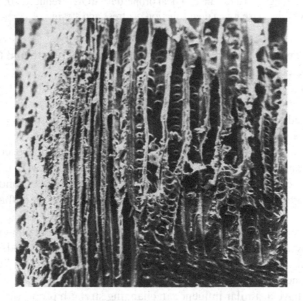

FIGURE 11.19   Scanning electron micrograph of sodium chloride in southern pine cell (×40).

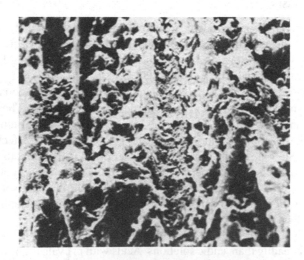

FIGURE 11.20   Scanning electron micrograph of sodium chloride in southern pine cell (×400).

decomposition of the cellulose (Koenigs, 1974). This is also true for copper, chromium, tin, zinc, and other similar reactive metals.

11.6.3 SWELLING SOLVENTS

Solutions that swell wood tend to plasticize it and reduce its strength properties. Water, for example, swells the intrapolymeric spaces, reduces cross-linking and, thus, reduces strength. In general, the greater the swelling, the greater the strength loss. Nonswelling liquids generally do not decrease strength properties. For example, oven-dry wood and wood saturated with water-free benzene have virtually the same strength (Erickson and Rees, 1940; Siau, 1969).

11.6.4 ULTRAVIOLET DEGRADATION

Wood exposed to the outdoors undergoes chemical reactions due to UV radiation. UV radiation causes photochemical degradation primarily in the lignin component which gives rise to characteristic color changes (see Chapter 7). Southern pine, for example changes from a light-yellow natural color to brown and evenly to gray. As the lignin degrades, the wood surface becomes richer in cellulose content. Although the cellulose is much less susceptible to UV degradation (Kalnins, 1966), it is eventually washed off the surface with water during rain, which exposes new lignin-rich surfaces that then start to degrade. As this process continues, the wood surface is said to weather.

Because UV radiation does not penetrate wood more than a few cells deep, weathering is considered a surface phenomenon. Over time it can account for a significant loss in surface fiber (see Figure 7.1, Chapter 7). As the degradative process continues, the loss in fiber may eventually cause a reduction in the material's load-carrying capacity.

11.6.5 THERMAL DEGRADATION

Wood strength is inversely related to temperature (see Chapter 6). A nearly linear decrease in strength is observed on increasing the temperature from $-200°C$ to $+160°C$ (Figure 11.4; derived from Kollman and Cote, 1968; Gerhards, 1982; Green, 1999). Heat has two types of effects on wood, immediate effects that occur only as long as the increased temperature is maintained and permanent effects that result from thermal degradation of wood polymers. The immediate effects of heat are recoverable, but permanent effects are not. The combination of immediate and permanent effects is multiplicative rather than additive.

In an environment without adequate humidity, the initial effect of heating wood is dehydration. As temperatures approach $55-65°C$ for extended periods (2–3 months), hemicellulose and cellulose depolymerization slowly begins (Feist et al., 1973; LeVan et al., 1990). This progressively escalates to pyrolysis and volatilization of cell wall polymers which rapidly occurs at about $250°C$ followed by char formation in the absence of air and combustion in the presence of air.

Heating dry Douglas-fir in an oven at $102°C$ for 335 days reduced MOE by 17%, MOR by 45%, and fiber stress at proportional limit by 33% (MacLean, 1945, 1953; Millett and Gerhards, 1972). Similar losses might be observed in a week or so at $160°C$. In the absence of air, heating softwood at $210°C$ for 10 min reduced MOR by 2%, hardness by 5%, and toughness by 5% (Stamm et al., 1946). Under the same conditions at $280°C$ MOR was reduced 17%, hardness was reduced 21%, and toughness was reduced 40%. Both examples illustrate the compound effect of heat, air, and time.

Comparison of photomicrographs of southern pine at $25°C$ (Figure 11.21a) and the same sample after heating from $20°C$ to $295°C$ under nitrogen over a period of 15 min (Figure 11.21b) shows the cell structure still intact, but the cell wall components have been darkened by pyrolysis. LeVan et al. (1990) noted an on going darkening of pinewood exposed at $82°C$, which corresponded to a loss in arabinose and to a lesser degree xylose. They, then, attributed the darkening brown color at $82°C$ to

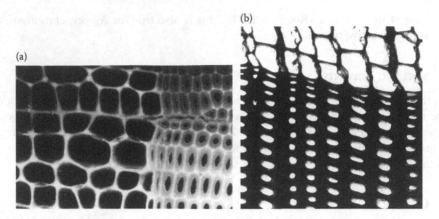

FIGURE 11.21 Southern pine at 25°C (×400) (a) and southern pine after heating in a nitrogen environment from 20°C to 295°C over 15 min (×400) (b).

hydrolysis of furan-ringed arabinose and xylose to formation of chocolate-brown colored furfural. Over the last ±25 years, the permanent effect of extended high-temperature and cyclic exposure on wood strength has been extensively studied (LeVan et al., 1990; Winandy et al., 1991; Winandy, 1994, 1995a; LeVan et al., 1996; Green et al., 2003; Green and Evans, 2008) and has been fairly thoroughly reviewed (Winandy, 2001). Predictive kinetic-based models have also been developed

TABLE 11.5
Relationship between Loss of Strength and Chemical Composition for Untreated Southern Pine

Temp (°C)	Time (days)	MOE (GPa)	MOR (MPa)	WML (kJ/m³)	Lig (%)	Glu (%)	Xyl (%)	Gal (%)	Arab (%)	Mann (%)
23	3	13.9	117.4	101.2	29.4	48.1	8.0	3.0	1.3	12.9
23	16	14.5	127.9	116.6	27.6	44.9	6.1	2.1	1.2	11.2
54	3	13.9	124.1	110.2	29.6	47.0	7.5	3.4	1.2	12.6
54	7	13.8	116.9	98.5	28.4	48.0	7.0	2.4	1.3	13.2
54	21	13.5	119.5	101.2	29.6	47.1	7.1	2.6	1.2	12.4
54	60	13.8	121.3	101.5	29.4	45.9	6.6	2.5	1.0	12.0
54	160	13.9	120.1	99.3	27.8	52.9	7.1	2.2	1.3	13.6
54	7	14.6	115.9	95.8	28.6	42.9	6.1	1.9	1.0	10.8
66	21	14.0	113.8	96.7	28.5	42.7	5.9	1.8	1.0	11.1
66	60	14.8	122.2	97.3	28.6	42.8	5.9	1.9	0.9	11.1
66	160	13.5	116.0	79.7	28.2	43.1	5.9	2.0	0.8	11.0
66	290	14.3	113.4	74.5	29.3	43.4	5.9	2.2	0.5	10.8
66	560	14.7	96.5	51.3	29.4	42.9	5.7	2.2	0.4	11.0
66	1095	14.6	81.3	32.7	29.7	43.0	5.4	1.7	0.2	10.9
66	1460	11.2	42.5	10.8	32.8	43.0	3.8	0.7	0.1	8.3
82	3	13.5	119.0	112.0	29.2	46.7	7.3	2.6	1.2	12.4
82	7	14.0	125.0	105.3	29.3	46.5	6.9	2.5	1.2	12.3
82	21	14.7	124.2	101.4	29.7	48.0	6.8	2.4	0.9	12.7
82	60	14.0	118.7	84.4	29.1	46.9	6.8	2.6	0.7	12.7
82	160	13.8	104.7	58.0	27.6	53.1	6.9	2.3	0.6	13.4

Note: Arab = arabinose, Gal = galactose, Glu = glucose, Lig = Klasson lignin, Mann = mannose, Xyl = xylose.

(Woo, 1981; Pasek and McIntyre, 1990; Winandy and Lebow, 1996; Lebow and Winandy, 1999a; Green et al., 2003). The reasons for these permanent thermal effects on strength relate to changes in the wood polymeric substance and structure and predictive models have been developed (Winandy and Lebow, 2001). The comprehensive analyses of almost 10,000 specimens systematically exposed to various high-temperature regimes, followed by their development of kinetic models, debunked one long-held misconception that a thermal threshold existed below which permanent effects did not occur (Lebow and Winandy, 1999a). That work concluded that thermal degradation of wood was a continuum, but at most ambient temperature exposures below 40–50°C, the rate of degrade was so slow as to be unnoticeable, and thus virtually negligible.

Tables 11.5 through 11.9 show loss in mechanical properties as a function of sugar analysis when southern pine is heated at different temperatures for different times either untreated or treated with some commercial fire retardants. Table 11.5 shows the effects of these variables on untreated wood and it can be seen that heat alone results in a decrease in MOR and WML as the time and temperature increase. As the loss of mechanical properties, there is an accompanying loss of xylose, galactose and arabinose. The greatest loss is in arabinose. The loss of arabinose may be the causative event leading to initial strength losses.

Table 11.6 shows the effects of these variables on wood that has been treated with phosphoric acid. MOE, MOR, and WML all decrease with increasing temperature and time. Significant losses in glucose, xylose, galactose, arabinose, and mannose also occur as the time and temperature increase. Table 11.7 shows a similar trend for southern pine treated with monoammonium phosphate; Table 11.8 for guanylurea phosphate/boric acid; and Table 11.9 for borax/boric acid. The most sensitive and consistent hemicellulose sugar lost in untreated and fire retardant treated wood is

TABLE 11.6
Relationship between Loss of Strength and Sugar Analysis of Southern Pine Treated with Phosphoric Acid (58.2 kg/m³)

Temp (°C)	Time (days)	MOE (GPa)	MOR (MPa)	WML (kJ/m³)	Lig (%)	Glu (%)	Xyl (%)	Gal (%)	Arab (%)	Mann (%)
23	3	11.5	58.3	22.5	26.7	40.9	5.9	2.6	0.7	10.8
23	160	11.4	56.9	21.6	24.8	44.3	5.8	1.8	1.2	11.3
54	3	11.8	55.8	18.5	27.1	41.5	5.9	2.4	0.8	11.9
54	7	11.2	52.9	17.7	27.0	41.6	5.9	2.7	1.1	10.7
54	21	13.2	55.4	19.1	26.8	43.6	6.1	2.7	1.1	12.2
54	60	10.7	50.8	19.0	27.4	43.2	5.7	2.8	0.8	10.8
54	160	11.3	52.4	17.7	26.7	44.1	6.4	1.5	0.6	10.3
66	7	11.3	49.1	17.2	26.2	41.8	5.6	1.8	0.9	10.9
66	21	10.2	41.1	13.9	27.2	40.2	4.7	1.7	0.7	9.6
66	60	10.5	43.8	13.0	29.3	41.2	3.5	0.9	0.3	8.4
66	160	8.2	27.5	5.5	33.0	40.5	2.2	0.2	0.3	5.3
66	290	7.0	19.2	3.2	36.6	36.6	2.1	0.4	0.1	4.1
66	560	3.3	6.9	0.8	44.4	32.2	1.0	0.2	0.0	1.8
66	1095	0.0	0.0	0.0	0.0	0.0	0.0	0.0	0.0	0.0
66	1460	0.0	0.0	0.0	0.0	0.0	0.0	0.0	0.0	0.0
82	3	10.6	41.0	12.6	28.4	5.4	5.4	2.5	0.7	10.2
82	7	10.4	43.6	12.5	29.3	4.4	4.4	1.8	0.6	10.1
82	21	8.6	30.9	6.5	34.0	2.6	2.6	1.7	0.2	6.2
82	60	7.6	20.4	3.4	37.9	2.0	2.0	0.8	0.0	4.2

Note: Arab = arabinose, Gal = galactose, Glu = glucose, Lig = Klasson lignin, Mann = mannose, Xyl = xylose.

TABLE 11.7
Relationship between Loss of Strength and Sugar Analysis of Southern Pine Treated with Monoammonium Phosphate (55.5 kg/m³)

Temp (°C)	Time (days)	MOE (GPa)	MOR (MPa)	WML (kJ/m³)	Lig (%)	Glu (%)	Xyl (%)	Gal (%)	Arab (%)	Mann (%)
23	3	12.5	101.8	81.3	27.3	43.8	5.9	2.1	0.7	11.3
23	160	12.7	103.5	79.3	23.6	42.5	6.2	2.1	1.0	11.8
54	3	12.5	96.1	76.6	27.5	43.6	6.6	2.4	1.2	11.7
54	7	12.2	99.2	77.7	27.2	41.8	6.4	2.2	0.9	11.0
54	21	15.6	98.6	62.6	26.6	42.6	6.1	1.7	0.7	11.3
54	60	12.6	95.9	65.1	27.0	41.5	6.2	2.6	0.9	11.1
54	160	13.2	90.9	44.3	26.3	44.2	6.4	2.0	0.6	11.6
66	7	13.4	99.5	69.5	26.6	40.8	5.6	1.8	0.8	10.6
66	21	12.3	90.7	55.1	26.5	40.5	5.7	1.7	0.3	10.2
66	60	12.8	79.2	34.7	27.6	42.1	5.8	2.2	0.4	11.0
66	160	11.7	66.1	25.6	27.8	41.2	5.0	1.6	0.1	10.0
66	290	11.7	57.5	19.0	29.3	40.6	4.0	0.8	0.0	9.1
66	560	10.5	32.7	6.7	33.6	43.0	2.8	0.3	0.0	6.6
66	1095	7.6	18.8	2.6	39.7	41.2	2.2	0.2	0.0	3.7
66	1460	2.2	5.8	0.6	43.4	41.3	1.9	0.1	0.0	3.3
82	3	13.0	89.2	45.7	27.1	41.8	6.2	2.1	0.6	10.9
82	7	12.3	77.5	33.5	27.3	41.1	5.7	2.0	0.5	10.6
82	21	13.3	76.0	29.0	27.7	42.7	6.0	1.7	0.6	10.5
82	60	12.5	61.1	21.1	30.2	43.5	4.9	1.2	0.0	9.8
82	160	10.5	42.7	11.2	26.3	44.7	2.6	0.2	0.0	5.6

Note: Arab = arabinose, Gal = galactose, Glu = glucose, Lig = Klasson lignin, Mann = mannose, Xyl = xylose.

arabinose. The loss of arabinose can be used as an approximate indication of strength loss without having to do a bending test to determine strength properties.

Analysis of these materials and results indicated that strength losses from external exposure of untreated or fire retardant treated wood to elevated temperatures were a result of loss of hemicelluloses or cell wall matrix structure not a loss in the degree of polymerization of cellulose (Sweet and Winandy, 1999).

11.6.6 MICROBIAL DEGRADATION

When certain organisms come into contact with wood, several types of degradation occur (see Chapter 5). The mechanical damage caused by metabolic action can result in significant losses in strength. Microbial activity via enzymatic pathways induces wood fiber degradation by chemical reactions such as hydrolysis, dehydration, and oxidation. Brown-rot fungi preferentially metabolize holocellulose, especially the strength-critical cellulose fraction (Cowling, 1961). White-rot fungi metabolize nearly all fractions of the wood, but strength is not affected to the same degree as with a brown-rot fungus. An initial 10% weight loss for sweetgum (Liquidambar styraciflua L.) attacked by a brown-rot fungus reduced the degree of polymerization of the holocellulose from 1500 to 300 (Figure 11.22) (Cowling, 1961). As the side chains of the hemicellulose (e.g., arabinose, galactose) in the wood fiber are degraded, mechanical properties begin to wood decrease (Winandy and Morrell, 1993; Curling et al., 2002). This work has shown that the initial 5–20% of strength loss seems to be directly related to this initial hemicellulose degradation. Later, as the main-chain backbones of the hemicellulose (e.g., xylose, mannose) is degraded, additional strength loss occurs.

TABLE 11.8

Relationship between Loss of Strength and Sugar Analysis of Southern Pine Treated with 70% Guanylurea Phosphate/30% Boric Acid (55.5 kg/m³)

Temp (°C)	Time (days)	MOE (GPa)	MOR (MPa)	WML (kJ/m³)	Lig (%)	Glu (%)	Xyl (%)	Gal (%)	Arab (%)	Mann (%)
23	3	12.2	108.1	82.8	28.3	41.4	5.7	2.3	1.2	11.2
23	160	12.6	107.8	80.7	25.9	44.1	6.0	2.4	1.1	12.4
54	3	12.6	107.4	84.4	27.9	43.0	6.2	1.9	1.1	11.4
54	7	13.4	108.5	77.6	27.5	43.1	5.9	1.8	1.1	11.4
54	21	12.5	107.6	75.6	28.1	42.6	6.3	2.3	1.2	11.6
54	60	12.9	104.2	72.0	27.9	42.4	5.8	1.9	1.3	11.3
54	160	13.0	104.6	65.1	25.6	46.2	6.2	2.3	0.8	12.0
66	7	14.1	105.9	73.8	26.9	40.5	5.9	1.7	1.0	10.6
66	21	13.3	104.9	72.5	28.1	40.3	5.4	1.8	0.8	10.3
66	60	13.3	98.8	57.5	28.2	40.8	5.7	1.6	0.6	10.4
66	160	12.2	78.3	33.6	29.1	—	5.7	1.7	0.4	10.7
66	290	13.3	69.8	24.6	28.9	41.7	5.6	1.4	0.1	10.4
66	560	12.7	56.2	16.8	30.3	42.8	5.2	1.1	0.1	9.9
66	1095	11.7	38.5	7.6	32.8	41.0	3.7	0.7	0.1	7.3
66	1460	8.0	17.9	3.3	37.6	42.1	2.8	0.4	0.1	5.8
82	3	13.5	111.1	70.1	27.9	42.2	5.7	2.1	1.5	10.9
82	7	12.8	110.3	69.3	28.1	42.3	6.2	2.3	1.4	10.9
82	21	13.3	105.0	55.1	29.5	44.2	6.6	2.4	1.2	11.1
82	60	13.4	83.7	37.2	29.3	44.2	6.5	2.6	0.9	10.4
82	160	12.5	66.3	25.2	26.9	48.4	5.3	1.6	0.3	9.8

Note: Arab = arabinose, Gal = galactose, Glu = glucose, Lig = Klasson lignin, Mann = mannose, Xyl = xylose.

Finally, after strength has been reduced about 40–60% from the virgin strength level, noticeable degradation of glucose and lignin becomes evident (Winandy and Lebow, 2001).

Using isolated polymers derived from wood, previous research has indicated a large drop in degree of polymerization which was then related to the large drop in wood strength properties without corresponding weight loss (Ifju, 1964; Cowling, 1961; Kennedy, 1958). This work indicated that, in the initial stages of biological attack, hydrolytic chemical reactions play an important part. Yet, more recent work that carefully approached this same subject using less invasive isolation techniques to study the polymers, found that hydrolytic hemicellulose depolymerization preceded cellulose depolymerization (Sweet and Winandy, 1999). In either case, in the initial phase(s) of degradation, large polymers are broken into smaller, more digestible units, but the initial degradation products are not actually consumed by the organisms. Hydrogen peroxide and iron have been proposed as being involved in initial depolymerization (Koenigs, 1974). The initial enzymatic attack by microorganisms is probably not only hydrolytic but also oxidative in nature.

Curling et al. (2002) recently evaluated the effect of brown-rot decay on Southern pine (*Pinus* spp.) exposed in a laboratory test for periods ranging from 3 days to 12 weeks. The relative effects of decay on strength, weight and chemical composition can be seen by comparing the loss in bending strength (MOR), loss in work to maximum load (WML), loss in stiffness (MOE) to loss in dry weight (Figure 11.23).

These results demonstrate that considerable bending strength loss occurs before measurable weight loss. At 10% weight loss, strength loss was approximately 40% and the loss in the energy properties, like work to maximum load, was reduced by 70–80%. The decay–strength relationship

TABLE 11.9
Relationship between Loss of Strength and Sugar Analysis of Southern Pine Treated with a 50/50% Borax/Boric Acid (56.3 kg/m³)

Temp (°C)	Time (days)	MOE (GPa)	MOR (MPa)	WML (kJ/m³)	Lig (%)	Glu (%)	Xyl (%)	Gal (%)	Arab (%)	Mann (%)
23	3	12.4	115.7	69.3	27.8	42.2	6.7	2.6	1.21	14.8
23	160	13.3	115.3	65.4	26.3	39.3	5.9	2.4	1.2	10.5
54	3	12.9	106.1	57.6	27.1	42.8	6.5	2.3	1.0	14.7
54	7	13.9	115.4	62.2	26.8	44.4	7.0	2.7	1.4	15.5
54	21	16.0	117.1	59.8	26.7	43.9	6.5	1.9	1.0	15.7
54	60	13.5	126.5	86.3	27.0	43.5	6.8	1.9	1.2	14.9
54	160	13.7	122.7	70.9	25.7	44.9	6.5	2.0	1.0	12.4
66	7	13.8	116.5	69.0	27.4	42.2	6.1	1.9	1.0	11.2
66	21	13.1	116.0	74.7	27.5	41.8	6.4	2.0	1.0	11.1
66	60	13.1	108.5	66.1	28.2	42.0	6.2	1.8	1.0	10.7
66	160	13.4	115.0	77.8	28.9	43.5	6.0	2.0	0.8	11.2
66	290	13.3	108.7	76.3	28.6	44.0	6.0	1.7	0.7	11.1
66	560	13.4	99.2	65.8	29.3	44.1	6.3	1.9	0.8	10.9
66	1095	12.8	73.5	32.3	27.4	42.7	5.4	1.4	0.3	9.7
66	1460	11.7	50.3	16.2	31.6	45.2	4.3	0.8	0.1	8.9
82	3	14.6	123.1	66.9	27.0	43.1	6.7	2.7	1.2	15.7
82	7	14.5	131.1	76.9	27.1	43.5	6.3	1.4	0.6	14.8
82	21	14.4	132.7	73.9	27.0	46.0	6.8	2.3	0.9	15.4
82	60	14.8	137.9	89.1	27.0	42.3	6.5	2.5	0.9	14.5
82	160	14.3	124.0	69.1	26.3	43.8	6.1	2.3	0.7	11.4

Note: Arab = arabinose, Gal = galactose, Glu = glucose, Lig = Klasson lignin, Mann = mannose, Xyl = xylose.

appears very consistent as decay initiates and progresses. It seems not only qualitative, but may also be quantitative.

Curling et al. (2002) also found that changes in chemical composition were directly related to strength loss (Figure 11.24). As decay progressed it sequentially affected different chemical components. A virtually similar effect had been previously noted by Winandy and Morrell (1993). Initial

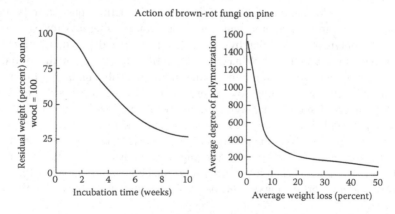

Action of brown-rot fungi on pine

FIGURE 11.22 Action of brown rot fungus on holocellulose. (Adapted from Cowling, E.B. 1961. *Comparative Biochemistry of the Decay of Sweetgum Sapwood by White-rot and Brownrot Fungi.* USDA, Forest Service, Tech. Bull. No. 1258, 50 p.)

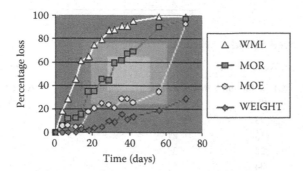

FIGURE 11.23 Effect of decay by *G. trabeum* on weight loss and mechanical properties. (Adapted from Curling, S.F, Clausen, C.M., and Winandy, J.E. 2002. *Forest Prod. J.*, 52(7/8):34–39.)

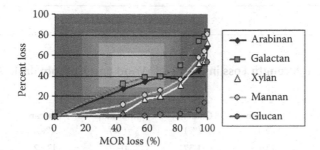

FIGURE 11.24 Comparison of loss of carbohydrate components with loss in bending strength (MOR) caused by *G. trabeum*. (Adapted from Curling, S.F, Clausen, C.M., and Winandy, J.E. 2002. *Forest Prod. J.*, 52(7/8):34–39.)

loss in the mannan and xylan components correlated with the start of measurable weight loss, both occurred at about 40% strength loss. Because mannan/xylan components make up approximately 18% of the wood, compared to 8% for galactan and arabinan, it is understandable why measurable weight loss is not evident in incipient decay. Final stage of brown-rot decay occurs once the glucan-rich cellulose is broken down, at a strength loss of about 80%. It is also at this stage (80% loss in MOR) that loss in stiffness (MOE) increases rapidly suggesting that the stiffness of the wood is related to the cellulose rather than the hemicellulose composition.

Table 11.10 shows the loss in MOE and maximum compression strength (MCS) in wood exposed to the brown-rot fungus *Postia placenta* in standard ASTM soil block and in ground tests. In the wood blocks, at 2 weeks into the standard 12 week test, there is only a 3% weight loss but already a 13% loss in MOE and a 21% loss in MCS. At five weeks, there is a 26% weight loss but a 53% loss in MOE and a 71% loss in MCS. In the wood stakes, the losses are less since the specimen size is larger and requires more time for the fungus to degrade the wood. Even so, the wood stakes at 5 weeks show a weight loss of 7% and a loss of 12% in MOE and 22% in MCS.

Table 11.11 shows the correlation between time in test and changes in sugar analysis that has occurred during that time. The greatest loss is in the arabinose sugar which was also the case in fire retardant treated wood.

11.6.7 NATURALLY OCCURRING CHEMICALS

Some woods have a higher acidic extractives content that can cause greater strength loss due to hydrolysis. This may be a problem in some of the tropical species coming into the market. These

TABLE 11.10
Strength Loss in Decayed Wood

	Wood Blocks			Wood Stakes		
Time (Weeks)	Weight Loss (%)	MOE Reduction (%)	MCS Reduction (%)	Weight Loss (%)	MOE Reduction (%)	MCS Reduction (%)
1	0	3	6	0	5	2
2	3	13	21	0	8	2
3	9	22	37	1	7	8
4	18	34	53	3	9	19
5	26	53	71	7	12	22

Source: Adapted from Clausen, C.A. and Kartal, S.N. 2003. *Forest Prod. J.*, 53(11/12):90–94.

Note: Fungus used in test: *Postia placenta,* MOE = modulus of elasticity, MCS = maximum compression strength.

TABLE 11.11
Correlation between Strength Loss in Decayed Wood and Carbohydrate Analysis

		Carbohydrate Composition					
Sample	Time (Weeks)	Arabinan (%)	Galactan (%)	Rhamnan (%)	Glucan (%)	Xylan (%)	Mannan (%)
Blocks	0	1.01	1.37	0.07	42.82	5.88	11.32
	1	0.87	1.40	0.07	42.56	5.05	12.15
	2	0.78	1.10	0.06	42.46	5.60	10.32
	3	0.68	1.02	0.05	42.23	4.82	9.65
	4	0.55	1.07	0.04	41.89	3.91	10.73
	5	0.61	0.86	0.04	39.13	4.36	7.87
Stakes	0	1.02	2.63	0.09	41.97	6.44	9.84
	1	0.93	2.20	0.09	42.80	6.39	10.43
	2	0.94	2.13	0.09	42.07	6.61	10.53
	3	0.86	2.05	0.08	42.23	6.21	10.62
	4	0.82	1.69	0.07	41.25	6.13	9.84
	5	0.77	2.04	0.07	40.48	5.93	8.98

Source: Adapted from Clausen, C.A. and Kartal, S.N. 2003. *Forest Prod. J.*, 53(11/12):90–94.

more acidic woods will not only be more susceptible to strength loss but will also increase the corrosion potential of iron fasteners used with the wood. It is not uncommon to find silica and calcium salt crystals in wood fiber (Figure 11.25), and they can lead to strength losses by abrasion of the fibers during machining. Naturally occurring chemicals may also cause increased hygroscopicity and hydrolysis when salts dissolve in water.

11.7   TREATMENT EFFECTS

From a quantitative viewpoint, the effects of preservatives have been investigated by many researchers. A previous review identified and then classified several treatment-related issues that controlled the treatment effect with waterborne preservatives (Winandy, 1996a). In that review, waterborne-preservative treatments were shown to generally reduce the mechanical properties of wood. The

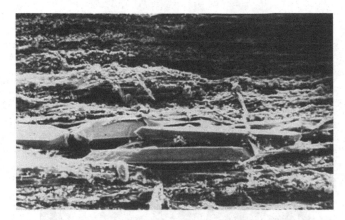

FIGURE 11.25   Scanning electron micrograph of calcium carbonate crystals in lumens of southern pine (×800).

effects of waterborne preservatives on mechanical properties are related to several key wood material factors and pretreatment, treatment, and post-treatment processing factors. That review was the basis for recent categorical statements in the 2010 edition of the *Wood Handbook*. These key factors include preservative chemistry or chemical type (Wood et al., 1980; Bendtsen et al., 1983; Winandy et al., 1983), retention (Winandy et al., 1985, 1989), posttreatment drying temperature (Winandy and Boone, 1988; Winandy 1989), initial kiln-drying temperature (Winandy and Barnes, 1991), grade of material (Winandy and Boone, 1988; Winandy, 1989), and incising (if required) (Lam and Morris, 1991; Morrell et al., 1998; Winandy and Morrell, 1998). The effects of treatments on time-dependent relationship previously shown in Figure 11.5 have also been studied, but while they seemed to have little practical significance related to long-term duration-of-load (Soltis and Winandy, 1989) they did have significant effects on short-term impact loading (Winandy, 1995b).

In summary, each factor has the potential of influencing the effects of the preservative treatment on mechanical properties. As might be expected, the interactive nature of the processing and chemical factors is critical and these factors must always be considered when attempting to define the effects of waterborne preservatives on mechanical properties. A cumulative-damage model to predict these effects has also been developed (Winandy, 1996b).

From a qualitative viewpoint, metallic salts such as chromated copper arsenate (CCA), amine- or ammoniacal-copper quat (ACQ), Copper Azol (CA), and other metallic salts of copper, arsenic, zinc, and tin are used to increase the service life of wood in use against biological attack. One major advantage of many waterborne preservatives is their resistance to leaching by water. Some organo-metallic salt preservatives (e.g., ACQ, CA) become insoluble as the treated wood dries which dehydrates the preservative complex within the wood. In this process, known as immobilization, the cell wall is not directly affected and subsequently strength virtually unaffected. Other metal salts (e.g., CCA) undergo hydrolytic reduction upon contact with the reducing sugars found in wood. In this process, known as fixation, the cell wall is oxidized and subsequently strength is affected. At most usable concentrations these preservatives are sufficiently acidic or alkaline to cause some cell wall hydrolysis (Hatt, 1906; Betts and Newlin 1915; Wood et al., 1980). The effects on strength are greater in kiln-dried waterborne preservative-treated wood than in air-dried salt-treated wood (Bendtsen et al., 1983; Winandy et al., 1985). Most water-borne preservative salts increase the hygroscopicity of the wood (Bendtsen, 1966; Bendtsen et al., 1983). This causes an increased EMC, which further influences strength.

After treatment and either immobilization or fixation, precipitated preservative salts can be seen in the wood structure. Comparison of photomicrographs of untreated southern pine (Figure 11.26 top) and southern pine treated with 2.5 lb CCA/ft³ (Figure 11.26 bottom) reveals that the salts in the treated wood form a rough coating on the lumen walls. The exact final location of these fixed salts

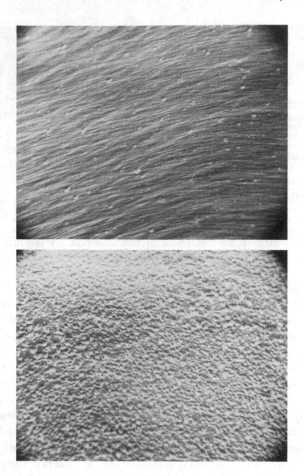

FIGURE 11.26   Scanning electron micrograph of CCA in southern pine.

is still subject to discussion. Undoubtedly, most of these injected hydrolytic preservatives remain in the cell lumen, but the extent to which these preservatives enter the cell wall and react with the cell wall substance dictates the preservative's effect on strength.

Preservative formulations that contain copper and chromium salts reportedly promote afterglow in treated wood subjected to fire. Once the treated wood starts to burn or glow, the wood may continue to glow until the entire member is consumed, even when no flame is present (Dale, 1966; McCarthy et al., 1972). This characteristic can cause serious problems utility poles, fence posts, and highway signs, structures that might be subjected to accidental fires or controlled ground fire, which is used as a forest or agricultural management tool.

Petroleum-based chemicals such as creosote and pentachlorophenol in oil are also used as wood preservatives. These organic preservatives are inert toward the cell wall substance and do not seem to cause any appreciable strength losses (MacLean, 1951; Koukal et al., 1960; Gillwald, 1961; USDOD, 1979; Barnes and Winandy, 1986).

Salts such as sodium tetraborate, diammonium phosphate, trisodium phosphate, diammonium sulfate, and salts of boric acid have long been used as fire retardants. Hygroscopicity, corrosion of fasteners, and increased acidity are also problems with these salts. Like the preservative salts, these salts also precipitate in the cell cavity and the cell wall (Figure 11.27). They appear on the fiber surface of pine treated with 4.2 lb of ammonium dihydrogen phosphate ($NH_4H_2PO_4$) per cubic foot of wood. The initial effects of FR treatments on strength are greater in kiln-dried salt-treated wood than in air-dried salt-treated wood (Gerhards, 1970; Brazier and Laidlaw, 1974; Adams et al., 1979;

FIGURE 11.27 Scanning electron micrograph of southern pine with FRT crystal deposition on surface.

Johnson, 1967). Like the waterborne preservatives, this is due to hydrolytic degradation of the fiber, caused by the combination of high moisture content, high temperature, and acid salts or alkali's during the posttreatment redrying process.

Many fire-retardant salt treatments are very hygroscopic salts causing increased moisture sorption and irreversible swelling (Bendtsen, 1966). Most interior FR formulations are not recommended for use where relative humidity is over 92% (AWPA, 2010).

FR-treated plywood roof sheathing is often required by U.S. Building Codes in roof systems for multifamily dwellings having common property walls. In the mid-1980s, users of some FR-treated materials, especially plywood roof sheathing exposed to elevated in-service temperatures, were experiencing significant thermal degrade (Figure 11.28). From the late 1980s until about 2001, a major +10-year research program was carried on at the U.S. Forest Products Laboratory to identify the factors and mechanisms of fire-retardant (FR)-treatment-induced strength loss. Qualitatively,

FIGURE 11.28 Example of fully serviceable untreated plywood roof sheathing (top left) adjacent to thermally degraded FR-treated plywood sheathing (center) used next to a gypsum-sheathed 60-min-rated fire wall (lower right).

the mechanism of thermal degrade in FR-treated plywood was acid hydrolysis. The magnitude of strength loss could be cumulatively related to FR chemistry, thermal exposure during pretreatment, treatment, and post-treatment processing and in-service exposure. The effects of FR chemistry could be mitigated by use of pH buffers (Winandy, 1997; Lebow and Winandy, 1999b). The strength effects were similar for many quality levels of plywood (Lebow and Winandy, 1998). Quantitatively, a kinetics-based approach could be used to predict strength loss based on its time–temperature history (Lebow and Winandy, 1999a).

From a mechanistic approach to FR thermal degradation, the effects of FR treatments on strength properties were shown to depend on FR chemistry and thermal processing. The research confirmed that field problems with FR-treated plywood roof sheathing and roof-truss lumber resulted from thermal-induced acid degradation. In one comprehensive study, over 6000 specimens of density matched southern pine ($16 \times 35 \times 250$ mm^3) were systematically exposed at one of four temperatures: 25°C, 54°C, 66°C, and 82°C for exposures up to 4 years (LeVan et al., 1990; Winandy, 1995a; Lebow and Winandy, 1999a). Data on the rates and magnitudes of thermal-induced strength loss at 27°C, 54°C, 66°C, and 82°C were obtained for specimens treated with one of six FR-model chemicals or untreated. The influence of temperature and treatment pH was progressive, as shown by this example at 66°C in which each treatment (going from left-to-right on z-axis) has a progressively higher pH (Figure 11.29).

Kinetic-based models for predicting strength loss as a function of exposure temperature and duration of exposure were then developed from this data obtained at four temperatures. These kinetic models can be used predict thermal degradation at other temperatures (Lebow and Winandy, 1999a). A single-stage approach quantitatively based on time–temperature superposition was used to model reaction rates, such that

$$Y_{ij} = b_j * \exp(-X * A * \left(H_i / H_o\right) * e^{(-E_a / RTij)})$$
(11.10)

where

i = Temperature of exposure
j = FR chemical
Y_{ij} = bending strength (MPa) at temperature (T_i) for FR$_j$
X = time (days) at temperature (T_i) for FR$_j$
b_{ij} = initial bending strength (MPa) at time ($X_i = 0$)
H_i = relative humidity at test
H_o = normalized relative humidity (67% R.H. (per ASTM D5516)
A = preexponential factor
E_a = activation energy
R = gas constant (J/°K*mol)
T_{ij} = temperature (°K) for FR$_j$

This research also proved that the use of pH buffers in FR chemicals, such as borates, can partially mitigate the initial effect of the FR treatment on strength and then significantly enhance resistance to subsequent thermal degradation. For modulus of elasticity there appeared to be few real benefits derived from adding borate-based pH buffers to the FR-mixture on the subsequent thermal degrade of FR-treated plywood exposed to high temperatures (Winandy, 1997). However, after 290 days of exposure at 66°C there were significant ($p < 0.05$) benefits derived with respect to limiting strength loss and loss in energy-related properties, such as work to maximum load, by the addition of borate to the FR-chemical mixture (Figure 11.30). Remedial treatments based on surface

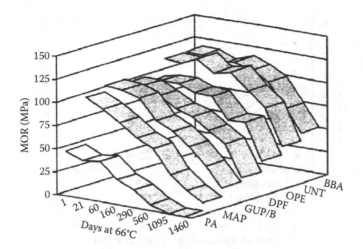

FIGURE 11.29 Strength loss over time of exposed to 66°C. Key: PA, phosphoric acid; MAP, monoammonium phosphate; GUP/B, guanylurea phosphate/boric acid; DPF, dicyandiamide-PA-formaldehyde; OPE, organophosphonate ester; UNT, untreated; and BBA, borax/boric acid.

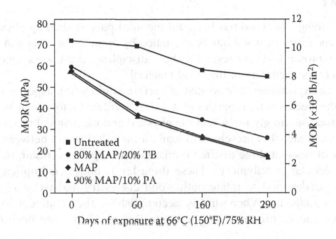

FIGURE 11.30 Effect of pH buffers on the rate of thermal degrade. Key: MAP, monoammonium phosphate; PA, phosphoric acid; and TB, sodium tetraoctaborate. (Adapted from Winandy, J.E. 1997. *Forest Prod. J.*, 47(6):79–86.)

application of pH-buffered borate/glycol solutions were also developed to protect against additional in-service strength loss (Winandy and Schmidt, 1995).

Other works also found that the rate of strength loss was largely independent of plywood quality or grade (Lebow and Winandy, 1998) and could be directly related to changes in treated wood pH (Lebow and Winandy, 1999b). Further, variation in redrying temperatures from 49°C to 88°C had little differential effect on the subsequent rate of thermal degradation when the treated plywood was exposed at 66°C for up to 290 days (Winandy, 1997; Figure 11.31). This was related to the shorter kiln-residence times required at higher temperatures yielding similar states of entropy via differing, but thermodynamically comparable, temperature–duration histories.

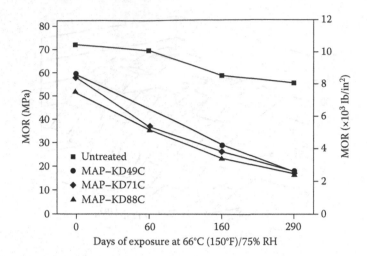

FIGURE 11.31 Effects of kiln drying after MAP treatment at 49°C, 71°C, and 88°C on thermal degrade. (Adapted from Winandy, J.E. 1997. *Forest Prod. J.*, 47(6):79–86.)

11.8 SUMMARY

Over the years the strength of wood has been, for the most part, studied by physical chemists and engineers and the chemistry of wood has been studied by organic chemists and biochemists. In a materials science approach to wood research, these disciplines must work together to relate the physical properties to the chemistry of the wood material.

In this chapter an explanation is presented of certain engineering aspects that are important in understanding the mechanical properties of wood. Individual factors such as growth, environment, chemicals, and use can greatly affect the physical and mechanical properties of the wood material. A theoretical model is presented to explain the relationship between physical properties and chemistry of wood at three distinct levels: macroscopic or cellular, microscopic or cell wall, and nano-molecular or polymeric. These three levels and their implications on material properties must be understood to relate both wood chemistry and wood engineering from a materials science standpoint. When this is accomplished, the treatment and processing of wood and wood composite products can be controlled to yield more desirable and uniform properties.

While considerable work has been undertaken in the last 20–30 years, many of the underlying governing principles are yet unknown. The theories presented in this paper are offered as a starting point for consideration and discussion. Hopefully they may some day become a point from which, through mutual cooperation between the fields of engineering and chemistry, the chemistry of wood strength may be truly explained.

REFERENCES

Adams, E.H., Moore, G.L., and Brazier, J.D. 1979. The effect of flame-retardant treatments on some mechanical properties of wood. BRE Infor. Paper IP 24/79 Princes Risborough Lab., Aylesbury, Bucks., UK, 4 p.

Alliott, E.A. 1926. Effect of acids on the mechanical strength of timber. A preliminary study. *J. Soc. Chem. Ind.* 45:463T–466T.

American Society for Testing and Materials. 2009a. *Standard Test Method for Small Clear Specimens of Timber, ASTM Stand. Desig. D 143–94.* ASTM Annual Book of Standards. West Conshohocken, PA.

American Society for Testing and Materials. 2009b. *Standard Test Method for Static Tests of Lumber in Structural Sizes. ASTM Stand. Desig. D 198–08.* ASTM Annual Book of Standards. West Conshohocken, PA.

American Society for Testing and Materials. 2009c. *Standard Practice for Establishing Structural Grades and Related Allowed Properties for Visually Graded Lumber. ASTM Stand. Desig. D 245–06*. ASTM Annual Book of Standards. West Conshohocken, PA.

American Society for Testing and Materials. 2009d. *Standard Test Methods for Establishing Clear Wood Strength Values. ASTM Stand. Desig. D 2555–06*. ASTM Annual Book of Standards. West Conshohocken, PA.

American Society for Testing and Materials. 2009e. *Standard Test Methods for Mechanical Properties of Lumber and Wood Based Structural Material. ASTM Stand. Desig. D 4761–05*. ASTM Annual Book of Standards. West Conshohocken, PA.

American Wood Preservers Association. 2003. *Standard C20: Structural Lumber: Fire-Retardant Treatment by Pressure Processes*. AWPA Annual Book of Standards. Selma, AL.

American Wood Protection Association. 2010. *Standard U-1 Specification for Treated Wood: Commodity Specification H: Fire Retardants*. AWPA Annual Book of Standards, Birmingham, AL.

Barnes, H. M. and Winandy, J.E. 1986. Effects of seasoning and preservatives on the properties of treated wood. *Am. Wood Pres. Assoc. Proc.* 82:95–104.

Barrett, J.D., Green, D.W., and Evans, J.W. 1989. Temperature adjustments for the North American in-grade testing program. In: *Proceedings of the Workshop on In-grade Testing of Structural Lumber*, Forest Products Research Society, Madison, WI, 27–38.

Bendtsen, B.A. 1966. *Sorption and Swelling Characteristics of Salt-Treated Wood*. USDA, For. Serv. Res. Pap., FPL-RP-60. Madison, WI.

Bendtsen, B.A., Gjovik, L.R., and Verrill, S.P. 1983. *Mechanical Properties of Longleaf Pine Treated with Waterborne Salt Preservatives*. USDA, For. Serv. Res. Paper, FPL 434.

Betts, H.S. and Newlin, J.A. 1915. *Strength Tests of Structural Timbers*. USDA Bull. No. 286. Washington, D.C. 15 p.

Brazier, J.D. and Laidlaw, R.A. 1974. *The Implications of using Inorganic Salt Flame-retardant Treatments with Timber*. BRE Infor. Supp. IS 13/74 Princes Risborough Lab., Aylesbury, Bucks. U.K., 3 p.

Clausen, C.A. and Kartal, S.N. 2003. Accelerated detection of brown-rot decay: Comparison of soil block test, chemical analysis, mechanical properties and immunodetection. *Forest Prod. J.*, 53(11/12):90–94.

Cowling, E.B. 1961. *Comparative Biochemistry of the Decay of Sweetgum Sapwood by White-rot and Brown-rot Fungi*. USDA, For. Serv., Tech. Bull. No. 1258, Washington, DC, 50 p.

Curling, S.F, Clausen, C.M., and Winandy, J.E. 2002. Solid wood products relationships between mechanical properties, weight loss, and chemical composition of wood during incipient brown-rot decay. *Forest Prod. J.*, 52(7/8):34–39.

Dale, F.A. 1966. Fence posts and fire. *Forest Products Newsletter*, 328:1–4.

Erickson, H.D. and Rees, L.W. 1940. Effect of several chemicals on the swelling and crushing strength of wood. *Agric. Res.* 60(1):593–604.

Feist, W.C., Hajny, G.J., and Springer, E.L. 1973. Effect of storing green wood chips at elevated temperatures. *Tappi.*, 56(8):91–95.

Freudenberg, K. 1932. The relation of cellulose to lignin in wood. *Chem. Ed.*, 9(7):1171–1180.

Gerhards, C. C. 1977. *Effect of Duration and Rate of Loading on Strength of Wood and Wood-Based Materials*. USDA, Forest Service, Forest Products Laboratory, Res. Paper, FPL 283, Madison, WI.

Gerhards, C. C. 1982. Effect of moisture and temperature on the mechanical properties of wood: An analysis of immediate effects. *Wood Fiber*, 14(1):4–36.

Gerhards, C.C. 1970. *Effect of Fire-Retardant Treatment on Bending Strength of Wood*. USDA, For. Serv. Res. Pap., FPL 145, Madison, WI.

Gillwald, W. 1961. Influence of different impregnation means on the physical and strength properties of wood. *Holztechnologie*, 2:4–16.

Green, D.W. 1999. Adjusting modulus of elasticity of lumber for changes in temperature. *Forest Prod. J.*, 49(10):82–94.

Green, D.W. and Evans, J.W. 2008. Effect of cyclic long-term temperature exposure on the bending strength of lumber. *Wood an Fiber Science*, 40(2):288–300.

Green, D.W., Evans, J.W., and Craig, B. 2003. Durability of structural lumber products at high temperature. Part 1:66°C at 75% RH and 82°C at 30% RH. *Wood Fib. Sci.*, 35(4):499–523.

Green, D.W., Evans, J.W., Hatfield, C.A., and Byrd, P.J. 2005. *Durability of Structural Lumber Products after Exposure at 82°C and 80% Relative Humidity*. USDA, For. Serv. Res. Pap. FPL-RP-631, Madison, WI.

Hatt, W.K. 1906. *Experiments on the Strength of Treated Timber*. USDA, For. Serv., Circ. 39. Washington, D.C., 31 p.

Ifju, G. 1964. Tensile strength behavior as a function of cellulose in wood. *Forest Prod. J.*, 74(8):366–72.

Johnson, J.W. 1967. *Bending Strength for Small Joists of Douglas-Fir Treated with Fire Retardants.* Oregon State University Rep. T-23. Forest Res. Laboratory, School of Forestry, Oregon State University, Corvallis, OR, 12 p.

Kalnins, M.A. 1966. *Surface Characteristics of Wood as they Affect Curability of Finishes.* Part II. Photochemical degradation of wood. USDA, For. Serv. Res. Pap. FPL 57.

Kennedy, R.W. 1958. Strength retention in wood decayed to small weight losses. *Forest Prod. J.*, 8:308–14.

Kerr, A.J. and Goring D.A.I. 1975. The ultrastructural arrangement of the wood cell wall. *Cellul. Chem. Technol.* 9:563–573.

Koenigs, J.W. 1974. Hydrogen peroxide and iron: A proposed system for decompodition of wood by brown-rot basidiomycetes. *Wood Fiber*, 6(1):66–80.

Kollmann, F. and Cote, W.A. Jr. 1968. *Principles of Wood Science and Technology. Vol. 1. Solid Wood*, Springer-Verlag, Berlin.

Koukal, M., Bednarcik, V., and Medricka, S. 1960. The effect of PCP and NaPCP on increasing the resistance of waterproof plywood to decay and on its physical and mechanical properties. *Drevarsky Vyskum*, 5(1):43–60.

Lagergren, S., Rydholm, S., and Stockman, L. 1957. Studies on the interfibre bonds of wood. Part 1. Tensile strength of wood after heating, swelling and delignification. *Sven. Papperstidn.*, 60, 632–44.

Lam, F. and Morris, P.I., 1991. Effect of double-density incising on bending strength. *Forest Prod. J.*, 41(6):43–47.

Larson, P.G. and Cox, W.T. 1938. *Mechanics of Material*, Wiley and Sons, New York, 408 p.

Larsen, M.J., Winandy, J.E., and Green, F. III. 1995. A proposed model of the tracheid cell wall having an inherent radial ultrastructure in the S2 layer. *Materials and Organism*, 29(3):197–210.

Lebow, S. T. and Winandy, J.E. 1998. The role of grade and thickness in the degradation of fire-retardant-treated plywood. *Forest Prod. J.*, 48(6):88–94.

Lebow, P.K. and Winandy, J.E. 1999a. Verification of the kinetics-based model for long-term effects of fire retardants on bending strength at elevated temperatures *Wood Fib. Sci.*, 31(1):49–61.

Lebow, S.T. and Winandy, J.E. 1999b. Effect of fire-retardant treatment on plywood pH and the relationship of pH to strength properties. *Wood Science and Technology*, 33, 285–298.

LeVan. S.M., Ross, R.J., and Winandy, J.E. 1990. *Effects of Fire Retardant Chemicals on the Bending Properties of Wood at Elevated Temperatures.* U.S.D.A. Research Paper FPL-RP-498, Madison, WI.

LeVan, S.M., Kim, J.M., Nagel, R.J., and Evans, J.W. 1996. Mechanical properties of fire-retardant treated plywood after cyclic temperature exposure. *Forest Prod. J.*, 46(5):64–71.

Mark, R. E. 1967. *Cell Wall Mechanics of Tracheids.* Yale Press: New Haven, CT, 310 p.

McCarthy, W.G., Seaman, E.W., DaCosta, B., and Bezemer, L.D. 1972. Development and evaluation of a leach resistance fire retardant preservative for pine fence posts. Inst. *Wood Set.*, 6(1):24–31.

MacLean, J.D. 1945. *Effect of Heat on the Properties and Serviceability of Wood.* Experiments on thin wood specimens. USDA, Forest Service, Forest Products Laboratory, Report R1471.

MacLean, J.D. 1953. Effect of steaming on the strength of wood. *Proceedings Annual Meeting Am. Wood-Preserv. Assoc.* Cleveland, OH, 49, 88–112.

MacLean, J.D. 1951. *Preservative Treatment of Wood by Pressure Processes.* USDA Agric. Handbook No. 40. Washington, DC, 160 p.

Millett, M A. and Gerhards, C.C. 1972. Accelerated aging: Residual weight and flexural properties of wood heated in air at 115 to 175 C. *Wood Sci.*, 4(4):193–201.

Morrell, J.J., Gupta, R., Winandy, J.E., and Riyanto, D.S. 1998. Effects of incising on torsional shear strength of lumber. *Wood Fib. Sci.*, 30(4):374–381.

Pasek, E. and McIntyre C.R. 1990. Heat effects on fire-retardant treated wood. *J. Fire Sci*, 8:405–420.

Prentice, I.W. 1949. Annual Report. East Mailing Res. Station. Kent, UK, 122–25.

Rowell, R.M. 1980. *How the Environment Affects Lumbar Design*, In: Lyon, D E., and Galligan, W.L. Eds: USDA, Forest Service, For. Prod. Lab. Report, Madison, WI.

Siau, J.F. 1969. The swelling of basswood by vinyl monomers. *Wood Sci.*, 1(4):250–253.

Soltis, L.A. and Winandy, J.E. 1989. Long-term strength of CCA-treated lumber. *Forest Prod. J.*, 39(5):64–68.

Sweet, M. and Winandy, J.E. 1999. Influence of degree of polymerization of cellulose and hemicellulose on strength loss in fire-retardant-treated southern pine. *Holzforschung*, 53(3):311–317.

Stamm, A.J. 1964. *Wood and Cellulose Science.* The Ronald Press Co. New York.

Stamm, A.J., Burr, A.K., and Kline, H.A. 1946. Staybwood: Heat-stabilized wood. *Ind. Eng. Chem.*, 38, 630–37.

Thomas, R.J. 1981. *Wood: Its Structure and Properties*, In: Wangaard, F. F., Ed: Penn State Univ. Press: University Park, PA, 101–46.

U.S. Department of Agriculture, 2010. *Forest Service. Wood Handbook*. USDA General Technical Report GTR-190, Madison, WI, 508 p.

U.S. Department of Defense, 1979. *Department of the Navy*. Civil Engineering Laboratory. GEL Tech. Data Sheet 79–07. USN, GEL: Port Hueneme, CA, 4 p.

Wilson, T.R.C. 1932. *Strength-Moisture Relations for Wood*. USDA Tech. Bull. No. 282, Washington, D.C., 88 p.

Wangaard, F.F. 1950. *The Mechanical Properties of Wood*, John Wiley and Sons: New York, 1950, 377 p.

Winandy, J.E. 1989. The effects of CCA preservative treatment and redrying on the bending properties of 2 by 4 Southern Pine lumber. *Forest Prod. J.*, 39(9):14–21.

Winandy, J.E. 1994. Effects of long-term elevated temperature on CCA-treated Southern Pine lumber. *Forest Prod. J.*, 44(6):49–55.

Winandy, J.E. 1995a. *Effects of Fire Retardant Treatments After 18 Months of Exposure at* 150°F (66°C). USDA, Research Note FPL– RN– 0264. Madison, WI.

Winandy, J.E. 1995b. The Influence of time-to-failure on the strength of CCA-treated lumber. *Forest Prod. J.*, 45(3):82–85.

Winandy, J.E. 1996a. Effects of treatement, incising and drying on mechanical properties of Timber. In: Ritter, M. Ed: *Proceedings: National conf. On wood in Transportation Structures*. USDA. General Technical Report GTR-94. 371–378.

Winandy, J.E. 1996b. Treatment-processing Effects Model for WBP-treated Wood. In: Gopu, V.K.A. Ed. *Proceedings: International Wood Engineering Conference*. Louisiana State University: Baton Rouge, LA, 3:125–133.

Winandy, J.E. 1997. Effects of fire retardant retention, borate buffers, and re-drying temperature after treatment on thermal-induced degradation. *Forest Prod. J.*, 47(6):79–86.

Winandy, J.E. 2001. Thermal degradation of fire-retardant-treated wood: Predicting residual service-life. *Forest Prod. J.*, 51(2):47–54.

Winandy, J.E. and Barnes, H.M. 1991. The influence of initial kiln-drying temperature on CCA-treatment effects on strength. *Proc. Am. Wood Pres. Assn.*, 87:147–152.

Winandy, J.E., Bendtsen, B.A., and Boone, R S. 1983. The effect of time-delay between CCA preservative treatment and redrying on the toughness of small clear specimens of Southern Pine. *Forest Prod. J.*, 33(6):53–58.

Winandy, J.E. and Boone, R.S. 1988. The effects of CCA preservative treatment and redrying on the bending properties of 2 by 6 Southern Pine lumber. *Wood Fib. Sci.*, 20(3):50–64.

Winandy, J.E., Boone, R.S., and Bendtsen, B.A. 1985. The interaction of CCA preservative treatment and redrying: effects on the mechanical properties of Southern Pine. *Forest Prod. J.*, 35(10):62–68.

Winandy, J.E., Boone, R.S., Gjovik, L.R., and Plantinga, P.L. 1989. The effects of ACA and CCA preservative treatment and redrying: Effects on the mechanical properties of small clear specimens of Douglas Fir. *Proc. Am. Wood Pres. Assn.*, 85:106–118.

Winandy, J.E. and Lebow, P.K. 1996. Kinetic models for thermal degradation of strength of fire-retardant-treated wood. *Wood Fib. Sci.*, 28(1):39–52.

Winandy, J.E. and Lebow, P.K. 2001. Modeling wood strength as a function of chemical compostion: An individual effects model. *Wood Fib. Sci.*, 33(2):239–254.

Winandy, J.E., LeVan. S.M., and Ross, R.J. 1991. Thermal Degradation of Fire-Retardant-Treated Plywood— Development and Evaluation of a Test Protocol. USDA, Research Paper FPL-RP-501, Madison, WI.

Winandy, J.E. and Morrell, J.J. 1993. Relationship between incipient decay, strength, and chemical composition of Douglas-fir heartwood. *Wood Fib. Sci.*, 25(3):278–288.

Winandy, J.E. and Morrell, J.J. 1998. Effects of incising on lumber strength and stiffness: Relationship between incision density and depth, species and MSR-grade. *Wood Fib. Sci.*, 30(2):185–197.

Winandy, J.E. and Rowell, R. M. 1984. The chemistry of wood strength. In: Rowell, R.M. Ed: *The Chemistry of Solid-wood*. ACS Sym Series #208, Washington, DC, 211–255.

Winandy, J.E. and Schmidt, E.L. 1995. Preliminary development of remedial treatments for thermally degraded fire-retardant-treated wood. *Forest Prod. J.*, 45(2):51–52.

Woo, J.K. 1981. *Effect of Thermal Exposure on Strength of Wood Treated with Fire Retardants*. Dept. Wood Science. Univ. of California PhD thesis, Berkeley, CA.

Wood, L.W. 1951. *Relation of Strength of Wood to Duration of Load*. USDA, For. Serv., For. Prod. Lab. Rep. No. R-1916, Madison, WI.

Wood, M.W., Kelso K.C. Jr., Barnes, H.M., and Parikh, S. 1980. Effects of the MSU process and high preservative retentions on Southern Pine treated with CCA–Type C. *Proc. Am. Wood Pres. Assn.*, 76:22–37.

12 Fiber Webs

Roger M. Rowell

CONTENTS

Wood fibers can be used to produce a wide variety of low-density, three-dimensional webs, mats, and fiber-molded products. Short wood fiber can be blended with a long fibers can be formed into flexible fiber mats, which can be made by physical entanglement, nonwoven needling or thermoplastic fiber melt matrix technologies. The three most common types of flexible mats are carded, air-laid, needle punched, and thermobonded mats. In carding, the fibers are combed, mixed and physically entangled into a felted mat. These are usually of high density but can be made at almost any density. Air-laid webs are made by laying down layers of wood fibers combined with a low melting thermoplastic fiber that is then passed through a heated chamber that melts the thermoplastic in the web. The heated web is then passed through calender rolls that press the melted fibers together with the wood fibers holding the web together. A needle punched mat is produced in a machine which passes a randomly formed machine made web through a needle board that produces a mat in which the fibers are mechanically entangled. The density of air-laid webs and needled mats can be controlled by the amount of fiber going through the processes or by overlapping webs or mats to give the desired density. A thermobonded mat is made by combining natural fibers with a thermoplastic fiber in the needled mat technology that is then melted in a heated press holding the mat together. The webs and mats can be used as filters, geotextiles, sorbents, and mulch mats.

Wood fibers can also be formed into fiber-based products using air or water as a carrier. Fibers can be sprayed in an air stream and used as insulation or ground cover. Fibers can be slurried in water, molded into wide variety of shapes (pulp molding) and dewatered to form the final dry product.

12.1   WEBS AND MATS

Early information indicates that the Russians experimented with air forming during the 1930s. During this period, a patent was issued to two Russians describing a method for the production of a dry web using synthetic fiber and air (Pusyrev and Dimitriev 1960). In the early 1960s, a patent was issued to James Clark for the air forming of fibrous material and consolidating it into a web or sheet (Clark 1960). Later in the decade, a Finnish inventor named H. J. Hieldt developed an air forming method that involved the use of an electrostatic current to help guide the fibers.

In the mid-1960s, the Rando Corporation in the United States developed a different system in order to process long synthetic fibers for use in medium density fiberboard (Curlator Corp. 1967, Figure 12.1). Section A in Figure 12.1 is where the fiber is fed into the system. Section B is a fiber opener where fiber bundles are separated and mixed with other fibers. Between A and B, the fibers are formed into a continuous mat which is fully formed at C. There are several options at C. A liquid or powdered adhesive can be added if the final product is a web to be thermally formed into a three-dimensional composite. Another option is to place a seed applicator here to incorporate different types of seeds into the mat to be used as seeding geotextiles. At D, the web can go on through a needle board where the web is "needled" together in a nonwoven process. Some of the Rando Systems were modified to make lighter weight webs that include wood pulp (Rando 1993). Both of these early systems were relatively speed limited. Figure 12.2 shows a web that has been made using the Rando system and a needle board.

In the early 1970s, a Japanese firm, Honshu, developed a process for making a variety of nonwovens using wood pulp and synthetic fibers. Most recently, Danweb Forming International, Ltd. developed a drum former capable of utilizing various length synthetic or natural fibers (Danweb 2003, Figure 12.3). In addition, this firm has made use of a new and simpler horizontal machine layout (Wolff and Byrd 1990, Byrd 1990a,b).

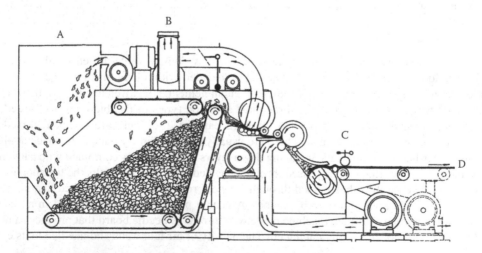

FIGURE 12.1   Schematic of Rando fiber mat forming system.

FIGURE 12.2   Fiber mat made using the Rando system.

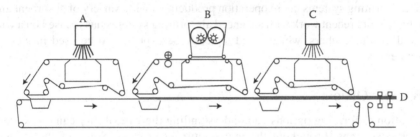

FIGURE 12.3   Schematic of Danweb fiber mat forming system.

The essential features of the drum system comprise two perforated counter-rotating drums (Figure 12.3b) located transversely above the forming wire within a square-section box. These drums are connected to a fixed pipe, such that the drums and their pipes form a track. A series of brush rolls are located inside the drums, transverse to the wire. The box itself is sealed in the transverse direction by means of seal rolls, and in the longitudinal direction with side plates. Figure 12.4 shows a typical mat made using the Danweb system.

In operation, the fibers, dispersed in air, are fed into the rotating drums via the fixed pipes. As the fiber stream passes into the drums, the brush rolls will force the fibers—partly as a result of a centrifugal effect and partly as a result of turbulence—through the perforated walls of the drum. The air from the suction box then draws the fibers onto the forming wire. The supply of single or mixed fibers is connected to the "horse-track" in such a way that a circular movement is insured within the system. This guarantees a completely uniform distribution.

The drum system also has the advantage that it allows partially opened fibers to be separated out of the flow, because, as a result of their inertia, they remain on the outer wall of the track. This feature of the system makes it particularly useful in handling recycled fibers.

An additional advantage of the drum-type former is its ability to handle fibers up to 25 mm in length. This permits the addition of regular staple fibers or bonding fibers without modifying the former. Forming capacity remains fairly constant with fiber lengths up to 12 mm and then declines somewhat with fiber length from 12 to 25 mm.

One of the unique features of modern air forming systems, unlike those of a decade ago, is their flexibility with respect to feedstock. Much of the early work on air forming centered around recycled and virgin cellulose fibers. This work has advanced to the point where a number of full-scale

FIGURE 12.4   Fiber mat made using the Danweb system.

commercial air forming systems are in operation producing a wide variety of absorbent and decorative disposables. More recent work with advanced air forming systems has focused on a wide variety of natural and synthetic fibers, which could logically be expected to be used in more advanced composite materials.

12.1.1 FORMING OPTIONS

The identification of forming options and understanding their flexibility can conserve materials as well as process steps throughout the manufacturing process. Some of the composite web or molded product options that would be available with forming unit layouts include the following.

12.1.1.1 Layering

By placing different feedstocks in different forming heads, a manufacturer can produce a composite material with high-performance/high-cost materials on the exterior of the web and low-performance/low-cost material on the interior of the web. Figure 12.3 shows a Danweb system with several different options. Section A and/or C can be used to add a top or bottom layer of fiber or other film to the fiber mat that is formed in B.

12.1.1.2 Fiber Mixing

The inherent ability of the air former to handle streams of mixed fibers of different fiber types, deniers, and lengths permits in-line forming of a composite. Also, the fiber mixture can be varied for each forming head which provides additional product flexibility.

12.1.1.3 Use of Additives

Work has been done which demonstrates how additives, such as superabsorbent powders and binders can be added to the web during the forming process (Figure 12.1, section C and Figure 12.3, section A or C). In the case of super absorbents, one advantage of this approach is that the super absorbent powder when near the area of maximum void space in the web can absorb liquids faster and in greater quantity than if added to a finished web as part of a laminate in an off-line process. Also, because of their uniform dispersion, powdered binders can perform in much the same manner to insure maximum strength with a minimum add-on.

12.1.1.4 Scrim Addition

Composite materials can be further enhanced through the addition of an open net or scrim between two of the air forming heads. This can simplify the process and possibly result in the use of less total raw material (Figure 12.3, section A or C).

12.1.1.5 Card Combined with Air Forming

The first technique discussed is a means of combining a card with an air forming unit. From the standpoint of size and speed, these two processes are quite compatible.

The advantages of a process combination such as this for composites includes: combination of long and short fibers, increased uniformity and bulk of composite materials, the ability to adjust the quantity of different fibers to be used in the process and process simplicity; both are proven processes and compatible in terms of speed and machine trim.

Economic advantages obtained from a combination of air forming techniques include (a) the ability to make a thermally bondable composite for either single or multi-stage bonding; (b) the ability to substitute low-cost raw materials such as wood pulp and waste synthetic fiber in place of higher cost fibers; and (c) the ability to limit capital expenditures by upgrading an existing card line versus purchasing an entirely new nonwoven line.

12.1.1.6 Melt-Blown Polymer Unit Combined with Air Forming

Another technique is a combination of melt blown and air forming processes. This approach permits a composite to be produced in-line as a single process. If required, the process could include a provision for blanking out the center wood pulp core where the top and bottom layers of melt blown are thermally bonded.

This technology represents options to produce such products as (a) oil absorbent pads, (b) backing, laminating composites, and (c) wipes.

Air forming systems offer many product and process advantages in the production of both flexible and rigid composite materials. The units are of manageable size and can be combined with other process equipment to offer significant materials flexibility.

From a commercial standpoint, air forming is a relatively young technology. We can expect current and improved systems to play an increasing role in future production of composite materials.

12.2 PULP MOLDING

Composites can be made using wood fibers in a wet process that forms a composite by dewatering the slurry (Laufenberg 1996). This is used today to make such products as egg cartons and nesting packaging where products are kept apart with a thin wall of molded pulp. Pulp molding is done using a forming slurry of pulp fibers with a consistency of 0.5–5% (dry fiber weight/water weight). Pulp molding is done in two steps: forming a dense fiber network from a wet fiber slurry onto a configured surface or mold and drying. Most of the pulp molding is done using a drainable surface (fine mesh screen), initial dewatering by gravity, vacuum dewatering and finally, heat drying. The pulp slurry is either poured into the mold or the mold is dipped into the pulp slurry and withdrawn or the slurry is pumped into the mold. For highly uniform surfaces, the pulp slurry consistency should be very low (see Figure 12.5).

Drying the wet formed product is carried out in one of several ways. Densification and drying can be done applying pressure to the molded product. Minimal density and strength will result without the use of pressure in the drying step. The partially dewatered product can be dried in an oven. The strength of the final product comes from inter-fiber bonding similar to the bonds formed in the paper making process. Starch can be added for increased bonding strength, however, since these products were formed in a wet media, the products have very little wet strength or wet stiffness.

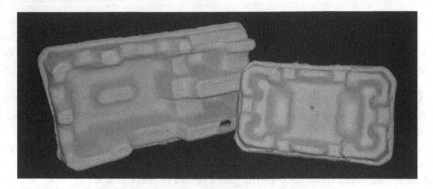

FIGURE 12.5 Products made by pulp molding.

FIGURE 12.6 Spaceboard—a structural pulp molded product.

Structural pulp molded products can also be made using a similar process. Setterholm (1985) developed a method of forming a three-dimensional, waffle-like structure using a hard flexible rubber forming head. The product was called "Spaceboard" and could be used for structural applications. Further advancement were made by Hunt and Gunderson (1998) and Scott and Laufenberg (1994) (see Figure 12.6).

12.3 GEOTEXTILES

Geotextiles derive their name from the two words, geo and textile and, therefore, mean the use of fabrics in association with the earth. Erosion control is mainly done today using geotextiles made using synthetic materials such as polypropylene and polyethylene. Wood and other agricultural fibers can be made into geotextiles that can be used to control erosion by using either the Rando or the Danweb process.

12.3.1 EROSION CONTROL

Soil type and vegetation coverage are critical factors in the ability of the land to sorb water. With a healthy forest where, at least, 60–75% of the ground is covered with vegetation, only about 2% or less of rainfall becomes surface runoff and erosion is low (<0.05 tons loss per acre). When the

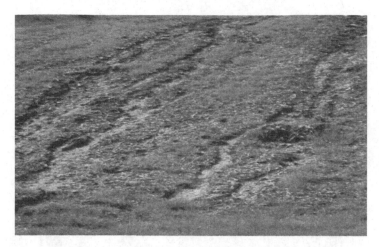

FIGURE 12.7   Example of severe soil erosion.

vegetation cover is between 35% and 50%, the surface runoff increases to 14% with a soil loss of 0.5 tons per acre. Finally, when the vegetation cover is less than 10%, the surface runoff increases to 73% with a soil loss of 5.55 tons per acre (Sedell 2000). From these simple statistics, it is easy to see the effect catastrophic fires in our forests can have on our fresh water supply.

When there is a severe fire and little ground vegetation remains, surface runoff can increase over 70% and erosion can increase by three orders of magnitude (Bailey and Copeland 1961). There are three components to erosion: detachment, transport, and deposition. The rate of erosion will depend on the geology, topography, vegetation, and climate. In flat terrain, erosion may be minimal after a fire, however, in steep terrain, surface soil loss can be severe following a fire (Figure 12.7).

After the fire, burned land usually sorbs water more slowly than unburned land (Anderson and Brooks 1975). A severe fire in a forest can easily create a ground condition where surface runoff can lead to "flash floods" and erosion can result in not only loss of soil but badly contaminated water. Increases in water flow after a fire can result in more solids and dissolved materials in the water (DeBano et al. 1998). Water soluble and insoluble nutrients can increase aquatic plant growth that may decrease water flow. Inorganic compounds leached into the water increase the soluble ions that may increase both turbidity and toxicity of the water (Robichaud et al. 2000).

The addition of such chemicals could be based on silvicultural prescriptions to ensure seedling survival and early development on planting sites where severe nutritional deficiencies, animal and fire damage, insect attack, and weed problems are anticipated. Medium density fiber mats can also be used to replace dirt or sod for grass seeding around new homesites, along highway embankments or stream beds. Grass or other type of seed can be incorporated in the fiber mat. Fiber mats promote seed germination and good moisture retention. Low and medium density fiber mats can be used for soil stabilization around new or existing construction sites. Steep slopes, without root stabilization, lead to erosion and loss of top soil.

In one type of geotextile, seeds are added to the geotextiles while the web is being formed. Grass and wild flower seeds can be added so that the geotextiles not only prevents erosion by forming a surface physical barrier but allows grass to grow establishing a new layer of plant growth with a root system to stabilize the soil after the geotextiles has degraded. Seeds can also be planted under the geotextile so that the seeds can germinate and grow above the geotextile. This type of geotextile should have enough physical strength to endure strong wind and at the same time about 30% of sunlight should pass through the geotextiles. Wood fiber-based geotextiles can hold moisture to help germinate the seeds. Other chemicals, such as fertilizers, can also be added to the web during its formation. Figure 12.8 shows a geotextiles application on a very steep embankment beside a highway.

FIGURE 12.8   Application of fiber-based geotextiles.

Medium- and high-density fiber mats can also be used below ground in road and other types of construction as a natural separator between different materials in the layering of the back fill. It is important to restrain slippage and mixing of the different layers by placing separators between the various layers. Jute and kenaf geotextiles have been shown to work very well in these applications but the potential exists for any of the long agro-based fibers.

Geotextiles in general are expected to be biodegraded within given period of time and this timing is very much depending upon the materials use. Lignin contents contribute the biodegradability with other factors such as density, hydrophobicity, extractive contents, and so on. Usually, the geotextiles are expected to last until the germination of seeds—between four and six weeks. The biodegradability can be controlled by addition of preservatives to prolong the decay and addition of fungi to speedup the decay.

It has been estimated that the global market for geotextiles is about 800 million square meters, but this estimate has not been broken down into use categories so it is impossible to determine the portion that is available for natural geotextiles.

12.4   FILTERS

12.4.1   Types

12.4.1.1   Physical Type

Fiber-based filters can be used to remove suspended solids from both air and water (Boving and Rowell 2010). The physical types of wood fiber filters can be in several forms. Fibers can be made into webs, mats, packed into a column or chamber. Webs or mats increase the surface area of the filter and stabilize hydraulic pressure. The suspended solids are physically captured and held in the webs until the filters are cleaned.

12.4.1.2   Chemical Type

Fiber-based filters can also remove dissolved inorganic ions, organic chemicals and other soluble contaminates from water. Most of the wood fiber-based mats have limited capacities for removing soluble contaminates from water but their capacity can be greatly improved with chemical or plasma modification.

12.4.2 APPLICATIONS

We live in a water-based world. Water sculpts our landscape, provides navigational opportunities, transports our goods, and is the medium of life. It is the basis of all life on earth so it is not surprising that one of our high priorities is to insure a long-term supply of clean water.

Seventy percent of the earth's surface is covered with water. Most of this water, 97.5%, is in the oceans and seas and is too salty to drink or grow crops. Of the remaining 2.5%, 1.73% is in the form of glaciers and icecaps leaving only about 0.77% available for our fresh water supply. Said another way, of the total water on earth, only 0.0008% is available and renewable in rivers and lakes for human and agricultural use. It is the water that falls as rain or snow or that has been accumulated and stored as groundwater that we depend on for our "clean" water resource.

For 1.5–2.5 billion people in the world, clean water is a critical issue (Lepkowski 1999). It is estimated that by the year 2025, there will be an additional 2.5 billion people on the earth that will live in regions already lacking sufficient clean water. In the United States, it is estimated that 90% of all Americans live within 10 miles of a body of contaminated water (Hogue 2000b). The U.S. Environmental Protection Agency (EPA) is working on guidelines and regulations to establish total maximum daily load (TMDL) for each pollutant that "continues to be a problem in a particular impaired water" (Hogue 2000a, pp. 31–33). The materials that EPA have listed as water impairments include sediments, nutrients, pathogens, dissolved oxygen, metals, suspended solids, pesticides, turbidity, fish contamination, and ammonia. Other conditions to be considered for clean water on the list include pH, temperature, habitat, and noxious plants. Of these, sediments, nutrients, pathogens, and dissolved oxygen contribute the greatest to our contaminated water (EPA data, 1998 from Hogue 2000a).

On one specific issue of arsenic in drinking water, the EPA has proposed lowering the maximum allowed level of arsenic from 50 ppb to 5 ppb due to concerns of bladder, lung, and skin cancer (Hileman 2000). Meeting these targets will not be easy. Arsenic in water is a global concern especially in countries like Bangladesh where most of their water wells are contaminated with arsenic (Lepkowski 1999).

About 80% of the fresh water in the United States originates on the 650 million acres of forestlands that cover about 1/3 of the Nation's land area. The nearly 192 million acres of National Forest and Grasslands are the largest single source of fresh water in the United States In many cases, the headwaters of large river basins originate in our National forests. In 1999, the EPA estimated that 3400 public drinking-water systems were located in watersheds contained in national forests and about 60 million people lived in these 3400 communities (Sedell 2000).

The wood and bark structure is very porous and has a very high free surface volume that should allow accessibility of aqueous solutions to the cell wall components. One cubic inch of a lignocellulosic material, for example, with a specific gravity of 0.4, has a surface area of 15 square feet. However, it has been shown that breaking wood down into finer and finer particles does increase sorption of heavy metal ions.

Lignocellulosics are hygroscopic and have an affinity for water. Water is able to permeate the noncrystalline portion of cellulose and all of the hemicellulose and lignin. Thus, through absorption and adsorption, aqueous solutions come into contact with a very large surface area of different cell wall components (Browning 1967).

Laszlo and Dintzis (1994) have shown that wood has ion-exchange capacity and general sorptive characteristics, which are derived from their constituent polymers and structure. The polymers include extractives, cellulose, hemicelluloses, pectin, lignin, and protein. These are adsorbents for a wide range of solutes, particularly divalent metal cations (Laszlo and Dintzis 1994). Wood contains, as a common property, polyphenolic compounds, such as tannin and lignin, which are believed to be the active sites for attachment of heavy metal cations (Waiss et al. 1973, Masri et al. 1974, Randall et al. 1974, Bhattacharyya and Venkobachar 1984, Phalman and Khalafalla 1988, Verma et al. 1990, Shukla and Sakhardande 1991, Maranon and Sastre 1992, Lalvani et al. 1997,

Vaughan et al. 2001). Sawdust has been used to remove cadmium and nickel (Basso et al. 2002) and several types of barks have been used to remove heavy metal ions from water (Randall 1977, Randall et al. 1974, Kumar and Dara 1980, Pawan and Dara 1980, Vazquez et al. 1994, Seki et al. 1997, Tiwari et al. 1997, Gaballah and Kibertus 1998, Bailey et al. 1999) from aqueous solution. Cellulose can also sorb heavy metals from solution (Acemioglu and Alma 2001). Isolated kraft lignin has been used to remove copper and cadmium (Verma et al. 1990) and organosolv lignin has been used to remove copper (Acemioglu et al. unpublished data) from aqueous solutions.

Acemioglu et al. postulate that metal ions compete with hydrogen ions for the active sorption sites on the lignin molecule (Acemioglu et al. unpublished data). They also conclude that metal sorption onto lignin is dependent on both sorption time and metal concentration. Basso et al. (2002) studied the correlation between lignin content of several woods and their ability to remove heavy metals from aqueous solutions. The efficiency of removing Cd(II) and Ni(II) from aqueous solutions was measured and they found a direct correlation between heavy metal sorption and lignin content. Reddad et al. (2002) showed that the anionic phenolic sites in lignin had a high affinity for heavy metals. Mykola et al. (1999) also showed that the galacturonic acid groups in pectins were strong binding sites for cations.

Extracting fibers with different solvents will change both the chemical and physical properties of the fibers. It is known, for example, that during the hot water and 1% sodium hydroxide extraction of fibers, the cell walls delaminate (Kubinsky 1971). A simple base treatment has been shown to greatly increase the sorption capacity of wood fibers (Tiemann et al. 1999, Reddad et al. 2002). At the same time, some of the amorphous matrix and part of the extractives, which have a bulking effect, are removed (Kubinsky and Ifju 1973), so that the individual microfibrils become more closely packed and shrunken (Kubinsky and Ifju 1974). Therefore, delamination and shrinkage may also change the amount of exposed cell wall components that may affect the heavy metal ions sorption capacity of the fibers.

Han et al. (2005) have shown that phosphorus can be removed from water using a juniper-fiber-based web that is first saturated with a heavy metal. Figure 12.9 shows a plot of phosphorus uptake versus time with webs made of juniper fiber, base treated juniper fiber, and juniper fiber that has been saturated with iron. The filter made using the heavy metal loaded fiber removed much more phosphorus than the webs without the heavy metal.

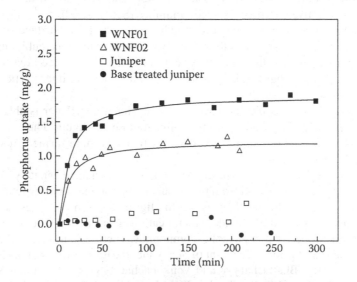

FIGURE 12.9 Plot of phosphorus removal from solution using juniper fiber filters.

12.4.3 Testing Protocols for Filters

12.4.3.1 Kinetic Tests

A kinetic tests determines the time it takes to remove a given concentration of a contaminate. Two approaches are used. One uses a constant weight of sorbent that is applied to various concentrations of solution, and the other uses a constant concentration of solution applied to variable weights of the sorbent (Eaton-Dikeman 1960).

12.4.3.2 Isotherms

Adsorption occurs on the surface of the fibers and when the rate of sorption equals the rate of desorption, equilibrium has been achieved and the capacity of the fiber has been reached. The theoretical adsorption capacity of the fiber for a particular containment can be determined by calculating its adsorption isotherm. Wood samples are carefully weighed (M) and placed in a known volume (V) of standard solution of and shaken at 150 rpm. After 24 h the mixture of the standard solution and fiber sample are filtered and the final concentration of the solution is determined by ICP.

The quantity of adsorbate that can be taken up by an adsorbent is a function of both the characteristics and concentration of adsorbate and the pH (some authors use various temperature, but pH has more impact than temperature in wood fiber filtration). Generally, the amount of material adsorbed is determined as a function of the concentration at a constant pH, and the resulting function is called adsorption isotherm. One equation that is often used to describe the experimental isotherm data was developed by Freundrich et al. The original Freundrich isotherms were used to describe the adsorption characteristic of the activated carbon used in wastewater treatment. Applying this principle in wood fiber, the empirically derived Freundrich isotherm is defined as follows:

$$\frac{x}{m} = K_f C_e^{1/n} \tag{12.1}$$

where

x/m = amount adsorbate adsorbed per unit weight of adsorbent (fiber)
C_e = equilibrium concentration of adsorbate in solution after adsorption
K_f, n = empirical constants

$$\log\left(\frac{x}{m}\right) = \log K_f + \frac{1}{n}\log C_e \tag{12.2}$$

The constants in Freundrich isotherm can be determined by plotting (x/m) versus C and making use of Equation 12.1

$$\log\left(\frac{x}{m}\right) = \log K_f + \frac{1}{n}\log C_e \tag{12.3}$$

12.4.4 Biofilters for Organic Compounds

A biofilter is a filter that contains a microorganism capable of degrading the organic compound that is trapped by the filter. The filter first sorbs the compound by physical or chemical entrapment and then the microorganism decomposes the chemical into smaller chemicals such as carbon dioxide and water. Langseth and Pflum (1994) reported that biofilters could be used to remove organic compounds with an estimated 95% efficiency. The chemicals removed were alcohols, aldehydes, organic acids, and small amounts of low-molecular weight volatile organics such as benzene and toluene. The retention times for small simple volatile compounds is very short with a 90% reduction in low-molecular weight at around 10 s. Higher molecular weight organics decompose at a slower rate (English 1996).

12.5 SORBENTS

Wood-based sorbents have been used for years as cleaning aids. For example, bark has been used to sorb oil for sea water and treated sawdust and wood shavings has been used to clean industrial floors. There are several important factors that influence sorbtive capacity such as density, porosity, surface area, selectivity, and retention (English 1996).

12.5.1 DENSITY

Bulk density is simply a measure of wood mass per unit volume. Wood bulk density varies from a lot of about 0.2 g/cm³ to greater than 1 g/cm³. In general, the lower the bulk density, the greater the sorption capacity. The true density of the wood cell wall is between 1.3 and 1.5 and does not take into consideration any internal voids.

12.5.2 POROSITY AND SURFACE AREA

Porosity can be defined in several ways. It can be a measure of the size of the voids in the wood or can be a measure of how quickly and easily a liquid or gas can penetrate through a piece wood of a given size. Porosity can be measured as the time it takes for a gas or liquid to travel through a piece of wood or it can be a measurement of the sizes of internal voids in the wood. Woods with a low bulk density have a very large internal void space. One cubic inch of wood with a specific gravity of 0.4, has a surface area of 15 square feet.

12.5.3 SELECTIVITY

Wood is a hydrophilic resource so it strongly hydrogen bonds with polar liquids such as water. Selectivity is a measurement of the ability of wood to preferentially sorb one gas, liquid, or a chemical in a gas or liquid over another. Selectivity is also influenced by the wood pore size, ability to wet the wood surface, and capillary pressure.

12.5.4 RETENTION

Retention is the ability of a saturated wood sample after it has been used as a sorbent to retain the sorbed gas, liquid, or chemical that was removed from a gas or a liquid. Retention can be a critical factor in choosing a sorbent system. If the wood has the ability to remove a given contaminate but quickly releases it upon, for example, exposure to air, the sorbed contaminate will be released into the environment before it can be disposed of.

12.6 MULCH MATS

Mulching materials have been used around plants for many years (Waggoner et al., 1960) and are generally one of two types: particle mulches and sheet mulches. Particle mulches, like bark, wood shavings, or sawdust are used to heat and hold moisture close to the plant. Sheet mulches are similar to geotextiles but are usually thicker and last longer in the environment. They are used to enhance seedling survival in wood species where the seedling remains in a "grass like" state for several years, that is, the southern pines (Figure 12.10). The mulch mats, like the loose mulches

FIGURE 12.10   Fiber-based mulch mat.

hold moisture and heat as well as reducing weed growth around the seedling. Mulch mats must remain intact for as long as competition from unwanted plants remains. For this reason, many mulch mats are made from plastics such as polyethylene or polypropylene.

REFERENCES

Acemioglu, B. and Alma, M.H. 2001. Equilibrium studies on the adsorption of Cu(II) from aqueous solution onto cellulose. *J. Colloid Interf. Sci.*, 243:81.

Acemioglu, B., Samil, A., Alma, M.H., and Gundogan, R. Copper(II) removal from aqueous solution by organosolv lignin and its recovery. Unpublished data.

Anderson, W.E. and Brooks, L.E. 1975. Reducing erosion hazard on a burned forest in Organ by seeding. *J. Range Manage.*, 28(5):394–398.

Bailey, S.E., Olin, T.J., Bricka, R.M., and Adrian, D.D. 1999. A review of potentially low cost sorbents for heavy metals. *Water Res.*, 33:2469–2479.

Bailey, R.W. and Copeland, O.L. 1961. Vegetation and engineering structures in flood and erosion control. In: *Proceedings, 13th Congress, International Union of Forest Research Organization*. September, 1961, Vienna, Austria. Paper 11–1:23.

Basso, M.C., Cerrella, E.G., and Cukierman, A.L. 2002. Lignocellulosic materials as potential biosorbents of trace toxic metals from wastewater. *Ind. Eng. Chem. Res.*, 3580–3585.

Bhattacharyya, A.K. and C. Venkobachar, C. 1984. Removal of cadmium (II) by low cost adsorbents. *J. Environ. Eng.*, Division (ASCE), 110:110.

Boving, T. and R. Rowell. 2010. Water filtration using biomaterials. In: *Sustainable Development in the Forest Products Industry*. R.M. Rowell, F. Caldera, and J.K. Rowell, eds. Fernando Pessoa, Oporto, Portugal, Chapter 9, pp. 213–252.

Browning B.L. 1967. The extraneous components of wood, In: *Methods of Wood Chemistry*, Chapter 5, pp. 75–90, John Wiley & Sons, Inc., New York, N.Y.

Byrd, V.L. 1990a. Dan-Web process revolutionizes dry-forming. *Nonwovens Ind.* 21(5):74.

Byrd, V.L. 1990b. New air-laid process revolutionizes industry. *The Business Edition* (March, pp. 1,3,5).

Clark, J.A. 1960. Apparatus and method for producing fibrous structures. U.S. Patent 2,931,076.

Curlator Corp. 1967. Method of making a random fiber web. British Patent 1,088,991.

Danweb, Research and Development, 2003. Danweb, Aarhus, Denmark.

DeBano, L.F., Neary, D.G., and Folliott, P.F. 1998. *Fire's Effects on Ecosystems*. John Wiley and Sons, New York, 333pp.

Eaton-Dikeman Company, 1960. *Handbook of Filtration*, The Eaton-Dikeman Company, Mt. Holly Springs, PA.

English, B. 1996. Filters, sorbents and geotextiles. In: *Paper and Composites from Agro-Based Resources*, R.M. Rowell, R.A. Young, and J.K. Rowell, eds., Chapter 13, 403–426, CRC Lewis Publishers, Boca Raton, FL.

Gaballah, I. and Kibertus, G. 1998. Recovery of heavy metal ions through decontamination of synthetic solutions and industrial effluents using modified barks. *J. Geochem. Exploration.* 62:241–286.

Han, J.S., Min, S-H., and Kim, Y-K. 2005. Removal of phosphorus using AMD-treated lignocellulosic material. *Forest Prod. J.* 55(11):48–53.

Hogue, C. 2000a. Clearing the water. *Chem. Eng. News*, 78(11):31–33.

Hogue, C. 2000b. Muddied waters. *Chem. Eng. News*, 78(35):19–20.

Hileman, B. 2000. Rules in the fast lane. *Chem. Eng. News*, 78(41):43–44.

Hunt, J.F. and Gunderson, D.E. 1998. FPL spaceboard development. In: Tappi *Proceedings of the 198th Corrugated Containers Conference*, Tappi Press, Atlanta, GA, 11.

Kubinsky, E. 1971. Influence of steaming on the properties of *Quercus rubra* L. Wood. *Holzforschung*, 25(3):78–83.

Kubinsky, E. and Ifju, G. 1973. Influence of steaming on the properties of red oak. Part I. Structural and chemical changes. *Wood Sci.*, 6(1):87–94.

Kubinsky, E. and Ifju, G. 1974. Influence of steaming on the properties of red oak. Part I. Changes of structural and related properties. *Wood Sci.*, 7(2):103–110.

Kumar, P. and Dara, S.S. 1980. Modified barks for scavenging toxic heavy metal ions. *Indian J. Envir. Health*, 22:196.

Laufenberg, T.L. 1996. Packaging and light weight structural composites. In: *Paper and Composites from Agro-Based Resources*, R.M. Rowell, R.A. Young, and J.K. Rowell, eds., Chapter 10, 337–350, CRC Lewis Publishers, Boca Raton, FL.

Lalvani, S.B., Wiltowski, T.S., Murphy, D., and Lalvani, L.S. 1997. Metal removal from process water by lignin. *Environ. Technol.* 18(11):1163–1168.

Langseth, S. and Pflum, D. 1994. Weyerhauser tests large pilot biofilters for VOCs removal. *Panel World*, March.

Laszlo, J.A. and Dintzis, F.R. 1994. Crop residues as ion-exchange materials. Treatment of soybean hull and sugar beet fiber (pulp) with epichlorohydrin to improve cation-exchange capacity and physical stability. *J. Appl. Polymer Sci.*, 52:521–528.

Lepkowski, W. 1999. Science meets policy in shaping water's future. *Chemical and Engineering News*, 77(51): 127–134.

Maranon, E. and Sastre, H. 1992. Behaviour of lignocellulosic apple residues in the sorption of trace metals in packed beds. *Reactive Polymers*, 18:172–176.

Masri, M.S., Reuter, F.W., and Friedman, M. 1974. Binding of metal cations by natural substances. *J. Appl. Polymer Sci.*, 18:675–681.

Mykola, T.K., Kupchik, L.A., and Veisoc, B.K. 1999. Evaluation of pectin binding of heavy metal ions in aqueous solutions. *Chemosphere*, 38(11):2591–2596.

Pawan, K. and Dara, S.S. 1980. Modified barks for scavenging toxic heavy metal ions. *Indian J. Environ. Health*, 22:196.

Phalman, J.E. and Khalafalla, J.E. 1988. Use of ligochemicals and humic acids to remove heavy metals from process waste streams. U.S. Department of Interior, Bureau of Mines: RI 9200.

Pusyrev, S.A. and Dimitriev, M. 1960. Production of paper and board by the dry process. *Pulp Paper Mag. Can.* 61(1):T3–6.

Randall, J.M. 1977. Variations in effectiveness of barks as scavengers for heavy metal ions. *Forest Products J.*, 27(11):51.

Randall, J.M., Bermann, R.L., Garrett, V., and Waiss, A.C. 1974. Use of bark to remove heavy metal ions from waste solutions. *Forest Product J.*, 24(9):80–84.

Rando, Product improvement bulletin, 1993. Rando Machine Corporation, Macedon, NY, PIB-018.

Rando Machine Corp. Installation, Operation and Maintenance Manual No.631 for the Rando-Webber, The Commons, Macedon, NY 14502.

Reddad, Z., Gerente, C., Andres, Y., Ralet, M-C., Thibault, J-F., and Cloirec, P.L. 2002. Ni(II) and Cu(II) binding properties of native and modified sugar beet pulp. *Carbohydrate Polymers* 49:23–31.

Robichaud, P.R., Beyers, J.L., and Neary, D.G. 2000. Evaluating the effectiveness of postfire rehabilitation treatments. USDA, Forest Service, Rocky Mountain Research Station, General Technical Report RMRS-GTR-63, September.

Rowell, R.M. and Han, J.S. 1999. Changes in kenaf properties and chemistry as a function of growing time. In: *Kenaf Properties, Processing and Products*, T. Sellers, Jr. and N.A. Reichert, eds, Mississippi State University Press, Mississippi State, MS, Chapter 3, 33–41.

Scott, C.T. and Laufenberg, T.L. 1994. Spaceboard II panels: Preliminary evaluation of mechanical properties. In: *PTEC 94 Timber Shaping the Future, Proceedings of the Pacific Timber Engineering Conference*, July, Gold Coast, Fortitude Valley, Queensland, Australia, TRADA 632, 2.

Sedell, J. S. 2000. Water and the Forest Service, USDA, Forest Service Washington Office, ES 660, Jan.

Seki, K., Saito, M., and Aoyama, M. 1997. Removal of heavy metals from solutions by coniferous barks. *Wood Sci. Technol.*, 31:441–447.

Setterholm, V.C. 1985. FPL spaceboard new structural sandwich concept. *Tappi*, 68(6):40.

Shukla, S.R. and Sakhardande, V.D. 1991. Metal ion removal by dyed cellulosic materials. *J. Appl. Polymer Sci.*, 42:825–829.

Tiemann, K.J., Gardea-Torresdey, J.L., Gamez, G., Dokken, K., Sias, S., Renneer, M.W., and Furenlid, L.R. 1999. Use of x-ray absorption spectroscopy and esterification to investigate Cr (III) and NI(II) ligands in alfalfa biomass. *Environ. Sci. Technol.*, 33:150–154.

Tiwari, D.P., Saksena, D.N., and Singh, D.K. 1997. Kinetics of adsorption of Pb(II) on used tea leaves and Cr(VI) on *Acacia Arabica* bark. *Dev. Chem. Eng. Miner. Process*, 5:79.

Vaughan, T., Seo, C.W., and Marshall, W.E. 2001. Removal of selected metal ions from aqueous solution using modified corncobs. *Bioresource Tech.*, 78:133–139.

Vazquez, G., Antorrena, G. Gonzalez, J., and Doval, M.D. 1994. Adsorption of heavy metal ions by chemically modified *Pinus pinaster* bark. *Bioresource Technol.* 48:251–255.

Verma, K.V.R., Swaminathan T., and Subrahmnyam, P.V.R. 1990. Heavy metal removal with lignin. *J. Environ. Sci. Health*, A25(2):242–265.

Waggoner, P. E., Miller, P. M., and DeRoo, H. C., 1960. *Plastic Mulching—Principles and Benefits*, Bull. No. 634, New Haven: Connecticut Agricultural Experiment Station.

Waiss, A.C., Wiley, M.E., Kuhnle, J.A., Potter, A.L., and McCready, R.M. 1973. Adsorption of mercuric cation by tannins in agricultural residues. *J. Environ. Quality*, 2:369–371.

Wolff, H. and Byrd, V.L. 1990. Flexible composite and multicomponent air-formed webs. *Tappi J.* 73(9):159.

13 Wood/Nonwood Thermoplastic Composites

Craig M. Clemons, Roger M. Rowell, David Plackett, and B. Kristoffer Segerholm

CONTENTS

Composites made from wood, other biomass resources and polymers have existed for a long time but the nature of many of these composites has changed in recent decades. Wood-thermoset composites date to the early 1900s. "Thermosets" or thermosetting polymers are plastics that, once cured, cannot be remelted by heating. These include cured resins such as epoxies and phenolics, plastics used as wood adhesives with which the forest products industry is traditionally most familiar (see Chapter 9). For example, an early commercial composite marketed under the trade name Bakelite was composed of phenol–formaldehyde and wood flour. Its first commercial use was reportedly as a gearshift knob for Rolls Royce in 1916 (Gordon 1988). "Thermoplastics" are plastics that can be repeatedly melted, such as polyethylene (PE), polypropylene (PP), and polyvinyl chloride (PVC). Thermoplastics are used to make many diverse commercial products such as milk jugs, grocery bags, and siding for homes. In contrast to the wood–thermoset composites, wood–thermoplastic composites have seen large growth in recent decades. Wood–thermoplastic composites are now most often simply referred to as wood-plastic composites (WPCs) with the common understanding that plastic refers to a thermoplastic.

While wood is the most used filler in WPCs, other biomass resources have also been used both in PE and PP. More recently, biomass resources have been used in bioplastics such as polylactic acid. This chapter will review all of these thermoplastic materials.

13.1   WOOD THERMOPLASTICS

The performance of a WPC largely depends on its constituent materials, how they are assembled, and how they interact. Therefore, to adequately understand (WPCs), we must first understand the advantages and limitations of its main constituents.

The plastics industry often uses fillers and reinforcements to modify the performance of plastic. Fillers (e.g., talc and calcium carbonate) are typically equidimensional and do not generally improve properties such as strength but can provide other benefits such as reduced cost, increased stiffness, and reduced thermal expansion, for example. Reinforcements (e.g., glass and carbon fibers) are fibrous, having one dimension much longer than the other, and are well-bonded to the plastic matrix (Carley 1993). Reinforcements markedly improve strength by transferring applied stress to the stronger reinforcing fiber. Wood-derived materials can be used as both a filler or a reinforcement.

There are a number of reasons that manufacturers use wood as a filler or reinforcement in plastics. As with many other fillers and reinforcements, wood is added to modify mechanical performance. However, wood has lower density than inorganic fillers, which can lead to benefits such as fuel savings when composites made with it are used in transportation and packaging applications. With changing consumer perceptions, some manufacturers have used the natural look of composites made with wood as a marketing tool. Others add wood to increase bio-based material content. Though not specifically added to plastics to impart biodegradability, wood can be used as a filler or reinforcement in biodegradable polymers where its biodegradability is an attribute rather than the detriment it is sometimes considered in more durable composites.

The wood used in WPCs is most often in particulate form (e.g., wood flour) or very short fibers or fiber bundles, rather than long individual wood fibers (i.e., it is used as a filler). A wide range of wood flour contents in WPCs are used depending on the processing method used, required performance, and economics. Extruded WPCs (e.g., deckboards) typically contain approximately 50–60% wood but injection molded WPCs usually contain less so that the melt viscosity is not too high. The relatively high bulk density and free-flowing nature of wood flour compared with wood fibers or other natural fibers, as well as its low cost, familiarity, and availability, is attractive to WPC manufacturers and users. Common species used include pine, maple, and oak but others are used as well. Typical particle sizes are roughly 2–0.2 mm (10–70 mesh), although there are exceptions. WPC manufacturers obtain wood flour either directly from forest products companies such as lumber mills and furniture, millwork, or window and door manufacturers that produce it as a by-product or buy it from companies that specialize in wood flour production.

Wood fibers are an order of magnitude stronger than the wood from which they derive (Rowell 1992). In a WPC, this higher strength, as well as their higher aspect ratio compared to wood flour can improve the efficiency of stress transfer from the plastic to the stronger wood fiber when well bonded to the plastic. This leads to improved mechanical performance. Because of the potential for improved mechanical properties, there has been a continuing interest in the use of individual wood, pulp, or paper fibers rather than wood flour as reinforcement in WPC. Though lower in mechanical performance than glass, the balance of properties that wood fibers yield, along with other advantages such as lower density, aesthetics, and low abrasiveness during processing offer advantages in some applications. Fiber preparation methods have a large effect on reinforcing ability. For example, high quality pulp fibers are more effective than lower cost, thermomechanical pulp fibers and can offer other benefits such as a lighter color. Processing difficulties, such as feeding and metering low-bulk-density fibers, have limited their use in WPCs. While there have been some developments in densification, handling, and processing of these fibers (e.g., Jacobson et al. 2002), these approaches usually add cost. Wood-derived fibers also have to compete on economics and performance with other natural fibers such as flax, depending on regional availability. Natural variability in wood and other natural fibers is typically greater than with inorganic fibers and is also a consideration.

Wood and other plant-derived flours and fibers are unusual fillers/reinforcements in that they are compressible. The high pressures during plastics processing can collapse the hollow fibers or fill them with low molecular weight additives and polymers. The degree of collapsing or filling will depend on such variables as particle size, processing method, and polymer/additive viscosity, but wood densities in WPCs that approach wood cell wall density can be found in high-pressure processes such as injection molding. Consequently, adding wood to commodity plastics such as PP, PE, and polystyrene increases their density. However, even these higher densities are considerably lower than those of common inorganic fillers and reinforcements such as talc, calcium carbonate, or glass fibers. This density advantage is important in applications where weight is important such as in automotive or packaging applications.

Wood's low thermal stability limits the plastics used in WPCs to those that can be processed at low temperatures, typically lower than about 200°C, although high purity cellulose pulps with higher thermal stability have been added to plastics (e.g., some nylons) that are processed at higher temperatures than most commodity thermoplastics (Caulfield et al. 2001). The thermal expansion of wood is less than that of the commodity plastics commonly used as matrices and adding wood can reduce part shrinkage and warping due to wide temperature changes during processing.

Wood's hygroscopicity must be considered both in composite fabrication and product performance. Wood usually contains at least 4% moisture when delivered, which is much higher than most thermoplastics. This moisture must be removed before or during processing with thermoplastics. Though moisture could potentially be used as a foaming agent to reduce density, this approach is difficult to control and is not used commercially. Even compounded material (i.e., blended pellets of wood, plastic, and additives) often needs to be dried prior to further processing, especially if high weight percentages of wood flour are used. The hygroscopicity of wood flour can also affect the performance of WPC products. If the composite is not formulated and processed correctly for exterior applications, moisture sorption can lead to mold or fungal attack and volume changes due to moisture sorption, especially repeated moisture cycling, can lead to loss of adhesion between the wood and plastic components or matrix cracking.

13.1.1 THERMOPLASTIC MATRIX MATERIALS

Much of how a polymer processes and performs is determined by its molecular structure, which is developed during the polymerization process. The number-average molecular weights of many commercial synthetic polymers are typically about 10,000–100,000 (Billmeyer 1984). Table 13.1 shows the basic chemical structural units of several common synthetic. Polymers can contain one type of monomer (homopolymer) or multiple monomers (copolymers, terpolymers, etc.). Tacticity is important in the arrangement of repeat units in polymers with asymmetrical repeat units. For example, PP contains a methyl group (CH_3) attached to a carbon chain (Table 13.1) that can be attached to one side of the chain (isotactic), alternating sides of the chain (syndiotactic), or lack a consistent arrangement (atactic). Polymers can have a few branches, such as high density polyethylene, HDPE or long branches, such as low-density polyethylene, LDPE.

Polymers are often categorized by their behavior, which is influenced by their molecular organization. Unlike thermosets, thermoplastic polymers can be repeatedly softened by heating. When cooled, they harden as motion of the long molecules is restricted. If the polymer molecules remain disordered as they are cooled from the melt, they are considered amorphous thermoplastics and the temperature at which the polymer molecules cease to rotate and slip past one another is called its glass transition temperature (T_g). Below its glass transition, amorphous plastics (or amorphous

TABLE 13.1

Structural Units for Selected Polymers with Approximate Glass Transition (T_g) and Melting (T_m) Temperatures

Structural Unit	Polymer	T_g (°C)	T_m (°C)
$-CH_2-CH_2-$	Polyethylene (PE)	−125	135
$-CH_2-CH_2-$ $\quad\quad\quad\vert$ $\quad\quad\quad CH_3$	Polypropylene (PP)	−20	170
$-CH_2-CH-$ $\quad\quad\vert$ $\quad\quad C_6H_5$	Polystyrene (PS)	100	—
$-CH_2-CH-$ $\quad\quad\vert$ $\quad\quad C1$	Polyvinylchloride (PVC)	80	—
$-\overset{\overset{O}{\|\|}}{C}-\bigcirc-\overset{\overset{O}{\|\|}}{C}-O-CH_2-CH_2-O-$	Polyethylene-terephthalate (PET)	75	280

Source: Condensed from Osswald, T.A. and G. Menges. 1996. *Materials Science of Polymers for Engineers.* Carl Hanser Verlag, New York.

regions of semicrystalline plastics) become glassy, stiff, and sometimes brittle. Polystyrene is an amorphous thermoplastic, for example.

Some thermoplastics form regions of highly ordered and repetitive molecular arrangements on cooling. In these semicrystalline plastics much, though not all, of its molecular structure is in an ordered state. A crystallinity of 40–80% is typical for common semicrystalline plastics such as PP and PE but depends on molecular architecture as well as processing history. In addition to a glass transition temperature (T_g), semicrystalline plastics have crystalline melting points (T_m), above which temperature the crystal order disappears and flow is greatly enhanced. This ability of plastics to flow is a great advantage in processing complex shapes.

Though molecular architecture such as tacticity, polymer branching, and molecular weight are important parameters affecting crystallization rates and crystal structure, processing also plays a large role and affects characteristics such as molecular orientation and crystallinity that also influence performance. The presence of fillers, reinforcements, and additives can influence crystallization processes. For example, Figure 13.1 shows a PP melt that is slowly being cooled and whose crystallization is being partially nucleated by a cellulose fiber.

The properties of thermoplastics are often highly dependent on the temperature at which they are measured and the speed at which they are tested. Plastics are viscoelastic, behaving as it were a combination of a viscous liquid and an elastic solid. Most plastics have higher moduli when stress is rapidly applied versus when it is applied slowly. Also, some plastics have a much greater tendency than wood to sag over time (i.e., creep) when bearing sustained loads, an important consideration in structural applications.

Typical room temperature properties of plastics commonly used in WPCs are summarized in Table 13.2. These values are provided to give a general indication of the properties. However, the exact performance of these plastics are difficult to summarize since various grades are produced, whose performance has been tailored by controlling polymerization and additive content. PVC in particular often contains a considerable amount of additives such as heat stabilizers and plasticizers resulting in a wide range of processability and performance. However, some general comments can be made.

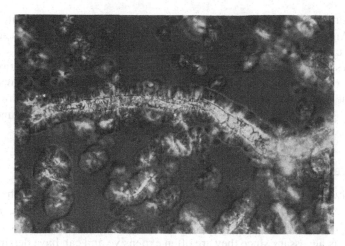

FIGURE 13.1 Cellulose fiber crystallizing a PP melt.

PE, PP, and PVC are the most common plastics used in WPCs. Though they tend to have considerably lower mechanical performance than the so-called engineering plastics, these commodity plastics have reasonably good mechanical performance for many applications, low price, and tend to have lower melting or softening temperatures. Generally WPCs are processed at temperatures lower than about 200°C to avoid significant degradation of the wood. The large use of PE is due, in part, to that fact that much of the early WPCs were developed as an outlet for recycled film as well as the low cost and availability of recycled sources of PE. They absorb little moisture and can act as effective moisture barriers. This is important since moisture sorption in WPCs can negatively affect the performance of the composite.

Though PE and PP are largely impervious to moisture, they are susceptible to degradation by UV radiation, and the use of light stabilizing additives is common in exterior applications. The thermal expansion and contraction of PE and PP are significant and they tend to creep (or sag over time), especially under load or at high temperatures, limiting their structural performance. However thermal expansion and creep can be reduced with fillers and reinforcements.

Polyvinyl chloride is also used in WPCs. PVC can be much stiffer than PE or PP and, with appropriate additives, have been commonly used in exterior applications such as siding. However,

TABLE 13.2
Typical Room Temperature Properties of Common Polymers

Polymer	Density (g/cm³)	Tensile Strength (MPa)	Tensile Modulus (GPa)	Elongation at Break (%)	Water Absorption in 24 h (%)	Coefficient of Thermal Expansion ($K^{-1} \times 10^6$)
Low-density polyethylene	0.91–0.93	8–23	0.2–0.5	300–1000	<0.01	250
High-density polyethylene	0.94–0.96	18–35	0.7–1.4	100–1000	<0.01	200
Polypropylene	0.90–0.92	21–37	1.1–1.3	20–800	0.01–0.03	150
Rigid polyvinyl chloride	1.4–1.6	50–75	1.0–3.5	10–50	3–18	70–80

Source: Condensed from Osswald, T.A. and G. Menges. 1996. *Materials Science of Polymers for Engineers.* Carl Hanser Verlag, New York.

complexities in formulating and processing as well as patent issues have limited broader use in WPCs in North America. Also, the negative perceptions of chlorinated plastics such as PVC have also limited their use in some parts of the world such as Japan.

Different grades of a particular plastic have been tailored for a specific application and processing method. For example, a bottle grade of HDPE has high molecular weight to provide the toughness needed for a bottle application and high melt strength necessary for the melt blowing process. An injection molding grade might have a lower molecular weight yielding lower melt viscosity and good flow properties.

13.1.2 ADDITIVES

The term "additive" refers to a broad class of materials added to affect the processing and performance of polymers. Once additives such as stabilizers, plasticizers, etc. are incorporated into a polymer, it is usually referred to as a "plastic" (Carley 1993). Additives are only added in as small amounts as necessary since they are often expensive and can have detrimental effects on properties other than those for which they are added. Though the plastics themselves contain additives, more are often added during the processing of WPCs to overcome the limitations of the constituent materials, improve their interaction, or improve the processing or the performance of the end product. It is difficult to fully describe all of the additives used in WPC production because of the variety of additives, the proprietary nature of the formulations used, and the range of WPC applications. However, several of the major additive types used in WPCs are briefly described below.

Adding wood increases the viscosity of plastics, especially at high loadings such as those used in exterior building applications (e.g., extruded deckboards). Additionally, composite melts at these high wood contents have low strength and can stick and tear (especially at the edges) as it exits the extruder die (Figure 13.2a). Adding lubricants can reduce the viscosity of the composite melt, increase output, and lubricate the interface between the composite melt and the die to prevent edge-tearing (Figure 13.2b). Lubricants for WPCs are usually added at about 1–5% by weight and are made from materials such as metal stearates, aliphatic carboxylic acid salts, mono and diamides and modified fatty acid esters (Anonymous 2002). However, they can negatively impact some mechanical properties and interfere with other additives such as coupling agents if not chosen carefully.

 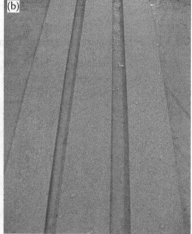

FIGURE 13.2 Extruded WPC: (a) without lubricant showing edge tearing and (b) with lubricant.

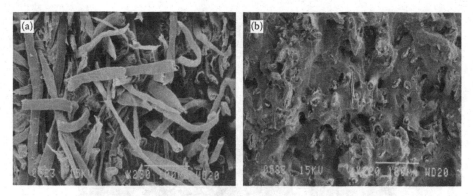

FIGURE 13.3   PP with 40% by weight of pulp fibers (a) without and (b) with a MAPP coupling agent.

The chemical incompatibility of the wood and many plastics, especially nonpolar plastics such as PP and PE can be problematic. For example, if wood fiber is used to reinforce plastic, a coupling agent is necessary to improve the adhesion between the wood and plastic so that applied stress can be transferred to the stronger fibers. Many different coupling agents have been investigated for use in WPCs and are reviewed elsewhere (Lu et al. 2000). However, one common coupling agent is maleic anhydride grafted polypropylene, often simply referred to as maleated polypropylene (MAPP). With MAPP, the anhydride portion of the molecule reacts with the wood's hydroxyl groups to form an ester bond and the long polymer chain incorporates itself into the bulk PP or PE network, thereby bonding the two together (see next section for mechanism). Figure 13.3 shows the effect of MAPP on the fracture surfaces of a composite of PP and 40% by weight of pulp fibers. Without MAPP the fibers were pulled out of the PP as the specimen failed. When 3% of a MAPP was added, the adhesion between PP and the pulp fibers was improved sufficiently that the fibers were broken during composite failure. Coupling agents can improve properties such as strength and unnotched Izod impact energy and can also decrease moisture sorption.

Adding wood to most plastics increases its density because of partial or complete collapse of wood fibers due to the high pressures during processing or filling of the lumen with low molecular weight additives and polymers. These densities are also well above those of solid wood. Additives have been used to decrease density. One approach uses foaming agents that react and evolve gas at critical temperatures (i.e., chemical foaming). Another approach uses special processing equipment that can inject nitrogen or carbon dioxide in a supercritical state that creates a very fine cell (i.e., microcellular) structure that has some advantages in performance. Other additives such as hollow glass spheres or heat-expandable polymer microspheres have also been investigated to reduce WPC density (Guo et al. 2008).

A variety of additives have been used to improve the durability of WPCs, particularly in exterior applications. Moisture sorption can be reduced with some additives. For example, a common secondary effect of adding coupling agents such as MAPP is reduced moisture sorption. Colorants and light stabilizers can improve resistance to color fade and UV degradation. Biocides such as zinc borate are sometimes added to WPCs to improve fungal resistance. Flame retardants have also been investigated for certain applications.

13.1.3   Processing

Although there are a wide variety of methods for preparing WPCs, all involve melting the plastic, mixing in the wood and additives, and forming a product. "Compounding" is the feeding and dispersing of fillers and additives in the molten polymer. Many options are available for compounding, using either batch or continuous mixers. The compounded material can be immediately pressed or shaped into an end product or formed into pellets for future processing. Some product manufacturing

FIGURE 13.4   WPC exiting the extruder die (left) and entering cooling tank (right).

options for WPCs force molten material through a die (sheet or profile extrusion), into a cold mold (injection molding), between calenders (calendering), or between mold halves (thermoforming and compression molding) (Youngquist 1999). Combining the compounding and product manufacturing steps is called in-line processing.

The majority of WPCs are manufactured by profile extrusion, in which molten composite material is forced through a die to make a continuous profile of the desired shape (Figure 13.4). Extrusion lends itself to processing the high viscosity of the molten WPC blends and to shaping the long, continuous profiles common to building materials. These profiles can be a simple solid shape, or highly engineered and hollow.

Although extrusion is by far the most common processing method for WPCs, the processors use a variety of extruder types and processing strategies (Mapleston 2001a). Some processors run compounded pellets through single-screw extruders to form the final shape. Others compound and extrude final shapes in one step using twin-screw extruders. Some processors use several extruders in tandem, one for compounding and the other for profiling (Mapleston 2001a). Moisture can be removed from the wood component before processing, during a separate compounding step (or in the first extruder in a tandem process), or by using the first part of an extruder as a dryer in some in-line process. Equipment has been developed for many aspects of WPC processing, including materials handling, drying and feeding systems, extruder design, die design, and downstream equipment (i.e., equipment needed after extrusion, such as cooling tanks, pullers, and cut-off saws). Equipment manufacturers have partnered to develop complete processing lines specifically for WPCs. Some manufacturers are licensing new extrusion technologies that are very different from conventional extrusion processing (Mapleston 2001a,b).

Compounders specializing in wood and other natural fibers mixed with thermoplastics have fueled growth in several markets. These compounders supply preblended, free-flowing pellets of high bulk density that can be reheated and formed into products by a variety of processing methods. The pellets are advantageous to manufacturers who do not typically do their own compounding or do not wish to compound in-line (e.g., most single-screw profilers or injection molding companies).

Other processing technologies such as injection molding, thermoforming, and compression molding are also used to produce WPCs. These alternative processing methods have advantages when processing of a continuous piece is not desired or a more complicated shape is needed. Composite formulations must be adjusted to meet processing requirements (e.g., the low viscosity needed for injection molding can limit wood content).

13.1.4 Performance

The wide variety of WPC formulations and markets makes it difficult to discuss the performance of these composites. Performance depends on the inherent properties of the constituent materials, interactions between these materials, additives used, processing, product design, and service environment. Because formulations can vary greatly and are proprietary, performance data should be obtained directly from the manufacturer or supplier. However, some generalizations regarding performance can be made.

13.1.5 Mechanical Properties

Wood flour is often added to thermoplastics as a low cost filler to alter mechanical performance, such as the stiffness and heat deflection temperature without increasing density excessively. Tensile and flexural strengths are, at best, maintained and more often decreased in the absence of a coupling agent. Adding wood fiber (rather than wood flour) and a coupling agent can improve strength properties and somewhat mitigate losses in other properties such as unnotched Izod impact energy.

Table 13.3 demonstrates some of these effects of wood on the mechanical performance of PP. Both tensile and flexural moduli increase with wood flour since wood is much stiffer than commodity thermoplastics such as PP. However, the increase in modulus with addition of wood flour comes at the expense of elongation, a drastic reduction in unnotched impact energy, and a general decrease in tensile strength. Not surprisingly, the wood fibers are more effective reinforcements than the wood flour when a coupling agent (MAPP) is added, nearly doubling the strength unfilled PP. Wood fibers are an order of magnitude stronger than the wood that they come from Rowell (1992) and their higher aspect ratio (ratio of length to diameter) results in increased stress transfer area and stress transfer efficiency (relative to wood flour) when a coupling agent is used.

13.1.6 Durability

Since the majority of WPCs are used in exterior applications, durability is often important. Rather than relying on painting, staining, or chemical treatments to protect it from moisture intrusion, the durability of WPCs largely lies in the ability of the thermoplastic to at least partially encapsulate the wood and retard moisture sorption. This results in a low maintenance product, which is a common selling point for WPCs in outdoor applications. Moisture management is critical since volume

TABLE 13.3
Selected Properties of PP and Its Composites with Wood

Composite	Specific Gravity	Tensile Properties		Flexural Properties		Izod Impact Energy	
		Strength (MPa)	Modulus (GPa)	Strength (MPa)	Modulus (GPa)	Notched (J/m)	Unnotched (J/m)
PP	0.90	28.5	1.53	38.3	1.19	20.9	656
PP + 40% wood flour	1.05	25.4	3.87	44.2	3.03	22.2	73
PP + 40% wood flour + MAPP	1.05	32.3	4.10	53.1	3.08	21.2	78
PP + 40% wood fiber	1.03	28.2	4.20	47.9	3.25	23.2	91
PP + 40% wood fiber + MAPP	1.03	52.3	4.23	72.4	3.22	21.6	162

Source: Condensed from Stark, N.M. and R.E. Rowlands. 2003. *Wood and Fiber Science.* 35(2), 167–174.

TABLE 13.4
Weight Gain in Aspen-Polypropylene Composites
at 90% Relative Humidity after *D* days

	Weight Gain (%)					
ASPEN/PP/MAPP[a]	25 D	50 D	75 D	100 D	150 D	200 D
0/100/0	0	0	0	0	0.2	0.4
30/70/0	0.7	1.4	1.7	2.1	2.4	2.8
30/68/2	0.7	0.7	1.1	1.5	1.5	2.2
40/60/0	0.7	1.4	1.7	2.0	2.4	3.0
40/58/2	0.4	1.2	1.5	1.9	2.7	3.5
50/50/0	1.3	2.0	2.6	3.6	4.3	5.3
50/48/2	1.5	1.8	2.2	2.9	4.0	5.1
60/40/0	3.7	4.5	5.6	6.0	6.3	6.7
60/38/2	1.6	2.2	3.5	4.4	5.1	6.0

[a] Compositions are weight percent.

changes due to moisture sorption, especially repeated moisture cycling, can lead to warping, buckling, interfacial damage, matrix cracking, and biological attack.

Table 13.4 shows the moisture absorbed by wood fiber–PP composites that were subjected to 90% relative humidity (RH) for an extended period of time. Even after 200 days, all of the composites continued to gain weight and equilibrium was not yet reached. The higher the wood fiber content, the more moisture was picked up by the specimen. Adding MAPP helped to reduce moisture sorption. Table 13.5 shows data on a cyclic humidity test where the specimens were subjected to 30% RH for 60 days, measured and then subjected to 90% RH for an additional 60 days. This cycle was repeated four times. As with the static 90% RH tests, the specimens continued to gain weight with each 90% RH cycle. As the percentage of hydrophilic wood fiber increases in the wood–thermoplastic composites, there is a corresponding increase in moisture gain. The moisture gain is slow but continues over a very long period of time. The data also suggests that at about 50% wood, the rate and

TABLE 13.5
Weight Changes in Repeated Humidity Tests on Aspen-Polypropylene
Composites Cycled between 30% and 90% Relative Humidity

	Weight Gain (%)							
ASPEN/PP/MAPP[a]	30%	90%	30%	90%	30%	90%	30%	90%
0/100/0	0	0	0	0	0	0.2	0	0.4
30/70/0	0.6	0.9	0.7	1.4	0.7	1.4	0.9	1.9
30/68/2	0.4	0.9	0.7	1.2	0.7	1.3	0.9	1.8
40/60/0	0.4	0.9	0.7	1.4	0.7	1.6	0.7	2.1
40/58/2	0.2	1.2	0.8	1.6	1.0	2.0	1.2	2.2
50/50/0	0.5	1.5	1.2	2.5	1.2	2.7	1.2	3.2
50/48/2	0.6	1.3	1.1	2.2	1.3	2.2	1.3	2.6
60/40/0	0.7	2.5	1.5	4.1	1.6	4.1	1.3	4.8
60/38/2	0.2	1.5	1.0	2.6	1.3	2.6	1.3	3.3

[a] Compositions are weight percent.

extent of moisture pickup increases. At this fiber content, there more connectivity (percolation) between the fibers and moisture can wick faster and further into the composite.

Since WPCs are not usually protected by these means, those containing no pigments usually fade to a light gray when exposed to sunlight. Photostabilizers or pigments are commonly added to help reduce this color fade when used in exterior environments. Because WPCs absorb less moisture and do so more slowly than unprotected solid wood, they have better fungal resistance and dimensional stability when exposed to moisture. For composites with high wood contents, some manufacturers incorporate additives such as zinc borate to improve fungal resistance. Mold can form on surfaces of WPCs and can be caused by moisture sorption by the wood flour or buildup of organic matter on the composite surfaces. Although mold does not reduce the structural performance of the composite, it is an aesthetic issue and WPCs usually have to be periodically cleaned. Manufacturers ensure a useful level of durability by limiting wood content, careful processing, and judicious use of additives. To further improve durability as well as stain and scratch resistance some manufacturers cover the WPC with a capstock or protective layer.

13.1.7 MARKETS AND FUTURE TRENDS

By far the greatest use of WPCs is in building products that have limited structural requirements. Customers and builders have a certain familiarity with wood in applications such as decking (Figure 13.5) and railings (the largest WPC market) and often desire an alternative that may have similar attributes. Mixing wood flour with plastic is seen as a way to use wood in these applications yet improve its durability without chemical treatment or the need for painting or staining. Although more expensive than wood, many consumers have been willing to pay for the lower maintenance required when WPCs are used. Mechanical properties such as creep resistance, stiffness, and strength are lower than those of solid wood. Hence, these composites are not currently used in applications that require considerable structural performance unless carefully formulated, designed, and tested. For example, WPCs are used for deckboards but not the substructure. Development of codes and standards for these materials are also important for their acceptance in building applications, for example. Solid, rectangular profiles are manufactured as well as more complex hollow or ribbed profiles. However, WPCs have expanded beyond decking and railing to other exterior building products including roofing (Figure 13.6), fencing, siding, and window/door profiles.

Automotive components are another market for WPCs, although plastic composites are also made with natural fibers other than wood (e.g., flax). These products take advantage of the balance

FIGURE 13.5   Examples of extruded WPCs.

FIGURE 13.6   WPC roofing tiles.

of performance and low density of natural fillers and fibers relative to their inorganic counterparts. Some components made from WPCs are interior substrates thermoformed from wood-PP sheets and injection-molded glove boxes, fixing hooks, and sound system components (Carus et al. 2008).

The development and use of WPCs varies by region. While there has been maturation of some major markets and some consolidation due, in part, to a recent slow-down of the building industry, wider acceptance of WPCs beyond North America has fueled growth in China, Japan, and southeast Asia, for example (Toloken 2010). There is increasing diversification of WPC markets beyond building components and automotive components as well. Flooring, benches, shelving, chairs, and other furniture components are currently produced in Europe, for example, Anonymous (2011).

The future of WPCs is very difficult to predict and depends on many factors including the economy, legislation, societal trends, and the technological advancements of WPCs and competing materials. A continued increase in familiarity and acceptance of WPCs by consumers, designers, and manufacturers as well as a desire for more sustainably sourced materials will likely continue to increase demand. As the use of biopolymers increases, wood and other plant-derived materials are a logical consideration for fillers and reinforcements.

Research and development efforts will likely have a large influence on the future of WPCs as well. Through better understanding of material behavior and more rigorous testing and engineering, WPCs are being used in more structural applications such as deckboards, chocks, and whales of naval piers and pedestrian bridges, for example (Wolcott et al. 2009). Advanced processing techniques such as reactive extrusion to produce cross-linked WPCs (Bengtsson et al. 2005), micro-cellular processing to produced lighter WPCs with a fine (microcellular) foam structure (Guo et al. 2008), and co-injection or co-extrusion techniques that uses multiple materials to their best advantage (Stark and Matuana 2009) are just a few examples of new approaches to processing that are being investigated.

Many different materials are also being investigated for use in WPCs. For example, use of wood with higher performing plastics (e.g., the so-called "engineering polymers" such as nylon, Caulfield et al. 2001) or new nano-scale additives such as nano-sized titanium dioxide (Stark and Matuana 2009) may greatly improve performance. Although somewhat beyond what is typically considered WPCs, wood-derived materials such as carbon fibers from lignin (Kadla et al. 2002), and nano-fibrillated cellulose (Siró and Plackett 2010) are being investigated as reinforcements in high-performance polymer composites. If the remaining technical and economic hurdles can be overcome and these reinforcements are accepted by industry, these new composites may look and perform very differently from previous generations of composites from plastics and wood.

13.2 NONWOOD FIBERS IN THERMOPLASTIC COMPOSITES

13.2.1 AGRICULTURAL FIBERS

While wood fiber and flour are used the most in wood plastic composites, there are many other natural fibers that can and have been used to make these composites (Rowell et al. 1997). Chapter 18 presents a table of the world inventory of wood and agricultural biomass and it can be seen that collectively, the nonwood biomass inventory is as large as that of wood. In some countries, wood is very scarce and is not used as a building material. These countries must resort to nonwood sources of fiber and flour in order the produce composites.

There has been a lot of research done on agricultural fibers in thermoplastic composites (Sanadi et al. 1994a,b, 1996, Rowell et al. 1999). Some of this research was on kenaf due to a large USDA research project in the 1990s to promote the growing of kenaf in the United States. Table 13.6 shows the physical properties of kenaf at two different levels and jute in polypropylene composites compared to pure PP, glass, mica and calcium carbonate filled PP (Sanadi et al. 1995a,b, 1996). The fibers and fillers were blended in a kinetic mixer with PP and a coupling or compatibilizing agent (maleic anhydride grafted polypropylene, MAPP) and discharges at 190°C. There are many different types of MAPP's differing in molecular weight and the degree of maleic anhydride substitution. One used most often had a number average molecular weight of 20,000, a weight average molecular weight of 40,000 and about had 6% by weight of maleic anhydride in the polymer. Specimens were injected molded and tested according to ASTM standards.

Some of the earliest research on the effect of coupling agents was done using kenaf (Sandi et al. 1994b). Figure 13.7 shows the results of using a coupling agent in kenaf-PP composites. The figure

TABLE 13.6
Properties of Kenaf Polypropylene Composites

Property	Pure PP	Kenaf	Kenaf	Jute	Mica	CaCO$_3$	Glass
% filler by weight	0	40	50	50	40	40	40
% filler by volume	0	30	39	39	18	18	19
Tensile modulus, GPa	1.7	6.0	8.3	8.8	7.6	3.5	9
Specific tensile modulus, GPa	1.9	5.9	7.2	8.1	6.0	2.8	7.3
Tensile strength, MPa	33	56	65	74	39	25	110
Specific tensile strength, MPa	37	55	61	69	31	20	89
Elongation at break, %	≫10	1.9	2.2	2.3	2.3	—	2.5
Flexural strength, MPa	41	82	98	97	62	48	131
Specific flexural strength, MPa	46	80	92	90	49	38	107
Flexural modulus, GPa	1.4	5.9	7.3	7.3	6.9	3.1	6.2
Specific flexural modulus, GPa	1.6	5.8	6.8	6.8	5.5	2.5	5.0
Notched Izod impact, J/m	24	28	32	43	27	32	107
Specific gravity	0.9	1.02	1.07	1.08	1.26	1.25	1.23
Water absorption, % in 24 h	0.02	—	1.05	—	0.03	0.02	0.06
Mold (linear) shrinkage, cm/cm	0.028	0.004	0.003	—	—	0.01	0.004

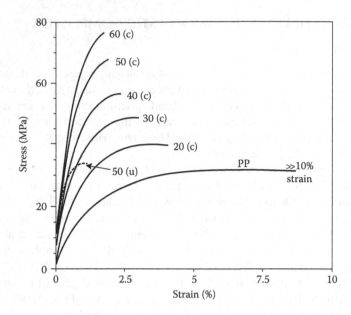

FIGURE 13.7   Stress–strain curve for kenaf-PP composites.

shows how the strength and stiffness increases as the percentage of kenaf is increased in the composite using a coupling agent.

The strength in the noncompatibilized blend is much lower as compared to the compatibilized composite.

The mechanism of compatibilization or coupling is suspected to be as follows: First, the anhydride reacts with a cell wall polymer hydroxyl group to form an ester bond the then the PP polymer attached to the anhydride intertangles into the PP or PE network in the melt (see Figures 13.8 and 13.9). Figure 13.10 shows a composite where a compatibilizer has been used. It can be seen that when the composite was broken in two pieces, the fiber shown broke rather than be pulled out of the thermoplastic matrix. There is a small gap at the base of the fiber showing that there was some slippage but the fiber broke rather than be pulled out of the matrix due to weak interfacial bonding between the fiber and the matrix.

Figures 13.11 through 13.21 show the mechanical properties of several fibers other than wood in a PP matrix (Jacobson et al. 1995a,b, 1996).

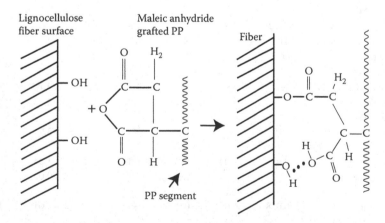

FIGURE 13.8   Reaction of maleic anhydride grafted PP with cell wall hydroxyl groups.

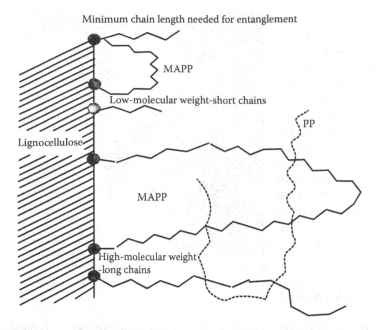

Minimum chain length needed for entanglement

MAPP

Low-molecular weight-short chains

PP

Lignocellulose

MAPP

High-molecular weight -long chains

FIGURE 13.9 Entanglement of MAPP with PP.

13.2.2 OTHER FIBERS

Another option for the biomass fraction is to utilize the solids in animal manures (Rowell et al. 2008).

Animal agriculture is under increasing pressure to produce more and more meat, milk and eggs giving rise to an increasing amount of manures. In the past, manures have been viewed as a waste by-product used mainly as a fertilizer that has a value of 2–4 cents/dry pound. We need to change our view of manures from waste to asset. Destroying manures by burning or lagooning may solve the environmental problem but it does nothing to add to animal income.

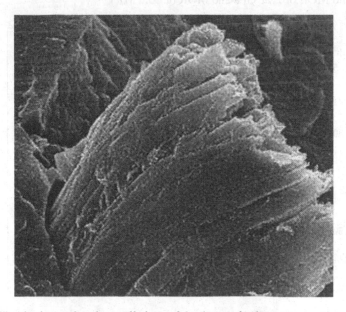

FIGURE 13.10 Fiber broken rather than pulled out of the thermoplastic matrix.

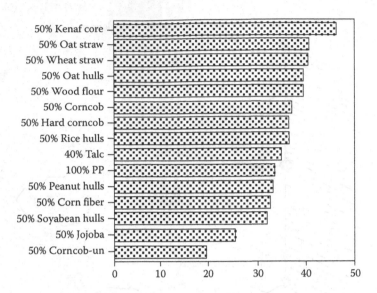

FIGURE 13.11   Tensile strength (MPa).

One of the alternatives is to use animal manures in industrial products. It is possible to use swine and cow manures and residue from a methane digester plant as reinforcing fillers in HDPE and HDPP. This is a win-win situation as it increases the value of the animal manures, decreases the cost and improves mechanical properties of the thermoplastic composites.

A 40% blend of dry dairy manure with 58% HDPE and 2% MAPE gives a composite with an MOE in bending of 2.18 GPa and MOR 34.7 MPa as compared to 40% pine flour with 58% HDPE and 2% MAPE of MOE 2.98 and MOR 33.4 MPa.

A 40% blend of dry swine manure with HDPE and 2% MAPE gives a composite with MOE in bending of 1.31 GPa and MOR of 34.7 MPa as compared to unfilled HDPE MOE of 0.75 GPa and MOR of 15.1 MPa.

A 40% blend of dry residue left after methane digestion with 58% HDPE and 2% MAPE gives a composite with and MOE of 2.21 GPa and MOR of 26.9 MPa.

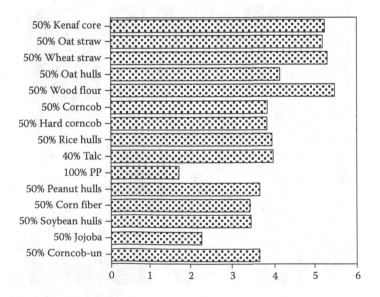

FIGURE 13.12   Tensile modulus (GPa).

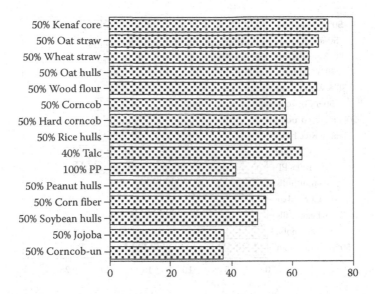

FIGURE 13.13   Flexural strength (MPa).

13.3   BIOPLASTICS

13.3.1   INTRODUCTION

The opportunity to manufacture plastics from renewable resources, especially from biomass, has been the subject of research and development and considerable commercial interest for a number of decades (Belgacem and Gandini 2008). Interest in these materials in academe as well as in industry has grown further in recent years for a number of key reasons. First, there is an increasing awareness of the limited nature of fossil fuels and the need to investigate renewable resources for our future energy and material needs. Second, the use of materials such as plastics obtained from biomass offers a way to reduce overall greenhouse gas (GHG) emissions and reduce carbon footprints. Third, most but not all bio-derived plastics are biodegradable or compostable and, providing

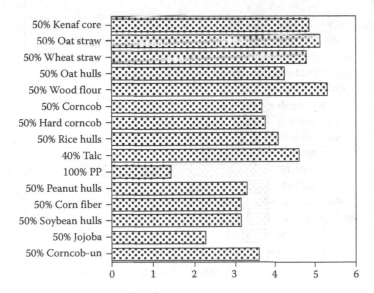

FIGURE 13.14   Flexural modulus (GPa).

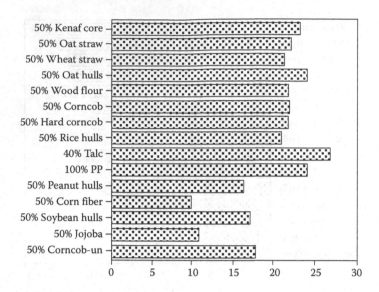

FIGURE 13.15 Notched Izod impact (J/M).

suitable materials sorting and other infrastructure can be put in place, can make a contribution to reducing the plastic waste disposal problem in land and marine environments around the world.

Bio-derived polymers can be broadly divided into three categories as illustrated in Figure 13.22. In the first category there are those polymers which can be extracted more or less directly from biomass, including starch and cellulose from plants, chitin from the shells of marine organisms and proteins or lipids derived from plant or animal sources. The second category encompasses polymers which are composed of and synthesized from bio-derived monomers, prime examples being poly-lactide (PLA) and related polyesters. In the third category there are polymers which are synthesized by natural or genetically modified (GM) organisms and, in this case, the best-known examples from a bioplastics perspective are the polyhydroxyalkanoates (PHAs) which are produced in-situ by certain strains of bacteria under nutrient-limited growth conditions (Figure 13.23). Celluloses synthesized by bacteria and algae as well as by tunicates, a group of oceanic filter feeders, can also be

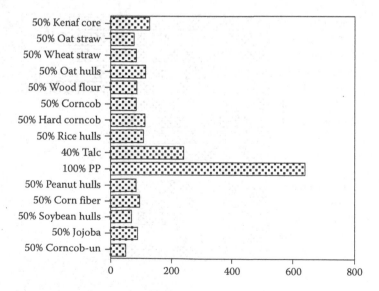

FIGURE 13.16 Unnotched Izod impact (J/M).

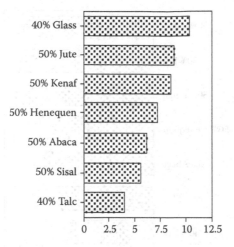

FIGURE 13.17   Tensile modulus (GPa).

included in the third category of bio-derived polymers. As far as is known, tunicates are the only members of the animal kingdom which naturally produce cellulose (Kimura and Itoh 2004).

Although there has been extensive research over the years aimed at deriving plastics or fibers for plastic reinforcement from each of the various types of biopolymer identified in Figure 13.22, significant commercial development has been limited to only a few types. Historically, derivatives of cellulose such as esters and ethers have been important commodities and some of these derivatives are the oldest thermoplastic materials prepared by man, dating back well into the nineteenth century (Belgacem and Gandini 2011). Products such as cellulose acetate and cellulose acetate butyrate (CAB) are still widely used today in the form of various films, coatings and printing inks. Of the other bio-derived polymers which have been commercialized, thermoplastic starch, PHAs and PLAs have been particularly important.

Starch is the most abundant reserve polysaccharide in plants and, as a source of bioplastics, has the advantage of being both inexpensive and biodegradable in soil or water. Starch can be plasticized by the addition of glycols, polyethers, urea or water and converted to thermoplastic starch (TPS) by the application of heat and mechanical energy during extrusion processing. Starch-based plastics have a number of limitations, including water sensitivity and poor mechanical properties

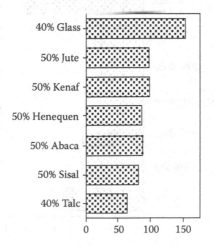

FIGURE 13.18   Flexural strength (MPa).

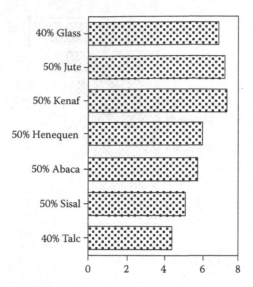

FIGURE 13.19   Flexural modulus (GPa).

and, for that reason, there have been numerous studies on the blending of plasticized starch with other biodegradable polymers to obtain inexpensive, biodegradable or compostable materials with improved properties (Vazquez et al. 2011). Such blends offer possibilities for reductions in manufacturing costs, tunable rates of degradation, and materials with properties which are a combination of those of the individual monomers. At the present time, Novamont based in Italy is one of the main suppliers of starch-based bioplastics with its Mater-Bi® range of products, which find application as agricultural mulching films and in disposable tableware. In addition, Mater-Bi products are used in biodegradable or compostable bags for shopping or waste collection, thermoformed trays for foodstuffs, transparent film for packaging of fruit and vegetables, and extruded or woven nets for food products such as potatoes and onions. There are also a variety of non-food related products based on Mater-Bi such as cotton buds, sanitary towels, nappies, and chewable items for pets.

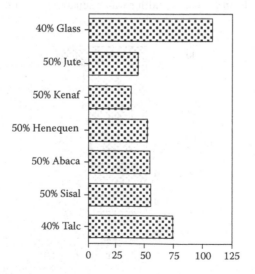

FIGURE 13.20   Notched Izod impact (J/M).

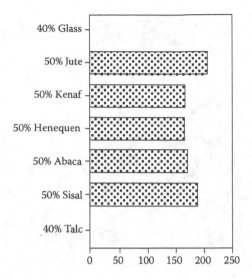

FIGURE 13.21 Unnotched Izod impact (J/M).

PHAs or, as they are otherwise known, bacterial polyesters, were first observed in detail during the 1920s in the laboratory of Maurice Lemoigne, a French microbiologist (Lemoigne 1923, 1924a, 1924b). However, it was not until the 1950s that the structure and mechanism of synthesis of the PHAs was understood and there has been considerable research and development as well as industrial interest ever since (Macrae and Wilkinson 1958, Chodak 2008, Pollet and Averous 2011). In one of the first commercial developments in the 1960s, ICI introduced Biopol®, a polyhydroxy-butyrate-*co*-hydroxyvalerate (PHBV) copolymer, but this product was not widely adopted, probably for cost reasons, and was taken over by Zeneca BioProducts when ICI was split up in 1993. Subsequently, the technology was sold to Monsanto and then acquired by the American company Metabolix in 1998. At the time of writing, Metabolix is working with Archer Daniels Midland (ADM) in a joint venture (Telles) to manufacture PHAs under the Mirel™ trade name. This joint venture is currently establishing a plant in Iowa with a 50,000 tonnes/year capacity which will be one of the biggest, if not the biggest, such facilities worldwide. A recent survey suggests that there are more than 15 companies around the world at various stages and scales of PHA production. These companies are either raw material suppliers or are targeting application of PHAs in packaging or medicine (e.g., drug delivery).

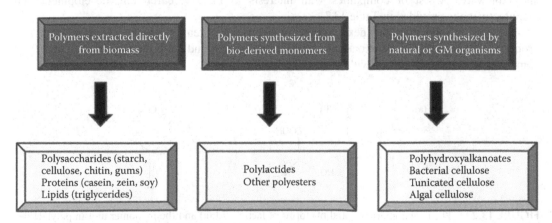

FIGURE 13.22 Biopolymer categories.

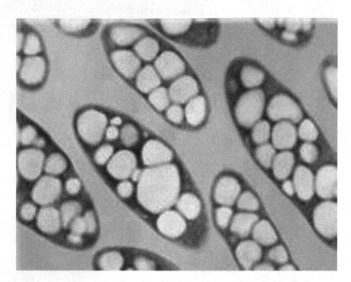

FIGURE 13.23 An optical microscopy image showing polyhydroxyalkanoate (PHA) granules produced in bacterial cells under nutrient-limited conditions.

PLA thermoplastics are at present the most advanced and widely used of the commercial bioplastics with interest driven by their potential uses in packaging, medicine, electronics, and composite materials (Auras et al. 2004, Kricheldorf 2001, Søndergård and Stolt 2002). The most widely available PLA is the L-form (Figure 13.24) which, although strictly not biodegradable in a technical sense, is compostable in the presence of high humidity and at temperatures above the glass transition (normally in the 50–60°C range). PLA is attractive for applications such as food packaging because of its processability, transparency, and other properties, which can be similar to those of polystyrene. Wider adoption of PLA in this large commodity market has, however, been hindered so far by cost factors as well as by limitations in some physical characteristics, notably thermal stability and barrier properties. Ongoing research and development and innovations which have already been introduced commercially are going some way to improving these characteristics. PLA is manufactured by companies in Europe, North America, and Asia; however, NatureWorks LLC, which was originally created out of a joint venture between Cargill and Dow, has by far the largest share of PLA production capacity. NatureWorks has a production plant in Blair, Nebraska with a capacity of 140,000 metric tonnes PLA per year under the Ingeo™ trade name. Major applications for Ingeo™ include food packaging, bottles, films, nonwovens, textiles, and tableware. A list of companies with interests in PLA research and development and manufacturing is provided in Table 13.7.

With this introduction, the next section of this chapter deals with the manufacturing and properties of PLAs as a prelude to a summary of past research and development aimed at new PLA composite materials reinforced with wood or agricultural fibers.

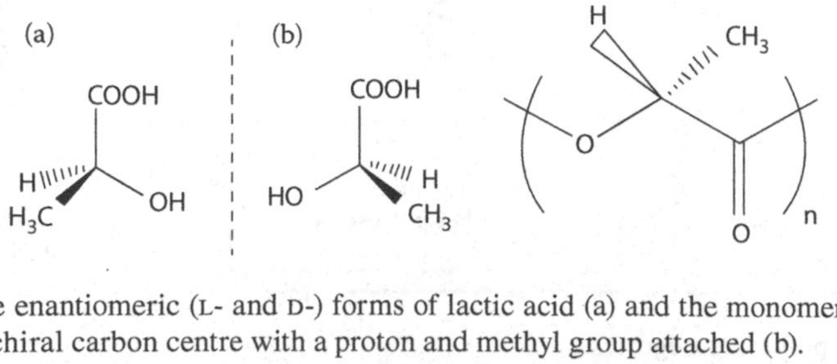

FIGURE 13.24 The enantiomeric (L- and D-) forms of lactic acid (a) and the monomer unit in polylactides (PLAs) showing the chiral carbon centre with a proton and methyl group attached (b).

TABLE 13.7

Companies Currently Manufacturing Lactic Acid-Based Polymers

Company	Location	Main Products	Trade Name
Birmingham Polymers	USA	Lactide polymers for medical applications	Lactel
Boeringer Ingelheim		Lactide polymers for medical applications	Resomer
Durect	USA	Range of lactide, glycolide and caprolactone polymers for medical applications	Lactel
FKuR	Germany	PLA blends for extrusion and injection molding	
Galactic	Belgium	Lactic acid and latates	Galacid, Galaflow, etc.
Mitsui Chemicals	Japan	PLAs for film and fiber uses	Lacea
Nature works	USA	PLA raw materials for film, fiber and other diverse applications	Ingeo
Phusis	France	Medical-grade lactide polymers	Phusiline
Purac	The Netherlands	Monomers and lactide polymers for diverse applications	Purasorb
Shimadzu Corporation	Japan	PLA raw materials	Lacty

13.3.2 PLA MANUFACTURING

The starting raw material for production of PLA is lactic acid in all cases. This naturally occurring α-hydroxy acid, which exists in two enantiomeric forms (L- and D-) (Figure 13.24) is widely used in the food, pharmaceutical, textile, leather and chemical industries. The L-form of lactic acid is found in all living organisms but the D-form is not commonly found in nature. Microbial fermentation of carbohydrates is the favored method for manufacturing lactic acid and allows the production of optically pure forms. The use of optically pure lactic acid is critically important in respect to PLA production, since even small amounts of enantiomeric impurities can cause very significant changes in polymer crystallinity and biodegradability.

The synthesis of PLA can in principle follow three routes: (1) condensation polymerization, (2) azeotropic dehydrative condensation, and (3) ring opening polymerization from the lactide dimer. The condensation polymerization approach is the least expensive, but high-molecular weight PLA is difficult to obtain when using this method. One possibility is to introduce chain-extending agents; however, the final product may contain unreacted chain-extending agents, or residual impurities from the catalyst. Coupling or esterification-promoting agents may also be used to increase chain length, albeit with increased cost and process complexity. The second approach, involving azeotropic condensation polymerization, can be used to produce PLAs with long chain lengths and without the use of chain extenders and adjuvants. A process of this type has been commercialized by Mitsui Chemicals (Japan) in which lactic acid and a catalyst are azeotropically dehydrated in a refluxing, high boiling, aprotic solvent under reduced pressure to obtain high molecular weight PLA (Mw ≥ 300,000). A drawback of this approach is the need for relatively high catalyst concentrations in order to achieve acceptable reaction rates, with the result that polymer hydrolysis and degradation occur during processing. Catalyst deactivation has been proposed as one solution to this problem as residual catalyst concentrations can then be reduced to ppm levels. The third method, based on ring-opening polymerization (ROP) of lactide (Figure 13.25) is the only practical technique for producing pure high molecular weight PLA (Mw ≥ 100,000) and the one that has been developed most widely for industrial-scale production. ROP also has the advantage that the chemistry and therefore the properties of the final polymer can be accurately controlled and tuned to requirements. As shown in Figure 13.25, the first step in ROP is polycondensation of lactic acid, which is then followed by depolymerizaton to obtain the lactide dimer. The latter step is typically carried out by raising the reaction temperature, lowering the pressure and distilling off the resulting lactide. The

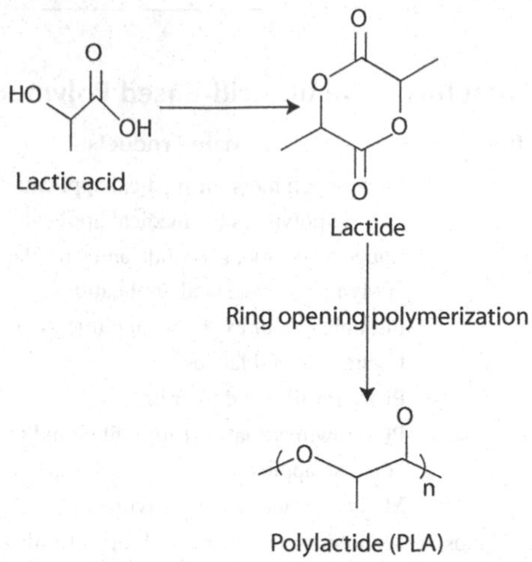

FIGURE 13.25 Schematic showing synthesis of PLA by ring-opening polymerization from the lactide dimer.

use of ROP to produce PLAs has been carried out by solution polymerization, bulk polymerization, melt polymerization and suspension polymerization techniques and, depending upon which type of catalyst is used, the reaction mechamism is of ionic, coordination or free-radical type. Reports indicate that ROP of lactide can be catalyzed by many different metal compounds but tin(II) 2-ethylhexanoate (tin octanoate) is the most frequently used. Just as lactic acid is available in L- and D-forms, so lactide also has two different versions, namely D,D- and L,L- and, in addition, can be formed from one D- and one L-lactic acid molecule, yielding D,L-lactide (meso-lactide). On this basis, ROP can be used to create PLA copolymers of different stereoforms, with the various combinations having a large effect on the final product properties.

13.3.3 MELT PROCESSING OF PLAs

Melt processing is the main method used to convert PLA raw materials (i.e, granulates) into useful materials such as bottles, cups, tableware, fibers, films, and coatings, using either extrusion or injection molding or both. As with other polymers containing ester linkages, an important processing consideration is the tendency towards degradation on exposure to heat for even relatively short periods. Melt extrusion of PLAs is often linked to a second processing step (e.g., injection molding, fiber drawing, film blowing) and the properties of the final product are determined by melt processing factors including the temperature and temperature gradient, residence time, atmosphere and moisture content. The moisture level is particularly important given the hydrolytic tendency of PLAs and, as an example, the processing recommendations for NatureWorks Ingeo™ 2003D extrusion grade suggest that the polymer granulate should be dried to a maximum of 250 ppm moisture as determined by the Karl Fischer method. Furthermore, processes which use longer residence times and/or higher temperatures will benefit from pre-drying of granulate to ~50 ppm water content. Typically NatureWorks supplies products in sealed bags dried to 400 ppm moisture level and the company recommends drying and re-sealing of unused material. The recommended drying conditions for Ingeo™ 2003 D and other PLA granulates are shown in Table 13.8.

13.3.4 PROPERTIES OF PLAs

The physical properties of PLAs are determined by the molecular structure (e.g., stereochemistry, copolymers) as well as by the processing conditions. Methods for characterization and analysis of lactide polymers and their precursors have recently been reviewed (Inkinen et al. 2011). The key

TABLE 13.8

Typical Drying Conditions Recommended for Nature Works LLC Ingeo™ PLA Granulates

Drying Parameter	Settings for Dry Pellets Received in Packaging with a Barrier Liner[a]	Settings for Pellets with 2000 ppm Moisture Content
Residence time (h)	2	4
Air temperature (°C)	90	90
Air dew point (°C)	−40	−40
Air flow rate (m³/h · kg resin)	1.85	1.85

[a] Material is dried to less than 400 ppm moisture content before shipping in foil-lined containers.

properties of PLA in relation to its use in composite materials are discussed in the following sections.

13.3.4.1 Thermal Properties

As mentioned earlier, the limited thermal stability of PLA is of concern when melt processing. Thermal degradation of PLA is a complex process, reported to have first-order kinetics, and involves both radical and nonradical reactions (Sødergård and Stolt 2002). Racemization, the process by which an enantiomerically pure form is converted into a mixture of L- and D-forms, can also take place at high temperatures. Thermal degradation of PLA is accelerated by high amounts of reactive end groups (i.e., low molecular weight) and high polydispersity (Mw/Mn), as well as by the presence of residual catalyst, lactide or lactic acid, moisture and various impurities. PLA stereocomplexes based on 50:50 blends of L- and D-PLA were first reported by Ikada et al. (1987), who also discovered that the melting point of the stereocomplex is about 50°C higher than that of either the pure L- or pure D-forms. PLA stereocomplexes have also been shown to have higher thermal resistance. For example, this approach was recently used to produce a cup which could withstand boiling oil (US Patent 2008207840 2008). The T_g of PLA increases with increasing polymer molecular weight up to a threshold value and usually lies in the 50–80°C range, although lower values have been reported for oligomers or low molecular weight polymers. The melting point of PLA typically lies in the 170–180°C range and, like the T_g, increases as the molecular weight increases up to a specific threshold.

PLA exists in semi-crystalline or amorphous states depending upon stereochemistry, composition, and thermal history. As with other polymers, the solid state structure of PLA influences its mechanical characteristics, electrical and solvent resistance, and biodegradation properties and the polymer may therefore be modified in order to tune these properties for particular needs. As an example in the field of food packaging, higher crystallinity generally means higher gas barrier properties, which is desirable for some of these packaging applications. Differential scanning calorimetry (DSC) is a very useful method for analyzing thermal transitions and crystallinity in PLA and, as an example, the second heating curve in the DSC thermogram for PLA with 5% clay nanofiller is shown in Figure 13.26. It is not uncommon to observe multiple melting point behavior in the DSC of PLAs, which may be due to the presence of crystallites of different morphologies and dimensions in the initial sample but can also arise from formation of imperfect crystals as a result of annealing during the DSC scan (Sarasua et al. 1998).

13.3.4.2 Hydrolytic Stability

Hydrolysis of PLA, leading to molecular fragmentation, is influenced by a variety of factors, including chemical structure, molar mass and its distribution, purity, morphology, specimen shape and history, and the environmental conditions. The mechanism of PLA hydrolysis involves random attack at the ester linkages as well as chain-end scission in the presence of water. Amorphous parts of the polymer generally exhibit faster water uptake than the crystalline regions and therefore

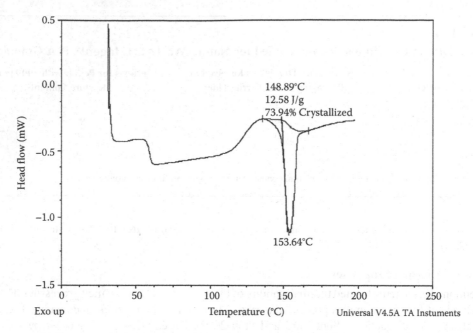

FIGURE 13.26 Second heating curve in the DSC thermogram of PLA with 5% clay nanofiller, showing the glass transition in the 50–60°C range, a pre-melting exotherm and a peak melting temparture close to 153°C.

will be the first to undergo hydrolysis. This being the case, the polymer chains in the remaining nondegraded material will have more space and greater mobility with consequent reorganization and increased crystallinity. This provides an explanation for the greater opacity of PLA test films when subjected to hydrolytic conditions over a short period of time. After the initial phase, crystalline regions of PLA are also hydrolyzed, which produces increased mass loss and eventually complete resorption of the material. It has been suggested that hydrolysis takes place more rapidly in the core of PLA materials due to autocatalysis in the presence of compounds with carboxylic end groups, which are unable to permeate the outer shell of the material (Li et al. 1990). Meanwhile, degradation products in the surface layer are continuously dissolved in the surrounding aqueous medium. As would be expected, hydrolytic degradation of PLA is accelerated at higher temperatures.

13.3.4.3 Biodegradability

The biodegradation of aliphatic biopolyesters has been extensively reported in the literature (Amass et al. 1998, Sødergård and Stolt 2002, Zhang and Sun 2005). Biodegradation of PLA occurs when the polymer is hydrolyzed at or above the T_g in a composting environment. When the oligomeric breakdown products reach a threshold molecular weight, microorganisms start to digest these lower molecular weight compounds, releasing carbon dioxide and water (Figure 13.27). Specific factors which will influence the rate of biodegradation of articles made from PLA include chemical structure, molar mass and molar mass distribution. At a higher level, the influence of the T_g, T_m, crystallinity and modulus will also be important. Material surface area and surface hydrophilicity or hydrophobicity may also play a role. In general, as the molar mass and T_m increase, the rate of biodegradation decreases. The biodegradation of PLA also depends upon the environment to which it is exposed. For example, tests in soil have shown that it can take a long time for degradation to start. For instance, in one study, PLA films showed no degradation after six weeks of soil burial. On the other hand, in a composting environment at 50–60°C PLA can be substantially degraded to smaller molecules within 45–60 days.

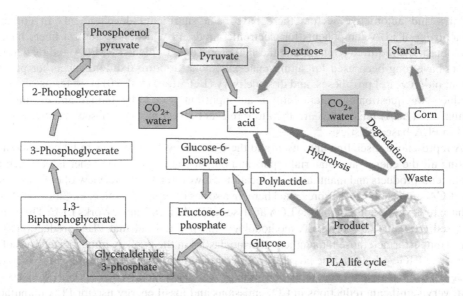

FIGURE 13.27 Schematic showing the pathways for synthesis of lactic acid in nature and the cycles for production and breakdown of PLA.

13.3.4.4 Mechanical Properties

The mechanical properties of PLA can be varied to a large extent, providing materials which range in character from soft and elastic to stiff and strong, as a complex function of parameters such as polymer structure, molecular weight, crystallinity, formulation and processing. It has been reported that semi-crystalline PLA has a tensile strength of 50–60 MPa, tensile modulus of 3 GPa and an elongation at break of 4% (Sødergård and Stolt 2002). In terms of dependence on molar mass, significant changes in tensile strength and modulus have been observed when the molar mass is raised from 50 to 100 kDa; however, further increases in molar mass to 300 kDa appear not to have such a significant effect. Fiber spinning has been used as a way to increase the tensile strength and modulus of L PLA and its blends or copolymers with D-PLA or polycaprolactone (PCL), a synthetic biodegradable polyester. Solution spinning processes may offer an advantage in this respect since melt spinning is often associated with very significant thermal degradation of PLAs (Penning et al. 1993). As a result, the L-PLA fiber modulus increases from a range of 7–9 GPa when using melt spinning to a range of 10–16 GPa when solution spinning is used (Sødergård and Stolt 2002).

Brittle, high modulus PLAs can be modified by copolymerization with polymers having lower T_g values. For example, after copolymerization with ε-caprolactone, products are obtained which exhibit lower tensile strength, increased elongation at break and higher impact strength (Grijpma et al. 1992, 1991, Hiljanen-Vainio et al. 1997).

13.3.5 Sustainability

PLA as a raw material for products in packaging, electronics and automotive, medicine and the broad field of composite materials offers the advantages of compostability, biocompatibility, and nontoxicity and can, in principle, be manufactured starting from a wide range of different biomass residues (Sodergaard and Inkinen 2011). As a bioplastic, it therefore fits well with the cradle-to-cradle principle espoused by McDonough and Braungart (2002). In the commodity packaging field, PLA has attracted considerable interest and is already used in some countries for certain types of food packaging. Wider adoption in this sector is feasible as and when the price of PLA comes down relative to that of synthetic petroleum-derived plastics and when certain technical properties are improved, notably the heat distortion temperature and gas barrier properties. In the electronics and automotive sectors there have been significant developments in commercial manufacturing of PLA components for electronic goods, in which case natural fibers have been used to obtain the required strength and thermal properties. In medicine, PLA and related lactide polymers have long been used for suture materials, bone fracture fixation, sheets for preventing adhesion, blood vessel prostheses, and drug delivery (Letchford et al. 2011). In terms of commodity products, the question of sustainability is an important issue to be defined and evaluated. Life cycle analysis (LCA) techniques are therefore relevant and a number of such studies have been applied to PLA-based plastics.

LCA represents a discipline unto itself and there are many complexities to be considered when evaluating all the energy and material inputs and outputs, not to mention other issues such as the particular end-products and manufacturing location. However, a brief overview of conclusions from various LCA studies on PLAs is presented here. In a series of reports, Vink et al. (2003, 2004, 2007, 2010) have reviewed studies involving LCA analyses of PLA from NatureWorks LLC. These authors emphasized the value of using LCA methodology and pointed out that PLA production systems generally outperformed those for synthetic thermoplastics in the two key impact categories of fossil fuel use and GHG emissions. The possibility for continued improvement in the processing systems for PLA as part of an overall cost reduction strategy was also noted. With continued process development, very significant reductions in CO_2 emissions and fossil energy use for PLA manufactured in 2006 relative to data from 2003 were reported. The values of 0.27 kg CO_2 eq./kg emitted and 27.2 MJ/kg fossil energy used compare favorably with data for commercial thermoplastics published by Plastics Europe (Figure 13.28). In a study on biodegradable food packaging published in 2004, PLA was compared with polypropylene (PP) for the specific application of thermoformed yoghurt cups. The authors concluded in this case that the use of PLA was a more energy-efficient option; however, the differences were considered marginal when uncertainties in the various estimates were taken into account. Furthermore, uncertainties concerning PLA biodegradation in landfills complicated the GHG emission estimates. More recently, Uihlein et al. (2008) compared PLA with polystyrene for production of drinking cups and concluded that life-cycle assessments did not unequivocally support decision-making either for or against the use of materials from renewable

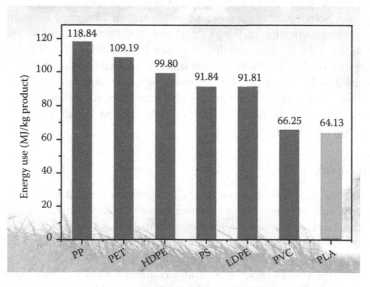

FIGURE 13.28   A comparison of energy used for production of PLA with comparable values for a range of commodity thermoplastics. (Data from: http://www.plasticseurope.org, The contribution of plastic products to resource efficiency—Final report (2005), p. 46.)

resources. The substantial further reductions in GHG emissions and fossil energy utilization in PLA manufacturing by NatureWorks are detailed in the most recent report by Vink et al. (2010).

The assessment of PLA production and the use of PLA-based products from a sustainability perspective using LCA methodologies is likely to continue developing in parallel with the evolution of commercial PLA manufacturing. Similarly, PLA use in combination with natural fiber reinforcements in biocomposites or nano-biocomposites will also be increasingly evaluated as these new materials are adopted more widely in industry. In this respect, there will be a need to ensure the basic requirements for LCA, such as data transparency, uncertainties, sensitivity analysis for key data and methodological choices are all taken into consideration if, as seems probable, sustainability assessments become increasingly relevant in the future.

13.4 POLYLACTIDE FIBER COMPOSITES

Wood or other bio-fiber composites using PLA as matrix material have been widely studied the last decade. The mechanical performance of this type of composites can, as with PLA alone, be varied to a large extent. Apart from the inherent properties of the constituents used, variation depending on the manufacturing method is also to expect. Moisture can, if present during processing hydrolyze the PLA, it is therefore of outmost importance to keep both the fibers and the PLA dry before processing. The properties of the composites can also be controlled by cooling rates and annealing steps after processing. Table 13.9 demonstrates mechanical performance of a variety of injection molded wood/agro fiber PLA composites. The addition of fibers to PLA gives a higher Young's modulus, the tensile strength is also possible to improve by the addition of fibers but Table 13.9 shows that the resulting strength reported in some studies could also be lower compared to neat PLA. The notched impact strength is increased or similar when fibers are added to the PLA, the unnotched impact strength is decreased except for a study with man-made cellulose where the unnotched impact strength is increased almost 4 times when the fiber was added to composite (Ganster et al. 2006).

The biological durability of these composites has mainly been studied regarding the ability to degrade the composite at the easiest way after end of use. In order to degrade the material it is needed for the PLA to be hydrolyzed and composting is the most preferred disposal route (Mathew et al. 2005). Mathew et al. (2005) also studied the biodegradability in a 58°C, 60% relative humidity of PLA composites with 25% microcrystalline cellulose, wood flour and wood fiber. Initially the neat PLA started to degrade first, after 75 days all samples shoved a rapid increase in degradation, where the PLA with wood flour was the most rapid. For the in-use biological durability there is not many studies performed, Table 13.10 demonstrates resulting weight losses from a 10 week soil block test (ASTM D1413), there the composites first are subjected to a preconditioning step by water soaking for 2 weeks followed by 10 weeks of exposure to a brown rot fungus. The mass losses obtained from these types of laboratory tests normally are very low for WPCs, this is due to the slow moisture transport and relative short duration of the test, therefore it could be valuable to combine durability tests with mechanical tests before and after decay testing. Tables 13.11 and 13.12 present strength and stiffness properties of the PLA fiber composites before and after soil block testing. There it is seen that the losses in strength and stiffness is similar for both sterile and fungus soil, so the brown rot fungus used had no effect on the mechanical properties of the composites. However, the strength and stiffness of the PLA unmodified fiber composite show dramatically lower values after decay testing, this would be related to the moisture induced movements which causes cracks and destroyed interfaces between the fiber and the PLA. Figure 13.29 shows scanning electron micrographs of the surface of a PLA unmodified composite after water soaking and after a decay test in soil, the water soaking does not create visible cracks but after decay testing there are surface cracks formed over the entire surfaces. The PLA acetylated fiber composite show almost no loss in mechanical properties after decay testing, so dimensional stabilization by acetylation proved to be a good route to ensure a long term performance of these PLA composites.

TABLE 13.9
Comparison of the Mechanical Properties of Some Injection Molded PLA Wood/Agro Fiber Composites

Fiber	PLA Polymer	Fiber Wt%	Tensile Strength		Young's Modulus		Unnotched Charpy Impact Strength		Notched Impact Strength		Source
			Neat PLA (MPa)	Composite (MPa)	Neat PLA (GPa)	Composite (GPa)	Neat PLA (kJ/m²)	Composite (kJ/m²)	Neat PLA	Composite	
Abaca fibers	PLA 4042D, Nature Works	30	63	74	3.4	8.0			2.2[a]	5.3[a]	Bledzki et al. (2009)
Cellulose fibers	POLLAIT	25	50	45	3.6	6.0					Mathew et al. (2005)
Cellulose nanofibers	ESUN, Shenzhen Bright China Co	5	59	71	2.9	3.6					Jonoobi et al. (2010)
Flax	PLA 6202D, Nature Works	10	44	43	3.1	3.9	16.1	10.0			Bax and Müssig (2008)
Flax	PLA 6202D, Nature Works	20	44	49	3.1	5.1	16.1	10.5			Bax and Müssig (2008)
Flax	PLA 6202D, Nature Works	30	44	54	3.1	6.3	16.1	11.1			Bax and Müssig (2008)
Hemp	PLA 4042D, Nature Works	10	51	52	3.5	4.1					Sawpan et al. (2007)
Hemp	PLA 4042D, Nature Works	20	51	60	3.5	5.5					Sawpan et al. (2007)
Hemp	PLA 4042D, Nature Works	30	51	66	3.5	7.0					Sawpan et al. (2007)
Kenaf fiber	Biomer L 9000	30	61	53	3.3	5.5	76	52	7.4[a]	12.2[a]	García et al. (2008)
Kenaf fiber	LACTY 9020, Shimadzu	30	47	36	1.4	1.9					Pan et al. (2007)
Man-made cellulose fibers	PLA 2002D, Nature Works	25	71	108	2.8	4.2	14.4	69.0	2.8[a]	8.4[a]	Ganster et al. (2006)
Man-made cellulose fibers	PLA 4042D, Nature Works	30	63	92	3.4	5.8			2.2[a]	7.9[a]	Bledzki et al. (2009)

Microcrystalline cellulose	POLLAIT	25	50	36	3.6	5.0				Mathew et al. (2005)
Rice husk fiber	Biomer L 9000	30	61	36	3.3	4.5	76	7.4[a]	6.8[a]	García et al. (2008)
Silkworm silk fibers	Eask Link Degradable Materials Ltd.	5	65	62	1.8	2.5	36			Cheung et al. (2008)
Sugar beet pulp	PLA, Dow Cargill	45	70	30	1.9	2.6				Finkenstadt et al. (2007)
Wood flour	PLA 3001D, Nature Works	10	56	53	0.6	2.0				Pilla et al. (2009)
Wood flour	Biomer L 9000	20	63	66	2.7	4.8		25.7[b]	23.9[b]	Huda et al. (2006)
Wood flour	PLA 3001D, Nature Works	20	56	56	0.6	2.7				Pilla et al. (2009)
Wood flour	POLLAIT	25	50	45	3.6	6.3				Mathew et al. (2005)
Wood flour	Biomer L 9000	30	63	63	2.7	5.3		25.7[b]	23.2[b]	Huda et al. (2006)
Wood flour	Biomer L 9000	40	63	59	2.7	6.3		25.7[b]	21.9[b]	Huda et al. (2006)
Wood flour	PLA 7000D, Nature Works	50	69	55	3.6	8.3				Sykacek et al. (2009)

[a] Charpy (kJ/m^2).
[b] Izod (J/m).

TABLE 13.10
Mass Loss of PLA Fiber Composite after Soil Block Test

Matrix	Wood	Wt% wood	Mass Loss %
PLA 4042D, Nature Works	—	0	0.1
PLA 4042D, Nature Works	Unmodified fiber	50	1.6
PLA 4042D, Nature Works	Acetylated fiber	50	0.2

TABLE 13.11
Strength of the PLA Fiber Composites before and after Soil Block Test

Matrix	Wood	Wt% wood	Reference (MPa)	After Water Soaking (MPa)	After Soil Block Test Sterile Soil (MPa)	After Soil Block Test Fungus Soil (MPa)
PLA 4042D, Nature Works	—	0	103	102	94	95
PLA 4042D, Nature Works	Unmodified fiber	50	102	72	59	55
PLA 4042D, Nature Works	Acetylated fiber	50	94	89	89	92

TABLE 13.12
Stiffness of the PLA Fiber Composites before and after Soil Block Test

Matrix	Wood	Wt% wood	Reference (GPa)	After Water Soaking (GPa)	After Soil Block Test Sterile Soil (GPa)	After Soil Block Test Fungus Soil (GPa)
PLA 4042D, Nature Works	—	0	3.17	4.02	3.72	3.80
PLA 4042D, Nature Works	Unmodified fiber	50	7.00		4.00	3.67
PLA 4042D, Nature Works	Acetylated fiber	50	6.37	7.17	6.88	6.83

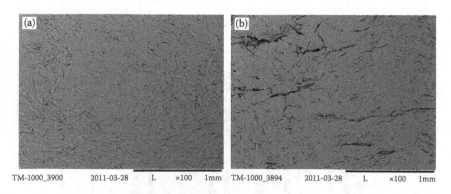

TM-1000_3900   2011-03-28   L   ×100   1mm   TM-1000_3894   2011-03-28   L   ×100   1mm

FIGURE 13.29 PLA-unmodified fiber, (a) subjected to preconditioning with moisture, (b) after testing in compost soil.

REFERENCES

Amass, W., A. Amass, and B. Tighe. 1998. A review of biodegradable polymers: Uses, current developments in the synthesis and characterization of biodegradable polyesters, blends of biodegradable polymers and recent advances in biodegradation studies. *Polym Int* 47, 89–144.

Anonymous. 2002. Want to improve your plastic composite? Struktol knew you wood." Technical brochure, Struktol Company of America. Stow, OH.

Anonymous. 2011. *iBIB2011: International Business Directory for Innovative Bio-Based Plastics and Composites*. Dominik Vogt ed., nova-Institute GmbH. Huerth, Germany and Polymedia Publisher GmbH, Moenchengladbach, Germany.

Auras, R., B. Harte., and S. Selke. 2004. An overview of polylactides as packaging materials. *Macromolecular Bioscience* 4, 835–864.

Bax, B. and J. Müssig. 2008. Impact and tensile properties of PLA/cordenka and PLA/flax composites. *Comp. Sci. Tech.* 68, 1601–1607.

Belgacem, M.N. and A. Gandini. (eds). 2008. *Monomers, Polymers and Composites from Biomass*. Elsevier Science and Technology, Oxford, UK.

Belgacem, M. N. and A. Gandini. 2011. Production, chemistry and properties of cellulose-based materials. Chapter 8 in Plackett D. (ed). *Biopolymers—New Materials for Sustainable Films and Coatings*. John Wiley and Sons, Ltd. Chichester, West Sussex, England.

Bengtsson, M.P., P. Gatenholm, and K. Oksman. 2005. The effect of crosslinking on the properties of polyethylene/wood flour composites. *Composites Science and Technology* 65(10), 1468–1479.

Billmeyer, F.W. Jr. 1984. *Textbook of Polymer Science*. John Wiley and Sons, New York.

Bledzki, A.K., A. Jaszkiewicz, and D. Scherzer. 2009. Mechanical properties of PLA composites with manmade cellulose and abaca fibres. *Composites: Part A* 40, 404–412.

Carley, J.F. (ed.) 1993. *Whittington's Dictionary of Plastics*. 1993. Technomic Publishing Company, Inc. Lancaster, PA.

Carus, M., C. Gahle, and H. Korte. 2008. Market and future trends for wood-polymer composites in Europe: The example of Germany. In Oksman Niska, K. and Sain, M. Woodhead (eds). *Wood-Polymer Composites*. Publishing and CRC Press LLC, Boca Raton, FL.

Caulfield, D., R. Jacobson, K. Sears, and J. Underwood. 2001. Fiber reinforced engineering plastics. In *Proceedings: 2nd International Conference on Advanced Wood Composites*, Bethel, ME.

Cheung, H.-Y., K.-T. Lau, X.-M. Tao, and D. Hui. 2008. A potential material for tissue engineering: Silkworm silk/PLA biocomposite. *Composites Part B* 39, 1026–1033.

Chodak, I. 2008. Polyhydroxyalkanoates: Origin, properties and applications. In Belgacem, M. and Gandini, A. (eds). *Monomers, Polymers and Composites from Renewable Resources*. Elsevier, Amsterdam.

Finkenstadt, V.L., L.S. Liu, and J.L. Willett. 2007. Evaluation of poly(lactic acid) and sugar beet pulp green composites. *J Polym Environ* 15, 1–6.

García, M., I. Garmendia, and J. García. 2008. Influence of natural fiber type in eco-composites. *J. Appl. Polym. Sci.* 107, 2994–3004.

Ganster, J., H.-P. Fink, and M. Pinnow. 2006. High-tenacity man-mad cellulose fibre reinforced thermoplastics—injection moulding compounds with polypropylene and alternative matrices. *Composites: Part A* 37, 1796–1804.

Gordon, J.E. 1988. *The New Science of Strong Materials (or why you don't Fall Through the Floor)*. Princeton University Press, Princeton, NJ.

Grijpma, D.W, A.J. Nijenhuis, P.G.T. van Wijk, and A.J. Pennings. 1992. High impact strength as-polymerized PLA. *Polym Bull* 29, 571–578.

Grijpma D.W., G.J. Zondervan, and A.J. Pennings. 1991. High molecular weight polymers of L-lactide and ε-caprolactone as biodegradable elastomeric implants. *Polym Bull* 25, 327–333.

Guo, G., G.M. Rizvi, and C.B. Park. 2008. Wood-polymer composite foams. In Oksman Niska, K. and Sain, M. (ed). *Wood-Polymer Composites*. Woodhead Publishing and CRC Press LLC, Boca Raton, FL.

Hiljanen-Vainio, M., P.A. Orava, and Seppala, J.V. 1997. Properties of ε-caprolactone/DL-lactide (ε-CL/DL-LA) copolymers with a minor ε-CL content. *J Biomed Mater Res* 34, 39–46.

Huda, M.S., L.T. Drzal, M. Misra, and A.K. Mohanty. 2006. Wood-fiber-reinforced poly(lactic acid) composites: Evaluation of the physicomechanical and morphological properties. *J. Appl. Polym. Sci.* 102, 4856–4869.

Inkinen, S., M. Hakkarainen, A-C. Albertsson, and A. Södergård. 2011. From lactic acid to poly(lactic acid) (PLA): Characterization and analysis of PLA and its precursors. *Biomacromolecules* 12, 523–532.

Ikada, Y., K. Jamshidi, H. Tsuji, and S.-H. Hyon. 1987. Stereocomplex formation between enantiomeric poly(lactides). *Macromolecules* 20, 904–906.

Jacobson, R.E., R.M. Rowell, D.F. Caulfield, and A.R. Sanadi. 1996. Property improvement effects of agricultural fibers and wastes as reinforcing fillers in polypropylene-based composites. Woodfiber-Plastic Composite: Virgin and recycled wood fiber and polymers for composites, D.F. Caulfield, R.M. Rowell and J.A. Youngquist, eds. pp 211–219, Forest Products Society, Madison, WI.

Jacobson, R.E., R. Rowell, D. Caulfield, and A.R. Sanadi. 1995a. United States based agricultural "waste products" as fillers in a polypropylene homopolymer. *Proceedings, Second Biomass Conference of the Americas: Energy, Environment, Agriculture, and Industry*. August, 1995, Portland, OR, NREL/CP-200–8098, DE95009230, 1219–1227.

Jacobson, R.E., R.M. Rowell, D.F. Caulfield, and A.R. Sanadi. 1995b. Mechanical properties of natural fiber-polyolefins composites: Dependence on fiber type. In *Proceedings, Inside Automotives 95'*, Dearborn, MI.

Jacobson, R., D. Caulfield, K. Sears, and J. Underwood. 2002. Low Temperature Processing (LTP) of Ultra-Pure Cellulose Fibers into Nylon 6 and other Thermoplastics, In *Proceedings: Sixth International Conference on Woodfiber/Plastic Composites*. Madison, WI, April.

Jonoobi, M., J. Harun, A.P. Mathew, and K. Oksman. 2010. Mechanical properties of cellulose nanofiber (CNF) reinforced polylactic acid (PLA) prepared by twin screw extrusion. *Comp. Sci. Tech.* 70, 1742–1747.

Kadla, J.F., S. Kubo, R.A. Venditti, R.D. Gilbert, A.L. Compere, and W. Griffith. 2002. Lignin-based carbon fibers for composite fiber applications. *Carbon* 40(15), 2913–2920.

Kimura, S. and Itoh, T. 2004 Cellulose synthesizing terminal complexes in the Ascidians. *Cellulose* 11, 377–383.

Kricheldorf, H.R. 2001. Synthesis and applications of polylactides. *Chemosphere* 43, 49–54.

Lemoigne, R. 1923. Production of β-hydroxybutyric acid by certain bacteria of the B. subtilis group. *Compt Rend* 176, 1761–1763.

Lemoigne, R. 1924a. Mechanism of the production of β-hydroxybutyric acid by the biochemical method. *Compt Rend* 178, 1093–1095.

Lemoigne, R. 1924b. Production of β-hydroxybutyric acid by a bacterial process. *Compt Rend* 179, 253–256.

Letchford, K., A. Sodegraad, D. Plackett, S. Gilchrist, and H. Burt. 2011. Lactide and glycolide polymers. Chapter 9 in (A. Domb, N. Kumar, A. Azra eds.) *Biodegradable Polymers in Clinical Use and Clinical Development*. John Wiley and Sons, Chichester, West Sussex, England.

Li, S.M., H. Garreau, and M. Vert. 1990. Structure-property relationships in the case of the degradation of massive aliphatic poly(α-hydroxy acids) in aqueous media. Part 1. Poly(D,L-lactic acid). *J Mater Sci, Mater Med.* 1, 123–130.

Lu, J.Z., Q. Wu, and H.S. McNabb Jr. 2000. Chemical coupling in wood fiber and polymer composites: A review of coupling agents and treatments. *Wood and Fiber Science*, 32(1): 88–104.

McDonough, W. and M. Braungart. 2002. *Cradle to Cradle—Rethinking the Way We Make Things*. North Point Press, New York.

Macrae, R.M. and J.F. Wilkinson. 1958. Poly-β-hydroxybutyrate metabolism in washed suspensions of Bacillus cereus and Bacillus megaterium. *J Gen Microbiol* 19, 210–222.

Mapleston, P. 2001a. Processing technology: A wealth of options exist. *Modern Plastics* June, 56–60.

Mapleston, P. 2001b. Wood composite suppliers are poised for growth in Europe. *Modern Plastics* October, 41.

Mathew, A.J., K. Oksman, and M. Sain. 2005. Mechanical properties of biodegradable composites from poly lactic acid (PLA) and microcrystalline cellulose (MCC). *J. Appl. Polym. Sci.* 97, 2014–2025.

Osswald, T.A. and G. Menges. 1996. *Materials Science of Polymers for Engineers*. Carl Hanser Verlag, New York.

Pan, P., B. Zhu, W. Kai, S. Serizawa, M. Iji, and Y. Inoue. 2007. Crystallization behavior and mechanical properties of bio-based green composites based on poly(L-lactide) and kenaf fiber. *J. Appl. Polym. Sci.* 105, 1511–1520.

Penning, J.P., H. Dijkstra, and A.J. Pennings. 1993. Preparation and properties of adsorbable fibres from L-lactide copolymers. *Polymer* 34, 942–951.

Pilla, S., S. Gong, E. ÓNeill, L. Yang, and R.M. Rowell. 2009. Polylactide-recycled wood fiber composites. *J. Appl. Polum. Sci.* 111, 37–47.

Pollet, E. and L. Averous. 2011. Production, chemistry and properties of polyhydroxyalkanoates. In: Plackett (ed) *Biopolymers—New Materials for Sustainable Films and Coatings*. John Wiley & Sons Ltd.,Chichester, UK.

Rowell, R.M. 1992. Opportunities for lignocellulosic materials and composites, In R.M. Rowell, T.P. Schultz, and R. Narayan, eds. *Emerging Technologies for Materials and Chemicals from Biomass*, American Chemical Society, Washington, DC.

Rowell, R.M., E. O'Neill, A. Krzysik, A.D. Bossman., D.F. Galloway, and M. Hemenover. 2008. Industrial applications of animal manure filled thermoplastics. *Mol. Cryst. Liq. Cryst.* 484:616–622.

Rowell, R.M., A. Sanadi, R. Jacobson, and D. Caulfield. 1999. Properties of kenaf/polypropylene composites. *Kenaf Properties, Processing and Products*, T. Sellers, Jr. and N.A. Reichert, Eds, Mississippi State University Press, Mississippi State, MS, Chapter 32, 381–392.

Rowell, R.M., A.R. Sanadi, D.F. Caulfield, and R.E. Jacobson. 1997. Utilization of natural fibers in plastic composites: Problems and opportunities. *Lignocellulosic-Plastics Composites*, A.L. Leao, F.X. Carvalho, and E. Frollini, Eds., pp. 23–52, Universidade de Sao Paulo Press, Sao Paulo, Brazil.

Sanadi, A.R., D.F. Caulfield, K. Walz, L. Wieloch, R.E. Jacobson, and R.M. Rowell. 1994a. Kenaf fibers-Potentially outstanding reinforcing fillers in thermoplastics, In *Proceedings, Sixth Annual International Kenaf Conference*, New Orleans, LA, pp. 155–160.

Sanadi, A.R., D.F. Caulfield, and Rowell, R.M. 1994b. Reinforcing polypropylene with natural fibers. *Plastic Engineering* L(4), 27–28.

Sanadi, A.R., R.E. Jacobson, D.F. Caulfield, and R.M. Rowell. 1995a. Mechanical properties of natural fiber-polyolefins composites: An overview. In *Proceedings, Seventh Annual International Kenaf Conference*, Irving, TX, March, 1995, pp. 47–59.

Sanadi, A.R., D.F. Caulfield, R.E. Jacobson, and R.M. Rowell. 1995b. Renewable agricultural fibers as reinforcing fillers in plastics: Mechanical properties of kenaf fiber-polypropylene composites. *Ind. Eng. Chem. Research* 34, 1889–1896.

Sanadi, A.R., R.M. Rowell, and D.F. Caulfield. 1996a. Agro-based fiber/polymer composites, blends and alloys. *Polymer News* 21(1), 7–17.

Sanadi, A.R., K. Walz, L. Wieloch, R.E. Jacobson, D.F. Caulfield., and R.M. Rowell. 1996b. Effect of Matrix Modification on Lignocellulosic Composite Properties, D.F. Caulfield, R.M. Rowell and J.A. Youngquist, eds. pp 166–172, Forest Products Society, Madison, WI.

Sarasua, J-R., R.E. Prud'homme, M. Wisniewski, A. Le Borgne, and N. Spassky. 1998 Crystallization and melting behavior of polylactides. *Macromolecules* 31, 3895–3905.

Sawpan, M.A., K.L. Pickering, and A. Fernyhough. 2007. Hemp fibre reinforced poly(lactic acid) composites. *Adv. Mater. Res.* 29–30, 337–340.

Siró, I. and D. Plackett. 2010. Microfibrillated cellulose and new nanocomposite materials: a review. *Cellulose* 17(3): 459–494.

Søbergård, A. and S. Inkinen. 2011. Production, chemistry and properties of polylactides. Chapter 3 in Plackett D (ed) *Biopolymers—New Materials for Sustainable Films and Coatings*. John Wiley and Sons, Ltd. Chichester, West Sussex, England.

Søbergård, A. and M. Stolt. 2002. Properties of lactic acid based polymers and their correlation with composition. *Progress in Polymer Science* 27, 1123–1163.

Stark, N.M. and R.E. Rowlands. 2003. Effects of wood fiber characteristics on mechanical properties of wood/polypropylene composites. *Wood and Fiber Science.* 35(2), 167–174.

Stark, N.M. and L.M. Matuana. 2009. Co-Extrusion of WPCs with a Clear Cap Layer to Improve Color Stability. In *The Proceedings of 4th Wood Fibre Polymer Composites International Symposium*. Bordeaux, France, 1–13.

Sykacek, E., M. Hrabalova, H. French, and N. Mundigler. 2009. Extrusion of five biopolymers reinforced with increasing wood flour concentration on a production machine, injection moulding and mechanical performance. *Composites*: Part A 40, 1272–1282.

Toloken, S. 2010. China's WPC firms see growth from Europe and China. *PlasticsNews.com/China*. Crain Communications Inc.

Uihlein, A., S. Ehrenberger, and L. Schebeck. 2008. Utilisation options of renewable resources: A life cycle assessment of selected products. *J Cleaner Prod* 16, 1306–1320.

US Patent 2008207840, Assignees: Tate & Lyle PLC. Inventors: Sodergard, N.D.A., Stolt, E.M., Siistonen, H.K. et al., 2008.

Vazquez, A., M.L. Foresti, and V. Cyras. 2011. Production, chemistry and degradation of starch-based polymers. In: Plackett D. (ed). *Biopolymers—New Materials for Sustainable Films and Coatings*. John Wiley and Sons, Ltd. Chichester, West Sussex, England.

Vink, E.T.H., K.R. Rabago, D.A. Glassner, and P.R. Gruber, P.R. 2003. Applications of life cycle assessment to NatureWorks™ polylactide (PLA) production. *Polym Degrad Stab* 80, 403–419.

Vink, E.T.H., K.R. Rabago, D.A. Glassner, R. Springs, R.P. O'Connor, J.J. Kolstad, and P.R. Gruber. 2004. The sustainability of NatureWorks™ polylactide polymers and Ingeo™ polylactide fibers: Aan update of the future. *Macromol Biosci* 4, 551–564.

Vink, E.T.H., D.A. Glassner, J.J. Kolstad, R.J. Wooley, and R.P. O'Connor. 2007. The eco-profiles for current and near-future NatureWorks® polylactide (PLA) production. *Ind Biotech* 3, 58–81.

Vink, E.T.H., S. Davies., and J.J. Kolstad. 2010. The eco-profile for current Ingeo® polylactide production. *Ind Biotech* 6, 212–224.

Wolcott, M.P., P.M. Smith, and D.A. Bender. 2009. Natural fiber thermoplastic composites for Naval facilities. In *The Proceedings of the International SAMPE Symposium and Exhibition*. Baltimore Convention Center, Baltimore, Maryland, p. 54.

Youngquist, J.A. 1999. Wood-based composites and panel products. *Wood Handbook: Wood as an Engineering Material*. USDA Forest Serv. Forest Prod. Lab. Forest Prod. Soc., Madison, WI. 27–28.

Zhang, J.F. and Sun, X. 2005. Poly(lactic acid) based bioplastics. Chapter 10 in Smith R (ed) *Biodegradable Polymers for Industrial Applications*. CRC, Woodhead Publishing Ltd, Cambridge, England, pp. 251–288.

Part IV

Property Improvements

14 Heat Treatment

Roger M. Rowell, Ingeborga Andersone,
and Bruno Andersons

CONTENTS

14.1 INTRODUCTION

With an increased awareness of the fragility of our environment and the need for durability in wood products, new technologies have been developed to increase the service life of wood materials without the use of toxic chemicals. Issues of sustainability, carbon sequestration, and improved performance come together in a search for environmentally friendly methods of wood preservation.

Heat treatments of wood have been studied for many years and are now commercial (Esteves and Pereira 2009). Heating wood results in increased decay resistance and improved dimensional stability; however, strength properties of the wood are, in general, reduced.

14.2 PROCESSES

Heating wood to improve performance dates back to thousands of years. The Kalvträsk wooden ski was bent using heat by the Saami and used in northern Sweden over 5200 years ago (Åström 1993, Insulander 1999). From hieroglyphic pictures of furniture in Egypt between 2500 and 3000 BC, there are chairs made using heat-bent wood members (Rivers and Umney 2005). The Egyptians also bent wood for bows using hot water around 1900 BC as shown on tomb paintings (Ostergard 1987). In the eleventh century, the Viking shipbuilders also bent wood using heat for parts of the ship (Olsen and Crumlin-Pedersen 1967). In the Norwegian stave churches, large beams were bent using heat and joined (Bugge 1953, Holan 1990). Finally, the wooden cask makers knew how to heat-bend wood 2000 years ago (Twede 2005). A picture of a wooden cask is shown on the wall of a 2690 BC Egyptian tomb (Kilby 1971). Wooden casks were also made by the Romans to store wine (Twede 2005).

In the early part of the twentieth century, it was found that drying wood at high temperature increased dimensional stability and a reduction in hygroscopicity (Tiemann 1915, Stamm and Hansen 1937). Later, it was found that heating wood in molten metal at temperatures between 140°C and 320°C reduced swelling in Sitka spruce by 60% and also increased resistance to microbiological attack (Stamm et al. 1946). They also found that the increase in stability and durability also increases brittleness and loss in some strength properties including impact toughness, modulus of rupture, and work to failure. The treatments usually cause a darkening of the wood and the wood has a tendency to crack and split.

Burmester studied the effect of temperature, pressure and moisture on wood properties (1973). He found that the optimum conditions for pine were 160°C, 20–30% moisture and 0.7 MPa pressure. He reported high-dimensional stability and resistance to brown-rot fungi attack and minimal loss of strength. He named his process as FWD (Feuchte-Wärme-Druck).

More recently, Norimoto et al. (1993) and Ito et al. (1998) heated wet wood to improve dimensional stability. Inoue et al. (1993) heated wet wood to fix compression set in wood.

Several historic names have been given to the various heat-treated products produced in the past including Staypak and Stabwood in the United States (using dry wood); Lignostone and Lignofol in Germany (using dry wood), Jicwood and Jablo in the United Kingdom. New industrial processes are Thermowood from Finland (using water vapor), Plato from The Netherlands (using water), Perdure from Canada (using steam), Oil Heat Treated (OHT) from Germany (using oil) and New Option Wood/Retification from France (using nitrogen) (Hill 2006). These will be described in detail later in this chapter.

14.3 CHEMISTRY

There are a variety of thermal modification processes that have been developed in the laboratory and several have been taken to commercialization. The results of the process depend on several variables including time, temperature, treatment atmosphere, wood species, moisture content, wood dimensions, and the use of a catalyst (Stamm 1956, Millett and Gerhards 1972). Temperature and time of treatment are the most critical elements and treatments done in air result in oxidation reactions not leading to the desired properties of the treated wood. Wood degrades faster when heated in either steam or water as compared with dry conditions (MacLean 1951, 1953, Millett and Gerhards 1972, Hillis 1975, 1984). Stamm and Hansen (1937) and Stamm (1956, 1964) showed that wood heated in the presence of oxygen degraded more rapidly than wood heated in an oxygen-free atmosphere. As the treatment time increases, embrittlement of the wood increases (Yao and Taylor 1979, Edlund 2004). Generally, weight loss occurs to a higher extent in hardwoods as compared with softwoods which may be due to the higher content of acetyl groups in hardwoods releasing more acetic acid during heat treatment contributing to the acid hydrolysis (MacLean 1951, 1953, Millett and Gerhards 1972, Hillis 1975).

14.3.1 WEIGHT LOSS

As the wood is heated, the first weight loss is due to the loss of water, followed by a variety of chemistries that produce degradation products and volatile gasses (Shafizadeh and Chin 1977). As the temperature increases, wood cell wall polymers start to degrade. Degradation of the hemicelluloses starts to take place at a lower temperature as compared to the degradation of cellulose. Lignin is much more stable to high temperature (Stamm 1956). In the early stages of degradation, 83% of the weight loss is water (Mitchell et al. 1953). The hemicelluloses polymers are the first to degrade followed by cellulose resulting in the formation of furans such as furfural and hydroxymethyl furfural and other degradation products such as acetic acid, formic acid, 3-vanilpropanol, catechol, coniferaldehyde, vanillic acid, vanillin, and levoglucosan (Dunlop 1948, Esteves et al. 2007, Rowell et al. 2009). Waste water after thermal treatment of wood shows an increase of both acids and ester compounds and anhydrides and ketones decrease (Biziks et al. 2010).

TABLE 14.1
Weight Loss Due to Heating Wood in the Presence of Moisture[a]

Heating Temperature (°C)	Time (h)	Weight Loss (%)
120	2	2–3
120	10	5
140	1	1
160	1	6
160	3	10
170	1	12
180	1	16
180	10	10–12
190	2	4–7
190	12	5–10
200	10	10–17
220	1	6–7
235	4	25
240	0.5	7

[a] Data taken from several sources.

The acetyl groups in the hemicelluloses side chains are cleaved and the released acetic acid reduces the pH of the wood (Tjeerdsma et al. 1998, Tjeerdsma and Militz 2005, Boonstra and Tjeerdsma 2006). The heating process reduces the pH of the heated wood to 3.5–4.0 due to the production of acetic and formic acids when the starting wood pH was 5.0–5.5 (Boonstra et al. 2007a).

The hemicelluloses and cellulose are also degraded to internal ethers and other rearrangement products (Bächle et al. 2007). The optimum amount of water that can be lost is three molecules of water from two anhydroglucose units or 16.7% of the carbohydrate polymers or 12% of the wood (Stamm and Baechler 1960).

Cellulose is very stable up to about 165°C and when cellulose starts to degrade upon heating, the first chemistry is thought to be one of crystallization of the amorphous cellulose (Bhuiyan and Hirai 2000). Some of the amorphous cellulose does degrade when heated for long periods of time at high temperatures (above 180°C, Boonstra and Tjeerdsma 2006).

Table 14.1 shows the weight loss resulting from heating wood at different temperatures in the presence of moisture. As the temperature increases, so does the weight loss.

Table 14.2 shows the sugar analysis before and after heating aspen with high-pressure steam (220°C) for 8 min (Rowell et al. 2009). The data show that in the very early stages of weight loss, the hemicellulose polymers are breaking down and cellulose remains unchanged. Over half of the arabinan and rhamnans are lost during this early heating.

Table 14.3 shows that even at lower temperatures (82°C), arabinose is lost during heating (Winandy and Rowell 2005). The data indicate that arabinose is the most unstable sugar in the hemicelluloses polymers.

TABLE 14.2
Sugar Analysis of Aspen before and after Heating at 220°C for 8 min

Wt Loss (%)	Total Carbohydrate (%)	Arabian (%)	Galactan (%)	Rhamnan (%)	Glucan (%)	Xylan (%)	Mannan (%)
0	61.8	0.56	0.68	0.30	42.4	16.3	1.6
2.2	59.2	0.20	0.58	0.14	42.2	14.6	1.5

TABLE 14.3

Arabinose Analysis after Heating

Wood at 82°C for up to 160 h

Time (h)	Arabinose Content (%)
0	1.3
7	1.2
2	0.9
60	0.7
160	0.6

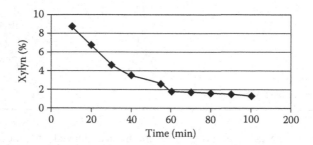

FIGURE 14.1   Loss of xylans from the cell wall as wood is heated at 200°C.

Figure 14.1 shows the loss of xylans as pine wood is heated at 200°C for 0 to 100 h. The loss of xylans, in the early stages of heating, is nearly directly proportional to the heat time. At approximately 60 min, the proportional relationship does not continue. This may indicate that the first xylans lost are much noncrystalline and the later xylans lost are more crystalline in nature.

Similar results were found when heating beech wood at 240°C for 8 h (Dumarçay et al. 2007). The total weight loss after 8 h was 21%. Table 14.4 shows the results of these experiments. The xylose content was reduced over 60%.

During the heating process, the lignin not only starts to degrade but also undergoes condensation (Tjeerdsma et al. 1998, Sivonen et al. 2002, Hakkou et al. 2005). The covalent bonds between lignin and hemicelluloses are broken and lignin fragments are released. Demethylation or more likely demethoxylation of the methoxy groups of the aromatic nuclei of lignin takes place (Tjeerdsma et al. 1998).

Heating both coniferous and deciduous woods from 140°C to 160°C for 1 h showed a decrease in hemicelluloses content (Biziks et al. 2010). Beech wood lost the highest amount of hemicelluloses at all temperatures tested going from a hemicelluloses content of 27% in the control to approximately 4% at 180°C. At 180°C for 1 h, the hemicelluloses content of aspen was 8.6% and for alder 11.9%.

TABLE 14.4

Lignin, Glucose and Xylose Analysis of Control and Beech Heated

for 8 h at 240°C

Sample	Klason Lignin (%)	Glucose (%)	Xylose (%)
Control	22.7 (±0.4)	44.0 (±0.4)	14.2 (±0.3)
Heated (weight loss = 21%)	38.0 (±2.0)	35.5 (±0.6)	5.5 (±0.1)

TABLE 14.5
Changes in Wood Components as a Result of Modification Temperature

Species	Temperature (°C)	Acetone Extractables (%)	Cellulose (%)	Lignin (%)	Other Comp (%)	Cellulose/ Lignin (ratio)
Alder	Control	4.0	47.3	24.7	28.0	1.9
	140	3.4	48.4	25.5	26.1	1.9
	160	9.2	48.2	28.9	22.9	1.7
	170	11.3	51.8	36.4	11.8	1.4
	180	12.8	50.5	37.7	11.8	1.3
Aspen	Control	1.8	53.7	19.7	26.6	2.7
	140	2.7	53.5	19.2	28.3	2.9
	160	10.4	61.3	21.1	17.6	2.9
	170	13.6	60.1	28.9	11.0	2.1
	180	12.8	61.0	30.4	8.6	2.0
Birch	Control	1.7	51.9	20.1	28.0	2.6
	140	2.0	52.8	20.2	27.0	2.6
	160	13.2	63.2	22.5	14.3	2.8
	170	14.1	61.8	30.8	7.5	2.0
	180	14.3	61.7	34.3	4.0	1.8

Andersons et al. studied the change in the acetone-soluble extractives, cellulose, lignin, and the cellulose/lignin ratio in alder, aspen, and birch (2009). Table 14.5 shows that as the temperature goes up, there is an increase in acetone extractables and a relative increase in both cellulose and lignin and a decrease in other compounds. The heating in this table was done for 1 h at each temperature. The ratio of cellulose to lignin stays about the same within each species except at the highest temperature.

Essential changes in the chemical composition of soft deciduous wood start at the modification temperature 160°C and are especially pronounced at 180°C, when relative amounts of cellulose and lignin grow; the cellulose/lignin content decreases.

FTIR spectroscopy is a suitable tool for characterizing lignin after thermal treatments, especially in the case of hardwoods (Windeisen and Wegener 2008).

FTIR spectra for soft deciduous wood (alder, aspen, birch) are similar, but differ quantitatively in terms of the aromatic compounds (lignin) and polysaccharides ratio (Grinins et al. 2009) (Figure 14.2).

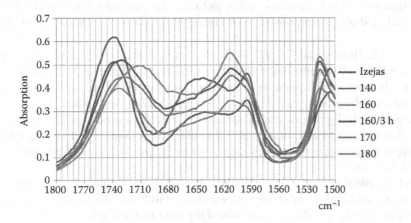

FIGURE 14.2 Example. FTIR spectra of hydrothermally modified birch wood in the range of 1800–1500 cm⁻¹.

The higher is the wood treatment temperature, the more acetate groups (variations in the range 1750–1735 cm⁻¹) are split out and the more wood is oxidized. Oxidation products such as ketones and aldehydes are formed (Tjeerdsma and Militz 2005, Tjeerdsma et al. 1998). After the thermal treatment at 180°C, hemicelluloses acetate groups are split, and the absorption maximum disappears at 1740 cm⁻¹. As a result of hydrothermal modification, double-bond configurations change—the content of isolated double bonds (1680–1620 cm⁻¹) in wood decreases; hence, conjugated double-bond systems are possibly formed. The peak fluctuations typical for lignin in the range from 1520 to 1505 cm⁻¹ are somewhat shifted and become sharper with increasing treatment temperature, probably, as a result of the decomposition of less thermally stable compounds. The higher intensity of the aromatic skeletal vibration and the shift from 1596 to 1612 cm⁻¹ testify the splitting out of the lignin aliphatic side chains and/or crosslinking in condensation reactions. The observed decrease of intensity at 1330 cm⁻¹ and simultaneous increase at 1320 to 1316 cm⁻¹, as well as shift from 1245 to 1220 cm⁻¹, with increasing modification intensity, according to Faix (1992) are also explained by the presence of more condensed structures.

Also the TGA, DTA, and DTG data testify that, with increasing hydrothermal modification (HTM) treatment intensity, more thermally stable (thermally less degradable) structures are formed in wood. The exothermal degradation temperature of the crystalline part of the cellulose isolated from HTM wood grows. The most thermostable lignin is isolated from the wood, modified for 3 h at 160°C and for 1 h at 170°C. The lignin from the wood modified 1 h at 180°C has the maximal oxidation rate temperature; correspondingly, it has become thermally more unstable at this treatment, obviously due to the chemical transformations. Because the HTM wood at 180°C/1 h is most thermally stable, then this could be explained by the growth in the cellulose ordering or also the most ordered common matrix, which is cleaved in the processes of the chemical isolation of individual components of wood.

The effect of the HTM on lignin on the cell level is studied for aspen, birch and alder samples, applying scanning microspectrophotometry (UMSP) (Irbe et al. 2006). The UV points' measurements for untreated wood cell walls show a typical hardwood lignin curve with a pronounced maximum at 278 nm, and a local minimum at about 250 nm, which agrees with other earlier documented hardwood UV microspectrophotometry analyses, for example, for birch (Fergus and Goring 1970) and beech (Takabe et al. 1992).

The applied UV microscopy scanning method enables characterizing the lignin distribution in separate individual cell wall layers. The control samples and those treated at 140°C show a similar clear differentiation of the cell wall layer and do not show any essential difference between the UV absorption in individual cell wall layers.

For hydrothermally treated samples, with increasing temperature from 140°C to 180°C, absorption values grow both in the secondary wall S_2 and the compound middle lamella CML; and for the samples treated at 180°C, absorption is much higher, in comparison with the case of the untreated wood.

According to Kollmann and Fengel (1965), the lignin content grows due to polycondensates. Such compounds for all species under study are formed only between 160°C and 180°C, and a greater growth is observed for aspen, in comparison with the case of gray alder and birch.

The appearance of a pronounced shoulder at 300–310 nm for the samples treated at 180°C testifies that reactions in the lignin side chains occur. Such effects could be caused by the appearance of conjugated carbonyl groups and conjugated α-β-double bonds (Sander and Koch 2001). The HTM microspectrophotometry scans for aspen are shown in Figure 14.3, for alder Figure 14.4, and for birch, Figure 14.5.

The effect of hydrothermal treatment on the cell wall layers for the soft deciduous wood under study is different. Alder, in comparison with aspen and birch, has the highest lignin content in S_2 and CML, which is reflected also in the results of the chemical analysis of lignin. The character of the lignin absorption changes in CML for alder and birch as well as in the secondary wall for birch, is similar, namely, absorption grows linearly in the temperature range of 140–180°C. For aspen, the

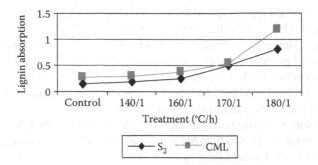

FIGURE 14.3 Hydrothermal modification microspectrophotometry scan of aspen.

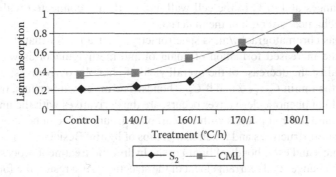

FIGURE 14.4 Hydrothermal modification microspectrophotometry scan of alder.

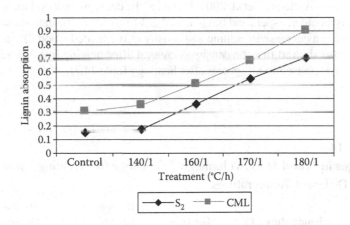

FIGURE 14.5 Hydrothermal modification microspectrophotometry scan of birch.

growth of the lignin sorption in CML in the temperature range 140–170°C is gradual, but increases dramatically, when carrying out HTM at 180°C. For alder, the lignin absorption in the secondary wall grows more dramatically in the temperature range 160–170°C, but, with increasing HTM temperature till 180°C, it does not change any more.

The distinctions between the lignin changes under study for HT-modified hardwoods can be explained by both the distinctions in the secondary walls and the CML components' composition, and the lignin syringyl and guaiacyl structures, respectively, their different thermal stability.

According to Fergus and Goring (1970a) the lignin content in the birch (*Betula verrucosa*) fiber's S_2 layer, middle lamella (ML), and cell corners are 16–19%, 30–40%, and 72–85%, respectively.

It is found that the different anatomical elements of wood contain lignins with a different content of guaiacyl and syringyl groups. According to the data by Fergus and Goring (1970b) already mentioned above, the secondary wall of vessels for birch *Betula papyrifera* contains mainly guaiacyl structures, but the S_2 of libriform fibers and wood rays—syringyl structures; in its turn, the middle lamella between both the fibres contains the lignin of both types.

According to the data by Whiting and Goring (1982), the black spruce S_2 lignin contains more than 1.7 times more methoxyl groups than ML; in its turn, the ML lignin contains three times more carbonyl groups and carboxyl groups than S_2.

Although the data on the lignin content in the ML of the wood of different tree species are different, it is clear that ML contains 60–90% of lignin, and the lignin in middle lamella differs both chemically and morphologically.

The notable changes at 180°C in the cell wall and CML are connected with the considerable decrease in the mechanical strength of the material.

The pyrolytic gas chromatography/mass spectrometry spectra testify that the modified wood is characterized by the increased formation of guaiacol and methylguiacol as well as syringol and methylsyringol against the decrease of the overall yield of phenolic derivatives, which is explained by the cleavage of the lignin C_{arom}- α and ß-C_{aliph} bonds. The main components, owing to which the decrease in the yield of phenolic derivatives occurs, are the derivatives with the unsaturated bond in the aliphatic chain and the compounds with the carbonyl group. This can be explained by the high reactivity of the similar structures and functional groups of lignin. Besides, the hydrolytic activation of the carbonyl groups and ester bonds of lignin in the hydrolytic treatment process probably results in the condensation changes and rearrangement of the structure. As a result, the formation of derivatives without the aliphatic chain (guaiacol and syringol) and their methyl derivatives increases. The yield of the rest of the guaiacyl and syringyl derivatives decreases, because these compounds participate in condensation reactions.

These same authors (Andersons et al. 2009) looked at the decrease in wood mass and volume and decreases in density in alder, aspen, and birch heated at 140–160°C. The data show that as the mass loss increases, there is a decrease in volume and density of the heated wood (Table 14.6).

Chaouch et al. also showed that the density decreased after heating five different wood species (2010). Density was decreased in all species after heating (Table 14.7).

TABLE 14.6

Changes in Wood Mass, Volume, and Density After Heating 1 h at Three Different Temperatures

Species	Temperature (°C)	Mass Losses	Decrease in Volume (%)	Decrease in Density (%)
Alder	140	1.9	0	4.5
	160	6.3	1.9	8.1
	180	14.7	4.9	13.0
Aspen	140	0.8	1.5	1.4
	160	5.0	3.2	5.2
	180	14.2	5.9	11.0
Birch	140	0.6	2.2	1.2
	160	5.2	5.4	4.7
	180	18.0	9.6	11.4

TABLE 14.7

Change in Density Heating Different Species at 230°C

Species	Density Before (kg/m³)	Density After (kg/m³)
Beech	653	591
Poplar	437	394
Ash	675	589
Pine	461	439
Fir	447	410

Zauer and Pfriem (2009) found an increase in porosity and pour diameter and a decrease in density after heating spruce and maple to 180–200°C.

Taking into account the mass loss, loss of hemicelluloses, decrease in density and the increase in porosity on heating wood, Figure 14.6 presents some idea of what the cell wall looks might like as the wood is degraded by heat based on the cell wall model as described by Henriksson in Ek et al., *Ljungberg Textbook* Vol. 1, p. 143 (2006).

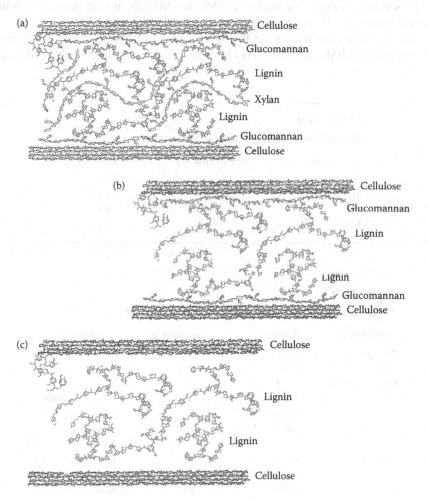

FIGURE 14.6 Schematic of the wood cell wall (a), with the xylans removed (b) and the glucomannans removed (c).

In Figure 14.6a, the control wood has a complete and intact cell wall. As the wood is heated, the hemicelluloses start to decompose and Figure 14.6b shows the cell wall with all the xylans removed. Finally, Figure 14.6c shows the cell wall with all of the xylans and glucomannans removed. Since not all the hemicelluloses are removed during heating, Figure 14.6c depicts an extreme case. The models, however, do show that losing part of the hemicelluloses would increase mass loss, decrease density, and increase porosity. The loss of the hemicelluloses would also decrease moisture sorption, decrease the equilibrium moisture content, and improve dimensional stability.

14.4 MOISTURE AND DIMENSIONAL STABILITY

Because of the loss of hygroscopic hemicelluloses polymers during heat treatment, the hydroxyl content is reduced by 20–40% (Andersons et al. 2007), equilibrium moisture content (EMC) is reduced (Rowell et al. 2009). Dimensional stability of heat-treated wood was first thought to be due to a cross-linking of cellulose chains (Stamm and Hansen 1937). The furan intermediates that form from the decomposition of the hemicelluloses have poor hygroscopicity and have little effect on the EMC.

Table 14.8 shows the reduction in EMC for pine that has been heated to different temperatures for different times.

Table 14.9 shows a summary of results on EMC resulting from heat treatments from various authors. In most cases, a 50% reduction in EMC is observed (Figure 14.7).

Equilibrium moisture content at the air W_{rel} 98% for HTM deciduous wood modifying at 160°C and higher temperatures EMC decreases more than twice, and is similar to that for all species under study.

TABLE 14.8
Reduction in EMC as Wood Is Heated

Temperature (°C)	Time of Heating	EMC at 90% RH
25	0	18.9
220	8 min	13.1
220	1 h	9.3
220	2 h	9.1

TABLE 14.9
Reduction in EMC as a Result of Heating Wood

Wood Species	Conditions	Reduction in	Reference
Pinus pinaster	190°C 8–24 h	50	Esteves et al. (2006)
Pinus sylvestris	220°C 1–3 h	50	Thermowood Handbook (2003)
	Plato process	50-60	Tjeerdsma (2006)
	190°C 8–24 h	50	Kegel (2006) Militz (2002)
Fagus sylvatica	190°C 8–24 h	50	Esteves et al. (2006)
Eucalyptus Globules	190°C 2–24 h	50 2–24 h	Kegel (2006)

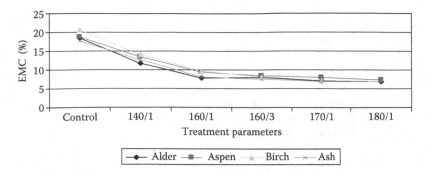

FIGURE 14.7   Equilibrium moisture content (EMC) of heat-treated wood.

Sorption isotherms are also changed as a result of heating wood (Figure 14.8) (Pfriem and Zauer 2009). The adsorption and desorption for heated maple and spruce are reduced as compared to untreated wood. Wood heated to 200°C showed the lowest adsorption and desorption curves. The hysteresis between adsorption and desorption was greatly reduced in spruce when heated at 200°C.

To characterize the microstructure and hygroscopicity in the water vapor medium (Thermowood Finnforest, temperature 180°C and extra 2 days, 220°C) and nitrogen medium (New Option, temperature 200°C) for thermally modified pine (*Pinus sylvestris*), the water vapor sorption method is applied, and sorption–desorption isotherms are measured in three cycles (Chirkova et al. 2005).

A considerable effect of hydrophobization (decrease in the mass concentration of hydrophilic centers by 40–55% depending on the type and temperature of the treatment) has been established. After the second sorption–desorption cycle, the hydrophobization effect decreases 1.5–2 times, then the structure and properties of the samples get stabilized. In our opinion, the distinctions in the structural characteristics of untreated and thermally treated wood show that the primary effect of hydrophobization is caused mainly by the decrease of the surface accessible for water, connected mainly with the inaccessibility of the inner wood surface molecules, and not by the change in their hydrophilic properties. This is indicated also by wide loops of hysteresis for thermally treated samples in comparison with the case of untreated wood. The decrease of accessible specific surface upon heating up to high temperatures proceeds owing to the closing of the entries in the cell wall pores, with the formation of interstructural bonds. At high humidity (close to the pressure of saturated vapors) during the water vapor sorption experiment, these bonds disrupt, and the greater

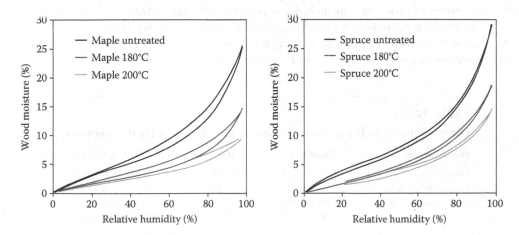

FIGURE 14.8   Sorption isotherms for control and heated maple and spruce. (Reproduced with approval of the authors.)

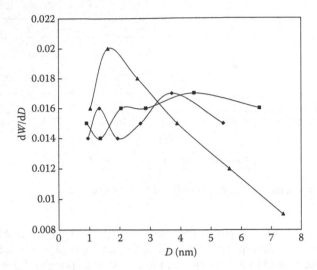

FIGURE 14.9 Effective pore size of the thermo-wood samples: ◆ *Pinus sylvestris*, NOW, 200°C; ■ *Pinus sylvestris*, Finnforest, 180°C; ▲ *Pinus sylvestris*, Finnforest, 220°C.

surface becomes accessible for the sorption of water. Therefore, the sorption in the second cycle grows, and the hysteresis loop narrows.

Similarly, it has been shown by Obataya and Tomita (2002) that thermal treatment has a reversible and irreversible effect on the wood hygroscopicity. They also believe that the modification of the lignin and lignocellulose complex during thermal treatment is mainly responsible for the irreversible decrease in the wood hygroscopicity but the reversible effect is based on the cell wall matrix softening and relaxation at high temperatures, which causes the matrix structure's folding when cooling. As a result, the polymer chains' mobility decreases. At a low relative moisture, the folded structure hampers the water adsorption, while, at a high relative moisture, the material gets moistened, the polymer chains recover the mobility, and water adsorption proceeds again. Irreversible effects are caused by the material's chemical modification.

Figure 14.9 shows the distribution of pore volumes in their sizes for thermotreated samples. A notable change in the microstructure is observed in the case of treatment at a higher temperature (220°C), namely, the wood becomes microporous and has rather homogenous porosity. This result agrees with the conclusion about the changing in the fractal sizes of wood (from 2.5 to about 3) at high temperatures of treatment (Takato Nakano and Junko Miyazaki 2003).

Table 14.10 shows the reduction in swelling of wood that has been heat treated.

As a result of the thermal treatment, treating for 3 h at 160°C or for 1 h at 170°C or 180°C, ASE for deciduous wood grows by 40–60%, in comparison with that for unmodified wood.

TABLE 14.10
Dimensional Stability of Various Woods as a Result of Heat Treatment

Wood Species	Conditions	ASE (%)	Reference
Pinus pinaster	170°C WL[a] 2%	50	Esteves et al. (2006)
Eucalyptus globulus	190°C WL 2%	77	Esteves et al. (2006)
Eucalyptus globulus	200°C WL 10%	60	Esteves et al. (2006)
Pinus sylvestris	220°C	40	Rapp and Sailer (2001)

[a] WL = Weight loss, ASE = antishrink efficiency.

TABLE 14.11

Change in Contact Angle of Aspen and Alder after Heating at Various Temperatures

Sample	Temperature	Contact Angle
Aspen	140	69.3
	160	81.2
	170	85.1
	180	99.0
Alder	140	64.4
	160	91.0
	170	89.4
	180	99.0
Birch	140	46.0
	160	72.7
	170	70.3
	180	81.3

Table 14.11 shows the change in contact angle in water when wood is heated (Sansonetti et al. 2010). Aspen, alder, and birch were heated at 140°C to 180°C for 1 h and the contact angle was measured with water. In general, the heating temperature goes up so does the contact angle showing that the wood surface is becoming more hydrophilic.

14.5 DECAY RESISTANCE

As dimensional stability increases and hygroscopicity decreases in heat-treated wood, resistance to decay also increases (Stamm et al. 1946, Stamm and Bacchler 1960, Rowell 2010). Table 14.12 shows the data of Stamm et al. (1946). A reduction in swelling over 40% results in no weight loss in the decay test. Similar results were obtained when the test fungus was *Lenzites trabea* 517 (Stamm and Bacchler 1960).

Table 14.13 shows fungal resistance of two types of heat-treated wood (Rapp and Sailer 2001). Pine (*Pinus sylvestris*) and spruce (*Picea abies*) were heated in oil or air at three different temperatures and the treated blocks were tested for decay for 19 weeks in the EN 113 (1996) decay test using *Coniophora puteana* fungus. The data show that heating in oil is more effective in reducing attack by fungi than heating in air. This may be due to the presence of the oil in the samples.

Welzbacher and Rapp compared different types of industrially heat-treated products using several different fungi in laboratory tests and in different field and compost conditions. Table 14.14

TABLE 14.12

Relationship between Reduced Swelling and Decay Resistance of White Pine Using *Trametes serialis* Fungus

Reduction in Swelling (%)	Weight Loss due to Decay (%)
30–33	12
33–38	4.5
40 or more	0

Source: Adapted from Stamm, A.J., Burr, H.K., and Kline, A.A. 1946. *Industrial and Engineering Chemistry*, 38(6):630–634.

TABLE 14.13
Fungal Resistance of Pine and Spruce Heated in Oil or Air[a]

Temperature (°C)	Oil Heated		Air Heated	
	Pine Wt Loss (%)	Spruce Wt Loss (%)	Pine Wt Loss (%)	Spruce Wt Loss (%)
Control	40	48		
180	13	15	25.0	31.2
200	1.9	13.1	15.8	26.7
220	2.0	0.0	11.0	5.5

[a] Data taken from several sources.

TABLE 14.14
Weight Loss (%) of Different Woods and Different Heat-Treating Processes[a] Due to Attack by *Poria placenta, Coriolus versicolor,* and *Coniophora puteana*

Material	P. placenta	C. versicolor	C. puteana
Pine (control)	31.0	5.1	47.5
Douglas fir (control)	14.0	2.6	27.4
Oak (ontrol)	0.8	14.3	3.9
Plato (heat treated)	10.0	6.8	3.7
Premium (heat treated)	16.0	9.0	1.9
NOW (heat treated)	13.3	7.8	12.2
OHT (heat treated)	7.4	5.6	3.4

[a] Plato = Heat treated (The Netherlands), Premium = Heat treated (Finland), NOW = New Option Wood, OHT = Oil-heated wood.

shows the weight loss during an EN 113 test using three different fungi. The heat treatment in oil was the most effective but the effect of the oil in the decay test is not known (Welzbacher and Rapp 2007).

Table 14.15 shows the resistance of pine sapwood and heartwood heated to different temperatures and then exposed to several decay fungi in a standard 10-week laboratory test (Metsa-Kortelainen

TABLE 14.15
Changes in Mass of Heat-Treated Pine Sapwood and Heartwood after 10 Weeks in Test

Wood	Temperature (°C)	Coniophora puteana	Poria placenta
Pine sapwood	Untreated	33.6	31.7
	170	24.9	27.2
	190	19.1	23.1
	210	2.3	7.9
	230	0.0	1.7
Pine heartwood	Untreated	32.8	23.8
	170	21.7	20.7
	190	16.1	19.9
	210	2.4	6.6
	230	0.0	0.1

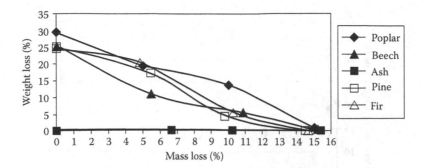

FIGURE 14.10 Weight loss due to heating wood and the weight loss due to attack by *Poria placenta*. (Adapted from Chaouch, M. et al. 2010. In: *Proceedings, The Fifth European Conference on Wood Modification*, Riga, Latvia, pp. 135–142.)

and Viitanen 2010). The data show that both sapwood and heartwood of pine are resistant to decay if the wood was heated above 210°C. There have been other laboratory tests for durability of heat-treated wood using many different wood species (Alfredsen and Westin 2009, Ohnesorge et al. 2009).

Figure 14.10 shows the relationship between weight loss due to heat treatment and mass lost in a laboratory decay test using *Poria placenta* as the test fungi on poplar, beech, ash, pine, and fir (Chaouch et al. 2010). This shows the correlation between the mass lost due to heat wood and the weight lost due to fungal attack as a result of that weight loss. The higher the mass loss, the higher the weight loss due to fungal attack.

Also unpublished results on hydrothermally modified soft deciduous species show a similar relationship between the mass losses as a result of thermal modification and the mass losses after the action of the fungi (EN 113). However, although the mass losses, modifying the samples at the same parameters, are similar, their resistance to rot fungi is different. For example, performing the modification at 160°C for 3 h, the mass losses for alder, aspen, and birch are <2%, but the alder mass losses after the action of rot fungi, according to EN 113 and EN 84, do not exceed 5%, whereas those for aspen and birch reach 10–20%. One of the reasons could be the different composition of the wood components, and its changes as a result of modification.

If the aim of modification is to reach the improved biodurability of wood, then it should be taken into account that the biodurability of the HT-modified soft deciduous wood against different rot fungi is different, which can be well seen from Figures 14.11 through 14.14. The HTM deciduous wood is relatively durable against the brown rot fungus *C. puteana*, but is more intensively degraded by *P. placenta*, as well as the white rot fungus *C. versicolor*. The mass losses after the

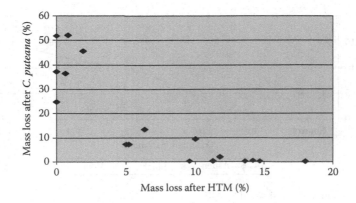

FIGURE 14.11 Mass loss of HTM soft deciduous wood after HTM and *C. puteana*.

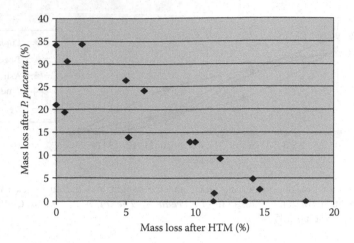

FIGURE 14.12 Mass loss of soft deciduous wood after HTM and *P. placenta*.

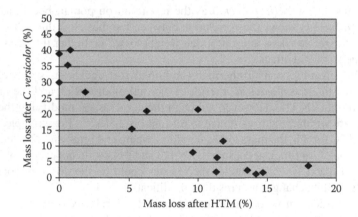

FIGURE 14.13 Mass loss of soft deciduous wood after HTM and *C. versicolor*.

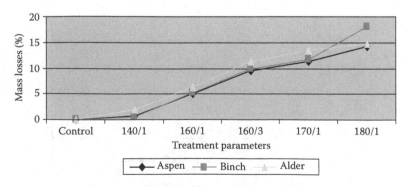

FIGURE 14.14 Mass loss of HTM wood.

TABLE 14.16

Results of In-Ground Tests on Heat-Treated Wood

Wood	Years in Test	Rating[a]
Pine	0	0.0
	1	0.5
	3	1.0
	5	2.1
Spruce	1	0.0
	3	0.1
	5	0.4

[a] Rating: 0 = No attack; 1 = Slight attack; 2 = Moderate attack; 3 = Severe attack; 4 = Destroyed.

---

action of *C. versicolor* and *P. pacenta* are <5%, if the mass losses after modification are 13–15%, correspondingly, the chemical composition of wood is considerably changed. Such mass losses are reached, hydrothermally modifying the wood for 1 h at the temperature 170–180°C. One of the reasons could be the different fungal enzymatic systems and microorganisms' reaction to the modified wood substance. In the studies on the effect of the thermal treatment of beach on the extracellular enzymes of the white-rot fungus *Trametes versicolor* involved in wood degradation, it has been found that the chemical changes in the wood components in the thermal treatment process change the fungal enzymatic system involved in wood degradation and that the improvement of wood durability can be reached only at a sufficient level of chemical changes in wood (Lekounougou et al. 2009).

There have been many reports of heat-treated wood in in-ground tests EN 252 (Alfredsen and Westin 2009, Stingl et al. 2009, Plaschkies et al. 2010). Table 14.16 gives an overview of the results of these tests.

14.6 STRENGTH PROPERTIES

It is well known that heating wood to high temperatures reduces mechanical properties (Stamm et al. 1946, Seborg et al. 1953, Davis and Thompson 1964, Stamm 1964, Rusche 1973, Tjeerdsma et al. 1998, Kubojima et al. 2000, Bekhta and Niemz 2003, Boonstra et al. 2007b). Stamm et al. (1946) heated wood at 320°C for 1 min or 150°C for a week and found a 17% reduction in modulus of rupture (MOR) when heated in molten metal and 50% when heated in air. Under the same conditions, there was less loss in modulus of elasticity (MOE).

Mitchell (1988) found that losses in strength properties were much less when wood was heated dry as opposed to wet, with a greater loss in MOR than in MOE. Rusche (1973) found that strength losses were related to rate and extent of mass loss when heating beech and pine in the presence and absence of air at temperatures ranging from 100°C to 200°C. Losses in maximum strength and work to maximum load were greater for compression than for tension loads. Loss in MOE was only significant when the mass loss was greater than 8–10% and was the same for both species.

Rapp and Sailer (2001) heated pine and spruce at 180–220°C for various times in air and in oil and determined MOR and MOE in a three-point bending test. The highest MOE was 11,000 N/mm^2 in oil heating and little loss in MOE in both air and oil heating. MOR, however, decreased 30% in oil heating and impact bending strength decreased and the wood became brittle. Oil-heated wood lost about 50% and air-heating lost over 70% of the impact strength compared to controls.

TABLE 14.17

Change in MOR and MOE as a Result of Heating Beach Wood for 3 Hours at 190°C

	MOE (N/mm²) Mean	MOR (N/mm²) Maximum	Minimum	Mean	Maximum	Minimum
Sapwood	11359	18470	7177	94	132	66
Heartwood	10765	18250	7200	89	139	49

TABLE 14.18

Change in MOR as a Result of Heat Treatment of Various Woods

Species	Treatment	MOR (%)	Reference
Betula pendula	Vapor, 200°C	−43	Johansson and Moren (2005)
Beech	Vapor 200°C	−50	Yidiz et al. (2006)
Pinus sylvestris	OHT 220°C	−30	Rapp and Sailer (2001)
Pinus pinaster	O₂ 180–200°C	−6–25	Esteves et al. (2006)

Table 14.17 shows the average values for MOR and MOE for beech wood heated for 3 h at 190°C (Todorović et al. 2010). The air-dry density of the sapwood was 0.68 and that of the heartwood was 0.69.

Table 14.18 gives a summary of other data on strength changes in heat-treated wood. In all cases, there was only a small decrease in MOE but major changes in MOR depending on temperature, time, and atmosphere. In some cases, MOE actually increases with heat treatment (Borysiuk et al. 2007, González-Pena and Hale 2007, Kocaefe et al. 2010). In some cases, there is a slight increase in MOE in the early stages of heating which may be due to a relative increase in crystalline cellulose (Boonstra et al. 2007).

Other strength and hardness properties are changed when wood is heated. Tensile strength parallel to the grain and perpendicular to the grain decreases upon heating as well as radial and tangential shear strength (Heräjärvi 2007). Impact strength, compression strength parallel and perpendicular to the grain, work to maximum load, and fiber stress at proportional limit are all reduced (González-Pena and Hale 2007). Finally, hardness is also reduced (González-Pena and Hale 2007, Welzbacher et al. 2010).

The physical damage resulting for heating wood can be clearly seen in Figure 14.15 (Birkinshaw and Dolan 2009). It can be seen that there is extensive damage to the cell wall matrix as a result of

(a)   (b)

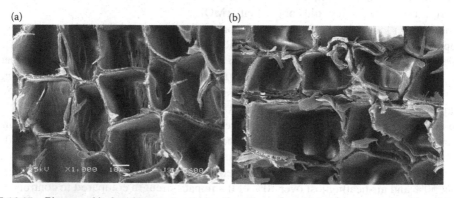

FIGURE 14.15 Pine wood before (a) and after (b) heating. (Reproduced with permission from the authors.)

heating. Matrix integrity is essential for efficient stress transfer. The cell wall fragmentation evident in the micrographs, suggests failure of the binding polymers with loss of stress transfer mechanisms.

Finally, it is possible to approximate the degree of mechanical property loss by measuring the color change in the heated wood (Johansson and Moren 2005, Welzbacher et al. 2010). In general, the darker the color of the wood after heating, the greater the loss of mechanical properties. The color is due to the formation of quinones (Tjeerdsma et al. 1998). The color change also depends on the wood species and is correlated to density as the color gets darker with increasing wood density.

14.7 INDUSTRIAL PROCESSES

14.7.1 THERMOWOOD®

ThermoWood is a registered trademark name that is owned by the Finnish ThermoWood Association which was formed in 2000 (www.thermowood.fi). The process was developed by VVT in Finland and the first commercial plant was built in Mänttä Finland in the early 1990s. The ThermoWood process heats wood in the presence of steam. The steam limits the amount of oxygen that comes in contact with the wood during the process so oxidative degradation is kept to a minimum. The process is broken down in three time/temperature phases. There are two processes that vary only in the temperature in the second phase (Figure 14.16). In both processed, in Phase 1, the wood is heated to 100°C over 6 h then the temperature is increased steadily to 130°C over the next 12 h. In Phase 2, Thermo-S wood is heated to 190°C over 6 h and held there for 7 h. Phase 2 for Thermo-D wood, the temperature is increased to 212°C over 6 h and held for 4 h. In both processes, Phase 3, the heated wood is cooled to 100°C over 6 h, held at that temperature for 6 h and then cooled to room temperature over then next 3 h. The total time for bother processes is just over 42 h (ThermoWood handbook).

Table 14.19 shows the bending and compression strength of Scots pine treated with the Thermo-S and Thermo-D processes as compared with nonheated wood (Rikala et al. 2010).

Figure 14.17 shows the color change in pine as a result of the ThermoWood process.

14.7.2 PLATO® WOOD

Plato Wood (Proving Lasting Advanced Timber Option) was originally developed by Royal Dutch Shell which was part of a research program by them to convert biomass into liquid fuels. The process has four stages: Stage 1 hydro-thermolysis where the wood is heated to 150–180°C in steam under 6–8 bar (0.6–0.8 PMa) for 4–5 h; Stage 2 the heated wood is dried in a conventional kiln for from 5

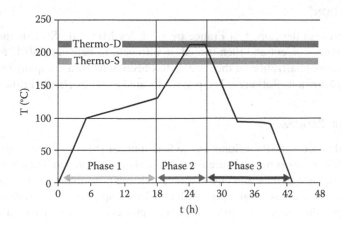

FIGURE 14.16 The Thermalwood® process.

TABLE 14.19
Bending and Compression Strength of ThermoWood®
(Standard Deviation)

Treatment	Bending Strength (MPa)	Compression Strength (MPa)	Wood Density (kg/m³)
Thermo-S	77.4 (13.8)	48.2 (6.0)	407.0 (24.7)
Thermo-D	65.9 (15.3)	50.0 (6.8)	419.6 (38.8)
Control	109.4 (9.7)	57.8 (4.8)	514.3 (24.2)

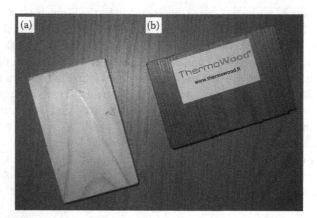

FIGURE 14.17   Control pine (a) and ThermoWood® (b).

days to 3 weeks to a moisture content of 8–10%; Stage 3 the wood is heated again to 150–190°C in dry conditions that leaves the wood at less than 1% moisture; and, Phase 4 a kiln is used to increase the moisture content to about 4–6% over 3 days (www.platowood.nl, Thermowood Handbook 2003).

The Plato Wood plant is located in Arnhem, the Netherlands right next to the Titan Wood acetylation plant and has a plant capacity of 35,000 m³ per year. They mainly treat Spruce, Scots Pine, Douglas fir, poplar, and birch.

14.7.3 RETIFICATION®

Retification is a process developed in France by Écol des Mines de St. Etienne. Wood is dried to about 12% moisture content and then heated, in nitrogen, to between 180°C and 250°C (www.retiwood.com). Commercialization of the process has been done by a company called New Option Wood (NOW, also known as Retitech). The plant is located in La Rochelle, France.

14.7.4 MOLDRUP PROCESS

The hydro-thermal process (SSP for Superheated Steam under Pressure) is executed in an autoclave in stainless steel. It has a working pressure from 0 to 10 bar (abs) (vacuum and pressure). Inside the autoclave fan(s) have been mounted to circulate steam through heating coils to heat the water vapor used for heat transfer as well as a water pipe with nozzles for spraying water into the autoclave various moments during the process (Figure 14.18). The process consists of the following main elements: (a) Initial heating to 80°C at normal atmospheric pressure and with an air/vapor mix circulated by the fan(s) and heated by the heating coils. (b) Vacuum at 80°C for 1 h to suck out

FIGURE 14.18   Wood going into the Moldrup process.

sufficient oxygen of the wood and the plant to avoid self-ignition during the rest of the process. (c) The heating and the vacuum will dry down the wood to a lower moisture content (as in vacuum drying). The transfer of heat from the heating coils to the wood is done by circulating the water vapors either from the wood moisture resulting from the drying or water sprayed into the autoclave in this part of the process. (d) The wood is heated gradually through at least four steps to the working temperature between 160°C and 205°C. The wood moisture will be released from the wood and increase the working pressure up to 8–10 bar pressure, where it leaves the autoclave through a valve system to prevent the pressure from increasing higher. The wood moisture drops to 0–2% moisture content. Once the working temperature has been kept for the time necessary to transfer the heat to the center of the wood pieces and keeping it there for the time necessary to decompose the hemicellulose, the pressure is released. (e) A climatization/cooling of the wood is then carried out by spraying water as a mist through nozzles in the autoclave and circulating the mist with the fans. The pressure is regulated to follow the dropping wood-temperature down to app. 120°C, when the process is ended by letting the pressure drop to atmospheric pressure and doing a vacuum of 30 min to limit the odors from the wood. Process times are generally from 8 to 18 h for poplar, ash, aspen up to 27 mm thickness. Working temperatures are between 160°C and 185°C. Softwoods have process times between 12 and 18 h again for 27 mm material with working temperatures between 185°C and 205°C. The process times depend on the initial moisture content (IMC) and the above cycles are based on wood with IMC between 6% and 12%. Oak is treated at temperatures between 160°C and 185°C with a process time of up to 36 h, again depending on the IMC.

14.7.5   OTHER PROCESSES

The Perdure™ process based in Quebec Canada involves first drying the wood and then heating to 200–230°C in a steam atmosphere (www.perdure.com). The Menz Holz process consists of heating wood with vegetable oil at temperatures in the range of 180–220°C for up to 18 h. The oil excludes oxygen so that there is no or little oxidative degradation. Figure 14.19 shows wood coming out of a completed Men Holz process in Reulbach, Germany (thermomodifiedwood.com/Worldwide/Germany.pdf).

The Royal or Royale Process also uses hot oil but at lower temperatures (60–90°C). The process starts with wet wood, oil is added to a treatment vessel and the temperature is raised to the desired temperature while a vacuum is applied. When the wood has reached the desired moisture content, the oil is removed.

FIGURE 14.19 Wood coming out of a Menz Holz process.

REFERENCES

Alfredsen, G. and Westin, M. 2009. Durability of modified wood—laboratory and field performance. In: *Proceedings, The Forth European Conference on Wood Modification*, Staockholm, Sweden, pp. 515–522.

Andersons, B., Andersone, I., Biziks, V., Irbe, I., Grininsh, J., and Zudrags, K. 2009. Pecularities of the thermal modification of hardwood. In: *Proceedings, The Forth European Conference on Wood Modification*, Staockholm, Sweden, pp. 141–145.

Andersons, B., Andersone, I., Chirkova, J., Suttie, E., and Jones, D. 2007. Changes in the wood cell wall microstructure as a result of modification processes. In: *Proceedings, The Third European Conference on Wood Modification*, Cardiff, UK, pp. 393–400.

Åström, K. 1993. The ski from Kalvträsk. *Västerbotten*, 74(3):129–131.

Bächle, F., Niemz, P., and Schneider, T. 2007. Physical–mechanical properties of hard- and softwood heat treated in an autoclave. In: *Proceedings, The Third European Conference on Wood Modification*, Cardiff, UK, pp. 177–182.

Bekhta, P. and Niemz, P. 2003. Effect of high temperature on the change in color, dimensional stability and mechanical properties of spruce wood. *Holzforschung* 57:539–546.

Bhuiyan, M.T.R. and Hirai, N. 2000. Changes in crystallinity in wood cellulose by heat treatment under dried and moist conditions. *Journal of Wood Science* 46:431–436.

Birkinshaw, C. and Dolan, S. 2009. Mechanism of strength loss in heat treated solftwoods. In: *Proceedings, The Forth European Conference on Wood Modification*, Stockholm, Sweden, pp. 337–343.

Biziks,V., Andersons, B., Andersone, I., Grinins, J., Irbe, I., Kurnosova, N., and Militz, H. 2010. Hydrothermal modification of soft deciduous wood: Bending Strength Properties. In: *Proceedings, The Fifth European Conference on Wood Modification*, Riga, Latvia, pp. 99–106.

Biziks, V., Grinins, J., Andersons, B., Andersone, I., and Dobele, G. 2010. Chemical composition of waste water after hydrothermal treatment of wood. In: *Proceedings, The Forth International Conference on Environmentally-Compatible Forest Products*, Porto, Portugal.

Boonstra, M.J. and Tjeerdsma, B.F. 2006. Chemical analysis of heat treated softwoods. *Holz als Roh und Werkstoff*, 64:204–211.

Boonstra, M.J, Van Acker, J., kegel, E., and Stevens, M. 2007a. Optimization of a two-stage heat treatment process: Durability aspects. *Wood Science and Technology*, 41:31–57.

Boonstra, M.J, Van Acker, J., and Pizzi, A. 2007b. Anatomical and molecular reasons for property changes of wood after full-scale industrial heat treatment. In: Proceedings, *The Third European Conference on Wood Modification*, Cardiff, UK, pp. 343–358.

Borysiuk, P., Mamiński, M., Grześkiewicz, Parzuchowski, P., and Mazurek, A. 2007. Thermally modified wood as raw material for particleboard manufacture. In: *Proceedings, The Third European Conference on Wood Modification*, Cardiff, UK, pp. 227–230.

Burmester, V.A. 1973. Effect of heat-pressure treatments of semi-dry wood on its dimensional stability. *Holz als Roh und Werkstoff*, 31(6):237–243.

Bugge, A. 1953. *Norwegian Stave Churches*. Dreyers Forlag, Oslo.

Chaouch, M., Pétrissans, M., Pétrissans, A., and Gérardin, P. 2010. Prediction of decay resistance of different softwood and hardwood species after heat treatment based on analysis of wood elemental composition. In: *Proceedings, The Fifth European Conference on Wood Modification*, Riga, Latvia, pp. 135–142.

Chirkova, J., Andersons, B., and Andersone, I. 2005. Water sorption properties of thermo-modified wood. *The Second European Conference on Wood Modification*, October 6–7, 2005, Göttingen, Germany, pp. 65–69.

Davis, W.H. and Thompson, W.S. 1964. Influence of thermal treatments of short duration on the toughness and chemical composition of wood. *Forest Products Journal* 14:350–356.

Dumarçay, S., Nguila Inari, G., Mounguengui, S., Pétrissans, M., and Gérardin, P. 2007 Reinvestigations of the wood polymers behavior during heat treatment. In: *Proceedings, The Third European Conference on Wood Modification*, Cardiff, UK, pp. 165–168.

Dunlop, A.P. 1948. Furfural formation and behavior. *Industrial and Engineering Chemistry* 40:204–209.

Edlund, M.-L. 2004. Durability of some alternatives to preservative treated wood. *International Research Group on Wood Preservation*. Doc. No. IRG/WP 04-30353.

EN 113 1996. European Committee for Standardization (CEN), Wood Preservatives—Test method for determining the protective effectiveness against wood destroying basidiomycetes—Determination of toxic values.

EN 252 1989. European Committee for Standardization (CEN), Field test method for determining the relative protective effectiveness of a wood preservative in ground contact.

Esteves, B., Domingos, I., and Pereira, H. 2006. Variation of dimensional stability and durability of eucalypt wood by heat treatment. In: *Proceedings of ECOWOOD 2006, 2nd International Conference on Environmentally Compatible Forest Products*, Oporto, Portugal, pp. 20–22.

Esteves, B.M. and Pereira, H.M. 2009. Wood modification by heat treatment: A Review. *Bioresources* 4(1):370–404.

Esteves, B., Videira, R., and Pereira, H. Composition and ecotoxicity of heat treated pine wood extractives. 2007. In: *Proceedings, The Third European Conference on Wood Modification*, Cardiff, UK, pp. 325–332.

Faix, O. 1992. Fourier Transform Infrared Spectroscopy. In: *Methods in Lignin Chemistry*, S.Y. Lin and C.W.Dence (Eds.), Springer, Berlin, 1992, pp. 83–109.

Fergus B.J. and Goring D.A.I. 1970a. The distribution of lignin in birch wood as determined by ultraviolet microscopy. *Holzforschung*, 24:118–124.

Fergus, B.J. and Goring, D.A.I. 1970b. The location of guiacyl and syringyl lignins in birch xylem tissue. *Holzforschung*, 24:113–117.

González-Pena, M.M. and Hale, M.D.C. 2007. The relationship between mechanical performance and chemical changes in thermally modified wood. In: *Proceedings, The Third European Conference on Wood Modification*, Cardiff, UK, pp. 169–172.

Grinins, J., Biziks, V., Andersons, B., and Andersone, I. 2009. Changes in the chemical composition of soft deciduous wood after thermal treatment. *Proc. of the 5th Meeting of the Nordic Baltic Network in Wood Materials Sciences & Engineering, (WSE)*, October 1–2, 2009, Copenhagen, Denmark, pp. 41–47.

Hakkou, M., Pétrissans, M., Zoulalian, A., and Géradin, P. 2005. Investigations of the reasons for the increase of wood durability after heat treatment based on changes of wettability and chemical composition. In: *Proceedings, The Second European Conference on Wood Modification*, Gottingen, Germany, pp. 38–46.

Heräjärvi, H. 2007. Shear and tensile strength of conventionally dried, press dried and heat treated aspen. In: *Proceedings, The Third European Conference on Wood Modification*, Cardiff, UK, pp. 173–176.

Hill, C.A.S. 2006. *Wood Modification: Chemical, Thermal and other Processes*. John Wiley & Sons, Chichester, England, 239 pp.

Hillis, W.E. 1975) The role of wood characteristics in high temperature drying. *Journal of the Institute of Wood Science*, 7(2):60–67.

Hillis, W.E. 1984. High temperature and chemical effects on wood durability. *Wood Science Technology*, 18:281–293.

Holan, J. 1990. *Norwegian Wood: A Tradition of Building*. Rizzoli International Publications, Inc., New York, ISBN 0-8478-0954-4.

Inoue, M., Norimoto, M., Tanahashi, M., and Rowell, R.M. 1993) Steam or heat fixation of compressed wood. *Wood and Fiber Science*, 25(3):224–235.

Insulander, R. 1999. Reconstruction of the saami bow: A comparison between findings in Sweden, Norway and Finland. *Fornvanen*, 94(2):73–87.

Irbe, I., Noldt, G., Koch, G., Andersone, I., and Andersons, B. 2006. Application of scanning UV microspectrophotometry for the topochemical detection of lignin in individual cell wall layers of brown rotted Scots pine sapwood. *Holzforschung*, 60:601–607.

Ito, Y., Tanahashi, M., Shigematsu, M., and Shinoda, Y. 1998. Compressive-molding of wood by high pressure-steam treatment: Part 2. Mechanism of permanent fixation. *Holzforschung*, 52(2):217–221.

Johansson, D. and Moren, T. 2005. The potential of colour measurement for strength prediction of thermally treated wood. *Holz als Roh und Werkstoff*, 64:104–110.

Kilby, K. 1971. *The Cooper and his Trade*. John Baker Ltd., London, ISBN 0-212-98399-7.

Kegel, E. 2006. *The Plato Technology—A Novel Wood Upgrading Technology*, Plato International BV, www.platowood.nl.

Lekounougou, S., Pétrissans, M., Jacquot, J.P., Gelhaye, E., and Gérardin, P. 2009. Effect of heat treatment on extracellular enzymatic activities involved in beech wood degradation by *Trametes versicolor. Wood Science and Technology*, 43 (3–4):331–341.

Kocaefe, D., Kocaefe, Y., Poncsak, S., Lekounougou, S.T., Younsi, R., and Oumarou, N. 2010. In: *Proceedings, The Forth International Conference on Environmentally-Compatible Forest Products*, Porto, Portugal.

Kollmann, F. and Fengel, D. 1965. Änderungen der chemischen Zusammensetzung von Holz durch thermische Behandlung. *Holz als Roh und Werkstoff*, 23:461–468.

Kubojima, Y., Okano, T., and Ohta, M. (2000). Bending strength and toughness of heat treated wood. *Journal of Wood Science*, 46:8–15.

MacLean, J.D. 1951. Rate of disintegration of wood under different heating conditions. *American Wood Preservers' Association Proceedings*, 47:155–168.

MacLean, J.D. 1953. Effect of steaming on the strength of wood. *American Wood Preservers' Association Proceedings*, 49:88–112.

Metsä-Kortelainen, S. and Viitanen, H. 2010. Decay resistance of sapwood and heartwood of thermally modified Scots pine and Norway spruce. In: *Proceedings, The Fifth European Conference on Wood Modification*, Riga, Latvia, pp. 171–174.

Militz, H. 2002. Heat treatment technologies in Europe: Scientific background and technological state-of-art. In: *Proceedings, Conference on Enhancing the Durability of Lumber and Engineering Wood Products*, Kissimmee, Orlando, FL, pp. 11–13.

Millett, M.A. and Gerhards, C.C. 1972. Accelerated aging: Residual weight and flexural properties of wood heated in air at 115°C to 175°C. *Wood Science*, 4(4):193–201.

Mitchell, P.H. 1988. Irreversible property changes of small loblolly pine specimens heated in air, nitrogen, or oxygen. *Wood and Fiber Science*, 20(3):320–355.

Mitchell, R.L., Seborg, R.M., and Millett, M.A. 1953. Effect of heat on the properties and chemical composition of Douglas-fir wood and its major components. *Journal of the Forest Products Research Society*, 3(4):38–42.

Norimoto, M., Ota, C., Akitsu, H., and Yamada, T. 1993. Permanent fixation of bending deformation in wood by heat treatment. *Wood Research*, 79:23–33.

Obataya, E. and Tomita, B. 2002. Hygroscopicity of heat-treated wood II: Reversible and irreversible reductions in the hygroscopicity of wood due to heating. *Mokuzai Gakkaishi*, 48(4): 288–295.

Ohnesorge, D., Tausch, A., Krowas, I., Huber, C., Becker, G., and Fink, S. 2009. Labaoratory tests on the natural durability of six different wood species after hygrothermal treatment. In: *Proceedings, The Forth European Conference on Wood Modification*, Stockholm, Sweden, pp. 159–164.

Olsen, O. and Crumlin-Pedersen, O. 1967. A report of the final underwater excavation in 1959 and the salvaging operation in 1962. *Acta Archaeologicas*, 38:73–174.

Ostergard, D.E. 1987. *Bent Wood and Metal Furniture 1850–1946*. The American Federation of Arts, New York, ISBN 0-295-96409-X.

Pfriem, A. and Zauer, M. 2009. In: *Proceedings, The Forth European Conference on Wood Modification*, Stockholm, Sweden, pp. 363–370.

Plaschkies, K., Scheiding, W., Jacobs, K., and Weiss, B. 2010. Durability of thermally modified timber assortments against fungi—Results of a 6 year field test in comparison with results from lab tests. In: *Proceedings, The Fifth European Conference on Wood Modification*, Riga, Latvia, pp. 119–126.

Rapp, A.O. and Sailer, M. 2001. Oil heat treatment of wood in Germany—State of the art. Environmental optimization of wood protection. Cost Action E22. In: *Proceedings of Special Seminar*, Antibes, France.

Rikala, J. Havimo, M., and Sipi, M. 2010. Variation in strength properties of Scots pine wood after two different heat treatments. In: *Proceedings, The Sixth Nordic-Baltic Network in Wood Material Science and Engineering*, Tallinn, Estonia, pp. 126–132.

Rivers, S. and Umney, N. 2005. *Conservation of Furniture*. Elsevier, Butterworth-Heinemann, Oxford, ISBN 0-7506-09583.

Rowell, R.M. 2010. Heat treatments of wood to improve decay resistance. In: *Proceedings, The Sixth Nordic-Baltic Network in Wood Material Science and Engineering*, Tallinn, Estonia, pp. 23–33.

Rowell, R.M., Ibach, R.E., McSweeny, J., and Nilsson, T. 2009. Understanding decay resistance, dimensional stability and strength changes in heat treated and acetylated wood. In: *Proceedings, The Forth European Conference on Wood Modification*, Stockholm, Sweden, pp. 489–502.

Rusche, V.H. 1973. Thermal degradation of wood at temperatures up to 200°C—Part I: Strength properties of dried wood after heat treatment. *Roh- und Werkstoff*, 31:273–281.

Sander, C. and Koch, G. 2001. Effects of acetylation and hydrothermal treatment on lignin as revealed by cellular UV-spectroscopy in Norway spruce (*Picea abies L. Karst.*). *Holzforschung*, 55:1–6.

Sansonetti, E., Andersons, B., Biziks, V., Grinins, J., and Chircova, J. 2010. Surface properties of the hydrothermally modified soft deciduous wood. In: *Proceedings, The Fifth European Conference on Wood Modification*, Riga, Latvia, pp. 183–186.

Seborg, R.M., Tarkow, H., and Stamm, A.J. 1953. Effects of heat upon the dimensional stabilization of wood. *Journal Forest Products Research Society*, 56:648–654.

Shafizadeh, F. and Chin, P.P.S. 1977. Thermal degradation of wood. In: *Wood Technology: Chemical Aspects*, I.S. Goldstein (Ed.), ACS Symposium Series 43:57–81.

Sivonen, H., Maunu, S.L., Sundholm, F., Jämsä, S., and Viitaniemi, P. 2002. Magnetic resonance studies of thermally modified wood. *Holzforschung*, 56:648–654.

Stamm, A.J. 1956. Thermal degradation of wood and cellulose. *Industrial and Engineering Chemistry*, 48(3):413–417.

Stamm, A.J. 1964. *Wood and Cellulose Science*. Ronald Press, New York, NY.

Stamm, A.J. and Hansen, I.A. 1937. Minimizing wood shrinkage and swelling: Effects of heating in various gasses. *Industrial and Engineering Chemistry*, 29:831–833.

Stamm, A.J. and Baechler. R.H. 1960. Decay resistance and dimensional stability of five modified woods. *Forest Products Journal*, 10:22–26.

Stamm, A.J., Burr, H.K., and Kline, A.A. 1946. Staybwood. heat stabilized wood. *Industrial and Engineering Chemistry*, 38(6):630–634.

Stamm, A.J. and Hansen, L.A. 1937. Minimizing wood shrinkage and swelling. Effect of heating in various gasses. *Industrial and Engineering Chemistry*, 29(7):831–833.

Stingl, R., Weigl, M., Teischinger, A., and Hansmann, C. 2009. Moderate thermal treated Norway spruce (*Picea abies*(L) exposed to ground contact in Austria for five years. In: *Proceedings, The Forth European Conference on Wood Modification*, Stockholm, Sweden, pp. 165–168.

Takabe, K.S., Miyauchi, R., Tsunoda, R., and Fukazaw, K. 1992. Distribution of guiacyl and syringyl lignins in Japanese beech (*Fagus crenata*): Variation within annual ring. *IAWA Bulletin* 13(1):105–112.

Takato Nakano and Junko Miyazaki. 2003. Surface fractal dimensionality and hygroscopicity for heated wood. *Holzforschung*, 57(3):289–294.

Tiemann, H.D. 1915. The effect of different methods of drying on the strength of wood. *Lumber World Review*, 28(7):19–20.

Thermowood Handbook 2003. Finnish thermowood association, c/o Wood Focus Oy, P.O. Box 284, Snellmaninkatu 13, Helsinki, Finland, FIN-00171.

Tjeerdsma, B.F. 2006. *Heat Treatment of Wood—Thermal Modification*. SHR Timber Research, Wageningen, The Netherlands.

Tjeerdsma, B.F., Boonstra, M., Pizzi, A., Tekely, P., and Militz, H. 1998. Characterization of thermally modified wood: Molecular reasons for wood performance improvement. *Holz als Roh- und Werstoff*, 56:149–153.

Tjeerdsma, B.F. and Militz, H. 2005. Chemical changes in hydrothermal treated wood: FTIR analysis of combined hydrothermal and dry heat-treated wood. *Holz als Roh- und Werstoff*, 63:102–111.

Todorović, N., Schdwanninger, M., and Popović, Z. 2010. Prediction of mechanical properties of thermally modified beech wood by use of near infred spectroscopy. In: *Proceedings, The Fifth European Conference on Wood Modification*, Riga, Latvia, pp. 187–190.

Twede, D. 2005. The cask age: The technology and history of wooden barrels. *Packing Technology and Science*, 18(5):253–264.

Welzbacher, C.R., Meyer, L., and Brischke, C. 2010. Prediction of flooring-relevant properties based on colour values of thermally modified timber (TMT). In: *Proceedings, The Fifth European Conference on Wood Modification*, Riga, Latvia, pp. 143–150.

Welzbacher, C.R. and Rapp, A. 2007. Durability of thermally modified timber from industrial-scale processes in different use classes: Results from laboratory and field tests. *Wood Material Science and Engineering*, 2:4–14.

Winandy, J. and Rowell, R.M. 2005. Chemistry of wood strength. *Handbook of Wood Chemistry and Wood Composites*, Ed, R.M. Rowell. Taylor & Francis, Chapter 11, pp. 303–349.

Windeisen, E. and Wegener, G. 2008. Behaviour of lignin during thermal treatments of wood. *Ind. Crops & Products*, 27 (2):157–162.

Whiting, P. and Goring, D.A.I 1982. Chemical characterization of tissue fractions from the middle lamella and secondary wall of black spruce tracheids. *Wood Science and Technology*, 16(4):261–267.

Yao, J. and Taylor, F. 1979. Effect of high-temperature drying on the strength of southern pine dimension lumber. *Forest Products Journal*, 29(8):49–51.

Yidiz, S, Gezer, E.D., and Yildiz, U.C. 2006. Mechanical and chemical behavior of spruce wood modified by heat. *Building and Environment*, 41:1762–1766.

Zauer, M. and Pfriem, A. 2009. Alteration of the pore structure of spruce and maple due to thermal treatment. In: *Proceedings, The Forth European Conference on Wood Modification*, Stockholm, Sweden, pp. 169–172.

15 Chemical Modification of Wood

Roger M. Rowell

CONTENTS

15.1 DEGRADATION OF WOOD

For the most part, we have designed and used wood putting up with the "natural defects" that nature has given to us, such as dimensional instability and the degradations due to weathering, fire, and decay. Wood is a hygroscopic resource that was designed to perform, in nature, in a wet environment. Nature is programmed to recycle wood in a timely way through biological, thermal, aqueous, photochemical, chemical, and mechanical degradations. In simple terms, nature builds wood from carbon dioxide and water and has all the tools to recycle it back to the starting chemicals. We harvest a green tree and convert it into dry products, and nature, with its arsenal of degrading chemistries, starts to reclaim it at its first opportunity.

The properties of any resource are, in general, a result of the chemistry of the components of that resource. In the case of wood, the cell wall polymers (cellulose, hemicelluloses, and lignin) are the components that, if modified, would change the properties of the resource. If the properties of the wood are modified, the performance of the modified wood would be changed. This is the basis of chemical modification of wood to change properties and improve performance. This idea is applied to both solid wood and wood composites.

In order to produce wood-based materials with a long service life, it is necessary to interfere with the natural degradation processes for as long as possible (Figure 15.1). This can be done in several ways. Traditional methods for decay resistance and fire retardancy, for example, are based on treating the product with toxic or corrosive chemicals which are effective in providing decay and fire resistance but can result in environmental concerns. In order to make property changes, you must first understand the chemistry of the components and the contributions each play in the properties of the resource. Following this understanding, you must then devise a way to modify what needs to be changed to get the desired change in property.

Properties of wood, such as dimensional instability, flammability, biodegradability, and degradation caused by acids, bases, and ultraviolet radiation are all a result of chemical degradation reactions

Biological degradation	Fungi, bacteria, insects, termites
Enzymatic reactions	Oxidation, hydrolysis, reduction
Chemical reactions	Oxidation, hydrolysis, reduction
Mechanical	Chewing
Fire degradation	Lightning, sun
Pyrolysis reactions	Dehydration, hydrolysis, oxidation
Water degradation	Rain, sea, ice, acid rain, dew
Water interactions	Swelling, shrinking, freezing, cracking
Weather degradation	Ultraviolet radiation, water, heat, wind
Chemical reactions	Oxidation, hydrolysis
Mechanical	Erosion
Chemical degradation	Acids, bases, salts
Chemical reactions	Oxidation, reduction, dehydration, hydrolysis
Mechanical degradation	Dust, wind, hail, snow, sand
Mechanical	Stress, cracks, fracture, abrasion

FIGURE 15.1 Degradation reactions which occur when wood is exposed to nature.

Biological degradation

    Hemicelluloses → Accessible cellulose → Noncrystalline cellulose → Crystalline cellulose → Lignin

Moisture sorption

    Hemicelluloses → Accessible cellulose → Noncrystalline cellulose → Lignin → Crystalline cellulose

Ultraviolet degradation

    Lignin → Hemicelluloses → Accessible cellulose → Noncrystalline cellulose → Crystalline cellulose

Thermal degradation

    Hemicelluloses → Cellulose → Lignin

Strength

    Crystalline cellulose → Matrix (Noncrystalline cellulose + Hemicelluloses + Lignin) → Lignin

FIGURE 15.2   Cell wall polymers responsible for the properties of wood.

which can be prevented or, at least, slowed down if the cell wall chemistry is altered (Rowell 1975a, 1983, 1992, Rowell and Youngs 1981, Rowell and Konkol 1987, Rowell et al. 1988a, Hon 1992, Kumar 1994, Banks and Lawther 1994, Hill 2006).

Figure 15.2 shows the cell wall polymers involved in each property as we understand it today (Rowell 1990). Wood changes dimensions with changing moisture content because the cell wall polymers contain hydroxyl and other oxygen-containing groups that attract moisture through hydrogen bonding (Stamm 1964, Rowell and Banks 1985) (see Chapter 4). The hemicelluloses are mainly responsible for moisture sorption, but the accessible cellulose, noncrystalline cellulose, lignin, and surface of crystalline cellulose also play major roles. Moisture swells the cell wall, and the fiber expands until the cell wall is saturated with water (fiber saturation point, FSP). Beyond this saturation point, moisture exists as free water in the void structure and does not contribute to further expansion. This process is reversible, and the fiber shrinks as it loses moisture below the FSP.

Wood is degraded biologically because organisms recognize the carbohydrate polymers (mainly the hemicelluloses) in the cell wall and have very specific enzyme systems capable of hydrolyzing these polymers into digestible units (Figure 15.3, see Chapter 5). Biodegradation of the cell wall matrix and the high molecular weight cellulose weakens the fiber cell (Rowell et al. 1988b). Strength is lost as the cell wall polymers and matrix undergoes degradation through oxidation, hydrolysis, and dehydration reactions.

Wood exposed outdoors undergoes photochemical degradation caused by ultraviolet radiation (Figure 15.4, see Chapter 7). This degradation takes place primarily in the lignin component, which is responsible for the characteristic color changes (Rowell 1984). The lignin acts as an adhesive in the cell walls, holding the cellulose fibers together. The surface becomes richer in cellulose content as the lignin degrades. In comparison to lignin, cellulose is much less susceptible to ultraviolet light degradation. After the lignin has been degraded, the poorly bonded carbohydrate-rich fibers erode easily from the surface, which exposes new lignin to further degradative reactions. In time, this "weathering" process causes the surface of the composite to become rough and can account for a significant loss in surface fibers.

Wood burn because the cell wall polymers undergo pyrolysis reactions with increasing temperature to give off volatile, flammable gases (Figure 15.5, see Chapter 6). The gases are ignited by some external source and combust. The hemicelluloses and cellulose polymers are degraded by heat much before the lignin (Rowell 1984). The lignin component contributes to char formation, and the charred layer helps insulate the composite from further thermal degradation.

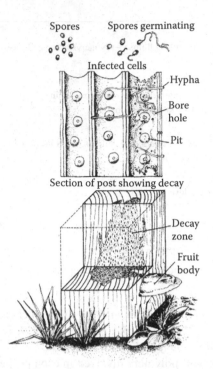

FIGURE 15.3 Wood, in ground contact, attacked by microorganisms.

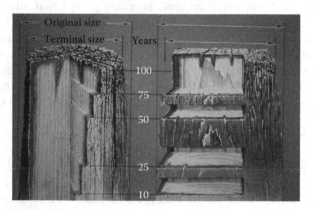

FIGURE 15.4 Wood loss in the weathering of wood.

15.2 CHEMICAL REACTION SYSTEMS

The term "chemical modification" has been used to mean different things by different authors over the years. For this chapter, chemical modification will be defined as a chemical reaction between some reactive part of wood and a simple single chemical reagent, with or without catalyst, to form a covalent bond between the two. This excludes all simple chemical impregnation treatments, which do not form covalent bonds, monomer impregnation's that polymerize *in situ* but do not bond with the cell wall, polymer inclusions, coatings, heat treatments, and so on.

There are several approaches to chemically modifying the wood cell wall polymers. The most abundant single site for reactivity in these polymers is the hydroxyl group and most reaction schemes have been based on the reaction of hydroxyl groups. Sites of unsaturation in the lignin structure can also be used as a point of reactivity as well as free radical additions and grafting. However, the most studied class of chemical reactions are those involving hydroxyl substitutions.

FIGURE 15.5 Wood burning.

In modifying wood for property improvement, there are several basic principles that must be considered in selecting a reagent and a reaction system. Of the thousands of chemicals available, either commercially or by synthetic means, most can be eliminated because they fail to meet the requirements or properties listed below.

If hydroxyl reactivity is selected as the preferred modification site, the chemical must contain functional groups, which will react with the hydroxyl groups of the wood components. This may seem obvious but there are several failed reaction systems in the literature using a chemical that could not react with a hydroxyl group.

The overall toxicity of the chemicals must be carefully considered. The chemicals must not be toxic or carcinogenic to humans in the finished product, and should be as nontoxic as possible in the treating stage. The chemical should be as noncorrosive as possible to eliminate the need for special stainless steel or glass-lined treating equipment.

In considering the ease with which excess reagents can be removed after treatment, a liquid treating chemical with a low boiling point is advantageous. Likewise, if the boiling point of a liquid reagent is too high, it will be very difficult to remove the chemical after treatment. It is generally true that the lowest member of a homologous series is the most reactive and will have the lowest boiling point. The boiling point range for liquids to be considered is 90–150°C. In some cases, the lowest member of a homologous series is a gas. It is possible to treat wood with a gas system, however, there may be processing challenges in handling a pressurized gas in a continuous reactor. Gases do not penetrate deeply or quickly into the wood cell wall so penetration of the reaction system may be limited.

Accessibility of the reagent to the reactive chemical sites is a major consideration. In some cases, this may be the rate limiting step in a reaction system. To increase accessibility to the reaction site, the chemical should be able to swell the wood structure. If the reagents do not swell the structure, then another chemical or co-solvent can be added to aid the penetration of chemicals. Accessibility to the reactive site is a major consideration in a gas system unless there is a condensation step in the procedure.

If the chemical reaction system requires a catalyst, a strong acid or base catalyst cannot be used as they cause extensive degradation. The most favorable catalyst from the standpoint of wood degradation is a weakly alkaline one. The alkaline medium is also favored as in many cases these chemicals swell the cell wall matrix structure and give better penetration. The properties of the catalyst parallel those of reagents, that is, low boiling point liquid, nontoxic, effective at low temperatures, and so on. In most cases, the organic tertiary amines or weak organic acids are best suited.

The experimental reaction condition which must be met in order for a given reaction to go is another important consideration. If the reaction time is long, the temperature required for complete reaction must be low enough so there is little or no fiber degradation. The reaction must also have a relatively fast rate of reaction with the cell wall components. It is important to get as fast a reaction as possible at the lowest temperature without wood degradation. High-temperature reactions are possible (up to 170°C) if the reaction time is very short and no strong acid or base catalysts are used. Wood degrades rather quickly at temperatures above 175°C (Stamm 1964) (see Chapter 6).

The moisture present in the wood is another consideration in the reaction conditions. It is costly to dry wood to less than 1% moisture, but it must be remembered that the –OH group in water is more reactive than the −OH group available in the wood components, that is, hydrolysis is faster than substitution. The most favorable condition is a reaction which requires a trace of moisture and the rate of hydrolysis is relatively slow.

Another consideration in this area is to keep the reaction system as simple as possible. Multicomponent systems will require complex separation after reaction for chemical recovery. The optimum would be a reactive chemical that swells the wood structure and acts as the solvent and catalyst as well.

If possible, avoid byproducts during the reaction that have to be removed. If there is not a 100% reagent skeleton add-on, then the chemical cost is higher and will require recovery of the by-product for economic and environmental reasons.

The chemical bond formed between the reagent and the wood components is of major importance. For permanence, this bond should have great stability to withstand natures recycling system if the product is used out doors. In order of stability, the types of covalent chemical bonds that may be formed are: ethers > acetals > esters. The ether bond is the most desirable covalent carbon–oxygen bond that can be formed. These bonds are more stable than the glycosidic bonds between sugar units in the wood polysaccharides so the polymers would degrade before the grafted ether. It may be desired, however, to have the bonded chemical released by hydrolysis or enzyme action in the final product so that an unstable bond may be required from the modification.

The hydrophobic nature of the reagent needs to be considered. The chemical added to the wood should not increase the hydrophilic nature of the wood components unless that is a desired property. If the hydrophilicity is increased, the susceptibility to microorganism attack increases. The more hydrophobic the component can be made, the better the moisture exclusion properties of the substituted wood will be.

Single-site substitution versus polymer formation is another consideration. For the most part, a single reagent molecule that reacts with a single hydroxyl group is the most desirable. Cross-linking can occur when the reagent contains more than one reactive group or results in a group, which can further react with a hydroxyl group. Cross-linking can cause the wood to become more brittle. Polymer formation within the cell wall after initial reaction with the hydroxyl groups of the wood components gives, through bulking action, dimensional stabilization. The disadvantage of polymer formation is that a higher level of chemical add-on is required for biological resistance than is required in the single-site reactions.

The treated wood must still possess the desirable properties of wood. That is, strength should not be reduced, no change in color, good electrical insulation properties retained, final product not dangerous to handle, no lingering chemical smells, still gluable and finishable unless one or more of these properties are the object of change in the product.

A final consideration is, of course, the cost of chemicals and processing. In laboratory scale experimental reactions, the high cost of chemicals is not a major factor. For commercialization of a process, however, the chemical and processing costs are very important factors. Laboratory scale research is generally done using small batch processing, however, rapid, continuous processes should always be studied for scale up. Economy of scale can make an expensive laboratory process economical.

In summary, the chemicals to be laboratory tested must be capable of reacting with wood hydroxyls under neutral, mildly alkaline or acid conditions at temperatures below 170°C. The chemical system should be simple and capable of swelling the structure to facilitate penetration. The complete molecule should react quickly with wood components yielding stable chemical bonds, and the treated wood must still possess the desirable properties of untreated wood.

15.3 CHEMICAL REACTIONS WITH WOOD

As was stated before, because the properties of wood result from the chemistry of the cell wall components, the basic properties of wood can be changed by modifying the basic chemistry of the cell wall polymers. There are several approaches to cell wall modification depending on what property is to be modified. For example, if the objective is water repellency, then the approach might be to reduce the hydrophilic nature of the cell wall by bonding on hydrophobic groups. If dimensional stability is to be improved, the cell wall can be bulked with bonded chemicals, or cell wall polymer components cross-linked together to restrict cell wall expansion, or groups can be bonded that reduce hydrogen bonding or increase hydrophobicity. If fire retardancy is desired, chemical groups can be bonded onto cell wall polymers that contain retardants or flame suppressants. If resistance to ultraviolet radiation is desired, UV blockers or absorbers could be bonded to lignin. The chemical modification system selected must, therefore, perform the desired chemical change to result in the desired change in performance.

Many chemical reaction systems have been published for the modification of wood and these systems have been reviewed in the literature several times in the past (Rowell 1975, 1983, 1991, 1999, 2006; Kumar 1994; Hon 1996; Hill 2006). These chemicals include anhydrides such as, phthalic, succinic, malaic, propionic and butyric anhydride, acid chlorides, ketene carboxylic acids, many different types of isocyanates, formaldehyde, acetaldehyde, difunctional aldehydes, chloral, phthaldehydic acid, dimethyl sulfate, alkyl chlorides, β-propiolactone, acrylonitrile, epoxides, such as, ethylene, propylene, and butylene oxide, and difunctional epoxides.

15.3.1 ACETYLATION

The acetylation of wood was first performed in Germany in 1928 by Fuchs, using acetic anhydride and sulfuric acid as a catalyst (Fuchs 1928). Fuchs found an acetyl weight gain of over 40%, which meant that he decrystallized the cellulose in the process. He used the reaction to isolate lignin from pine wood. In the same year, Horn (1928) and Suida and Titsch (1928) acetylated beech wood to remove hemicelluloses in a similar lignin isolation procedure. A year later, Suida and Titsch (1929) acetylated powdered beech and pine using pyridine or dimethylaniline as a catalyst to yield an acetyl weight gain of 30–35% after 15–35 days at 100°C. In 1945, Tarkow first demonstrated that acetylated balsa was resistant to decay (Tarkow 1945). In 1946, Tarkow, Stamm, and Erickson (Tarkow 1946, Tarkow et al. 1950) first described the use of wood acetylation to stabilize wood from swelling in water.

Acetylation of wood using acetic anhydride has mainly been done as a liquid phase reaction. The early work was done using acetic anhydride catalyzed with either zinc chloride (Ridgway and Wallington 1946) or pyridine (Stamm and Tarkow 1947). Through the years, many other catalysts have been tried both with liquid and vapor systems. Some of the catalyst used include urea–ammonium sulfate (Clermont and Bender 1957), dimethylformamide (Clermont and Bender 1957, Risi and Arseneau 1957a, Baird 1969), sodium acetate (Tarkow 1959), magnesium persulfate (Arni et al. 1961a,b, Ozolina and Svalbe 1972, Truksne and Svalbe 1977), trifluoroacetic acid (Arni et al. 1961a,b), boron trifluoride (Risi and Arseneau 1957a) and g-rays (Svalbe and Ozolina 1970). The reaction has also been done without catalyst (Rowell et al. 1986b) and by using an organic cosolvent (Goldstein et al. 1961). Gas phase reactions have also been reported using acetic anhydride, however, the diffusion rate is very slow so this technology has only been applied to thin veneers (Tarkow et al. 1950, Tarkow 1959, Baird 1969, Avora et al. 1979, Rowell et al. 1986a).

The reaction with acetic anhydride results in esterification of the accessible hydroxyl groups in the cell wall with the formation of byproduct acetic acid (Rowell et al. 1975a).

$$WOOD-OH + CH_3-C(=O)-O-C(C=O)-CH_3 \rightarrow WOOD-O-C(=O)-CH_3 + CH_3-C(=O)-OH$$

Acetic anhydride   Acetylated wood   Acetic acid

Acetylation is a single-site reaction which means that one acetyl group is on one hydroxyl group with no polymerization. This means that all of the weight gain in acetyl can be directly converted into units of hydroxyl groups blocked. This is not true for a reaction where polymer chains are formed (epoxides and isocyanates, for example). In these cases, the weight gain can not be converted into units of blocked hydroxyl groups.

Acetylation has also been done using ketene gas (Tarkow 1945, Karlson and Svalbe 1972, 1977, Rowell et al. 1986c). In this case, esterification of the cell wall hydroxyl groups takes place but there is no formation of byproduct acetic acid.

$$WOOD-OH + CH_2=C=O \rightarrow WOOD-O-C(=O)-CH_3$$

Ketene   Acetylated wood

While this is interesting chemistry and eliminates a by-product, it has been shown that reactions with ketene gas results in poor penetration of reactive chemical and the properties of the reacted wood are less desirable than those of wood reacted with acetic anhydride (Rowell et al. 1986c).

The preferred method of acetylating wood today is using acetic anhydride without a catalyst or cosolvent. The method depends on the size of the wood to be acetylated (Hill 2006). Large thick wood (thinker than 2 cm) is best acetylated by first saturating the wood with cold acetic anhydride followed by heating (Militz 1991a). Because of the by-product acetic acid generated during the reaction, if the thick wood is contacted first with hot acetic anhydride, the by-product acetic acid dilutes the reaction site and the reaction either slows down or stops. With the thick wood pre-saturated with cold acetic anhydride, when the reaction starts there is a sufficient amount of anhydride to keep the reaction going. For wood veneer and wood up to about 1.5 cm in thickness it can be acetylated using hot acetic anhydride (Rowell et al. 1986b, 1986d, 1991a). Variations of this procedure have been used to modify fibers, particles, flakes, and chips. The fact that only a limited quantity of acetic anhydride is used means that less chemical has to be heated during the reaction and less chemical has to be cleaned up after the reaction. A small amount of acetic acid seems to be needed in the reaction mixture to swell the cell wall.

Fiber can be acetylated in a continuous process but it is hard to feed fiber into a reactor and energy consuming to remove the excess acid and anhydride in the reaction clean up. The most recent method of producing acetylated fiber is to use a refiner as part of the process (Rowell 2010). Chips can first be acetylated and then refined, or acetylated chips can be exploded and then refined or the chips can be acetylated in a closed refiner where the chips are refined in the presence of hot acetic anhydride.

Many different types of wood have been acetylated using a variety of procedures including several types of wood (Narayanamurti and Handa 1953, Rowell 1983, Rowell et al. 1986b, Rowell and Plackett 1988, Imamura et al. 1989, Plackett et al. 1990, Beckers and Militz 1994, Chow et al. 1996, Hill 2006), as well as other lignocellulosic resources such as bamboo (Rowell and Norimoto 1987, 1988), bagasse (Rowell and Keany 1991), jute (Callow 1951, Andersson and Tillman 1989, Rowell et al. 1991a), kenaf (Rowell 1993, Rowell and Harrison 1993), wheat straw (Gomez-Bueso et al. 1999, 2000), pennywort, and water hyacinth (Rowell and Rowell 1989).

15.3.2 ACID CHLORIDES

Acid chlorides can also be used to esterify wood (Suida 1930). The product is the ester of the related acid chloride with hydrochloric acid as a byproduct. Using lead acetate as a

$$WOOD-OH + R-C(=O)-Cl \rightarrow WOOD-O-C(=O)-R + HCl$$

catalyst with acetyl chloride, Singh et al. (1981) found a lower acetyl content than with acetic anhydride. Using a 20% lead acetate solution in the reaction reduced the amount of free hydrochloric acid released in the reaction. The very strong acid released as a byproduct in this reaction causes extensive degradation of the wood and because of this, very little work has been done in this area.

15.3.3 OTHER ANHYDRIDES

Other anhydrides have been used to react with wood. Risi and Arseneau (1958) reacted wood with phthalic anhydride resulting in high dimensional stability. Popper and Bariska (1972), however, found that chemical weight gain was lost after soaking in water showing that the phthaly group was lost to hydrolysis. Phthaly groups have a greater affinity for water than hydroxyl groups in wood, so phthalylated wood is more hydroscopic than untreated wood (Popper and Bariska 1972, 1973). While dimensional stability resulting from acetylation is due to bulking the cell wall, dimensional stability from phthalylation seems to be due to mechanical bulking of the submicroscopic pores in the wood cell wall (Popper and Bariska 1975). Very high weight gains are achieved by phthalylation (40–130%) which may result as a result of polymerization (Risi and Arseneau 1958, Popper 1973).

Goldstein et al. (1961) reacted ponderosa pine with propionic and butyric anhydrides in xylene without a catalyst. After 10 h at 125°C, the reaction weight gain was 4% for propionylation and zero for butyrylation. After 30 h reaction time, propionylation only resulted in a weight gain of 10%. Papadopoulos and Hill (2002) reacted Corsican pine with acetic, propionic, butric, valeric and hexanoic anhydrides. Butylation has also been done using microwave heating to improve dimensional stability and lightfastness (Chang and Chang 2003).

Succinic and malaic anhydrides have also been reacted with wood fiber (Clemons et al. 1992, Rowell and Clemons 1992). Reaction of these two anhydrides with wood resulted in making the wood thermoplastic and it was possible to thermoform the wood fiber into a high-density composite under pressure.

Propionic, butyric, valeric, hexanoic, and heptanoic anhydrides have also been used in the esterification of wood (Hill and Jones 1996a, Hill and Kwon 2009). As the molecular weight increases, reactivity decreases.

The reaction of wood with a cyclic anhydride does not result in a by-product acid and it has been shown that cross-linking of cell wall polymers is possible (Matsuda 1987).

15.3.4 CARBOXYLIC ACIDS

Carboxylic acids have been esterified to wood catalyzed with trifluoroacetic anhydride (Arni et al. 1961a,b, Nakagami et al. 1974).

$$WOOD-OH + (CH_3)_2-C=CH-COOH \rightarrow WOOD-O-C(=O)-CH-C(CH_3)_2$$

Several unsaturated carboxylic acids were found to react with wood by the "impelling" method to increase the oven dry volume without a change in color, a decrease in crystallinity and moisture content (Nakagami and Yokata 1975). Reaction with β-methylcrotonic acid gave a degree of substitution high enough to make the reacted wood soluble in acetone and chloroform to the extent of 30% (Nakagami et al. 1976). Further esterification increased the solubility but was accompanied by considerable degradation of wood components. Solubilization seems to be hindered by both lignin and hemicelluloses (Nakagami 1978, Nakagami and Yokata 1978).

15.3.5 ISOCYANATES

In the reaction of wood with hydroxyls with isocyanates, a nitrogen-containing ester is formed. Clermont and Bender (1957) exposed wood veneer swollen in dimethylformamide to vapors of phenyl isocyanate at 100–125°C. The resulting wood was very dimensionally stable and showed increased mechanical strength with little change in color. Baird (1969) reacted dimethylformamide-soaked cross-sections of white

$$WOOD-OH + R-N=C=O \rightarrow WOOD-O-C(=O)-NH-R$$

Isocyanate

pine and Englemann spruce with ethyl, allyl, butyl, *t*-butyl, and phenyl isocyanate. Vapor phase reactions of butyl isocyanate in dimethylformamide gave the best results.

White cedar was reacted with 2,4-tolylene diisocyanate with and without a pyridine catalyst to a maximum nitrogen content of 3.5 and 1.2, respectively (Wakita et al. 1977). Compressive strength and bending modulus increased with increasing nitrogen content. Beech wood was reacted with a diisocyanate that gave the resulting wood very high decay resistance (Lutomski 1975).

Methyl isocyanate reacts very quickly with wood without a catalyst to give high add-on weight gains (Rowell and Ellis 1979). Ethyl, *n*-propyl, and phenyl isocyanates react with wood without the need for a catalyst, but *p*-tolyl-1,6-diisocyanate and tolylene-2,4-diisocyanate require either dimethylformamide or triethylamine as a catalyst (Rowell and Ellis 1981).

15.3.6 FORMALDEHYDE

The reaction between wood hydroxyls and formaldehyde occurs in two steps. Because the bonding is with two hydroxyl groups, the reaction is called cross-linking. The two

$$WOOD-OH + H-C(=O)-H \rightarrow WOOD-O-C(OH)-H_2 + WOOD-OH \rightarrow WOOD-O-CH_2-O-WOOD$$

Formaldehyde   Hemiacetal   Acetal

hydroxyl groups can come from (1) hydroxyls within a single sugar unit, (2) hydroxyls on different sugar residues within a single cellulose chain, (3) hydroxyls between two different cellulose chains, (4) same as 1,2, and 3 except reaction occurring on the hemicelluloses, (5) hydroxyl groups on different lignin residues, and (6) interaction between cellulose, hemicelluloses and lignin hydroxyls. The possible cross-linking combinations are large and theoretically all of them are possible. Since the reaction is a two-step mechanism, some of the added formaldehyde will be in a non-cross-linked hemiacetal form. These chemical bonds are very unstable and would not survive long after reaction.

The reaction is best catalyzed by strong acids such as hydrochloric acid (Tarkow and Stamm 1953, Burmeister 1970, Ueyama et al. 1961, Minato and Mizukami 1982), nitric acid (Tarkow and Stamm 1953), sulfur dioxide (Dewispelaere et al. 1977, Stevens et al. 1979), p-toluene sulfonic acid and zinc chloride (Stamm 1959, Ueyama et al. 1961). Weaker acids such as sulfurous and formic acid do not work (Tarkow and Stamm 1953). Schurch (1968) speculated that bases such as lime water and tertiary amines can initiate the reaction.

15.3.7 OTHER ALDEHYDES

Acetaldehyde (Tarkow and Stamm 1953) and benzaldehyde (Tarkow and Stamm 1953, Weaver et al. 1960) reacted wood using either nitric acid or zinc chloride as catalyst. Glyoxal, glutaraldehyde and α-hydroxyadipaldehyde have also been catalyzed with zinc chloride, magnesium chloride, phenyldimethylammonium chloride, and pyridinium chloride (Weaver et al. 1960). Chloral (trichloroacetaldehyde) without catalyst and phthaldehydic acid in acetone catalyzed with p-toluenesulfonic acid have also been used as reaction systems with wood (Kenaga 1957, Weaver et al. 1960). Other aldehydes and related compounds have also been tried either alone or catalyzed with sulfuric acid, zinc chloride, magnesium chloride, ammonium chloride or diammonium phosphate (Weaver et al. 1960). Compounds such as N,N'-dimethylol-ethylene urea, glycol acetate, acrolein, chloroacetaldehyde, heptaldehyde, o- and p-chlorobenzaldehyde, furfural, p-hydroxybenzaldehyde and m-nitrobenzaldehyde all bulk the cell wall but no cross-linking seems to occur.

15.3.8 METHYLATION

The simplest ether that can be formed is the methyl ether. Reactions of wood with dimethyl sulfate and sodium hydroxyl (Rudkin 1950, Narayanamurti and Handa 1953) or methyl iodide and silver oxide (Narayanamurti and Handa 1953) are two systems that have been reported.

$$WOOD-OH + CH_3I \rightarrow WOOD-O-CH_3$$

Methylation up to 15% weight gain did not affect the adhesive properties of casein glues.

15.3.9 ALKYL CHLORIDES

In the reaction of alkyl chlorides with wood, hydrochloric acid is generated as a by-product. Because of this, strength properties of the treated wood are poor. Reaction of allyl chloride in pyridine (Kenaga et al. 1950, Kenaga and Sproull 1951) or aluminum chloride resulted in good dimensional stability but on soaking in water, dimensional stability was lost (Kenaga and Sproull 1951). In the case of allyl chloride-pyridine case,

$$WOOD-OH + R-Cl \rightarrow WOOD-O-R + HCl$$

Alkyl chloride

The dimensional stability was not due to the bulking of the cell wall with bonded chemical but to the formation of allyl pyridinium chloride polymers which are water soluble and easily leached out (Risi and Arseneau 1957d).

Other alkyl chlorides reported are crotyl chloride (Risi and Arseneau 1957b) and *n*- and *t*-butyl chlorides (Risi and Arseneau 1957c) catalyzed with pyridine.

15.3.10 β-PROPIOLACTONE

The reaction of β-propiolactone with wood is interesting in that different products are possible depending on the pH of the reaction. Under acid conditions, an ether bond is formed with the hydroxyl group on wood along with the formation of a free acid end group.

$$
\begin{array}{l}
\quad\quad\quad\quad\quad\quad\quad\quad\quad H+ \\
WOOD-OH+CH_2-CH_2 \ \rightarrow WOOD-O-CH_2-CH_2-COOH \\
\quad\quad\quad\quad | \quad | \quad\quad\quad\quad\quad OH- \\
\quad\quad\quad\quad O-C=O \quad\quad\quad \rightarrow WOOD\,O-C(=O)-CH_2-CH_2-OH \\
\quad\quad\quad\beta-propiolactone
\end{array}
$$

Under basic conditions, an ester bond is formed with the wood hydroxyl giving a primary alcohol end group.

Southern yellow pine was reacted with β-propiolactone under acid conditions to give a carboxy-ethyl derivative (Goldstein et al. 1959, Goldstein 1960). High concentrations of β-propiolactone caused delamination and splitting of the wood due to very high swelling (Rowell unpublished data).

β-Propiolactone has now been labeled as a very active carcinogen. For this reason, this very interesting chemical reaction system will probably not be studied again.

15.3.11 ACRYLONITRILE

When acrylonitrile reacts with wood in the presence of an alkaline catalyst, cyanoethylation occurs.

$$WOOD-OH + CH_2=CH-CN \rightarrow WOOD-O-CH_2-CH_2-CN$$

Acrylonitrile

With sodium hydroxide, weight gains up to 30% have been reported (Goldstein et al. 1959, Baechler 1959b, Fuse and Nishimoto 1961, Kenaga 1963). Ammonium hydroxide has also been used giving lower weight gains as compared to sodium hydroxide.

15.3.12 EPOXIDES

The reaction between epoxides and the hydroxyl groups is an acid or base catalyzed reaction; however, all work in the wood field has been alkali catalyzed.

$$WOOD-OH + R-CH(-O-)CH_2 \rightarrow WOOD-O-CH_2-CH(OH)-R$$

Epoxide

The simplest epoxide, ethylene oxide, catalyzed with trimethylamine has been used as a vapor phase reaction (McMillin 1963, Liu and McMillin 1965, Mo and Domsjo 1965, Rowell and Gutzmer 1975). Barnes et al. (1969) showed that an oscillating pressure was better than a constant pressure in this reaction. Sodium hydroxide has also been used as a catalyst with ethylene oxide (Zimakov and Pokrovskii 1954). Other epoxides that have been studied include 1,2-epoxy-3,3,3-trichloropropane, 1,2-epoxy-4,4,4-trichlorobutane, 1-allyloxy-2,3-epoxypropane, p-chlorophenyl-

2,3-epoxypropyl ether, 1,2:3,4-diepoxybutane, 1,2:7,8-diepoxyoctane, 1,4-butanediol diglycidal ether, 3-eposxyethyl-7-oxabicyclo heptane (Rowell and Ellis 1984a,b).

Propylene and butylenes oxides have also been reacted with wood along with epichlorohydrin and mixtures of epichlorohydrin and propylene oxide (Rowell 1975a,b, Rowell and Ellis 1984a,b). It is theoretically possible for cross-linking to occur with epichlorohydrin resulting in the splitting out of hydrochloric acid but this has never been observed (Rowell and Ellis 1981, Rowell and Chen 1994).

In the case of the epoxy system, after the initial reaction with a cell wall hydroxyl group, a new hydroxyl group originating from the epoxide is formed. From this new hydroxyl, a polymer begins to form. Because of the ionic nature of the reaction and the availability of alkoxyl ions in the wood components, the chain length of the new polymer is probably short owing to chain transfer.

15.3.13 Furfural Alcohol

Reaction of wood with furfural alcohol began in the 1950's at the Forest Products Laboratory in Madison, WI (Stamm 1977). In the 1960s, Goldstein and Dreher investigated this chemistry further (Goldstein and Dreher 1960). This chemistry was further developed by Anaya et al. (1984). In the 1980's, the University of New Brunswick developed this chemistry further (Schneider and Witt 2004). They postulated that the reaction that occurred with furfural alcohol and wood was a polymerization forming a furan polymer in the wood (Choura et al. 1996). They developed a two stage process where the wood was first impregnated with an initiator and then treated with furfural alcohol. The product was trademarked as "WISTIwood" and licensed to a company in the 1990s called Wood Polymer Products Inc. (Hill 2006). The commercialization failed but improvements were made to the process in the late 1990s in Norway and a new process was developed. The products are sold under the trade names of "Kebony" or "VisorWood" (Lande et al. 2008, Lande 2008). The products are known today as Kebony XXX where the XXX stands for the wood species with a general weight gain of about 30%. For example, Kebony Southern yellow pine, Kebony ash, and Kebony beech.

More recently, the mechanism of the chemistry involved has been modified (Lande et al. 2008, Lande 2008). The reaction starts with a dehydration condensation between two furfural alcohol molecules (1) or (2a). There can then be a reaction eliminating formaldehyde (2b). Cross-linking also takes place (3 and 4). All of these reactions form oligomers and polymers but no bonding to wood. More recently, it has been shown that furfural alcohol can react with lignin model compounds (5) so that bonding to a hydroxyl on lignin may be possible (Nordstierna et al. 2007). This reaction is less understood. The industrial process is shown in a later section of this chapter.

$$2 \quad \text{(furan ring)} + HCHO \longrightarrow \quad \text{(product)} \qquad (4)$$

Lignin guaiacyl unit

$$\text{(furan-CH}_2\text{OH}) + \text{(Lignin guaiacyl unit)} \xrightarrow{-H_2O} \text{(product)} \qquad (5)$$

15.3.14 DIMETHYLOLDIHYDROXYETHYLENEUREA (DMDHEU)

The reaction between wood and 1,3-dimethylol-4,5-dihydroxyethyleneurea (DMDHEU) was first carried out in the 1960s by Weaver and coworkers (Weaver et al. 1960). Later, Nicholas and Williams reacted wood with DMDHEU to improve dimensional stability (Nicholas and Williams 1987). The reaction is acid catalyzed and both inorganic acids (HCl) and Lewis acids (aluminum chloride, manganese chloride). Methanesulfonic acid has also been used as a catalyst with both sweetgum and southern yellow pine (Ashaari et al. 1990a,b). The predominate reaction seems to be polymerization with some evidence of cross-linking (Simonson 1998). It has also been used in combination with other chemicals including vinyl polymers (Ahmed Kabir et al. 1992) and polyethylene glycol (Simonson 1998). A maximum weight gain of 31% shows that 39% of the available –OH groups have been substituted (Dieste et al. 2009).

15.3.15 REACTION RATES

Table 15.1 shows the rate of reaction of southern pine reacted with propylene and butylene oxide, methyl and butyl isocyanate, and acetylated under different reaction conditions using liquid acetic anhydride, acetic anhydride vapor and ketene gas. The propylene and butylenes oxide reactions were done using 5% triethyl amine as a catalyst. The methyl and butyl isocyanates were uncatalyzed. In the case of acetic anhydride vapor reactions, the pine veneer was suspended above a supply of acetic acid anhydride (Rowell et al. 1986a). In the case of ketene gas, the pine chips were subjected to ketene gas in batches (Rowell et al. 1986c).

In the reactions of the two epoxides and two isocyanates, the longer the reaction is run, the higher the weight percent gain (WPG). Acetylation, however, reaches a maximum WPG of about 22 no mater how it is done and further reaction time does not increase this value. The acetylation of aspen follows a similar pattern except a maximum WPG of about 17–18 is reached. Other softwood and hardwood that have been acetylated using these procedures have followed a similar pattern.

It has been shown that the acetic anhydride reaction solution could contain up to 30% acetic acid without any detrimental effect on the reaction rate (Rowell et al. 1990). There is actually an increase in reaction rate at 10–20% acetic acid in the impregnation solution due to the swelling effect of the acetic acid and, possibility, as an effect of the increased acidity.

TABLE 15.1
Reaction Rates for Pine Using Different Chemicals

Chemical	Temperature (°C)	Time (min)	Weight Percent Gain
Propylene oxide-TEA	120	40	35.5
	120	120	45.5
	120	240	52.2
Butylene oxide-TEA	120	40	24.6
	120	180	36.9
	120	360	42.2
Methyl isocyanate	120	10	25.7
	120	20	40.4
	120	60	51.8
Butyl isocyanate	120	120	5.0
	120	180	16.0
	120	360	24.3
Acetic anhydride (liquid)	100	60	7.8
	100	120	11.2
	100	360	19.9
	120	60	17.2
	120	180	21.4
	140	60	17.9
	140	180	22.1
	160	60	21.4
Acetic anhydride (vapor)	120	480	7.2
	120	1440	22.1
Ketene	50–60	60	6.8
	50–60	120	21.7

Since the reaction of wood with acetic anhydride is an exothermic reaction, once the acetylation starts to take place, the temperature of the system goes up. This heating can be taken advantage of to reduce the external heat supplied to the system.

15.3.16 EFFECT OF MOISTURE

Since all of the chemicals that react with wood cell wall hydroxyl groups can also react with water, the amount of water in the wood before reaction is very important (Rowell and Ellis 1984a). Hydrolysis with water is much faster than the reaction with a cell wall hydroxyl group. There is a trade off between the cost and energy to remove water below about 10% moisture content and the cost of chemical lost to hydrolysis. In the case of epoxides, reaction with water leads to the formation of nonbonded polymers in the cell wall. In the reaction with anhydrides, free acid is generated, which in the case of acetylation can be an advantage so long as the amount of acetic acid generated is not too large.

15.4 PROPERTIES OF CHEMICALLY MODIFIED WOOD

It is not possible to describe all of the properties of chemically modified wood in a short chapter. An entire book would be needed to describe all of the changes in properties from all of the reaction systems in the published literature. The next section will select some of the major property changes with a few examples from several of the chemical reactions systems. Since most of the historical and recent research in chemical modification of wood has been in the area of acetylation, much of the property and performance data will focus on this technology.

15.4.1 CHANGES IN WOOD VOLUME RESULTING FROM REACTION

Table 15.2 shows the increase in volume of pine wood after reaction with several chemicals and the calculated volume of chemical added to the cell wall after redrying the wood. Since the volume increase due to reactions with propylene and butylenes oxide, methyl and butyl isocyanates and acetic anhydride is equal to the calculated volume of chemical added, this shows that the reaction has taken place in the cell wall and not in the void spaces of the wood (Rowell 1983). This correlation is not true for reactions of wood with, for example, acrylonitrile (Rowell 1983). In this case, the chemical is located in the voids in the wood structure.

15.4.2 STABILITY OF BONDED CHEMICAL TO CHEMICAL LEACHING

One of the indirect ways to tell if the reaction has resulted in bonding with the cell wall is to leach the reacted wood with several solvents and determine loss of weight in the reacted sample. At least two different solvents are used. One is a solvent that the starting chemicals are soluble in and a second in which any polymers that might be formed in the reaction are soluble.

Table 15.3 shows the weight loss due to leaching in benzene/ethanol, water in a Soxhlet extractor for 24 h and water soaking for 7 days. The data show that reactions with methyl isocyanate, butylene oxide, and acetic anhydride form chemical bonds that are stable to solvent extraction while propylene oxide reacted wood loses some chemical in water leaching. The reaction with acrylonitrile catalyzed with either ammonium or sodium hydroxyl forms bonds that are not stable to solvent extraction.

Table 15.4 shows the stability of acetyl groups on acetylated pine and aspen to various pHs and temperatures and Table 15.5 shows the activation energy for the deacetylation of pine and aspen (Rowell et al. 1992a,b). These results show that acetylated wood is much more stable under slightly acid conditions as opposed to slightly alkaline conditions. The data can be used to predict the stability of acetylated wood or other wood at any pH and temperature combinations to estimate the life expectancy of the product in its service environment. However, it must be considered that the data from this study were collected using either finely ground powder or small fiber. As a result, a very

---

TABLE 15.2

Changes in Pine Volume and Volume of Chemical Added as a Result of Chemical Reactions

Chemical	WPG	Increase in Wood Volume[a] (cm³)	Calculated Volume of Added Chemical[b] (cm³)
Propylene oxide	26.5	7.1	7.5
	36.2	8.9	9.0
Butylene oxide	25.3	6.9	6.9
Methyl isocyanate	12.4	0.16	0.14
	25.7	0.21	0.27
	47.7	0.46	0.54
Acetic anhydride	17.5	3.0	2.9
	22.8	3.9	4.0
Acrylonitrile	25.7	0.46	0.77
	36.0	0.74	1.2

[a] Difference in oven-dry volume between reacted and nonreacted wood.

[b] Density used in volume calculations: Propylene and butylenes oxides—1.01, methyl isocycnate—0.967, acetic anhydride—1.049, and acrylonitrile—0.806.

TABLE 15.3
Ovendry Weight Loss of Chemically Reacted Wood Extracted with Several Solvents

Chemical	WPG	Benzene/Ethanol	Water	Water
		4 h Soxhlet 20 mesh (% wt loss)	24 h Soxhlet 40 mesh (% wt loss)	7 days Soxhlet 20 mesh (% wt loss)
None	0	2.3	11.2	0.6
Methyl isocyanate	23.5	6.5	11.6	1.0
Propylene oxide	29.2	5.2	10.7	4.0
Butylene oxide	27.0	3.8	11.7	1.6
Acetic anhydride	22.5	2.8	12.2	1.2
Acrylonitrile (NH₄OH)	26.1	22.3	20.4	21.7
Acrylonitrile (NaOH)	25.7	17.6	18.7	13.5

TABLE 15.4
Effects of Various pH and Temperature Conditions on the Stability of Acetyl Groups in Acetylated Pine and Aspen (Acetyl Content: Pine 20.2 WPG, Aspen 20.6 WPG)

Wood	Temperature (°C)	pH	Rate Constant ($k \times 10^3$)	Half Life (days)
Pine	24	2	0.26	2640
		4	0.15	4630
		6	0.06	10,900
		8	1.40	500
Aspen	24	2	0.23	3083
		4	0.15	4697
		6	0.06	10,756
		8	1.11	623
Pine	50	2	2.07	340
		4	1.06	650
		6	1.71	410
		8	7.31	95
Aspen	50	2	1.18	590
		4	0.66	1050
		6	1.29	540
		8	8.51	80
Pine	75	2	11.3	61
		4	8.41	82
		6	15.5	45
		8	32.2	22
Aspen	75	2	7.33	95
		4	4.78	145
		6	12.6	55
		8	32.5	21

TABLE 15.5
Activation Energy for Deacetylation of Pine and Aspen over a Temperature Range 24–75°C

Wood	pH	Activation Energy (kJ/mol)
Pine	2	58
	4	58
	6	89
	8	57
Aspen	2	63
	4	68
	6	93
	8	53

large surface area came into contact with the various pH and temperature environments. Therefore, this data represents the fastest possible degradation rates. It is well documented that in archaeological wood, acetyl groups are stable over thousands of years and little loss of acetyl has been observed.

Table 15.6 shows the stability of acetyl groups in pine and aspen flakes to cyclic exposure to 90% and 30% relative humidity (RH) (Rowell et al. 1992a,b). Each cycle represents exposing the flakes for three months at 30% RH and then three months at 90% RH. Within experimental error, there is no loss of acetyl over 41 cycles of humidity changes. This data was collected in 1992 and this experiment has continued. After almost 10 years of cycling these chips from 30% to 90% RH, recent acetyl analysis shows that there is still no loss of acetyl resulting from humidity cycling.

15.4.3 ACCESSIBILITY OF REACTION SITE

The rate controlling step in the chemical modification of wood is the rate of penetration of the reagent into the cell wall (Rowell et al. 1991b). In the reaction of liquid acetic anhydride with wood, at an acetyl weight percent gain of about 4, there is more bonded acetyl in the S_2 layer than in the middle lamella. At a weight gain of about 10, there is an equal distribution of acetyl throughout the S2 layer and the middle lamella. At a WPG over 20, there is a slightly higher concentration of acetyl in the middle lamella than in the rest of the cell wall. This was found using chloroacetic anhydride and following the fate of the chlorine by energy-dispersive x-ray analysis (Rowell et al. 1991b).

When larger pieces of wood are used, the time for the anhydride to penetrate increased in proportion to the size of wood used. For larger timber, it is important to use cold acetic anhydride to start the procedure so that there is sufficient reactive chemical distributed throughout the timber before the reaction starts. If hot anhydride is used on large timbers, a "shell" acetylation may occur since the by-product acetic acid produced during the reaction will dilute the penetrating anhydride to the point where the reaction slows down or stops. It has been shown that a concentration of greater than 15% acetic acid in the anhydride mixture, slows the rate of acetylation (Rowell et al. 1990).

TABLE 15.6
Stability of Acetyl Groups in Pine and Aspen Flakes after Cyclic Exposure between 90% Relative Humidity (RH) and 30% RH

Wood	Acetyl Content (%) after Cycle (Number)				
	0	13	21	33	41
Pine	18.6	18.2	16.2	18.0	16.5
Aspen	17.9	18.1	17.1	17.8	17.1

TABLE 15.7
Effects of Size of Spruce Specimen on Acetyl Content

Specimen	WPG	Acetyl Content
Chips acetylated	14.2	15.6
Acetylated chips to fiber	14.2	15.4
Chips to fiber and then acetylated	22.5	19.2
Acetylated chips to fiber	22.5	19.4
Acetylated chips to fiber and again acetylated	20.4	20.5

Of course, the critical dimension is the size in the longitudinal dimension as chemical penetrates much faster in this direction as compared to the radial and tangential directions.

There is a difference in the rate and extent of acetylation and probably in the distribution of acetyl groups within a given species and between species. Heartwood, sapwood, juvenile wood, mature wood, hardwood, softwood, springwood and latewood all react at different rates and to different levels though the difference may be small. In the acetylation of loblolly pine flakes, it was found that the reactivity increased with an increased content of juvenile wood (Hon and Bangi 1996). However, no difference was found in the reactivity of heartwood and sapwood of *Pinus radiate* flakes (Rowell and Plackett 1988). In larger wood samples, pine sapwood reacted faster than heartwood and reached a higher level of acetylation (Beckers and Militz 1994). This reduction in reaction rate may be a function of heartwood permeability. In a pilot-scale study, it was found that there was a difference in reaction rates between beech, eucalyptus, poplar, pine Douglas fir and spruce (Beckers and Militz 1994).

Table 15.7 shows that if spruce chips are reacted for 30 min at 120°C, the WPG is 14.2 with an acetyl content of 15.6. If the chips are first reduced to fiber and then acetylated under the same reaction conditions, the WPG is 22.5 and the acetyl content is 19.2. If the acetylated chips were reduced to fiber and then reacetylated under the same conditions, the WPG is 20.4 with an acetyl content of 20.5. Since no acetyl groups were lost in the refining step, this shows that new –OH sites become available when the acetylated chips are reduced to fiber (Rowell and Rowell 1989). This also shows that some –OH groups are not accessible in a chip that becomes accessible in a fiber.

15.4.4 ACETYL BALANCE IN ACETYLATED WOOD

The mass balance in the acetylation reaction shows that all of the acetic anhydride going into the acetylation of hardwood and softwood could be accounted for as increased acetyl content in the wood, acetic acid resulting from hydrolysis by moisture in the wood or as unreacted acetic anhydride (Rowell et al. 1990). The consumption of acetic anhydride can be calculated stoichiometrically based on the degree of acetylation and the moisture content of the wood. This is true for all wood acetylated to date.

Table 15.8 shows the comparison of the weight gain due to acetylation with the acetyl content found by chemical analysis. At the lower weight gain levels, there is always a higher acetyl content as compared to the WPG. This may be due to the removal of extractives and some cell wall polymers into the acetic anhydride solution resulting in an initial specimen weight loss. At WPGs greater than about 15, the acetyl content and the WPG values are almost the same.

15.4.5 DISTRIBUTION OF BONDED CHEMICAL

Figure 15.6 shows the reaction of uncatalyzed acetic anhydride with various isolated wood cell wall polymers and whole wood at 120°C. The data show that lignin is the fastest to react but the

TABLE 15.8
Weight Percent Gain and Acetyl Analysis of Acetylated Pine and Aspen

Pine		Aspen	
WPG	Acetyl Content	WPG	Acetyl Content
0	1.4	0	3.9
6.0	7.0	7.3	10.1
14.8	15.1	14.2	16.9
21.1	20.1	17.9	19.1

hemicelluloses react to a higher level (Rowell et al. 1991b). Isolated cellulose did not react under these conditions.

Table 15.9 shows distribution of bonded chemical and the degree of substitution (DS) of hydroxyl groups in methyl isocyanate reacted pine (Rowell 1982, Rowell and Ellis 1981, Rowell et al. 1991b). This analysis is based on many assumption: (1) pine has a holocellulose content of 67% and a lignin content of 27%, (2) the holocellulose content is 87% hexosans, 13% pentosans, (3) the cellulose content of the holocellulose is 71.8%, 14.8% hemicellulose hexosans, and 13.4% hemicellulose pentosans, (4) whole pine wood is 67% holocellulose of which 48.1% is cellulose, 9.0% pentosans and 9.9% hexosans, (5) the calculated theoretical acetyl content of the holocellulose is 77.7% and 25.7% for the lignin, and (6) there are 1.1 hydroxyl groups for each nine carbon unit of lignin.

Based on this analysis, the lignin is completely substituted at a WPG of about 47 and at that point, only about 20% of the hydroxyls in the holocellulose are substituted. These calculations are also based on the added assumption that all of the theoretical hydroxyl groups are accessible during the acetylation reaction. Assuming 100% accessibility of the hemicellulose fraction but only 35% accessibility of the cellulose (based on the degree of crystallinity), the degree of substitution

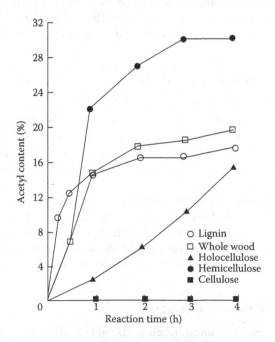

FIGURE 15.6 Acetylation of isolated cell wall polymers and whole wood.

TABLE 15.9
Distribution of Methyl Isocyanate Nitrogen and Degree of
Substitution (DS) of Hydroxyl Groups in Modified Pine

WPG	MeIsoN in Lignin		MeIsoN in Holo	
	%	DS	%	DS
5.5	1.42	0.17	0.59	0.03
10.0	2.36	0.28	1.19	0.05
17.7	3.44	0.41	2.11	0.08
23.5	4.90	0.59	2.94	0.12
47.2	7.46	0.89	5.24	0.21

of the accessible holocellulose would be 0.48. In experiments acetylating isolated cell wall components, lignin reacts faster than the hemicelluloses and whole wood and cellulose did not react at all (Rowell et al. 1991b). It is possible that the cellulose is modified during the isolation procedure so this result does not necessarily mean that no cellulose modification takes place during the acetylation of whole wood.

The theoretical hydroxyl content of pine softwood is 14.9 mmoles/g which corresponds to an accessible −OH hydroxyl of 8.6 mmoles/g assuming that none of the crystalline cellulose are substituted (Hill and Jones 1996b, 2006). Based on these calculations, at a weight percent gain of 25, 5.9 mmoles of −OH groups would be substituted per gram of oven dried pine wood. The difference between the maximum level of acetylation possible without using a catalyst, there is a significant number of −OH groups left unsubstituted in a "fully acetylated" wood specimen. This indicates that not all −OH groups are involved in a modified wood that shows a high degree of dimensional stability and decay resistance. Only the most accessible –OH are substituted in the noncatalyzed acetylation process.

15.4.6 MOISTURE SORPTION

By replacing some of the hydroxyl groups on the cell wall polymers with bonded acetyl groups, the hygroscopicity of the wood material is reduced. Table 15.10 shows the fiber saturation point for acetylated pine and aspen (Rowell 1991). As the level of acetylation increases the fiber saturation point decreases in both the soft and hard wood

TABLE 15.10
Fiber Saturation Point for Acetylated
Pine and Aspen

WPG	Pine (%)	Aspen (%)
0	45	46
6	24	—
8.7	—	29
10.4	16	—
13.0	—	20
17.6	—	15
18.4	14	—
21.1	10	—

TABLE 15.11
Equilibrium Moisture Content of Control and Chemically Modified Pine

Chemical	WPG	Equilibrium Moisture Content at 27°C		
		35% RH	60% RH	85% RH
Control	0	5.0	8.5	16.4
Propylene oxide	21.9	3.9	6.1	13.1
Butylene oxide	18.7	3.5	5.7	10.7
Formaldehyde	3.9	3.0	4.2	6.2
Acetic anhydride	20.4	2.4	4.3	8.4

Table 15.11 shows the equilibrium moisture content (EMC) of control and several types of chemically modified pine at three levels of relative humidities. In all cases, as the level of chemical weight gain increases, the EMC of the resulting wood goes down (Rowell et al. 1986b). Reaction of formaldehyde with wood is the most effective in reducing EMC as a function of chemical weight gain. Of the other chemicals, acetylation is the most effective followed by butylene oxide reaction. Propylene oxide reacted wood has only a slight decrease in EMC as compared to nonreacted wood. Table 15.12 shows the EMC of acetylated pine and aspen at several levels of acetylation.

In the case of acetylated wood, if the reductions in EMC at 65% RH of many different types of acetylated woods referenced to unacetylated fiber is plotted as a function of the bonded acetyl content, a straight line plot results (Rowell and Rowell 1989). Even though the points represent many different types of wood, they all fit a common curve. A maximum reduction in EMC is achieved at about 20% bonded acetyl. Extrapolation of the plot to 100% reduction in EMC would occur at about 30% bonded acetyl. This represents a value not too different from the water fiber saturation point in these fibers. Because the acetate group is larger than the water molecule, not all hygroscopic hydrogen-bonding sites are covered so it would be expected that the acetyl saturation point would be lower than that of water.

The fact that EMC reduction as a function of acetyl content is the same for many different types of wood indicates that reducing moisture sorption may be controlled by a common factor. The lignin, hemicellulose, and cellulose contents of all the woods are different.

TABLE 15.12
Equilibrium Moisture Content of Acetylated Pine and Aspen

Specimen	WPG	Equilibrium Moisture Content at 27°C		
		30% RH	65% RH	90% RH
Pine	0	5.8	12.0	21.7
	6.0	4.1	9.2	17.5
	10.4	3.3	7.5	14.4
	14.8	2.8	6.0	11.6
	18.4	2.3	5.0	9.2
	20.4	2.4	4.3	8.4
Aspen	0	4.9	11.1	21.5
	7.3	3.2	7.8	15.0
	11.5	2.7	6.9	12.9
	14.2	2.3	5.9	11.4
	17.9	1.6	4.8	9.4

TABLE 15.13

Equilibrium Moisture Content of Furfural Alcohol-Treated Pine Wood

	Sapwood			Heartwood		
Sample	EMC (35%)	EMC (65%)	EMC (85%)	EMC (35%)	EMC (65%)	EMC (85%)
Control	8.9	12.9	17	7.4	10.9	14.5
Kebony 70	0.1	0.1	0.1	0.2	0.4	0.4

Table 15.13 shows the equilibrium moisture content of furfural alcohol-treated wood (Esteves et al. 2009). The measurements were done on Kebony 70 which is pine treated with furfural alcohol. EMC was measured at three different relative humidity conditions, 35%, 65%, and 85%.

Figure 15.7 shows the sorption–desorption isotherms for acetylated spruce fibers (Stromdahl 2000). The 10 min acetylation curve represents a WPG of 13.2 and the 4 h curve represents a WPG of 19.2. The untreated spruce reaches an adsorption/desorption maximum at about 35% moisture content, the 13.2 WPG a maximum of about 30%, and the 19.2 WPG a maximum of about 10%. There is a very large difference between the adsorption and desorption curves for both the control and the 13.2 WPG fibers but much less difference in the 19.2 WPG fibers. The sorption of moisture is presumed to be sorbed either as primary water or secondary water. Primary water is the water that is sorbed to primary sites with high binding energies such as the hydroxyl groups. Secondary water is that water sorbed to less binding energy sites which is water molecules sorbed on top of the primary layer. Since some of the hydroxyl sites are esterified with acetyl groups, there are fewer primary sites to which water sorbs. And since the fiber is more hydrophilic due to acetylation, there may also be less secondary binding sites.

In a recent study, Wålinder et al. found a contact angle, using the Wilhelmy method for acetylated Scots pine sapwood at 20% weight gain, to be 81.5 while the control was 59.7 (Wålinder et al. 2010).

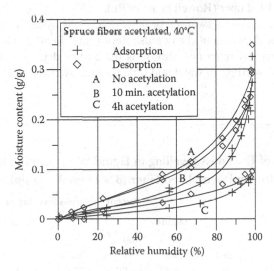

FIGURE 15.7 Sorption/desorption isotherms for control and acetylated spruce fiber.

TABLE 15.14

Thickness Swelling (TS) of Aspen Fiberboards
Made from Control and Acetylated Fiber

WPG	Phenolic Resin Content (%)	TS at 27°C		
		30% RH (%)	65% RH (%)	90% RH (%)
0	5	0.7	3.0	12.6
	8	1.0	3.1	11.2
	12	0.8	2.5	9.7
17.9	5	0.2	1.8	3.2
	8	0.2	1.7	3.1
	12	0.1	1.7	2.9

15.4.7 Dimensional Stability

Changes in dimensions in tangential and radial direction in solid wood and in thickness and in linear expansion for composites are a great problem in wood composites (see Chapter 4). In composites, they not only undergo normal swelling (reversible swelling) but also swelling caused by the release of residual compressive stresses imparted to the board during the composite pressing process (irreversible swelling). Water sorption causes both reversible and irreversible swelling with some of the reversible shrinkage occurring when the board dries.

Thickness swelling at three levels of relative humidity is greatly reduced as a result of acetylation (Table 15.14). Linear expansion is also greatly reduced as a result of acetylation (Krzysik et al. 1992, 1993). Increasing the adhesive content can reduce the thickness swelling but not to the extent that acetylation does.

The rate and extent of thickness swelling in liquid water of fiberboards made from control and acetylated fiber is shown in Table 15.15. Both the rate and extent of swelling are greatly reduced as a result of acetylation. At the end of 5 days of water soaking, control boards swelled 36% whereas boards made from acetylated fiber swelled less than 5%. After drying the boards at the end of the test showed that control boards exhibit a greater degree of irreversible swelling as compared to boards made from acetylated fiber (Rowell et al. 1991b).

Table 15.16 shows the thickness swelling of pine, beech and wheat straw fiberboards after 24 h of water swelling and the thickness of the boards after drying. In all cases, the control fiberboards swelled on water soaking and remained almost as thick after drying while the acetylated fiberboards showed very little swelling in water and little residual swelling after drying.

TABLE 15.15

Rate and Extent of Thickness Swelling in Liquid Water of Pine Fiberboards
Made from Control and Acetylated Fiber (8% Phenolic Resin)

Fiberboard	Percent Thickness Swelling at						
	Minutes			Hours			Days
	15	30	60	3	6	24	5
Control	25.7	29.8	33.5	33.8	34.0	34.0	36.2
Acetylated (21.6 WPG)	0.6	0.9	1.2	1.9	2.5	3.7	4.5

TABLE 15.16

**Thickness Swelling of Fiberboards in Water
and Thickness after Drying (10% Phenolic Resin)**

Fiber	WPG (%)	Thickness Swelling (24 h in Water)	Residual Thickness Swelling (Oven Dry) (%)
Pine	0	21.3	19.7
	21.5	2.1	1.0
Beech	0	17.0	13.0
	19.7	2.2	0.9

TABLE 15.17

**Antishrink Efficiencies of Chemically Modified Solid
Pine Wood**

Chemical	WPG	ASE
None	0	—
Propylene oxide	29.2	62
Butylene oxide	27.0	74.3
Acetic anhydride	22.5	70.3
Methyl isocyanate	26.0	69.7
Acrylonitrile		
NH_4OH	26.1	80.9
NaOH	25.7	48.3

TABLE 15.18

**Repeated Antishrink Efficiency (ASE) of Chemically
Modified Solid Pine**

Chemical	WPG	ASE_1	ASE_2	ASE_3	ASE_4	Weight Loss after Test
Propylene oxide	29.2	62.0	43.8	50.9	50.3	5.7
Butylene oxide	26.7	74.3	55.6	59.7	48.1	4.6
Acetic anhydride	22.5	70.3	71.4	70.6	69.2	<0.2
Methyl isocyanate	26.0	69.7	62.8	65.0	60.7	4.3
Acrylonitrile						
NH_4OH	26.1	80.9	0	0	0	22.6
NaOH	25.7	48.3	0	0	0	14.7

Table 15.17 shows the antishrink efficiencies (ASE, see Chapter 4 for definition and calculations) of several different types of chemically modified solid pine wood. All chemically reacted pine wood shows an ASE of approximately 70 at weight gains between 22 and 26 WPG.

If the water swelling test is continued through several cycles of water soaking and reoven drying, the ASE may change if chemical is lost in the leaching process. Table 15.18 shows the ASE for several types of chemically modified solid pine. ASE_1 is calculated from the wood going from oven dry to water soaked (Rowell and Ellis 1978). ASE_2 is then calculated from the wood going from water soaked back to the oven dry state. ASE_3 and ASE_4 are the values on the second complete cycle.

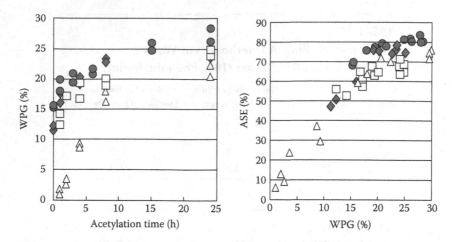

FIGURE 15.8 Weight gain of acetyl on sugi wood as a function of reaction time (left), antishrink efficiency as a function of weight gain (right). ▲ = supercritical CO_2 (120°C, 10–12 MPa), ● = supercritical CO_2 (130°C, 10–12 MPa), □ = liquid phase(120°C, △ = vapor phase (120°C). (Reproduced with permission from the authors.)

These data in Table 15.18 show that the highest ASE is for acrylonitrile catalyzed with ammonium in the first wetting cycle but redrying results in a loss of much of the chemical so ASE_2 is zero with a final weight loss of 22.6% which is almost a complete loss of reacted chemical. Propylene and butylene oxides, methyl isocyanate, and acetic anhydride reacted wood is more stable to swelling and shrinking cycles. Acetylation is the most stable treatment with less than 0.2% weight loss after the two cycle test.

Acetylation of sugi wood using supercritical carbon dioxide also produced a board with high dimensional stability (Matsunnaga et al. 2010). Figure 15.8 (left) shows the percent weight gain due to acetylation using super critical CO_2 at two different reaction temperatures, acetylation using liquid acetic anhydride and acetylation using acetic anhydride vapor. Figure 15.8 (right) shows the antishrink efficiency of the acetylated samples. Super critical CO_2 produces acetylated samples with very high weight gains and also show the highest ASE. The super critical CO_2 started to form a single phase with the acetic anhydride at about 90°C.

Treating wood with furfural using zinc chloride, citric acid or formic acid as polymerization catalysts resulted in an ASE of 50–70%. Highter weight gains (up to 120%) of resin did not increase the ASE value (Stamm 1977). Furfural alcolol treatment of *Pinus pinaster* (Kebony FA 70) decreased the equilibrium moisture content (EMC) at 35% relative humidity (RH) from 9% to 5% and from 17% to 9% at 85% RH (Esteves et al. 2010). Larsson Brelid and Westin determined the EMC of furfural alcohol treated wood and found that at a 10% weight gain, the ASE was 20, at 20% weight gain, the ASE was 37, 30% weight gain, ASE 42% and at a 35% loading, the ASE was 50% (Larsson Brelid and Westin 2007).

DMDHEU modified wood has also been shown to improve dimensional stability (Zee et al. 1998). Using $MgCl_2$, $AlCl_3$ and citric acid as catalysts, an ASE of approximately 40% was achieved witrh a resin loading of 20%. Curing at higher temperatures, increasing the catalyst concentration and treating to a resin loading of 30%, the ASE value went up to 80%.

15.4.8 LOSS OF DIMENSIONAL STABILITY

At very high WPG with isocyanates and epoxide-modified wood, the ASE values started to drop (Rowell and Ellis 1979, 1981, 1984a,b). Figure 15.9 shows a plot of ASE versus WPG for propylene and butylene oxide reacted wood. As the WPG increases up to about 25–30, the ASE increases. At

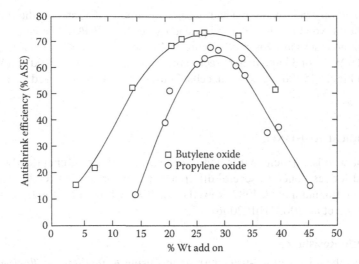

FIGURE 15.9 Loss of antishrink efficiency at high chemical weight gains.

WPGs higher than 25–30, ASE starts to decrease. The same effect is also observed in methyl isocyanate reacted wood.

Explanation of this phenomenon is observed in an electron microscopic examination of the epoxide and isocyanate reacted wood at different WPG levels. Figure 15.10 shows the electron

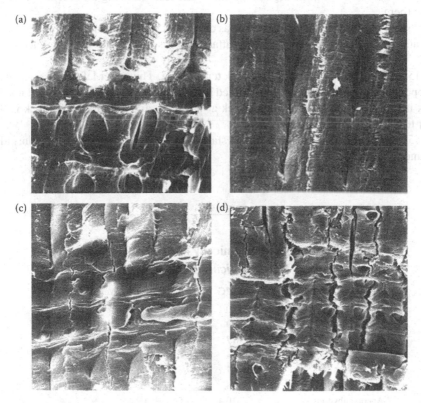

FIGURE 15.10 Scanning electron micrographs of radial-split southern pine showing swelling of wood when reacted with propylene oxide-triethylamine. (a) Control (1100×), (b) WPG 25 (1100×), (c) WPG 32.6 (600×), and (d) WPG 45.3 (550×).

micrographs for propylene oxide reacted wood. Figure 15.10a shows the unreacted wood. Figure 15.10b shows wood reacted with propylene oxide to 25 WPG. The cell wall is swollen in Figure 15.10b but no cracks are observed. Figure 15.10c shows small cracks starting to form in the tracheid walls at a WPG of 33 and major cracks are observed in Figure 15.10d at a WPG of 45. The largest cracks in Figure 15.10d are between cells. Similar results are observed in methyl isocyanate reacted wood.

15.4.9   BIOLOGICAL RESISTANCE

Various types of woods, particleboards and flakeboards made from chemically modified wood have been tested for resistance to several different types of organisms (Goldstein 1960, Stamm and Baecher 1960, Imamura et al. 1987, Rowell et al. 1989, Chow et al. 1994, Beckers et al. 1994, Militz 1991b, Wang et al. 2002, Hill 2006).

15.4.9.1   Termite Resistance

Table 15.19 shows the results of a 2-week termite test using *Reticulitermes flavipes* (subterranean termites) on several types of chemically modified pine (Rowell et al. 1979, 1988a). Propylene and butylene oxide and acetic anhydride modified wood becomes somewhat resistant to termite attack at about 30% for propylene oxide and about 20–25% for butylene oxide and acetic anhydride reacted wood. There was not complete resistance to attack and this may be attributed to the severity of the test. However, since termites can live on acetic acid and decompose cellulose to mainly acetate, perhaps it is not surprising that acetylated wood is not completely resistant to termite attack. Termite survival was quite high at the end of the tests showing that the modified wood was not toxic to them.

The increased hardness of acetylated wood as compared to controls may be part of the resistance to termite attack since termites are known to attack softer spring wood in preference to the harder late wood.

Figure 15.11 shows the results of a two week termite test. The upper left is the control before the test; the upper right is a propylene oxide modified block at 34 WPG showing some attack. The lower right block in a propylene oxide modified block at 17 WPG, and the lower left block is the control block after test.

Figure 15.12 shows the end of an in-ground stake that has been attacked by termites after a year in the ground in Gulf Port, Mississippi.

TABLE 15.19

Wood Weight Loss in Chemically Modified Pine a Two-Week Exposure Test to *Reticulitermes flavipes*

Chemical	WPG	Wood Weight Loss (%)
Control	0	31
Propylene oxide	9	21
	17	14
	34	6
Butylene oxide	27	4
	34	3
Acetic anhydride	10.4	9
	17.8	6
	21.6	5

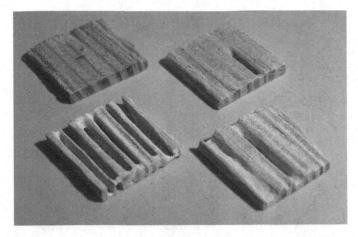

FIGURE 15.11 Small blocks in a termite test. Upper left, control; upper right, propylene oxide modified block at 25 WPG; lower right, propylene oxide modified block at 5 WPG; lower left, control block after test.

15.4.9.2 Decay Resistance

Chemically modified wood has been tested with several types of decay fungi in an ASTM standard 12 week soil block test using the brown-rot fungus *Gloeophyllum trabeum* or the white-rot fungus *Trametes versicolor* (Figure 15.13) (Nilsson et al. 1988, Rowell et al. 1987, 1988a). Table 15.20 shows that all of the chemical modifications show good resistance to white-rot fungi and all but propylene oxide shows good resistance to brown-rot fungi.

Figure 15.14a shows a control pine sample before the soil block test. Figure 15.14b shows the early stages of fungal attach with evidence of fungal mycelia developing and Figure 15.14c shows the surface of the block at the 16 week test. The block is badly attacked with about a 60 percent loss in weight. Figure 15.14d shows an acetylated block at 18 WPG after the 16-week test. There are fungal hyphae visible but no attack is evident.

Weight loss resulting from fungal attack is the method most used to determine the effectiveness of a preservative treatment to protect wood composites from decaying. In some cases, especially for brown-rot fungal attack, strength loss may be a more important measure of attack since large

FIGURE 15.12 Wood stake destroyed by termites.

FIGURE 15.13 Standard ASTM fungal test for decay. (a) Control block after 16 weeks test; (b) an acetylated block with 7% weight gain; (c) acetylated block at 18% weight gain.

strength losses are known to occur in solid wood at very low wood weight loss (Figure 15.15) (Cowling 1961).

Figure 15.14 shows the results of a brown-rot fungal test. The left graph shows the residual weight left during the fungal test and the right graph shows the average degree of polymerization left at different sample weight losses. Comparing the two graphs, it is obvious that there is substantial strength loss at a very low weight loss which occurs early in the 16 week Standard ASTM soil block test.

A dynamic bending-creep test has been developed to determine strength losses when wood composites are exposed to a brown- or white-rot fungus (Figure 15.16) (Imamura and Nishimoto 1985, Norimoto et al. 1987, 1992). A sample is placed between two upright holders and a standard load is applied to the center of the sample (~10% of its breaking strength. A pan of water is placed under the sample, the sample is inoculated with a test fungi and the entire test rig is placed in a closed polyethylene bag. Using this bending-creep test on aspen flakeboards, control boards made with phenol-formaldehyde adhesive failed in an average of 71 days using the brown-rot fungus *Tyromyces palustris* and 212 days using the white-rot fungus *Traetes versicolor* (Figure 15.17, Imamura et al. 1988, Rowell et al. 1988b). At failure, weight losses averaged 7.8% for *T. palustris* and 31.6% for

TABLE 15.20
Resistance of Chemically Modified Pine against Brown- and White-Rot Fungi

| | | Weight Loss After 12 Weeks | |
| | | Brown-Rot | White-Rot |
Chemical	WPG	Fungus (%)	Fungus (%)
Control	0	61.3	7.8
Propylene oxide	25.3	14.2	1.7
Butylene oxide	22.1	2.7	0.8
Methyl isocyanate	20.4	2.8	0.7
Acetic anhydride	17.8	1.7	1.1
Formaldehyde	5.2	2.9	0.9
β-propiolacetone	25.7	1.7	1.5
Acrylonitrile	25.2	1.9	1.9

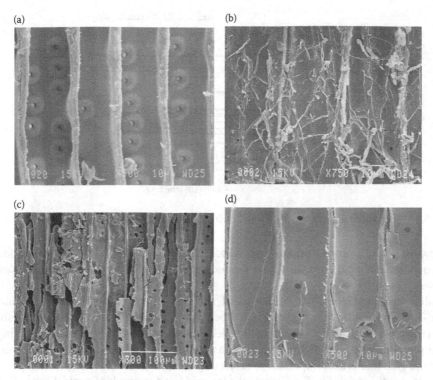

FIGURE 15.14 Pine sample before soil block test (a), pine sample after 1 week in the soil block test (b), control block after 16 weeks in test (c), and an acetylated block at 18% weight gain after the 16-week test (d).

T. versicolor. Isocyanate-bonded control flakeboards failed in an average of 20 days with *T. palustris* and 118 days with *T. versicolor*, with an average weight loss at failure of 5.5% and 34.4%, respectively (Rowell et al. 1988b). Very little or no weight loss occurred with both fungi in flakeboards made using either phenol–formaldehyde or isocyanate adhesive with acetylated flakes. None of these specimens failed during the 300-day test period.

Mycelium fully covered the surfaces of isocyanate-bonded control flakeboards within 1 week, but mycelial development was significantly slower in phenol-formaldehyde-bonded control flakeboards.

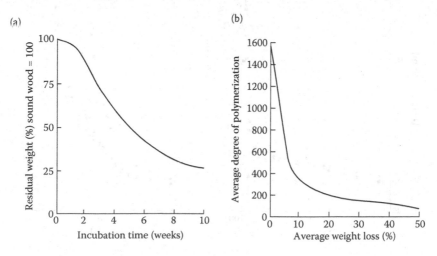

FIGURE 15.15 Residual weight left after zero to ten weeks (a) and average degree of polymerization vs. average weight loss (b) in a brown-rot decay test.

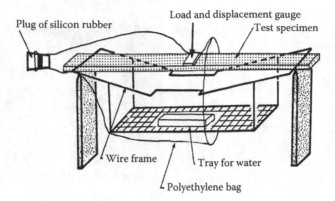

FIGURE 15.16   Set up for the dynamic bending-creep test.

Both isocyanate– and phenol–formaldehyde-bonded acetylated flakeboards showed surface myce-lium colonization during the test time, but the fungus did not attack the acetylated flakes so little strength was lost.

In similar bending-creep tests, both control and acetylated pine particleboards made using melamine–urea–formaldehyde adhesive failed because *T. palustris* attacked the adhesive in the glue line (Imamura et al. 1988). Mycelium invaded the inner part of all boards, colonizing in both wood and glue line in control boards but only in the glue line in acetylated boards. These results show that the glue line is also important in protecting composites from biological attack.

After a 16-week exposure to *T. palustris*, the internal bond strength of control aspen flakeboards made with phenol–formaldehyde adhesive was reduced over 90% and that of flakeboards made with isocyanate adhesive was reduced 85% (Imamura et al. 1987). After 6 months of exposure in moist unsterile soil, the same control flakeboards made with phenol–formaldehyde adhesive lost 65% of their internal bond strength and those made with isocyanate adhesive lost 64% internal bond strength. Failure was due mainly to great strength reductions in the wood caused by fungal attack.

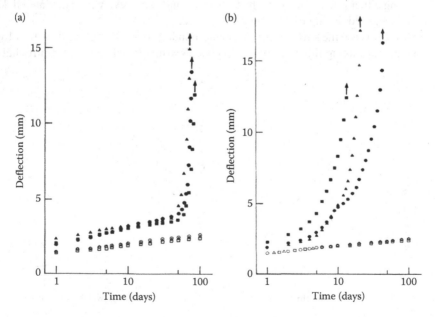

FIGURE 15.17   Deflection-time curve for flakeboards. (a) Phenol-formaldehyde bonded flake board; (b) isocyanate bonded flake board.

Acetylated aspen flakeboards lost much less internal bond strength during the 16-week exposure to *T. palustris* or 6-month soil burial. The isocyanate adhesive was somewhat more resistant to fungal attack than the phenol–formaldehyde adhesive. In the case of acetylated composites, loss in internal bond strength was mainly due to fungal attack in the adhesive and moisture, which caused a small amount of swelling in the boards.

Wood modified with furfural alcohol has also been shown to be resistant to attack by fungi (Venås and Felby 2009). These authors followed the release of carbohydrates during enzymatic attack and showed that at 20–25% weight gain of furfural alcohol reacted wood, the release of carbohydrates was reduced 55–60%. *Pinus pinaster* Aiton treated with furfural alcohol reduced decay due to the brown-rot *Postia placenta* from 29% for the control to 1% for the furfural-treated wood and reduced decay due to the white-rot *Coniophora puteana* from 6% for the control to 1% for the furfural-treated wood (Esteves et al. 2010).

DMDHEU treatment of wood has also been shown to be resistance to attack by white- and brown-rot fungi (Mai et al. 2009). Complete protection against attack by *T. versicolor* and *C. puteana* was achieved at a weight percent gain of 15% for beech and 10% for pine. Beech wood (*Fagus silvatica*) treated with different concentrations (1.3 and 2.3 M) of DMDHEU showed resistance to attach by several fungi (Table 15.21) (Bollmus et al. 2010).

The mechanism of resistance to fungal attack by chemical modification has been suggested to be due to the specific enzymatic reactions that can take place due to a change in configuration and conformation of the modified wood and/or when the reduction in equilibrium moisture content becomes too low to support fungal growth. In the specific case of brown-rot fungal attack, a mechanism has been suggested that explains the loss of strength before there is very much weight loss (Nilsson 1986, Winandy and Rowell 1984). This mechanism is consistent with the data from the soil block weight loss tests and the strength loss tests on acetylated wood.

Enzymes → Hemicelluloses
          (Energy source for
          generation of
          chemical oxidation
          system) → Hemicellulose matrix
          (H_2O_2, Fe^{3+})   (Strength losses)
                       (Energy source for generation
                       of β-glucosidases)   → Weight loss

Another test for fugal and bacterial resistance that has been done on acetylated composites is with brown-, white-, and soft-rot fungi and tunneling bacteria in a fungal cellar (Figure 15.18, Table 15.22). Control blocks were destroyed in less than 6 months while flakeboards made from acetylated furnish above 16 WPG showed no attack after 1 year (Nilsson et al. 1988, Rowell et al. 1988a). This data shows that no attack occurs until swelling of the wood occurs (Rowell and Ellis 1984, Rowell et al. 1988a). This fungal cellar test was continued for an additional 5 years with no attack at 17.9 WPG.

TABLE 15.21

Mass Loss of DMDHEU Treated Beech and Controls with Different Fungi

Fungus	Mass Loss % Control	Mass Loss % DMDHEU 1.3 M	Mass Loss % DMCHEU 2.3 M
C. puteana	36.5	8.5	0.6
P. placenta	21.1	0.4	0.7
T. versicolor	25.5	1.2	0.8

FIGURE 15.18   Fungal cellar with small sticks in test. (BAM Federal Institute for Materials Research and Testing, Berlin, Germany.)

This is more evidence that the moisture content of the cell wall is critical before attack can take place (Ibach and Rowell 2000, Ibach et al. 2000).

In the reaction of ponderosas pine with butylenes oxide and then exposed to unsterile soil in a fungal cellar, at low weight gains there was extensive surface attack by soft-rot and tunneling bacteria (Nilsson and Rowell 1983). This attach was reduced at 15% weight gain and completely stopped at 24% weight gain. The same modified wood was exposed to the brown-rot fungi *Fomitopsis pinicola* and at lower weight gains, attack was evident. At higher weight gains, there was no attack but micro cracks were observed in the ML regions and cell wall corners of latewood tracheids. The fungus entered through these micro cracks.

Table 15.23 shows the weight loss and the percent acetyl before and after a brown rot, soil block test for both solid pine and pine fiber. Both the solid wood and the wood fiber show very little loss in acetyl content during the test.

Table 15.24 shows the sugar analysis of solid wood and wood fiber after the brown rot, soil block test. The fiber control sample lost 86% of the carbohydrate polymers while the acetylated sample only showed a 13% loss of carbohydrate polymers. No glucan polymer was lost during the decay test with arabinose the largest single sugar loss. The solid wood control sample lost 88% of the carbohydrate polymers while the acetylated showed only an 18% loss of these polymers. Again arabinose showed the greatest loss in the acetylated solid wood sample.

TABLE 15.22

Fungal Cellar Tests of Aspen Flakeboards Made from Control and Acetylated Flakes[a,b]

	Rating at Intervals (Months)[c]							
WPG	**2**	**3**	**4**	**5**	**6**	**12**	**24**	**36**
0	S/2	S/3	S/3	S/3	S/4	—	—	—
7.3	S/0	S/1	S/1	S/2	S/3	S/4	—	—
11.5	0	0	S/0	S/1	S/2	S/3	S/4	
13.6	0	0	0	0	S/0	S/1	S/2	S/3
16.3	0	0	0	0	0	0	0	0
17.9	0	0	0	0	0	0	0	0

[a] Nonsterile soil containing brown-, white-, and soft-rot fungi and tunneling bacteria.

[b] Flakeboards bonded with 5% phenol–formaldehyde adhesive.

[c] Rating system: 0 = no attack; 1 = slight attack; 2 = moderate attack; 3 = heavy attack; 4 = destroyed; S = swollen.

TABLE 15.23

Equilibrium Moisture Content of Southern Pine before and Acetyl Content before and after Attack by Brown-Rot Fungus

Solid Wood	Acetyl Before %	Acetyl After %	Weight Loss %
		WPG	
0	0.63	0.22	65.8
19	18.60	17.67	5.0
		Fiber	
0	0.81	0.39	51.7
13	14.96	14.75	1.4

Arabinose is the only sugar in wood that is in a strained five-membered ring. It is possible that this easily hydrolyzed sugar is the recognition site for the fungal enzymes that starts the entire decay process in brown-rot fungi.

In-ground tests have also been done on acetylated solid wood and flakeboards (Figure 15.19) (Hadi et al. 1995, Rowell et al. 1997, Larsson Brelid et al. 2000). Specimens have been tested in the United States, Sweden, New Zealand, and Indonesia (Chow et al. 1994). The specimens in the United States, Sweden, and New Zealand are showing little or no attack after 15 years while the specimens in Indonesia failed in less than three years (Hadi et al. 1996). The failure in Indonesia was mainly due to termite attack. Figure 15.20 shows an in-ground stake that has been pulled for inspection.

Recent results show that acetylated pine at a WPG of 21.2 is outperforming CCA (copper–chromium–arsenic) at 10.3 kg/m^3 after 18 years in test in Sweden (Larsson Brelid and Westin 2010).

In ground stake tests run in Finland showed that out of 10 stakes placed in the ground, three failed in 15 years with an average acetyl weight gain of 16.6% while none failed with weight gains of 19.8% and 22.0% (Larsson Brelid and Westin 2007). These data combined with a lot of other data suggest that the threshold of acetyl weight gain for in ground contact is approximately 18–19%.

Finally, resistance to biological attack by chemical modified wood is due, in part to stabilizing the hemicelluloses against enzyme attack. Fungal resistance may also be due to changes conformation and configuration for enzymatic reactions to take place. There is very strong evidence that the moisture content too low for fungal attack.

15.4.9.3 Marine Resistance

Table 15.25 shows the data for chemically modified pine in a marine environment (Johnson and Rowell 1988). As with the termite test, all types of chemical modifications of wood help resist attack

TABLE 15.24

Carbohydrate Analysis after Brown-Rot Degradation

Wood Fiber WPG	Weight Loss (%)	Total Carbon Lost (%)	Araban Lost (%)	Galactan Lost (%)	Rhamnan Lost (%)	Glucan Lost (%)	Xylan Lost (%)	Mannan Lost (%)
0	51.7	85.8	87.9	71.9	90.0	83.8	90.6	92.5
13	1.4	13.2	89.0	55.2	70.0	0	38.3	42.0
Solid Wood WPG	**Weight Loss (%)**	**Total Carbon Lost (%)**	**Araban Lost (%)**	**Galactan Lost (%)**	**Rhamnan Lost (%)**	**Glucan Lost (%)**	**Xylan Lost (%)**	**Mannan Lost (%)**
0	65.8	88.0	86.7	66.8	75.0	89.0	84.3	93.5
19	5.0	18.0	27.4	17.5	16.7	18.0	11.1	20.9

FIGURE 15.19   In-ground stake test for preservative effectiveness.

FIGURE 15.20   Inspection of an in-ground stake.

TABLE 15.25

Ratings of Chemically Modified Southern Pine Exposed to a Marine Environment[a]

			Mean Rating due to Attack by	
Chemical	WPG	Years of Exposure	*Limnoriid and Teredinid borers[b]*	*Shaeroma terebrans[c]*
Control	0	1	2–4	3.4
Propylene oxide	26	11.5	10	—
		3	—	3.8
Butylene oxide	28	8.5	9.9	—
		3	—	8.0
Butyl isocyanate	29	6.5	10	—
Acetic anhydride	22	3	8	8.8

[a] Rating system—10 = no attack; 9 = slight attack; 7 = some attack; 4 = heavy attack; 0 = destroyed.
[b] Installed in Key West, FL. 1975 to 1987.
[c] Installed in Tarpon Springs, FL 1984 to 1987.

by marine organisms. Control specimens were destroyed in 6 months to 1 year, mainly because of attack by *Limnoria tripunctata*, while propylene and butylene oxide, butyl isocyanate and acetic anhydride reacted wood showed good resistance. Similar tests were run in Sweden on acetylated wood and the modified wood failed in marine tests after 2 years (Larsson Brelid et al. 2000). The failure was due to attack by crustaceans and molluscs.

Figure 15.21 shows the test rack as it is just pulled up from the sea test area near Key West Florida (Left) and samples before and after test. The sample in the left picture, left side is a control specimen before the test began. The center specimen is a control sample after one year in test and the specimen on the right is an acetylated sample at 20 WPG.

Figure 12.22 shows a wood sample destroyed by marine borers after one year in test.

Acetylated wood with an acetyl content of 22% showed moderate attack in test in the Swedish west coast after 10 years exposure (Larsson Brelid and Westin 2010). There is borer activity (*Limnoria lignorum*, *Teredo navalis* L., and *Nototeredo norvagica*) in these waters all year long. The service of control pine samples in this test was less than 1 year.

Furfural alcohol treated has been tested for resistance to marine borers (Westin et al. 2006, 2007). After seven years in both the North Sea and the Baltic Sea, there was no significant attack by marine borers. The most active borer in these waters was the common shipworm *Teredo navalis*.

(a) (b)

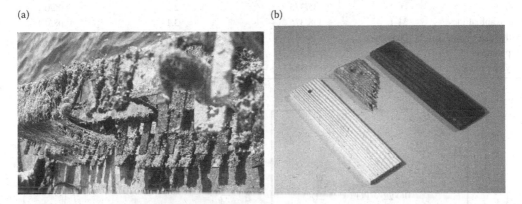

FIGURE 15.21 Samples on test rack as it is pulled from the sea (a) and (b), (from a to b) control specimen before test, control specimen after one year in test and acetylated (20 WPG) specimen after one year in test.

FIGURE 15.22 Control wood sample destroyed by marine borers.

DMDHEU treated wood has also been exposed to a marine environment (Borges et al. 2005, Klüppel et al. 2010). *Pinus sylvestris* sapwood treated above 15% weight gain showed no attack after one year in test in the North Sea and the Baltic Sea (Klüppel et al. 2010).

15.4.10   THERMAL PROPERTIES

Table 15.26, Figures 15.23 and 15.24 show the results of thermogravimetric and evolved gas analysis of chemically modified pine (see Chapter 6). The unreacted, acetylated and methyl isocyanate samples show two peaks in the thermogravimetric runs while propylene and butylene oxide only show one peak. The lower temperature peak represents the hemicellulose fraction and the higher peak represents the cellulose in the fiber. Acetylated pine fibers pyrolyze at about the same temperature and rate (Rowell et al. 1984). The heat of combustion and rate of oxygen consumption are approximately the

TABLE 15.26
Thermal Properties of Control and Acetylated Pine Fiber

Chemical	WPG	Temperature of Maximum Weight Loss (°C)	Heat of Combustion (kcal/g)	Rate of Oxygen Consumption (mm/g s)
Control	0	335/375	2.9	0.06/0.13
Acetic anhydride	21.1	338/375	3.1	0.08/0.14
Methyl isocyanate	24.0	315/375	2.6	0.07/0.12
Propylene oxide	32.0	380	4.3	0.23
Butylene oxide	22.0	385	4.1	0.24

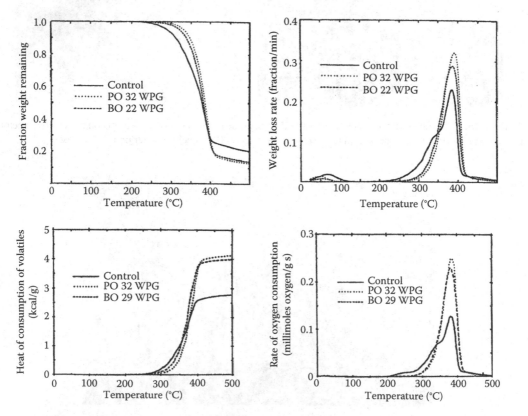

FIGURE 15.23   Thermogravimetric and evolved gas analysis of propylene and butylene oxide reacted pine.

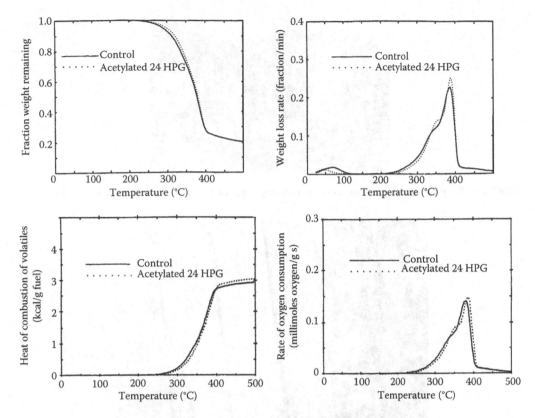

FIGURE 15.24 Thermogravimetric and evolved gas analysis of acetylated pine.

same for control and acetylated fibers which means that the acetyl groups added have approximately the same carbon, hydrogen and oxygen content as the cell wall polymers. The two peaks in control, acetylated and methyl isocyanate samples are combined into only one large peak in the epoxide reacted pine (Rowell et al. 1984). The hemicelluloses seems to become more thermally stable when epoxidized. The heat of combustion and rate of oxygen consumption is almost double for the epoxide reacted samples as compared to control, acetylated and methyl isocyanate-modified pine.

Total smoke release from pine wood was slightly higher than for controls because of the small difference in density (Morozovs and Buksans 2010). In general, acetylated and control hardwood species showed lower smoke production levels as compared to softwood species. Acetylation decreases smoke optical density by 84% for aspen, by 64% for alder, 53% for oak, and 41% for birch as compared to only 3% for pine. The heat release was higher for acetylated wood as compared to controls.

Reactive fire retardants could be bonded to the cell wall hydroxyl groups in reactions similar to this technology. The effect would be an improvement in dimensional stability, biological resistance as well as fire retardancy (Rowell et al. 1984, Ellis and Rowell 1989, Ellis et al. 1987, Lee et al. 2000).

15.4.11 ULTRAVIOLET RADIATION AND WEATHERING RESISTANCE

Figure 15.25 shows a typical outdoor exposure site for testing wood for UV resistance. Samples are places on racks and the samples can be oriented to the sun in several ways. The samples shown in the figure are vertical and facing east. Samples are also placed at 45° to the earth and some are on racks that maintain direct contact with the sun during the entire day.

Figure 15.26 shows control and acetylated samples after 13 years of outdoor weathering. The control samples on the left were coated with the same alkyd primer followed by acrylic top coating showing that acetylation is effective in extending the life of a surface coating.

FIGURE 15.25   Field area for testing wood and coatings against UV degradation in Garston, UK.

FIGURE 15.26   Performance of control and acetylated coated panels after 13 years of exposure.

Reaction of wood with epoxides and acetic anhydride has also been shown to improve ultraviolet resistance of wood (Feist et al. 1991a,b). Table 15.27 shows the weight loss, erosion rate, and depth of penetration resulting from 700 h of accelerated weathering for acetylated aspen. Control specimens erode at about 0.12 μm/h or about 0.02 %/h. Acetylation reduces surface erosion by 50%. The depth of the effects of weathering is about 200 μm into the fiber surface for the unmodified boards and about half that of the acetylated boards.

Table 15.28 shows the acetyl and lignin content of the outer 0.5 mm surface and of the remaining specimen after the surface had been removed before and after accelerated weathering. The acetyl content is reduced in the surface after weathering that shows that the acetyl blocking group is removed during weathering (Hon 1995). UV radiation does not remove all the blocking acetyl group so some stabilizing effect to photochemical degradation still is in effect. The loss of acetate is confined to the outer 0.5 mm since the remaining wood has the same acetyl content before and after

---

TABLE 15.27

Weight Loss and Erosion of Acetylated Aspen after 700 h of Accelerated Weathering

WPG	Weight Loss in Erosion (%/h)	Erosion Rate (μm/h)	Reduction in Erosion (%)	Depth of Weathering (μm)
0	0.019	0.121	—	199–210
21.2	0.010	0.059	51	85–105

TABLE 15.28

Acetyl and Lignin Analysis before and after 700 h of Accelerated Weathering of Aspen Fiberboards Made from Control and Acetylated Fiber

	Before Weathering		After Weathering	
WPG	Surface (%)	Remainder (%)	Surface (%)	Remainder (%)
		Acetyl		
0	4.5	4.5	1.9	3.9
19.7	17.5	18.5	12.8	18.3
		Lignin		
0	19.8	20.5	1.9	17.9
19.7	18.5	19.2	5.5	18.1

accelerated weathering. The lignin content is also greatly reduced in the surface as a result of weathering which is the main cell wall polymer degraded by UV radiation. Cellulose and the hemicelluloses are much more stable to photochemical degradation.

It is known that ultraviolet energy can remove the –R group as shown below to form an intermediate quinone-methide that is in a reversible tautomeric equilibrium with the free phenol. Because of this, it is not surprising that a substituted phenol on lignin that is intended to protect lignin from photochemical degradation offers little real protection.

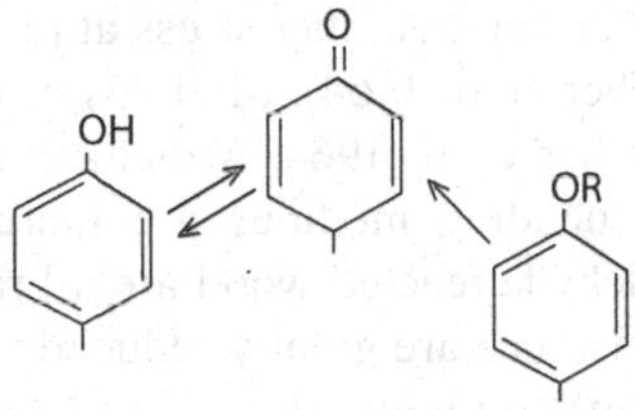

In outdoor tests, flakeboards made from acetylated pine wood maintain a light yellow color after one year while control boards turn dark orange to light gray during this time (Feist et al. 1991b). Within two years the acetylated pine had started to turn gray. Acetylated pine in test indoors retains its bright color after 10 years while control pine turns light orange after a few months.

Figure 15.27 shows that if a acetylated wood sample is subjected to artificial weathering in a weatherometer, the rate of weight loss due to weathering is greatly reduced but if the acetylated wood is subsequently treated with methyl methacrylate and the acrylic polymerized in the wood, UV resistance is greatly increased (Feist et al. 1991a). This is probably due to the new acrylic matrix holding degraded lignin in place so that erosion of the cell wall is greatly reduced.

UV stabilizers can be bonded to the wood cell wall hydroxyl groups (Kiguchi and Evans 1998, Kiguchi et al. 2001, Olsson et al. 2010). The chemistry shown the reaction of 2,4-dihydroxybenzophenone with wood for form the 2-hydroxy-4-(2,3-epoxypropoxy)benzophenone. Wood reacted with 2,4-dihydroxybenzophenone has been shown to be more resistant to UV degradation as compared to untreated wood (Olsson et al. 2010).

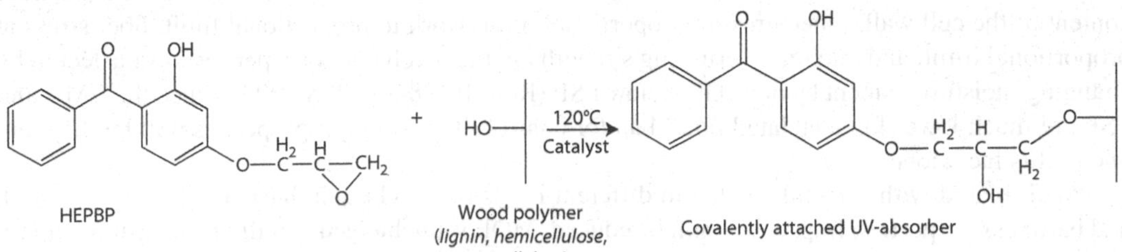

HEPBP

+

Wood polymer
(lignin, hemicellulose, or cellulose)

120°C
Catalyst

Covalently attached UV-absorber

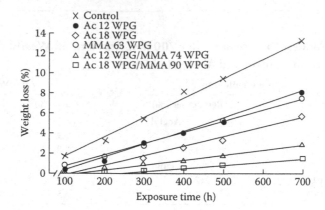

FIGURE 15.27   Weight loss due to weathering vs. exposure time.

15.4.12 MECHANICAL PROPERTIES

Strength properties are modified as a result of many of the chemical reactions done on wood. For example, shear strength parallel to the grain is decreased in acetylated wood (Dreher et al. 1964), a slight decrease in the modulus of elasticity (Narayanamurti and Handa 1953) but no change in impact strength (Koppers 1961) or stiffness (Dreher et al. 1964). Wet and dry compressive strength (Koppers 1961, Dreher et al. 1964), hardness, fiber stress at proportional limit, and work to proportional limit are increased (Dreher et al. 1964). Modulus of rupture is increased in softwoods but decreased for hardwoods (Dreher et al. 1964, Minato et al. 2003). For isocyanate reacted wood, compressive strength and bending modulus are increased (Wakita et al. 1977). The mechanical properties for formaldehyde reacted wood are all reduced as compared to unreacted wood. Toughness and abrasion resistance are greatly reduced (Tarkow and Stamm 1953, Stamm 1959), crushing and bending strength are reduced 20%, and impact bending strength is reduced up to 50% (Burmeister 1967). A definite embrittlement is observed in formaldehyde reacted wood. This may be due to the short inflexible cross-linking unit of –O–C–O– type. Some of the strength loss may also be polymer hydrolysis due to the strong acid catalyst. Strength properties of methylated wood are also reduced as a result of the strong acid used as catalyst (Narayanamurti and Handa 1953). Cyanoethylated wood, using sodium hydroxide as catalyst, has a lower impact strength (Goldstein et al. 1959, Kenaga 1963). Many of the mechanical properties of propylene oxide reacted wood are reduced. Modulus of elasticity is reduced 14%, modulus of rupture reduced 17%, fiber stress at proportional limit reduced 9% and maximum crushing strength reduced 10% (Rowell et al. 1982).

Acetylated wood does not significantly lose mechanical properties as compared to nonreacted wood (Table 15.29). It has been shown that there is very little effect on tensile strength properties of thin wood strips in a zero span test as a result of acetylation (Rowell and Banks 1987). Acetylated pine has both higher dry and wet strength as compared to control pine but the acetylated pine has also greater wet stiffness.

It should also be pointed out that strength properties of wood are very dependent on the moisture content of the cell wall. Fiber stress at proportional limit, work to proportional limit, fiber stress at proportional limit, and maximum crushing strength are the mechanical properties most affected by changing moisture content by only ±1% below FSP (Rowell 1984, USDA 1999). Since the EMC and FSP are much lower for acetylated fiber than for control fiber, strength properties will be different due to this fact alone.

Wood treated with furfuryl alcohol to different levels of weight gain have an effect on strength and hardness properties. Impact strength is reduced, MOE is unchanged, bending strength is slightly higher than controls, and Brinell, radial and tangential hardness is greatly increased (Esteves et al.

TABLE 15.29

Wet and Dry Strength and Stiffness of Control and Acetylated Pine

Sample	Dry Strength MOR (N/mm²)	Dry Stiffness MOE (N/mm²)	Wet Strength MOR (N/mm²)	Difference (%)	Wet Stiffness MOE (N/mm²)	Difference (%)
Radiata pine	63.6	10,540	39.4	−62	6760	−36
Acetylated radiata pine	64.4	10,602	58.0	−10	9690	−8.6

TABLE 15.30

Mechanical Properties of Control and Furfural Alcohol-Treated Pine Wood

Sample	MOE (MPa)	Bending Strength (MPa)	Radial Hardness (N)	Tangential Hardness (N)
Control	10,924	166	4505	4363
Kebony 70	10,833	176	7013	6534

2009). Table 15.30 shows the mechanical properties of pine treated with furfural alcohol (Kebony 70) (Esteves et al. 2009).

Pine treated with DMDHEU with $MgCl_2$ as catalyst, lost 50% of its tensile strength (Mai et al. 2007). Strength loss due to $MgCl_2$ accounted for half of the strength loss. Treatment with DMDHEU alone also showed a strength loss of about 20%.

15.4.13 ADHESION OF CHEMICALLY MODIFIED WOOD

Acetylated wood is more hydrophobic than natural wood so studies have been done to determine which adhesives might work best to make composites (Rowell et al. 1987, 1996, Youngquist et al. 1988, Vick and Rowell 1990, Youngquist and Rowell 1990, Larsson et al. 1992, Vick et al. 1993, Gomez-Bueso et al. 1999). Shear strength and wood failure was measured using acetylated yellow poplar at 0, 8, 14, and 20 WPG and adhesives including emulsion polymer isocyanate cold set, polyurethane cold set, polyurethane hot-melt, polyvinyl acetate emulsion, polyvinyl acetate cold set, polyvinyl acetate cross-link cold set, rubber-based contact-bond, neoprene contact-bond cold set, waterborne contact-bond cold set, casein, epoxy-polyamide cold set, amino resin, melamine-formaldehyde hot set, urea–formaldehyde hot set, urea–formaldehyde cold set, resorcinol–formaldehyde cold set, phenol–resorcinol–formaldehyde cold set, phenol–resorcinol–formaldehyde hot set, phenol–formaldehyde hot set, and acid-catalyzed phenol–formaldehyde. In all cases, adhesive strength was reduced by the level of acetylation—some adhesives to a minor degree and others to a severe degree.

Many adhesives were capable of strong and durable bonds at the 8 WPG level of acetylation, but not at 14 and 20 WPG levels. Most of the adhesives tested contained polar polymers, and all but four were aqueous systems, so that their adhesion was diminished in proportion to the presence of the nonpolar and hydrophobic acetate groups in acetylated wood (Rudkin 1950). Thermosetting adhesives produced the strongest bonds in both dry and wet conditions, but thermoplastic adhesives were capable of high shear strengths in the dry condition. With the exception of the acid-catalyzed phenol–formaldehyde adhesive, thermosetting adhesives that were hot pressed became highly mobile and tended to over-penetrate the wood because of the limited capacity of the acetylated wood to sorb water from the curing bond-line. The abundance of hydroxyl groups in the highly reactive resorcinol adhesive permitted excellent adhesion at room temperature, despite the limited availability of hydroxyl groups in the acetylated wood.

An emulsion polymer-isocyanate, a cross-linking polyvinyl acetate, a resorcinol–formaldehyde, a phenol–resorcinol–formaldehyde and an acid-catalyzed phenolic-formaldehyde adhesive developed bonds of high shear strength and wood failure at all levels of acetylation in the dry condition. A neo-prene contact-bond adhesive and a moisture-curing polyurethane hot-melt adhesive performed as well on acetylated wood as untreated wood in tests of dry strength. Only the cold-setting resorcinol–formaldehyde and the phenol–resorcinol–formaldehyde adhesive, alone with the hot-setting acid-catalyzed phenolic adhesive, developed bonds of high strength and wood failure at all levels of acetylation when tested in a water-saturated condition.

15.4.14 ACOUSTICAL PROPERTIES

Wood is an anisotropic, viscoelastic and hydroscopic material. All three of these properties have a profound effect on the acoustical properties of wood (Ono and Norimoto 1983, 1984, Norimoto et al. 1984, 1986, Yano et al. 1986, Tanaka et al. 1987, Zhao et al. 1987, Sasaki et al. 1988, Ono 1993, 1999). As an anisotropic composite, the properties of wood differ in all three growing directions of the tree: lengthwise (longitudinal or transverse), from the center out (radial), and along the annual rings (tangential). The density of wood varies from lower values in the earlywood or springwood (cells added in the spring) to higher values in the latewood or summerwood (cells added in the summer and early fall). The higher the density, the more swelling will occur when moisture is added to the cell wall. In the presence of moisture, the tangential grain direction in wood swells about twice as much as does grain in the radial direction. In most wood species, there is very little, if any, swelling in the longitudinal or transverse direction.

As a viscoelastic composite, the wood cell wall is made up of an elastic phase, which is the crystalline cellulose polymer, and a viscous phase, which is the noncrystalline cellulose phase, the amorphous hemicelluloses polymers and the lignin polymer. What this means in acoustical terms is that sound travels faster through the more elastic or crystalline portion of wood and more slowly through the more viscous or noncrystalline portion. When a vibration is propagated in wood, the vibration quickly passes through the elastic crystalline cellulose but since the crystalline cellulose is buried a viscous phase, the vibration is "held up" or delayed in the viscous phase. This mixed pattern of variation in vibrations is what gives wooden instruments their "mellow" or "soft" tone. Instruments made of, for example, silver, have only one elastic phase and vibrations travel faster through the metal resulting in a "bright" sound. Instruments made of, for example, plastic, have only one viscose phase and the sound from these instruments is often referred to as "muddy" or "dark." Since the crystalline cellulose is oriented somewhat (fibril angle) parallel to the longitudinal or transverse direction in the wood, sound travels faster in this direction as compared to the tangential or radial directions.

Viscoelasticity in wood is studied using dynamic mechanical analysis where an oscillatory force (stress) is applied to the wood and the resulting displacement (strain) is measured. In purely elastic materials, the stress and strain occur in phase so that the response of one occurs simultaneously with the other. In purely viscous materials, there is a phase difference between stress and strain, where strain lags stress by a 90° phase lag. Wood exhibits behavior somewhere in between that of a purely viscous and purely elastic material resulting in some phase lag in strain.

As a hydroscopic composite, the moisture content in wood depends on the relative humidity of the environment in which the wood is exposed. The sorption of water molecules in the wood cell-wall polymers acts as a plasticizer to loosen the cell-wall microstructure, which increases the viscose properties of wood. This affects the tone quality of wooden musical instruments because, as the moisture content increases, the acoustic properties of wood are reduced or dulled. The higher the moisture content, the greater the effect on the viscose phase. Because of this, someone playing a musical instrument made of wood tries to keep the instrument as dry as possible and may change instruments during a long piece of music. As mentioned before, water molecules in the cell wall also cause the wood to swell, which increases the deformation of wooden parts. This is particularly bad from a durability and sound quality perspective for parts that are under stress, for example, the strings of guitars and violins, the pin block in a piano, and in joints of instruments.

One way to study the viscoelastic properties of wood as they relate to acoustic properties is through vibrational analysis (Moore et al. 1983, Norimoto et al. 1988, Akitsu et al. 1991, 1993, Yano et al. 1993). A simple harmonic stress results in a phase difference between stress and strain. The ration of dynamic Young's modulus (E') to specific gravity (γ) (E'/γ = specific modulus) and internal friction (tangent of the phase angle, tan δ) are important properties to measure in relationship to the acoustic properties of wood. The E'/γ is related to sound velocity and tan δ to sound absorption or damping within the wood.

In a dynamic vibrational analysis, using a simple free-free vibrational beam test using acetylated and non-acetylated Sitka spruce, sound velocity and sound absorption were determined at several different relative humidities (Yano et al. 1993). The conclusions of this research showed that acetylation of wood slightly reduced both sound velocity and sound absorption when compared to unreacted wood. Acetylation greatly reduced variability in the moisture content of the cell-wall polymers thereby stabilizing the physical dimensions of the wood, and therefore, its acoustic properties. These tests did not determine whether acetylation enhanced sound quality.

Several wooden instruments and parts for musical instruments have been made from acetylated wood and have been evaluated by both amateur and professional musicians. Two violins were made: one in Japan and one in Sweden from acetylated wood. Both violins were played by professional violinists. Based on their experience, the musical quality of both violins was recorded as excellent. A guitar was also made in Japan and played by several amateur guitarists. The guitar was very responsive with good note separation. A piano soundboard was made by Yahama and evaluated at their factory in Japan. The quality of the sound was excellent. Also, a thin wooden diaphragm speaker system was made in Japan (Ono et al. 1988). Clarinet reeds have also been made using acetylated wood and it was recorded that the vibrational properties and the anticreep properties of the reeds were enhanced by the acetylation process (Obataya et al. 2000).

As a result of several years of experimentation on the effects of acetylation on the acoustical properties of wood, it can be concluded that acetylation of wood slightly increases density, and slightly (about 5%) reduces both sound velocity and sound absorption when compared to unreaceted wood. Acetylation does not change the acoustic converting efficiency. In short, acetylation slightly increases the viscose properties of the wood. Acetylation reduces the amount of moisture in the cell wall decreasing the effect of moisture on the viscose properties of wood. This alloys a wooded musical instrument to be played longer without having to let it dry out. This gives an instrument made from acetylated wood a greater range of moisture conditions it can be played in without losing tone quality. Acetylation also greatly stabilizes the physical dimensions of the wood. The major effect of acetylation of wood, therefore, is to stabilize its acoustic properties.

15.5 COMMERCIALIZATION OF CHEMICALLY MODIFIED WOOD

15.5.1 ACETYLATION

Early attempts to commercialize acetylation of wood failed. One in the United States (Koppers 1961) and one in Russia (Otlesnow and Nikitina 1977, Nikitina 1977) came close to commercialization but were discontinued presumably because they were not cost-effective. There was a commercial acetylation plant for solid wood in Japan by the Daiken Company with a product name of alpha-wood (α-Wood) but production has stopped.

15.5.1.1 Accsys Technologies

In 2008, Accsys Technologies PLC started commercial production of acetylated wood in Arnhem, The Netherlands under the trade name Accoya® wood (Kattenbroek 2005, 2007, Tjeerdsma et al. 2007). The new 30,000 m^2 production facility at Arnhem acetylates plantation-grown radiata pine from New Zealand and Chile. The process uses high temperature liquid acetic anhydride in a vacuum-pressure cycle in a large retort followed by a closed-loop chemical removal and recovery

FIGURE 15.28 Loading the acetylation cylinder at titan wood in Arnhem, the Netherlands.

system (Figure 15.28). There is very little color change in the acetylated wood (Figure 15.29. It has been noted that darker woods become lighter with acetylation and light woods become slightly darker. The acetylated wood is being used for window frames, doors, conservatories, siding, decking, shutters and louvers. In 2009, the first timber traffic bridge of Accoya wood was built and in the city Sneek in the Netherlands (Tjeerdsma and Bongers 2009) (Figure 15.30). A second bridge is planned. In 2009, a new product called Tricoya™ which is technology to produce acetylated wood fibers, chips, and particles for use in the fabrication of wood-based composites including panel products such as MDF, OSB, and particle board.

15.5.1.2 Acetylation Process Development

Two new processes are presently under way in Sweden to commercialize the acetylation of wood. One is a fiber process and the second a process to acetylate wood of large dimensions using microwave technology. One of the concerns about the acetylation of wood, using acetic anhydride as the reagent, has been the by-product acetic acid. Many attempts have been made for the "complete

FIGURE 15.29 Logo for Accoya and a sample of Accoya next to a non-acetylated pine sample.

FIGURE 15.30 Bridge made from Accoya in the Netherlands.

removal" of the acid to eliminate the smell, make the process more cost effective, and to remove a potential ester hydrolysis causing chemical. Complete removal of by-product acetic acid has now been achieved in both the fiber process and the solid wood microwave process.

There is a pilot plant in Sweden with a capacity of approximately 4000 tones of acetylated fiber a year. Figure 15.31 shows the schematic of the new continuous fiber acetylation process (Nelson et al. 1994, 1995a, 1995b, 1999, Simonson and Rowell 2000). The fiber is first dried in an optional dryer section to reduce the moisture content to as low a moisture content as is economically feasible realizing that the anhydride will react with water to form acetic acid and that a certain amount of acetic acid is needed to swell the fiber wall for chemical access.

The dried fiber is then introduced, by a screw feeder, into the reactor section and the acetylating agent is added. The temperature in this section is within the range of 110–140°C so the acetylating agent is in the form of a vapor/liquid mixture. Back flow of the acetylating agent is prevented by a fiber plug formed in the screw feeder. A screw-conveyor or similar devise is used to move the material through the reactor and to mix the fiber–reagent mixture. During the acetylation reaction, which is exothermic, the reaction temperature can be maintained substantially constant by several conventional methods. The contact time in the reactor section is from 6 to 30 min. The bulk of the acetylation reaction takes place in this first reactor.

The resultant acetylated fiber from the first reactor contains excess acetylating agent and formed acetic acid as it is fed by a star feeder into the second reactor, designed as a long tube and working as an anhydride stripper. The fiber is transported through the stripper by a stream of superheated vapor of anhydride and acetic acid. The temperature in the stripper is preferably in the range of 185°C to 195°C. The primary function of this second step is to reduce the content of the unreacted acetylating medium remaining in the fiber emerging from the first reactor. An additional acetylation of the fiber is, however, also achieved in this step. The residence in this step is relatively short and normally less than 1 min. After the second reactor (stripper), superheated vapor and fiber are separated in a cyclone and part of the superheated vapor is recirculated after heating to the stripper fiber inlet and part is transferred to the system for chemical recovery.

The acetylated fiber from the second reactor may still contain some anhydride and acetic acid that is sorbed or occluded in the fiber. In order to remove remaining chemicals and the odor from them, the acetylated fiber is introduced into a second stripper step also acting as a hydrolysis step. The transporting medium in this step is superheated steam, and any remaining anhydride is rapidly hydrolyzed to acetic acid, which is evaporated. The acetylated fiber emerging from the second stripper is essentially odor free and is completely dry. The acetylated fiber can, as a final treatment be resinated for fiberboard production or conditioned and baled for other uses as desired. The steam and acetic acid removed overhead from this step is processed in the chemical recovery step.

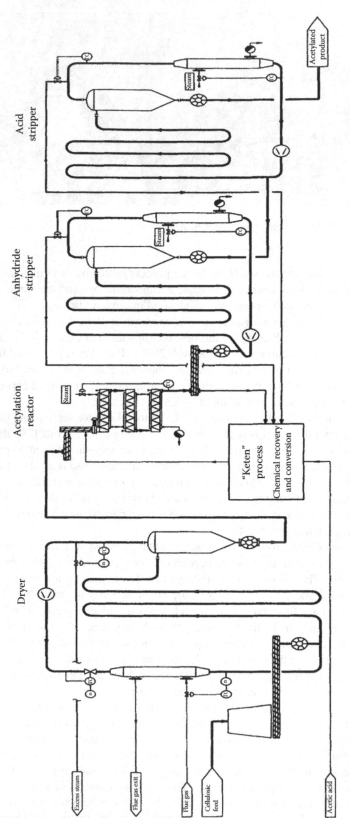

FIGURE 15.31   Schematic diagram of the fiber acetylation process.

FIGURE 15.32 Pilot plant for producing acetylated fiber.

The preferred recovery of chemicals includes separation of acetic anhydride from acetic acid by distillation, and conversion of acetic acid, recovered as well as purchased, by the ketene process into anhydride. The raw materials entering the production site is thus fiber and acetic acid to cover the acetyl groups introduced in the fiber. This minimizes the transportation costs and the chemical costs and makes the process much more cost effective.

Figure 15.32 shows the pilot plant being assembled. The plant was built during the spring of 2000, taken apart, and reassembled in Kvarntorp, Sweden. The designated production rate is 500 kg per hour or 12 tons per day or about 4000 tons per year of acetylated wood fiber. The process can be applied to any wood fiber and fiber other than wood will be used. Due to environmental issues (pilot plant too close to a school), the pilot plant has been disassembled waiting rebuilding at a new location.

Microwave energy has been shown to heat acetic anhydride and acetic anhydride-impregnated wood (Larsson Brelid and Simonson 1997, Larsson Brelid et al. 1999, Risman et al. 2001). Figure 15.33 shows the microwave reactor that was located at Chalmers University in Göteborg, Sweden. The absorption of microwave energy in acetic anhydride impregnated wood is preferred over other

FIGURE 15.33 Microwave reactor to acetylated wood.

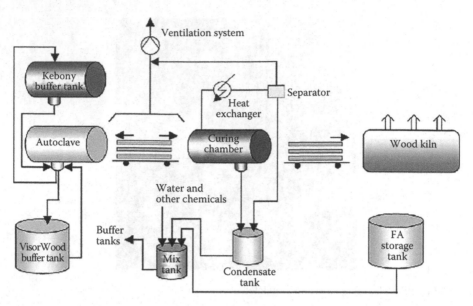

FIGURE 15.34 Schematic of the process for producing Kebony.

FIGURE 15.35 Kebony Wood coming out of the curing chamber.

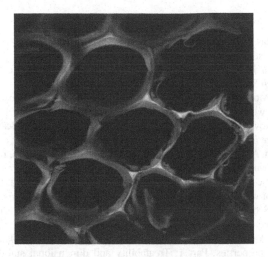

FIGURE 15.36 Furfural alcohol polymer in the wood cell wall.

methods of heating since it heats only specific parts of the wood and provides some self-regulation of the overall temperature rise and promotes a more uniform heating pattern. Acetic anhydride is supplied to the reactor, under vacuum, then pressure is applied for a short time, and following another vacuum step to remove excess anhydride. Microwave energy is then applied to heat the anhydride soaked wood. This equipment has now been moved to SP Trätek in Borås, Sweden.

The penetration depth of the microwaves at 2450 MHz is approximately 10 cm, which means this technology can be used to acetylate large wood members. The variation in acetyl content, both within and between samples is less than 2%. Microwave energy can also be used to remove the excess acetic anhydride and by-product acetic acid after acetylation.

15.5.2 FURFURYLATION

Figure 15.34 shows a schematic diagram of a Kebony plant (www.kebony.com). A factory was built in 2004 near Porsgrum, Norway. It is now being used to produce products made of southern yellow pine, beech and maple (Brynildsen and Myhre 2007, Brynildsen and Bendiktsen 2009, Figures 15.35 through 15.37).

FIGURE 15.37 Control pine sample on left and kebony sample on right.

REFERENCES

Ahmed Kabir, F.R., Nicholas, D.D., Vasishth, R.C., and Barnes, H.M. 1992. Laboratory methods to predict the weathering characteristics of wood. *Holzforschung*, 46(5): 395–401.

Akitsu, H., Norimoto, M., and T. Morooka. 1991. Vibrational properties of chemically modified wood. *Journal, Japan Wood Research Society* 37(7): 590–597.

Akitsu, H., Norimoto, M., Morooka, T., and Rowell, R.M. 1993. Effect of humidity on vibrational properties of chemically modified wood. *Wood and Fiber Science* 25(3): 250–260.

Anaya, M., Alvarez, A., Novoa, J., González, M., and Mora, M. 1984. Modification of wood with furfuryl alcohol. *Revista sobre los Derivados de la Cana de Azucar* 18: 49–53.

Andersson, M. and Tillman, A.-M. 1989. Acetylation of Jute. Effects on strength, rot resistance and hydrophobicity. *J. Applied Polymer Sci.* 37: 3437–3447.

Arni, P.C., Gray, J.D., and Scougall, R.K. 1961a. Chemical modification of wood. I. Use of trifluoroacetic acid in the esterification of wood by carboxylic acids. *Journal of Applied Chemistry* 11: 157–163.

Arni, P.C., Gray, J.D., and Scougall, R.K. 1961b. Chemical modification of wood. II. Use of trifluoroacetic acid as catalyst for the acetylation of wood. *J. Applied Chem.* 11: 163–170.

Ashaari, Z., Barnes, H.M., Vasishth, R.C., Nicholas, D.D., and Lyon, D.E. 1990a. Effect of aqueous polymer treatments on wood properties, Part 1: Treatability and dimensional stability International Research Group on Wood Preservation, Doc. No. IRG/WP 3610.

Ashaari, Z., Barnes, H.M., Vasishth, R.C., Nicholas, D.D., and Lyon, D.E. 1990b. Effect of aqueous polymer treatments on wood properties, Part 2: Mechanical properties. International Research Group on Wood Preservation, Doc. No. IRG/WP.

Avora, M., Rajawat, M.S., and Gupta, R.C. 1979. Studies on the acetylation of wood. *Holzforschung und Holzverwertung* 31(6): 138–141.

Baechler, R.H. 1959a. Improving wood's durability through chemical modification. *Forest Products Journal* 9: 166–171.

Baechler, R.H. 1959b. Fungus-resistant wood prepared by cyanoethylation. U.S. Patent 2,959, 496.

Baird, B.R. 1969. Dimensional stabilization of wood by vapor phase chemical treatments. *Wood and Fiber* 1(1): 54–63.

Banks, W.B. and Lawther, J.M. 1994. Derivation of wood in composites. In *Cellulosic Polymers, Blends and Composites*, R.G. Gilbert, ed., Hanser Publishers, New York, NY, 131.

Barnes, H.M., Choong, E.T., and McIlhenny, R.L. 1969. An evaluation of several vapor phase chemical treatments for dimensional stabilization of wood. *Forest Products Journal* 19(3): 35–39.

Beckers, E.P.J., Militz, H., and Stevens, M. 1994. Resistance of acetylated wood to basidiomycetes, soft rot and blue stain. International Research Group on Wood Preservation, 25th Annual Meeting, Bali, Indonesia, Document No: WP 94-40021.

Beckers, E.P.J. and Militz, H. 1994. Acetylation of solid wood—Initial trials on lab and semi-industrial scale. *Proceedings: Second Pacific Rim Bio-Based Composite Symposium*, Vancouver, Canasa, 125–134.

Bollmus, S., Rademacher, P., Krause, A., and Militz, H. 2010. In: *Proceedings, The Fifth European Conference on Wood Modification*. Hill, C.A.S, Militz, H., and Andersons, B. Eds. Riga, Latvia September 15–22.

Borges, L.M.S., Cragg, S., Zee, S.V.D., and Homan, W.J. 2005. Laboratory and field tests of antimarine borer potential of wood modified with dimethyloldihydroxylenurea (DMDHEU) and phosphobutane tricarboxylic acid (PBTC). In: *Proceedings, The Second European Conference on Wood Modification*, Göttingen, 198–201.

Brynildsen, P. and Bendiktsen, R. 2009. State-of-the-art Kebony factory and its main products. In: *Proceedings, 4th European conference on wood modification*, Stockholm Sweden 37–42.

Burmeister, A. 1967. Tests for wood treatment with monomeric gas of formaldehyde using gamma rays. *Holzforschung* 21(1): 13–20.

Burmeister, A. 1970. Wood of dimensional stability to moisture. German Patent 1, 812,409.

Brynildsen, P. and Myhre, E. 2007. Commercial Development of VisorWood® and Kebony® Furfurylated Wood. In: *Proceedings, The Third European Conference on Wood Modifcation*, Cardiff, UK.

Callow, H.J. 1951. Acetylation of cellulose and lignin in jute fiber. *Journal of the Indian Chemical Society* 43: 605–610.

Clemons, C., Young, R.A., and Rowell, R.M. 1992. Moisture sorption properties of composite boards from esterified aspen fiber. *Wood and Fiber Sci.* 24(3): 353–363.

Chang, H-T. and Chang, S-T. 2003. Improvements in dimensional stability and lightfastness of wood by butylation using microwave heating. *J. Wood Sci.* 49: 455–469.

Choura, M., Belgacem, N.M., and Gandini, A. 1996. Acid-catalyst polycondensation of furfuryl alcohol: Mechanisms of chromophore formation and cross-linking. *Macromolecules* 29(11): 3839–3850.

Chow, P., Bao, Z., Youngquist, J.A., Rowell, R.M., Muehl, J.H., and Krzysik, A.M. 1996. Properties of hardboards made from acetylataed aspen and southern pine. *Wood and Fiber Sci.* 28(2): 252–258.

Chow, P., Harp, T., Meimban, R., Youngquist, J.A., and Rowell, R.M. 1994. Biodegradation of acetylated southern pine and aspen composition board. *Proceedings: International Research Group on Wood Preservation 25th Annual Meeting*, IRG/WP94-40020, May-June, 1994, Bali, Indonesia.

Clermont, L.P. and Bender, F. 1957. Effect of swelling agents and catalysts on acetylation of wood. *Forest Prod. Journal* 7(5): 167–170.

Cowling, E.B. 1961. Comparative biochemistry of the decay of Sweetgum sapwood by white-rot and brown-rot fungus, U.S. Department of Agriculture, *Forest Serv. Technol. Bull.*, No. 1258, p. 50.

Dewispelaere, W., Raemedondk, J., and van Stevens, M. 1977. Decay resistance of wood treated for dimensional stabilization with monomers and formaldehyde. *Material und Organismen* 12(3): 211–222.

Dieste, A., Krause, A., Mai, C., and Militz, H. 2009. The determination of EMC and its effect on the analysis of moisture sorption in wood modified with DMDHEU. In: *Proceedings, 4th European Conference on Wood Modification*, F. Englund, C.A.S Hill, H. Militz, and B.K Segerholm, eds. 85–91.

Dreher, W.A., Goldstein, I.S., and Cramer, G.R. 1964. Mechanical properties of acetyolated wood. *Forest Products Journal* 14(2): 66–68.

Ellis, W.D. and Rowell, R.M. 1989. Flame-retardant treatment of wood with a diisocyanate and an oligomer phosphonate. *Wood and Fiber Sci.* 21(14): 367–375.

Ellis, W.D., Rowell, R.M., LeVan, S.L., and Susott, R.A. 1987. Thermal degradation properties of wood reacted with diethylchlorophosphate or phenylphosphonic dichloride as potential flame retardants. *Wood and Fiber Sci.* 19(4): 439–445.

Esteves, B., Nunes, L., and Percira, H. 2009. Furfurlation of *Pinus pinastere* wood. In: *Proceedings, 4th European Conference on Wood Modification*, Stockholm, Sweden, 415–418.

Esteves, B., Nunes, L., and Pereira, H. 2010. Durability and stability improvement of *Pinus pinastere* wood by furfurylation. In: *Proceedings, 4th International Conference on Environmental Compatible Forest Products*. Porto, Portugal, September.

Feist, W.C., Rowell, R.M., and Ellis, W.D. 1991a. Moisture sorption and accelerated weathering of acetylated and/or methyl methacrylate treated aspen. *Wood and Fiber Sci.* 23(1): 128–136.

Feist, W.C., Rowell, R.M., and Youngquist, J.A.1991b. Weathering and finish performance of acetylated aspen fiberboard. *Wood and Fiber Sci.* 23(2): 260–272.

Fuchs, W. Genuine lignin. I. 1928. Acetylation of pine wood. *Ber.* 61B: 948–51.

Fuse, G. and Nishimoto, K. 1961. Preservation of wood by chemical modification. I. Preservative properties of cyanoethylated wood. *Journal of the Japan Wood Research Society* 7(4): 157–161.

Goldstein, I.S. 1960. Improving fungus resistance and dimensional stability of wood by treatment with β-propiolactone. U.S. Patent 2,931,741.

Goldstein, I.S., Dreher, W.A., and Jeroski, E.B. 1959. Wood processing inhibition against swelling and decay. *Industrial and Engineering Chemistry* 5(10): 1313–1317.

Goldstein, I.S. and Drehrer, W.A. 1960. Stable furfuryl alcohol impregnating solutions. *Industrial and Engineering Chemistry* 52(1): 57–58.

Goldstein, I.S., Jeroski, E.B., Lund, A.E., Nielson, J.F., and Weater, J.M. 1961. Acetylation of wood in lumber thickness. *Forest Products Journal* 11(8): 363–370.

Gomez-Bueso, J., Westin, M., Torgilsson, R., Olesen, P.O., and Simonson, R. 1999. Composites made from acetylated wood fibers of different origin, Part II. The effect of nonwoven fiber mat composition upon molding ability. *Holz als Roh- und Werkstoff* 57: 178–184.

Gomez-Bueso, J., Westin, M., Torgilsson, R., Olesen, P.O., and Simonson, R. 2000. Composites made from acetylated wood fibers of different origin, Part I. Properties of dry-formed fiberboards. *Holz als Roh- und Werkstoff* 58: 9–14.

Hadi, Y.S., Darma, I.G.K.T., Febrianto, F., and Herliyana, E.N. 1995. Acetylated rubberwood flakeboard resistance to bio-deterioration. *Forest Products Journal* 45(10): 64–66.

Hadi, Y.S., Rowell, R.M., Nelsson, T., Plackett, D.V., Simonson, R., Dawson, B., and Qi, Z.-J. 1996. In-ground testing of three acetylated wood composites in Indonesia. Toward the new generation of bio-based composite products, In: *Proceedings: Third Pacific Rim Bio-Based Composites Symposium*, Kyoto, Japan, Dec.

Hill, C.A.S. 2006. *Wood Modification: Chemical, Thermal and Other Processes*. John Wiley & Sons, Chichester, England, 239 pp.

Hill, C. and Jones, D. 1996a. The dimensional stabilization of Corsican pine sapwood by reaction with carboxylic acid anhydrides. *Holzforschung* 50(5): 457–462.

Hill, C. and Jones, D. 1996b. A chemical kinetics study of the propionic anhydride modification of Corsican pine. Determination of activation energies. *Journal of Wood Chemistry and Technology* 16(3): 235–247.

Hill, C. and Kwon, J.H. 2009. The influence of wood species upon the decay protection mechanisms exhibited by anhydride modified woods. In: *Proceedings, 4th European Conference on Wood Modification*, Stockholm Sweden, 95–102.

Hon, D.N.-S. 1992. Chemical modification of wood materials: Old chemistry, new approaches. *Polymer News* (17): 102.

Hon, D.N.-S. 1995. Stabilization of wood color: Is acetylation blocking effective? *Wood and Fiber Sci.* 27(4): 360–367.

Hon, D.N.-S. 1996. *Chemical Modification of Wood Materials*. Marcel Dekker, New York, 370 pp.

Hon, D.N.-S. and Bangi, A.P. 1996. Chemical modification of juvenile wood. Part 1: Juvenility and response of southern pine OSB flakes to acetylation. *Forest Products Journal* 46(7/8): 73–78.

Horn, O. 1928. Acetylation of beech wood. *Ber.* 61B: 2542–2545.

Ibach, R.E. and Rowell, R.M. 2000. Improvements in decay resistance based on moisture exclusion. *Mol. Cryst. and Liq. Cryst.* 353: 23–33.

Ibach, R.E., Rowell, R.M., and Lee, B.-G. 2000. Decay protection based on moisture exclusion resulting from chemical modification of wood. *Proceedings: 5th Pacific Rim Bio-Based Composites Symposium*, Canberra, Australia, 197–204.

Imamura, Y. and Nishimoto, K. 1985. Bending creep test of wood-based materials under fungal attack. *J. Soc. Materials Sci.* 34(38): 985–989.

Imamura, Y., Nishimoto, K., and Rowell, R.M. 1987. Internal bond strength of acetylated flakeboard exposed to decay hazard. *Mokuzai Gakkaishi.* 33(12): 986–991.

Imamura, Y., Rowell, R.M., Simonson, R., and Tillman, A.-M. 1988. Bending-creep tests on acetylated pine and birch particleboards during white- and brown-rot fungal attack. *Paperi ja Puu* 9: 816–820.

Imamura, Y., Subiyanto, B., Rowell, R.M., and Nelsson, T. 1989. Dimensional stability and biological resistance of particleboard made from acetylated albizzia wood particles. *Japan J. Wood Res.* 76: 49–58.

Johnson, B.R. and Rowell, R.M. 1988. Resistance of chemically-modified wood to marine borers. *Material und Organismen* 23(2): 147–156.

Karlson, I. and Svalbe, K. 1972. Method of acetylating wood with gaseous ketene. *Uchen. Zap. Latv. Univ.* 166: 98–104.

Karlson, I. and Svalbe, K. 1977. Method of acetylating wood with gaseous ketene. *Latv. Lauksiamn. Akad. Raksti.* 130: 10–21.

Kattenbroek, B. 2005. How to introduce acetylated wood from the first commercial production into Europe. In: *Proceedings of the Second European Conference on Wood Modification*. Göttingen, Germany, pp. 398–403.

Kattenbroek, B. 2007. The Commercialisation of Wood Acetylation Technology on a Large Scale. In: *Proceedings of the Third European Conference on Wood Modification*. Cardiff, UK, pp. 19–22.

Kenaga, D.L. 1957. Stabilized wood. U.S. Patent 2,811,470.

Kenaga, D.L. 1963. Dimenstioal stabilization of wood and wood products. U.S. Patent 3,077,417.

Kenaga, D.L. and Sproull, R.C. 1951. Further experiments on dimensional stabilization of wood by allylation. *J. Forest Products Research Soc.* 1: 28–32.

Kenaga, D.L., Sproull, R.C., and Esslinger, J. 1950. Preliminary experiments on dimensional stabilization of wood by allylation. *Southern Lumberman* 180(2252): 45.

Kiguchi, M. and Evans, P.D. 1998. Photostabilization of wood surfaces using grafted UV absorber. *Polymer Degradation and Stability* 61: 33–45.

Kiguchi, M., Evans, P.D., Ekstedt, J., Williams, R.S., and Kataoka, Y. 2001. Improvement of the durability of clear coatings by grafting of UV-absorbers on to wood. *Surface Coatings International*. Part B: Coatings Transactions 84(4): 263–270.

Klüppel, A., Militz, H., Cragg, S., and Mai, C. 2010. Resistance of modified wood to marine borers. In: *Proceedings, The Fifth European Conference on Wood Modification*. Riga, Latvia September 389–396.

Koppers' Acetylated Wood. 1961. Dimensionally stabilized wood. New Materials Technical Information No. (RDW-400) E-106, 23 pp.

Krzysik, A.M., Youngquist, J.A., Muehl, J.M., Rowell, R.M., Chow, P., and Shook, S.R. 1992. Dry-process hardboards from recycled newsprint paper fibers. In, *Materials Interactions Relevant to Recycling of wood-Based Materials*, R.M. Rowell, R, T.L. Laufenberg, and J.K. Rowell, eds., Materials Research Society, Pittsburgh, PA, 266, pp. 73–79.

Krzysik, A.M., Youngquist, J.A., Muehl, J.M., Rowell, R.M., Chow, P., and Shook, S.R. 1993. Feasibility of using recycled newspaper as a fiber source for dry-process hardboards. *Forest Products Journal* 43(7/8): 53–58.

Kumar, S. 1994. Chemical modification of wood. *Wood and Fiber Sci* 26(2): 270–280.

Lande, S. 2008. Furfurylation of wood—Wood modification by the use of furfuryl alcohol. PhD Thesis Norwegian University of Life Sciences, Department of Ecology and Natural Resources Management, Ås, Norway.

Lande, S., Westin, M., and Schneider, M. 2008. Development of modified wood products based on furan chemistry. *Molecular Crystals and Liquid Crystals* 484: 367–378.

Larsson, P., Mahlberg, R., Vick, C., Simonson, R., and Rowell, R. 1992. Adhesive bonding of acetylated pine and spruce. *Forest Research Institute Bulletin* No. 176, D.V. Plackett, and E.A. Dunningham, ed., Rotorua, New Zealand, pp. 16–24.

Larsson Brelid, P. and Simonson, R. 1997. Chemical modification of wood using microwave technology. *Proceedings of the Forth International Conference of Frontiers of Polymers and Advanced Materials*, Cairo, Egypt.

Larsson Brelid, P. and Simonson, R. 1999. Acetylation of solid wood using microwave heating: Part II: Experiments in laboratory scale. *Holz als Roh- und Werkoff* 57: 383–389.

Larsson Brelid, P., Simonson, R., Bergman, O., and Nilsson, T. 2000. Resistance of acetylated wood to biological degradation. *Holz als Roh- und Werkstoff* 58: 331–337.

Larsson Brelid, P., Simonson, R., and Risman, P.O. 1999. Acetylation of solid wood using microwave heating: Part I: Studies of dielectric properties. *Holz als Roh- und Werkstoff* 57: 259–263.

Larsson Brelid, P. and Westin, M. 2007. Acetylated wood—Results from long-term field tests. In: *Proceedings, The Third European Conference on Wood Modification*, Cardiff, UK, 11–18, May.

Larsson Brelid, P. and Westin, M. 2010. Biological degradation of acetylated wood after 18 years in ground contact and 10 years in marine water. In: *Proceedings, International Research Group on Wood Protection, 41st Annual Meeting*, Biarritz, France, May, IRG/WP 10-40522.

Lee, H-L., Chen, G.C., and Rowell, R.M. 2000. Chemical modification of wood to improve decay ajnd thermal resistance. *Proceedings: 5th Pacific Rim Bio-Based Composite Symposium*, Evans, P.D., ed., Canberra, Australia 179–189.

Liu, C. and McMillin, C.W. 1965. Treatment of wood with ethylene oxide gas and propylene oxide gas. U.S. Patent 3,183,114.

Lutomski, K. 1975. Resistance of beechwood modified with styrene, methyl methacrylate and diisocyanates against the action of fungi. *Material und Organismen* 10(4): 255–262.

Mai, C., Verma, P., Xie, Y., Dyckmans, J., and Militz, H. 2009. Mode of action of DMDHEU treatment against wood decay by white and brown rot fungi. In: *Proceedings, 4th European Conference on Wood Modification*, Stockholm, Sweden, 45–52.

Mai, C., Xie, Y., Xiao, Z., Bollmus, S., Vetter, G., Krause, A., and Militz, H. 2007. Influence of the Modification with Different Aldehyde-based Agents on the Tensile Strength of Wood. In: *Proceedings, The Third European Conference on Wood Modification*, Cardiff, UK 49–56, May.

Matsuda, H. 1987. Preparation and utilization of esterified woods bearing carboxyl groups. *Wood Science and Technology* 21(1): 75–88.

Matsunnaga, M., Kataoka, Y., Matsunnaga, H., and Matsui, H. 2010. A noval method of acetylation of wood using supercritical carbon dioxide. *J. Wood Sci.* 56: 293–298.

McMillin, C.W. 1963. Dimensional stabilization with polymerizable vapor of ethylene oxide. *Forest Products Journal* 13(2): 56–61.

Militz. H. 1991a. The improvement of dimensional stability and durability of wood through treatment with non-catalysed acetic acid anhydride. *Holz als Roh- und Werstoff* 49(4): 147–152.

Militz, H. 1991b. Improvements of stability and durability of beechwood (*Fagus sylvatica*) by means of treatment with acetic anhydride. *Proceedings: International Research Group on Wood Preservation*, 22nd Annual Meeting, Kyoto, Japan, Document No: WP 3645.

Minato, K. and Mizukami, F. 1982. A kinetic study of the reaction between wood and vaporus formaldehyde. *Journal of the Japan Wood Research Society* 28(6): 346–354.

Minato, K., Takazawa, R., and Ogura, K. 2003. Dependence of reaction kinetics and physical and mechanical properties on the reaction systems of acetylation II: Physical and mechanical properties. *J. Wood Sci.* 49: 519–524.

Mo, D. and Domsjo Aktiebolag. 1965. Wood impregnated with gaseous ethylene oxide French Patent 1,408,170.

Moore, G.R., Kline, D.E., and Blackenhorn, P.R. 1983. Dynamic mechanical properties of epoxy-poplar composite materials. *Wood and Fiber Science* 15(4): 358–375.

Morozovs, A. and Buksans, S. 2010. Influence of acetylation, moisture and heat flux on reaction to fire in cone calorimeter tests on pine wood. In: *Proceedings 5th European Conference on Wood Modification*, Riga, Latvia, 57–64.

Nakagami, T. 1978. Esterification of wood with unsaturated carboxylic acids. V. Effect of delignification treatments on dissolution of wood esterified by the TFAA method. *Journal of the Japan Wood Research Society* 24(5): 318–323.

Nakagami, T., Amimoto, H., and Yokata, T. 1974. Esterification of wood with unsaturated carboxylic acids. I. Preparation of several wood esters by the TFAA method. Bulletin, Kyoto University Forests No. 46, 217–224.

Nakagami, T., Ohta, M., and Yokata, T. 1976. Esterification of wood using unsaturated carboxylic acids. III. Dissolution of wood by the TFAA esterification method. Bulletin, Kyoto University Forests No. 48, 198–205.

Nakagami, T. and Yokata, T. 1975. Esterification of wood using unsaturated carboxylic acids. II. Reaction conditions of esterification and properties of the prepared esters of wood. Bulletin, Kyoto University Forests No. 47, 178–183.

Nakagami, T. and Yokata, T. 1978. Esterification of wood using unsaturated carboxylic acids. IV. Chemical composition and molecular weight of the acetone soluble fraction of β-methylcrotonylated woods. *Journal of the Japan Wood Research Society* 24(5): 311–317.

Narayanamurti, V.D. and Handa, B.K. 1953. Acetylated wood. *Das Papier* 7: 87–92.

Nelson, H.L., Richards, D.I., and Simonson, R. 1994. Acetylation of lignocellulosic materials. European patent 650,998.

Nelson, H.L., Richards, D.I., and Simonson, R. 1995a. Acetylation of lignocellulosic materials. European patent 746,570.

Nelson, H.L., Richards, D.I., and Simonson, R. 1995b. Acetylation of lignocellulosic fibers. European patent 799,272.

Nelson, H.L., Richards, D.I., and Simonson, R. 1999. Acetylation of wood materials. European patent EP 0 650 998 B1.

Nicholas, D.D. and Williams, A.D. 1987. Dimensional stabilization of wood with dimethylol compounds. International Research Group of Wood Preserevation, Doc. No. IRG/WP 3412.

Nikitina, N. 1977. Quality control of acetylated wood. *Latvijas Lauksaimniecibas Akademijas Raksi* 130: 54–55.

Nilsson, T. 1986. Personal Communication, Upsalla, Sweden.

Nilsson, T. and Rowell, R.M. 1983. Decay patterns observed in butylenes oxide modified ponderosa pine attacked by *Fomitopsis pinicola*. International Research Group on Wood Preservation, Doc. No. IRG/WP/1183.

Nilsson, T., Rowell, R.M., Simonson, R., and Tillman, A.-M. 1988. Fungal resistance of pine particle boards made from various types of acetylated chips. *Holzforschung* 42(2): 123–126.

Nordstierna, L., Lande, S., Westin, M., Furó, I., and Brynildsen, P. 2007. ^{1}H NMR demonstration of chemical bonds between lignin-like model molecules and poly(furfural alcohol): Relevance to wood furfurylation. Third European Conference on Wood Modification, Oct, Cardiff, Wales, UK.

Norimoto, M., Grill, J., Minato, K., Okamura, K., Mukudai, J., and Rowell, R.M. 1987. Suppression of creep of wood under humidity change through chemical modification. *Wood Industry* (Japan) 42(18): 14–18.

Norimoto, M., Grill, J., and Rowell, R.M. 1992. Rheological properties of chemically modified wood: Relationship between dimensional stability and creep stability. *Wood and Fiber Sci.* 24(1): 25–35.

Norimoto. M., Grill, J., Sasaki, T., and Rowell, R.M. 1988. Improvement of acoustical properties of wood through chemical modification. In: *Proceedings of the European Scientific Colloquium, Mechanical Behavior of Wood*, Bordeaux, France, June 8–9, p. 37–44.

Norimoto, M., Ono, T., and Watanabe, Y. 1984. Selection of wood used for piano soundboards. *Journal society of rheology*, Japan 12: 115.

Norimoto, M., Tanaka, F., Ohogama, T., and Ikimune, R. 1986. Specific cynamic Young's modulus and internal friction of wood in the longitudinal direction. *Wood Research and Technical Notes*, No. 22, 53.

Nordstierna, L., Lande, S., Westin, M., Karlsson, O., and Furó, I. 2008. Towards novel wood-based materials: Chemical bonds between lignin-like model molecules and poly(furfuryl alcohol) studied by NMR. *Holzforschung* 62(6): 709–713.

Obataya, E., Ono, T., and Norimoto. M. 2000. Vibrational properties of wood along the grain. *Journal of Materials Science* 35(12): 2993–3001.

Olsson, S., Johansson, M., Westin, M., and Östmark, E. 2010. Covalently attached UV-absorbers for improved UV-protection of wood. In: *Proceedings 5th European Conference on Wood Modification*, Riga, Latvia, 263–266.

Ono, T. 1993. Effects of varnishing on acoustical characteristics of wood used for musical instrument soundboards. *Journal of Acoustical Society of Japan* 14(6): 397–407.

Ono, T. 1999. Transient response of wood for musical instruments and its mechanism in vibrational property. *Journal of Acoustical Society of Japan* 20(2): 117–124.

Ono, T., Katoh, Y., and Norimoto, M. 1988. Humidity-proof, quasi-isotropic wood diaphragm for loudspeakers. *Journal of Acoustical Society of Japan* 9(1): 25–33.

Ono, T. and Norimoto, M. 1983. Study on Young's modulus and internal friction on wood in relation to the evaluation of wood for musical instruments. *Japanese Journal of Applied Physics* 22: 611.

Ono, T. and Norimoto, M. 1984. On physical criteria for the selection of wood for soundboards of musical instruments. *Rheology Acta* 23: 652.

Otlesnow, Y. and Nikitina, N. 1977. Trial operation of a commercial installation for modification of wood by acetylation. *Latvijas Lauksaimniecibas Akademijas Raksi* 130: 50–53.

Ozolina, I. and Svalbe, K. 1972. Acetylation of wood by an anhydrone catalyst. *Latvijas Lauksaimniecibas Akademijas Raksi* 65: 47–50.

Papadopoulos, A.N. and Hill, C.A.S. 2002. The biological effectiveness of wood modified with linear chain carboxylic acid anhydrides against *Coniophora puteana. Holz als Roh- und Werkstoff* 60: 329–332.

Plackett, D.V., Rowell, R.M., and Close, E.A. 1990. Acetylation and the development of new products from radiata pine *Proceedings, Composite Products Symposium*, Rotorua, New Zealand, November, 1988, *Forest Research Institute Bulletin* No. 153, p. 68–72.

Popper, R. 1973. Treatments to increase dimensional stability of wood. SAH Bulletin. *Schweizerische Arbeitsgemeinschaft fur Holzforschung* 1(1): 3–12.

Popper, R. and Bariska, M. 1972. Acylation of wood. I. Water vapour sorption properties. *Holz als Roh- and Werkstoff* 30(8): 289–294.

Popper, R. and Bariska, M. 1973. Acylation of wood. II. Thermodynamics of water vapour sorption properties. *Holz als Roh- and Werkstoff* 31(2): 65–70.

Popper, R. and Bariska, M. 1975. Acylation of wood. III. Swelling and shrinkage behaviour. *Holz als Roh- and Werkstoff* 33(11): 415–419.

Ridgway, W.B. and Wallington, H.T. 1946. Esterification of wood. British Patent 579, 255.

Risi, J. and Arseneau, D.F. 1957a. Dimensional stabilization of wood. I. Acetylation. *Forest Products Journal* 7(6): 210–213.

Risi, J. and Arseneau, D.F. 1957b. Dimensional stabilization of wood. II. Crotonylation and crotylation. *Forest Products Journal* 7(7): 245–246.

Risi, J. and Arseneau, D.F. 1957c. Dimensional stabilization of wood. III. Butylation. *Forest Products Journal* 7(8): 261–265.

Risi, J. and Arseneau, D.F. 1957d. Dimensional stabilization of wood. IV. Allylation. *Forest Products Journal* 7(9): 293–295.

Risi, J. and Arseneau, D.F. 1958. Dimensional stabilization of wood. V. Phthaloylation. *Forest Products Journal* 8(9): 252–255.

Risman, P.O., Simonson, R., and Laarsson-Bredil, P. 2001. Microwave system for heating voluminous elongated loads. Swedish Patent 521,315.

Rowell, R.M. 1975a. Chemical modification of wood: advantages and disadvantages, *Proceedings, Am. Wood Preservers' Assoc.*, San Francisco, CA, April 28–30, pp. 1–10.

Rowell, R.M. 1975b. Chemical modification of wood: Reactions of alkylene oxides with southern yellow pine. *Wood Science* 7(3): 240–246.

Rowell, R.M. 1982. Distribution of reacted chemicals in southern pine modified with acetic anhydride, *Wood Sci.* 15(2): 172–182.

Rowell, R.M. 1983. Chemical modification of wood: A review, Commonwealth Forestry Bureau, Oxford, England, 6(12): 363–382.

Rowell, R.M. 1984. The Chemistry of Solid Wood, Advances in Chemistry Series No. 207, American Chemical Society, Washington, DC, 458 pp.

Rowell, R.M. 1990. Chemical modification of wood: Its application to composite wood products, *Proceedings, Composite Products Symposium*, Rotorua, New Zealand, November, 1988, *Forest Research Institute Bulletin*, No. 153, 57–67.

Rowell, R.M. 1991. Chemical modification of wood. In, *Handbook on Wood and Cellulosic Materials*, Hon, D. N.-S. and Shiraishi, N.,eds., Marcel Dekker, Inc., New York, NY, Chapter 15, pp. 703–756.

Rowell, R.M. 1992. Property enhancement of wood composites, Composites Applications: The role of matrix, fiber, and interface, Vigo, T. L. and Kinzig, B.J., eds., VCH Publishers, Inc, New York, NY, 365–382.

Rowell, R.M. 1993. Opportunities for composite materials from jute and kenaf, International consultation of jute and the environment, Food and Agricultural Organization of the United Nations, ESC:JU/IC 93/15, 1–12.

Rowell, R.M. 1999. Chemical modification of wood. In: *Proceedings, International Workshop on Frontiers of Surface Modification and Characterization of Wood Fibers*, P. Gatenholm and T. Chihani, eds. Fiskebackskil, Sweden, May 30–31, 1996. ISBN 91-7197-593-4, 31–47.

Rowell, R.M. 2006. Acetylation of wood: A journey from analytical technique to commercial reality. *Forest Products Journal* 56(9): 4–12.

Rowell, R.M. 2010. Production of acetylated wood fiber: Part 1: Refining acetylated chips; Part 2: exploding acetyated chips; Part 3: Acetylation of chips in a refiner. In: *Proceedings of Fifth European Conference on Wood Modification*, Sept., Riga, Latvia, p.73–80.

Rowell, R.M. and Banks, W.B. 1985. Water repellency and dimensional stability of wood, USDA Forest Service General Technical Report FPL 50, Forest Products Laboratory, Madison, WI, 24 pp.

Rowell, R.M. and Banks, W.B. 1987. Tensile strength and work to failure of acetylated pine and lime flakes. *British Polymer J.* 19: 479–482.

Rowell, R.M. and Chen, G.C. 1994. Epichlorohydrin coupling reactions with wood. I. Reaction with biologically active alcohols. *Wood Science and Technology* 28: 371–376.

Rowell, R.M. and Clemons, C.M. 1992. Chemical modification of wood fiber for thermoplasticity, compatibilization with plastics and dimensional stability, In: *Proceedings, International Particleboard/Composite Materials Symposium*, Pullman, WA, p. 251–259.

Rowell, R.M., Dawson, B.S., Hadi, Y.S., Nicholas, D.D., Nilsson, T., Plackett, D.V., Simonson, R., and Westin, M. 1997. Worldwide in-ground stake test of acetylated composite boards. International Research Group on Wood Preservation, Section 4, Processes, Document no. IRG/WP 97-40088, Stockhold, Sweden, p. 1–7.

Rowell, R.M. and Ellis, W.D. 1978. Determination of dimensional stabilization of wood using the water-soak method. *Wood and Fiber* 10(2): 104–111.

Rowell, R.M. and Ellis, W.D. 1979. Chemical modification of wood: Reaction of methyl isocyanate with southern yellow pine. *Wood and Fiber* 12(1): 52–58.

Rowell, R.M. and Ellis, W.D. 1981. Bonding of isocyanates to wood. *Am. Chem. Soc. Symposium Series* 172: 263–284.

Rowell, R.M. and Ellis, W.D. 1984a. Effects of moisture on the chemical modification of wood with epoxides and isocyanates. *Wood and Fiber Sci.* 16(2): 257–276.

Rowell, R.M. and Ellis,W.D. 1984b. Reaction of epoxides with wood, USDA Forest Service Research Paper, FPL 451, Forest Products Laboratory, Madison, WI, 41 pp.

Rowell, R.M., Esenther, G.R., Nicholas. D.D., and Nilsson, T. 1987. Biological resistance of southern pine and aspen flakeboards made from acetylated flakes. *J. Wood Chem. Tech.* 7(3): 427–440.

Rowell, R.M., Esenther, G.R., Youngquist, J.A., Nicholas, D.D., Nilsson, T., Imamura, Y., Kerner-Gang, W., Trong, L., and Deon, G. 1988a. Wood modification in the protection of wood composites, In: *Proceedings: IUFRO Wood Protection Subject Group*, Honey Harbor, Ontario, Canada. Canadian Forestry Service, 238–266.

Rowell, R.M. and Gutzmer, D.L. 1975. Chemical modification of wood: Reactions of alkylene oxides with southern yellow pine. *Wood Science* 7(3): 240–246.

Rowell, R.M. and Harrison, S.E. 1993. Property enhanced kenaf fiber composites, *Proceedings, Fifth Annual International Kenaf Conference*, Bhangoo, M. S., ed., California State University Press, Fresno, CA, 129–136.

Rowell, R.M., Hart, S.V., and Esenther, G.R. 1979. Resistance of alkylene-oxide modified southern pine to attack by subterranean termites. *Wood Sci* 11(4): 271–274.

Rowell, R.M., Imamura, Y., Kawai, S., and Norimoto, M. 1989. Dimensional stability, decay resistance and mechanical properties of veneer-faced low-density particleboards made from acetylated wood. *Wood and Fiber Sci.* 21(1): 67–79.

Rowell, R.M. and Keany, F. 1991. Fiberboards made from acetylated bagasse fiber. *Wood and Fiber Sci.* 23(1): 15–22.

Rowell, R.M. and Konkol, P. 1987. Treatments that enhance physical properties of wood, USDA, Forest Service, Forest Products Laboratory Gen. Technical Report FPL-GTR-55, Madison, WI, 12 pp.

Rowell, R.M., Lichtenberg, R.S., and Larsson, P. 1992a. Stability of acetyl groups in acetylated wood to changes in pH, temperature, and moisture. Forest Research Institute Bulletin No. 176, Rotorua, New Zealand, p. 33–40.

Rowell, R.M., Lichtenberg, R.S., and Larsson, P. 1992b. Stability of acetylated wood to environmental changes. *Wood and Fiber Sci.* 25(4): 359–364.

Rowell, R.M., Moisuk, R., and Meyer, J.A. 1982. Wood polymer composites: Cell wall grafting of alkylene oxides and lumen treatments with methyl methacrylate. *Wood Science* 15(2): 90–96.

Rowell, R.M. and Norimoto, M. 1987. Acetylation of bamboo fiber. *J. Jap. Wood Res. Soc.* 33(11): 907–910.

Rowell, R.M. and Norimoto, M. 1988. Dimensional stability of bamboo particleboards made from acetylated particles. *Mokuzai Gakkaishi* 34(7): 627–629.

Rowell, R.M. and Plackett, D. 1988. Dimensional stability of flakeboards made from acetylated *Pinus radiata* heartwood and sapwood flakes. *New Zealand J. For. Sci.* 18(1): 124–131.

Rowell, R.M. and Rowell, J.S. 1989. Moisture sorption of various types of acetylated wood fibers. In *Cellulose and Wood*, C. Schuerch, ed., John Wiley and Sons, New York, NY, pp. 343–356.

Rowell, R.M., Simonson, R., and Tillman, A.-M. 1986d. A simplified procedure for the acetylation of chips for dimensionally stable particle board products. *Paperi ja Puu*. 68(10): 740–744.

Rowell, R.M., Simonson, R., and Tillman, A.-M. 1990. Acetyl balance for the acetylation of wood particles by a simplified procedure. *Holzforschung* 44(4): 263–269.

Rowell, R.M., Simonson, R., and Tillman, A.-M. 1991a. A process for improving dimensional stability and biological resistance of wood materials, European Patent 0213252.

Rowell, R.M., Simonson, R., Hess, S., Plackett, D.V., Cronshaw, D., and Dunningham, E. 1991b. Acetyl distribution in acetylated whole wood and reactivity of isolated wood cell wall components to acetic anhydride. *Wood and Fiber Sci.* 26(1): 11–18.

Rowell, R.M., Susott, R.A., De Groot, W.G., and Shafizadeh, F. 1984. Bonding fire retardants to wood. Part I, *Wood and Fiber Sci.* 16(2): 214–223.

Rowell, R.M., Tillman, A.-M., and Simonson, R. 1986a. Vapor phase acetylation of southern pine, Douglas-fir, and aspen wood flakes. *J. Wood Chem. and Tech.* 6(2): 293–309.

Rowell, R.M., Tillman, A.-M., and Simonson, R.A. 1986b. simplified procedure for the acetylation of hardwood and softwood flakes for flakeboard production. *J. Wood Chem. and Tech.* 6(3): 427–448.

Rowell, R.M., Wang, R.H.S., and Hyatt, J.A. 1986c. Flakeboards made from aspen and southern pine wood flakes reacted with gaseous ketene. *J. Wood Chem. and Tech.* 6(3): 449–471.

Rowell, R.M., Young, R.A., and Rowell, J.K. 1996. *Paper and Composites from Agro-Based Resources*. CRC Lewis Publishers, Baco Rato, FL, 446 pp.

Rowell, R.M., Youngquist, J.A., and Imamura, Y. 1988b. Strength tests on acetylated flakeboards exposed to a brown rot fungus. *Wood and Fiber Sci.* 20(2): 266–271.

Rowell, R.M., Youngquist, J.A., Rowell, J.S., and Hyatt, J.A. 1991b. Dimensional stability of aspen fiberboards made from acetylated fiber. *Wood and Fiber Sci.* 23(4): 558–566.

Rowell, R.M., Youngquist, J.A., and Sachs, I.B. 1987. Adhesive bonding of acetylated aspen flakes. Part I. surface changes, hydrophobicity, adhesive penetration, and strength. *International J. Adhesion Adhesives* 7(4): 183–188.

Rowell, R. M. and Youngs, R.L. 1981. Dimensional stabilization of wood in use, USDA, Forest Serv. Res. Note. FPL-0243, Forest Products Laboratory, Madison, WI.

Rudkin, A W. 1950. The role of hydroxyl group in the gluing of wood. *Austral. J. Appl. Sci.* 1: 270–283.

Sasaki, T., Noromoto, M., Yamada, T., and Rowell, R.M. 1988. Effect of moisture on the acoustical properties of wood. *Journal, Japan Wood Research Society* 34(10): 794–803.

Schneider, M.H. and Witt, A.E. 2004. History of wood polymer composite commercialization. *Forest Products Journal* 54(2): 19–24.

Schurch, C. 1968. Treatment of wood with gaseous reagents. *Forest Products Journal* 18(3): 47–53.

Simonson, J. 1998. Lack of dimensional stability in cross-linked wood-polymer composites. *Holzforschung* 52(1): 102–104.

Simonson, R. and Rowell, R.M. 2000. A new process for the continuous acetylation of wood fiber. In, *Proceedings of Fifth Pacific Rim Bio-Based Composite Symposium*, P.D. Evans, ed. Canberra, Australia, 190–196.

Singh, S.P., Dev, I., and Kumar, S. 1981. Chemical modification of wood II. Vapour phase aceylation with acetyl chloride. *International Journal of Wood Preservation* 1(4): 169–171.

Stamm, A.J. 1959. Dimensional stabilization of wood by thermal reactions and formaldehyde crosslinking. *Tappi* 42: 39–44.

Stamm, A.J. 1964. *Wood and Cellulose Science*. The Ronald Press Co., New York, 549 pp.

Stamm, A.J. 1977. Dimensional stabilization of wood with furfuryl alcohol. In, *Wood Technology; Chemical aspects*. I. Goldstein (Ed.). ACS Symposium series 43, American Chemical Society, Washington DC, pp.141–149.

Stamm, A.J. and Baecher, R.H. 1960. Decay resistance and dimensional stability of five modified woods. *Forest Products Journal* 10(1): 22–26.

Stamm, A.J. and Tarkow, H. 1947. Acetylation of wood and boards. U.S. Patent 2417995.

Stevens, M., Schalck, J., and Raemdonck, J.V. 1979. Chemical modification of wood by vapour-phase treatment with formaldehyde and sulfur dioxide. *International Journal of Wood Preservation* 1(2): 57–68.

Stromdahl, K. 2000. Water sorption in wood and plant fibers, PhD thesis, The Technical University of Demark, Department of Structural Engineering and Materials, Copenhagen, Denmark.

Suida, H. 1930. Acetylating wood. Austrian Patent 122, 499.

Suida, H. and Titsch, H. 1928. Chemistry of beech wood: Acetylation of beech wood and cleavage of the acetyl-beech wood. *Ber* 61B: 1599–1604.

Suida, H. and Titsch, H. 1929. Acetylated wood: The combination of the incrustation and a method of separation of the constituents of wood. *Monatsh.* 53/54: 687–706.

Svalbe, K. and Ozolina, I. 1970. Modification of wood by acetylation. *Plast. Modif. Drev.* 145–146.

Tanaka, C., Najia, T., and Takahashi, A. 1987. Acoustical property of wood. *Journal, Japan Wood Research Society* 33(10): 811–817.

Tarkow, H. 1945. Acetylation of wood with ketene. Office Report, Forest Products Laboratory, USDA, Forest Service, 2 pp.

Tarkow, H. 1946. *A New Approach to the Acetylation of Wood.* USDA Forest Service, Forest Products Laboratory, Madison, WI, Internal Report, 9 pp.

Tarkow, H. 1959. A new approach to the acetylation of wood. Office Report, Forest Products Laboratory, USDA, Forest Service, 9 pp.

Tarkow, H., Stamm, A.J., and Erickson, E.C.O. 1950. *Acetylated Wood.* USDA, Forest Service, Forest Products Laboratory Report 1593, 29 pp.

Tarkow, H. and Stamm, A.J. 1953. Effect of formaldehyde treatments upon the dimensional stabilization of wood, *Journal Forest Products Research Society* 3: 33–37.

Tjeerdsma, B.F. and Bongers, F. 2009. The making of a traffic timber bridge of acetylated radiata pine. In: *Proceedings, The Forth European Conference on Wood Modification,* Stockholm, Sweden 15–22.

Tjeerdsma, B.F., Kattenbroek, B., and Jorissen, A. 2007. Acetylated wood in exterior and heavy load-bearing constructions: building of two timber traffic bridges of acetylated Radiata pine. In: *Proceedings of the Third European Conference on Wood Modification.* Cardiff, UK, pp. 403–412.

Truksne, D. and Svalbe, K. 1977. Water-repellent properties and dimensional stability of acetylated pine wood in relation to the degree and method of acetylation. *Latvijas Lauksaimniecibas Akademijas Raksi* 130: 26–31.

Ueyama, A., Araki, M., and Goto, T. 1961. Dimensional stability of woods. X. Decay resistance of formaldehyde treated wood. *Wood Research* (26): 67–73.

United States Department of Agriculture. 1999. Forest Service, Wood Handbook,. USDA Agri. Handbook 72, General Technical Reort FPL-GTR-113, Washington, D.C.

Venås, T.M. and Felby, C. 2009. Enzymatic hydrolysis of furfurlated scots pine sapwood. In: *Proceedings, 4th European Conference on Wood Modification,* F. Englund, C.A.S Hill, H. Militz and B.K Segerholm, eds. 53–60.

Vick, C.B. and Rowell, R.M. 1990. Adhesive bonding of acetylated wood. *Internat. J. Adhesion and Adhesives* 10(4): 263–272.

Vick, C.B., Larsson, P.Ch., Mahlberg, R.L., Simonson, R., and Rowell, R.M. 1993. Structural bonding of acetylated Scandinavian softwood for exterior lumber laminates. *Internat. J. Adhesion and Adhesives* 13(3): 139–149.

Wakita, H., Onishi, H., Jodai, S., and Goto, T. 1977. Studies on the improvement of wood materials. XVIII. Vapor phase reactions of 2,4-tolyene diisocyanate in wood. *Zairyo* 26(284): 460–464.

Wålinder, M., Segerholm, K., Larsson-Brelid, P., and Westin, M. 2010. Liquids and coatings wettability and penetrability of acetylated scots pine sapwood. In: *Proceedings 5th European Conference on Wood Modification,* Riga, Latvia, 381–388.

Wang, C.-L., Lin, T.-S., and Li, M.-H. 2002. Decay and termite resistance of planted tree sapwood modified by acetylation. *Taiwan Journal of Forest Sci.* 17(4): 483–490.

Weaver, J.W., Nielson, J.F., and Goldstein, I.S. 1960. Dimensional stabilization of wood with aldehydes and related compounds. *Forest Products Journal* 10(6): 306–310.

Westin, M., Rapp, A., and Nilsson, T. 2006. Field test of resistance of modified wood to marine borers. *Wood Material Science and Engineering* 1(11): 34–38.

Westin, M., Rapp, A.O., and Nilsson, T. 2007. Marine borer resistance of modified wood- Results from seven years in field. Secretariat of the International Research Group on Wood Protection. IRG Document No. IRG/WP 07–40375. pp. 1–7.

Winandy, J.E. and Rowell, R.M. 1984. The chemistry of wood strength. In, R.M. Rowell, ed., American Chemical Society Advances in Chemistry Series No. 207. Washington, DC, Chapter 4, 211–255.

Yano, H., Norimoto, M., and Rowell, R.M. 1993. Stabilization of acoustical properties of wooden musical instruments by acetylation. *Wood and Fiber Sci.* 25(4): 395–403.

Yano, H., Norimoto, M., and Yamada, T. 1986. Changes in acoustical properties of Sitka spruce due to acetylation. *Journal, Japan Wood Research Society* 32(12): 990–995.

Yano, H., Norimoto, M., and Rowell, R.M. 1993. Stabilization of acoustical properties of wooden musical
instruments by acetylation. *Wood and Fiber Science* 25(4): 395–403.

Youngquist, J.A. and Rowell, R.M. 1990. Adhesive bonding of acetylated aspen flakes. Part III. Adhesion with
isocyanates. *Internat. J. Adhesion and Adhesives* 10(4): 273–276.

Youngquist, J.A., Sachs, I.B., and Rowell, R.M. 1988. Adhesive bonding of acetylated aspen flakes, Part II.
Effects of emulsifiers on phenolic resin bonding. *Internat. J. Adhesion and Adhesives* 8(4): 197–200.

Zee, M.E., Beckers, E.P.J., and Militz, H. 1998. Influence of concentration, catalyst and temperature on dimensional stability of DMDHEU modified Scots pine. International Research Group on Wood Preservation,
Doc. No. IRG/WP 98-40119.

Zhao, J., Norimoto, M., Tanaka, F., Yamada, T., and Rowell, R.M. 1987. Structure and properties of acetylated
wood I. Changes in degree of crystallinity and dielectric properties by acetylation. *J. Japan Wood Res.
Soc.* 33(2): 136–142.

Zimakov, P.V. and Pokrovskii, V.A. 1954. A peculiarity in the reaction of ethylene oxide with wood. *Zhurnal
Prikladnoi Khimii* 27: 346–348.

16 Lumen Modifications

Rebecca E. Ibach and Roger M. Rowell

CONTENTS

16.1 METHODS

16.1.1 *In situ* Polymerization of Liquid Monomers in the Lumens

When wood is vacuum impregnated with liquid vinyl monomers that do not swell wood, and then *in situ* polymerized either by chemical catalyst-heat, or gamma radiation, the polymer is located almost solely in the lumens of the wood. Figure 16.1 is a scanning electron microscopy (SEM) micrograph of unmodified wood showing open cells that are susceptible to indentation and wear. In contrast, Figure 16.2 is a micrograph of wood after impregnation and polymerization showing the voids filled with polymer that will resist indentation and wear.

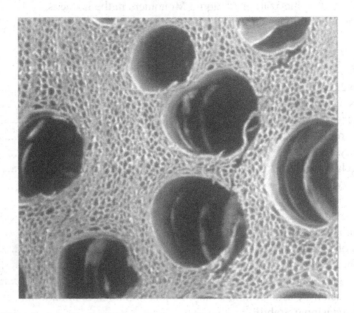

FIGURE 16.1   SEM micrograph of solid wood before polymer impregnation with open lumens.

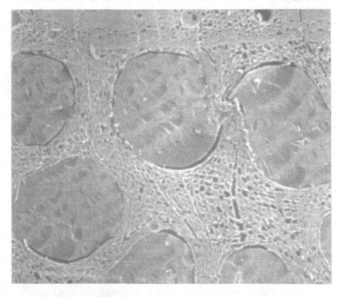

FIGURE 16.2   SEM micrograph of solid wood after polymer impregnation with filled lumens.

The process for impregnating wood with acrylics involves drying the wood (usually at 105°C) overnight to remove moisture, and then weighing. The wood is placed in a container (large enough for the wood, and an equal volume of solution), and a weight is placed on top of the wood to hold it under the solution. A vacuum (0.7–1.3 kPa) is applied to the wood for 30 min. (It can be longer depending on the size of the wood to be treated.) The acrylic monomer solution (containing the acrylic monomer, a catalyst [such as azo compounds] and perhaps a cross linking agent) is introduced into the container. The vacuum is maintained for 5–10 min to remove air from the monomer. The vacuum is then released, and the chamber returned to atmospheric conditions. The wood and solution are allowed to stand for usually 30 min. If the wood specimens have large dimensions, or are hard to penetrate species, then pressure is applied. The amount of pressure and time under pressure depends again on the size of the wood to be treated. Pressure is applied for 30 min, released, and again the wood is allowed to soak in the solution for 30 min. The treated wood is removed from the solution, drained, and wiped to remove excess chemical from the outside of the specimens. The monomer in the wood can be cured by either heat or irradiation. With heat, the chamber itself can be heated, or the wood can be removed from the chamber and heat cured in an oven or heated press. If polymerizing by heat, then the temperature is described by the catalyst, that is, Vazo 67 is heated at around 67°C. Those monomers that do not polymerize in the presence of air require a curing environment without air present. Heat is applied usually overnight or until the monomer has polymerized in the wood. Samples are weighed again, and percentage weight gain calculated. Some polymer will be on the surface of the wood and is sanded off.

There are many sources of acrylics and many different types of acrylics. Some monomers will be discussed in this chapter. The general one is methyl methacrylate. The thickness of the piece of wood being treated will determine the amount of pressure and/or vacuum needed. Small thin pieces of wood may not need any vacuum.

16.2 POLYMERIZATION METHODS

Free radicals used to initiate polymerization can be generated in two ways—by temperature sensitive catalysts or radiation curing. Chemical curing is a cheaper method with small-scale productions, while gamma radiation is more economical on a larger scale (Lee 1969).

A free radical catalyst or gamma-irradiated monomer generates the free radicals ($R\bullet + R\bullet$)

Initiation Step: $R\bullet + M$ (monomer) $\rightarrow R\text{-}M\bullet$
Propagation Step: $R\text{-}(M)_n\text{-}M\bullet + M \rightarrow R\text{-}(M)_{n-1}\text{-}M\bullet$
Termination Step: $R\text{-}(M)_n\text{-}M\bullet + R\text{-}(M)_n\text{-}M\bullet \rightarrow R\text{-}(M)_n\text{-}M\text{-}M\text{-}(M)_n\text{-}R$

16.2.1 CHEMICAL INITIATORS

16.2.1.1 Peroxides

Peroxides form free radicals when thermally decomposed. These radicals initiate polymerization of vinyl monomers. Some peroxides used to initiate polymerization of monomers in wood include t-butyl hydroperoxide, methyl ethyl ketone peroxide, lauroyl peroxide, isopropyl hydroperoxide, cyclohexanone peroxide, hydrogen peroxide, and benzoyl peroxide. Each of the radicals generated from these peroxides has a different reactivity. The phenyl radical is more reactive than the benzyl radical, and the allyl radical is unreactive. Benzoyl peroxide is the most commonly used initiator. Usually the amount of peroxide added ranges from 0.2% to 3% by weight of monomer.

16.2.1.2 Vazo Catalysts

Dupont manufactures a series of catalysts with the trade name Vazo® that are substituted azonitrile compounds. The catalysts are white crystalline solids that are soluble in most vinyl monomers.

Upon thermal decomposition, the catalysts decompose to generate two free radicals per molecule. Nitrogen gas is also generated. The grade number is the Celsius temperature at which the half-life in solution is 10 h. The series consists of:

Vazo® 52, the low-temperature polymerization initiator,

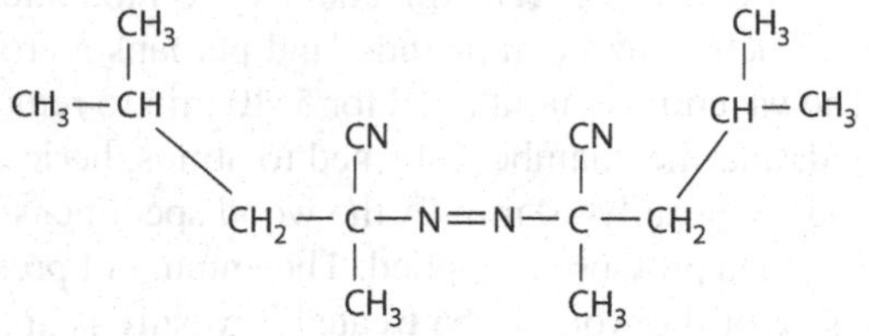

2,2'-azobis-2,4-dimethylvaleronitrile

Vazo® 64, also known as AIBN, (toxic tetramethylsuccinonitrile (TMSN) is produced, therefore better to substitute Vazo 67),

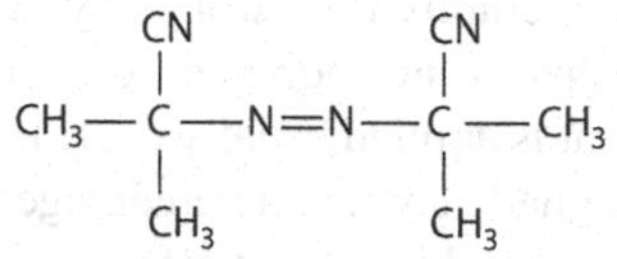

2,2'-azobisisobutyronitrile

Vazo® 67, best solubility in organic solvents and monomers,

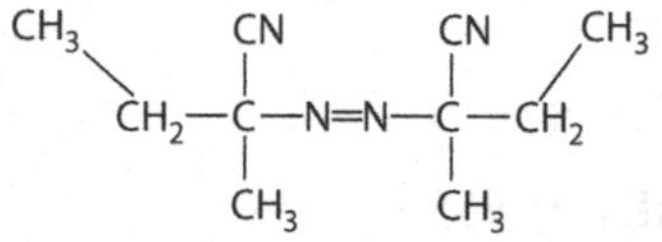

2,2'-azobis-(2-methylbutyronitrile)

and Vazo® 88,

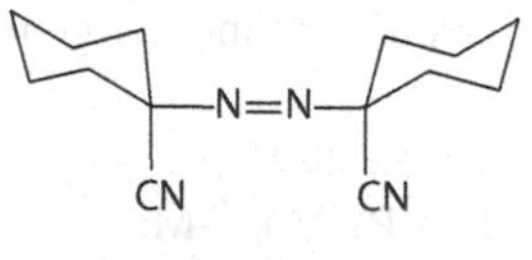

1,1'-azobis-(cyclohexanecarbonitrile)

Vazo® free radical initiators are solvent soluble and have a number of advantages over organic peroxides. They are more stable than most peroxides, so they can be stored under milder conditions, and are not shock-sensitive. They decompose with first-order kinetics, are not sensitive to metals, acids, and bases, and they are not susceptible to radical-induced decompositions. The Vazo catalysts produce less energetic radicals than peroxides, so there is less branching and cross-linking. They are a weak oxidizing agent, which lets them be used to polymerize unsaturated amines, mercaptans, and aldehydes without affecting pigments and dyes.

Catalysts are most frequently used in concentrations of 1% or less by weight of the monomers. The rate of free radical formation is dependent on the catalyst used and is controlled by regulating the temperature. For Vazo 52, the temperature range is 35°C to 80°C; for Vazo 64 and Vaso 67, 45°C to 90°C; and for Vazo 88, 80°C to 120°C.

AIBN and cyclohexanone peroxide were used to initiate the polymerization of styrene in birch-wood (Okonov and Grinberg 1983). Benzoyl peroxide or AIBN were used as initiators for beech wood impregnated with trimethylolpropane trimethacrylate and polyethylene glycol dimethacrylate (Nobashi et al. 1986). Buna sapwood was impregnated with tetraethylene glycol dimethacrylate containing AIBN as initiator (Nobashi et al. 1986). WPC were prepared from wood by impregnating with a mixture of unsaturated polyester, MMA, styrene and AIBN or benzoyl peroxide followed by heat-curing (Pesek 1984). Wood materials such as birch wood, basswood, and oak wood were impregnated with MMA or unsaturated polyester-styrene containing AIBN or benzoyl peroxide and polymerized. The polymerization was faster in the presence of AIBN than with benzoyl peroxide (Kawakami and Taneda 1973). Free radical copolymerization of glycidyl methacrylate (GMA) and N-vinyl-2-pyrrolidone was carried out using AIBN, in chloroform at 60°C (Soundararaian and Reddy 1991).

16.2.1.3 Radiation

There are two main radiation-initiated polymerization methods used to cure monomers in wood: gamma radiation and electron beam.

16.2.1.3.1 Gamma Radiation

Wood is a mixture of high molecular weight polymers; therefore, exposure to high-energy radiation will depolymerize the polymers, creating free radicals to initiate polymerization. With gamma radiation, polymerization rate and extent of polymerization are dependent on the type of monomer, other chemical additives, wood species, and radiation dose rate (Aagaard 1967). An example of radiation polymerization of the vinyl monomer MMA using cobalt 60 gamma ray dose rates of 56, 30, and 9 rad/s produced exotherms at 120°C, 90°C, and 70°C, respectively, with reaction times of 5, 7, and 12 h, respectively, produced 70–80% wood weight gain (Glukhov and Shiryaeya 1973). A 1.5–2.5 megarad-dose of gamma irradiation from a cobalt 60 source of isotope activity 20,000 Ci can be used to polymerize MMA in wood. Addition of a solid organic halogen compound with a high content of Cl or Br, accelerates the polymerization (Pesek et al. 1969). Addition of tributyl phosphate accelerates the polymerization rate of MMA 2.5 times and decreases the required radiation dosage. Addition of alkenyl phosphonates or alkenyl esters of phosphorus acids increases the polymerization rate and imparts fire resistance and bioresistance to the resultant WPC (Schneider, Phillips et al. 1990). Pietrzyk reports the optimum irradiation conditions for MMA in wood are: irradiation dose 1.5 Mrad and dose strength approximately 0.06 Mrad/h (Pietrzyk 1983). It is best if the irradiation is done in a closed container without turning the samples in order to minimize the escape of the monomer from the wood. Beech wood impregnated with MMA alone or in carbon tetrachloride or methanol solutions, can be cured with cobalt 60 gamma-radiation giving polymer loadings of up to 70% by weight. Radiation doses of 2–4 Mrad are necessary for complete conversion (Proksch 1969).

Moisture in wood accelerates polymerization (Pesek et al. 1969). A small amount of water in the wood or monomer improves the properties of the WPC (Pietrzyk 1983). The polymerization rate of MMA in beech wood is increased by using aqueous emulsions containing 30% MMA and 0.2% oxyethylated fatty alcohol mixture instead of 100% MMA. The complete conversion of MMA required ~5 kJ/kg radiation dose when the aqueous emulsions were used, in comparison to >16 kJ/kg when 100% MMA was used. The radiation polymerization of MMA in wood is inhibited by lignin (Pullmann et al. 1978).

Polymerization rate of vinyl compounds in wood, by gamma-ray irradiation, decreases in the presence of oxygen giving 50–90% conversion for styrene, methyl-, ethyl-, propyl-, and butyl methacrylates, and 4–8% conversion of vinyl acetate. Toluene diisocyanate addition increases monomer conversion, and decreases benzene extractives from the composite (Kawase and Hayakawa 1974).

The U.S. Atomic Energy Commission sponsored research that used gamma radiation to make WPC's in the early 1960s, but one drawback is the safety concerns and regulations needed when using radiation. Some advantages are that the monomer can be stored at ambient conditions, as long as inhibitor is included, and the rate of free radical generation is constant for cobalt-60 and does not increase with temperature (Meyer 1984).

16.2.1.3.2   Electron Beam

High-energy electrons are another way of generating free radicals to initiate polymerization, and have been used with some success. Electron beam irradiation was used to make WPC's of beech sapwood veneers with styrene, MMA, acrylonitrile, butyl acrylate, acrylic acid, and unsaturated polyesters (Handa et al. 1973; Handa et al. 1983). Gotoda used electron beam to polymerize several different monomers and monomer combinations in wood (Gotoda et al. 1970a,b; Gotoda and Takeshita 1971; Gotoda et al. 1971, 1974, 1975; Gotoda and Kitada 1975). Increasing the wood moisture content has a positive effect on electron curing. For example, the monomer conversion in the electron beam-induced polymerization of MMA pre-impregnated in beech veneer increases with increases of moisture content in the wood up to 20–30% moisture, and is proportional to the square root of the electron dosage. The polymerization of styrene and acrylonitrile in veneer is also affected similarly by moisture content (Handa et al. 1973).

Some studies have indicated that curing of monomer systems in wood causes some interaction of the polymer with the wood. WPCs made with MMA, MMA-5% dioxane, and vinyl acetate impregnation into the wood cellular structure, followed by electron-beam irradiation show an increase in the compressive and bending strength, indicating some interaction at the wood–polymer interface (Boey et al. 1985). The dynamic modulus of WPC made from beech veneer impregnated with acrylic acid and acrylonitrile containing unsaturated polyester or polyethylene glycol methacrylate by electron beam irradiation, increased logarithmically as the weight polymer fraction increased, suggesting an interaction between the polymer and cell wall surface. The temperature dispersion of the dynamic viscoelasticity of composites also indicate an interaction between polymer and wood cell walls (Handa et al. 1981).

16.3 MONOMERS

16.3.1 ACRYLIC MONOMERS

Methyl methacrylate (MMA) is the most commonly used monomer for WPCs (Meyer 1965). It is one of the least expensive and readily available monomers and is used alone or in combination with other monomers to cross-link the polymer system. MMA has a low boiling point (101°C) that can result in significant loss of monomer during curing and it must be cured in an inert atmosphere, or at least in the absence of oxygen. MMA shrinks about 21% by volume after polymerization, which results in some void space at the interface between the cell wall of the wood and the polymer. Adding cross-linking monomers such as di- and tri-methacrylates increases the shrinkage of the polymer which results in larger void spaces between the polymer and cell walls (Kawakami et al. 1981). Polymerization of MMA is exothermic and a lot of heat is generated during the polymerization that must be controlled.

Methyl methacrylate   Poly(methyl methacrylate)

MMA can be polymerized in wood using catalysts (Vazo or peroxides) and heat, or radiation. Curing of MMA using cobalt-60 gamma radiation requires a longer period of time (8–10 h depending upon the radiation flux) while catalyst-heat-initiated reactions are much faster (30 min or less at 60°C) (Meyer 1981).

Hardness modulus values determined for untreated and poly methyl methacrylate-treated red oak, aspen, and sugar maple, on both flat and edge-grained faces show untreated wood hardness values are related to sample density. There are significant relationships between treated wood hardness modulus, wood density, and loading. Large variations in hardness modulus of treated aspen and maple are related to their diffuse-porous structure. In contrast, the hardness modulus of treated red oak is predictable on the basis of density or polymer loading (Beall et al. 1973).

The compressive and bending strengths of a tropical wood (*Kapur-Dryobalanops* sp.) are improved significantly by impregnation of MMA (Boey et al. 1985). Using a gamma irradiation method, some tropical wood-poly methyl methacrylate and—poly(vinyl acetate) composites are produced which exhibit a significant improvement in uniaxial compressive strength (Boey et al. 1987). Samples with an average polymer content of 63% (based on dry wood) show increases in compressive strength, toughness, radial hardness, compressive strength parallel to the grain, and tangential sphere strength (Bull et al. 1985). Hardness and mechanical properties of poplar wood are improved by impregnation with MMA and polymerization of the monomer by exposure to gamma-irradiation, the hardness of the product increases with impregnation pressure and weight of polymer (Bull et al. 1985; Ellis 1994).

$$H_2C=C\!\!\begin{array}{c}H\\ \\ C-O-\!\!\left[CH_2\right]_6\!\!-O-C\end{array}\!\!\begin{array}{c}H\\ \\ \end{array}C=CH_2$$

1,6-Hexanediol diacrylate (HDDA)

$$H_2C=C\!\!\begin{array}{c}CH_3\\ \\ C-O-CH_2-CH_2-OH\\ \| \\ O\end{array}$$

2-Hydroxyethyl methacrylate (HEMA)

Various other acrylic monomers have been investigated (Ellis and O'Dell 1999). WPCs were made with different chemical combinations and evaluated for dimensional stability, ability to exclude water vapor and liquid water, and hardness. Different combinations of hexanediol diacrylate (HDDA), hydroxyethyl methacrylate (HEMA), hexamethylene diisocyanate (Desmodur N75, DesN75), and maleic anhydride (MAn) were *in situ* polymerized in solid pine, maple and oak wood. The rate of water vapor and liquid water absorption was slowed, and the rate of swelling was less than that of unmodified wood specimens, but the dimensional stability was not permanent (see Figure 16.3). The WPCs were much harder than unmodified wood (see Table 16.1). Wetting and penetration of water into the wood was greatly decreased, and hardness and dimensional stability increased with the chemical combination of hexanediol diacrylate, hydroxyethyl methacrylate, and hexamethylene diisocyanate. Treatments containing hydroxyethyl methacrylate were harder and excluded water and moisture more effectively. This is probably due to the increased interfacial adhesion between the polymer and wood, due to the polarity of HEMA monomer.

16.3.2 STYRENE

$$H_2C=CH$$

Styrene is another monomer that is commonly used for WPCs. It can be polymerized in wood using catalysts (Vazo or peroxides) and heat, or radiation. Other monomers are commonly added to

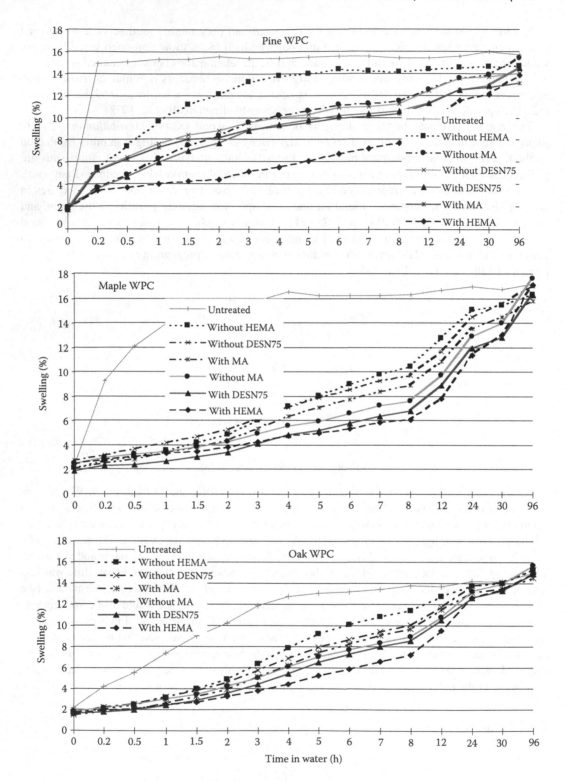

FIGURE 16.3 Volumetric swelling of WPC specimens in water.

TABLE 16.1
The Rockwell Hardness of WPC Specimens

Treatment	Earlywood	Latewood
Pine		
HDDA/DesN75 (3:1)	37.3	45.0
HDDA	32.2	41.9
HDDA/MA (3:1)	31.7	26.0
HDDA/HEMA/MA (1:2:1)	61.7	67.8
HEMA/DesN75/MA (2:1:1)	61.0	70.1
HDDA/DesN75/MA (2:1:1)	31.7	49.3
HDDA/HEMA (1:1)	47.0	55.3
HDDA/HEMA/DesN75 (1:2:1)	63.2	74.2
CONTROL (untreated)	−15.5	−10.6
Maple		
HDDA/DesN75 (3:1)	44.8	
HDDA	46.8	
HDDA/MA (3:1)	49.2	
HDDA/HEMA/MA (1:2:1)	60.0	
HEMA/DesN75/MA (2:1:1)	56.4	
HDDA/DesN75/MA (2:1:1)	46.6	
HDDA/HEMA (1:1)	49.5	
HDDA/HEMA/DesN75 (1:2:1)	65.6	
CONTROL (untreated)	−9.4	
Red Oak		
HDDA/DesN75 (3:1)	23.3	22.7
HDDA	27.9	25.1
HDDA/MA (3:1)	23.1	20.5
HDDA/HEMA/MA (1:2:1)	38.6	46.3
HEMA/DesN75/MA (2:1:1)	26.8	35.4
HDDA/DesN75/MA (2:1:1)	23.0	13.6
HDDA/HEMA (1:1)	29.6	25.5
HDDA/HEMA/DesN75 (1:2:1)	39.3	40.8
CONTROL (untreated)	−17.1	−25.1

Note: Rockwell hardness of the longitudinal face of $25 \times 25 \times 0.6$ mm^3 specimens. 1/4 inch ball indenter and 60 Kgf (Rockwell scale L). Maple measured without regard to earlywood or latewood.

control the polymerization rate, extent of polymerization, and to cross-link the styrene for improved physical properties of the WPCs.

Hardness, impact strength, compression and shear strength, and bending and cleavage strengths of styrene treated wood are better than for untreated samples and the same as, or better than, those for samples impregnated with MMA. The treated wood is sometimes unevenly colored and more yellow than the original samples (Autio and Miettinen 1970).

Modification of several types of hard- and softwoods with polystyrene improves their resistance to wear. Wood–polystyrene composites made from the softwood species birch, gray and black alder, and spruce exhibits abrasion resistance comparable to that of natural oakwood (Dolacis 1983). The flexural strength, hardness, and density of alderwood are increased by impregnating it with styrene and heating to obtain the polystyrene saturated wood (Lawniczak 1979). Poplar wood modified with polystyrene has increased hardness, static bending strength and toughness; the increases in toughness depend on the polymer content to a certain limit (Lawniczak 1973).

WPC can be prepared from a mixture of acrylonitrile–styrene–unsaturated polyester in wood. This mixture gives a tough cross-linked polymer, and is more favorable for radiation polymerization than the systems of MMA, MMA unsaturated polyester, and acrylonitrile-styrene (Czvikovszky 1977, 1981). Composite materials obtained by evacuation of wood (beech, spruce, ash, and tropical wood *Pterocarpus vermalis*) followed by its impregnation with an unsaturated polyester–MMA–styrene mixture or unsaturated polyester–acrylonitrile–styrene mixture and gamma-irradiation-induced curing exhibit decreased water vapor absorption and improved dimensional stability, hardness, compression strength, and wear resistance, compared to untreated wood (Czvikovszky 1982).

Curing of unsaturated polyester–styrene mixture can be affected by the initiator-heat technique by either using 0.1–0.2% benzoyl peroxide or 1% methyl ethyl ketone peroxide (Doss et al. 1991). Polymer-reinforced alderwood can be prepared by impregnating it with styrene and peroxide catalyst, followed by thermal polymerization for 3–7 h. The addition of 1.0% divinylbenzene, triallyl phosphate, or trimethylolpropane trimethacrylate cross-linking agent to styrene results in an increased polymerization rate, with divinylbenzene having the most pronounced effect on the polymerization rate (Lawniczak and Szwarc 1987). Gamma-ray-induced polymerization of styrene in impregnated samples of beech wood in the presence of carbon tetrachloride require a minimum dose of 159 kGy for full monomer hardening. The polymer content in the resulting samples is 53% at a monomer conversion of >90%. A modified sample has ~50% increase in density, ~90% increase in hardness, and ~125% decrease in absorptivity, compared to unmodified wood (Raj and Kokta 1991).

The impregnation of beech wood with ternary mixtures of styrene, dioxane, acetone, or ethanol, and water then curing by ionizing radiation gives a product with some dimensional stability due to chemical fixation of the polymer on the lignocellulosic material. This change is accompanied by a marked change in the structure of the cell wall. Pure styrene or styrene in aqueous solution gives a composite with low-dimensional stability (Guillemain et al. 1969). Wood polymers based on aqueous emulsion polyester–styrene mixtures are dimensionally more stable than those produced with a pure polyester–styrene mixture (Jokel 1972). Impregnation of poplar wood with styrene–ethanol–water followed by polymerization at 70°C gives 50% increase in dimensional stability with 30–40% polystyrene content in the wood. Use of styrene alone increased wood dimensional stability by only 10% even with >100% styrene retention. Dimensional stability of poplar wood is also significantly increased (~40%) by a 90:5 styrene–ethanol system (Katuscak et al. 1972). The use of a mixture of polar solvents with styrene to make WPC seems to improve the dimensional stability of the composites. The use of styrene alone for the modification of wood was not as favorable as using a styrene–methanol–water system which gives greater bending strength and better dimensional stability. Hardness increases with increasing polystyrene in the wood (Varga and Piatrik 1974).

Untreated woods of ash, birch, elm, and maple, absorb about 4 times more water than woods containing acrylonitrile–styrene copolymer (Spindler et al. 1973). Addition of acrylonitrile and butyl methacrylate to styrene does not affect significantly the maximum amount of water sorbed by the composites but decreases their swelling rate and increases their dimensional stability and bending strength (Lawniczak and Pawlak 1983). WPC prepared using styrene–acrylonitrile have increased hardness, substantially improved dimensional stability, and give no difficulties in machining and gluing (Singer et al. 1969).

Monomer- and polyester prepolymer-impregnated beech wood veneer irradiated with 3–6 Mrads and cured at 80°C has improved shrinkage resistance and water repellency and provides laminates suitable for flooring and siding. Various mixtures of unsaturated polyester and styrene, as well as the individual monomers MMA, ethyl acrylate, butyl acrylate, acrylonitrile, and vinyl acetate have been used to treat veneers. A variety of tests of physical properties show the styrene–polyester system to be superior (Handa et al. 1972).

WPCs can be prepared by gamma irradiation of hardwood impregnated with a styrene-unsaturated polyester mixture, MMA, or acrylonitrile–styrene mixture. The addition of chlorinated paraffin oil to any of these monomer systems imparts fire resistance to the composites and reduces the gamma-ray dosage needed for total polymerization of the monomers. Styrene-unsaturated polyester mixtures containing about 30% chlorinated paraffin oil are suitable systems for large-scale preparation of composites (Iya and Majali 1978).

Modification of wood samples with polystyrene increases the resistance of the composites to degradation in contact with rusting steel (Helinska-Raczkowska and Molinski 1983). Conversion of unsaturated polyester–styrene mixture and dimensional stability of the wood–styrene–unsaturated polyester composites decreases with an increase in moisture content of the wood to be treated (Yamashina et al. 1978).

Polymerization of styrene in wood can result in the grafting of styrene to cellulose, lignin, and pentosans (Lawniczak et al. 1987). The treatment of wood with diluted hydrogen peroxide solution leads to an increase in the viscosity-average molecular weight of the polystyrene, and to the graft polymerization of the monomer, which, in turn, enhances the stress properties of wood–polystyrene composites (Manrich and Marcondes 1989).

Wood impregnated with a styrene–ethylene glycol dimethacrylate mixture under full vacuum (0.64 kPa), has higher densities and hardness in the early wood than that of late wood, and early wood shows hardness increases roughly double those in late wood at the same density, indicating styrene uptake is predominantly in early wood (Brebner et al. 1985).

Kenaga (1970) researched high boiling styrene-type monomers including vinyltoluene, tert-butylstyrene, and o-chlorostyrene. In the preparation of WPCs the cure rate, monomer loss, and composite physical properties can be varied by appropriate selection and concentration of catalyst, comonomers, and cross-linking agents. The composite can be bonded to untreated crossbanded veneers simultaneously with polymerization in a press because these three styrene-type monomers have boiling points from 27°C to 74°C higher than styrene's boiling point. The monomer tert-butylstyrene has the highest boiling point at 219°C and the least shrinkage, 7%, on polymerization. Cross-linking agents increase reaction rate and improve the WPC physical properties. Effects of the cross-linking agents trimethylolpropane triacrylate, trivinyl isocyanurate, trimethylolpropane trimethacrylate, ethylene glycol dimethacrylate, trimethylene glycol dimethacrylate, tetraethylene glycol dimethacrylate, poly-ethylene glycol dimethacrylate, and divinylbenzene were studied. Generally 10% or more cross-linking agent is needed to give the best improvement in abrasion resistance. Copolymers of tert-butylstyrene with di-ethyl maleate, di ethyl fumarate, and acrylonitrile were studied in basswood and birch wood blocks. All the copolymers except acrylonitrile improved the abrasion resistance of the composite. Polyesters lowered cure time and styrene monomer loss during cure but increased the exotherm temperature to a level that could be unacceptable for larger pieces of wood.

16.3.3 POLYESTERS

Unsaturated polyester resins are most often used in combination with other monomers, making them less expensive and improving their properties. Many polyester resins are available as commercial products. Polyester, MMA, and styrene were polymerized individually and in combinations by gamma radiation or benzoyl peroxide (Miettinen et al. 1968; Miettinen 1969). MMA composites had higher tensile strength and abrasion resistance, but lower bending strength and impact strength compared to the polyester composites. Styrene is frequently mixed with polyester resins to reduce viscosity, thus enabling better penetration into the wood. Polyesters decrease the loss of styrene monomer, and the time to heat cure (Kenaga 1970).

16.3.4 MELAMINE RESINS

Melamine resins have many uses with paper and wood products. Paper can be impregnated with melamine resins, and then laminated to the surface of wood veneers, fiber boards, or other panel products resulting in a hard, smooth, and water resistant surface. Wood veneers can also be impregnated with melamine resins to improve dimensional stability, water resistance, and hardness (Inoue et al. 1993; Takasu and Matsuda 1993). The melamine resin-modified wood is 1.5–4 times harder than untreated wood (Inoue et al. 1993). Yet, the maximum hardness achieved by melamine modified wood is less than that of wood modified with acrylate, methacrylate, and other vinyl monomers which increase hardness 7–10 times that of unmodified wood (Mizumachi 1975). But, the hardness of melamine resin-impregnated wood can be increased by compressing the wood (Inoue et al. 1993). The melamine resins decrease the abrasion resistance of wood (Inoue et al. 1993; Takasu and Matsuda 1993).

16.3.5 ACRYLONITRILE

$$H_2C=CH$$
$$|$$
$$CN$$

Acrylonitrile is used in the production of WPCs mostly in combination with other monomers because the polymer does not improve properties by itself. It is most frequently used with styrene, and less frequently with MMA, methyl acrylate, unsaturated polyester, diallyl phthalate, and vinylidene chloride. WPCs made with MMA–acrylonitrile or styrene–acrylonitrile mixtures were cured using either gamma radiation or catalyst, and the resultant composites were found to be very similar (Yap et al. 1990, 1991).

Styrene–acrylonitrile WPCs show high-dimensional stability which is probably due to swelling of the wood by the acrylonitrile during treatment creating a bulking action (Loos 1968). Addition of acrylonitrile to styrene gives substantial improvement in hardness and compressibility of the wood (Rao et al. 1968). Ratios of acrylonitrile to styrene between 7:3 and 4:1 give the most substantial improvements in dimensional stability, compressibility, and hardness. The antiswell efficiencies of wood–styrene–acrylonitrile combinations are 60–70% (no swelling is 100%) (Ellwood et al. 1969). Moisture absorption increases with increase of acrylonitrile in WPCs made with various ratios of acrylonitrile and methyl acrylate (Gotoda et al. 1970b).

Copolymerization of bis(2-chloroethyl) vinylphosphonate with vinyl acetate or acrylonitrile in beechwood improves the dimensional stability of the WPCs (Ahmed et al. 1971). A ternary resin mixture of styrene, acrylonitrile, and unsaturated polyester that can be cured in the wood with a low dose of gamma radiation also has favorable properties (Czvikovszky 1977, 1981, 1982).

The addition of acrylonitrile to a diallyl phthalate prepolymer improves the glueability of the composite against a substrate, such as plywood or particle board. The weatherability of a wood composite laminate containing diallyl phthalate prepolymer and acrylonitrile is improved by incorporating polyethylene glycol dimethacrylate (Gotoda et al. 1971).

Acrylonitrile is highly toxic and is a carcinogen therefore attempts have been made to find chemicals that can be substituted for acrylonitrile in the treating solutions. These attempts have been only partially successful. N-vinyl carbazol can be used as a partial replacement of acrylonitrile. Several other compounds including acryloamide, N-hydroxy acryloamide and 1-vinyl-2-pyrrolidone were tried unsuccessfully (Schaudy and Proksch 1982).

16.4 CROSS-LINKING AGENTS

Some of the cross-linking agents frequently used with MMA, styrene, or other vinyl monomers are trimethylolpropane triacrylate, trivinyl isocyanurate, trimethylolpropane trimethacrylate, ethylene glycol dimethacrylate, trimethylene glycol dimethacrylate, tetraethylene glycol dimethacrylate, polyethylene glycol dimethacrylate, and divinylbenzene. Cross-linking agents generally increase reaction rate and improve the WPC's physical properties (Kenaga 1970).

Several cross-linking monomers, including 1,3-butylene dimethacrylate ethylene dimethacrylate and trimethylolpropane trimethacrylate and the polar monomers 2-hydroxyethyl methacrylate and glycidyl methacrylate, were added at 5–20% concentration to MMA, and their effects upon the polymerization and properties of the composites were examined (Kawakami et al. 1977). WPCs with only MMA show a void space at the interface between cell wall and polymer. With addition of cross-linking esters such as di- and tri-methacrylate, the shrinkage (and hence void spaces) of the polymer during polymerization increases. On the other hand, in the WPCs containing polar esters having hydroxyethyl and glycidyl groups, the voids due to the shrinkage of polymer was found to form inside the polymer itself, suggesting better adhesion of the polymer to the inner surface of cell wall (Kawakami et al. 1981).

Impregnation with ethyl α-hydroxymethylacrylate (EHMA) plus another multifunctional monomer 2 vinyl-4,4-dimethyl 2-oxazolin-5-one (vinyl azlactone) results in improved mechanical properties of wood samples. Improvements of 38–54% in impact strength and 27–44% in compression modulus are achieved depending on the relative amount of vinyl azlactone incorporated (Mathias et al. 1991).

16.4.1 ISOCYANATES

The addition of isocyanate compounds with acrylic monomers reduces the brittleness of WPCs consisting only of acrylic compounds (Schaudy and Proksch 1981). WPC properties improve by adding a blocked isocyanate to a mixture of MMA and 2-hydroxyethyl methacrylate (Fujimura et al. 1990). The isocyanate compound cross-links the copolymer.

The mechanical properties of a wood–polystyrene composite are improved by the addition of an isocyanate compound to the styrene treating mixture. Polymethylene (polyphenyl isocyanate) forms a bridge between wood and polymer on the interfaces. The isocyanate compound then becomes instrumental in efficient stress transfer between the wood and polymer (Maldas et al. 1989).

16.4.2 ANHYDRIDES

A maleic anhydride and styrene mixture has been used to make WPCs (Ge, Peng et al. 1983). Also, a mixture of tetraethylene glycol dimethacrylate and chlorendic anhydride have been used to increase fire, chemical and abrasion resistance as well as hardness (Paszner et al. 1975). A process has been developed that is designed to prepare cross-linked oligoesterified wood with improved dimensional stability and surface properties. Maleic, phthalic, and succinic anhydrides are used. The wood is reacted with the anhydride then impregnated with glycidyl methacrylate then heated to cause cross-linking. In a one-step process the anhydride and glycidyl methacrylate are impregnated into the wood together, then polymerized and reacted with the wood simultaneously. The resulting wood is hard and has smooth surfaces. As the anhydride in the anhydride:glycidyl methacrylate ratio increases the dimensional stability increases (Ueda et al. 1992).

16.5 PROPERTIES OF WOOD–POLYMER COMPOSITES

WPCs can improve many properties of solid wood, and therefore be tailored for a specific application. Some of these properties are surface hardness, toughness, abrasion resistance, dimensional stability, moisture exclusion, and fire, decay and weather resistance. Table 16.2 is a summary of some of the properties of woods modified by five different treatments.

16.5.1 HARDNESS

Hardness is the property that resists crushing of wood and the formation of permanent dents. Hardness is, for the most part, a function of density, that is, the more dense the wood, the harder it is. The density of dry wood varies widely based on the volume of void (lumens and vessels) space in the wood. For example, the density of dry balsa wood ranges from 100 to 200 kg/m³ with a typical density of about 140–170 kg/m³ (about one-third the density of other hard woods) while a wood like lignum vitae has a density of 1280–1370 kg/m³ (this wood does not float).

TABLE 16.2
Properties of Wood after Five Different Modifications

Property	Water-Soluble Polymers and Synthetic Resins		Compression		Heat	Organic Chemicals or Cross-Linking Agents		Liquid Monomers	
	Polyethylene Glycol	Impreg	Staypak	Compreg	Staybwood	Bulking	Cross-Linking	Methyl Methacrylate	Epoxy Resin
Specific gravity	Slightly increased	15–20 pct greater than normal wood	1.2–1.4	1.0–1.4	Unchanged	Slightly increased	Unchanged	Increased	Increased
Permeability to water vapor	Hygroscopic	Better than normal	Better than normal	Greatly improved	Better than normal	Unchanged	Unchanged	Greatly improved	Greatly improved
Liquid water repellency	Hygroscopic	Better than normal	Better than normal	Greatly improved	Better than normal	Better than normal	Better than normal	Greatly improved	Greatly improved
Dimensional stability	80 pct	60–70 pct	Slightly improved	80–85 pct	40 pct	65–75 pct	80–90 pct	10 pct	Slightly improved
Decay resistance	Better than normal	Better than normal	Unchanged	Much better than normal	Better than normal	Much better than normal	Better than normal	Somewhat increased	Somewhat increased
Heat resistance	No data	Greatly increased	No data	Greatly increased	No data	No data	No data	Increased	No data
Fire resistance	No data	Unchanged	Unchanged	Unchanged	Unchanged	Unchanged	Unchanged	Unchanged	No data
Chemical resistance	No data	Better than normal	Slightly better than normal	Much better than normal	Better than normal	No data	No data	Much better than normal	Much better than normal
Compression strength	Slightly increased	Increased	Increased	Greatly increased	Reduced	Slightly reduced	Slightly reduced	Greatly increased	Greatly increased
Hardness	Unchanged	Increased	Increased	10–20 times greater	Reduced	Slightly reduced	Slightly reduced	Greatly increased	Greatly increased
Abrasion resistance	Slightly reduced	Reduced	Increased	Increased	Greatly reduced	Slightly reduced	Greatly reduced	Greatly increased	Greatly increased
Machinability	Unchanged	Better than normal but dulls tools	Metalworking tools required	Metalworking tools required	Unchanged	Unchanged	Unchanged	Metalworking tools preferred	Metalworking tools preferred
Glueability	Special glues required	Unchanged	Unchanged	Same as normal after sanding	Unchanged	Unchanged	Unchanged	Special glues required	Epoxy used as adhesive
Finishability	Requires poly, urethane, oil, or 2 parts polymer	Unchanged	Unchanged	Plastic-like surface (can be polished without finish)	Unchanged	Unchanged	Unchanged	Plastic-like Surface (no finish required)	Plastic-like surface (no finish required)
Color change	Little change	Reddish brown	Little change	Reddish brown	Darkened	Little change	Little change	Little change	Little change

The Janka Scale of Hardness measures the force required to embed a 1.11 cm (0.444 in.) steel ball to half its diameter in wood. It is the industry standard for determining the ability of various species to resist denting and wear. The Janka hardness of lignum vitae ranks highest of the trade woods, with a Janka hardness of 4500. Table 16.3 shows the Janka hardness of several domestic and foreign woods.

The hardness or indent resistance of WPC is measured by any of several methods. The test method used depends on the WPC and the expected final product. Measurement can be made using a hand-held Shore Durometer tester, ball indenters such as Brinell and Rockwell hardness, the Janka ball indenter or the Gardner Impact tester that uses a falling dart to make dents that can be measured (Miettinen et al. 1968; Beall et al. 1973; Schneider 1994).

Hardness of a WPC depends on the polymer loading and the hardness of the polymer. Polymer loading is affected by wood porosity and density. For example, a more porous and lower density wood will require a higher polymer loading. Generally, a higher polymer loading will give a greater WPC hardness. Figure 16.4 is an SEM micrograph of a WPC with no polymer attachment and the lumens incompletely filled. Figure 16.5 is an SEM micrograph of a WPC with filled lumens and some interaction of the polymer with the wood. The hardness of a WPC is improved when the cells are completely filled and there is attachment of the polymer to the wood.

TABLE 16.3
Janka Hardness of Hardwoods and Softwoods

Wood Species	Janka Hardness
Eastern white pine	380
Basswood	410
Chestnut	540
Douglas fir	660
Southern yellow pine	690
Sycamore	770
Cedar	900
Black cherry	950
Teak	1000
Black walnut	1010
Yellow birch	1260
Red oak	1290
American beech	1300
Ash	1320
White oak	1360
Hard maple	1450
Birch	1470
Brazilian oak	1650
Locust	1700
Rosewood	1780
Hickory	1820
Purple heart	1860
African rosewood	1980
Mesquite	2345
Brazilian cherry	2350
Brazilian rosewood	3000
Ebony	3220
Brazilian teak	3540
Brazilian walnut	3680

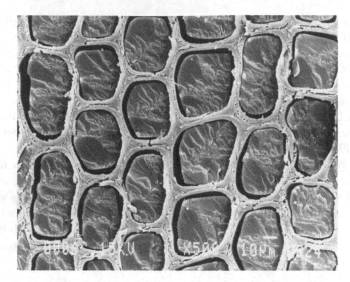

FIGURE 16.4   SEM micrograph of a WPC having the wood cells incompletely filled with polymer and having no attachment of polymer to the wood.

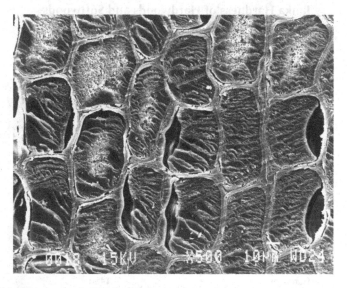

FIGURE 16.5   SEM micrograph of a WPC with the wood cells filled with polymer and having some interaction of the polymer with the wood.

The type of polymer, cross-linking chemicals and method and extent of polymerization affect polymer hardness. A 7–10-fold increase in hardness can be expected by most treatments for example, MMA-impregnated alderwood has more than a 10-fold increase in hardness of the sides and more than a 7-fold increase in hardness of the cross-cut areas (Miettinen et al. 1968).

16.5.2   TOUGHNESS

Increasing the toughness of wood with polymer increases the crack resistance and brittleness at room temperature. Impact strength and toughness are closely related; both refer to the WPCs ability to resist fracturing. Measurements of impact strength are made using the Izod and the Charpy impact test instruments. The test involves striking the specimen with a pendulum and measuring the impact energy necessary to initiate fracture. Treating sugar maple wood with a vinyl polymer increases toughness in both radial and tangential impact directions compared to untreated wood. Increased polymer load results in increased toughness. Microscopy indicates brittle polymer fracture extends across lumens but stops at the polymer and cell wall interface (Schneider et al. 1989). Brittleness of a composite can be severely increased by increasing the amount of a cross-linker such as ethyleneglycol dimethacrylate even to as little as 1.5% in MMA (Schaudy and Proksch 1982). WPC's with high toughness (low brittleness), have been prepared by using a treating mixture consisting of MMA and an isocyanate that has an acrylic functionality. This treating mixture increased the impact bending strength of the WPC by about 100% (Schaudy et al. 1982).

16.5.3 ABRASION RESISTANCE

Abrasion resistance is determined by the Taber wear index, which is the weight loss (mg/1000 cycles) caused by an abrasive wheel turning on a specimen. The lower the weight loss value, the better the resistance to wear. In general, abrasion resistance increases with increasing polymer content in the wood (Kawakami and Taneda 1973). Softwood species such as birch, gray and black alder, and spruce when made into a composite with polystyrene have abrasion resistance comparable to that of natural oak wood (Dolacis 1983). Alder wood and birch wood impregnated with MMA had up to 85% less weight loss than untreated wood (Miettinen et al. 1968).

16.5.4 DIMENSIONAL STABILITY

Dimensional stability is the property of wood that allows it to resist changes in dimensions when exposed to various moisture conditions. Dimensional stability is reported as percent volumetric swelling or as antishrink efficiency (ASE). ASE is the percent reduction in volumetric swelling of treated wood compared to untreated wood at equilibrium water or moisture saturated conditions (see Chapter 4). Many WPCs are not dimensionally stable so that with time in water or high humidity, most WPCs will swell to the same amount as untreated wood.

There are two approaches to improve the dimensional stability of WPCs. One approach is to direct the penetration into the wood cell walls to bulk the wood at or near its wet or green dimensions. Aqueous and nonaqueous solvents have been used to swell the wood and carry the monomers into the cell walls, and polar monomers have been used to increase the swelling of the wood and penetration of the monomers into the wood. The second approach is to react the chemicals with the cell wall hydroxyl groups, therefore decreasing its affinity for moisture (Loos 1968; Rowell et al. 1982). WPCs with polymer located just within the lumen do not make a significant contribution to dimensional stability compared with chemical modifications of the cell wall (Fujimura and Inoue 1991).

16.5.5 MOISTURE EXCLUSION

Moisture exclusion efficiency (MEE) is the property of a WPC to exclude moisture and is related to the rate at which the composite absorbs moisture and swells and not to the maximum extent of swelling or moisture uptake (see Chapter 4). If the WPCs are not allowed to reach equilibrium with respect to moisture or water, then MEE can be mistaken for dimensional stability or ASE values (Loos 1968). Many WPCs will absorb water and swell at a slower rate than untreated wood, but in most cases the maximum swelling is nearly the same as that of untreated wood.

16.5.6 FIRE RESISTANCE

There are several methods of measuring different aspects of the property of fire retardancy (see Chapter 6). Thermogravimetry measures char formation and decomposition temperatures by heating small specimens in an inert atmosphere. More char generally indicates greater fire retardancy. The oxygen index test measures the minimum concentration of oxygen, in an oxygen and nitrogen atmosphere, that will just support flaming combustion. Highly flammable materials are likely to have a low oxygen index. Flame spread tests are those in which the duration of flaming and extent of flame spread are measured. The results of any of the test methods that use small specimens often do not correlate with the actual performance of materials in a real fire situation. The surface burning characteristics of WPCs used as building materials are best measured by flame spread tests that use large specimens, such as the ASTM E84 test that requires specimens approximately 514 mm wide by 7.3 m long. The test chamber in this test also has a photometer system built in to measure smoke and particulate density. Smoke evolution is very important because many fire deaths are due to smoke inhalation.

Polymethyl methacrylate enhances the flammability of wood (Calleton et al. 1970; Lubke and Jokel 1983) but not styrene and acrylonitrile (Schaudy et al. 1982). Bis(2-chloroethyl) vinylphosphonate with vinyl acetate or acrylonitrile improves the fire retardancy, but is less effective than poly(dichlorovinyl phosphate) or poly(di-ethyl vinylphosphonate). Wood impregnated with dimethylaminoethyl methacrylate phosphate salt and then polymerized in the presence of cross-linking agents has high fire retardancy (Ahmed et al. 1971) as does trichloroethyl phosphate (Autio and Miettinen 1970). The addition of chlorinated paraffin oil to monomer systems imparts fire retardancy to composites (Iya and Majali 1978). The limiting oxygen index values of the MMA-bis(2-chloroethyl)vinyl phosphonate copolymer and MMA-bis(chloropropyl)-2-propene phosphonate copolymer wood composites are much higher than that of untreated wood and other composites, indicating the effectiveness of the phosphonates as fire retardants (Yap et al. 1991). WPC specimens made with MMA are smoke free, but styrene-type monomers create dense smoke (Siau et al. 1972). The presence of aromatic polymers, such as poly(chlorostyrene), and fire retardants having aromatic benzene rings in wood increase the smoke evolution, flame spread, and fuel contribution in a modified tunnel furnace test (Siau et al. 1975). In all specimens tested, the smoke evolution increased markedly after the flame is extinguished.

16.5.7 DECAY RESISTANCE

Most WPCs are not decay resistant because the polymer merely fills the lumens and does not enter the cell walls, which makes the cell walls accessible to moisture and decay organisms (see Chapter 5). WPCs prepared using MMA and several kinds of cross-linking monomers (1,3-butylene dimethacrylate, ethylene dimethacrylate, and trimethylolpropane trimethacrylate) and polar monomers (2-hydroxyethyl methacrylate and glycidyl methacrylate) added at 5–20% concentration have little resistance to brown rot decay (Kawakami et al. 1977). Using methanol with MMA or styrene allows the polymer to penetrate the cell walls. The amount of polymer in the cell wall is important for decay resistance. Some protection against biological degradation is possible at cell wall polymer contents of 10% or more (Rowell 1983). Acrylate monomers with various bioactive moieties were synthesized (Ibach and Rowell 2001). Pentachlorophenol acrylate and Fyrol 6 acrylate polymers provided no protection against decay, whereas tributyltin acrylate, 8-hydroxyquinolyl acrylate, and 5,7-dibromo-8-hydroxyquinolyl acrylate were found to be resistant to the brown-rot fungus *Gloeophyllum trabeum* at low polymer loading of 2–5% retention (Ibach and Rowell 2001).

16.5.8 WEATHERING RESISTANCE

WPCs made with birch and pine, impregnated with MMA or styrene–acrylonitrile were exposed in a weatherometer for 1000 h (Desai and Juneja 1972). The specimens were more resistant to surface checking than untreated wood and the styrene–acrylonitrile treatment performed better than MMA.

A combination of cell wall-modifying treatments (butylene oxide or methyl isocyanate) with MMA lumen-filled treatments results in a dual treatment that resists the degradative effects of accelerated weathering in a weatherometer (see Chapter 7). The use of MMA in addition to the cell wall-modifying chemical treatments provides added dimensional stability and lignin stabilization and has a significant effect on weatherability (Rowell et al. 1981).

16.5.9 MECHANICAL PROPERTIES

The strength properties of WPCs are enhanced compared to untreated wood. The hardness, compression and impact strength of wood composites increase with increasing monomer loading (Mohan and Iyer 1991). Cross-linking monomers increases static bending properties, compressive strength, and torsional modulus, but reduces dimensional stability, while polar monomers improve dimensional stability and static bending properties but have no significant effect on compressive strength and torsional modulus (Kawakami et al. 1977). Table 16.4 shows the increase in strength properties of acrylic lumen filled wood. All strength properties are greatly improved.

16.5.10 CHANGES IN COLOR

Dyes can be added to the monomer solution to change the color of the polymerized wood. Figure 16.6 shows natural oak on the left and a dark dyed oak on the right. Figure 16.7 shows a knife handle made from several layers of natural and dyed veneers.

TABLE 16.4
Strength Properties of Acrylic Lumen Filled Wood

Static Bending Properties of Acrylic Lumen Filled Wood

Property	Untreated MPa	Treated MPa
MOE	9.3	11.6
FSAPL	44.0	79.8
MOR	73.4	130.6
MCS	44.8	68.0

Note: MOE = modulus of elasticity, FSAPL = fiber stress at proportional limit, MOR = modulus of rupture, MCS = maximum crushing strength.

FIGURE 16.6 Various dyes can be added to change the color of the impregnated wood.

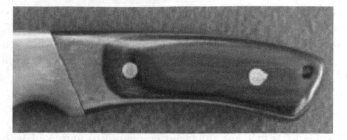

FIGURE 16.7   Knife handle made from acrylic impregnated veneers.

16.6   APPLICATIONS

The major uses of WPCs are for flooring, sports equipment, musical instruments, and furniture (Fuller et al. 1997). Flooring has the largest volume that includes solid plank flooring, top veneers of laminated flooring, and fillets for parquet flooring. As for sports equipment, patents have been issued for golf club heads (Katsurada and Kurahashi 1985), baseball bats, hockey sticks (Yamaguchi 1982), and parts of laminated skis. WPCs are used for wind instruments, mouthpieces of flutes and trumpets, and finger boards of stringed instruments (Knotik and Proksch 1971; Knotik et al. 1971). One area with potential is the use of veneer laminates for furniture, such as desk writing surfaces and tabletops (Maine 1971; Kakehi et al. 1985). A history of the commercialization of WPCs and future opportunities is covered by Schneider and Witt (2004).

16.7   POLYMER IMPREGNATION

Wood polymer composites are usually formed by impregnating the wood with a monomer which is polymerized *in situ* mainly in the lumen. Because the monomers are small, almost complete penetration of chemical is achieved. However, it is possible to impregnate wood with oligomers or polymers, however, penetration depends on the size of the oligomer or polymer. Chemical retention is often limited by the inability of the large polymers to penetrate into the wood structure. Several resin systems have been used to treat wood to improve performance properties. Figure 16.8 shows a very simple diagram of treating wood with a monomer.

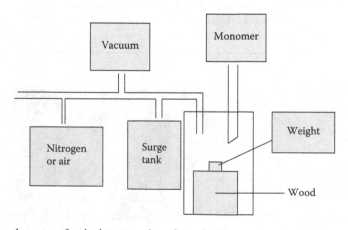

FIGURE 16.8   Simple system for the impregnation of wood with monomers.

16.7.1 Epoxy Resins

Epoxy resin is a partially polymerized, clear solution, with a consistency slightly thicker than varnish at room temperature (21°C). Just before treating wood, the epoxy resin is mixed with hardener. It is cured or hardened within the lumen structure from a few minutes to a few hours depending upon the hardener and temperature. Treatment with epoxy resin is usually performed on veneers because of its large molecular size and high viscosity which does not allow for deep penetration into larger specimens. Veneers are either vacuum treated or soaked in the epoxy resin-hardener solution and then cured. Mechanical properties are greatly increased with epoxy resin treatments, especially hardness (Rowell and Konkol 1987). Epoxy resins are used for wooden boat hulls, the outer ply of plywood, and strengthening softened or decayed wood.

16.7.2 Compression of Wood While Heating and Curing with Resin

Wood can be compressed using heat either with or without resin that improves strength, stiffness, and stability.

16.7.2.1 Staypak

Compressed wood containing no resin is called Staypak (Seborg et al. 1945). During compression with 1400–1600 lb/in² pressure, with temperatures of 170–177°C, lignin will flow relieving internal stresses, but also causing a darkening of the wood. The compression time varies with the thickness of the wood, but usually to a specific gravity of at least 1.3 (Rowell and Konkol 1987). The resultant wood product has a slower moisture absorption and hence a reduction in swelling. It is more dimensionally stable, but not necessarily more biological resistant. Tensile strength (both parallel and perpendicular to the grain), modulus of rupture, elasticity in bending, and impact bending strength of the wood are increased. Staypak is used for tool handles, mallet heads, and various tooling jigs and dies.

16.7.2.2 Compreg

Compreg is resin-treated, compressed wood, and made with layers of treated veneers (Stamm and Seborg 1951). The most common resin is phenol–formaldehyde. The veneers are treated to 25–30% weight gain based on oven-dried weight. The veneers are dried at 30°C or less to prevent the resin from curing. The resin is cured during the heating (140–150°C) and compression process (pressures of 1000–1200 lb/in²). Water absorption is greatly reduced; biological resistance to decay, termite, and marine borer is increased; electrical, acid, and fire resistance are increased. Strength properties of Compreg are increased, except for impact bending strength. The abrasion resistance and hardness are also increased compared to untreated wood. Compreg has many uses from knife handles and tools to musical instruments (Rowell and Konkol 1987). Figure 16.9 shows a clarinet made by Al Stamm from compreg in the 1950s.

FIGURE 16.9 Compreg clarinet made by Al Stamm.

To show that the instrument could be used even when wet, the clarinet was placed in a fish tank with fish the night before it was to be demonstrated. The next morning the clarinet was fine but the fish were dead probably from a small amount of phenol that may have leached from the instrument.

16.7.2.3 Staybwood

Staybwood is one product that is made by heating wood in a vacuum at high temperatures (93–160°C) in a bath of molten metal (Stamm et al. 1960). The high temperature causes the lignin to flow and the hemicelluloses to decompose, producing water-insoluble polymers. The process increases dimensional stability, but decreases strength and therefore has not been used commercially (Rowell and Konkol 1987).

16.8   WATER-SOLUBLE POLYMERS AND SYNTHETIC RESINS

Wood can be impregnated with water-soluble treatments such as PEG or resins (Impreg) that become insoluble after curing. Keeping the wood in a partially or completely swollen state increases dimensional stability, as well as strength and water repellency.

16.8.1   POLYETHYLENE GLYCOL

Polyethylene glycol (PEG) is a chemical with the following structure:

$$HO–CH_2CH_2–O–CH_2CH_2–(O–CH_2CH_2)_n–O–CH_2CH_2–OH$$

PEG 1000 is most commonly used when treating wood, usually undried, green wood (Mitchell 1972). It is a waxy, white solid that has an n value (average molecular weight) of 1000 and it can penetrate the cell wall because of its small size. It melts at 40°C, readily dissolves in warm water, is noncorrosive, odorless, and colorless, and has a very high fire point (305°C) (Rowell and Konkol 1987). Molecular weights up to 6000 are soluble in water.

To treat wood with PEG 1000, the wood is placed in a container and covered with a 30–50 wt.% solution dissolved in water. The treatment is based on diffusion and therefore soak time will vary depending upon the thickness of the specimen. PEG remains in the cell walls when the wood is dried because of its low vapor pressure. The rate of diffusion into the cell wall increases as water evaporates from the solution (Stamm 1964).

Treatment temperature is usually from 21°C to 60°C, but diffusion can be accelerated with increasing the temperature and/or the concentration of the solution. After treatment, the wood is air dried in a ventilated room. The drying time also depends upon specimen size.

PEG is not chemically attached to the wood, and because it is water soluble, it will leach out if it gets wet (see Chapter 4). Glycol attracts moisture, so if the relative humidity reaches above 70%, the wood becomes sticky.

PEG has many uses especially in prevention or reduction of cracking due to drying sound wood for tabletops, to partially decomposed wooden artifacts, or archeological waterlogged wood.

16.8.2   IMPREG

Impreg is wood that has been treated with a thermosetting, fiber-penetrating resin and is cured without compression (Stamm and Seborg 1962). Phenol–formaldehyde resin-forming systems with low molecular weights are the most successful thermosetting agents. The resins penetrate the cell wall (25–25% weight gain) and keep the wood in a swollen state, dried at 80–93°C for 30 min, and then are polymerized by heat (155°C for 30 min) to form a water-insoluble resin in the cell wall (Rowell and Konkol 1987). Treatments are usually done on thin veneers (<9 mm thick) due to time.

The cured product is usually reddish brown with reduced swelling, shrinkage, grain raising, and surface checking. It improves the compression strength, but reduces the impact bending strength. Impreg shows resistance to decay, termite, and marine-borer attack, and it has a high resistance to acid. It is suited for pattern and die models as well as electrical control equipment.

REFERENCES

Aagaard, P. 1967. Swedish studies on wood–polymer composites. *Svensk Kem. Tidskr.* 79(9): 501–510.

Ahmed, A. U., Takeshita, N., and Gotoda, M. 1971. Fire-retardant wood–polymer composite based on radiation-induced polymerization of phosphorous-containing vinyl monomers particularly bis(2-chloroethyl) vinylphosphonate. *Nippon Genshiryoku Kenkyusho Nempo* 5027: 82–90.

Autio, T. and Miettinen, J. K. 1970. Experiments in Finland on properties of wood–polymer composites. *Forest Products Journal* 20(3): 36–42.

Beall, F. C., Witt, A. E., and Bosco, L. R. 1973. Hardness and hardness modulus of wood–polymer composites. *Forest Products Journal* 23(1): 56–60.

Boey, F. Y. C., Chia, L. H. L., and Teoh, S. H. 1985. Compression, bend, and impact testing of some tropical wood-polymer composites. *Radiation Physics and Chemistry* 26(4): 415–421.

Boey, F. Y. C., Chia, L. H. L., and Teoh, S. H. 1987. Model for the compression failure of an irradiated tropical wood-polymer composite. *Radiation Physics and Chemistry* 29(5): 337–348.

Brebner, K. I., Schneider, M. H., and St-Pierre, L. E. 1985. Flexural strength of polymer-impregnated eastern white pine. *Forest Products Journal* 2: 22–27.

Bull, C., Espinoza, B. J., Figueroa, C. C., and Rosende, R. 1985. Production of wood–plastic composites with gamma radiation polymerization. *Nucleotecnica* 4: 61–70.

Calleton, R. L., Choong, E. T., and McIlhenny, R. C. 1970. Treatments of southern pine with vinyl chloride and methyl methacrylate for radiation-produced wood–plastic combinations. *Wood Science and Technology* 4(3): 216–225.

Czvikovszky, T. 1977. Pilot-scale experiments on radiation processing of wood–plastic composites. *Proc. Tihany Symp. Radiat. Chem.* 4: 551–560.

Czvikovszky, T. 1981. Wood-polyester composite materials. II. Dependence of the processing parameters on the initiation rate. *Angew. Makromol. Chem.* 96: 179–191.

Czvikovszky, T. 1982. Composite materials made of wood and a polyester resin. *Plast. Massy, 7, 37 and in (Plastics Manufacture and Processing)* 50: 1–43.

Desai, R. L. and Juneja, S. C. 1972. Weatherometer studies on wood–plastic composites. *Forest Products Journal* 22(9): 100–103.

Dolacis, J. 1983. Comparative abrasive wear of wood of various species modified radiochemically with polystyrene. *Modif. Svoistv Drev. Mater.* pp. 107–111. Edited by: Rotsens, K. A. Zinatne. Riga, USSR 37.

Doss, N. L., El-Awady, M. M., El-Awady, N. I., and Mansour, S. H. 1991. Impregnation of white pine wood with unsaturated polyesters to produce wood–plastic combinations. *Journal of Applied Polymer Science* 42: 2589–2594.

Ellis, W. D. 1994. Moisture sorption and swelling of wood polymer composites. *Wood and Fiber Science* 26(3): 333–341.

Ellis, W. D. and O'Dell, J. L. 1999. Wood-polymer composites made with acrylic monomers, isocyanate, and maleic anhydride. *Journal of Applied Polymer Science* 73(12): 2493–2505.

Ellwood, E., Gilmore, R., Merrill, J. A., and Poole, 1969. W. K. An investigation of certain physical and mechanical properties of wood–plastic combinations. *U. S. At. Energy Comm. ORO-638, 159pp. Avail. Dep.; CFSTIFrom: Nucl. Sci. Abstr.* 23(23), 48583.

Fujimura, T. and Inoue, M. 1991. Improvement of the durability of wood with acryl-high-polymer. III. Dimensional stability of wood with cross-linked epoxy-copolymer. *Mokuzai Gakkaishi* 37: 719–726.

Fujimura, T., Inoue, M., and Uemura, I. 1990. Durability of wood with acrylic-high-polymer. II. Dimensional stability with cross linked acrylic copolymer in wood. *Mokuzai Gakkaishi* 36(10): 851–859.

Fuller, B. S., Ellis, W. D., and Rowell, R. M. 1997. Hardened and fire retardant treatment of wood for flooring. U.S. Patent 5605767, February 25, 1997, U.S. Patent 5609915, March.

Ge, M., Peng, H., Dai, C., and Li, J. 1983. Heating wood–plastic composites. *Linye Kexue* 19(1): 64–72.

Glukhov, V. I. and Shiryaeva, G. V. 1973. Parameters of the radiation polymerization of vinyl monomers in wood. *Plasticheskie Massy* 6: 35–36.

Gotoda, M., Harada, O., Yagi, T., and Yoshizawa, I. 1975. Radiation curing of a mixture of diallyl phthalate prepolymer and vinyl monomer. X. Application of electron beam curing of low molecular weight diallyl phthalate prepolymer-vinyl monomer mixtures to the preparation of American hemlock-polymer composite laminated board. I. Thermal curing. *Nippon Genshiryoku Kenkyusho Nempo* 5030: 92–102.

Gotoda, M., Horiuchi, Y., and Urasugi, H. 1974. Radiation curing of mixtures of diallyl phthalate prepolymer and vinyl monomers. VIII. Application to the manufacture of resin composite-veneered plywood of electron beam curing within wood. 2. *Nippon Genshiryoku Kenkyusho Nempo* 5029: 79–91.

Gotoda, M. and Kitada, Y. 1975. Radiation curing of mixtures of diallyl phthalate prepolymer and vinyl monomer. IX. Fundamental examination of the application of low molecular weight diallyl phthalate prepolymer. *Nippon Genshiryoku Kenkyusho Nempo* 5030: 85–91.

Gotoda, M., Okugawa, H., Yagi, T., and Yoshizawa, I. 1975. Radiation curing of a mixture of diallyl phthalate prepolymer and vinyl monomer. XI. Application of electron beam curing of low molecular weight diallyl phthalate prepolymer-vinyl monomer mixtures to the preparation of American hemlock-polymer composite laminated board. 2. Electron beam curing. *Nippon Genshiryoku Kenkyusho Nempo* 5030: 103–113.

Gotoda, M. and Takeshita, N. 1971. Preparation of a wood-polymer composite by ionizing radiation. VII. Improvement of thermal stability of vinylidene chloride copolymers and fire retardancy of the wood-vinylidene chloride copolymer composites. *Nippon Genshiryoku Kenkyusho Nempo* 5027: 63–71.

Gotoda, M. and Takeshita, N. 1971. Preparation of a wood-polymer composite by ionizing radiation. VIII. Radiation curing of monomer solution of chlorinated allyl chloride oligomer and its application to the preparation of a flame-retardant wood-polymer composite. *Nippon Genshiryoku Kenkyusho Nempo* 5027: 72–81.

Gotoda, M., Tsuji, T., and Toyonishi, S. 1970a. Preparation of wood-polymer-composite by ionizing radiation. V. Effect of solvent extraction of wood as a pretreatment on the .gamma.-induced polymerization of vinylidene chloride in wood. *Nippon Genshiryoku Kenkyusho Nempo* 5026: 86–93.

Gotoda, M., Yokoyama, K., Takeshita, N., and Senzaki, Y. 1971. Radiation curing of mixture of diallyl phthalate prepolymer and vinyl monomer. VI. Electron-beam curing of diallyl phthalate prepolymer/monomer mixture in the preparation of wood-polymer composite piled board. *Nippon Genshiryoku Kenkyusho Nempo* 5027: 91–99.

Gotoda, M., Yokoyama, K., and Toyonishi, S. 1970b. Radiation curing of a mixture of diallyl phthalate prepolymer and vinyl monomer. IV. Radiation (especially electron-beam) curing of the mixture, impregnated in wood, for preparing wood-polymer composites. *Nippon Genshiryoku Kenkyusho Nempo* 5026: 108–120.

Guillemain, J., Laizier, J., Marchand, J., and Roland, J. C. 1969. Polymer-substrate bonding obtained by radiation in resin-treated woods. Large Radiat. Sources Ind. Processes, *Proc. Symp. Util. Large Radiat. Sources Accel. Ind. Process.*, pp. 417–33.

Handa, T., Otsuka, N., Akimoto, H., Ikeda, Y., and Saito, M. 1973. Characterization on the electron beam-induced polymerization of monomers in wood. I. Effect of moisture in the polymerization of vinyl monomer in presoaked beech veneer. *Mokuzai Gakkaishi.* 19(10): 493–498.

Handa, T., Seo, I., Akimoto, H., Saito, M., and Ikeda, Y. 1972. Physical properties of wood-polymer composite materials prepared by I. C. T-type electron-accelerator. *Proc. Jpn. Congr. Mater. Res.* 15: 158–163.

Handa, T., Seo, I., Ishii, T., and Hashizume, Y. 1983. Polymer-performance on the dimensional stability and the mechanical properties of wood-polymer composites prepared by an electron beam accelerator. *Polym. Sci. Technol.* 20(3): 167–190.

Handa, T., Yoshizawa, S., Seo, I., and Hashizume, Y. 1981. Polymer performance on the dimensional stability and the mechanical properties of wood-polymer composites evaluated by polymer-wood interaction mode. *Org. Coat. Plast. Chem.* 45: 375–381.

Helinska-Raczkowska, L. and Molinski, W. 1983. Effect of atmospheric corrosion in contact with rusting iron on the impact strength of lignomer. *Zesz. Probl. Postepow Nauk Roln.* 260: 199–210.

Ibach, R. E. and Rowell, R. M. 2001. Wood preservation based on *in situ* polymerization of bioactive monomers—Part 1. Synthesis of bioactive monomers, wood treatments and microscopic analysis. *Holzforschung* 55(4): 358–364.

Ibach, R. E. and Rowell, R. M. 2001. Wood preservation based on *in situ* polymerization of bioactive monomers—Part 2. Fungal resistance and thermal properties of treated wood. *Holzforschung* 55(4): 365–372.

Inoue, M., Ogata, S., Kawai, S., Rowell, R. M., and Norimoto, M. 1993. Fixation of compressed wood using melamine-formaldehyde resin. *Wood and Fiber Science* 25(3): 404–410.

Inoue, M., Ogata, S., Nishikawa, M., Otsuka, Y., Kawai, S., and Norimoto, M. 1993. Dimensional stability, mechanical properties, and color changes of a low molecular weight melamine-formaldehyde resin impregnated wood. *Mokuzai Gakkaishi* 39(2): 181–189.

Iya, V. K. and Majali, A. B. 1978. Development of radiation processed wood-polymer composites based on tropical hardwoods. *Radiation Physics and Chemistry* 12(3–4): 107–110.

Jokel, J. 1972. Wood plastics made from aqueous emulsions of a polyester-styrene mixture emulsion. *Drevarsky Vyskum* 17: 247–260.

Kakehi, M., Yoshida, Y., and Minami, K. 1985. Modified wood manufacture, Daiken Kogyo Co., Ltd.

Katsurada, S. and Kurahashi, K. 1985. Process and apparatus for producing wood heads of golf clubs, Sumitomo Rubber Industries Ltd.

Katuscak, S., Horsky, K., and Mahdalik, M. 1972. Phase diagrams of ternary monomer-solvent-water systems used for the preparation of wood-plastic combinations (WPC). *Drevarsky Vyskum* 17(3): 175–186.

Kawakami, H. and Taneda, K. 1973. Impregnation of sawn veneers with methyl methacrylate and unsaturated polyester-styrene mixture, and the polymerization by a catalyst-heat technique. Polymerization in several wood species and properties of treated veneers. *J. of the Hokkaido Forest Prod. Res. Inst.* 10: 22–27.

Kawakami, H., Taneda, K., Ishida, S., and Ohtani, J. 1981. Observation of the polymer in wood-polymer composite II On the polymer location in WPC prepared with methacrylic esters and the relationship between the polymer location and the properties of the composites. *Mokuzai Gakkaishi.* 27(3): 197–204.

Kawakami, H., Yamashina, H., and Taneda, K. 1977. Production of wood-plastic composites by functional resins. I. Effects of adding cross-linking and polar monomers to methyl methacrylate. *J. of the Hokkaido Forest Prod. Res. Inst.* 306: 10–17.

Kawase, K. and Hayakawa, K. 1974. Manufacturing of a wood-plastic combination by irradiation of microwave. Polymerization of impregnated monomer in wood with microwave irradiation. *Mokuzai Kogyo.* 29(32): 12–17.

Kenaga, D. L. 1970. Heat cure of high boiling styrene-type monomers in wood. *Wood and Fiber Science* 2(1): 40–51.

Knotik, K. and Proksch, E. 1971. Polymer-impregnated wood for musical instruments, Oesterreichische Studiengesellschaft fuer Atomenergie G.m.b.H.

Knotik, K., Proksch, E., and Bresimair, K. 1971. Polymer-impregnated wood mouthpieces for wind instruments, Oesterreichische Studiengesellschaft fuer Atomenergie m.b.H.

Lawniczak, M. 1973. Poplar wood modification. *Prace Komisji Technologii Drewna* 1: 3–42.

Lawniczak, M. 1979. Effect of temperature changes during styrene polymerization in wood on the quality of the lignomer. *Prace Komisji Technologii Drewna* 9: 69–82.

Lawniczak, M., Melcerova, A., and Melcer, I. 1987. Analytical characterization of pine and alder wood polymer composites. *Holzforschung und Holzverwertung* 39(5): 119–121.

Lawniczak, M. and Pawlak, H. 1983. Effect of wood saturation with styrene and addition of butyl methacrylate and acrylonitrile on the quality of the produced lignomer. *Zeszyty Problemowe Postepow Nauk Rolniczch* 260: 81–93.

Lawniczak, M. and Szwarc, S. 1987. Cross-linking of polystyrene in wood-polystyrene composite preparation. *Zeszyty Problemowe Postepow Nauk Rolniczch* 299(37): 115–125.

Lee, T. J. 1969. Wood plastic composites. *Hwahak Kwa Kongop Ui Chinbo.* 9(3): 220–224.

Loos, W. E. 1968. Dimensional stability of wood-plastic combinations to moisture changes. *Wood Science and Technology* 2(4): 308–312.

Lubke, H. and Jokel, J. 1983. Combustibility of lignoplastic materials. *Zesz. Probl. Postepow Nauk Roln.* 260: 281–292.

Maine, J. 1971. Composite wood-polymer product, Maine, C. W., and Sons, Inc.

Maldas, D., Kokta, B. V., and Daneault, C. 1989. Thermoplastic composites of polystyrene: Effect of different wood species on mechanical properties. *Journal of Applied Polymer Science* 38: 413–439.

Manrich, S. and Marcondes, J. A. 1989. The effect of chemical treatment of wood and polymer characteristics on the properties of wood polymer composites. *Journal of Applied Polymer Science* 37(7): 1777–1790.

Mathias, L. J., Kusefoglu, S. H., Kress, A. O., Lee, S., Wright, J. R., Culberson, D. A., Warren, S. C., Warren, R. M., Huang, S., Lopez, D. R., Ingram, J. E., Dickerson, C. W., Jeno, M., Halley, R. J., Colletti, R. F., Cei, G. and Geiger, C. C. 1991. Multifunctional acrylate monomers, dimers and oligomers—Applications from contact-lenses to wood-polymer composites. *Makromolekulare Chemie-Macromolecular Symposia* 51: 153–167.

Meyer, J. A. 1965. Treatment of wood-polymer systems using catalyst-heat techniques. *Forest Products Journal* 15(9): 362–364.

Meyer, J. A. 1981. Wood-polymer materials: State of the art. *Wood Science* 14(2): 49–54.

Meyer, J. A. 1984. Wood polymer materials. *Adv. Chem. Ser. 207 (Chem. Solid Wood)*: 257–289.

Miettinen, J. K. 1969. Production and properties of wood-plastic combinations based on polyester/styrene mixtures. *Amer. Chem. Soc., Div. Org. Coatings Plast. Chem., Pa.* 29(182–8).

Miettinen, J. K., Autio, T., Siimes, F. E., and Ollila, T. 1968. Mechanical properties of plastic-impregnated wood made from four Finnish wood species and methyl methacrylate or polyester by using irradiation. *Valtion Tek. Tutkimuslaitos, Julk.* 137: 58pp.

Mitchell, H. L. 1972. How PEG helps the hobbyist who works with wood. USDA, FS, Forest Products Laboratory, 20 pp.

Mizumachi, H. 1975. Interaction between the components in wood-polymer composite systems. *Nippon Setchaku Kyokai Shi.* 11(1): 17–23.

Mohan, H. and Iyer, R. M. 1991. Study of wood-polymer composites. *Conf. Proc., Rad Tech Int. North Am.* Northbrook, IL, pp. 93–97.

Nobashi, K., Koshiishi, H., Ikegami, M., Ooishi, K., and Kaminaga, K. 1986. WPC treatments of several woods with functional monomer. III. Hygroscopicities and dimensional stabilities of Buna WPC. *Shizuoka-ken Kogyo Gijutsu Senta Kenkyu Hokoku.* 31: 15–26.

Nobashi, K., Koshiishi, H., Ikegami, M., Ooishi, K., Oosawa, T., and Kaminaga, K. 1986. WPC treatments of wood with functional monomer. II. *Shizuoka-ken Kogyo Gijutsu Senta Kenkyu Hokoku* 30: 21–26.

Okonov, Z. V. and Grinberg, M. V. 1983. Modification of the properties of wood by polystyrene as applied to machine parts of the textile industry. *Modif. Svoistv Drev. Mater 102-6. Edited by: Rotsens; K. A. Zinatne: Riga, USSR* 38, 40.

Paszner, L., Szymani, R., and Micko, M. M. 1975. Accelerated curing and testing of some radiation curable methacrylate and polyester copolymer finishes on wood panelings. *Holzforschung und Holzverwertung* 27: 67–73.

Pesek, M. 1984. The possibilities of preparation of wood-plastic combinations based on unsaturated polyester resins and methyl methacrylate and styrene by combined radiation and chemical methods. *ZfI-Mitt.* 97: 359–363.

Pesek, M., Jarkovsky, J., and Pultar, F. 1969. Radiation polymerization of methyl methacrylate in wood. Effect of halogenated organic compounds on the polymerization rate. *Chem. Prum.* 19(11): 503–506.

Pietrzyk, C. 1983. Effect of selected factors on polymerization of methyl methacrylate by the radiation method in alder and birch wood. *Zesz. Probl. Postepow Nauk Roln* 260: 71–80.

Proksch, E. 1969. Wood-plastic combinations prepared from beechwood. *Holzforschung* 23: 93–98.

Pullmann, M., Jokei, J., and Manasek, Z. 1978. Dimensional stabilization of wood with a synthetic polymer in wood-plastic composites. II. Effect of components of the system wood + impregnant on the radiation polymerization of methyl methacrylate. *Drevarsky Vyskum* 22(4): 261–276.

Raj, R. G. and Kokta, B. V. 1991. Reinforcing high-density polyethylene with cellulosic fibers. I. The effect of additives on fiber dispersion and mechanical properties. *Polymer Engineering and Science* 31:1358–1362.

Rao, K. N., Moorthy, P. N., Rao, M. H., Vijaykumar, and Gopinathan, C. 1968. Wood-plastic combinations. II. Acrylic esters and their copolymers. *India, At. Energy Comm., Bhabha Atomic Research Cent.*, BARC-369: 11 pp.

Rowell, R. M. 1983. Bioactive polymer-wood composites. *Controlled Release Delivery Systems.* New York, New York and Basel: Marcel Dekker Inc. Chapter 23: 347–357.

Rowell, R. M., Feist, W. C., and Ellis, W. D. 1981. Weathering of chemically modified southern pine. *Wood Science* 13(4): 202–208.

Rowell, R. M. and Konkol, P. 1987. Treatments that enhance physical properties of wood. Madison, WI, USDA, Forest Service Forest Products Laboratory.

Rowell, R. M., Moisuk, R., and Meyer, J. A. 1982. Wood-polymer composites: Cell wall grafting with alkylene oxides and lumen treatments with methyl methacrylate. *Wood Science* 15: 290–296.

Schaudy, R. and Proksch, E. 1981. Wood-plastic combinations with high dimensional stability. *Oesterr. Forschungszent. Seibersdorf* Report No. 4113.

Schaudy, R. and Proksch, E. 1982. Wood-plastic combinations with high dimensional stability. *Ind. Eng. Chem. Prod. Res. Dev.* 21(3): 369–375.

Schaudy, R., Wendrinsky, J., and Proksch, E. 1982. Wood-plastic composites with high toughness and dimensional stability. *Holzforschung* 36(4): 197–206.

Schaudy, R., Wendrinsky, J., and Proksch, E. 1982. Wood-plastic composites with high toughness and dimensional stability under moisture. *Holzforschung* 36(4): 197–206.

Schneider, M. H. 1994. Wood polymer composites. *Wood and Fiber Science* 26(1): 142–151.

Schneider, M. H., Phillips, J. G., Brebner, K. I., and Tingley, D. A. 1989. Toughness of polymer impregnated sugar maple at two moisture contents. *Forest Products Journal* 39(6): 11–14.

Schneider, M. H., Phillips, J. G., Tingley, D. A., and Brebner, K. I. 1990. Mechanical properties of polymer-impregnated sugar maple. *Forest Products Journal* 40(1): 37–41.

Schneider, M. H. and Witt, A. E. 2004. History of wood polymer composite commercialization. *Forest Products Journal* 54(4): 19–24.

Seborg, R. M., Millett, M. A. and Stamm, A. J. 1945. Heat-stabilized compressed wood—Staypak. Mechanical Engineering 67(1):25–31.

Siau, J. F., Campos, G. S., and Meyer, J. A. 1975. Fire behavior of treated wood and wood-polymer composites. *Wood Science* 8(1): 375–383.

Siau, J. F., Meyer, J. A., and Kulik, R. S. 1972. Fire-tube tests of wood-polymer composites. *Forest Products Journal* 22(7): 31–36.

Singer, K., Vinther, A. and Thomassen, T. 1969. Some technological properties of wood-plastic materials. *Danish Atomic Energy Commission* Riso Rept. No. 211: 30 p.

Soundararaian, S. and Reddy, B. S. R. 1991. Glycidyl methacrylate and N-vinyl-2-pyrrolidone copolymers: synthesis, characterization, and reactivity ratios. *Journal of Applied Polymer Science* 43: 251–258.

Spindler, M. W., Pateman, R. and Hills, P. R. 1973. Polymer impregnated fibrous materials. Resistance of polymer-wood composites to chemical corrosion. *Composites* 4(6): 246–253.

Stamm, A. J. 1964. *Wood and Cellulose Science*. New York, The Ronald Press Co.

Stamm. A. J., Burr, H. K., and Kline, A. A. 1960. *Heat Stabilized Wood—Staybwood*. USDA, FS, Forest Products Laboratory, Report No. 1621.

Stamm, A. J. and Seborg, R. M. 1951. *Resin-Treated Laminated, Compressed Wood—Compreg*. USDA, FS, Forest Products Laboratory, Report No. 1381.

Stamm. A. J. and Seborg, R. M. 1962. *Resin Treated Wood—Impreg*. USDA, FS, Forest Products Laaboratory, Report No. 1380.

Takasu, Y. and Matsuda, K. 1993. Dimensional stability and mechanical properties of resin impregnated woods. *Aichi-ken Kogyo Gijutsu Senta Hokoku* 29: 57–60.

Ueda, M., Matsuda, H., and Matsumoto, Y. 1992. Dimensional stabilization of wood by simultaneous oligoes-terification and vinyl polymerization. *Mokuzai Gakkaishi* 38(5): 458–465.

Varga, S. and Piatrik, M. 1974. Contribution of the radiation preparation of wood-plastic materials [WPC]. VIII. Testing of the physical and mechanical properties of WPC. *Radiochem. Radio anal. Lett* 19: 255–261.

Yamaguchi, K. 1982. *Wood Hitting Parts for Sporting Goods*. Japan Patent JP 57203460.

Yamashina, H., Kawakami, H., Nakano, T., and Taneda, K. 1978. Effect of moisture content of wood on impregnation, polymerization and dimensional stability of wood-plastic composites. *J. of the Hokkaido Forest Prod. Res. Inst.* 316: 11–14.

Yap, M. G. S., Chia, L. H. L., and Teoh, S. H. 1990. Wood-polymer composites from tropical hardwoods I WPC properties. *Journal of Wood Chemistry and Technology* 10(1): 1–19.

Yap, M. G. S., Que, Y. T., and Chia, L. H. L. 1991. Dynamic mechanical analysis of tropical wood polymer composites. *Journal of Applied Polymer Science* 43(11): 1999–2004.

Yap, M. G. S., Que, Y. T., and Chia, L. H. L. 1991. FTIR characterization of tropical wood-polymer composites. *Journal of Applied Polymer Science* 43:2083–2090.

Yap, M. G. S., Que, Y. T., Chia, L. H. L., and Chan, H. S. O. 1991. Thermal-properties of tropical wood polymer composites. *Journal of Applied Polymer Science* 43(11): 2057–2065.

17 Plasma Treatment of Wood

Wolfgang Viöl, Georg Avramidis, and Holger Militz

CONTENTS

Plasma treatment of wood is a novel technology in the field of wood surface modification with a number of applications that have only been investigated for a few years (Denes et al., 2005). Some of the processes described in the literature are close to an industrial implementation.

There are many different methods to generate plasma discharges. Further, varying plasma sources can operate with different plasma parameters, which makes plasma treatment a topic of high complexity. Therefore, this chapter is limited to plasmas that can reasonably be used for a treatment of the material *wood* and which actually are suitable for industrial implementation.

17.1 THE PLASMA STATE

The *plasma* used for wood modification is an ionized gas with a characteristic optical spectrum. To give some examples, such a plasma is technically used for generating light in fluorescent lamps, in plasma displays for image presentation, or for pretreating plastic bags to assure the adherence which is necessary for printing shop logos on it.

Plasma is also called the "fourth state of matter." By applying energy, for example, in the form of heat, to a solid (i.e., the first state of matter), the solid melts and turns into a liquid (i.e., the second state of matter). With a further heating, the liquid can be vaporized and therefore be transferred to the third state of matter, gas. By a further supply of energy to the gas, single electrons disassociate

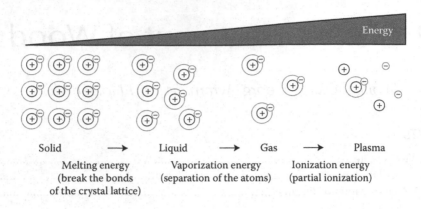

FIGURE 17.1   The four states of aggregation of matter.

from the gas atoms and molecules; the gas is ionized. Electrons and ions now move freely in this electrically conductive gas. This fourth state of matter is the most energy-rich one (see Figure 17.1).

Nature provides a lot of different forms of appearance for plasma, such as flashs, brush discharges or the *Aurora borealis*. Stars as the sun completely consist of plasma. Ninety-nine percent of matter in the cosmos is in the plasma state, the fourth state of matter.

17.1.1   GAS DISCHARGE

The easiest way to technologically generate plasma is the electrical gas discharge (see Figure 17.2).

Here, a voltage is applied in between two metal electrodes, generating an electric field in the gaseous gap. In common ambient air, cosmic radiation and natural radioactive radiation create approximately 1000 electrons/m^3. This implies that electrons already exist in the gaseous gap, being accelerated by the electric field and absorbing energy. The kinetic energy of the electrons can reach the range of electron volts (eV) before a collision with an atom or molecule. The kinetic energy of the electron can be used to stimulate atoms and molecules. Atoms and molecules emit this stimulating energy, for example, in the form of light. This process causes the light emission of a gas discharge or plasma.

The kinetic energy can increase to a threshold value, causing the emission of another electron from the atom or molecule. Thus, two electrons and a positively charged ion remain. The electrons are accelerated again by the electric field. Within further collisions with gas atoms and molecules, more electrons can be emitted in this vein: as a result, an electron avalanche occurs (see Figure 17.3) and the plasma state is created. Substantially, the plasma is electrically neutral to the outside, as many electrically negative as electrically positive charged particles are generated.

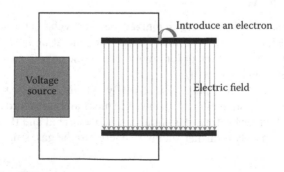

FIGURE 17.2   Electrical gas discharge.

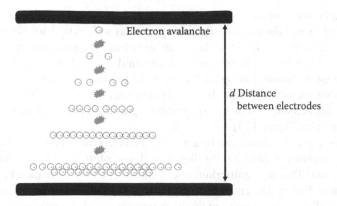

FIGURE 17.3 Generation of a gas discharge.

17.1.2 CLASSIFICATION OF PLASMAS

Whether plasma in an electric gas discharge setup ignites or not depend on the gas pressure p and the distance between the electrodes d (see Figure 17.3). The ignition behavior is described by the so-called *Paschen law* (see Figure 17.4).

Commonly, an electrode distance of 1 cm is chosen. Here, plasma ignition in gas at a low voltage can be realized at a pressure of approximately 100 Pa (= 1 hPa = 1 mbar = 0.75 Torr). At lower pressures, the gas particles, which are necessary as collision partners for ionization, are missing. At higher pressures, the electron cannot be accelerated up to an energy which is sufficient for an ionization between two collisions due to the shorter mean free path. Thus, a higher voltage is necessary for generating a gas discharge in both cases. Hence, for most plasma applications, a pressure of 1–10 hPa is chosen. A lot of research on the plasma treatment of wood has been executed in this range of pressure.

The vacuum methods used in wood technology operate at a negative or positive pressure of maximum 100 hPa, so they are not comparable to the vacuum chambers required for plasma technology.

Here, a negative pressure of 1000 hPa has to be achieved, causing extremely large forces on the vessel material by the ambient pressure, corresponding to approximately 10,000 kg/m². This shows the necessity of constructing vacuum vessels with a huge wall thickness and stability. In combination with the high demands for the associated vacuum pump to reach this low pressure, a high price level for the technology is reached. Nevertheless, only small wood samples with dimensions in the range of some centimeters can be modified in this way.

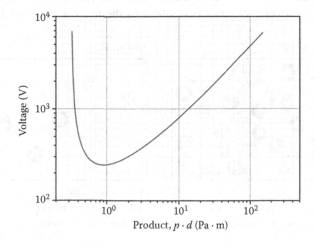

FIGURE 17.4 Required voltage for ignition of a gas discharge in air.

In addition, water is evaporated from the wood within the operating pressure range of 10 hPa, resulting in an elongation of the process of pumping down as well as both a deforming and shrinking of the wood sample. Considering these facts, an industrial implementation of the low-pressure plasma technology only seems to be practicable for a small range of individual cases.

In the recent past, substantial progress has been achieved in the development of atmospheric pressure plasma sources, which enable the user to omit vacuum systems. As well, the sources are capable of inline applications. Atmospheric pressure plasma technology is classified as a high-pressure technology (see Figure 17.5).

Another classification of plasmas can be accomplished referring to the equilibrium condition. In contrast to the three states of matter solid, liquid, and gas, plasma is able to exist in a thermodynamic nonequilibrium. This nonequilibrium initial state can easily be realized in the low-pressure range, but with some discharging arrangements it is as well possible to realize in the high-pressure range. The nonequilibrium state is caused by the extremely small electron mass which is 1/1860 of the atomic mass unit. When an electron elastically collides with an atom, the energy exchange $\Delta E/E$ per collision is calculated by applying both the principles of conservation of energy and the momentum conservation law:

$$\frac{\Delta E}{E} \approx \frac{4m_e}{m_a} \tag{17.1}$$

where m_e is the electron mass and m_a is the atomic mass.

Thus, 0.2% of the electron energy is transferred for an electron collision with a hydrogen atom, but only 0.005% is transferred for a collision with an argon atom. Thereby, the energy or velocity distribution of electrons in nonequilibrium plasma differs significantly from heavy atoms, molecules, and ions. Nevertheless, the electrons are at equilibrium with each other. The so-called electron temperature can then be assigned to these electrons.

So, different temperatures can be found in nonequilibrium plasmas:

Wall temperature
Neutral gas temperature
Rotational temperature
Vibration temperature
Electron temperature
Ion temperature

Not all the components in plasma necessarily have the same temperature. In most of the plasmas, the rotational temperature of molecules is identical to the neutral gas temperature. This is what

High pressure

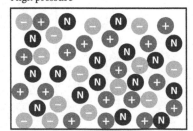

- Small free path
- Numerous collision processes
- Efficient energy exchange

Low pressure

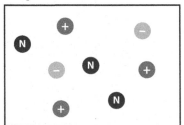

- Huge free path
- Rarely collision processes
- Different temperatures

FIGURE 17.5   Classification of plasmas referring to gas pressure.

makes nonequilibrium plasma attractive for an application to wood and wood products as basic materials. Both the neutral gas temperature and the ion temperature are barely higher than room temperature, so a thermal impact (in extreme cases even a burning) can easily be avoided. However, the vibration temperature can be up to some 1000 K and the electron temperature can rise up to 10,000 K in a nonequilibrium plasma. Since an interaction of these electrons with the wood molecules is inefficient as well (see Equation 17.1), a thermal effect to the wood caused by the electrons can be excluded here. The temperatures listed above can be measured with the help of spectroscopic methods (Viöl and Uhlenbusch 1996; Förster et al. 2005; Awakowicz et al. 2009; Kuchenbecker et al. 2009; Rajasekaran et al. 2009, 2010; Helmke et al. 2011).

These facts explain the efficiency of a wood treatment with non-equilibrium plasma. Only the needed energy for heating the electrons, which finally cause the molecular effect on the wood surface, has to be invested.

17.1.3 PLASMA PARAMETERS

The most important plasma parameters are the electron density n_e and the electron temperature T_e. The electron temperature corresponds to a certain electron velocity distribution. Numerous different microscopic processes among the atoms, molecules, ions and free electrons proceed in plasma. When the rates of all processes involved are as large as their particular inverse processes, the electron velocities v_e follow the Maxwell velocity distribution according to

$$f(v_e) = \frac{4}{\sqrt{\pi}} v^2 \left(\frac{m_e}{2\pi k T_e} \right)^{3/2} * \exp\left(-\frac{m_e v_e^2}{2 k T_e} \right) \tag{17.2}$$

with the Boltzmann constant k.

For a given electron temperature, the electron density n_e can be calculated with the help of the *Saha Eggert Equation* (a.k.a. *Saha Ionization Equation*). It describes the ionization recombination equilibrium:

$$\frac{n_{i+1} * n_e}{n_i} = \frac{2(2\pi m_e k T_e)^{3/2}}{h^3 \sum_{q=1}^{Q_i} g_q e^{-\frac{E_{q,i}}{kT_e}}} e^{-\frac{E_{ion,i}}{kT_e}} . \tag{17.3}$$

Here, n_i is the particle number density of ions at an ionization stage i (Griem, 1964). $E_{q,i}$ is the energy of a n_i state with the quantum number q, g_q is its degenerate energy level. Further,

$$Q_i = \frac{1}{2}(i+1)^{3/5}(a_0^3 n_e)^{-2/15} \tag{17.4}$$

is the maximum quantum number, where a_0 is the Bohr radius, which is given by

$$a_0 = \frac{4\pi\epsilon_0 h}{m_e e_0^2} = 5.29 * 10^{-11} \text{m}. \tag{17.5}$$

$E_{ion,i}$ is the ionization energy for the ionization stage i, the atomic number is represented by z. The electron density n_e is calculated as

$$n_e = \sum_{i=1}^{z} i\, n_i \tag{17.6}$$

The ionization state x of a single ionization is calculated by

$$x = \frac{n_e}{n_e + n_o} \tag{17.7}$$

with the neutral particle density n_o.

Figure 17.6 shows the ionization state x of caesium with an ionization energy of 3.9 eV, helium with an ionization energy of 24.6 eV and an additional notional ionization energy of 40 eV, all calculated by applying the Saha–Eggert equation.

It is quite remarkable that an ionization state can generally not be found at an average electron energy of 1/10 of the ionization energy, but even a complete ionization with $x \approx 1$ can be reached.

According to Equation 17.3, the dissociation equilibrium of molecules can be calculated with the help of the Saha–Eggert equation. To give an example, for temperatures above 3000 K, the calculation is provided for the dissociation of hydrogen (Artmann, 1963):

$$\frac{n_h^2}{n_{H_2}} = \frac{2g_0^2}{g_0(H_2)} * \frac{(\pi m_H)^{3/2}}{8\pi^2\theta} * \frac{v}{\sqrt{k_B T}} * \exp\left(-\frac{E_D}{k_B T_e}\right) \tag{17.8}$$

g_0: statistic weight of the hydrogen atom ground state
$g_0(H_2)$: statistic weight of the hydrogen molecule ground state
v: oscillator fundamental frequency
θ: moment of inertia
E_D: dissociation energy

With the dissociation degree:

$$D = \frac{n_H}{2n_{H_2} + n_H} \tag{17.9}$$

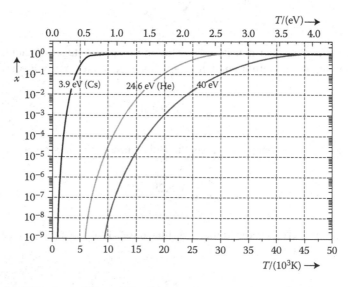

FIGURE 17.6 Ionization state x of Cs and He plasma and a gas with $E_{ion} = 40$ eV vs. temperature T at a gas pressure of approx. 0.1 MPa.

and the pressure p, Equation 17.8 can be written as

$$\frac{D^2}{1-D^2} = \frac{g_0^2}{2g_0(H_2)} * \frac{1}{p} * \frac{(\pi m_H)^{3/2}}{8\pi^2\theta} * v\sqrt{k_BT} \exp\left(-\frac{E_D}{k_BT_e}\right) \qquad (17.10)$$

For hydrogen, the following data are valid:

$g_0 = 2$
$g_0(H_2) = 1$
$\nu = 1.282 * 10^{14} Hz$
$E_D = 7.175 * 10^{-19} J = 4.5\ eV$
$\theta = 4.716 * 10^{-48}\ kg\ m^2$

In Figure 17.7, a complete hydrogen dissociation at an electron temperature of $T_e = 5000$ K and an electron energy of $k * T_e = 0.45$ eV, respectively (1/10 of dissociation energy) are shown.

This illustrates the effectiveness of plasma technology. Only about a tenth of the expected energy is needed to completely run a plasma physical or plasma chemical process. In nonequilibrium plasma, no energy is needed to heat the gas, an acceleration of the electrons is sufficient. The electrons are able to effect both the excitation process and the chemical reactions almost independently.

The rates for electron collision reactions such as

$$e + A \rightarrow e + A^* \qquad (17.11)$$

are calculated from the electron energy distribution function $f(E)$ and the cross section of the stimulation reaction q_{eA^*} corresponding to

$$\frac{dn_{A^*}}{dt} = n_e * n_a * k_{eA^*} = n_e * n_a * \int_0^\infty q_{eA^*}(E) * v_E(E) * f(E)dE. \qquad (17.12)$$

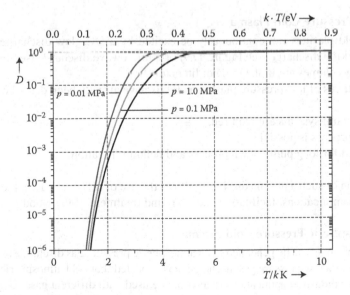

FIGURE 17.7 Dissociation state D vs. electron temperature T_e and electron energy kT_e at $p = 0.01$ MPa, $p = 0.1$ MPa (atmospheric pressure), and $p = 1.0$ MPa.

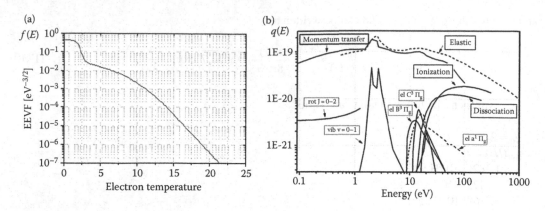

FIGURE 17.8   (a) Electron energy distribution function EEVF for air at atmospheric pressure of $p = 0.1$ MPa calculated with BOLSIG®; (b) electron collision cross section of nitrogen.

Figure 17.8 shows the numerically calculated energy distribution following the Boltzmann solver *BOLSIG®* for air in atmospheric pressure. As an example, the cross sections for nitrogen are displayed.

17.1.4   LOW-TEMPERATURE PLASMA

For a plasma treatment of wood, the application of thermic plasma (plasmas in the thermodynamic equlibrium) is evidently excluded, since the plasma features gas temperatures above 1000°C.

On the other hand, low-temperature plasma is in the nonequilibrium state as described above. Thus, the electron temperature is above 10,000 K and the gas temperature is below 400°C; preferably below 120°C in order to avoid a wood pyrolysis. In particular, it is at room temperature to exclude thermal effects completely. Such nonequilibrium plasmas can easily be generated in the low-pressure range ($p \approx 10$ hPa), since the energy exchange between electrons, atoms, molecules and ions is even more inefficient due to of the small gas particle density.

17.1.4.1   Low-Pressure Cold Plasma

Low-pressure cold plasmas can continuously be excited by a DC glow discharge, a capacitive or inductive coupled RF discharge (see Figure 17.9) or a microwave discharge (see Figure 17.10). The needed discharge volumes are in the scale of liters (dm³).

The disadvantages of low-pressure plasma methods have already been described:

Expensive vacuum systems are required.
Only a batch process is possible.
Time is needed for long pump down process and dehumidification.

The application of such a plasma discharge allows the use of different gases. It can be used without any further complications, facilitating an all-around treatment of the wood product.

17.1.4.2   Atmospheric Pressure Cold Plasma

Nonequilibrium plasmas can be generated in an atmospheric pressure gas discharge as well. Because of the disadvantages of a low-pressure gas discharge, it is expected that cold atmospheric pressure plasma is more suitable for industrial applications. It also can be used with different gases or gas mixtures.

Principally, there are two different types of plasma sources. On the one hand, the wood itself is used as a counter electrode, as in a corona discharge or dielectric barrier discharge. On the other

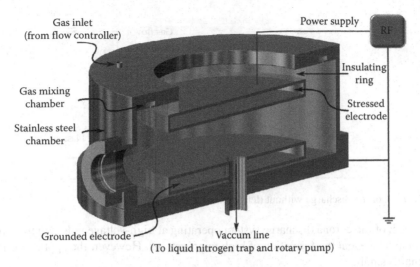

Gas inlet
(from flow controller)

Power supply

RF

Gas mixing
chamber

Insulating
ring

Stressed
electrode

Stainless steel
chamber

Grounded electrode

Vaccum line
(To liquid nitrogen trap and rotary pump)

FIGURE 17.9 Schematic of an RF-cold plasma reactor used to reduce weathering degradation of wood. (Adapted from Denes, A.R. and Young, R.A. 1999. *Holzforschung* 53:632–640.)

hand, plasma is generated between at least two electrodes in the plasma source, blowing the gas stream directly onto the wood surface simultaneously (e.g., the plasma jet).

The corona discharge is characterized by an ignition of a gas discharge in high field intensity at the electrode tip. The area near the pointy electrode is called a luminous area, followed by an extended dark drift zone up to the counter electrode (here: wood) (Figure 17.11).

The corona discharge can be generated with alternating current or with direct current. The power densities on the wood are quite low. So, the created effect is low as well.

By using higher power densities, the field which is physically defined as corona discharge is no longer valid, since there are no drift zones in this case. This kind of gas discharge leads to hot discharge channels on the wood. Hence, such a gas discharge is not suitable for the wood modification, too.

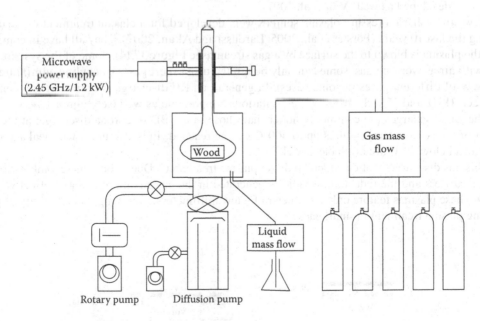

Microwave
power supply
(2.45 GHz/1.2 kW)

Wood

Gas mass
flow

Liquid
mass flow

Rotary pump   Diffusion pump

FIGURE 17.10 Schematic of the equipment used to modify wood surfaces by plasma polymerization. (Adapted from Podgorski, L. et al. 2001. *Pigment Resin Technol.* 31: 33–40.)

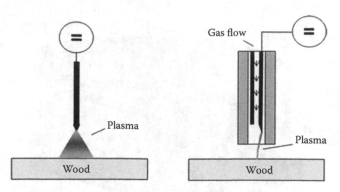

FIGURE 17.11   Corona discharge without (left) and with (right) gas flow.

An approach of the corona discharge system operating at high-voltage pulses in the nanosecond-range can help to prevent the formation of high temperatures. However, the applied average power density remains small.

The dielectric barrier discharge (DBD) has been proven successfully as an adequate plasma source setup for wood treatment (Viöl 2004). A characterizing attribute of the DBD is at least one dielectric between the electrodes (see Figure 17.12). It is used for electricity limitation. In general, a filament discharge can be found between a dielectric highly isolated high-voltage electrode and wood as a counter electrode.

The right picture in Figure 17.12 shows how the wood is electrically used as intermediate electrode. The filaments have a diameter of 0.1 mm; the gaseous gap typically features a thickness of 2 mm. In every filament, currents of some ampere flow for a few nanoseconds. Due to the dielectric, this gas discharge has to be conducted by AC voltage or pulsed voltage (Kogelschatz, 2003). Many different moving filaments are most suitable for a plasma treatment of wood (see Figure 17.13). Small gaseous gaps and a strong gas steam support this type of discharge. Commonly, the gas temperature is not exceeding the room temperature at a large scale.

Besides large industrial installations with treatment areas in the m² range, small and handy gadgets were developed as well (Viöl et al. 2007).

New atmopsheric pressure plasma sources were developed for a plasma treatment of surfaces during the last 10 years (Förster et al., 2005; Laroussi and Akan, 2007). They all have in common that the plasma is blown to the surface by a gas stream (see Figure 17.14). Some of the systems are run with air as working gas, some can only be ignited using inert gases (including admixtures of a few % of different gases in some cases). In general, an RF discharge is ignited at frequencies between 1kHz and 27MHz. Microwave excitations are possible as well (see Figure 17.15).

The gas discharges are commonly hotter than those in a DBD or corona discharge; at the gas outlet of the source, temperatures up to 400°C can be reached. Pulsed sources with leakage temperatures below 50°C are available as well.

This gas discharge is often called "indirect plasma treatment." Due to both its recombination of charge carriers and the reduction of radicals generated in the gas stream, it is not as effective as a DBD. These plasmas feature only a diameter of a few millimeters, providing the possibility of processing plasma treatment in small gaps.

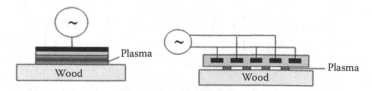

FIGURE 17.12   Dielectric barrier discharge on wood surfaces.

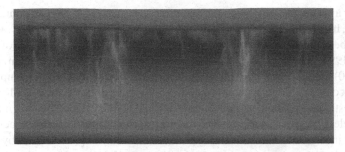

FIGURE 17.13   Dielectric barrier discharge in atmospheric air.

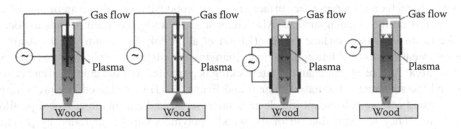

FIGURE 17.14   Different setups of atmospheric pressure plasma jets (APPJS).

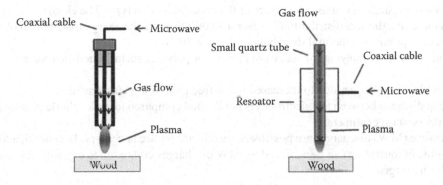

FIGURE 17.15   Atmospheric pressure plasma jets excited by microwaves.

17.2   PLASMA MODIFICATION OF WOOD

17.2.1   MAIN REACTIONS OF PLASMA TREATMENT ON WOOD

The interaction mechanisms between plasma and a solid surface are physical bombardment by energetic particles and ultraviolet photons, as well as chemical reactions at or near the surface, resulting in cleaning, ablation, cross-linking, and surface chemical modification (Liston et al., 1994). Whereas the highly energetic particles are limited to the surface, the ultraviolet photons can penetrate into the bulk material. Oxygen containing plasmas are capable of surface cleaning, that is, the removal of contaminants (e.g., oils or waxes) improving the chemical weak boundary layer. Thereby, a coating can get in direct contact with the substrate surface whereby adhesion increases, contributing to improved bonding results. A further important effect is the chemical modification by generating new functionalities at the surface (Liston et al., 1994). Oxygen or nitrogen-containing plasmas can alter a wide range of substrates and produce a variety of oxygen containing functionalities, for example, hydroxyl (OH), carbonyl (C=O), carboxyl (O=C–OH), ether (C–O–C) (Goldmann et al., 1985; Strobel et al., 1989; Liston et al., 1994; Jie-Rong et al., 1999; Drnovska et al., 2003). Belgacem et al. (1995) and others (Yuan et al., 2004; Klarhöfer et al., 2005; Avramidis et al., 2009a; Busnel et al., 2010) could verify the generation of the mentioned functional groups by x-ray photoelectron spectroscopy (XPS) on plasma-treated cellulose, lignin, and wood. Moreover, Klarhöfer et al. (2010) investigated plasma treated lignin and cellulose surfaces by a combination of XPS, ultraviolet photoelectron spectroscopy (UPS) and metastable impact electron spectroscopy (MIES). Plasma treatment in oxygen-containing atmospheres oxidizes the lignin surface by the generation of hydroxyl, carbonyl, and carboxyl groups, and reduces cellulose surfaces by the degradation of hydroxyl groups and the formation of double bonds between carbon and oxygen. Plasma treatment in argon leads to the reduction of both lignin and cellulose by the formation of double bonds under degradation of hydroxyl groups (Klarhöfer et al., 2010). These alterations in wood surface chemistry are of polar character and increase the surface energy of wood which results in an improved wetting of polar liquids on the wooden surface. Also chain scission and generation of radicals occur and contribute to an increased surface energy (Liston et al., 1994; Nussbaum, 2001). Additionally, plasma treatment results in ablation or etching, changing the surface morphology and can increase the surface area. As a result, mechanical interlocking is promoted and the contact area between the coating and the substrate is increased. Jamali and Evans (2011) investigated low-pressure plasma etching of wood cell walls by scanning electron microscopy and chromatic confocal profilometry (Figure 17.16). They revealed that all of the wood's polymers can be degraded by plasma even though lignin-rich cell wall layers are etched more slowly than other parts of the cell wall.

17.2.2 LOW-PRESSURE PLASMA MODIFICATION

Low-pressure plasmas as discussed here are of the nonequilibrium type. The electron energy distribution is close to a thermal distribution of several 10,000 K while the energy distribution of ions and neutrals corresponds to a thermal distribution of about 300 K.

The advantages of applying low-pressure plasma for polymer surface modification are

Higher electron energies due to increased mean-free path in low gas densities.
Gas gap distance between the electrodes is not limited compared to atmospheric plasma pressures (several centimeters).
Microwave (MW) discharges are possible, so no electrodes are necessary. In general, a higher degree of ionization can be achieved for MW discharges compared to capacitively coupled RF discharges.

17.2.2.1 Hydrophobic and Hydrophilic Wood Surfaces by Low-Pressure Plasma Treatment*

One of the first publications on the change of wettability of wood surfaces with plasma was by Bialski et al. (1975). They treated artificial fibers (Mica) and Aspen (*Populus tremula*) ground-wood fibers with MW-plasma. The fibers were considered as additives for extrusion-grade polypropylene composite preparations. The particle and fiber surface-modification reactions were performed in an MW installation at a pressure of approx. 67 Pa; an MW power of 1.5 kW and a treatment time of 90 s. Utilized plasma gases were argon, nitrogen, ethylene, sulfur dioxide, ammonia, and a mixture of ethylene and ammonia. Especially the ethylene-plasma treatment turned out to be effective in

---

* In the present and the subsequent chapters, only such studies will be described, that refer to pressure, frequency, voltage, power, and energy dissipated into the discharge respectively as the relevant experimental parameters for characterizing the applied plasma conditions. Other work on this subject was excluded for the sake of its missing repeatability.

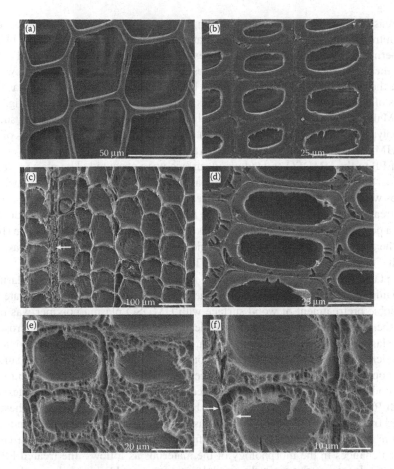

FIGURE 17.16 Transverse surfaces of redwood (Sequoia sempervirens) before and after plasma treatment: (a) untreated earlywood showing thin-walled tracheids; (b) untreated latewood showing thick-walled tracheids; (c) earlywood subjected to plasma treatment for 1333 s, note thinning of tracheid walls and resinous contents of a ray remaining at the surface (arrowed left); (d) latewood subjected to plasma treatment for 333 s showing radially and tangentially aligned voids and splits in cell wall; (e) latewood subjected to plasma treatment for 1333 s showing pronounced etching of secondary wall; (f) enlarged area of (e) showing middle lamella (arrowed) and thin lamellae radiating from the middle lamella to the tertiary wall layer (arrowed). (Adapted from Jamali, A. and Evans, P.D. 2011. *Wood Sci. Technol.* 45:169–182.)

increasing the mechanical strength of mica-based composites, whereas the plasma-treated wood fibers showed a less distinct effect.

In 1984, Legeay et al. (1984) treated oak (*Quercus*) samples in oxygen-, helium-, nitrogen-, CO_2-, and tetrafluoromethane-plasma at a power of 50–70 W, a pressure of 130–270 Pa and with a treatment duration of 5, 15, and 30 min. The plasma treatment led to decreased water contact-angle values. Bonding tests using polyvinylacetate emulsion indicated that most of the plasma treatments resulted in an increase of the rupture stress. The adhesive behavior of two different varnishes (glycerophtalic and polyurethane-monoconstituents) was also studied for four varieties of wood samples. Oxygen-plasma treatment increased the adhesion for the glycerophtalic varnish, while tests in the presence of polyurethane varnish exhibited decreased adhesion characteristics.

Chen and Zavarin (1990) exposed white fir (*Abies concolor*) and douglas fir (*Pseudotsuga menziesii*) heartwood and sapwood samples to oxygen-, nitrogen-, and helium-RF plasmas and evaluated alterations in wood permeability. The experiments were carried out with a commercial plasma device operating at 13.56 MHz, a power range of 0–600 W, and a pressure <130 Pa. The plasma treatment strongly increased wood permeability to nitrogen flow along the grain. Oxygen was most effective, followed by nitrogen and helium. It was found that wood extractives suppressed the plasma-induced increase of permeability, which was revealed by experiments with water and ethanol extracted wood samples. Furthermore, discs composed of compressed cellulose, xylan and lignin powder were used to investigate the rates of plasma-induced ablation by weight-loss measurements. It was found that the ablation rates of cellulose and xylan are two times higher than the ablation rates of lignin.

In 1998, Mahlberg et al. (1998) investigated the effect of oxygen plasma on the surface characteristics of polypropylene (PP) and birch (*Betula*). In addition, studies on the effects of hexamethyl-disiloxane (HMDSO) plasma on the surface characteristics of lignocellulosics and on the adhesion properties of PP to the HMDSO-treated wood surface were executed. The plasma treatment setup consisted of a 40-kHz RF parallel-plate reaction chamber. Oxygen-plasma treatments of PP and birch samples were carried out at pressures between 33 and 40 Pa at a power range between 60 and 100 W. The treatment duration was 15–90 s. The HMDSO plasma treatment of wood was executed at 27 Pa and a power of 200 W, the treatment duration was 2.5 and 8 min. It could be shown that the increase of the surface free energy of wood, which was effected by the oxygen plasma treatment, was due to the increase of the polar component (Table 17.1).

HMDSO-plasma treatments clearly hydrophobized wood sample surfaces. Plasma treatment duration of 5 min increased the contact angle values from 60° (for unmodified substrates) to 95–135°. The hydrophobic pretreatment of wood surface with HMDSO plasma polymer was not beneficial for adhesion between both wood and polypropylene film. AFM (atomic force microscopy) studies of HMDSO-plasma-treated wood surfaces indicated that the HMDSO-based macromolecular layer follows the microstructure of the substrate without forming an actual film on the surface. Atomic force microscopy was also used to study surfaces of oxygen plasma-treated PP, kraft pulp, filter paper, and wood (Mahlberg et al., 1999). The effect of plasma treatment on the adhesion properties between both PP film and wood was evaluated by means of a peel test. The highest adhesion to wood resulted from the shortest treatment times used. The effects of plasma on the adhesion properties were more pronounced when both the PP film and the wood surface were treated. Oxygen plasma caused changes in the morphology of the materials as follows: the treated PP surface was covered by a nodular structure, similarly a nodular structure could also be detected on the lignocellulosic materials. It was suggested that these nodules are responsible for a poor interaction between the surfaces. This could result in the generation of a weak interface between PP and wood.

Further, studies with plasma-generated, HMDSO-based hydrophobic layers on wooden substrates (southern yellow pine) were carried out by Denes et al. (1999). The wood was extracted by a Soxhlet-extraction process to remove extractives from wood surfaces. All depositions were performed in a parallel-plate RF plasma reactor (Denes and Young, 1999) at a pressure of 40 Pa; a RF-power in the range of 150–250 W and plasma treatment duration between 90 s and 600 s. A

TABLE 17.1
Disperse and Polar Component of the Surface Energy and Total Surface Energy for Untreated and Treated PP and Birch

Surface	Polar (mJ/m^2)	Disperse (mJ/m^2)	Total (mJ/m^2)
Untreated PP	1.2	27.6	28.8
PP treated for 30 s	19.4	32.3	51.7
PP treated for 60 s	25.4	26.3	51.7
Untreated birch	12.2	39.8	52.0
Birch treated 30 s	39.9	24.9	64.8
Birch treated 60 s	36.9	28.2	65.1

Source: Adapted from Mahlberg, R. et al. 1998. *Int. J. Adhesion Adhesives* 18:283–297.

plasma-generated layer was detected by AFM and DTA/TG on the treated wood surfaces. Water contact angles of approx. 130° indicated hydrophobic characteristics of the deposited films. ESCA, ATR-FTIR and GC-MS data indicated that the plasma-generated layers on the wood surfaces are based on highly cross-linked Si–C and Si–O–Si compounds. Denes and Young created a plasma-enhanced process to coat wood surfaces with a protective layer, which protected wood degradation due to artificial weathering (Denes and Young, 1999). An admixture of zinc oxid, electro-magnetic absorbents and polydimethylsiloxane (PDMSO) was applied onto wood surfaces (southern yellow pine) and consecutively cross-linked by low-pressure plasma-treatment at a pressure of 33 Pa, frequency of 30 kHz and 250 W power delivered to the electrodes. The plasma treatment duration was 10 min. The oxygen-plasma cross-linking resulted in surface layers that diminished the weathering degradation of wood (Figure 17.17). It could be proved that the Si–C, Si–O, and C–O bonds of the plasma-generated matrix-polymer were not degraded by artificial weathering.

A 2.450-GHz microwave plasma-enhanced surface treatment of scots pine (*Pinus sylvestris*) wood samples with an initial moisture content of 12% was carried out in gaseous- and liquid-phase environments (Podgorski et al., 2001). A variety of different plasma-gas environments was used for various purposes, including cleaning (Ar, He, H_2), oxidation (air, O_2, CO_2), nitriding (N_2, NH_3), fluorination (CF_4, fluorine-containing monomers, such as C_3F_6 and acrylate with a C_8F_{17} chain), and deposition (acetylene, propylene, hexamethyldisiloxane). Liquid-phase materials were also *ex situ* sprayed onto selected substrate surfaces and plasma-treated consecutively. Microwave power dissipated to the antenna was 700–1100 W, the treatment time was between 1 and 15 min. The "liquid/gas" line operates with liquid products featuring a vapor pressure, which is greater than the pressure used during the treatment. The authors indicate that both fluorine-containing compounds and hexamethyldisiloxane (HMDSO) generate wood surfaces with high contact angle values (76–120°) and that in most of the cases using fluorinated compounds is more effective in comparison with the use of an HMDSO-plasma environment. Improvement of coating adhesion using an oxygen plasma treatment at a pressure of 8 Pa, a power range of 400–1200 W, and a treatment duration of 300 s was also suggested (Podgorski and Roux, 1999), since surface free energy in the polar component was distinctly increased.

Macromolecular layers were deposited on pine (*Pinus*) wood by plasma polymerization using ethylene, acetylene, 1-butene, and vinyl acetate (Magalhaes and de Souza, 2002). The plasma treatments were carried out at a driving frequency of 60 Hz and a power of 10 W, the pressure was approx. 100 Pa and the treatment duration amounted to 15 min. The softwood samples were cut in radial, tangential, and longitudinal directions and oven-dried at 100°C for 12 h before plasma treatment. The plasma treatment generates hydrophobic layers, whereby the best hydrophobic results

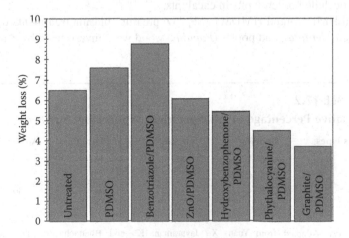

FIGURE 17.17 Percent weight loss of plasma-treated coated wood for the exposure trials under artificial weathering. (Adapted from Denes, A.R. and Young, R.A. 1999. *Holzforschung* 53:632–640.)

were achieved by 1-butene with water contact angle values of approx. 140°. Although the surface plasma treatment resulted in water repellence, permeability to water vapor remained. The hygroscopic properties of the wood could not be changed significantly by the plasma treatments. The plasma-deposited films were unable to prevent water vapor absorption and desorption. The authors attributed this to an insufficient coating of wood capillaries by the plasma-deposited layers.

Wistara et al. (2002) treated several tropic woods with a CF_4-RF plasma to generate hydrophobic surfaces in order to increase resistance against fungal attack. The treatments were carried out at a frequency of 13.56 MHz, a pressure 31 Pa and a power of 100 W. The resistance of the woods against decay due to attack of white rot (*Trametes versicolor*) could be increased significantly.

Yuan et al. (2004) exposed wood fibers to argon and air-plasma to improve the compatibility of both wood fibers and polypropylene (PP). The plasma treatments were carried out with a 13.56 MHz generator and a power output of 60 W. The wood fibers were treated for 30 s with argon plasma or air plasma at a pressure of approx. 270 Pa. It could be shown that the tensile strength and tensile modulus of the composite sheets were distinctly enhanced when using plasma treated wood fibers. In the process, air plasma shows even better results than the argon-plasma does. The improved bond between both plasma-treated wood fibres and PP was confirmed by scanning electron microscopy (SEM) imaging of fracture surfaces of the composite. Furthermore, SEM-analyses show an increased surface roughness and opened pits of wood fibers after plasma treatment with argon or air plasma. XPS-analysis revealed increased O/C ratios and the generation of oxygen-based functional groups due to argon-plasma and air plasma treatment, indicating oxidized fiber surfaces (Table 17.2). The authors ascribed the observed mechanical improvements to higher interfacial contact, enhanced mechanical interlocking, better viscoelastic dissipation at the interface and better adhesion due to etching and generation of oxygen functionalities.

Evans et al. (2007) proved the hypothesis that plasma modification could increase the surface energy of eucalypt wood (*Eucalyptus*) and improve its adhesive properties. For this purpose, a plasma reactor was built, generating a plasma with the working gas water (vapor). The driving frequency was 135 kHz, the pressure approx. 20 Pa and the power was in the range of 50–150 W. The treatment duration was between 3 and 10 min. The effectiveness of plasma treatment was evaluated by contact angle measurements and lap shear tests. The plasma treatment significantly increased the surface wettability of eucalypt wood even at low energy levels, but there was no corresponding increase in the glue-bond strength. At higher energy levels, the plasma modification increased the wettability and glue-bond strength of spotted gum and three other eucalypt species. The increase in glue-bond strength appeared to be associated with etching of the wood cell wall (which could be proved by Jamali and Evans 2011) and the removal of phenolic-rich materials and structures, including vestures that occlude bordered pits in eucalypts.

Effects of cold HMDSO and HMDSO+SF_6 low-pressure plasma treatments on spruce (*Picea*), chestnut (*Castaneta dentata*), and poplar (*Populus*) wood were investigated by Zanini et al. (2008).

TABLE 17.2
Relative Percentage of C1s-Spectra in Wood Fiber Surface

C1s Types	Bonding	Untreated	Ar Plasma	Air Plasma
C1	C–C, C–H	56	12	9
C2	C–O, C–O–C	29	42	47
C3	C=O, O–C–O	15	44	39
C4	O=C–O	0	2	6

Source: Adapted from Yuan, X., Jayaraman, K., and Bhattacharyya, D. 2004. *Composites Part A* 35:1363–1374.

The discharges were generated in a low-pressure, capacitively coupled radiofrequency (RF) plasma reactor. Radiofrequency power was delivered to the discharge by an RF antenna. The antenna was externally connected to a 13.56 MHz-RF power supply. Plasma depositions were effectuated at a fixed starting pressure of 20 Pa and an input power of 70 W. Treatment time was varied from 1 to 10 min. Plasma polymer films were deposited from hexamethyldisiloxane (HMDSO) and HMDSO + SF_6 on three different wood surfaces under different operating conditions. Structure and properties of the plasma polymerized layers were investigated by means of FT-IR spectroscopy, contact angle analysis, and LIBS analysis. The hydrophobicity of the treated samples was increased whereas no change of color and aspects of the wood occur by the treatment. The effect of the plasma treatment was not completely dissipated even after strong artificial ageing. The hydrophobic character was increased by consecutive plasma treatment of HMDSO + SF_6 compared to the pure HMDSO plasma-treated ones.

In order to study the effect of plasma treatments on wood using N_2, H_2, O_2, and Ar as process gases, two plasma sources at pressures between 13.3 and 665 Pa were used to functionalize sugar maple (*Acer saccharinum*) surfaces (Blanchard et al., 2009). One source delivered an inductively coupled plasma (ICP), and the other one delivered a capacitively coupled plasma (CCP). Both were driven at a frequency of 16.56 MHz at a power range from 150 to 200 W with treatment duration of 30 s to 60 min. For ICP treatments it was found that best improvements in adhesion of waterborne polyurethane acrylate coatings were obtained using Ar and N_2/H_2-admixtures at a pressure of 13.3 Pa. For N_2 plasma, best adhesion was achieved at 665 Pa. For CCP treatment at 133 Pa, best adhesion results were obtained with Ar. Water contact angle measurements and Raman spectroscopy showed that the adhesion improvement correlates with both wettability and coating penetration depth. XPS measurements indicated that nitrogen can be grafted to wood surfaces by ICP treatment.

Haase and Evans (2010) compared the penetration depth of adhesives into low-pressure plasma-treated and untreated wood (Figure 17.18). Samples of black spruce (*Picea mariana*) were exposed to glow discharge plasma for 0.5. 3, 10, or 20 min using the plasma reactor and reaction conditions described previously (Jamali and Evans 2011). The penetration of oil and water-borne clear coats into black spruce was measured by light microscopic methods. It could be proven that prolonged plasma treatment improved the penetration of both oil and waterborne coatings into subsurface wood layers.

In the study of Aydin and Demirkir (2010), low-pressure oxygen plasma treatment was applied to spruce (*Picea*) wood as an alternative to other possible surface treatment methods to reactivate the surfaces for glue bonding after long terms of natural surface inactivation. Plasma surface modification of spruce veneers was accomplished at a frequency of 40 kHz, a power of 200 W and a pressure of 80–150 Pa. The treatment duration was 15 min. Wettability, surface free energy, and pH-value of the spruce veneers were increased by plasma treatment. Similarly, bending strength and modulus of elasticity of plywood panels made from plasma treated veneers were found to be higher than those of inactivated samples. Marginal alterations in surface color of veneer sheets were detected after plasma treatment.

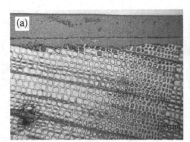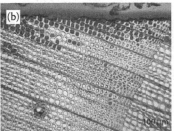

FIGURE 17.18 Effects of prolonged plasma treatment on the penetration of clear coatings into black spruce: (a) untreated and (b) plasma treated. (Adapted from Haase, J.G. and Evans, P.D. 2010. *Proceedings of the Fifth European Conference on Wood Modification* (Meeting date 2010), Riga, Latvia, 271–274.)

17.2.3 ATMOSPHERIC PRESSURE PLASMA MODIFICATION

Corona treatment in atmospheric pressure air was developed to a reliable process to increase the surface energy of polymer substrates in the 1960s. At that time, thin wires, threaded-rods or knife-edges were used as corona electrodes, which is why, nowadays, surface activation by plasma treatment is often misleadingly referred to as corona treatment, although most commercial plasma systems for surface activation now employ dielectric barrier discharges. DBD combine the advantageous properties of cold plasmas, that is, the feasibility of atmospheric pressure operation and easy scalability. A DBD's low operating temperatures make it well suited for the treatment of heat-sensitive materials, for example, wood, without compromising the material properties by excessive heat exposure. Operation at atmospheric pressure nullifies the requirement for extensive and expensive vacuum equipment and decreases apparatus complexity. Furthermore, a dielectric barrier discharge is more economical and much easier to transfer into industrial applications compared to low-pressure plasma processes. A treatment at low pressure in general requires cyclical processes whereas atmospheric pressure treatment can be conducted in a continuous process, without compromising product flow and performance. In particular for wood, the evacuation process can be complicated by evaporation of water and volatile wood extractives leading to extended evacuation times, contaminated process chambers or even negatively affected material properties which also militates in favor of atmospheric pressure treatment (Wolkenhauer, 2009a).

17.2.3.1 Hydrophobic Wood Surfaces by Atmospheric Pressure Plasma Treatment

Bente et al. (2004) hydrophobized spruce (*Picea abies*) surfaces with the help of gas discharges by using a DBD configuration. Methane, ethane, and silane/nitrogen (2% silane, 98% nitrogen) were injected into the plasma zone, creating thin water-proof layers. The wood surfaces treated with silane/nitrogen showed the best hydrophobic characteristics. The layer created in silane/nitrogen plasma had a water contact angle of 145° (Figure 17.19), whereas the contact angle of untreated wood was 72°. Both contact angles were taken 10 s after the application of the drop. The surface tension of silane/nitrogen-treated wood compared to untreated wood is fivefold lower. Using atomic force microscopy, it could be shown that the plasma generated surface layer had a regular topography with an enhanced surface roughness, which is assumed to contribute to the hydrophobic characteristics. The water uptake time of a 50 µL droplet for the untreated wood is about 165 s, whereas the absorbing time of methane- and ethane-treated surfaces is approx. 24,000 s and 27,000 s, respectively. The time measurement of silane/nitrogen-treated wood was aborted after 32,400 s since no absorption was noticeable. FT-IR-spectroscopy, EDX analysis, elementary analysis, and photometric methods gave information about the composition of the layer created in silane/nitrogen-plasma: carbon 0.6%, hydrogen 2.6%, nitrogen 4.3%, silicon 55.5%, and oxygen 37%.

Polymeric HMDSO-based layers were deposited on spruce (*Picea abies*) surfaces by Diffuse Coplanar Barrier Discharge (DCSBD, a special form of a DBD) for creating water-repellent surfaces (Odrášková et al. 2007). Plasma treatment of wooden samples was done at atmospheric pressure.

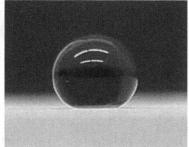

FIGURE 17.19   Water droplets on silane/nitrogen plasma-treated spruce surfaces.

The driving frequency was 14 kHz sinusoidal voltage, the plasma activation in the N_2/HMDSO atmosphere was done at an average plasma power density of 2.2 W/cm^2 and an exposure time of 90 s. The N_2/HMDSO mixture was prepared by mixing the pure N_2 with the N_2 gas bubbled through liquid HMDSO monomer. The optimum deposition rate was achieved at a relative volume concentration of 1.35% of HMDSO in N_2. The water droplet contact angle of water on spruce modified in these conditions was approx. 126° (untreated sample approx. 48°). The surface free energy of modified surfaces was 7 mJ/m^2 (untreated sample 56 mJ/m^2). A 50 μL water droplet took 172 ± 30 min to penetrate into the modified spruce. In comparison with the unmodified spruce, this is an increase by a factor of more than 20. The chemical composition of deposited layer was analyzed by ATR-FTIR confirming the presence of Si–O–Si and $Si(CH_3)$ functional groups, as shown in Figure 17.20.

Atmospheric pressure DBD treatments of wood were carried out to attain water repellency on wood surfaces (Toriz et al., 2008). Ethylene, methane, chlorotrifluoroethylene, and hexafluoropropylene were used as gases injected to the DBD. The driving frequency was in the range of 78–200 Hz, the maximum voltage delivered to the high-voltage electrode was up to 45 kV and the treatment time was between 30 s and 720 s. For methane and ethylene, XPS data showed an increased surface atomic concentration of carbon from 72.7% on untreated samples up to 80.7% and 96% on treated samples, respectively, whereas nearly 50% fluorine concentration was observed with fluorinated reagents. AFM images showed a deposit of a thin uniform film in the case of ethylene-DBD treatment, whereas the hexafluoropropylene-DBD treatment resulted in the nucleation of plasma-derived entities at the fiber surface and the subsequent growth of a film. Under optimized conditions, the water contact angle was in the range of 139°–145°. Water absorption was reduced at least 340 times and up to 530 times. Ethylene-DBD treatments showed that both the contact angle and the atomic concentration of carbon on the surface leveled off after 360 s. XPS analysis showed that DBD treatments with fluorinated reagents resulted in the deposition of fluorinated plasma polymers that resembled the parent monomers.

An atmospheric-pressure plasma jet was used to deposit HMDSO-based water repellent layers onto beech (*Fagus*) substrates (Avramidis et al., 2009b). The jet was driven at a frequency of 17 kHz (pulse duration 2 μs). The process gas was blown through the discharge gap toward the sample to be coated (distance between sample and jet outlet: 2 mm), the treatment time for all experiments was 18 s. Gas temperatures of max. 100°C were measured during plasma treatment. The process gas consisted of argon as carrier gas fed through liquid HMDSO acting as precursor. HMDSO-loaded argon was then mixed at different ratios with synthetic air. The electrical power dissipated into the plasma was in the

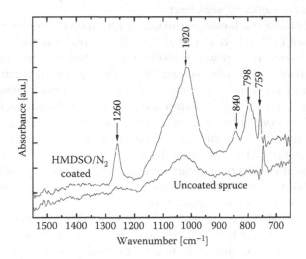

FIGURE 17.20   FTIR spectra of films prepared from N_2/HMDSO mixture. (Adapted from Odrášková, M. et al. 2007. *Proceedings of the 28th ICPIG*, 803–806.)

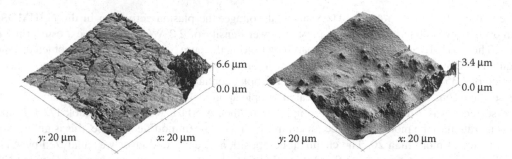

FIGURE 17.21 AFM images of untreated (left) and plasma-treated (right) beech. The image of the plasma-treated sample shows the wood surface coated with a silica-based layer. (Adapted from Avramidis, G. et al. 2009b. *Proceedings of the Fourth European Conference on Wood Modification*, 269–275.)

range of 67–102 W. Best water repellence was obtained for a power level of 102 W. As shown in Figure 17.21, AFM imaging revealed a compact layer completely covering the substrate surface. The chemical composition was determined by CHN analysis and FTIR spectrometry revealing that carbon, oxygen, silicon, and hydrogen are present in the form of $R–CH_3$, $Si–CH_3$, and $Si–O–Si$ groups.

17.2.3.2 Hydrophilic Wood Surfaces by Atmospheric Pressure Plasma Treatment

Pioneer work in the field of plasma modification of wood was done by Kim and Goring (1971). The authors treated birch (*Betula*), elm (*Ulmus*), poplar (*Populus*), pine (*Pinus*), polyethylene and polystyrene by plasma in order to increase the bond strength of thermally induced bonds between wood and synthetic polymers. For plasma treatment at atmospheric pressure with air, the authors used a corona cell, applying an electrical potential of 15 kV at a frequency of 60 Hz to the electrodes. The treatment times were in the range of 5 s and 30 min. It could be demonstrated that plasma treatment of only a few seconds prior to bonding led to distinctly enhanced bonding results, whereas treatment of the polymer has a higher influence on enhanced bonding than treatment of wood. For prolonged treatment duration, the effect was less distinct. Furthermore, only slight differences between the species of wood could be detected. Scanning electron microscopy revealed that the plasma treatment roughened the surface of the wood as well as the surface of the thermoplastics. In this work, it is already suggested that oxidation of the wood surface plays a role in increased bonding results. Moreover, the authors assumed that the wood is degraded by the plasma treatment and prolonged treatment generates a mechanically weak layer.

The improvement of adhesive bonding by oxidative activation of resinous wood surfaces applying a DBD was studied by Sakata et al. (1993). The plasma treatment was accomplished at a frequency of 5 kHz and a maximum power of 500 W, the gap thickness between the electrodes was 3.2 mm. The veneers (hardwoods, softwoods, and tropical woods) and their extractives were charged with an energy quantity between 0 and approx. 35 J/m². The plasma treatment of the surface of resinous wood veneers led to a considerable increase in terms of wettability of the surface and in the gluability of the veneers to water-based adhesives consequently. Negligible chemical effects of the plasma treatment on the surface modification of the wood's main components, such as cellulose, hemicellulose, and lignin, could be detected with the aid of the dye-adsorption method. The plasma treatment significantly affected certain extractives and oxidized them. The authors assumed that an increase of the wettability of corona-treated wood veneers resulted mainly from the oxidation of the high hydrophobic surface layer of some extractives.

A special DBD electrode system (Figure 17.22) was tested for air plasma treatment of wood surfaces (Rehn et al., 2003). A pulsed high-voltage (30 kV, pulse duration 2 μs; driving frequency: 15 kHz) was applied to the electrode in order to generate the discharge in air between the electrode and wood surface (1.2–3 mm gap) at atmospheric pressure and a gas temperature of 35°C. The treatment duration was between 0.25 s and 20 s. The wettability as well as the fracture strength of

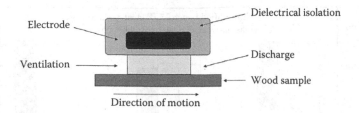

FIGURE 17.22 Plasma pretreatment setup.

plasma-treated wood samples were improved. Moisture content of wood does not have a significant influence on the uptake time of water droplets into the plasma-treated surfaces.

Furthermore, hydrophilic and hydrophobic wood surfaces were generated with a DBD setup using a parallel-plate electrode configuration, where both electrodes were insulated with quartz plates (Rehn and Viöl, 2003). In this case, the driving frequency for the pulsed high voltage was 17 kHz with a pulse duration of 2 μs. The amplitude of the alternating polarity was 30 kV. For creating hydrophilic surfaces, air, helium, nitrogen, and argon were used as plasma gases. When using air, the best hydrophilic surfaces were achieved. Plasma treatment durations from 1 s to 20 s significantly increased the wettability and water absorption of wood surfaces. Furthermore, the fracture strength of glued wood was also enhanced in comparison with unmodified substrates.

Mertens et al. (2006) compared laser- and plasma-induced surface activation of beech (*Fagus*) and thermowood beech by means of a water droplet test. Beyond ablation processes, the investigated lasers (excimer lasers with wavelengths of 157 nm, 193 nm, 248 nm, 308 nm) can also be used to influence surface energy. The same effect can be achieved by plasma treatment of wooden surfaces. The plasma treatment was carried out at atmospheric pressure with a dielectric barrier discharge configuration at an excitation frequency of 17 kHz, a pulse duration of 2 μs and a voltage applied to the electrodes of 40 kV. The treatment duration was 2 s; the distance between both electrode and wood amounted to 2 mm and the process gas used for surface treatment was ambient air. The wettability of the wooden surface could be improved by laser ablation as well as by plasma treatment, whereby the effect was more pronounced in the case of plasma treatment.

The surface free energies of both beech wood and heat-treated beech wood, untreated and plasma treated by a DBD at atmospheric pressure and ambient air, were investigated using the Lifshitz–van der Waals/acid–base approach (Wolkenhauer et al., 2008a). The plasma treatment was conducted with a DBD configuration. An alternating high voltage of 34 kV, a pulse duration of 2 μs and a frequency of 17 kHz was applied to one electrode. Ambient air at room temperature as process gas was blown through the discharge gap of 2 mm between the sample and the high-voltage electrode. The decreased wetting behavior of water on heat-treated beech wood could be improved by plasma treatment to a similar wetting behavior of unmodified beech wood. The basic component of heat-treated beech wood is significantly decreased compared to untreated beech. As shown in Figure 17.23, the determination of the surface free energy of beech, heat-treated beech, and plasma-treated heat-treated beech revealed a significant increase of the base component after plasma treatment of the heat-treated beech. This indicates the generation of functional groups to a comparable degree of unmodified beech wood. Furthermore, the drying behavior of a PVAc-glue and an EPI-glue on heat-treated beech wood in untreated and plasma-treated state was tested (Wolkenhauer et al., 2009b). Shear strength tests were carried out to monitor the increase in bonding strength during the drying process. In spite of increased surface energy and wettability of heat-treated beech wood by a plasma treatment, no significant effect on the drying behavior of the tested glues could be detected.

Samples of Pedunculate oak (*Quercus robur* L.) were activated by Diffuse Coplanar Surface Barrier Discharge (DCSBD) plasma (Odrášková et al., 2008). The electrodes were connected to a 14 kHz sinusoidal voltage, the plasma treatment was carried out at an average plasma power density of 2.2 W/cm² and an exposure time of 5 s. The distance between the sample and the electrode was in the

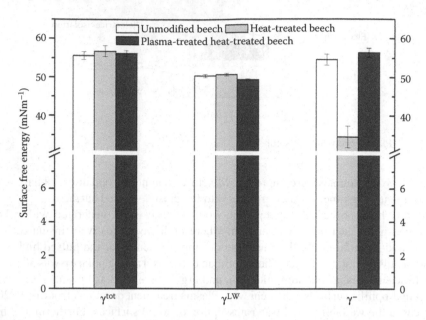

FIGURE 17.23 Surface free energy of beech, heat-treated beech, and plasma-treated heat-treated beech according to the Lifshitz–van der Waals/acid–base approach (γ^{tot}: total surface energy, γ^{LW}: Lifshitz—van der Waals component, γ^-: basic component).

range of 0 mm to 0.6 mm. A significant increase of the polar component of surface free energy and a reduction of water droplet uptake time was detected as a result of the treatment. FTIR analysis confirmed the generation of additional polar functional groups on the wood surface due to the plasma treatment. The authors suggested that this effect corresponds to the shift of wood surface pH level (Table 17.3) toward more acidic values, since the polar part of surface free energy is associated with the presence of acid/base forces. Furthermore, the aging of the treatment was tested revealing a distinct aging effect (within days). Nevertheless, the treated surface did not recover to its initial hydrophobic state.

Experimental work done by Lecoq et al. (2008) was on the treatment of wood (*Pinus pinaster*) samples using the afterglow of a nitrogen-driven DBD. For that purpose, wood samples were placed at a distance of 18 mm from the gas outlet of a DBD reactor. The DBD was operated with sinusoidal waveforms at a frequency of 130 kHz, the power was 900 W. The energy applied to the samples, varied by the duty cycle (voltage ON/voltage OFF phases) and treatment duration, was in the range of 0 kJ and 160 kJ. Soaking tests with an industrial fungicide showed that plasma treatment could enhance the absorption of fungicide into the wood. A very remarkable result was that nitrogen plasma-treated sample surfaces exhibit both a hydrophilic and a hydrophobic character depending

TABLE 17.3

Surface pH of Untreated and Plasma-Treated Radial Section of Oak

Sample	pH level
Oak heartwood	4.76 ± 0.11
Oak heartwood + plasma	3.29 ± 0.26

Source: Adapted from Odrášková, M. et al. 2008. *Plasma Chem. Plasma Process.* 28:203–211.

on the injected energy. Low energy results in increased values of water contact angle (110°–120°, untreated samples approx. 65°), applying high energies decreased the water contact angles down to 20°. XPS analyses revealed an increase of the O/C ratio and the presence of carboxyl groups on the surfaces after plasma treatments.

In order to compare sanding and plasma treatment by a (DBD) with respect to their effects on wood surface characteristics, beech (*Fagus sylvatica*), oak (*Quercus* sp.), spruce (*Picea abies*), and Oregon pine (*Pseudotsuga menziesii*) were investigated (Wolkenhauer et al., 2009c). The plasma treatments were performed using an alternating high-voltage pulse generator with a pulse duration of 2 µs and a repetition rate of 17 kHz. During plasma treatment, a voltage of 34 kV and a power dissipation of 60 W into the discharge zone were measured. The discharge gap between the sample and the high-voltage electrode had a thickness of 2 mm. Gas temperatures <40°C were measured during treatment. The surface energies of aged, freshly sanded, plasma-treated and the combination of both freshly sanded and plasma-treated surfaces were determined by contact angle measurement. As shown in Figure 17.24, both methods, sanding and plasma treatment, exhibit increased surface free energy in the polar part, whereas for the plasma-treated samples, the effect was more distinctive. This results in a higher work of adhesion between the wood surfaces and water after plasma treatment. A combination of both methods leads to no additional gain in surface energy. The authors assume that there is a limit for plasma treatment, which is independent from the initial surface energy. Besides, plasma treatment is not able to increase the surface energy, which might be interpreted as a saturation effect. For all tested wood species (beech, oak, spruce, and Oregon pine) and for all treatment methods, the dispersive part of surface energy was decreased.

Adhesion properties of freshly sanded sugar maple (*Acer saccharum*) and black spruce (*Picea mariana*) wood surfaces after treatment by a DBD were investigated (Busnel et al., 2010). The wood samples were plasma treated under the following experimental conditions: a driving frequency of the AC-generator of 9 kHz and a power of 2.2 kV were applied to the electrodes for Ar-plasma treatment. For plasma treatments using O_2, N_2, and CO_2, the applied voltage amounted to approx. 8 kV, the treatment duration was 1 s. After plasma treatment, the wood samples were coated with a waterborne urethane/acrylate coating. Pull-off tests and water contact angle measurements revealed that best improvements in

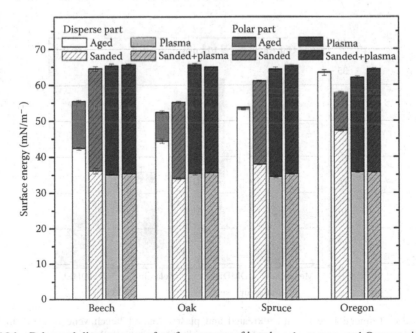

FIGURE 17.24 Polar and disperse part of surface energy of beech, oak, spruce, and Oregon pine samples: aged, freshly sanded, plasma treated, and freshly sanded + plasma treated.

adhesion (black spruce with N_2/O_2 (1:2) plasma and sugar maple Ar/O_2 (1:1) plasma) were correlated with the generation of hydrophobic surfaces. Further sugar maple surfaces were analyzed by XPS, indicating an increase of the O/C ratio due to the formation of functional groups after exposure to oxygen-containing plasmas. The authors suggested that a combination of structural and chemical change due to surface oxidation caused the observed surface modification of the plasma-treated wood samples.

In the study of Avramidis et al. (2010), the impact of DBD-treated wood veneers on their absorption characteristics for several liquids was investigated. For plasma treatment, the samples were positioned centrally between the two insulated electrodes, the discharge gap between the sample and the electrodes being 2 mm on each side. The experiments were carried out at 17 kHz driving frequency and an applied voltage of 35 kV (Wolkenhauer et al., 2009c). The treatment duration was between 1 s and 10 s. Ambient air at room temperature was blown through the discharge gaps at a velocity of ≈4 m/s, the gas temperature during the treatment remained below 50°C. The absorption characteristics of hardwood veneers (beech (*Fagus sylvatica*), oak (*Quercus* sp.)) were distinctly improved after plasma treatment. The plasma-treated hardwood species show a faster increase in weight induced by faster water absorption compared to the untreated ones. Plasma treatment of the softwood species (spruce (*Picea abies*), pine (*Pinus sylvestris*)) exhibited no impact concerning soaking ability, although the contact angle could be distinctly reduced. Only the beech veneers immersed in DMDHEU solution and melamine solution were found to benefit from a plasma treatment (Figure 17.25).

A DBD-plasma treatment was used to improve dye covering and impregnation of both oak and beech wood (Asandulesa et al., 2010). The utilized DBD system was operated at high-voltage square pulses (3 kV peak-to-peak, 2 kHz repetition frequency) and helium as process gas. The treatment duration was 5 s in all cases. Wetting characteristics of the surfaces were studied by dynamic water contact angle measurements. After plasma exposure, the adhesion of work between wood surfaces and water was significantly increased and the hydrophilic character of the wood surface was enhanced. The spreading area of water droplets on the sample surface was evaluated as a function of wetting time. Improved wettability of wood surfaces due to plasma treatment was also confirmed by the measurement of wetting areas of water and linseed oil drops.

The impact of DBD-treatment of wood veneers on the curing behavior of a PVAc adhesive was investigated (Avramidis et al., 2011a). The plasma treatment conditions were the same as described in

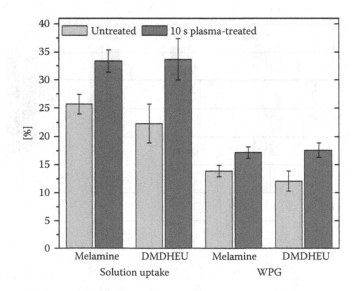

FIGURE 17.25 Immersion tests with untreated and plasma-treated beech veneers using melamine and DMDHEU as impregnation agents. (Adapted from Avramidis, G. et al. 2010. *Proceedings of the Fifth European Conference on Wood Modification*, 365–373.)

Avramidis et al. (2010), the treatment duration was 30 s for all experiments. Time-dependent shear bond strength tests on untreated and air plasma-treated maple (*Acer pseudoplanatus*), oak (*Quercus* sp.), beech (*Fagus sylvatica*), and teak (*Tectona grandis*) veneers showed that plasma treatment accelerated curing of the PVAc adhesive on the wood veneers as shown in Figure 17.26. Curing times were approximately halved. The authors ascribed the observed effect to faster water penetration into the wood.

Huang et al. (2011) exposed hybrid poplar (*Populus hybrids*) veneers to atmospheric pressure air plasma in order to improve wettability and plywood glue bond (shear) strength. The power output of the used plasma device was 0.65 kW; the operating voltage 100 V and the frequency 2.5 Hz. The air dried, oven dried, and overdried veneer samples were treated for 5 s. Compared to untreated references, the instantaneous contact angles of glycerin or urea formaldehyde (UF) droplets applied to the plasma-treated surfaces were distinctly decreased. 5 s after application of the droplets, overdried veneers showed complete wetting with both liquids. Compared to the air-dried and oven-dried poplar veneers, the overdried poplar veneers yielded the highest product glue bond strength after the plasma treatment.

17.2.3.3 Atmospheric Pressure Plasma Modification of Wood-Based Materials

The influence of an air-driven DBD at atmospheric pressure on wood plastic composites (WPC), particle board, and fiber board was evaluated by contact angle measurements and three different approaches of surface energy determination (Wolkenhauer et al., 2007a). Plasma treatment was conducted with alternating high voltage (pulse duration = 2 µs) of 34 kV at a frequency of 17 kHz. The power dissipated into the 2 mm gap between electrode and sample was 60 W and the treatment duration was in the range of 0.25 and 64 s. A significant increase in the polar part of surface energy due to plasma treatment could be validated for all tested samples (Figure 17.27). The work of adhesion between substrates and four different cured coatings was estimated revealing that the use of polar coating systems result in a greater work of adhesion than nonpolar coating systems after plasma treatment. Cross-cut tests demonstrate an increased adhesion of coatings after plasma treatment, corresponding to the calculated work of adhesion between cured coatings and substrates.

Particle boards and fiber boards were (DBD-)plasma treated and the alterations in soaking ability and shear strength were investigated (Wolkenhauer et al., 2007b). For plasma treatment the specimens were positioned between two insulated electrodes and placed onto the lower grounded electrode. An alternating high voltage pulse generator with pulse duration of 2 µs and frequency of 17 kHz drove the discharge. During the treatment with a duration of 60 s, a voltage of 34 kV (peak),

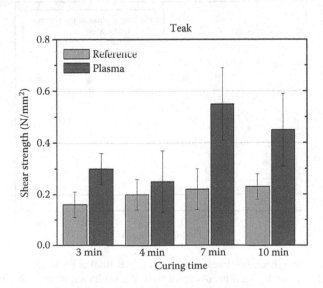

FIGURE 17.26 Time-dependent shear bond strength of untreated and plasma-treated teak veneers.

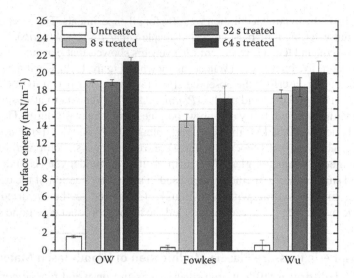

FIGURE 17.27 Change of surface energy (polar component) of WPC after different plasma treatment durations, according to the approaches of Owens-Wendt, Fowkes, and Wu.

a power dissipation of 60 W and a gas temperature <40°C were measured. The plasma pretreated specimens showed faster penetration of water compared to the untreated specimens. It could be demonstrated by shear strength tests that the increase in penetration speed led to a faster curing of PVA glued specimens (Figure 17.28).

Wolkenhauer et al. (2008b) investigated the adhesion of PVA-glue on fibre board and particle board in untreated and plasma-treated state by a force-sensitive peel test. The DBD plasma setup used for treatment of particle board and fiber board is described in Wolkenhauer et al. (2007a). It could be proven that the adhesion of PVA-glue on fiber board and particle board was enhanced by the plasma treatment. An increase in peel force of 4% for fiber board and 8% for particle board

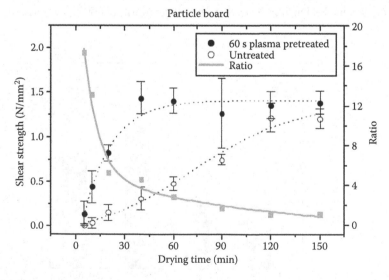

FIGURE 17.28 Shear strength (dashed lines) and the ratio (gray line) of PVAc-glued particleboard (untreated and plasma pretreated) against drying time. (Adapted from Wolkenhauer, A. et al. 2007b. *Proceedings of the third European conference on wood modification*, 271–274.)

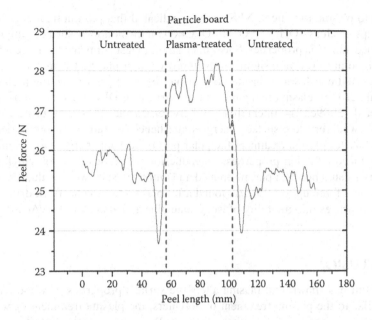

FIGURE 17.29   Peel force against peel length of textile glued on the particle board.

(Figure 17.29) was detected. The authors emphasized that with the used force-sensitive peel test, direct adjacent plasma-treated and untreated areas can be investigated and low differences in bonding strength are detectable.

Wolkenhauer et al. (2008c) investigated wood–plastic composites (WPCs) after DBD treatment at atmospheric pressure regarding their adhesion properties and surface characteristics. The plasma treatment conditions were the same as described in Wolkenhauer et al. (2007a). The treatment duration was 4 s for all experiments. Increased polar component of surface energy were detected by contact angle measurement. AFM micrographs revealed an increase in surface roughness after plasma treatment (Figure 17.30). Tensile bond strength tests, shear bond strength tests, and peel adhesion tests exhibited improved adhesion of waterborne, solventborne, and oil-based paints as well as for poly (vinyl acetate) and polyurethane adhesives bonded WPCs when WPCs were plasma pretreated.

Wood/PE composites were treated with air plasma at atmospheric pressure to improve adhesion properties of the low polar surfaces (Liu et al. 2010). The plasma treatment was carried out at a frequency of 100 kHz and a power range of 550–800 W. The samples were exposed for 30 s to the plasma. The altered surface characteristics of the treated and untreated composites were evaluated by contact angle, FTIR, SEM, AFM, and XPS-analysis. Contact angles decreased gradually with increasing discharge power. Polar groups (hydroxyl, carbonyl, and carboxyl) could be detected on the plasma-treated surfaces by FTIR analysis. SEM and AFM micrographs exhibit increased

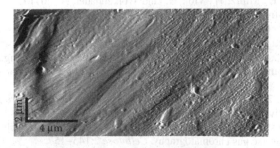

FIGURE 17.30   AFM image of untreated (left) and plasma-treated (right) WPC-PP surface.

roughness due to plasma treatment. XPS analysis indicated that plasma treatment introduce polar functional groups (such as –OH, –C=O, or –O–C=O) on the nonpolar surfaces improving the wettability and hence adhesion properties. Shear bond strength test confirmed that the adhesion properties of wood/PE composites were significantly improved after plasma treatment:

Wax-impregnated beech was subjected to plasma treatment prior to gluing (Scholz et al., 2010, Avramidis et al., 2011b). Plasma parameters of the air-driven DBD were the same as discussed in Wolkenhauer et al. (2009c). Treatment time was between 0 s and 120 s. The surfaces of wax-impregnated woods showed very low surface energies and hence low adhesion properties compared to untreated wood. After plasma treatment the polar part of surface energy was distinctly increased, pointing to enhanced adhesion properties. The adhesion properties of a polyvinyl acetate (PVAc) adhesive on wax-treated beech at plasma-treated and untreated state were evaluated with a peel test. An increase in peel force up to 60% was found after plasma treatment. Back-drying of the water-stored specimens proves the suitability of isocyanate glue and also that of PVAc after plasma treatment for 120 s.

17.3 CONCLUSION

This compendium has shown that plasma alters the surface properties of wood and wood-based materials. Similar to the plasma treatment of polymers, the plasma treatment of wood can be an efficient approach to improve surface characteristics. By a plasma treatment, it might be possible, for example, to decrease the use of primers and bonding agents, to facilitate the utilization of new, environmentally friendly or cheaper adhesives and paints or to create new materials and material combinations. The results of the presented studies and the multiplicity of possible applications demonstrate that plasma treatment represents a powerful and versatile tool and is a seminal technique for modern wood processing. Nevertheless, further research is needed to understand its mechanisms and to explore the full scope of its potential applications.

REFERENCES

Artmann, J. 1963. Untersuchung des Gleichgewichtszustandes hinter einer starken Stoßwelle in Wasserstoff. *Bericht des Physikalischen Institutes der TH Aachen*. Aachen, Germany.

Asandulesa, M., Topala, I., and Dumitrascu, N. 2010. Effect of helium DBD plasma treatment on the surface of wood samples. *Holzforschung* 64:223–227.

Avramidis, G., Hauswald, E., Lyapin, A. Militz, H., Viöl, W., and Wolkenhauer, A. 2009a. Plasma treatment of wood and wood-based materials to generate hydrophilic or hydrophobic surface characteristics. *Wood Material Science & Engineering* 4:52–60.

Avramidis, G., Wolkenhauer, A., Zhen, B., Militz, H., and Viöl, W. 2009b. Water-repellent coatings on wood surfaces generated by a dielectric barrier discharge plasma jet at atmospheric pressure. *Proceedings of the Fourth European Conference on Wood Modification*, Stockholm, Sweden, 269–275.

Avramidis, G., Tebbe, B., Nothnick, E., Militz, H., Viöl, W., and Wolkenhauer, A. 2010. Wood veneer modification by atmospheric pressure plasma treatment for improved absorption characteristics. *Proceedings of the Fifth European Conference on Wood Modification*, Riga, Latvia, 365–373.

Avramidis, G., Nothnick, E., Militz, H., Viöl, W., and Wolkenhauer, A. 2011a. Accelerated curing of PVAc adhesive on plasma-treated wood Veneers. *Eur. J. Wood Prod.* 69:329–332.

Avramidis, G., Scholz, G., Hauswald, E., Militz, H., Viöl, W., and Wolkenhauer, A. 2011b. Improved bondability of wax-treated wood following plasma treatment. *Wood Sci. Technol.* 45:359–368.

Awakowicz, P., Bibinov, N., Born, M., Busse, B., Gesche, R., Helmke, A., Kaemling, A. et al. 2009. Biological stimulation of the human skin applying health-promoting light and plasma sources. *Contrib. Plasma Phys.* 49:641–647.

Aydin, I. and Demirkir, C. 2010. Activation of spruce wood surfaces by plasma treatment after long terms of natural surface inactivation. *Plasma Chem. Plasma Process.* 30:697–706.

Belgacem, M.N., Czeremuszkin, G., Sapieha, S., and Gandini, A. 1995. Surface characterization of cellulose fibres by XPS and inverse gas chromatography. *Cellulose* 2:145–157.

Bente, M., Avramidis, G., Förster, S., Rohwer, E. G., and Viöl, W. 2004. Wood surface modification in dielectric barrier discharges at atmospheric pressure for creating water repellent characteristics. *Holz als Roh- und Werkstoff* 62:157–163.

Bialski, A., Manley, R.St.J., Wertheimer, M.R., and Schreiber, H.P. 1975. Composite materials with plasma-treated components: Treatment-property correlation. *Polym. Prepr.* 16:70–72.

Blanchard, V., Blanchet, P., and Riedl, B. 2009. Surface energy modification by radiofrequency inductive and capacitive plasmas at low pressures on sugar maple: An exploratory study. *Wood Fiber Sci.* 41:245–254.

Busnel, F., Blanchard, V., Pregent, J., Stafford, L., Riedl, B., Blanchet, P., and Sarkissian, A. 2010. Modification of sugar maple (Acer saccharum) and black spruce (Picea mariana) wood surfaces in a dielectric barrier discharge (DBD) at atmospheric pressure. *J. Adhes. Sci. Technol.* 24:1401–1413.

Chen, H.Y. and Zavarin, E. 1990. Interactions of cold radiofrequency plasma with solid wood, I: Nitrogen permeability along the grain. *J. Wood Chem. Technol.* 10:387–400.

Denes, A.R. and Young, R.A. 1999. Reduction of weathering degradation of wood through plasma-polymer coating. *Holzforschung* 53:632–640.

Denes, A.R., Tshabalala, M.A., Rowell, R.M., Denes, F.S., and Young, R.A. 1999. Hexamethyldisiloxaneplasma coating of wood surfaces for creating water repellent characteristics. *Holzforschung* 53:318–326.

Denes, F. S., Cruz-Barba, L.E., and Manolache, S. 2005. *Handbook of Wood Chemistry and Wood Composites*, Rowell, R. (ed.), Chap. 16, CRC Press, Boca Raton, Florida.

Drnovska, H., Lapcik, L., Bursikova, V., Zemek, J., and Barros-Timmons A.M. 2003. Surface properties of polyethylene after low-temperature plasma treatment. *Colloid Polym. Sci.* 281:1025–1033.

Evans, P.D., Ramos, M. and Senden, T. 2007. Modification of wood using a glowdischarge plasma derived from water. *Proceedings of the Third European Conference on Wood Modification*, 123–132.

Förster, S., Mohr, C., and Viöl, W. 2005. Investigations of an atmospheric pressure plasma jet by optical emission spectroscopy. *Surf. Coat. Technol.* 200:827–830.

Goldmann, M., Goldmann, A., and Sigmond, R.S. 1985. The corona discharge, its properties and specific uses. *Pure Appl. Chem.* 57:1353–1362.

Griem, H.R. 1964. *Plasma Spectroscopy*. McGraw-Hill Company, New York.

Haase, J.G. and Evans, P.D. 2010. Plasma modification of wood surfaces to improve the performance of clear coatings. *Proceedings of the Fifth European Conference on Wood Modification* (Meeting date 2010), Riga, Latvia, 271–274.

Helmke, A., Hoffmeister, D., Berge, F., Emmert, S., Laspe, P., Mertens, N., Viöl, W., and Weltmann, K.-D. 2011. Physical and microbiological characterization of *Staphylococcus epidermis* inactivation by dielectric barrier discharge plasma. *Plasma Process. Polym.* 8:278–286.

Huang, H., Wang, B.J., Dong, L., and Zhao, M. 2011. Wettability of hybrid poplar veneers with cold plasma treatments in relation to drying conditions. *Drying Technol.* 29:323–330.

Jamali, A. and Evans, P.D. 2011. Etching of wood surfaces by glow discharge plasma. *Wood Sci. Technol.* 45:169–182.

Jie-Rong, C., Xue-Yan, W., and Tomiji, W., 1999. Wettability of poly(ethylene terephthalate) film treated with low-temperature plasma and their surface analysis by ESCA. *J. Appl. Polym. Sci.* 72:1327–1333.

Kim, C.Y. and Goring, D.A.I. 1971. Corona-induced bonding of synthetic polymers to wood. *Pulp and Paper Magazine of Canada* 72:93–96.

Klarhöfer, L., Maus-Friedrichs, W., Kempter, V., and Viöl, W. 2005. Investigation of pure and plasma-treated spruce with surface analytical techniques. *Wood Modification: Processes, Properties and Commercialisation*, 339–345.

Klarhöfer, L., Viöl, W., and Maus-Friedrichs, W. 2010. Electron spectroscopy on plasma treated lignin and cellulose. *Holzforschung* 64:331–336.

Kogelschatz. U. 2003. Dielectric-barrier discharges: Their history, discharge physics, and industrial applications. *Plasma Chem. Plasma Process* 23:1–46.

Kuchenbecker, M., Bibinov, N., Kaemling, A., Wandke, D., Awakowicz, P., and Viöl, W. 2009. Characterization of DBD plasma source for biomedical applications. *J. Phys. D: Appl. Phys.* 42:045212.

Laroussi, M. and Akan, T. 2007. Arc-free atmospheric pressure cold plasma jets: A review. *Plasma Process. Polym.* 4:777–788.

Lecoq, E., Clément, F., Panousis, E., Loiseau, J.F., Held, B., Castetbon, A., and Guimon, C. 2008. *Pinus pinaster* surface treatment realized in spatial and temporal afterglow DBD conditions. *Eur. Phys. J. Appl. Phys.* 42:47–53.

Legeay, G., Epaillard, F., and Brosse, J.C.1984. Surface modification of natural or synthetic polymers by cold plasmas. *Proceedings from the 2nd Annual Int. Conf. Plasma Chem. Technol.*, 29–39.

Liston, E.M., Martinu, L., and Wertheimer, M.R. 1994. Plasma surface modification of polymers for improved adhesion: A critical review. In: Strobel, M., Lijons, C. and Mittal, K. (Eds.), *Plasma Surface Modification of Polymers—Relevance to Adhesion.* VSP, Utrecht, pp. 3–39.

Liu, Y., Tao, Y., Lu, X., Zhang, Y., and Di, M. 2010. Study on the surface properties of wood/polyethylene composites treated under plasma. *Appl. Surf. Sci.* 257:1112–1118.

Magalhaes, W.L.E. and de Souza, M.F. 2002. Solid wood coated with plasma-polymer for water repellence. *Surf. Coat. Technol.* 155:11–15.

Mahlberg, R., Niemi, H.E.-M., Denes, F.S., and Rowell, R.M. 1998. Effect of oxygen and hexamethyldisiloxane plasma on morphology, wettability and adhesion properties of polypropylene and lignocellulosics. *Int. J. Adhesion Adhesives* 18:283–297.

Mahlberg, R., Niemi, H.E.-M., Denes, F.S., and Rowell, R.M. 1999. Application of AFM on the adhesion studies of oxygen-plasma-treated polypropylene and lignocellulosics. *Langmuir* 15:2985–2992.

Mertens, N., Wolkenhauer, A., Leck, M., and Viöl, W. 2006. UV laser ablation and plasma treatment of wooden surfaces—a comparing investigation. *Laser Phys. Lett.* 3:380–384.

Nussbaum, R.M. 2001. Surface interactions of wood with adhesives and coatings. Doctoral thesis, Royal Institute of Technology, Stockholm.

Odrášková, M., Szalay, Z., Ráhel, J., Zahoranová, A., and Cernák, M. 2007. Diffuse Coplanar Surface Barrier Discharge assisted deposition of water repellent films from N2/HMDSO mixtures on wood surface. *Proceedings of the 28th ICPIG*, 803–806.

Odrášková, M., Ráheľ, J., Zahoranová, A., Tiňo, R., and Černák, M. 2008. Plasma activation of wood surface by diffuse coplanar surface barrier discharge. *Plasma Chem. Plasma Process.* 28:203–211.

Podgorski, L. and Roux, M. 1999. Wood modification to improve the durability of coatings. *Surf. Coat. Int.* 82:590–596.

Podgorski, L., Bousta, G., Schambourg, F., Maguin, J., and Chevet, B. 2001. Surface modification of wood by plasma polymerization. *Pigment Resin Technol.* 31:33–40.

Rajasekaran, P., Mertmann, P., Bibinov, N., Wandke, D., Viöl, W., and Awakowicz, P. 2009. DBD plasma source operated in single filamentary mode for therapeutic use in dermatology. *J. Phys. D: Appl. Phys.* 42:225201.

Rajasekaran, P., Mertmann, P., Bibinov, N., Wandke, D., Viöl, W., and Awakowicz, P. 2010. Filamentary and homogeneous modes of Dielectric Barrier Discharge (DBD) in air: Investigation through plasma characterization and simulation of surface irradiation. *Plasma Process. Polym.* 7:665–675.

Rehn, P. and Viöl, W. 2003. Dielectric barrier discharge treatments at atmospheric pressure for wood surface modification. *Holz als Roh- und Werkstoff* 61:145–150.

Rehn, P., Wolkenhauer, A., Bente, M., Forster, S., and Viöl, W. 2003. Wood surface modification in dielectric barrier discharges at atmospheric pressure. *Surf. Coat. Technol.* 174–175:515–518.

Sakata, I., Morita, M., Tsuruta, N., and Morita, K. 1993. Activation of wood surface by corona treatment to improve adhesive bonding. *J. Appl. Polym. Sci.* 49:1251–1258.

Scholz, G., Nothnick, E., Avramidis, G., Krause, A., Militz, H., Viöl, W., and Wolkenhauer, A. 2010. Adhesion of wax impregnated solid beech wood with different glues and by plasma treatment. *Eur. J. Wood Prod.* 68:315–321.

Strobel, M., Dunatov, C., Strobel, J.M., Lyons, C.S., Perron, S.J., and Morgen M.C. 1989. Low-molecular-weight materials on corona-treated polypropylene. *J. Adhes. Sci. Technol.* 3:321–335.

Toriz, G., Gutierrez, G.M., Gonzalez-Alvarez, V., Wendel, A., Gatenholm, P., and Martinez-Gomez, A. 2008. Highly hydrophobic wood surfaces prepared by treatment with atmospheric pressure dielectric barrier discharges. *J. Adhes. Sci. Technol.* 22:2059–2078.

Viöl, W. and Uhlenbusch, J. 1996. Generation of CO_2 laser pulses by Q-switching and cavity dumping and their amplification by a microwave excited CO_2 laser. *J. Phys. D: Appl. Phys.* 29:57–67.

Viöl, W. 2004. Method for modifying wooden surfaces by electrical discharges at atmospheric pressure. CZ PV 2002–1908, CA 2,393,952, EP 1 223 854.

Viöl, W., Kaemling, A., and Born, S. 2007. Device for Plasma treatment at atmospheric pressure. EP 07 723 181, US 12/207,585.

Wistara, N., Denes, F.S., and Young, R.A. 2002. The resistance of CF4 plasma treated tropical woods against white-rot (*Trametes versicolor* L. Fr. Pilat) attack. *Proceedings of the International Wood Science Symposium.* (Meeting date 2002), 157–163.

Wolkenhauer, A., Avramidis, G., Cai, Y., Militz, H., and Viöl, W. 2007a. Investigation of wood and timber surface modification by dielectric barrier discharge at atmospheric pressure. *Plasma Processes Polym.* 4:S470–S474.

Wolkenhauer, A., Avramidis, G., Militz, H., and Viöl, W. 2007b. Wood modification by atmospheric pressure plasma treatment. *Proceedings of the third European conference on wood modification*, Cardiff, Wales, 271–274.

Wolkenhauer, A., Avramidis, G., Militz, H., and Viöl, W. 2008a. Plasma treatment of heat treated beech wood—investigation on surface free energy. *Holzforschung* 62:472–474.

Wolkenhauer, A., Militz, H., and Viöl, W. 2008b. Increased PVA-glue adhesion on particle board and fibre board by plasma treatment. *Holz als Roh- und Werkstoff* 66:143–145.

Wolkenhauer A, Avramidis G, Hauswald E, Militz H., and Viöl W 2008c. Plasma treatment of wood-plastic composites to enhance their adhesion properties. *J. Adhes. Sci. Technol.* 22:2025–2037.

Wolkenhauer, A. 2009a. Plasma Treatment of Wood and Wood based Materials by Dielectric Barrier Discharge at Atmospheric Pressure. Doctoral Thesis, Sierke, Göttingen.

Wolkenhauer, A., Avramidis, G., Hauswald, E., Loose, S., Viöl, W., and Militz, H. 2009b. Investigations on the drying behaviour on plasma-treated wood materials. *Wood Res.* 54:59–66

Wolkenhauer, A., Avramidis, G., Hauswald, E., Militz, H., and Viöl, W. 2009c. Sanding vs. plasma treatment of aged wood: A comparison with respect to surface energy. *Int. J. Adhes. Adhes.* 29:18–22.

Yuan, X., Jayaraman, K., and Bhattacharyya, D. 2004. Effects of plasma treatment in enhancing the performance of woodfibre-polypropylene composites. *Composites Part A* 35:1363–1374.

Zanini, S., Riccardi, C., Orlandi, M., Fornara, V., Colombini, M.P., Donato, D.I., Legnaioli, S., and Palleschi, V. 2008. Wood coated with plasma-polymer for water repellence. *Wood Sci. Technol.* 42:149–160.

18 Sustainability of Wood and Other Biomass

Roger M. Rowell

CONTENTS

18.1   INTRODUCTION

The question can be asked, "why study wood chemistry?" Why would so many authors take their time to write a book on wood chemistry and wood composites? The simple answer is that wood and other biomass can be part of a sustainable materials future. However, wood and other biomass will not reach it highest potential until we understand its chemistry and materials properties and how these express themselves in performance. As our supply of nonrenewable resources runs out and becomes more expensive, we must turn out attention to renewable resources. Up until about 1920, almost all of our chemicals and materials were derived from wood. Then our dependence on petroleum, steel, aluminum, glass, plastics, and other synthetics started to grow. Today, we rely heavily on petroleum for our energy, metals, ceramics, and plastics for our materials but these resources are not sustainable.

The simplest definition of sustainability of wood and other biomass is to not use more today that you can replace tomorrow. Sustainable development can be achieved using a long-range proactive dynamic strategic plan that focuses not only on short-term profits and competitiveness without damage to the environment but maintains an equal emphasis on future maintenance and growth of the society. Sustainable development must fit certain criteria within a framework of economic, environmental, and social systems before it can last indefinitely. If we consume more than nature can replenish, we are in a state of environmental degradation that is not sustainable. If we only harvest and use what nature can replenish, we are in environmental equilibrium. If we consume less than what nature can replenish, then we have achieved an environmental renewable state and sustainable indefinitely.

We must create a closed loop in our use of forest resources. About 50% of the dry weight of wood is carbon. Growing trees take in carbon dioxide through the process of photosynthesis and convert it to tree biomass. The carbon dioxide is released again when the wood decays or burns. Thus, sustainable forest management systems study input carbon from photosynthesis, use carbon in products and carbon back into the atmosphere during combustion or decay.

We will have to live off of nature's income not its capital or we may find fewer choices and flexibility in our future lives. There will be resistance to change in the way we live but we must strive to become a sustainable society and integrate environmental, social, human, and economic needs into our policies and activities. We must set sustainable goals and achieve them both locally and globally.

In general, our present business economic model is focused on short-term profits. The "bottom line" for most companies is the black ink quarterly gains to stockholders. While this model will and must remain critical, a new economic model must be developed that gives credit for long-term survivability and sustainability of a business. It is interesting that historically the concept of sustainability comes from a German forest in the seventeenth century. They limited the number of trees that could be harvested for firewood to the number of new trees planted. The German word for sustainability is Nachhaltigkeit and was first mentioned in an eighteenth century treatise on forestry by Hans Carl von Carlowitz (Sustainable Times 2008). The English translation of this text is "sustainable yield."

Environmental sustainability is a direct result of this early German concept that is to use our natural resources only as fast as they can be regenerated. Sustainable forestry must integrate the elements of forest health, economic profitability, and social equity. In simpler terms, the concept can be described as the attainment of balance—balance between society's increasing demands for forest products and benefits, and the preservation of forest health and diversity. This balance is critical to the survival of forests, and to the prosperity of forest-dependent communities (Sathre 2007).

The modern concept of sustainability has it roots in a 1980 meeting of the International Union for the Conservation of Nature. They published the *World Conservation Strategy* and first used the term "sustainable development" (WCS 1980). Later in 1983, there was a meeting of the United Nations World Commission on Environment and Development (WCED 1983). The General Assembly passed Resolution 38/161 *"Process of preparation of the Environmental Perspective to the Year 2000 and Beyond."* It established a Commission to be chaired by the former Norwegian Prime Minister Gro Harlem Brundtland and its mission was to address growing concern "about the accelerating deterioration of the human environment and natural resources and the consequences of that deterioration for economic and social development." In establishing the commission, the UN General Assembly recognized that environmental problems were global and that it was in the best interest of all nations to establish policies for sustainable development. The new commission was to "(1) propose long-term environmental strategies for achieving sustainable development by the year 2000 and beyond; (2) recommend ways in which concern for the environment may be translated into greater co-operation among developing countries and between countries at different stages of economic and social development; (3) lead to the achievement of common and mutually supportive objectives which take account of the interrelationships between people, resources, environment and development; (4) consider ways and means by which the international community can deal more effectively with environmental concerns, in the light of the other recommendations in its report; and (5) help to define shared perceptions of long-term environmental issues and of the appropriate efforts needed to deal successfully with the problems of protecting and enhancing the environment, a long-term agenda for action during the coming decades, and aspirational goals for the world community" (WCED 1983, pp. 48–50).

In 1987, the report from the Brandtland Commission defined sustainable development as "development that meets the needs of the present without compromising the ability of future generations to meet their own needs" (BC 1987, p. 3).

In 1989, following the publication of the Brundtland Report, Swedish scientist, Karl-Henrik Robèrt published *The Natural Step* outlining the basic principals to help bring societies towards sustainability. The core message of *The Natural Step* is to help people from all over the world discuss environmental issues without getting bogged down in details or disputes (*The Natural Step* 1989). *The Natural Step* is derived from the first and second laws of thermodynamics: that energy is conserved and nothing disappears (all matter that will ever exist on Earth is now here) and that matter and energy tend to disperse over time (disorder increases in all closed systems). Applying these two laws into *The Natural Step* there are four system conditions: (1) "Nature cannot withstand a systematic buildup of dispersed matter mined from the Earth's crust; (2) Nature cannot withstand a systematic buildup of persistent compounds made by humans; (3) Nature cannot take a systematic deterioration of its capacity for renewal; and (4) if we want life to continue, we must (a) be efficient in our use of resources and (b) promote justice—because ignoring poverty will lead the poor, for short-term survival, to destroy resources that we all need for long-term survival."

18.2 ECONOMIC, ENVIRONMENTAL, AND SOCIAL FACTORS

The 2005 World Summit Outcome Document refers to the "independent and mutually reinforcing pillars" of sustainable development and must fit certain criteria within a framework of economic, environmental and social systems before it can last indefinitely (Figure 18.1, World Summit 2005). Each component of this framework is important and no component is more important than the other. Each element not only influences each other but also depends on each other. Using this model, Figure 18.1 shows how the three concepts fit together. The model does not rely on only one system but all three and the interface of the three systems shows areas that are bearable, viable, and equitable with a core intercept of sustainability. What is good for society may not be good for the environment and may not make long-term economic sense. It is within the core sector that sustainability can be achieved.

The model shows that there are areas that are bearable between the society and the environment, areas that are equitable between society and economic considerations and viable between the environment and the economics but only when the three are taken as a whole can the given situation become sustainable. The "bottom line" for this model requires a long-range proactive dynamic strategic plan that focuses not only on short-term profits and competitiveness without damage to the environment but maintains an equal emphasis on future maintenance and growth of the society. This new model also assumes that the long-range plan will maintain a quality of life indefinitely.

Western countries use approximately 30 times more resources than developing countries and countries such as China and India are striving to reach a western style of living. Taking these three factors into the future it is apparent that we cannot maintain the status quo and the present "business as usual" is not sustainable.

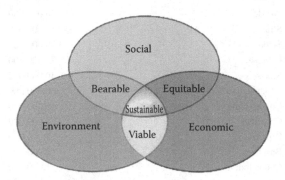

FIGURE 18.1   Model of the interaction between social, environmental and economic factors in sustainable development. (Adapted from World Summit 2005. United Nations 2005 World Summit. UN Headquarters, New York, NY (www.un.org/summit2005).)

18.3   GREEN AND SUSTAINABLE

The concept of green or environmentally friendly technologies can be in conflict with sustainable development in that green technologies gives priority to environmental considerations over those of economics and cultural considerations. An example of this might be a community that decides it will start a recycling program to reduce the amount of trash going into the local landfill. The cost of the program is expensive and there is no market for some of the recyclables so the program is environmentally sound but nonsustainable due to economics. Another example might be the desire of a society to convert corn into ethanol as a fuel supplement. This causes a higher price for corn which results in an increase in both animal feed and human food. In the long term, this may prove not to be sustainable.

18.4   ELEMENTS OF SUSTAINABILITY

There are several critical elements in a sustainable future. These include raw material supply, water, energy, emissions and waste products, product viability, human resources, and technology development (Rowell et al. 2010).

18.4.1   RAW MATERIAL SUPPLY

It has been said that the twenty-first century will be called the era of cellulosics. However, we have to remember that wood is not renewable; the tree it came from is renewable. This puts our focus on healthy ecosystems to maintain a renewable supply of wood. Wood is not the only source of lignocellulosics. The forest industries must look to other sources of biomass for composite products. Composites such as fiberboard and particleboard can also be made from agricultural resources. Table 18.1 shows an inventory of the world supply of agricultural resources that are available for utilization (Rowell 2002).

If we consume more than nature can replenish, we are in a state of environmental degradation that is not sustainable. We must work toward a stewardship and use of forests and forest lands in a way, and at a rate, that maintains their biodiversity, productivity, regeneration capacity, and vitality.

TABLE 18.1
Inventory of World Bio-Resources

Source	Metric Tons
Wood	1,750,000,000
Straw	1,145,000,000
Stalks	970,000,000
Bagasse	75,000,000
Reeds	30,000,000
Bamboo	30,000,000
Cotton staple	15,000,000
Core fiber	8,000,000
Papyrus	5,000,000
Bast fiber	2,900,000
Cotton linters	1,000,000
Grasses	700,000
Leaf	480,000
Total	4,033,080,000

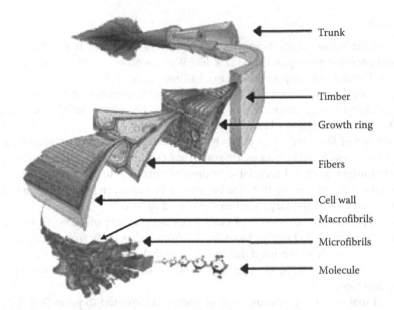

FIGURE 18.2   Possible product levels from a tree.

Figure 18.2 shows the possible product sizes that can come from a tree (see Chapters 10, 12, 13). With a Young's modulus of about 10 GPa for solid wood, 40 GPs for wood fibers, 70 GPa for microfibrils and 250 GPa for cellulose crystallites there are many possible biomaterials that can be produced from wood with a wide variety of furnish sizes and strengths.

18.4.2 WATER

Seventy percent of the Earth's surface is covered with water, however, most of this water, 97.5%, is in the oceans and seas and is too salty to drink or grow crops. Of the remaining 2.5%, 1.73% is in the form of glaciers and icecaps leaving only about 0.77% available for our fresh water supply. Of the total water on Earth, only 0.0008% is available and renewable in rivers and lakes for human and agricultural use. It is the water that falls as rain or snow or that has been accumulated and stored as groundwater that we depend on for our "clean" water resource (Rowell 2006).

For 1.5–2.5 billion people in the world, clean water is a critical issue. It is estimated that by the year 2025, there will be an additional 2.5 billion people on the Earth that will live in regions already lacking sufficient clean water. In the United States, it is estimated that 90% of all Americans live within 10 miles of a body of contaminated water. Contaminates in water include sediments, nutrients, pathogens, dissolved oxygen, heavy metal ions, suspended solids, pesticides, turbidity, fish contamination, and ammonia.

In 1990, the average water consumed per person was 1620 gallons per day and in 2000 this had decreased to 1430 gallons per day (USGS 2000). In 2000, 408 billion gallons per day were consumed. Sixty-nine percent of this water came from surface water the remainder from ground water. Thermoelectric-power plants consumed 195,000 million gallons per day (Mgal/d), 137,000 Mgal/d for irrigation, and 43,300 Mgal/d for household use. Of the household water consumed, 4% was for drinking, 20% for watering gardens, 32% for toilet flushing, 28% for bathing and cleaning, and 16% for laundry and dishes.

Future water use considerations must focus on reuse and closed-loop systems where waste water is kept to a minimum.

18.4.3 ENERGY

Since the start of the industrial revolution, the worldwide use of energy has had a steady growth. In 1900, the global energy consumption was 0.7×10^{12} Watts (watt = 1 joule per second) (KWE 2006). In 2005, the total world consumption of energy had increased to 5×10^{20} Joules with 85% of this energy coming from the combustion of fossil fuels (Energy Information 2006). From 1980 to 2004, the increase in global energy consumption increased 2% per year. The United States consumes approximately 25% of the total world energy.

The consumption of fossil resources puts what might be called historic or ancient carbon into the environment. Once these fossil resources are gone, they cannot be replaced. Burning of biomass for energy puts what might be called recyclable or modern carbon into the environment meaning that the small amount of carbon coming from the burning of biomass can be recycled into the photosynthetic pathways. Not so with the large amount of historic carbon.

Globally, on a percentage basis, 37% of our energy comes from oil, 25% from coal, 23% from gas, 6% from nuclear, 4% from burning biomass, 3% from hydro, 0.5% from solar, 0.3% from wind, and 0.2% for geothermal. It is estimated that there are global reserves of 290 zJ (1 zetta joule = 10^{20} joules) of coal, 19 zJ of oil, and 16 zJ of gas left. Coal is the most abundant fossil having a reserve of over 909 billion tons.

It is projected that our future consumption of energy is expected to jump 50% by 2030 and oil could cost between $113 and $186 a barrel (Hargreaves 2008). In 2012, this projected cost for oil seems very low and the actual cost may be much higher than this. Our future energy needs must be greatly reduced and may depend on sun-driven processes. The sun produces approximately YJ per year (1 YJ = 10^{24} J) or 8000 times the 2004 total energy usage.

18.4.4 EMISSIONS AND WASTE PRODUCTS

Emissions and waste products (pollutants) come from all human activity. Carbon dioxide, nitrogen oxides, carbon monoxide, volatile organic compounds, particulate matter, toxic metals, ammonia, chlorofluorocarbons, and radioactive pollution are the result of this activity. The gasses are causing an increase in global warming and sea levels rising. The solids are polluting our water, soil, and filling up our landfills. There is an advertisement from the Royal Dutch Shell Company that states, "Don't throw anything away. There is no away."

Table 18.2 shows the carbon footprint of bio-resources in the United States from 1910 to a projection on 2040 (USDA 2000). Part of the carbon is sequestered in products, part goes to the landfill, part to energy, and the rest to emission due to decay.

By 2000, the percentage of carbon going into products had come back to the level in the early part of the twentieth century and represents sequestered carbon. This was also true for burning

TABLE 18.2
Carbon Footprint of Bio-Resources in the United States in Trillion Tons

Year	Added to Products	Added to Landfills	Burned for Energy	Decay Nonenergy
1910	24.3	1.1	88.4	10.6
1920	22.9	3.1	51.9	14.7
1940	14.0	5.3	35.0	20.4
1960	9.0	7.1	34.6	30.6
1980	11.8	27.9	48.1	19.2
2000	25.0	32.5	88.1	14.3
2020	25.6	42.6	103.0	16.4
2040	22.9	50.8	119.0	17.5

biomass for energy. The largest increase in the carbon footprint is what the United States put in its landfills. The projections for the year 2020 and 2040 indicate a large increase in burning for energy and landfilling with little increases in carbon added to products or carbon lost due to decay. We must always work to minimize resource use and to minimize wastes.

18.4.5   PRODUCT VIABILITY

Along with strategies to remain sustainable in terms of raw material supply, water use, energy, and emissions/wastes is product viability. It is well known that at the turn of the nineteenth century the world had developed the best buggy whip ever made but there was no need for this tool since the era of the horse had passed. In order for a product to remain viable it is vital to make sure it is still needed and relevant in the modern society.

In general, the forest industry makes incremental improvements in the present technology, however, it is mainly universities and institutes that develop new technologies. It will be critical to continue research in the development of new technologies to insure a continuing supply of sustainable products.

18.4.6   HUMAN RESOURCES

There is no substitute for a well educated and competent human capital. New technologies must be developed and that comes from imagination, risk taking, experience, and vision. We cannot afford to lose skills and specialization. It is hoped that this book will help educate future generations in wood chemistry and wood composites.

Unfortunately, with the downturn in the forestry industries, fewer universities are teaching courses in wood science. If students cannot get a job in the industry, there are fewer students in that subject. This forces universities to reduce staff in those departments with few students. If forest resources make a comeback in a sustainable society, it will take years to rebuild the expertise that once existed in our universities and institutes.

Future generations must think new thoughts and new ways of doing things. Albert Einstein said that "Today's problems cannot be solved if we still think the way we thought when we created them." There is an old Chinese proverb that states "Unless we change direction, we are likely to end up where we are going."

18.4.7   TECHNOLOGY DEVELOPMENT

There must be constant and continuous improvements in technology to ensure that development remains sustainable. If any element in a sustainable project becomes unsustainable, then new technologies must be developed to get the project back in a sustainable mode. Life cycle management must be a dynamic part of all existing and new technologies. Within the life cycle assessment we must continue to analyze inventory, impact, and improvements of past, present, and future technologies. Innovation, research and development, and market integration must be part of all sustainable traditional and new businesses. Emphasis must be placed on carbon sequestration to reduce the carbon going into our atmosphere.

18.5   THE R'S OF SUSTAINABILITY

There are many "R's" in sustainable development: reduce, reuse, recycle, regulate, renovate, reinvent, reconsider, refresh, and respect. We must reduce our use of raw materials, energy, water, and emissions and wastes. We must find ways to reuse and recycle materials. There is a natural resistance to change so we must act by changing laws and regulations to force change. Instead of building new we must consider renovation of existing structures. We must reinvent, refresh, and update technologies to keep them relevant. Anthony Cortese, president of Second Nature added two more R's to this list, rethink (our use of materials) and refuse (to continue to consume). Finally, we must respect our environment so life on planet Earth can be sustained.

18.6　MAKING CHANGE HAPPEN

Sustainability is a concept that most support but few understand what it means in practice. Making the transition from our present economic model to one of sustainability is a process that will take time and great effort. It will be done in small realistic steps and it will be the responsibility of all involved in the system: workers, policymakers, researchers, companies, and consumers. It will be a dynamic process and we must be proactive not reactive to change.

It is easy to think that the other person will make the changes needed for us to become a sustainable global society. Do not wait for a grand scheme to come along to move society forward. Society is no more than a collection of "us." It is up to each of us to be involved. Margaret Mead said that "Never doubt that a small group of thoughtful, committed citizens can change the world. Indeed, it's the only thing that ever has." And, Mohandas Gandhi said "you must be the change you wish to see in the world."

We may find less choices and flexibility in our future lives. There will be resistance to change in the way we live but we must strive to become a sustainable society and integrate environmental, social, human, and economic needs into our policies and activities. We must set sustainable goals and achieve them both locally and globally.

18.7　CONCLUSIONS

The wood industry has the raw material supply, its own energy supply and a wide range of possible products from timbers, lumber, veneers, chips, flakes, particles, fibers, polymers, and chemicals that can be produced from a biorefinery concept, knowledge to produce adhesives and coatings from biomass and technology to increase stability and durability of biomaterials. New breakdown, separation, and synthesis technologies will be developed that will use less energy and water in conversion into products. We must make sure that our forests and agricultural lands are maintained and improved so that we can assure a sustainable supply of raw materials far into the future. So, why study wood chemistry? Wood and other biomass will be a major part of a global sustainable future.

Treat the Earth well, it is not inherited from our parents, it is borrowed from our children.

REFERENCES

BC 1987. Our Common Future, Report of the World Commission on Environment and Development, World Commission on Environment and Development. Published as Annex to General Assembly document A/42/427, Development and International Co-operation: Environment August, 1987. *Our Common Future* (1987), Oxford: Oxford University Press.

Brundtland Report 1989. *The Natural Step*, http://www.naturalstep.com.

Energy Information 2006. World consumption of primary energy by energy type and selected country groups, 1980–2004. Energy Information Administration, U.S. Department of Energy Report July 31, 2006 hwww.eia.ddoe.gov/pub/international/iealf/Table18.xls).

Hargreaves, S.http://money.cnn.com/2008/06/25/news/economy/eia_outlook/index.htm?postversion=200806 2510.

KWE 2006. Key world energy statistics. International Energy Agency (www.iea.org/textbase/nppdf/free/2006/key2006.pdf).

Rowell, R.M. 2002. Sustainable composites from natural resources. *High Performance Structures and Composites*. C.A. Brebbia and W.P. de Wilde, eds. WIT Press, Boston, MA, pp. 183–192.

Rowell, R.M. 2006. *Forest Water Contamination, McGraw-Hill Yearbook of Science and Technology*. McGraw-Hill, New York, NY, pp. 134–136.

Rowell, R.M., Caldeira, F., and Rowell, J.K. 2010. *Sustainable Development in the Forest Products Industry*. Fernando Pressoa University Press, Oporto, Portugal, 281pp.

Sathre, R. 2007. *Life-Cycle Energy and Carbon Implications of Wood-Based Products and Construction*. PhD thesis, Ecotechnology and Environmental Science, Department of Engineering, Physics and Mathematics, Mid Sweden University, Östersund, Sweden.

Sustainable Times. 2008. *Your Guide to a Natural Alternative*, Vol. IV, 4, January (www.sustainabletimes.net).

USDA 2000. United States Department of Agriculture, Forest Service, Forest Products Laboratory, Internal report.

USGS 2000. U.S. Geological Survey, estimated use of water in the United States in 2000. USGS Circular 1268.

WCS 1980. World Conservation Strategy: Living Resource Conservation for Sustainable Development. International Union for Conservation of Nature and Natural Resources, Gland, 1980.

WCED 1983. United Nations, Process of Preparation of the Environmental Perspective to the Year 2000 and Beyond. General Assembly Resolution 38/161, December 1983.

World Summit 2005. United Nations 2005 World Summit. UN Headquarters, New York, NY September, 2005 (www.un.org/summit2005).

Index

Printed in the United States
by Baker & Taylor Publisher Services